D0224233

Contents

ANDREW F. REX
The University of Puget Sound

MARTIN JACKSON
The University of Puget Sound

INTEGRATED PHYSICS AND CALCULUS

VOLUME II

ADDISON-WESLEY

An imprint of Addison Wesley Longman
San Francisco, California • Reading, Massachusetts • New York •
Harlow, England • Don Mills, Ontario • Sydney • Mexico City •
Madrid • Amsterdam

Acquisitions Editor: Sami Iwata
Marketing Manager: Jennifer Schmidt
Publishing Assistant: Nancy Gee
Production Coordination: Joan Marsh
Cover Design: Andrew Ogus ■ Book Design
Printer and Binder: Victor Graphics

Library of Congress Cataloging-in-Publication Data

Rex. Andrew F. [date]
Integrated physics and calculus volume II / by Andrew Rex and Martin Jackson.
p. cm.
Includes index.
ISBN 0-201-47397-6 (v.2)
1. Physics. 2. Calculus. I. Jackson, Martin. II. Title.
QC 21.2R49 1999
530—dc21 99-35366
CIP

ISBN: 0-201-47397-6

1 2 3 4 5 6 7 8 9 10--VG--02 01 00

Preface

The rules that describe nature seem to be mathematical.
Richard Feynman

Since the time of Galileo, and certainly Newton, physicists have relied upon mathematical models to help them understand natural phenomena. Newton developed calculus so that he could better understand motion and solve the particular problem of the motion of the planets under an inverse-square attractive force. For over 300 years, physicists have used calculus to solve a wide range of problems, including but not limited to those in Newtonian mechanics.

Mathematicians, and teachers of mathematics, have used physical phenomena as examples to help them understand and teach calculus. Calculus is not merely a tool for scientists but a fascinating and rich subject in its own right. Using physics and physical phenomena in the development of calculus helps students develop their intuition and deepen their understanding of the concepts of calculus.

The symbiotic relationship of physics and calculus is what we wish to exploit and enjoy with this book. The two subjects, each an exciting field of study in itself, benefit significantly by being connected. These benefits are often lost, or greatly reduced, when students take separate physics and calculus courses from different departments, with faculty speaking different languages and using different notation, and sequences of topics not synchronized.

Integrated Courses

This textbook contains a full year's course of both physics and calculus, taking the place of two textbooks that present the subjects separately. This unique approach comes just in time for the growing number of courses that emphasize the integration of subjects and connections between disciplines.

In recent years many colleges and universities have found ways to teach physics and calculus in an integrated course or in courses taught separately but somehow coordinated with each other. At some universities the integration or coordination also includes engineering, chemistry, or other subjects, and there are many excellent courses being taught that take advantage of the connections between disciplines. This textbook could be used for any such program that includes physics and calculus, regardless of what other subjects, if any, may be included. Because the physics and calculus topics are integrated in this textbook, it is important that the physics and calculus instructors coordinate the pacing of their classes to take advantage of the integration.

Since 1994, we have taught a successful integrated physics and calculus course at the University of Puget Sound. Our course is team-taught by a physicist and mathematician and meets eight hours per week for a full year. It is intended primarily for freshmen but may include upper-division students who are interested in both subjects. Our freshmen have interests in science, mathematics, engineering, computer science, or some combination thereof, and our course serves all such students. This textbook is based on our experiences teaching that course. In both our course and book we have included the topics that fit together naturally and can be reasonably covered in a full academic year (two semesters or three quarters).

The Integrated Textbook

This textbook assumes that students coming into the course have had some exposure to calculus, either in a high school course or in one semester or quarter of college calculus. Students should thus have had some experience with the concepts of limits and derivatives. We begin the textbook with a review of these topics, which then lead naturally into integral calculus (and simultaneously hang together with the study of kinematics in physics).

We have sought to make connections between physics and calculus wherever such connections could be fruitful, but we have not forced connections in every chapter. For example, the physics and calculus are fully integrated in Chapters 2 through 4, but then Chapter 5 and 7 are mostly physics, while Chapters 6 and 8 are mostly calculus. When there are chapters that are all physics or all calculus, they are close enough in the sequence that the physics and mathematics parts of the course can operate on separate tracks for a while and then come back together for the next integrated part. For example, Chapters 13 and 14 are almost all calculus, and Chapters 15 and 16 almost all physics, but then Chapters 17 and 18 contain both physics and calculus and have significant connections between the two. We leave it up to instructors to decide how best to use the integrated and non-integrated portions of the book.

We take advantage of recent innovations in teaching calculus, specifically by coordinating the use of graphical, numerical, and symbolic points of view. One advantage of this approach is that when calculus is applied to physics, it opens a variety of problems and solutions beyond the strictly symbolic methods used in traditional physics courses. Using graphical, numerical, and symbolic methods enhances students understanding and appreciation for both calculus and physics.

As noted above, we have included only what can be comfortably completed in a full-year course (that meets eight hours per week). Absent are some topics traditionally included in introductory physics, specifically heat and geometrical optics. We have found that these are subjects best studied in the laboratory portion of a course, with calorimeters, thermometers, lamps and thin lenses in hand. For those who find these indispensable in the classroom portion of their course, Addison Wesley Longman offers to custom-print those sections from one of their other calculus-based physics textbooks as a supplement to this book.

Contents Overview

Chapter 1 contains a review of functions, with which the entering student should be familiar. It goes on to introduce vector-output functions. This is a key feature of our text—moving the study of vector-output functions to the beginning of the course, where it can be applied to physics. The first chapter also contains a review of physical dimensions and units, with which most students should be familiar.

In Chapter 2 we review the basics of limit, continuity, and derivative. Although we assume students have seen these topics in their earlier calculus course, they will benefit from seeing them again and also seeing their relation to kinematics (position, velocity, and acceleration). This leads naturally into the development of definite integrals and antiderivatives in Chapter 3 (again with position, velocity, and acceleration in one dimension in mind). In Chapter 4 the discussion is extended to the calculus of vector-output functions along with the kinematics of more than one dimension.

Chapters 5 and 6 make a convenient package. Chapter 5 contains the study of dynamics and Newton's laws of motion. Newton's second law, expressed as a differential equation, leads naturally to the study of differential equations in Chapter 6. In Chapter 7 we consider work and energy. The definition of work, which involves a dot product, is more easily grasped by students who are familiar with vector-output functions. The need to calculate work for a variety of situations motivates our study of both symbolic and numerical antiderivative techniques in Chapter 8. This is a standard topic in most second-semester calculus courses.

We think of Chapters 9–11 as another unit, with Chapters 9 and 10 mostly physics (momentum and rotational motion), and Chapter 11 mostly calculus (sequences and series). These physics and calculus topics are considered essential by most faculty, and we give a full, though not completely integrated, treatment. Chapter 12 is another fully integrated chapter, combining harmonic motion and differential equations. The state space viewpoint is particularly useful in admitting graphical and numerical methods.

Chapters 13 and 14 are devoted mostly to mathematics, and they include much of the material one normally encounters in a third semester of calculus, including an introduction to multivariable functions and multiple integrals. These chapter can be covered simultaneously with Chapters 15 and 16, which are mostly physics. Gravitation (Chapter 15) and electrostatics (Chapter 16) are both governed by an inverse-square distance law. Coulomb's law leads to the introduction of electric fields in Chapter 17, which is integrated with the more general study of vector fields in calculus. Chapter 18 is also fully integrated, with the partial derivative and gradient applicable to electric potential.

The concept of electric potential is then used in the study of capacitors (Chapter 19) and DC circuits (Chapter 21). On the mathematics side Chapters 20 (extrema of multivariable functions) is separate from the physics but can be covered at this time to help prepare for what is to come later, when the subjects are integrated again.

Chapter 22 presents a thorough treatment of line integrals, motivated by their use throughout the course in the context of work and potential energy. Chapter 23 is then fully integrated once again. Gauss's law is presented first. This fundamental law of electromagnetism serves as motivation for the study of surface integrals (for which students have been prepared by the treatment of normal line integrals in Chapter 22).

Chapters 24-26 contain the fundamental laws of electromagnetism (an important part of most second-semester physics courses). In calculus the divergence and curl are introduced in Chapter 25. These vector operators are studied with the geometric interpretation of vector fields in mind, but they also serve to prepare students for Chapter 27, where Maxwell's equations are presented in integral form and then rewritten in differential form with the aid of Stokes's theorem and the divergence theorem from calculus. We stress the interpretation of these theorems as vector calculus versions of the second fundamental theorem of calculus.

We include Chapter 28 as a capstone to the first-year physics experience and a bridge to further study (many institutions follow the first-year course with a course in modern physics). The Bohr model is relatively straightforward (compared say to special relativity) and helps pull together a number of concepts used throughout the year: forces (especially electromagnetic), energy, potential energy, and conservation of energy. While these classical concepts are used, Bohr's radical assumptions give students the understanding that classical physics is not sufficient to explain atomic phenomena, and that a more complete theory is needed. This should help provide the inspiration for further study in physics.

Acknowledgments

The development of our course and textbook was supported by grant P116B0024 from the U.S. Department of Education's Fund for the Improvement of Post-Secondary Education (FIPSE).

We wish to thank Addison Wesley Longman for their support, and particularly Sami Iwata and Joan Marsh for their editorial and technical assistance. We also recognize Stuart Johnson and Steve Wuerz for their consideration and advice earlier in the project.

Our colleagues Matt Moelter and Bryan Smith, each of whom taught the course while we had sabbaticals, deserve recognition for their hard work and original ideas that contributed to the project. As we have taught the course and developed the textbook, we have benefited significantly from the input of students in the course and other colleagues in our university and elsewhere, who are too numerous to recognize individually.

A special word of thanks goes to Ruark Dreher, who created many of the electronic drawings that appear in the text.

We thank the following people for their reviews of a manuscript version of the textbook: F. R. Yeatts, *Colorado School of Mines*; Karl Zimmerman, *Union College*; Newton Greenberg, *Binghamton University*; Gary Reich,*Union College*; and Jennifer Quinn, *Occidental College*.

Last but not least, we acknowledge the support and patience of our families, especially our wives Sharon and Jill. They have helped make this all worthwhile and deserve our deepest gratitude.

Andrew Rex and Martin Jackson
University of Puget Sound
June 1999

Chapter 13

Functions of More Than One Variable

Thus far, we have studied two classes of functions. The first class comprises functions of the type $f : \mathbb{R} \to \mathbb{R}$. For these functions, each input is a single real number and each output is a single real number. Generally the first type of function to which we are introduced, these functions are the focus of the first pass through calculus. The second class comprises functions of the type $\vec{r} : \mathbb{R} \to \mathbb{R}^n$. For these functions, each input is a single real number, and each output is a vector of real numbers with n components. The calculus for functions of this type is very similar symbolically to that of the first class of functions, largely because both kinds of function have a single real number as input. Our geometric picture is quite different because the conventional plot for a function $\vec{r} : \mathbb{R} \to \mathbb{R}^2$ shows only outputs.

In this chapter we begin our study of functions for which each input is a vector of real numbers with n components and each output is a single real number. That is, we have functions of the type $f : \mathbb{R}^2 \to \mathbb{R}$ or $f : \mathbb{R}^3 \to \mathbb{R}$ or, most generally, $f : \mathbb{R}^n \to \mathbb{R}$ for some positive integer n. The focus of this chapter is visualizing functions of these types. This is inherently more difficult than visualizing functions $f : \mathbb{R} \to \mathbb{R}$ because we must use more than two dimensions. We start with functions of two variables $f : \mathbb{R}^2 \to \mathbb{R}$ in the first section. Visualizing the graph requires three dimensions, two for the input and one for the output. We also develop some tools, namely *level curves* and *cross sections*, that give us a two-dimensional graphical representation of these functions. In the second section we study some surfaces in $\mathbb{R}^3$. Some of these surfaces can be considered as graphs of functions of two variables, while others cannot. That is, some of these surfaces do not pass the "vertical line test." In the last section we examine functions of three variables. Visualization is difficult for this class because we need three dimensions for the input and one dimension for the output. We rely on a representation in three dimensions given by *level surfaces*, but even this requires substantial effort. For functions of more than three variables, we often "visualize by analogy" through thinking about generic lower-dimension situations.

13.1 Describing Planes in Space

In this section we examine the analytic description of planes by linear equations in three variables. We start with a brief review of the analytic description of lines in the plane by linear equations in three variables.

13.1.1 Lines in the Plane

For a line in the plane, we give an *analytic* description as a linear equation in two variables. The two variables correspond to the Cartesian coordinates of points in the plane. With x and y as the coordinates, the general linear equation in two variables has the form

$$Ax + By + C = 0$$

for constants A, B, and C. For example, the line that passes through the points $(3, 2)$ and $(5, -4)$ is described by the equation $3x + y - 11 = 0$. To be precise, this means if the ordered pair (x, y) satisfies the equation, then (x, y) is the coordinate pair for a point on the line *and* if (x, y) is the coordinate pair for a point on the line, then (x, y) satisfies the equation.

There are several convenient forms for the equation of a line. Given the slope m and the coordinates (x_0, y_0) for one point on the line, we have the *point-slope form*

$$y - y_0 = m(x - x_0).$$

Given the slope m and the coordinates $(0, b)$ of the y-intercept, we can write the equation of the line in the *slope-intercept form*

$$y = mx + b.$$

Each of these can be put into the general form $Ax + By + C = 0$ by suitable algebraic moves.

We now turn to a vector description of a line in the plane. This is the view we will generalize to describing planes in space. Putting the general form into the slope-intercept form results in

$$y = -\frac{A}{B}x - \frac{C}{B}.$$

The coefficients A and B are thus related to the slope by $m = -A/B$. Hence, the vector $\vec{d} = \langle B, -A\rangle$ points in the direction of the line. (Note that $\vec{d}$ has a "run" of B and a "rise" of $-A$.) Consider now the vector $\vec{N} = \langle A, B\rangle$. Since $\vec{d} \cdot \vec{N} = BA + (-AB) = 0$, we conclude that $\vec{N}$ is *perpendicular* to the line. We call this a **normal vector**. Note that $\vec{N}$ is generally not a unit vector. The normal vector has components given by the coefficients A and B on the two variables in the general form of the linear equation.

Given a normal vector $\vec{N}$ and a point on the line we can determine the equation of the line. We do this in a vector formulation. Let $\vec{r} = \langle x, y\rangle$ be a position vector for a generic point. Let $\vec{N} = \langle A, B\rangle$ denote the given normal vector and let $\vec{r}_0 = \langle x_0, y_0\rangle$ denote the position vector of the given point as shown in Figure 13.1. If the point $\vec{r}$ is on the line, then the difference vector $\vec{r} - \vec{r}_0$ lies along the line. Hence, $\vec{r} - \vec{r}_0$ is perpendicular to the normal vector $\vec{N}$, so we have

$$\vec{N} \cdot (\vec{r} - \vec{r}_0) = 0. \tag{13.1}$$

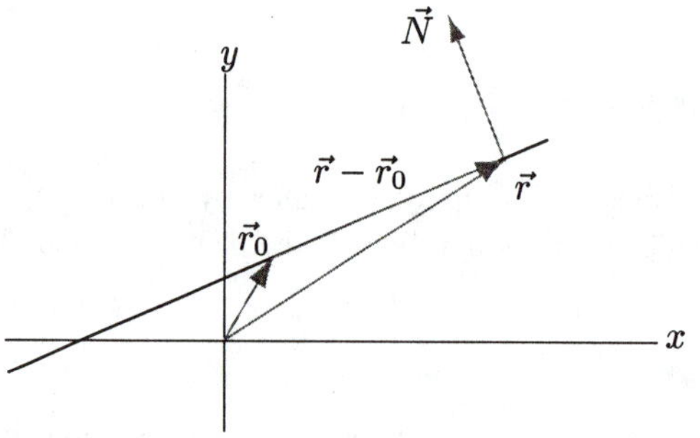

Figure 13.1. Line in the plane determined by a given normal $\vec{N}$ and point $\vec{r}_0$.

This is the **vector form** for the equation of the line. We can relate this to the general form by expressing the equation in terms of components as

$$0 = \vec{N} \cdot (\vec{r} - \vec{r}_0) = \langle A, B \rangle \cdot \langle x - x_0, y - y_0 \rangle = Ax - Ax_0 + By - By_0 = Ax + By + C$$

where we set $C = Ax_0 + By_0$. Note that we again see that the components of the normal vector are the coefficients on x and y in the general form of the linear equation.

Example 13.1. Find the slope-intercept form for the equation of the line that passes through the point $(-2, 5)$ and has a normal vector $\langle 3, 8 \rangle$.

Solution. We are given $\vec{r}_0 = \langle -2, 5 \rangle$ and $\vec{N} = \langle 3, 8 \rangle$. Substituting into Equation (13.1), we have

$$0 = \langle 3, 8 \rangle \cdot \langle x + 2, y - 5 \rangle = 3(x + 2) + 8(y - 5).$$

To get the slope-intercept form, we solve for y giving

$$y = -\frac{3}{8}x + \frac{17}{4}.$$

Thus, the slope is $-3/8$ and the y-intercept is $(0, 17/4)$. △

13.1.2 Planes in Space

Consider a plane in $\mathbb{R}^3$ specified by giving a vector $\vec{N} = \langle A, B, C \rangle$ normal to the plane and a position vector $\vec{r}_0 = \langle x_0, y_0, z_0 \rangle$ for a point on the plane. Our goal is to show that the plane is described by a general linear equation in three variables. Such an equation has the form

$$Ax + By + Cz + D = 0 \tag{13.2}$$

where A, B, C, and D are constants. We will argue that if a point with Cartesian coordinates (x, y, z) is on the plane, then the triple (x, y, z) satisfies a general linear equation in three variables.

Consider the position vector $\vec{r} = \langle x, y, z \rangle$ for a point in $\mathbb{R}^3$. If $\vec{r}$ corresponds to a point on the plane in question, then the difference vector $\vec{r} - \vec{r}_0$ lies in the plane, as shown in Figure 13.2. Hence, the difference vector $\vec{r} - \vec{r}_0$ is perpendicular to the given normal vector $\vec{N}$. Using the dot product, we express this statement as

$$\vec{N} \cdot (\vec{r} - \vec{r}_0) = 0. \tag{13.3}$$

To make an explicit connection to Equation (13.2), we express Equation (13.3) using components as

$$0 = \langle A, B, C \rangle \cdot \langle x - x_0, y - y_0, z - z_0 \rangle = Ax - Ax_0 + By - By_0 + Cz - Cz_0 = Ax + By + Cz + D$$

where $D = Ax_0 + By_0 + Cz_0$. We thus see that the plane is described by a general linear equation in three variables. Further note we can read off the components of a normal vector from the coefficients on the variables.

Example 13.2. Find the general form of the equation for the plane that contains the point $(2, -3, 6)$ and has a normal vector $\langle 5, 3, -8 \rangle$.

Solution. We are given $\vec{r}_0 = \langle 2, -3, 6 \rangle$ and $\vec{N} = \langle 5, 3, -8 \rangle$. Substituting into Equation (13.3), we have

$$0 = \langle 5, 3, -8 \rangle \cdot \langle x - 2, y + 3, z - 6 \rangle = 5(x - 2) + 3(y + 3) - 8(z - 6) = 5x + 3y - 8z + 47.$$

The general linear equation for the plane is thus $5x + 3y - 8z + 47 = 0$. △

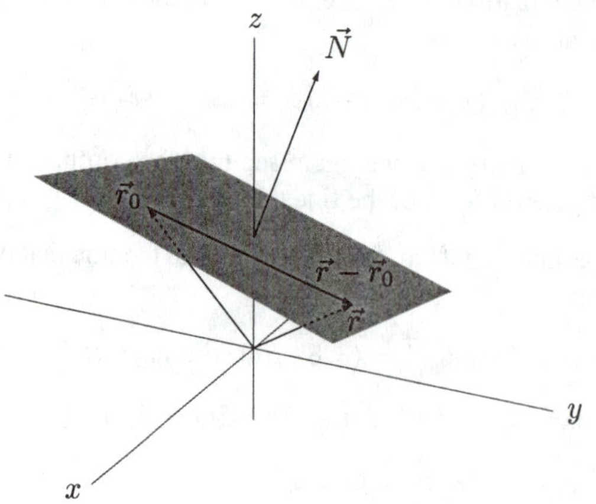

Figure 13.2. Line in the plane determined by a given normal $\vec{N}$ and point $\vec{r}_0$.

Example 13.3. Find the general form of the equation for the line that contains the three points $(6, 1, 4)$, $(-2, 3, 5)$, and $(1, 1, 0)$.

Solution. We have three position vectors $\vec{r}_1 = \langle 6, 1, 4\rangle$, $\vec{r}_2 = \langle -2, 3, 5\rangle$, and $\vec{r}_3 = \langle 1, 1, 0\rangle$. We can easily compute two difference vectors, say $\vec{r}_2 - \vec{r}_1 = \langle -8, 2, 1\rangle$ and $\vec{r}_3 - \vec{r}_1 = \langle -5, 0, -4\rangle$. These two difference vectors lie in the plane, so their cross product is perpendicular to the plane. We can thus use the cross product as a normal vector. Computing this gives us

$$\vec{N} = (\vec{r}_2 - \vec{r}_1) \times (\vec{r}_3 - \vec{r}_1) = \langle -8, -37, 10\rangle.$$

Using this together with $\vec{r}_1$ as the given point in Equation (13.3) gives us

$$0 = \langle -8, -37, 10\rangle \cdot \langle x - 6, y - 1, z - 4\rangle = -8x - 37y + 10z + 45.$$

The general equation of the plane is thus $8x + 37y - 10z - 45 = 0$. △

For a plane that is not vertical (with respect to the xy-plane), we can define *slopes* in analog to the slope of a line. There are two slopes to consider, one corresponding to a "run" parallel to the x-axis and the other corresponding to a "run" parallel to the y-axis. In each case, the slope is defined as the ratio of "rise" to "run." For a plane described by the general linear equation $Ax + By + Cz + D = 0$, we solve for z to get

$$z = -\frac{A}{C}x - \frac{B}{C}y - \frac{D}{C}.$$

The x-slope is given by $m_x = -A/C$, and the y-slope is given by $m_y = -B/C$. Note also that the quantity $d = -D/A$ gives the z-intercept of the plane. That is, the point $(0, 0, -D/A)$ is the point where the plane intersects the z-axis. The equation of a plane can thus be given in *slope-intercept form* as

$$z = m_x x + m_y y + d.$$

Example 13.4. Find the slope-intercept form of the equation for the plane of Example 13.3.

Solution. In the solution of Example 13.3, we found the general form of the equation for the plane to be $8x + 37y - 10z - 45 = 0$. Solving for z gives

$$z = \frac{4}{5}x + \frac{37}{10}y - \frac{9}{2}.$$

The plane has an x-slope of $4/5$, a y-slope of $37/10$ and a z-intercept of $-9/2$. △

13.2 Functions of Two Variables

A **function of two variables** has an input consisting of an ordered pair of real numbers and an output consisting of a single real number. With the input labeled (x, y), we write $f : \mathbb{R}^2 \to \mathbb{R}$ with $f : (x, y) \mapsto f(x, y)$. In many situations, it proves convenient to think of the input as a vector $\vec{r} = \langle x, y \rangle$, in which case we write $f : \vec{r} \mapsto f(\vec{r})$. We can represent a function of two variables in any of the ways we have represented functions of one variable: as an explicit formula for the outputs $f(x, y)$, as a list of numerical data for inputs and outputs, and as a graph. A given function may come to us in any one of these forms. Given an explicit formula for outputs, we can generate numerical data or a graphical representation. Given either a set of numerical data or a graph, we cannot necessarily produce an explicit formula.

Example 13.5. Produce numerical data and a graph for the function with outputs given by $f(x, y) = xy$.

Solution. We start with some numerical data. Given an input such as $(2, 3)$ we can easily compute the output $f(2, 3) = (2)(3) = 6$. We can arrange a list of input and outputs in a table such as Table 13.1. We can also present this same information in a grid, as shown in Table 13.2. The grid format is preferable because it is more compact. In either case, numerical data gives us limited understanding of the function. A graphical viewpoint can provide better understanding.

To produce a graph, we plot points $(x, y, f(x, y))$ in $\mathbb{R}^3$ with outputs $f(x, y)$ giving the third coordinate in the direction perpendicular to the xy-plane. For example, we can plot $(0.5, 0.5, f(0.5, 0.5) = (0.5, 0.5, 0.25)$ as shown in Figure 13.3(a). Plotting the numerical data of Table 13.2 results in Figure 13.3(b). (This figure is difficult to interpret because we are representing a three-dimensional plot in a two-dimensional page.) For "nice" functions, the points $(x, y, f(x, y))$ form a surface in $\mathbb{R}^3$ analogous to the curve formed in $\mathbb{R}^2$ by the points $(x, f(x))$ for $f : \mathbb{R} \to \mathbb{R}$. A portion of the surface given by the graph of $f(x, y) = xy$ is shown in Figure 13.3(c). Fundamental questions about the surface, such as "continuity," are precisely the questions we address with the calculus of functions of several variables. △

Table 13.1. Input/output data for $f(x, y) = xy$

(x, y)	$f(x, y)$
$(-1.0, -1.0)$	1.00
$(-1.0, -0.5)$	0.50
$(-1.0, 0.0)$	0.00
$(-1.0, 0.5)$	-0.50
$(-1.0, 1.0)$	-1.00
$(-0.5, -1.0)$	0.50
$(-0.5, -0.5)$	0.25
$(-0.5, 0.0)$	0.00
$(-0.5, 0.5)$	-0.25
$(-0.5, 1.0)$	-0.50

Table 13.2. Input/output data in grid form

$x \setminus y$	-1.00	-0.50	0.00	0.50	1.00
-1.00	1.00	0.50	0.00	-0.50	-1.00
-0.50	0.50	0.25	0.00	-0.25	-0.50
0.00	0.00	0.00	0.00	0.00	0.00
0.50	-0.50	-0.25	0.00	0.25	0.50
1.00	-1.00	-0.50	0.00	0.50	1.00

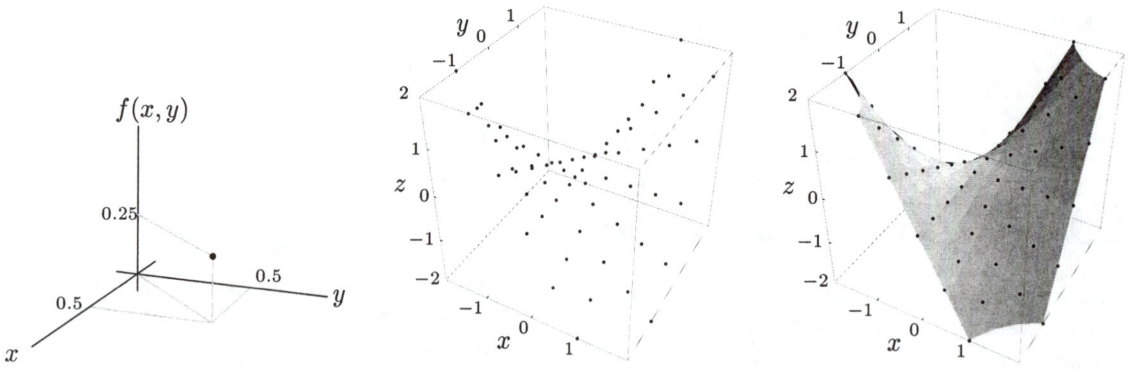

(a) Single point (b) Many points (c) Surface formed by plotting all points

Figure 13.3. Points plotted for the function $f(x, y) = xy$.

The terminology we use for functions of the type $f : \mathbb{R} \to \mathbb{R}$ is used also for functions of the type $f : \mathbb{R}^2 \to \mathbb{R}$. A function of two variables is a map or rule that assigns one output $f(x, y)$ to each input (x, y). The set of inputs is called the **domain** of f. A **source** of f is any set that contains the domain. For functions of two variables, we can use $\mathbb{R}^2$ as a source. The **range** of f is the set of all outputs; a **target** is any set that contains the range. We often use $\mathbb{R}$ as the target when we are not interested in the details of the range.

Drawing or visualizing a surface in three dimensions is difficult. We can give an alternate graphical view of a function $f : \mathbb{R}^2 \to \mathbb{R}$ by using **level curves**. A level curve is the set of points (x, y) in the domain of f that have the same output, say c. That is, a level curve is the set of points (x, y) that satisfy the equation $f(x, y) = c$.

Example 13.6. Plot level curves of $f(x, y) = xy$ for $c = -2, -1, 0, 1, 2$.

Solution. For $c = -2$, the level curve consists of the points that satisfy $xy = -2$. We can solve for y to get $y = -2/x$ and then plot the points given by this explicit relation for y in terms of x. Note that in this case the level curve for $c = -2$ consists of two disconnected pieces. We follow similar steps for $c = -1, 1, 2$. For $c = 0$, we have $xy = 0$. This equation is satisfied by $x = 0$ or $y = 0$. These correspond to the y- and x-axes, respectively. The level curves are shown in Figure 13.4(a). Note the value of c given with each level curve. Given a picture of the level curves, we can visualize the surface by imagining each level curve raised or lowered to the corresponding plane $z = c$. Figure 13.4(b) shows this for the level curves we constructed in this example. △

In the preceding example, we labeled the level curves with the relevant values of $f(x, y) = c$. It is also common to use shading or color to convey information about the output value for each level curve. The region between level curves is shaded or colored according to a scale that corresponds to the output value. Figure 13.4(c) shows the same level curves as in Figure 13.4(a) with shading added. The darkest shade corresponds to the lowest value of c, and the lightest shade corresponds to the highest value.

Functions of two variables arise often in modeling physical phenomena or measuring physical quantities. In many situations, the two input variables represent spatial coordinates; in other situations, the input variables represent other quantities. When the output variable represents a physical quantity, the level curves may have a special name. For example, if the output variable represents temperature, then the level curves are often called **isotherms**. Weather maps often include isotherms for the surface temperature. These are level curves for the functions with inputs (θ, ϕ) being the longitude and latitude coordinates for points on Earth's surface

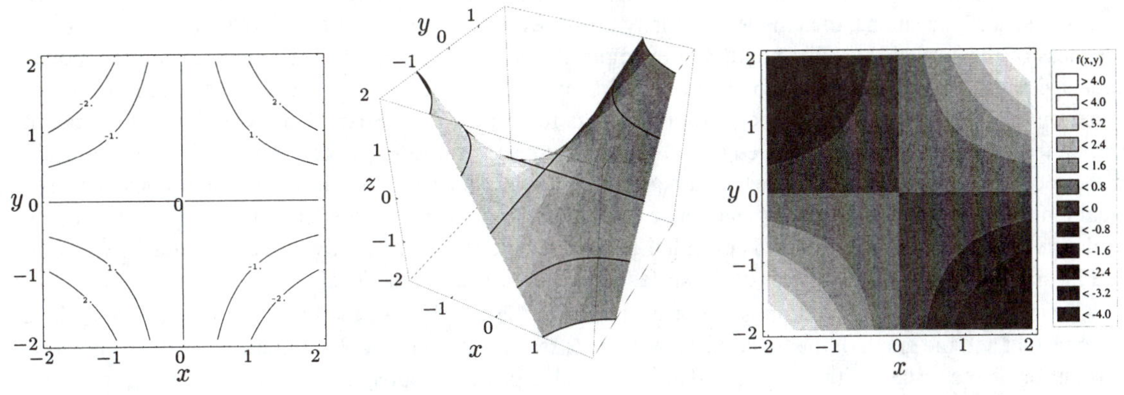

(a) Level curves in the xy-plane (b) Level curves on the graph (c) Shaded level curve plot

Figure 13.4. Level curves and surface for $f(x, y) = xy$.

and output $T(\theta, \phi)$ being the predicted air temperature near the surface. In this case, the level curves provide the most useful view of the function. It is unlikely we could find an explicit formula for $T(\theta, \phi)$.

A topographic map presents a situation for which we would be interested in constructing the graph of a function from knowledge of the level curves. A topographic map, such as the one shown in Figure 13.5, has level curves for the **elevation** of a given region of Earth's surface (or the surface of some other planet). We can think of a function with longitude and latitude as inputs and elevation above sea level as output. (We are assuming that each point on Earth's surface has an unambiguous elevation above sea level.) The graph of this function, a surface in $\mathbb{R}^3$, is a model for the real surface of Earth. It would be useful to visualize the surface corresponding to the given level curves if, for example, we were going to hike through the region. We can get a "global" view by imagining each level curve raised (or lowered for negative elevations) above the map by the amount corresponding to the elevation for that level curve.

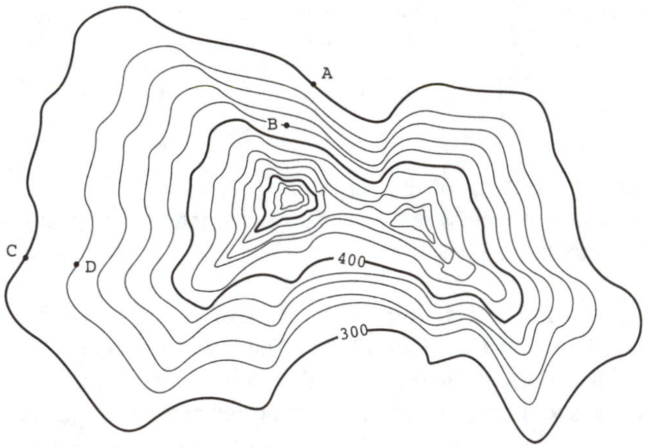

Figure 13.5. Topographic map showing level curves for elevation above sea level. Dark contours give elevation every 100 m. Lighter contours give elevations in 20 m increments.

We can also take a "local" view by examining the density of level curves. If the level curves are given for equal-spaced elevations, then the local density of the level curves at a particular point corresponds to the steepness of the surface over that point. For example consider the straight path on the map with endpoints labeled A and B shown in Figure 13.5. This path starts on the 300-m level curve and crosses three level curves to end on the 380-m level curve. The elevation change for the path is +80 m. Compare this to the straight path on the map with endpoints labeled C and D; this path is the same length as the first. The elevation gain on the second path is only 20 m since the path goes from one level curve to the next. In general, map paths of equal length cross more level curves and hence correspond to greater elevation change in regions where the level curves are denser. Note that it is essential to have level curves for equally spaced outputs if we are to read information about the steepness of the surface from the density of the level curves.

A level curve can be thought of as the projection onto the xy-plane of the intersection between a plane $z =$ constant and the graph of the function f. In some cases, it is useful to look at intersections of planes $x =$ constant or $y =$ constant with the graph. These are called **cross sections**. An example is shown in Figure 13.6(a) for the $y = 1$ plane. Figure 13.6(b) shows the curve formed by the intersection of the plane and the graph. As with level curves, it is convenient to display cross sections in a two-dimensional plot by projecting onto either the yz-plane or the xz-plane. Figure 13.6(c) shows the curve projected onto the xz-plane. We illustrate this in the following examples.

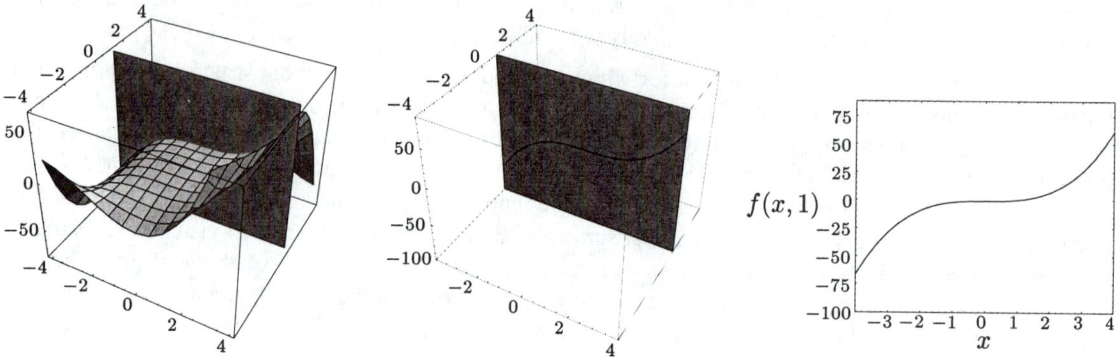

(a) Plane slicing the graph of a function

(b) Curve that is the intersection of the plane and the graph

(c) Projection of the curve into the xz-plane

Figure 13.6. Typical cross section.

Example 13.7. Construct cross sections of the graph of $f(x, y) = xy$ for $x = -2, -1, 0, 1, 2$ and $y = -2, -1, 0, 1, 2$. Graph the constant x cross sections as projected into the yz-plane and graph the constant y cross sections as projected into the xz-plane.

Solution. The cross section for $x = -2$ is given by the graph of $f(-2, y) = -2y$ considered as a function with one input y. This is the line shown in Figure 13.7(a). The cross section for $x = -1$ is given by the graph of $f(-1, y) = -y$. This is also a straight line. In fact, all of the constant x cross sections are lines. The same figure shows the cross sections for the other values of x projected onto the yz-plane.

The constant y cross sections are found in a similar manner. For example, the $y = -2$ cross section is given by the graph of $f(x, -2) = -2x$ considered as a function with one input x. The constant y cross sections are also straight lines for this example. These cross sections are shown in Figure 13.7(b). △

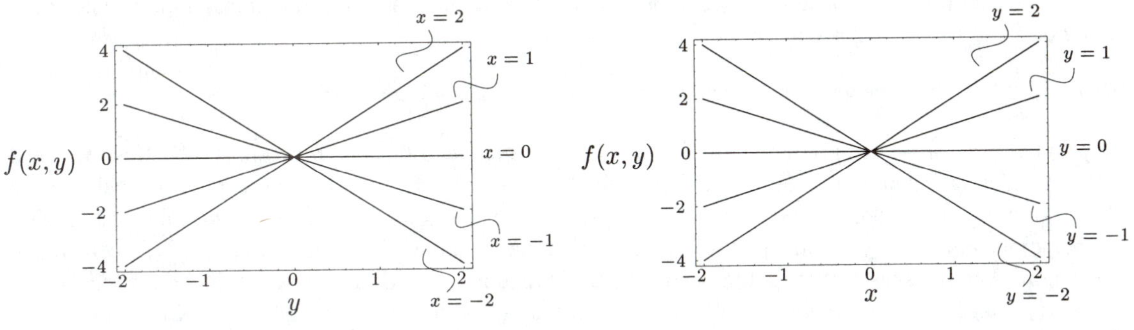

(a) Constant x cross sections (b) Constant y cross sections

Figure 13.7. Cross sections for Example 13.7

Example 13.8. Use both level curves and cross sections to visualize a graph of $f(x, y) = ye^{x-y^2}$.

Solution. Plots of the x = constant cross sections, and y = constant cross sections are shown in Figure 13.8(a) and 13.8(b). A level curve plot is shown in Figure 13.8(c). A surface plot of the graph is shown in Figure 13.8(d). △

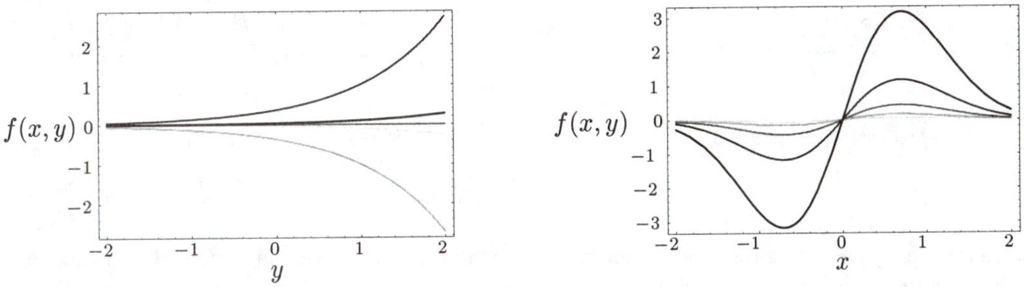

(a) Constant x cross sections (b) Constant y cross sections

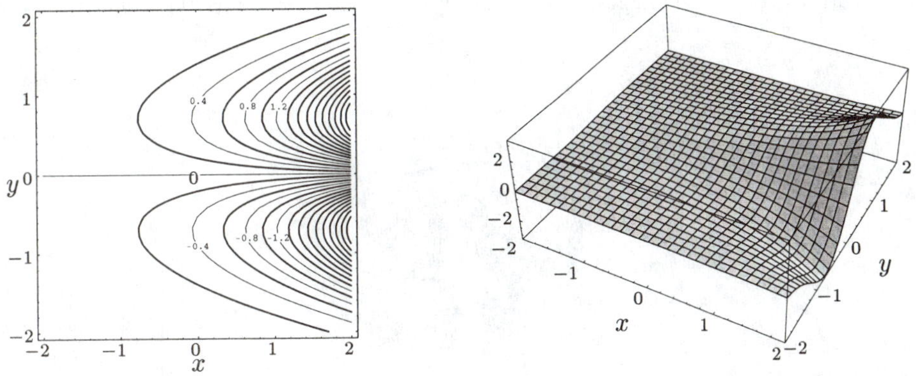

(c) Level curves (d) Graph of the surface

Figure 13.8. Cross sections, level curves, and graph for Example 13.8

The next example demonstrates a technique for visualizing the graph of a function f that depends on x and y only through the expression $x^2 + y^2$.

Example 13.9. Examine level curves and the graph of $f(x, y) = \sin(x^2 + y^2)$.

Solution. Level curves are given by $\sin(x^2 + y^2) = c$. The range of the sine function is $[-1, 1]$, so we need only consider values of c in that interval. For example, consider $c = 0.5$. The corresponding level curve is the set of points with coordinates (x, y) satisfying $\sin(x^2 + y^2) = 0.5$. One subset of the solutions consists of the points with $x^2 + y^2 = \arcsin 0.5 = \pi/6$. This is the equation of a circle of radius $\sqrt{\pi/6}$ centered at the origin. Another subset of the solutions for the 0.5 level curve equation consists of the points that satisfy $x^2 + y^2 = 5\pi/6$; this is the equation of a circle of radius $\sqrt{5\pi/6}$ centered at the origin. Since there are a infinite number of values for $\arcsin 0.5$, we have an infinite number of solution subsets each corresponding to a circle centered at the origin. Some of these are shown in Figure 13.9(a).

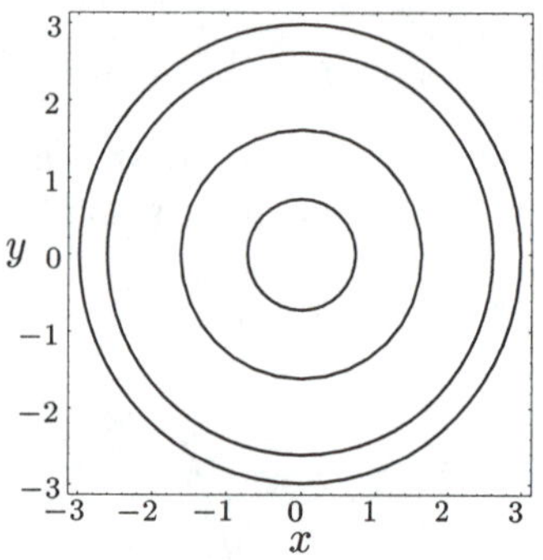

(a) Level curves for $c = 0$

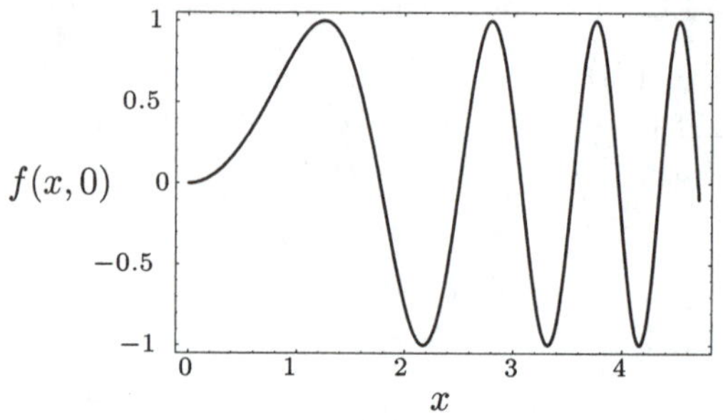

(b) The $y = 0$ cross section

Figure 13.9. Example 13.9.

Because the graph of this function is symmetric around the z-axis, we can understand the shape of the graph by looking at a cross section with a plane containing the z-axis. The graph can be generated by rotating that cross section around the z-axis. For this example, we first plot the $y = 0$ cross section $f(x, 0) = \sin x^2$ as shown in Figure 13.9(b). Now imagine rotating this graph around the z-axis to produce the surface shown in Figure 13.10. △

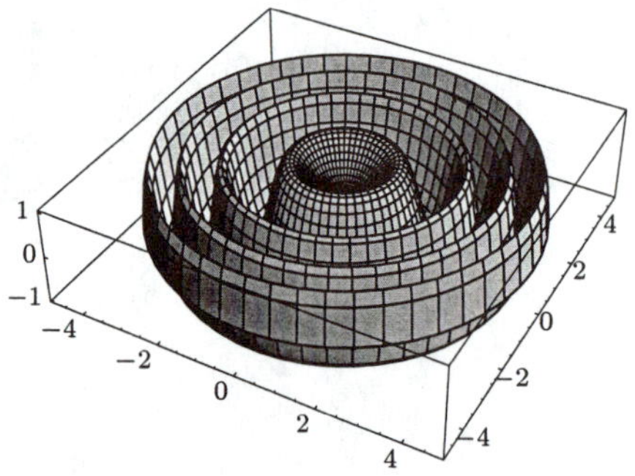

Figure 13.10. Graph of the function $f(x, y) = \sin(x^2 + y^2)$.

For functions of two variables that model physical quantities, the inputs often represent spatial positions, like those in the temperature and topographic map examples. This is not always the case, as we see in the following examples.

Example 13.10. Find an expression for the total energy of an object of mass m in simple harmonic motion with spring constant k. Analyze the energy E as a function of two input variables, the position x and the velocity v_x.

Solution. The object is moving in one dimension, so we can label positions with a single coordinate x. We label the corresponding velocity component v_x. As we saw in Chapter 12, the potential energy is given by $kx^2/2$. The kinetic energy is $mv_x^2/2$. Thus, the total energy is

$$E(x, v_x) = \frac{1}{2}kx^2 + \frac{1}{2}mv_x^2.$$

We think of this as a function of the two state variables x and v_x. Level curves of this function are ellipses in the xv_x-plane (i.e., in state space). A set of level curves is shown in Figure 13.11. Because energy is conserved in this system, the state-space trajectory of the system must remain on a level curve for the energy function. Recall that we found this to be the case in Section 12.2.

△

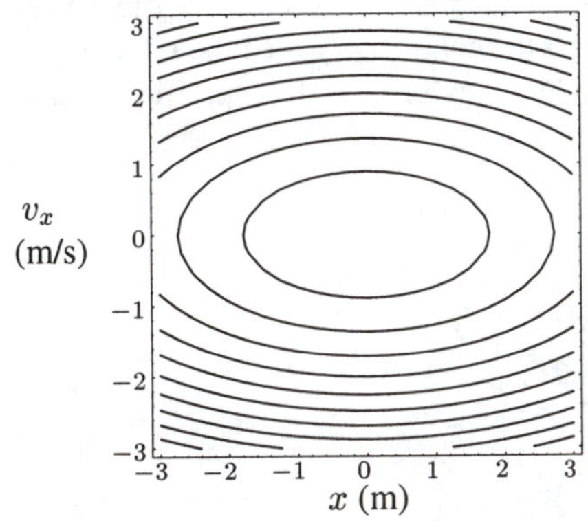

Figure 13.11. Level curves are constant energy curves in state space.

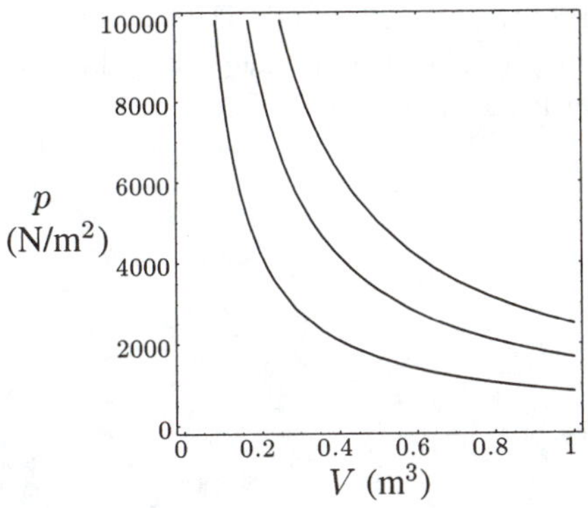

Figure 13.12. Isotherms in the pV-plane.

Example 13.11. Consider a 1 mol sample of gas in a balloon. The volume V and pressure p can change as can the temperature T. The **ideal gas law** models predicts a relationship between these quantities, namely $pV = RT$ for a constant $R \approx 8.314$ (N·m)/(mol·K). Examine isotherms in pressure-volume plane for this model of the sample of gas.

Solution. We think of the temperature as the output of a function with pressure and volume as inputs. The specific function given by the ideal gas laws is

$$T(p, V) = \frac{1}{R}pV.$$

We take the domain of this function to be the first quadrant of the pV-plane since only positive pressures and volumes are physically relevant. A graph showing the $T = 100$, 200, and 300 K isotherms is shown in Figure 13.12.

△

13.3 Examples of Surfaces

In the preceding section we saw that the graph of a function $f : \mathbb{R}^2 \to \mathbb{R}$ is a surface in $\mathbb{R}^3$. In this section we study other examples of surfaces in $\mathbb{R}^3$. Some of these surfaces are the graphs of functions $f : \mathbb{R}^2 \to \mathbb{R}$, while others are not. An analog here for plane curves is that a circle is not the graph of a function $f : \mathbb{R} \to \mathbb{R}$, while a parabola is. (A circle can, however, be split into pieces, each of which is the graph of a function.) We use the techniques of level curves and cross sections to understand these surfaces. We will make use of these surfaces in the next section for our study of functions of three variables.

From an analytic perspective, a surface in $\mathbb{R}^3$ corresponds to the solution set for an equation of the form $F(x, y, z) = c$ where $F(x, y, z)$ is an expression involving the three coordinate variables x, y, and z. A surface that is the graph of a function $f : \mathbb{R}^2 \to \mathbb{R}$ can be thought of from this perspective as corresponding to the solution set of $z - f(x, y) = 0$.

13.3.1 Cylinders

When we hear the word "cylinder" most often we think of something like the surface of a can. The mathematical term "cylinder" refers to something more general. The surface of a can is special kind of cylinder; we explain this in more detail below.

A **cylinder** is a surface generated by translating a plane curve along a line. An example is shown in Figure 13.13 where the plane curve is the portion of the parabola $y = x^2$ with x in $[-2, 2]$ in the xy-plane and the line has direction vector $\langle 1, 1, 1 \rangle$. The plane curve used to generate the cylinder is called the **base curve**, and the line along which it is translated is called the **axis** of the cylinder. A cylinder is thus specified by giving a base curve and an axis (or axis direction).

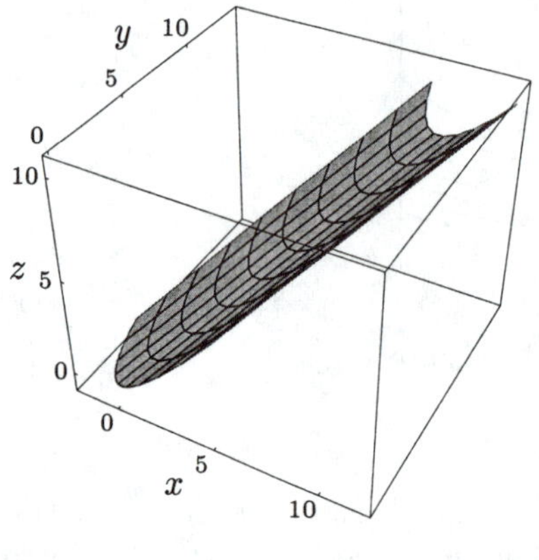

Figure 13.13. Cylinder generated by translating a parabola along a line.

Example 13.12. Sketch the cylinder with base curve given by the ellipse in the xy-plane with equation $3x^2 + y^2 = 1$ and axis given by the z-axis.

Solution. The ellipse is shown in Figure 13.14(a). Translating the ellipse parallel to the z-axis produces the surface shown in Figure 13.14(b).

△

The "standard" cylinder that comes to mind is a **right circular cylinder**. This means that the base curve is a circle and the axis is perpendicular (i.e., at a *right* angle) to the plane in which the circle lies.

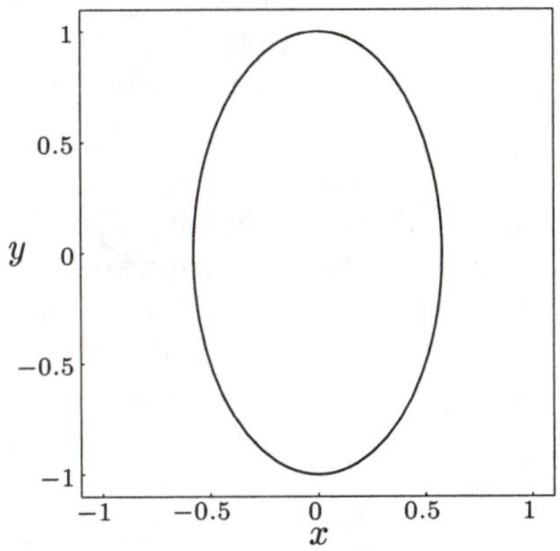

(a) Ellipse that generates a cylinder

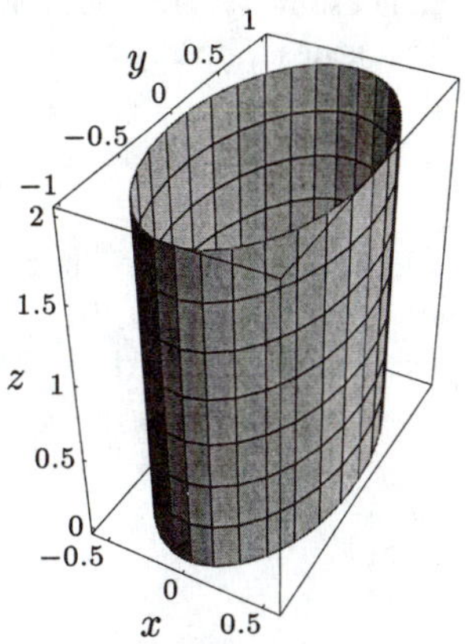

(b) Portion of the cylinder

Figure 13.14. Elliptic cylinder.

13.3.2 Quadric surfaces

The **conic sections** (parabolas, ellipses, hyperbolas) are a useful collection of plane curves. These curves are simple to describe and often arise in modeling applications. From an analytic point of view, we might call these **quadric curves** because each conic section is the graph of a second degree equation in two variables:

$$Ax^2 + By^2 + Cxy + Dx + Ey + F = 0.$$

Conversely, the graph of any second degree equation in two variables is a conic section. A **standard parabola** is described by

$$y = ax^2,$$

a **standard ellipse** by

$$\frac{x^2}{a^2} + \frac{y^2}{b^2} = 1,$$

and a **standard hyperbola** by

$$\frac{x^2}{a^2} - \frac{y^2}{b^2} = 1.$$

Other parabolas, ellipses, and hyperbolas can be generated by translating and rotating the standard ones.

A **quadric surface** is defined as the graph of a second degree equation in *three* variables:

$$Ax^2 + By^2 + Cz^2 + Dxy + Ezy + Fxz + Gx + Hy + Iz + J = 0.$$

As with the conic sections, there is a set of standard quadric surfaces from which all others can be generated by translation and rotation. We explore these in the following examples.

Example 13.13. Describe the quadric surfaces described by equations of the form

$$\frac{x^2}{a^2} + \frac{y^2}{b^2} + \frac{z^2}{c^2} = 1.$$

Solution. We can get some feel for the surface by examining cross sections. Let $x = 0$ to get the cross section in the yz-plane that is described by

$$\frac{y^2}{b^2} + \frac{z^2}{c^2} = 1.$$

This is the equation of an ellipse. The cross sections in the xz-plane and xy-plane are also ellipses. Figure 13.15(a) shows the ellipses in the cross sections for a specific choice of the parameters a, b, and c. Figure 13.15(b) shows the full surface. The surface is called an **ellipsoid**. A sphere is a special kind of ellipsoid with $a = b = c$. △

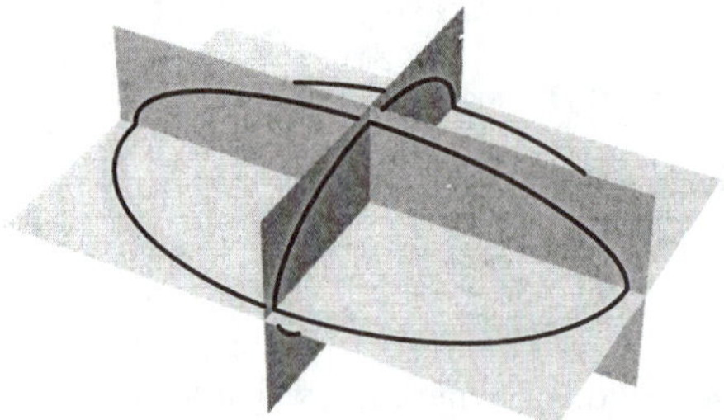

(a) Cross sections in the coordinate planes

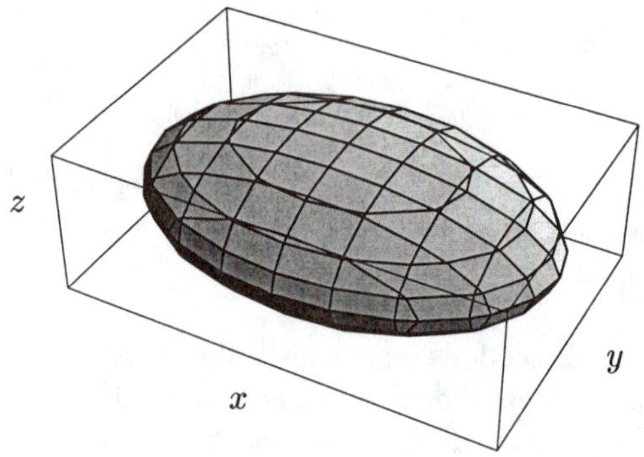

(b) Surface view of the ellipsoid

Figure 13.15. Ellipsoid.

Example 13.14. Describe the quadric surfaces described by equations of the form

$$\frac{x^2}{a^2} + \frac{y^2}{b^2} - \frac{z^2}{c^2} = 1.$$

Solution. Note that this class of equations differs from that of the previous example by the sign on the z term. We again get some feel for the surface by examining cross sections. Let $z = 0$ to get the cross section in the xy-plane that is described by

$$\frac{x^2}{a^2} + \frac{y^2}{b^2} = 1.$$

This is the equation of an ellipse. The cross sections in the xz-plane is described by

$$\frac{x^2}{a^2} - \frac{z^2}{c^2} = 1.$$

This gives a hyperbola for the cross section. Likewise, the cross section in the yz-plane is a hyperbola. Figure 13.16(a) shows the ellipse and hyperbolas in the cross sections for a specific choice of the parameters a, b, and c, and Figure 13.16(b) shows a portion of the full surface. The surface is called a **one-sheet hyperboloid**. The reason for the adjective "one-sheet" will become clear as our discussion unfolds.

△

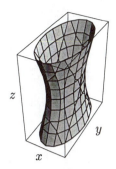

(a) Cross sections in the coordinate planes (b) Portion of the surface

Figure 13.16. Hyperboloid of one sheet.

Example 13.15. Describe the quadric surfaces described by equations of the form

$$\frac{x^2}{a^2} + \frac{y^2}{b^2} - \frac{z^2}{c^2} = 0.$$

Solution. As before, let $z = 0$ to get the cross section in the xy-plane that is described by

$$\frac{x^2}{a^2} + \frac{y^2}{b^2} = 0.$$

The only solution to this equation corresponds to a single point, the origin. The cross sections in the xz-plane is described by

$$\frac{x^2}{a^2} - \frac{z^2}{c^2} = 0$$

which is satisfied by the points on the two lines $z = \pm(c/a)x$. Likewise, the cross section in the yz-plane is a pair of lines $z = \pm(c/b)y$. We can also look at level curves for different values of z. Rewriting the original equation gives

$$\frac{x^2}{(\frac{z}{c}a)^2} + \frac{y^2}{(\frac{z}{c}b)^2} = 0.$$

Think now of z as a parameter. For a fixed choice of z, this equation desribes an ellipse with semi-axes $(z/c)a$ and $(z/c)b$. Note that both semi-axes scale linearly in z. Note also that we have the same ellipse for z and $-z$ since the factors in the denominator are squared. The overall surface is thus a cone with sections parallel to the xy-plane in the form of ellipses. Figure 13.17(a) shows some ellipses for a specific choice of the parameters a, b, and c, and Figure 13.17(b) shows a portion of the full surface. The surface is called an **elliptic cone**.

△

Example 13.16. Describe the quadric surfaces described by equations of the form

$$\frac{x^2}{a^2} + \frac{y^2}{b^2} - \frac{z^2}{c^2} = -1.$$

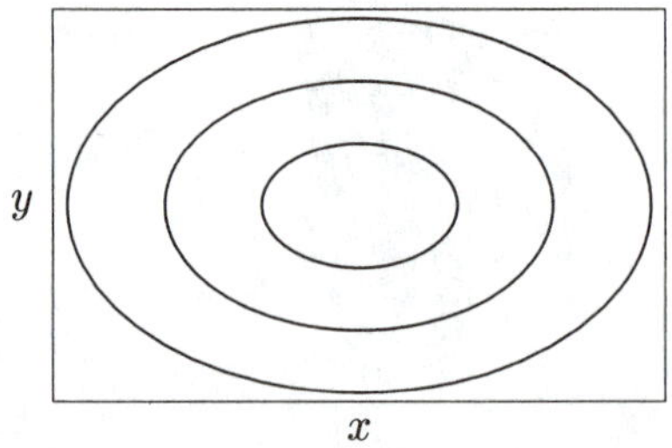

(a) Cross sections parallel to the xy-plane

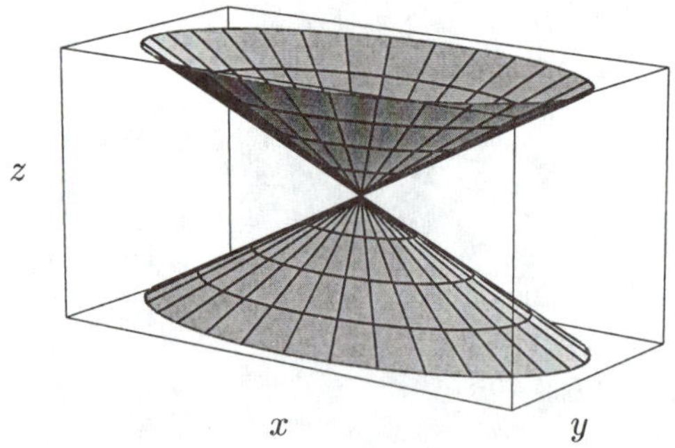

(b) Portion of the surface

Figure 13.17. Elliptic cone.

Solution. Note the similarity between this and equations in the previous two examples; the only difference comes in the value of the constant on the right side of the equation. Let $z = 0$ to get the cross section in the xy-plane that is described by

$$\frac{x^2}{a^2} + \frac{y^2}{b^2} = -1.$$

Since the expression on the left side is positive, there are no solutions. That is, the surface in question does not intersect the xy-plane. The cross section in the xz-plane is described by

$$\frac{x^2}{a^2} - \frac{z^2}{c^2} = -1.$$

This gives a hyperbola for the cross section. Likewise, the cross section in the yz-plane is a hyperbola. For a specific choice of the parameters a, b, and c, Figure 13.18(a) shows the ellipse and hyperbolas in the cross sections, and Figure 13.18(b) shows a portion of the full surface. The surface is called a **two-sheet hyperboloid**. △

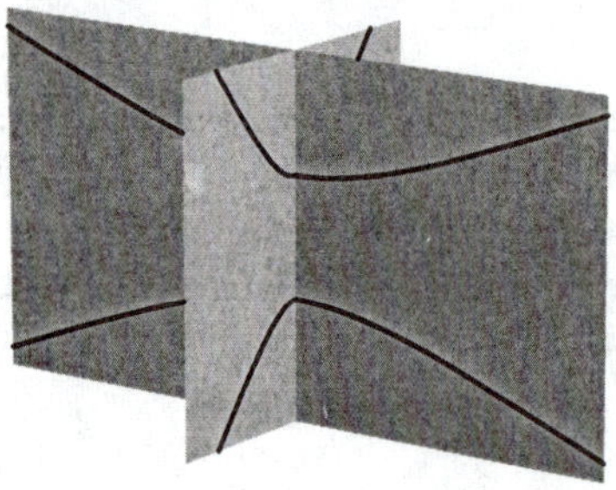

(a) Cross sections

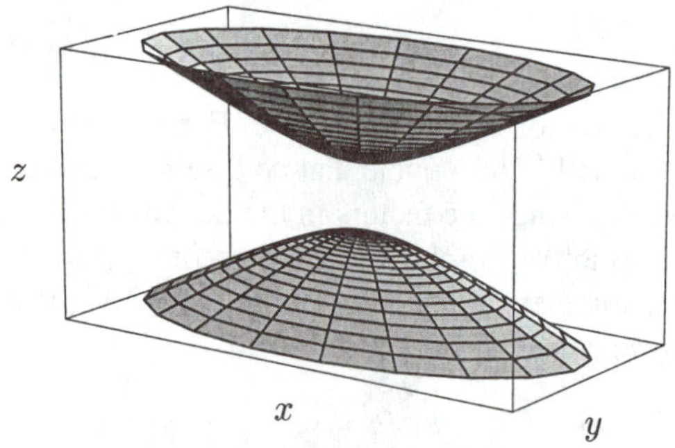

(b) Portion of the surface

Figure 13.18. Hyperboloid of two sheets.

The surfaces in the last three examples can be related in the following manner. Consider the family of equations

$$\frac{x^2}{a^2} + \frac{y^2}{b^2} - \frac{z^2}{c^2} = d.$$

Fix the parameters a, b, and c while considering d as varying from -1 to 1. Think of the corresponding surfaces in animation with d serving as the time. At $d = -1$, we see a one-sheet hyperboloid. As d increases, the neck of the one-sheet hyperboloid contracts in toward the origin. For $d = 0$, the neck pinches down to the origin itself and the surface is an elliptic cone. As d continues to increase (and is now positive), the surface splits into a two-sheet hyperboloid.

The following two examples explore quadric surfaces that are the graphs of functions of two variables. These graphs are important because, as we will see in Chapter 20, they represent the generic local nature of critical points for functions of two variables.

Example 13.17. Describe the quadric surfaces described by equations of the form

$$z = \frac{x^2}{a^2} + \frac{y^2}{b^2}.$$

Solution. It is straightforward to observe that the xz cross section is a parabola given by $z = x^2/a^2$ and the yz cross section is a parabola given by $z = y^2/b^2$. The level curves are all ellipses. A set of level curves is shown in Figure 13.19(a), and a portion of the surface in Figure 13.19(b). The surface is called an **elliptic paraboloid**. △

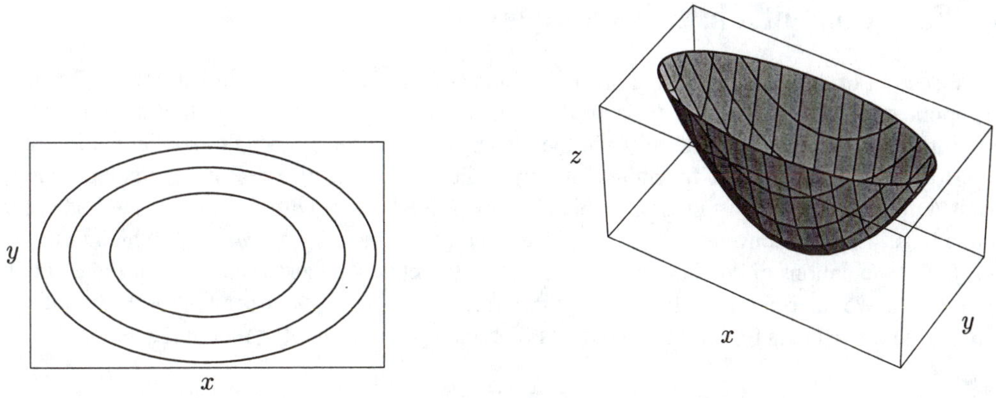

(a) Cross sections parallel to the xy-plane (b) Portion of the surface

Figure 13.19. Elliptic paraboloid.

Example 13.18. Describe the quadric surfaces described by equations of the form

$$z = \frac{x^2}{a^2} - \frac{y^2}{b^2}.$$

Solution. Note the sign difference in comparison to the previous example It is straightforward to observe that the xz cross section is a parabola given by $z = x^2/a^2$ and the yz cross section is a parabola given by

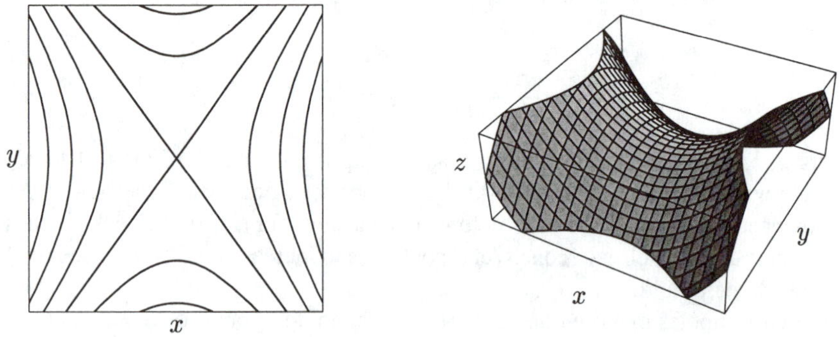

(a) Level curves (b) Portion of the surface

Figure 13.20. Hyperbolic paraboloid.

$z = -y^2/b^2$. The level curves with $z = 0$ is the pair of lines $y = \pm(b/a)x$. All other level curves are hyperbolas, as shown in Figure 13.20(a). A portion of the surface is shown in Figure 13.20(b). The surface is called a **hyperbolic paraboloid**.

△

The remaining quadric surfaces are cylinders with base curve one of the conic sections. These are explored in the exercises.

13.4 Functions of Three Variables

In the first section of this chapter we examined functions of two variables which we denote $f : \mathbb{R}^2 \to \mathbb{R}$. In many applications, such a function arises in modeling the physics of a planar configuration with the input variables representing coordinates of points in the plane. To model phenomena of spatial configurations, it is natural to consider the class of functions with inputs consisting of three real numbers and a single real number as output. We denote a function in this class by $f : \mathbb{R}^3 \to \mathbb{R}$ with $f : (x, y, z) \mapsto f(x, y, z)$. In many situations, it proves convenient to think of the input as a vector $\vec{r} = \langle x, y, z \rangle$, in which case we write $f : \vec{r} \mapsto f(\vec{r})$. The notions of domain, range, source, and target generalize to functions of three variables in the obvious way. We can represent a function of three variables as an explicit formula for the outputs $f(x, y)$ or as a list of numerical data for inputs and outputs. For example, consider the function given by

$$f(x, y, z) = x^2 + y^2 + z^2.$$

For any given input, we can compute an output. With the input $(2, -3, 1)$, we have $f(2, -3, 1) = (2)^2 + (-3)^2 + (1)^2 = 14$. In another situation, we might have a table of data for a function of three variables.

In thinking about a graph for a function $f : \mathbb{R}^3 \to \mathbb{R}$, we encounter a problem. We need three coordinates to represent an input and a fourth coordinate to represent the output. A point on the graph has coordinates $(x, y, z, f(x, y, z))$. That is, the graph is a set of points in $\mathbb{R}^4$. Since our visual world is $\mathbb{R}^3$, we have no direct sensory experience with $\mathbb{R}^4$, and we are somewhat at a loss in directly visualizing (or drawing) the graph. Let's take a cue from our experience with functions of two variables. For that class of functions, the graph requires three dimensions to visualize, two for the input variables and one for the output variable. However, to draw level curves, we need only two dimensions because the curves are planar. Let's look at an analogous idea for functions of three variables. In this class, we define a **level surface** to be the set of points in $\mathbb{R}^3$ corresponding to the inputs that have the same output. That is, a level surface is the set of points corresponding to the solutions of $f(x, y, z) = c$ for some constant c in the range of f.

Example 13.19. Describe the level surfaces of $f(x, y, z) = 3x + y - 4z$ for outputs $c = -10, 0, 10$.

Solution. This is a linear function, so the level surfaces $3x + y - 4z = c$ will be planes. The planes all have the same normal given by $\langle 3, 1, -4\rangle$. The plane for output $c = 0$ passes through the origin. The planes for $c = -10, 0, 10$ are shown in Figure 13.21. △

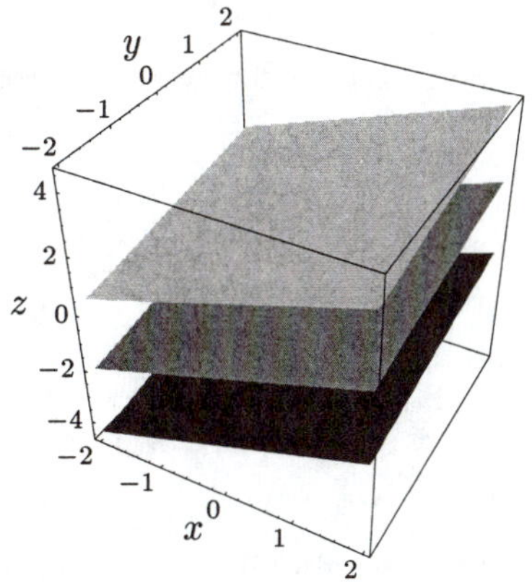

Figure 13.21. Level surfaces are planes for Example 13.19.

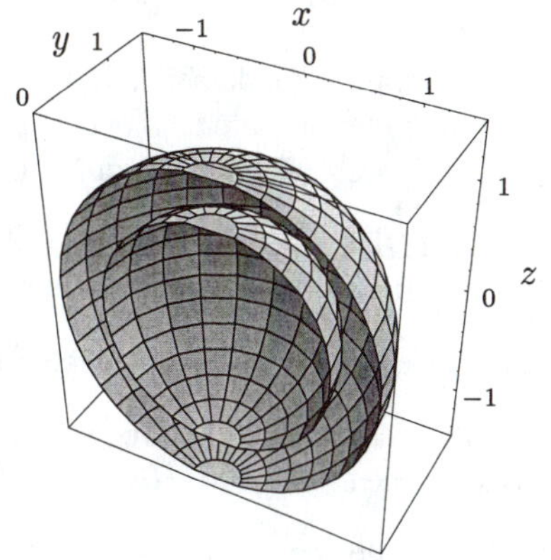

Figure 13.22. Level surfaces are spheres for Example 13.20.

Example 13.20. Describe the level surfaces of $f(x, y, z) = x^2 + y^2 + z^2$ for outputs $c = 0, 1, 2$.

Solution. Note that the range of this function is $[0, \infty)$ and that we have chosen output values c in this range. For $c = 0$, we have $x^2 + y^2 + z^2 = 0$ for which the only solution is $(0, 0, 0)$. The level "surface" is a single point, the origin. For $c = 1$, we have $x^2 + y^2 + z^2 = 1$, which describes the unit sphere centered at the origin. For $c = 2$, the level surface is the sphere of radius $\sqrt{2}$ centered at the origin. These surfaces are shown in Figure 13.22. △

Example 13.21. Describe the level surfaces of $f(x, y, z) = x^2/3 + 5y^2 - z^2$ for outputs $c = -1, 0, 1$.

Solution. Since the output is a second degree polynomial in three variables, the level surfaces will be quadric surfaces. (This is one reason we studied quadric surfaces in the preceding section.) For $c = -1$, we have the surface defined by $x^2/3 + 5y^2 - z^2 = -1$. We recognize this as a hyperboloid of two sheets. For $c = 0$, we have $x^2/3 + 5y^2 - z^2 = 0$, which gives an elliptic cone. For $c = 1$, we have $x^2/3 + 5y^2 - z^2 = 1$, which corresponds to a hyperboloid of one sheet. This function has output zero on the elliptic cone with negative outputs inside the cone and positive outputs outside the cone. △

In these examples, we have examined linear and quadratic functions because we have some experience with the corresponding surfaces. In general, it is quite difficult to visualize a specific level surface. It is often useful, however, to keep a generic picture of level surfaces in mind when dealing with functions of three variables.

Having examined functions of two variables and functions of three variables, it is not a big leap to consider functions of n variables where n is a postive integer. For $f : \mathbb{R}^n \to \mathbb{R}$, it is conventional to denote the input by $\vec{r} = (x_1, x_2, \dots, x_n)$ if the context does not provide a natural labeling. Thus, we write $f : \vec{r} \mapsto f(\vec{r})$

or $f : (x_1, x_2, \dots, x_n) \mapsto f(x_1, x_2, \dots, x_n)$. In applied situations, the context often provides a natural notation. For example, consider modeling the motion of an object moving in space with a known force acting. The position $\vec{r} = \langle x, y, z \rangle$ and velocity $\vec{v} = \langle v_x, v_y, v_z \rangle$ form the input for a function with output equal to the energy. This function has domain in $\mathbb{R}^6$ and range in $\mathbb{R}$. If we name the function E, we would write $E : \mathbb{R}^6 \to \mathbb{R}$ with $E : (\vec{r}, \vec{v}) \mapsto E(\vec{r}, \vec{v})$. The specific expression for the output $E(\vec{r}, \vec{v})$ depends on the nature of the force acting on the object.

13.5 Limits and Continuity

The concept of limit is fundamental in calculus. We review limits for functions of a single variable and then extend it to functions of more than one variable.

In Chapter 2, we gave the following somewhat informal definition.

Definition 13.1. The value L is the **limit of a function $f : \mathbb{R} \to \mathbb{R}$ as x approaches a** if outputs $f(x)$ can be made arbitrarily close to L by using inputs x sufficiently close, but not equal, to a.

We eliminated the ambiguity with the following precise definition.

Definition 13.2. The value L is the **limit of a function $f : \mathbb{R} \to \mathbb{R}$ as x approaches a** if for any $\epsilon > 0$ there exists a value δ such that $|f(x) - L| < \epsilon$ whenever $0 < |x - a| < \delta$.

Figure 13.23 illustrates the essential components of these definitions. In the formal definition, we can think of δ as a measure of how close the input x must be to the fixed value a to ensure that the output $f(x)$ is within ϵ of the limit value L. We say that L is the limit if a value of δ exists for any choice of ϵ.

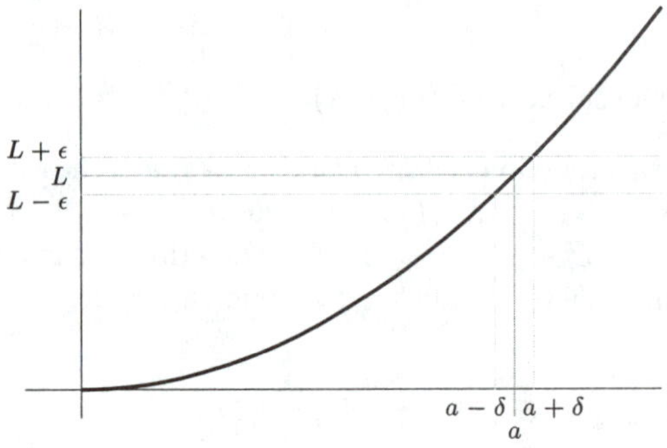

Figure 13.23. Generic limit for a function of one variable.

Our goal now is to generalize the concept of limit to functions of the type $f : \mathbb{R}^2 \to \mathbb{R}$. The basic idea remains the same: We say L is the limit of f if the outputs $f(x, y)$ are arbitrarily close to L for all inputs sufficiently close, but not equal, to (a, b). In a formal definition, we use the idea of an ϵ-neighborhood around L to make precise the notion of "arbitrarily close" and a *deleted δ-neighborhood* around (a, b) to make precise the notion of "sufficiently close." On the output side, no real change is needed; our "goal" is to get $f(x, y)$ within the ϵ-neighborhood $(L - \epsilon, L + \epsilon)$. From the graphical viewpoint, our goal is to have points on the graph of f lie with in the horizontal slab bounded by the planes $z = L - \epsilon$ and $z = L + \epsilon$ as shown in Figure 13.24(a). This slab is the analog of the horizontal band in Figure 13.23.

On the input side, we need to expand our definition of a neighborhood to $\mathbb{R}^2$. From the geometric perspective, a δ-neighborhood of a point (a, b) in the plane is a disk of radius δ centered at that point (see Figure 13.24(b)). The disk is the set of points (x, y) that satisfy the inequality $\sqrt{(x-a)^2 + (y-b)^2} < \delta$. Note

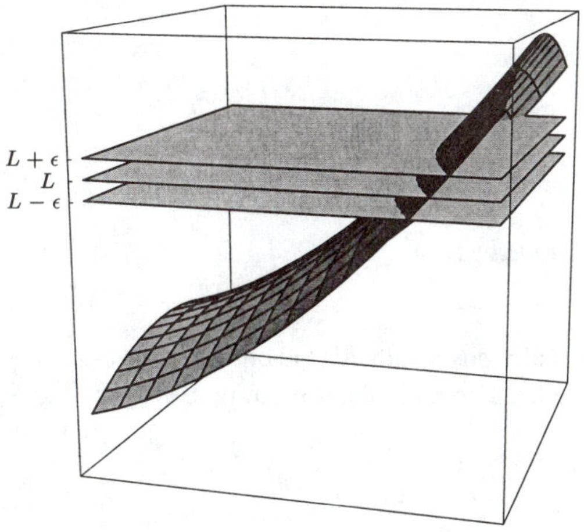

(a) An ϵ neighborhood of L

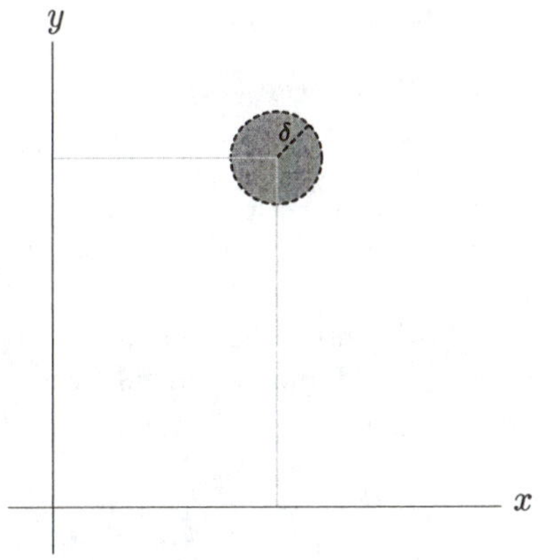

(b) A δ-neighborhood of (x, y)

Figure 13.24. Generic limit for a function of two variables.

that the neighborhood does not include the circle that constitutes the boundary of the disk. We can express the condition somewhat more compactly by using vector notation. Let $\vec{r} = \langle x, y \rangle$ and $\vec{r}_0 = \langle a, b \rangle$; the condition is then $||\vec{r} - \vec{r}_0|| < \delta$. This notation is particularly convenient when we consider generalizing to higher dimensions. A **deleted δ-neighborhood** of a point (a, b) is an open disk of radius δ centered at (a, b) with the center point (a, b) removed. A deleted δ-neighborhood can be described by $0 < \sqrt{(x-a)^2 + (y-b)^2} < \delta$ or $0 < ||\vec{r} - \vec{r}_0|| < \delta$.

With this notation, we can give a definition of limit for a function of the type $f : \mathbb{R}^2 \to \mathbb{R}$ as follows.

Definition 13.3. The value L is the **limit of a function $f : \mathbb{R}^2 \to \mathbb{R}$ as $\vec{r}$ approaches $\vec{r}_0$** if for any $\epsilon > 0$ there exists a value δ such that $|f(x) - L| < \epsilon$ whenever $0 < ||\vec{r} - \vec{r}_0|| < \delta$.

We denote the limit as

$$L = \lim_{(x,y)\to(a,b)} f(x, y) \qquad \text{or} \qquad L = \lim_{\vec{r}\to\vec{r}_0} f(\vec{r}).$$

The following example illustrates the use of the definition to prove a specific limit statement.

Example 13.22. Prove the limit $\displaystyle\lim_{(x,y)\to(0,0)} xy = 0$.

Solution. To show that an appropriate δ exists for any ϵ, we relate $\sqrt{(x-0)^2 + (y-0)^2} = \sqrt{x^2 + y^2}$ to $|xy - 0| = |xy|$. If $\sqrt{x^2 + y^2} < \delta$, then $|x| < \delta$ and $|y| < \delta$. Thus, $|x||y| < \delta^2$. With $\delta = \sqrt{\epsilon}$, the previous inequality implies $|xy| < \epsilon$. Thus, for any $\epsilon > 0$, let $\delta = \sqrt{\epsilon}$ to guarantee $|xy| < \epsilon$ whenever $\sqrt{x^2 + y^2} < \delta$. △

As with the limits of functions of one variable, we avoid direct ϵ-δ proofs by proving limit results. The limit results for functions of one variable carry over to the case of functions of two variables. Some of these are given in the following theorem.

Theorem 13.1. *If* $\lim_{(x,y)\to(a,b)} f(x, y) = K$ *and* $\lim_{(x,y)\to(a,b)} g(x, y) = L$, *then*

1. $\lim_{(x,y)\to(a,b)} (f(x,y) + g(x,y)) = K + L$

2. $\lim_{(x,y)\to(a,b)} \alpha f(x,y) = \alpha K$

3. $\lim_{(x,y)\to(a,b)} f(x,y)g(x,y) = KL$

4. $\lim_{(x,y)\to(a,b)} \frac{f(x,y)}{g(x,y)} = \frac{K}{L}$ *provided there are no divisions by zero.*

Proofs of these results are similar to the proofs of the analogous results for the one-variable case. We use these results to build limit statements from a few basic results. For example, it is straightforward to give ϵ-δ proofs of the limits

$$\lim_{(x,y)\to(a,b)} x = a \quad \text{and} \quad \lim_{(x,y)\to(a,b)} y = b. \tag{13.4}$$

Using these and result 3 of Theorem 13.1, we get the limit statement

$$\lim_{(x,y)\to(a,b)} xy = ab.$$

Note that we proved a special case of this directly in Example 13.22.

Consider the function $f(x,y) = \cos(xy)$. We can view this as a composition of two functions. On the "inside," we have $g(x,y) = xy$, and on the "outside," we have $h(u) = \cos(u)$. Generally, given a function $g : \mathbb{R}^2 \to \mathbb{R}$ and a function $h : \mathbb{R} \to \mathbb{R}$, we can look at the composition $h \circ g : \mathbb{R}^2 \to \mathbb{R}$. The following theorem gives a result about the limit of such a composition.

Theorem 13.2. *If* $\lim_{(x,y)\to(a,b)} g(x,y) = K$ *and* $\lim_{u\to K} h(u) = L$, *then*

$$lim_{(x,y)\to(a,b)} h(g(x,y)) = L.$$

Example 13.23. Prove that $\lim_{(x,y)\to(a,b)} \cos(xy) = \cos(ab)$.

Solution. We know that the cosine function is continuous, so

$$\lim_{u\to K} \cos(u) = K$$

for any K. Above we argued that

$$\lim_{(x,y)\to(a,b)} xy = ab.$$

We can thus apply Theorem 13.2 to conclude

$$\lim_{(x,y)\to(a,b)} \cos(xy) = \cos(ab).$$

To be explicit, note that we use $K = ab$ here. △

Our main use of limits is in giving a precise definition of continuity. The straightforward generalization of this concept from the one-variable case is given in the following definition.

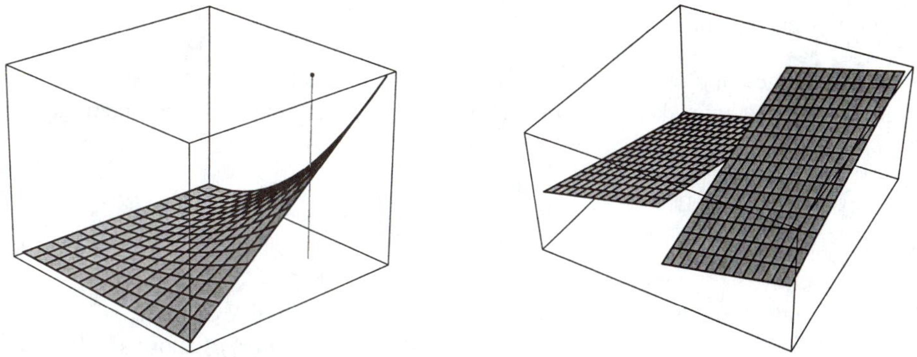

(a) Removable discontinuity (b) "Cliff" discontinuity

Figure 13.25. Two types of discontinuity.

Definition 13.4. A function $f : \mathbb{R}^2 \to \mathbb{R}$ is **continuous** for the input (a, b) if

$$\lim_{(x,y)\to(a,b)} f(x, y) = f(a, b).$$

Note that the existence of both the limit and the output $f(a, b)$ is implicit in the defining condition. As with functions of one variable, the idea of continuity is interpreted geometrically as an "unbroken" graph.

It is straightforward to show that any polynomial function of two variables is continuous for all (x, y) by applying Theorem 13.1 along with the two limits in Equation (13.4). Theorem 13.2 can be used to prove that a function such as $f(x, y) = \cos(xy)$ is continuous for all (x, y).

The possible types of discontinuities include analogs of the discontinuity types for functions of one variable. A function may have a single "misplaced" output as illustrated in Figure 13.25(a). A function may have an abrupt edge over some curve in the input plane, as shown in Figure 13.25(b). The discontinuity may be a vertical asymptote, at either a single input point as shown in Figure 13.26(a) or along a curve of inputs as shown in Figure 13.26(b).

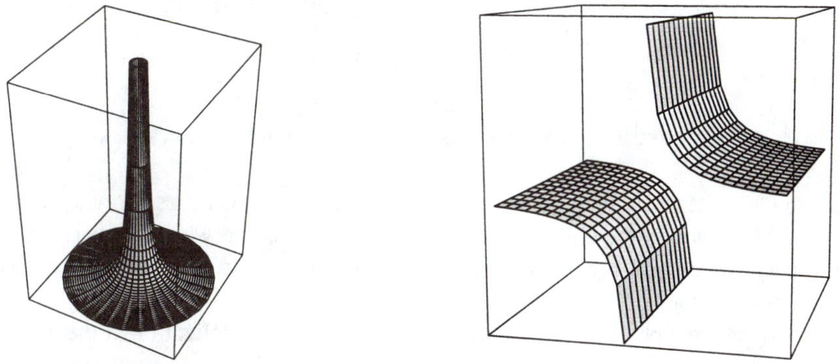

(a) Asymptote at a single point (b) Asymptote along a curve

Figure 13.26. Two types of asymptote.

There is also a somewhat more subtle type of discontinuity for functions of two variables which we illustrate in the following example.

Example 13.24. The function given by $f(x, y) = 2xy/(x^2 + y^2)$ is defined for all inputs except $(0, 0)$. Can $f(0, 0)$ be defined in a way that makes f continuous at the origin? Determine whether the limit

$$\lim_{(x,y)\to(0,0)} \frac{2xy}{x^2 + y^2}$$

exists.

Solution. We examine the limit of f as (x, y) approaches $(0, 0)$. If the limit exists, we can define $f(0, 0)$ to be that limit value. One way to get some feel for the possible limit is to let (x, y) approach $(0, 0)$ along some particular path. For example, we can examine a path along the x-axis by letting $y = 0$. This leads to the function of one variable $f(x, 0) = 0$, for which the limit as x approaches 0 is clearly 0. A similar analysis along the path consisting of the y-axis leads to the function $f(0, y) = 0$, for which the limit as y approaches 0 is also 0. However, if we look at the path $y = x$, we have $f(x, x) = 2x^2/(x^2 + x^2) = 1$. As x approaches 0, this function has a limit value of 1. More generally, we can look at the family of paths consisting of lines through the origin given by $y = mx$ with the slope m as a parameter. Along any line, the function has a constant value $f(x, mx) = 2mx^2/(x^2 + m^2x^2) = 2m/(1 + m^2)$. This constant value is the limit of f as x approaches 0. Clearly, there is no one value to which $f(x, y)$ is arbitrarily close for all inputs (x, y) sufficiently close to $(0, 0)$. The limit does not exist, and thus there is no value for $f(0, 0)$ that will make the function continuous at the origin. A plot of the surface near the origin in shown in Figure 13.27. △

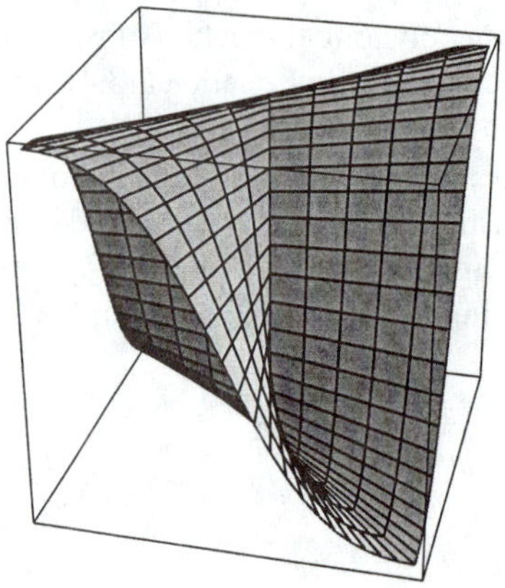

Figure 13.27. "Pinched surface" of Example 13.24.

One major lesson to be learned from this example is that we can sometimes show a limit does not exist by finding two paths along which the one-variable limits differ. Note that we can never prove a limit exists by examining approaches along different paths because we would have to examine all possible paths to the limit input, not just straight lines.

Limits for functions of the type $f : \mathbb{R}^3 \to \mathbb{R}$ are a straightforward generalization of the limits for functions of one and two variables. The idea of neighborhood generalizes from interval (in the one-variable case) and disk (in the two-variable case) to a ball. The δ-neighborhood around a point (a, b, c) is described by

$$\sqrt{(x - a)^2 + (y - b)^2 + (z - c)^2} < \delta$$

and the deleted δ-neighborhood by

$$0 < \sqrt{(x-a)^2 + (y-b)^2 + (z-c)^2} < \delta.$$

In vector notation, with $\vec{r} = \langle x, y, z\rangle$ and $\vec{r}_0 = \langle a, b, c\rangle$ we have $0 < ||\vec{r} - \vec{r}_0|| < \delta$. The defining condition for limit is the same statement as that given in Definition 13.3.

13.6 Problems

13.1 Describing Planes in Space

1. Sketch the plane with equation $2x + 5y + 9z - 90 = 0$ by finding the three coordinate axes intercepts and plotting those points.

2. Sketch the plane with equation $z = 4x - y + 3$ by finding the three coordinate axes intercepts and plotting those points.

3. Find the general form and the slope-intercept form of the equation for the plane that contains the point $(3, 4, 5)$ and has a normal vector $\langle 5, 4, 3\rangle$.

4. Find the general form and the slope-intercept form of the equation for the plane that contains the point $(-2, 6, -3)$ and has a normal vector $\langle 4, -2, 1\rangle$.

5. Find the general form and the slope-intercept form of the equation for the plane that contains the point $(3, 4, 5)$ and is parallel to the yz-plane.

6. Find the general form and the slope-intercept form of the equation for the plane that contains the point $(3, 4, 5)$ and is parallel to the xz-plane.

7. Find the general form and the slope-intercept form of the equation for the plane that contains the point $(3, 4, 5)$ and is parallel to the plane with equation $5x - 9y - 2z + 13 = 0$.

8. Find the general form and the slope-intercept form of the equation for the plane that contains the point $(2, 0, -1)$ and is parallel to the plane with equation $6x + y + 2z = 4$.

9. Find the general form and the slope-intercept form of the equation for the plane that contains the points $(3, 4, 5)$, $(1, -3, 5)$, and $(7, 2, -1)$.

10. Find the general form and the slope-intercept form of the equation for the plane that contains the points $(0, 1, 0)$, $(2, 4, -6)$, and $(1, 0, -1)$.

11. Find the equation of the osculating plane for the output curve of the vector-output function $\vec{r}(t) = \langle \cos t, \sin t, t\rangle$ with $t = \pi/2$. Recall that the osculating plane is defined by the unit tangent vector and the unit principal normal vector.

12. Find a parametrization for the line formed by the intersection of the planes with equations $3x + 7y - 2z + 4 = 0$ and $x - y + 2z - 1 = 0$.

13. Define **the angle between a plane and a curve at a point of intersection** as the angle between a normal vector for the plane and the tangent vector for the curve at the intersection point. Compute this angle between the plane given by $3x + y - 2z - 3 = 0$ and the curve given by $\vec{r}(t) = \langle 1, t^2, t\rangle$ for the intersection that occurs with $t = 2$.

13.2 Functions of Two Variables

For Problems 1–4, describe the major features of the graph corresponding to the given level curve plot.

1.

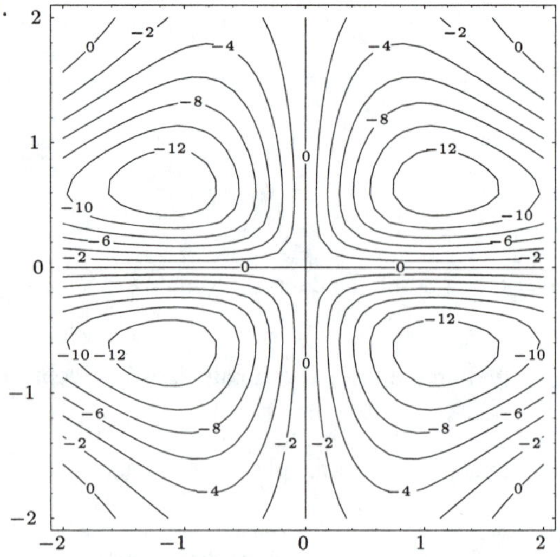

2.

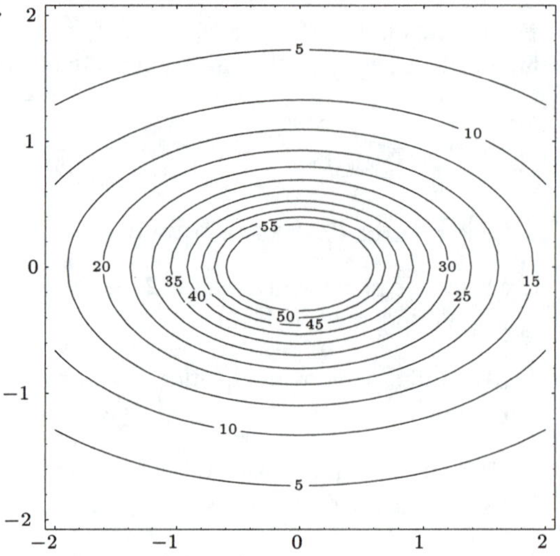

3.

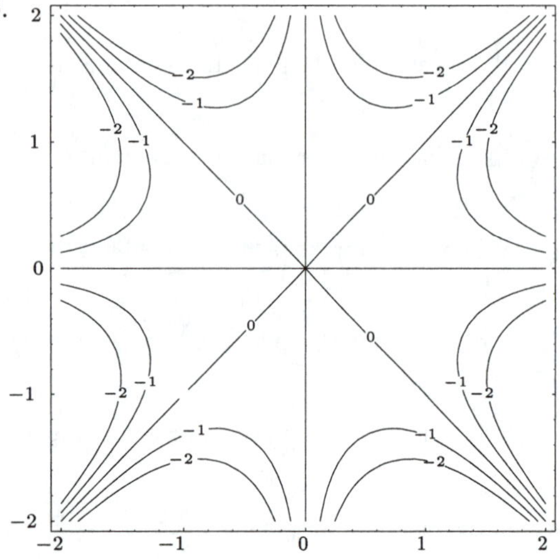

4.

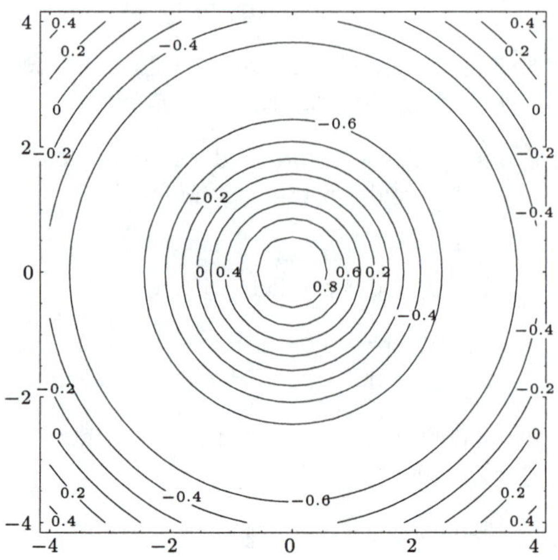

For Problems 5–12, plot the graph and level curves for the given function.

5. $f(x, y) = x^2 + 3y^2$

6. $f(x, y) = x^2 - 3y^2$

7. $f(x, y) = \dfrac{x + y}{x^2 + y^2 + 1}$

8. $f(x, y) = \arctan(xy)$

9. $f(x, y) = \cos(x^2 + y^2)$

10. $f(x, y) = \arctan x + y$

11. $f(x,y) = e^{-(x^2+y^2)}$

12. $f(x,y) = e^{-(x^2+3y^2)}$

13. Consider the topographic map shown below.

    (a) Sketch a cross section for the straight line from A to B.

    (b) Sketch a cross section for the straight line from C to D.

    (c) Describe the major topographic features in the region shown.

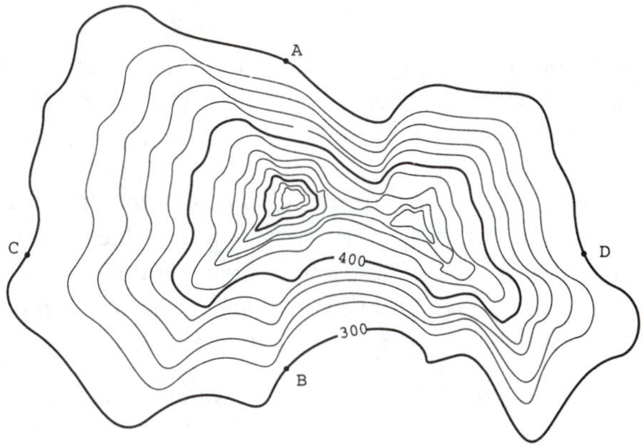

14. In Chapter 12, we studied a model for an object of mass m suspended from a pendulum of length L. We used the angular position θ and the angular velocity Ω as state variables.

    (a) Construct the energy function for the pendulum model with θ and Ω as inputs.

    (b) Sketch the level curves for the energy function you construct.

    (c) Develop a classification of the level curves. For each class, describe the generic motion of the pendulum.

15. A projectile is fired from the ground with an initial speed v_0 at an angle ϕ to the horizontal. Find an expression for the function with v_0 and ϕ as input and the range of the projectile as output. Assume the ground is flat for the extent of the range.

16. Consider the potential energy due to the gravitational force of two planets, one fixed at the position $(x,y) = (-1,0)$ and the other at the position $(x,y) = (1,0)$ where length is measured in some arbitrary unit. There is a function with the position of a point in the plane as input and the potential energy as output. Make a qualitatively correct sketch of level curves for this function.

17. Find an expression for the function with the base and height of a triangle as inputs and the area as output.

13.3 Examples of Surfaces

For Problems 1–3, sketch a picture of the cylinder with the given base curve and axis.

1. base curve $x = y^2$ in the xy-plane and axis direction $\langle 0,1,1 \rangle$

2. base curve $x^2 + z^2/9 = 1$ in the xz-plane and axis direction $\langle 0,1,0 \rangle$

3. base curve $y = e^x$ in the xy-plane and axis direction $\langle 0, 0, 1 \rangle$

4. The volume of a right circular cylindrical solid of radius r and height h is given by $V = \pi r^2 h$.

   (a) Generalize to right cylinder with base curve enclosing area A.

   (b) Generalize to a cylinder with base curve containing area A in xy-plane and with length l along axis direction $\langle a, b, c \rangle$.

   (c) Define "height" of a cylinder.

5. In this problem you will explore an example of how a quadric surface can be described as a rotation of one of the standard quadric surfaces. Consider the quadratic equation $z = xy$. Introduce a new pair of coordinates (u, v) to replace (x, y) by defining $u = (x + y)/2$ and $v = (x - y)/2$.

   (a) Draw a set of xy-axes. On that coordinate system draw the curves of constant u for values $u = -2, -1, 0, 1, 2$. Also draw the curves of constant v for values $v = -2, -1, 0, 1, 2$. (*Note*: the curves will be straight in this case.) These curves give a grid for the uv coordinate system. The $v = 0$ line is the u-axis and the $u = 0$ line is the v-axis (just as the $y = 0$ line is the x-axis and the $x = 0$ line is the y-axis in the original xy coordinate system).

   (b) Transform the second degree equation $z = xy$ in the variables x, y, z into a second degree equation in the variables u, v, z.

   (c) Identify the type of quadric surface corresponding to this quadric equation.

6. The **central axis** for the standard hyperbolic paraboloids discussed in the text is the z-axis. Find the second degree equations for the class of hyperbolic paraboloids with the x-axis as the central axis.

13.4 Functions of Three Variables

For Problems 1–4, describe some level surfaces of the given function of three variables.

1. $f(x, y, z) = -5x - 2y + 4z$
2. $f(x, y, z) = x^2 + y^2 + 4z^2$
3. $f(x, y, z) = \sin(x^2 + y^2 + z^2)$
4. $f(x, y, z) = e^{-(x^2+y^2+z^2)}$

5. Consider the function with $f(x, y, z) = x^2/3 + 5y^2 - z^2$ we examined in Example 13.21. Sketch the function given, and give a geometric interpretation of the plot in terms of the level surfaces.

   (a) $f(x, 0, 0)$ considered as a function of x

   (b) $f(0, y, 0)$ considered as a function of y

   (c) $f(0, 0, z)$ considered as a function of z

6. Find an expression for the function with the three side lengths of a rectangular solid as input and the volume as output.

7. Consider the potential energy due to the gravitational force of two planets, one fixed at the position $(x, y, z) = (-1, 0, 0)$ and the other at the position $(x, y, z) = (1, 0, 0)$ where length is measured in some arbitrary unit. There is a function with the position of a point in space as input and the potential energy as output. Make a qualitatively correct sketch of level surfaces for this function.

13.5 Limits and Continuity

For Problems 1–3, determine whether the given limit exists. If so give the value of the limit.

1. $\lim_{(x,y)\to(0,0)} \log\left(x^2+y^2\right)$

2. $\lim_{(x,y)\to(0,0)} \frac{x^2 y}{x^3+y^3}$

3. $\lim_{(x,y)\to(1,1)} \frac{x-y^4}{x^3-y^4}$

4. Prove result 1 of Theorem 13.1.

5. Prove result 2 of Theorem 13.1.

6. Prove result 3 of Theorem 13.1.

7. Prove result 4 of Theorem 13.1.

8. Use Theorem 13.1 to argue that

$$\lim_{(x,y)\to(a,b)} (x^2+5xy-6y^2) = a^2+5ab-6b^2.$$

9. Prove Theorem 13.2.

10. Use Theorems 13.1 and 13.2 to prove that $f(x,y)=\sin(x^2+y^2)$ is continuous for all (x,y).

11. Consider the function given by

$$f(x,y) = \begin{cases} x-y & \text{if } x<0 \\ x+y & \text{if } x\geq 0. \end{cases}$$

Determine whether this function is continuous at any point on the y-axis.

Chapter 14

Multiple Integrals

The definite integral is one of the major concepts in the calculus of functions $f : \mathbb{R} \to \mathbb{R}$. We have also seen many applications of definite integrals in physics. In this chapter we generalize the concept of definite integral to functions $f : \mathbb{R}^2 \to \mathbb{R}$. The generalization is called a *double definite integral* or *double integral* for short. A *triple definite integral* is the relevant generalization for functions $f : \mathbb{R}^3 \to \mathbb{R}$. Double and triple integrals are often most easily evaluated by using *iterated integrals*.

For $\mathbb{R}^2$, we can use *polar coordinates* as an alternative to Cartesian coordinates. Likewise, *cylindrical coordinates* and *spherical coordinates* are useful alternatives for describing points in $\mathbb{R}^3$. We define these coordinate systems and look at how to evaluated iterated integrals using them.

14.1 Double Integrals

14.1.1 Mass Density and Total Mass in One Dimension—Review

In Section 9.1, we introduced the notion of mass density for a straight one-dimensional (model) object. We review that idea here. We can think of the object as a thin rod of length L. The mass density λ is defined as the mass per unit length. That is, we determine the approximate mass density at some point of the rod by cutting out a small piece, measuring the length Δx and mass Δm, and computing the ratio $\Delta m/\Delta x$. The exact mass density is the limit of this (conceptual) ratio as $\Delta x \to 0$. The rod is not necessarily uniform in composition, so we think of the mass density as the output of a function with the location along the rod as input.

The total mass of the rod is related to the mass density through a definite integral. To construct this integral, we first introduce the coordinate system shown in Figure 14.1, with the rod extending from location

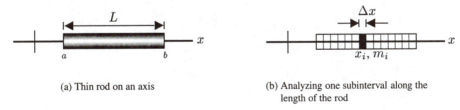

(a) Thin rod on an axis

(b) Analyzing one subinterval along the length of the rod

Figure 14.1. Center of mass for an extended object in one dimension.

$x = a$ to $x = b$ with $b - a = L$. Think of dividing the rod into n small pieces of length $\Delta x = (b-a)/n$, as shown in Figure 14.1. The mass m_i of a generic piece is given approximately by $\lambda(x_i)\Delta x$ where x_i is some location in that piece. The total mass is approximated by summing the masses of the n pieces:

$$M \approx \sum_{i=1}^{n} m_i = \sum_{i=1}^{n} \lambda(x_i)\Delta x.$$

As the number of pieces n increases, the sum on the right limits to a definite integral for the exact total mass so

$$M = \int_a^b \lambda(x)\, dx.$$

If we have a planar two-dimensional (model) object, the relevant mass density is mass per unit area. We can consider the two-dimensional mass density as a function of two variables with the input representing the location of a point in the object. Similarly, the mass density for a three-dimensional object, mass per unit volume, is the output of a function of three variables. The relationship between the mass density and the total mass is analogous to that of the one dimensional case. Our goal in this chapter is to generalize the concept of definite integral to functions of more than one variable. It is this generalized definite integral that appears in the relationship between mass density and total mass for two or three dimensions.

14.1.2 Mass Density and Total Mass in Two Dimensions

We now generalize the relationship between mass density and total mass to a flat two-dimensional (model) object. Think of the object as a thin plate. Introduce a coordinate system for the plane in which the plate lies and let R denote the region occupied by the plate (see Figure 14.2). The appropriate mass density is *mass per unit area*; we denote this by σ. We can consider the function that has coordinates (x, y) of a point in R as input and the mass density $\sigma(x, y)$ as output. To relate the total mass to the mass density, we follow a procedure analogous to constructing the definite integral for the one-dimensional case. Think of dividing the region R into small pieces each of area ΔA_{ij}, as shown in Figure 14.2. The double index ij labels the pieces. The mass m_{ij} of a generic piece is given approximately by $\sigma(x_{ij}, y_{ij})\Delta A_{ij}$ where (x_{ij}, y_{ij}) is some

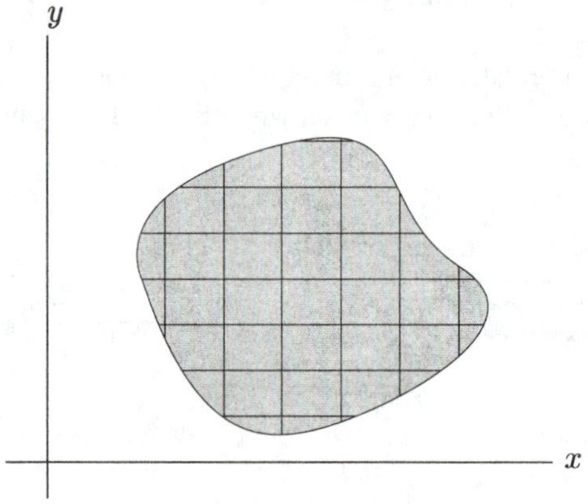

Figure 14.2. Planar region divided into subregions.

location in that piece. The total mass is approximated by summing the masses of the small pieces:

$$M \approx \sum_i \sum_j m_{ij} = \sum_i \sum_j \sigma(x_{ij}, y_{ij})\Delta A. \tag{14.1}$$

Note one difficulty in the procedure we just described: Some of the pieces are rectangles, while others are not. As a result, the pieces do not all have the same area. The areas of the rectangular pieces are simple to compute, but the areas of the nonrectangular pieces are not. To avoid this difficulty, we begin by limiting our attention to rectangular regions so that we can divide the region into pieces, all of which are also rectangles.

Example 14.1. A thin rectangular plate occupies the rectangular region $[0, 1] \times [0, 2]$ in the xy-plane with lengths measured in meters. The mass density is given by $\sigma(x, y) = 1 + xy$ in units of kg/m^2. Estimate the total mass of the plate.

Solution. We can divide the rectangular region into rectangular subregions by dividing the x-interval into m pieces of length Δx and the y-interval into n pieces of length Δy, as shown in Figure 14.3. The rectangular subregions will be of area $\Delta A = \Delta x \Delta y$. To start, we choose $m = 4$ and $n = 8$ for a total of $(4)(8) = 32$ subregions of area $\Delta A = (1/4)(2/8) = 1/16$. We must now choose a representive input from each subregion. Note that the function output $\sigma(x, y) = 1 + xy$ is increasing in both x and y. If we choose the smallest values of both x and y on each subregion, we have a lower bound on the mass of each subregion, and hence the sum will be a lower bound on the total mass. In a similar fashion, choosing the largest values of both x and y produces an upper bound on the total mass. Table 14.1 gives the data for the lower-bound outputs.

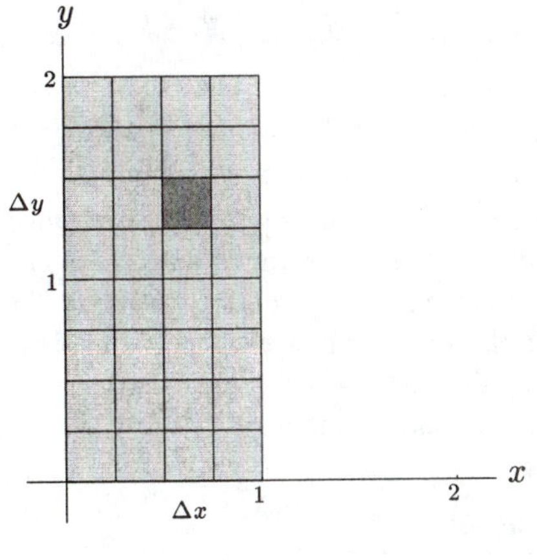

Figure 14.3. Rectangle of Example 14.1.

Table 14.1. Lower-bound output data for $\sigma(x, y) = 1 + xy$

$x \backslash y$	0	0.25	0.50	0.75	1.00	1.25	1.50	1.75
0	1.0	1.0	1.0	1.0	1.0	1.0	1.0	1.0
0.25	1.0	1.0625	1.125	1.1875	1.25	1.3125	1.375	1.4375
0.50	1.0	1.125	1.25	1.375	1.5	1.625	1.75	1.875
0.75	1.0	1.1875	1.375	1.5625	1.75	1.9375	2.125	2.3125

In a similar fashion, choosing the largest values of both x and y produces an upper bound on the total mass. Table 14.2 gives the data for the upper-bound outputs.

Table 14.2. Upper-bound output data for $\sigma(x, y) = 1 + xy$

$x \backslash y$	0.25	0.50	0.75	1.00	1.25	1.50	1.75	2.0
0.25	1.0625	1.125	1.1875	1.25	1.3125	1.375	1.4375	1.5
0.50	1.125	1.25	1.375	1.5	1.625	1.75	1.875	2.0
0.75	1.1875	1.375	1.5625	1.75	1.9375	2.125	2.3125	2.5
1.0	1.25	1.5	1.75	2.0	2.25	2.5	2.75	3.0

Note that each term in the sum has a factor of $\Delta A = 1/16$, so we can produce the lower sum by adding the outputs in Table 14.1 and then multiplying by $1/16$. The result is 2.65 kg. In a similar fashion, we find the upper bound to be 3.41 kg. Thus, the total mass is estimated by 2.65 kg $< M <$ 3.41 kg. We could also report this as $M = 3.03 \pm 0.38$ kg.

We could improve the estimate by using more subregions. For example, with $m = 10$ and $n = 20$, we find 2.855 kg $< M <$ 3.155 kg or $M = 3.005 \pm 0.15$ kg. It should be no surprise that the error decreases as the number of subregions increases. △

It is reasonable to think of the limit of the sum in Equation (14.1) as something analogous to a definite integral. For now, we do not worry about technical details and proceed on an intuitive basis. We assume the total mass is equal to the limit of this sum as the number of pieces increases and the area ΔA decreases. We denote this with

$$M = \iint\limits_R \sigma(x, y)\, dA.$$

The integral on the right side is an example of a **double definite integral**. The adjective "double" refers to the fact that the integrand is a function of two variables. It is common to drop the word "definite" and refer simply to a *double integral*. The notation here is analogous to our notation for definite integrals $\int_a^b f(x)\, dx$. We use a double integral sign to remind us that the region of integration is two dimensional. The R represents the region over which we are integrating and thus generalizes a and b. The dA generalizes dx and serves to remind us that the integral is a limit of a sum in which each term contains a factor of ΔA.

In the next section we define double definite integrals for generic functions of two variables, and in subsequent sections we examine how to evaluate these integrals.

14.1.3 Double Definite Integrals

We now turn to developing a definition of double definite integral. In doing so, we will think geometrically as we did in defining the single definite integral. In the case of a single definite integral for $f : \mathbb{R} \to \mathbb{R}$, we considered the area of the planar region bounded by the graph of a function, an interval on the input axis, and vertical lines at the boundaries of the interval. For $f : \mathbb{R}^2 \to \mathbb{R}$, the analog is the volume of a solid region bounded by the graph of the function, a rectangle in the input plane, and vertical planes at the boundaries of the rectangle.

Let's first introduce some notation. Consider a function $f : \mathbb{R}^2 \to \mathbb{R}$ with $f : (x, y) \mapsto f(x, y)$. Let R be a rectangular region $[a, b] \times [c, d]$ in the xy-plane. Consider the *volume* of the solid region bounded on the bottom by the rectangle R in the xy-plane, on the top by the graph of a function $f : \mathbb{R}^2 \to \mathbb{R}$, and on the sides by the planes $x = a$, $x = b$, $y = c$, and $y = d$. A typical example is shown in Figure 14.4.

The rectangle R can be divided into mn subrectangles by dividing the x-interval $[a, b]$ into m subintervals of length $\Delta x = (b - a)/m$ and the y-interval $[c, d]$ into n subintervals of length $\Delta y = (d - c)/n$. Each subrectangle has area $\Delta A = \Delta x \Delta y$. (It is possible to work with more general partitions in which the

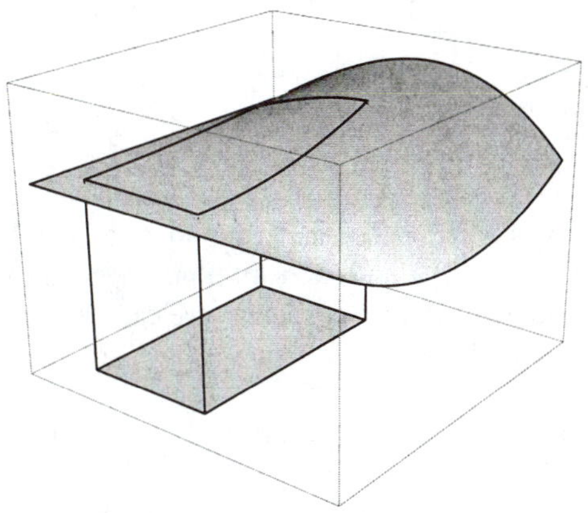

(a) View of the solid bounded above by the graph of f

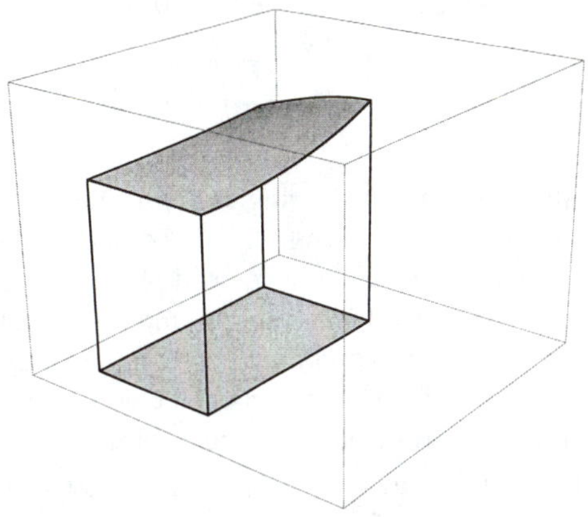

(b) The same solid region with the surface removed except for the portion bounded the top of the region

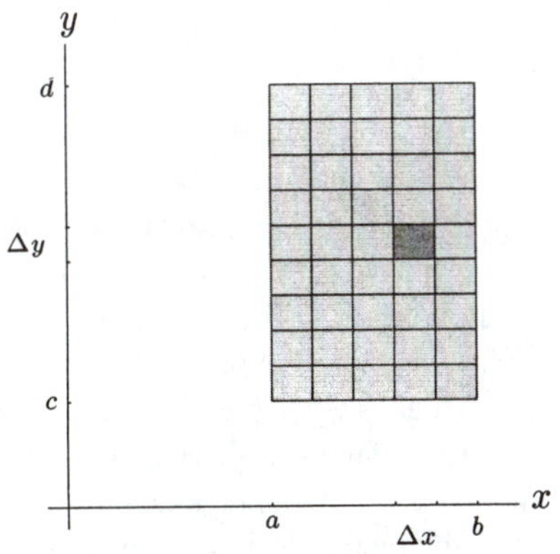

(c) Rectangle R divided into subrectangles

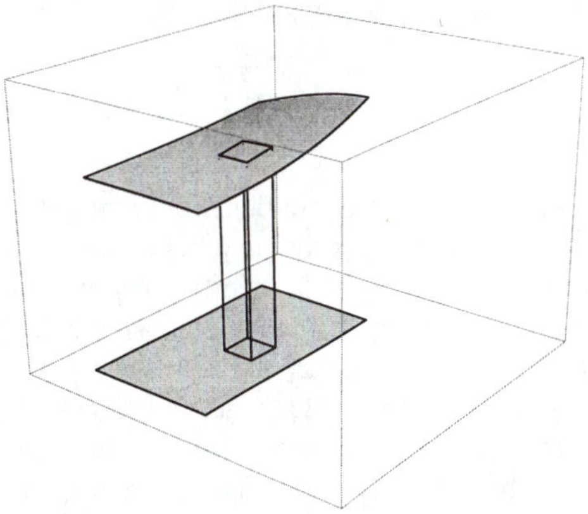

(d) Solid region over a typical subrectangle

Figure 14.4. Solid region bounded above by the graph of f, below by a rectangle R in the xy-plane, and on the sides by vertical planes.

subregions are not all of the same size. We use equal-size subregions for simplicity.) Let i be an index for the x subintervals and j be an index for the y subintervals. The subregions are labeled by a double index ij. We highlight a generic subregion kl in Figure 14.4(c). Think of constructing the rectangular solid defined by the boundaries of the subinterval on the sides, on the bottom by the xy-plane, and on top by a horizontal plane region at height $f(x_{kl}, y_{kl})$ where (x_{kl}, y_{kl}) is some point in the subregion. A typical example is shown in Figure 14.4(d).

The rectangular solid has a volume $f(x_{kl}, y_{kl})\Delta A$. The total volume of the solid region under the graph is approximated by a sum over both i and j:

$$V \cong \sum_{j=1}^{n} \sum_{i=1}^{m} f(x_{ij}, y_{ij})\Delta A. \tag{14.2}$$

We define the exact volume to be the limit of this sum as m and n increase without bound (presuming this limit exists). The limit is called a **double definite integral** or **double integral** for short. We denote it

$$\iint_R f(x,y)dA = \lim_{m,n\to\infty} \sum_{j=1}^{m}\sum_{i=1}^{n} f(x_{ij},y_{ij})\Delta A. \qquad (14.3)$$

As with the definition of a single definite integral, the limits involved here are quite subtle because the number of subregions is increasing without bound, while the area of the subregions is decreasing to zero. A more precise definition can be developed using lower and upper sums. With a precise definition, we can prove results giving conditions which guarantee the existence of a double integral. In particular, we can prove that if the integrand f is continuous on the rectangle R, then the limits in the definition exist and the double integral is well defined. We do not include these proofs in this text.

The basic properties of double integrals are analogous to those of single integrals. If a region R is composed of two nonoverlapping regions R_1 and R_2, then

$$\iint_R f(x,y)\,dA = \iint_{R_1} f(x,y)\,dA + \iint_{R_2} f(x,y)\,dA.$$

If the integrand f is a sum of two functions $f = g + h$, then

$$\iint_R f(x,y)\,dA = \iint_R g(x,y)\,dA + \iint_R h(x,y)\,dA.$$

Another feature of single definite integrals is the geometric interpretation that area above the x-axis is counted as positive, while area below the x-axis is counted as negative. What is the equivalent geometric interpretation for the volume associated with the double integral? If we think of approximating the volume associated with the double integral using a double Riemann sum [see Equation (14.3)], we can see that any term in the sum with $f(x_{ij}, y_{ij}) < 0$ is negative, and any term with $f(x_{ij}, y_{ij}) > 0$ is positive. Graphically, $z = f(x, y)$, so we conclude that any part of the graph where $f(x, y) > 0$ makes a positive contribution to the volume, and we associate a positive volume with solid regions above the xy-plane. Similarly, any part of the graph where $f(x, y) < 0$ makes a negative contribution to the volume, and we associate a negative volume with solid regions below the xy-plane. If $f(x, y)$ is positive over part of the rectangle $R = [a, b] \times [c, d]$ and negative for another part, the double integral is associated with the *net* volume above the xy-plane.

We can approximate double definite integrals using numerical techniques. We gave an example of this in the previous section. That example provides a clue on how we can evaluate a double integral by doing an iterated pair of single definite integrals. We develop this approach in the next section.

14.1.4 Iterated Integrals

In Example 14.1, we computed lower and upper sums that bounded the value of the double definite integral $\iint_R (1 + xy)\,dA$. To compute each sum, we added terms of the form $f(x_{ij}, y_{ij})\Delta A$ with i ranging from 1 to m and j from 1 to n. We can organize the calculation in many ways. One way is to first sum down the columns of Table 14.1 and then sum the column totals. If in summing down the columns we include a factor of $\Delta x = 0.25$ in each term, and in summing the column totals we include a factor of $\Delta y = 0.25$ in each term, then we will properly account for the factor of $\Delta A = \Delta x \Delta y$. The relevant sums for Example 14.1 are shown in Table 14.3.

Symbolically, we represent this method of summing by

$$\sum_{i=1}^{m}\sum_{j=1}^{n} f(x_{ij},y_{ij})\Delta A = \sum_{j=1}^{n}\left(\sum_{i=1}^{m} f(x_{ij},y_{ij})\Delta x\right)\Delta y$$

Table 14.3. Column sums for Example 14.1

y	(Column sum)$\times\Delta x$
0.00	1.00000
0.25	1.09375
0.50	1.18750
0.75	1.28125
1.00	1.37500
1.25	1.46875
1.50	1.56250
1.75	1.65625
(Total)$\times\Delta y$	2.65625

where the inner sum is done first with y_{ij} held constant. The inner sum should look familiar; it is a Riemann sum for a single definite integral with x as the variable of integration. In doing the sum, the y input is held constant. Likewise, the outer sum is a Riemann sum for a single definite integral with y as the variable of integration. It is plausible to conjecture that, by passing to the appropriate limits,

$$\iint_R f(x,y)\,dA = \int_c^d \left(\int_a^b f(x,y)\,dx \right) dy.$$

This statement is true; we do not include a rigorous proof in this text. We can use this to evaluate a double integral *exactly*, provided we can evaluate the single definite integrals on the right side using the second fundamental theorem. The pair of single definite integrals are called **iterated integrals**.

Example 14.2. Use iterated integrals to evaluate

$$\iint_R (1 + xy)\,dA.$$

where R is the rectangle $[0,1] \times [0,2]$.

Solution. We express the double integral as

$$\int_0^2 \left(\int_0^1 (1 + xy)\,dx \right) dy.$$

Now we evaluate the definite integral inside the parentheses (using the second fundamental theorem) holding y constant:

$$\int_0^2 \left(\int_0^1 (1 + xy)\,dx \right) dy = \int_0^2 \left(x + \frac{1}{2}x^2 y \right) \Big|_{x=0}^{1} dy = \int_0^2 \left(1 + \frac{1}{2}y \right) dy.$$

The remaining definite integral with respect to y can be evaluated easily:

$$\int_0^2 \left(\int_0^1 xy\,dx \right) dy = \left(y + \frac{1}{4}y^2 \right) \Big|_0^2 = 3.$$

Note that this exact result is consistent with the estimates we computed in Example 14.1 $\triangle$

Let's return to the task of computing the sum in Example 14.1. Rather than first summing down columns, we could first sum across rows and then sum the row totals. Table 14.4 gives the results with appropriate factors of Δy and Δx included. It should be no surprise that the final result is the same as before. Symbolically, we have

$$\sum_{i=1}^{m}\sum_{j}^{n} f(x_{ij}, y_{ij})\Delta A = \sum_{i=1}^{m}\left(\sum_{j=1}^{n} f(x_{ij}, y_{ij})\Delta y\right)\Delta x.$$

Passing to the limit gives

$$\iint_R f(x,y)\,dA = \int_a^b \left(\int_c^d f(x,y)\,dy\right)dx.$$

In words, we can evaluate a double integral by doing an iterated integral with the y integration first followed by the x integration.

Table 14.4. Row sums for Example 14.1

x	(Row sum)$\times\Delta y$
0.00	2.0000
0.25	2.4375
0.50	2.8750
0.75	3.3125
(Total)$\times\Delta x$	2.65625

Example 14.3. Use iterated integrals with the y integration first to evaluate

$$\iint_R (1+xy)\,dA.$$

where R is the rectangle $[0,1]\times[0,2]$.

Solution. We express the double integral as

$$\int_0^1 \left(\int_0^2 (1+xy)\,dy\right)dx.$$

Now we evaluate the definite integral inside the parentheses (using the second fundamental theorem) holding x constant:

$$\int_0^1 \left(\int_0^2 (1+xy)\,dy\right)dx = \int_0^1 \left(y+\frac{1}{2}xy^2\right)\Big|_{y=0}^{2} dx = \int_0^1 (2+2x)\,dx.$$

The remaining definite integral with respect to x can be evaluated easily:

$$\iint_R (1+xy)\,dA = (2x+x^2)\big|_0^1 = 3.$$

Note that this is the same result as the iterated integral with the x integration first. △

We can also view iterated integrals geometrically. Consider the function $f(x, y)$ with graph as shown in Figure 14.4 and the solid region whose volume is given by the double integral $\iint_R f(x, y)\, dA$. Look at a slice of that region for which y is a constant (called y_0), as shown in Figure 14.5(a). That slice is a planar region with area given by the single definite integral

$$\int_a^b f(x, y_0)\, dx.$$

The integrand in this expression is a function of just a single variable (x), because we have fixed y_0 to be a constant. We could compute such an area for any y_0 in $[c, d]$ we choose. We could just as well find the area of any such slice as a *function* of y by leaving y in the integrand (instead of y_0) but treating y as a constant. Evaluating the definite integral then produces a function $g(y)$ given by

$$g(y) = \int_a^b f(x, y)\, dx. \tag{14.4}$$

Now consider the "bread slice" formed by expanding the planar slice we have been considering to a solid region of width Δy as shown in Figure 14.5. As usual let m be the number of slices of equal width Δy on the interval $[c, d]$. The volume of a bread slice at $y = y_i$ is $g(y_i)\Delta y$. In the usual manner, the total volume of the region can be approximated by a Riemann sum

$$V \cong \sum_{j=1}^{m} g(y_i)\Delta y.$$

In the limit as m approaches infinity (or Δy approaches zero), the exact volume is the definite integral

$$V = \int_c^d g(y)\, dy. \tag{14.5}$$

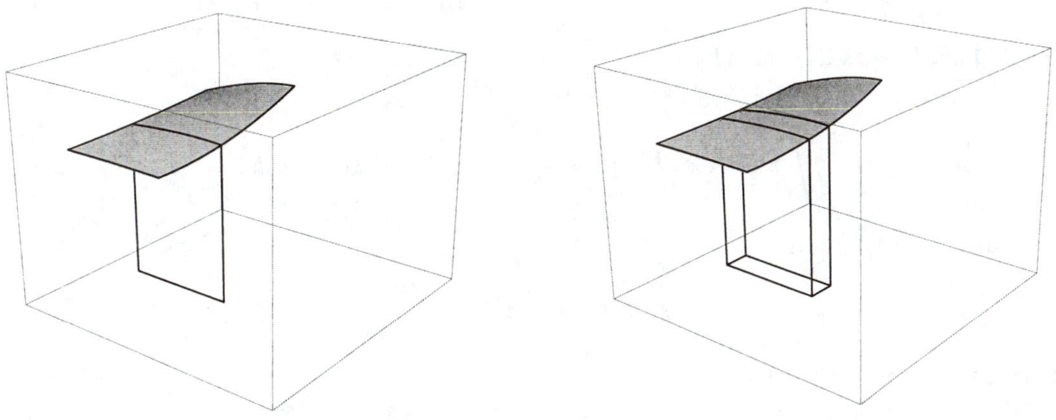

(a) Constant y slice defines a planar region

(b) "Bread slice" formed by expanding the planar slice into a solid region

Figure 14.5. Viewing iterated integrals by "slicing" solid region.

Combining Equations (14.4) and (14.5) we see that

$$V = \int_c^d \left(\int_a^b f(x,y)\,dx \right) dy \tag{14.6}$$

which is the form of the iterated double integral given previously. This view emphasizes the interpretation of a double integral as a volume.

Example 14.4. Evaluate the double integral

$$\iint_R x^2 y \, dA$$

where $R = [-2,2] \times [-2,1]$. Interpret the result as a net volume above or below the xy-plane.

Solution. We evaluate the double integral iteratively as usual. We no longer add parentheses when evaluating the first integral, since the order of iteration should be well understood now. We have

$$\iint_R x^2 y\,dA = \int_{-2}^{1}\int_{-2}^{2} x^2 y\,dx\,dy = \int_{-2}^{1} \left(\frac{1}{3}x^3 y\right)\Bigg|_{-2}^{2} dy = \frac{16}{3}\int_{-2}^{1} y\,dy = \frac{16}{3}\left(\frac{1}{2}y^2\right)\Bigg|_{-2}^{1} = -8.$$

The result is -8, which geometrically corresponds to a net 8 units of volume *below* the xy-plane. The negative result makes sense when we consider the nature of the integrand $f(x,y) = x^2 y$ over the region of integration $[-2,2] \times [-2,1]$. The factor x^2 is always greater than or equal to zero, but the factor y can be negative or positive. Thus, $x^2 y \geq 0$ when $y > 0$ and $x^2 y \leq 0$ when $y < 0$. Since a greater portion of the rectangular region of integration has $y < 0$, the integrand is less than zero over more of the region, and it makes sense that the definite integral is negative. △

In the previous example, the integrand $f(x,y) = x^2 y$ can be factored into a piece that depends only on x and a piece that depends only on y. That is, we can write $f(x,y) = F(x)G(y)$. In the example, $F(x) = x^2$ and $G(y) = y$. When the integrand can be separated in this way, we can "factor" the iterated integral into two factors, each of which is a single definite integral. In general, this has the form

$$\int_c^d \int_a^b f(x,y)\,dx\,dy = \int_c^d \int_a^b F(x)G(y)\,dx\,dy = \int_a^b F(x)\,dx \int_c^d G(y)\,dy. \tag{14.7}$$

Example 14.5. Rework Example 14.4 using the result in Equation (14.7).

Solution. We have

$$\iint_R x^2 y\,dA = \int_{-2}^{1}\int_{-2}^{2} x^2 y\,dx\,dy = \int_{-2}^{2} x^2\,dx \int_{-2}^{1} y\,dy.$$

The value of the first factor is

$$\int_{-2}^{2} x^2\,dx = \frac{1}{3}x^3\Bigg|_{-2}^{2} = \frac{16}{3}$$

and the value of the second factor is

$$\int_{-2}^{1} y\,dy = \frac{1}{2}y^2\Bigg|_{-2}^{1} = -\frac{3}{2}.$$

Thus the value of the double integral is $(16/3)(-3/2) = -8$, in agreement with the result of the previous example. △

14.2 Double Integrals Over Other Regions

In Section 14.1, we considered double integrals over rectangular regions. Here we apply the method of iterated integrals to nonrectangular regions. First, we study problems in which the double integrals can be evaluated using rectangular coordinates. Then we consider problems in which there is some advantage to evaluating the double integrals using polar coordinates.

Before we delve into the details of iterated integrals, let's comment on generalizing the definition of double integral to cover cases with nonrectangular regions. Looking at Figure 14.2, recall the issue of dividing a nonrectangular region into pieces. Some of the pieces will not be rectangles. We could define our generalization by dealing with the pieces that are not rectangles. An alternative approach, the conventional one, is outlined in the next paragraph.

Consider a function $f : \mathbb{R}^2 \to \mathbb{R}$ with $f : (x, y) \mapsto f(x, y)$ and a planar nonrectangular region R in the xy-plane. Since our definition of double integral is for rectangular regions, we introduce a rectangle $\tilde{R}$, which encloses the region R. We also introduce a new function $\tilde{f} : \mathbb{R}^2 \to \mathbb{R}$, which is defined in terms of f by

$$\tilde{f}(x, y) = \begin{cases} f(x, y) & \text{if } (x, y) \text{ is in } R \text{ but not in } \tilde{R} \\ 0 & \text{if } (x, y) \text{ is not in } R. \end{cases}$$

We now define a double integral over the nonrectangular region R by

$$\iint\limits_{R} f(x, y)\, dA = \iint\limits_{\tilde{R}} \tilde{f}(x, y)\, dA.$$

The previously undefined object on the left is defined by the known object on the right.

You may sense that we have not really done much in the definition of the preceding paragraph, and your sense is correct. We have failed to *prove* anything about the existence of a double integral over a nonrectangular region defined by this sleight of hand. In Section 14.1 we mentioned that we can show a double integral over a rectangular region exists if the integrand is continuous on the rectangle. However, note that the function $\tilde{f}$ will generally be discontinuous on the boundary of R since the outputs will be zero on one side and given by f on the other side. Nonetheless, we can prove that the double integral of a $\tilde{f}$ does exist if f is continuous and the boundary of R is not too "wild." Such a proof is technically detailed and not appropriate for this introductory text.

In the remainder of this section we develop techniques for evaluating double integrals over nonrectangular regions. We will not concern ourselves with proving that integrals exist using the above definition.

14.2.1 Cartesian Coordinates

Suppose we wish to evaluate the double integral of a function $f(x, y)$ over the region shown in Figure 14.6(a). Even though x varies from 0 to 2 and y from 0 to 1, this is *not* the rectangular region $R = [0, 2] \times [0, 1]$. Rather, it is a triangular region bounded by the lines $x = 0, y = 0$, and a third line that completes the triangle. That line has a slope of 1/2 and y-intercept of 0, and so we know the equation of the line is $y = x/2$. In this case, we write the double integral as

$$\int_0^2 \int_0^{x/2} f(x, y)\, dy\, dx.$$

Since we perform the integration with respect to y first (holding x constant), the limits of integration need to describe the fact that the region is bounded by $y = 0$ on the bottom and $y = x/2$ on top. This limits us to

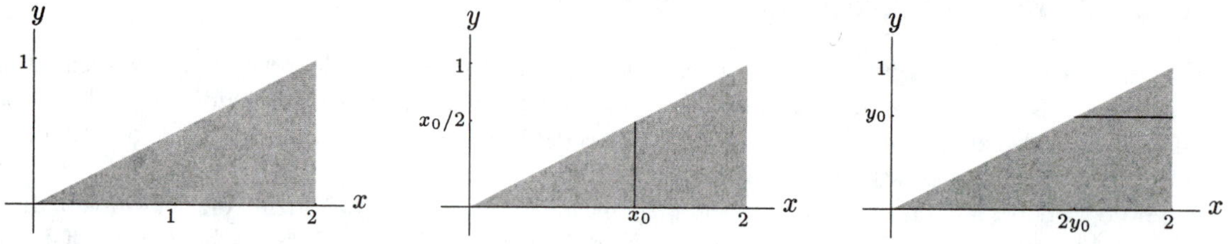

(a) Triangular region in the xy-plane (b) Constant x slice (c) Constant y slice

Figure 14.6. Region in the xy-plane.

the region between the lines $y = 0$ and $y = x/2$, and the limits on the second integral (0 to 2) then define the entire triangular region shown in Figure 14.6(a).

Why is this procedure (and the double integral written above) justified? We are really just employing the same "bread slice" method as described in Section 14.1, except that now the slices are not all the same width. In Figure 14.6(b), we show the cross section of an arbitrary slice parallel to the y-axis at $x = x_0$. When we do the iterated integral with respect to y, we are moving along this slice from the bottom ($y = 0$) to the top (at $y = x/2$). This gives the area of a region bounded by those two lines, the xy-plane, and the function $f(x_0, y)$. Then the second iterated integral (with respect to x) adds the volumes formed by taking each of those areas multiplied by a thickness dx. Thus, the double integral has a geometric representation as a *volume*, as before.

Example 14.6. Evaluate the definite integral of $f(x, y) = 2xy$ over the region shown in Figure 14.6(a).

Solution. Using the double integral constructed in the text,

$$\iint_R f(x,y)\,dA = \int_0^2 \int_0^{x/2} xy\,dy\,dx$$

where R is the triangular region shown. Evaluating iteratively,

$$\iint_R f(x,y)\,dA = \int_0^2 \int_0^{x/2} xy\,dy\,dx = \int_0^2 \left(\frac{1}{2}xy^2\right)\Big|_0^{x/2} dx = \frac{1}{8}\int_0^2 x^3\,dx = \frac{1}{2}.$$

△

Now, let's consider how to rework the previous example by interchanging the order of integration. If we want to integrate first with respect to x, notice that the region is bounded on the left by the diagonal line with equation $y = x/2$ (or $x = 2y$) and on the right by $x = 2$. In Figure 14.6(c), we have drawn an arbitrary slice of the region parallel to the x-axis to better illustrate this point. Thus, the double integral is

$$\int_0^1 \int_0^{2y} xy\,dx\,dy$$

where we have placed the limits 0 to 1 on y to complete the double integral. Evaluating iteratively,

$$\iint_R f(x,y)\,dA = \int_0^1 \int_{2y}^2 xy\,dx\,dy = \int_0^1 \left(\frac{1}{2}x^2y\right)\Big|_{2y}^2 dy = 2\int_0^1 (y - y^3)\,dy = \frac{1}{2}.$$

This produces the same result as before. Interchanging the order of integration has not changed the results. However, for this nonrectangular region we had to determine the limits of integration more carefully.

With the "slicing" method just demonstrated, we can compute double integrals of functions over other regions in the plane, as we show in the following examples.

Example 14.7. Find the double integral of $f(x, y) = 3x + 2y$ over the region that lies between the curves $y = x$ and $y = x^2$.

Solution. We choose to do the integration with respect to y first. Then from the boundaries of the region [Figure 14.7(a)], we see that the lower limit in this integration is the curve $y = x^2$ and the upper limit is $y = x$. From the intersection points, it follows that the limits on the second integration (in x) are 0 to 1. The first integral covers the vertical slice shown in Figure 14.7(a), and the second integral sums over all the slices from $x = 0$ to $x = 1$. Therefore, our double integral is equal to the iterated integral

$$\int_0^1 \int_{x^2}^{x} (3x + 2y)\, dy\, dx.$$

In performing the integral with respect to y, we treat x as a constant:

$$\begin{aligned}\int_0^1 \int_{x^2}^{x} (3x + 2y)\, dy\, dx &= \int_0^1 \left.(3xy + y^2)\right|_{x^2}^{x} dx = \int_0^1 \left[(3x(x) + x^2) - (3x(x^2) + (x^2)^2)\right] dx \\ &= \int_0^1 (4x^2 - 3x^3 - x^4)\, dx.\end{aligned}$$

Now for the second integration

$$\int_0^1 \int_{x^2}^{x} (3x + 2y)\, dy\, dx = \left.\left(\frac{4}{3}x^3 - \frac{3}{4}x^4 - \frac{1}{5}x^5\right)\right|_0^1 = \frac{23}{60}.$$

△

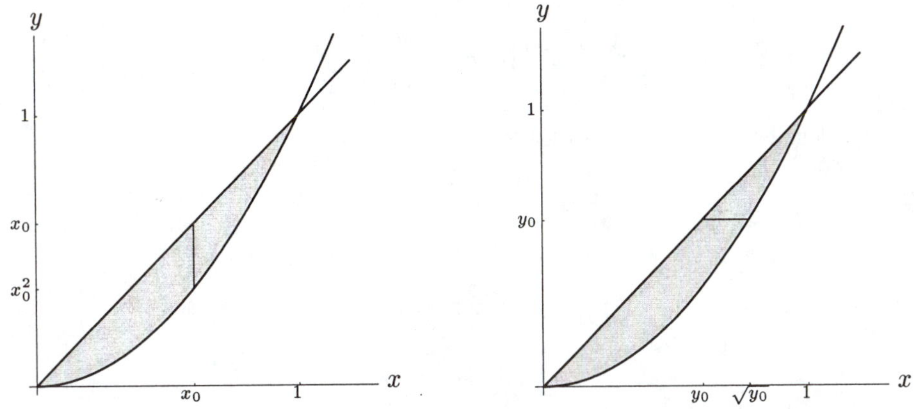

(a) Typical constant x slice

(b) Typical constant y slice

Figure 14.7. The slices used in Examples 14.7 and 14.8.

Example 14.8. Repeat the previous example, changing the order of integration.

Solution. The first integral covers the horizontal slice shown in Figure 14.7(b). This lower limit in x along an arbitrary slice is the left curve $x = y$, and the upper limit is the right curve $x = \sqrt{y}$. The limits on y are then 0 to 1, so the iterated integral is

$$\int_0^1 \int_y^{\sqrt{y}} (3x + 2y)\, dx\, dy.$$

In the integration with respect to x, we treat y as a constant:

$$\begin{aligned}\int_0^1 \int_y^{\sqrt{y}} (3x + 2y)\, dx\, dy &= \int_0^1 \left(\frac{3}{2}x^2 + 2xy\right)\bigg|_y^{\sqrt{y}} dy \\ &= \int_0^1 \left[\left(\frac{3}{2}(\sqrt{y})^2 + 2\sqrt{y}y\right) - \left(\frac{3}{2}y^2 + 2(y)y\right)\right] dy \\ &= \int_0^1 \left(-\frac{7}{2}y^2 + 2y^{3/2} + \frac{3}{2}y\right) dy.\end{aligned}$$

Now for the second integration,

$$\int_0^1 \int_y^{\sqrt{y}} (3x + 2y)\, dx\, dy = \left(-\frac{7}{6}y^3 + \frac{4}{5}y^{5/2} + \frac{3}{4}y^2\right)\bigg|_0^1 = \frac{23}{60}.$$

The result is unaffected by the order of integration. △

Example 14.9. Find the double integral of $f(x, y) = x^2 + 2y$ over the triangular region bounded by $y = |x|$ and $y = 3$.

Solution. This region is shown in Figure 14.8(a). We choose to integrate with respect to x first. The constant limits on y are clearly 0 to 3. This makes the lower and upper limits for the x integration $x = -y$ and $x = y$, respectively, as seen in Figure 14.8(b). The double integral is thus equal to the iterated integral

$$\int_0^3 \int_{-y}^{y} (x^2 + 2y)\, dx\, dy = \int_0^3 \left(\frac{1}{3}x^3 + 2xy\right)\bigg|_{-y}^{y} dy = \int_0^3 \left(\frac{2}{3}y^3 + 4y^2\right) dy = \left(\frac{1}{6}y^4 + \frac{4}{3}y^3\right)\bigg|_0^3 = \frac{99}{2}.$$

The value of the double integral is thus $99/2$. △

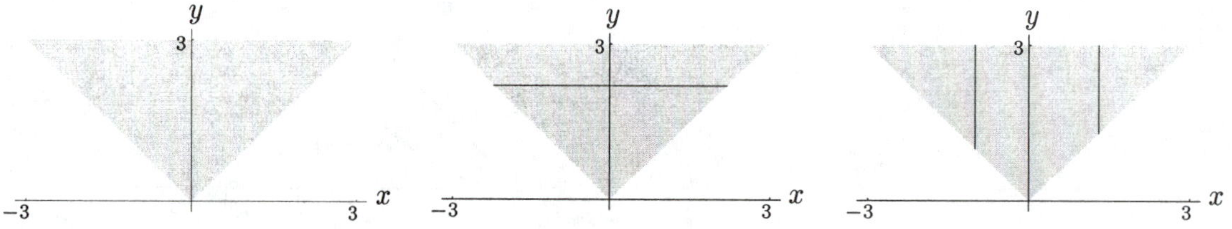

(a) Triangular region (b) Typical y-slice (c) Some typical x-slices

Figure 14.8. Region of Example 14.9.

In the preceding example, the order of integration was *not* chosen arbitrarily. Consider what would happen if we had tried to integrate over y first. The constant limits on x are -3 to 3. The lower limit for the y integration is $y = -x$ for the left half of the triangle (where $x < 0$) and $y = x$ for the right half of the triangle (where $x > 0$) as shown in Figure 14.8(c). The upper limit is $y = 3$ throughout. This necessitates breaking the y definite integral into two pieces. The process yields the same result, but with somewhat greater effort. It is worth putting some thought into which order of integration allows us to compute the value of a double integral most easily.

Example 14.10. Find the double integral of the function $f(x, y) = x^2y$ over the semicircle of radius 2 shown in Figure 14.9.

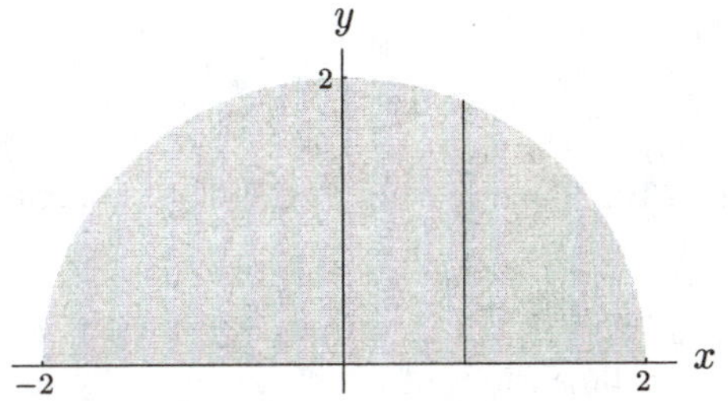

Figure 14.9. The region of integration for Example 14.10.

Solution. Because it is more difficult to define the limits of integration for a horizontal slice (i.e., integration with respect to x), we choose to integrate first with respect to y. The lower limit is simply $y = 0$, and the upper limit is the equation of the semicircle $y = \sqrt{4 - x^2}$. The limits on the second integral (with respect to x) are -2 and $+2$, so the double integral becomes

$$\int_{-2}^{2}\int_{0}^{\sqrt{4-x^2}} x^2y\,dydx = \int_{-2}^{2}\left(\frac{1}{2}x^2y^2\right)\bigg|_{0}^{\sqrt{4-x^2}} dx = \frac{1}{2}\int_{-2}^{2}\left(4x^2 - x^4\right)\,dx.$$

Evaluating the remaining definite integral gives

$$\int_{-2}^{2}\int_{0}^{\sqrt{4-x^2}} x^2y\,dy\,dx = \frac{1}{2}\left(\frac{4}{3}x^3 - \frac{1}{5}x^5\right)\bigg|_{-2}^{2} = \frac{64}{15}. \qquad \triangle$$

14.2.2 Polar Coordinates

In Example 14.10, the region of integration was not rectangular, but instead had a radial symmetry over the upper half-plane. The region is simply described using polar coordinates as $R = \{(r, \theta)|0 \le r \le 2, 0 \le \theta \le \pi\}$. This suggests that for this problem (and others having similar symmetry) polar coordinates may be a better choice for doing iterated integrals to evaluate double integrals.

To set up an iterated integral in polar coordinates, we need to be concerned with several things. First, using the standard transformation equations

$$x = r\,\cos\theta \qquad \text{and} \qquad y = r\,\sin\theta \tag{14.8}$$

it is possible to rewrite the integrand as $g(r, \theta) = f(x(r, \theta), y(r, \theta))$ in terms of the polar coordinates. But how do we handle the area element dA? We saw for rectangular coordinates that dA in a double integral

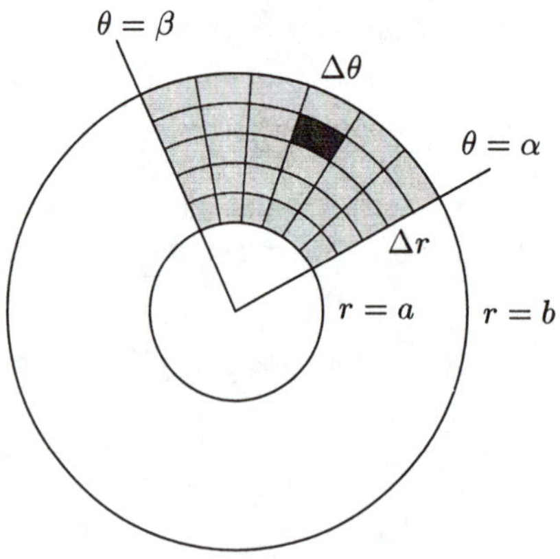

Figure 14.10. "Polar rectangle" split into subregions.

corresponded to $dx\,dy$ in an iterated integral, but the expression $dr\,d\theta$ is clearly incorrect on dimensional grounds alone: The quantity $drd\theta$ has dimensions of length but area has dimensions of length squared. To determine the correct correspondence, we return to first principles and look at the construction of an iterated integral as the limit of a double sum.

Consider a "polar rectangle" $R = \{(r,\theta) | a \le r \le b, \alpha \le \theta \le \beta\}$. To construct an approximating sum, we divide R into subregions by dividing the r interval into subintervals and the θ interval into subintervals. A generic subregion is shown in Figure 14.10. From the geometry of circles, we know that the length of the arc subtending that wedge is $r\Delta\theta$. Now look at the small shaded region at the end of the wedge, with length Δr in the radial direction. We approximate this region as a small rectangle with sides Δr and $r\Delta\theta$, so the area of the region is approximately $(\Delta r)(r\Delta\theta) = r\Delta r\Delta\theta$. The approximating sum that leads to the double integral thus has the form

$$\iint_R g(r,\theta)\,dA \approx \sum_{i=1}^{m}\sum_{j=1}^{n} g(r_{ij},\theta_{ij})r_{ij}\Delta r\Delta\theta.$$

This double sum can be rearranged to correspond to the iterated integral

$$\int_\alpha^\beta \int_a^b g(r,\theta)r\,dr\,d\theta.$$

The correct correspondence to dA is thus $r\,dr\,d\theta$. Note that this expression has the correct dimensions of length squared. We will refer to $dA = r\,dr\,d\theta$ as the **area element in polar coordinates**.

We illustrate the use of polar coordinates in iterated integrals in the following examples.

Example 14.11. Rework Example 14.10 using polar coordinates.

Solution. Using the transformations in Equation (14.8), the integrand becomes

$$g(r,\theta) = f(r\cos\theta, r\sin\theta) = (r\cos\theta)^2(r\sin\theta) = r^3\cos^2\theta\sin\theta.$$

From the figure it is clear that the limits on r are 0 to 2, and the limits on θ are 0 to π. (The fact that it is easier to establish the limits of integration for this region in polar coordinates shows the advantage of choosing polar coordinates for this problem.) Thus, the double integral becomes

$$\int_0^\pi \int_0^2 \left(r^3 \cos^2\theta \sin\theta\right) r\, dr\, d\theta = \int_0^\pi \int_0^2 r^4 \cos^2\theta \sin\theta\, dr\, d\theta.$$

Because the limits on both integrals are constant, the order of evaluation is irrelevant. We will evaluate in the order in which we have written the double integral above, that is, r first, giving

$$\int_0^\pi \int_0^2 r^4 \cos^2\theta \sin\theta\, dr d\theta = \int_0^\pi \frac{1}{5} r^5 \bigg|_0^2 \cos^2\theta \sin\theta\, d\theta = \frac{32}{5} \int_0^\pi \cos^2\theta \sin\theta\, d\theta.$$

Using a substitution, we let $u = \cos\theta$, so $du/d\theta = -\sin\theta$, or $\sin\theta\, d\theta = -du$, so

$$\int_0^\pi \int_0^2 r^4 \cos^2\theta \sin\theta\, dr\, d\theta = -\frac{32}{5} \int u^2 du = -\frac{32}{5}\frac{1}{3} \cos^3\theta \bigg|_0^\pi = \frac{64}{15}.$$

This is the same result we obtained previously. △

Example 14.12. Find the double integral of $f(r,\theta) = \sqrt{r} \sin\theta$ over the quarter-circle of radius 5 that lies in the second quadrant of the xy-plane.

Solution. The region of integration is shown in Figure 14.11. From the figure the limits on r are 0 to 5, and the limits on θ are $\pi/2$ to π. Thus, the double integral is

$$\int_{\pi/2}^\pi \int_0^5 (\sqrt{r} \sin\theta) r\, dr\, d\theta = \int_{\pi/2}^\pi \int_0^5 r^{3/2} \sin\theta\, dr\, d\theta.$$

Doing the r integration first,

$$\int_{\pi/2}^\pi \int_0^5 r^{3/2} \sin\theta\, dr\, d\theta = \int_{\pi/2}^\pi \frac{2}{5} r^{5/2} \bigg|_0^5 \sin\theta\, d\theta = \frac{2}{5} 5^{5/2} \int_{\pi/2}^\pi \sin\theta\, d\theta = \frac{2}{5} 5^{5/2} \left(-\cos\theta\right)\bigg|_{\pi/2}^\pi = \frac{2}{5} 5^{5/2}.$$

△

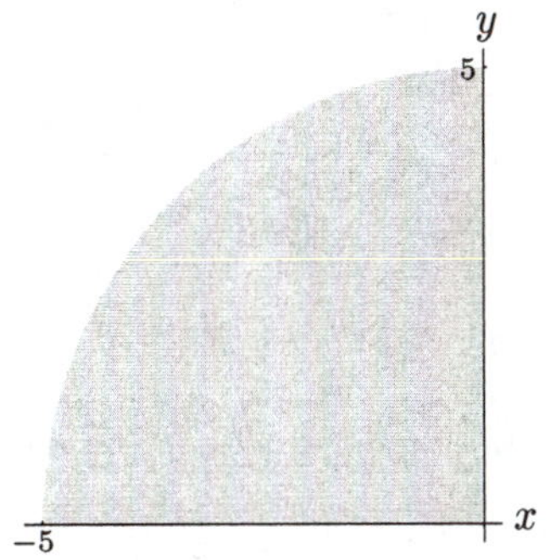

Figure 14.11. Region of integration for Example 14.12.

When the integrand can be factored in polar coordinates, the result is similiar to that of Equation (14.7). That is, if we can write $f(r,\theta) = F(r)G(\theta)$, the limits on both integrals are constants, and the double integral can be rewritten as a product of two single definite integrals:

$$\int_{\theta_a}^{\theta_b} \int_{r_a}^{r_b} f(r,\theta)\, dr\, d\theta = \int_{\theta_a}^{\theta_b} G(\theta)\, d\theta \int_{r_a}^{r_b} F(r)\, dr. \tag{14.9}$$

For example, in Example 14.12 we can apply Equation (14.9) to give

$$\int_{\pi/2}^{\pi}\int_0^5 r^{3/2}\sin\theta\,dr\,d\theta = \int_{\pi/2}^{\pi}\sin\theta\,d\theta\int_0^5 r^{3/2}\,dr = (1)\left(\frac{2}{5}5^{5/2}\right) = \frac{2}{5}5^{5/2}.$$

Be careful to apply Equation (14.9) only when the integrand is separable in this way and the limits of integration are constant.

As the following example shows, the ability to transform from rectangular to polar coordinates helps us solve problems that otherwise could not be solved in closed form.

Example 14.13. Consider the definite integral

$$I = \int_0^\infty e^{-x^2}\,dx.$$

Evaluate this definite integral by transforming it to a double integral in polar coordinates.

Solution. There is nothing special about the variable x, so we write

$$I = \int_0^\infty e^{-y^2}\,dy.$$

From Equation (14.9) we write

$$I^2 = I\cdot I = \int_0^\infty e^{-x^2}\,dx\int_0^\infty e^{-y^2}\,dy = \int_0^\infty\int_0^\infty e^{-x^2}e^{-y^2}\,dx\,dy = \int_0^\infty\int_0^\infty e^{-(x^2+y^2)}\,dx\,dy.$$

Now making the change to polar coordinates, we have $r^2 = x^2 + y^2$ and $dx\,dy = dA = r\,dr\,d\theta$. What about the limits of integration? With both x and y restricted to positive values, the region of integration includes only the first quadrant, so while r varies from 0 to ∞, the angle θ varies only from 0 to $\pi/2$. The double integral in polar coordinates is

$$I^2 = \int_0^\infty\int_0^{\pi/2} e^{-r^2}\,r\,dr\,d\theta = \int_0^\infty e^{-r^2}\,r\,dr\int_0^{\pi/2}d\theta.$$

The integral in θ is evaluated as

$$\int_0^{\pi/2}d\theta = \theta\Big|_0^{\pi/2} = \frac{\pi}{2}.$$

Thus,

$$I^2 = \frac{\pi}{2}\int_0^\infty e^{-r^2}\,r\,dr.$$

For the integration over r we can use substitution to evaluate the antiderivative (compare this integrand with the original integrand in I). Let $u = r^2$, so $du/dr = 2r$, and

$$I^2 = \frac{\pi}{2}\int_0^\infty e^{-u}\,r\,\frac{1}{2r}du = \frac{\pi}{4}\int_0^\infty e^{-u}\,du = \frac{\pi}{4}\left(-e^{-u}\right)\Big|_0^\infty = \frac{\pi}{4}[-(0-1)] = \frac{\pi}{4}.$$

Finally, we take the square root of I^2 to find

$$I = \sqrt{\frac{\pi}{4}} = \frac{\sqrt{\pi}}{2}.$$

We have selected only the positive root, because the integrand is positive over the entire region, so we know the result must be positive. △

14.3 Triple Integrals

The concepts and techniques we have developed for functions of two variables in the previous two sections are easily extended to functions of three variables. Double integrals generalize to triple integrals in which the region of integration is some solid region of space and the integrand is a function of three variables. The solid region can be either a rectangular solid or some nonrectangular region. Triple integrals can be equated to iterated integrals involving three nested single definite integrals, one for each of the input variables. Some solid regions are most easily described with coordinate systems other than Cartesian coordinates. Two such systems, which we introduce later in this section, are cylindrical coordinates and spherical coordinates.

14.3.1 Triple Integrals and Iterated Integrals

We begin by returning to the relation between mass density and total mass. For a three-dimensional object, the relevant mass density ρ is mass per unit volume (in SI units of kg/m^3). Let D be the solid region occupied by the object. We can think of a function with a location (x, y, z) in the object as input and mass density $\rho(x, y, z)$ at that location as output. To compute the total mass in terms of the mass density, we start by conceptually dividing D into pieces of volume ΔV. Let ijk be a triple index for the pieces. The mass m_{ijk} of a generic piece is approximately $\rho(x_{ijk}, y_{ijk}, z_{ijk})\Delta V$ where $(x_{ijk}, y_{ijk}, z_{ijk})$ is some location in that piece. The total mass is thus approximated by the triple sum

$$\sum_i \sum_j \sum_k m_{ijk} = \sum_i \sum_j \sum_k \rho(x_{ijk}, y_{ijk}, z_{ijk})\Delta V.$$

The total mass is equal to the limit of this sum as the number of pieces increases without bound (and ΔV approaches 0). That limit is a **triple integral**; we denote this

$$M = \iiint_D \rho(x, y, z)\, dV.$$

We can extract a general definition of triple integral from the applied example of the previous paragraph. Consider a function $f : \mathbb{R}^3 \to \mathbb{R}$ with $f : (x, y, z) \mapsto f(x, y, z)$. Let D be a rectangular solid $[a, b] \times [c, d] \times [u, v]$. Divide the x interval $[a, b]$ into m pieces of length $\Delta x = (b - a)/m$, the y interval $[c, d]$ into n pieces of length $\Delta y = (d - c)/n$,and the z interval into p pieces of length $\Delta z = (v - u)/p$. A generic subregion is shown in Figure 14.12. This induces a division of D into mnp subregions of volume $\Delta V = \Delta x \Delta y \Delta z$. In the generic ijk subregion, pick a point $(x_{ijk}, y_{ijk}, z_{ijk})$ and form the product $f(x_{ijk}, y_{ijk}, z_{ijk})\Delta V$.

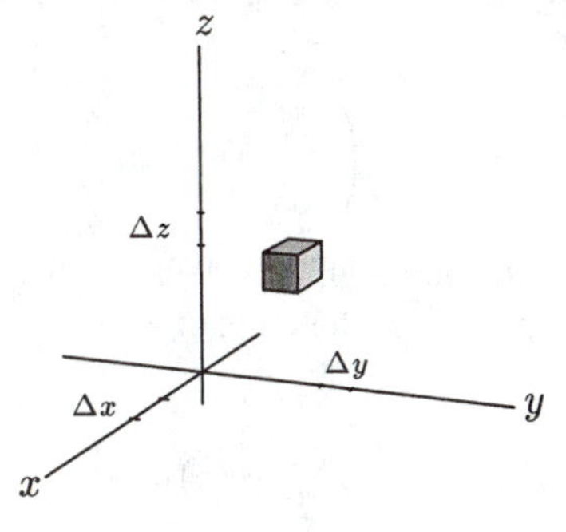

Figure 14.12. Typical subregion of a rectangular solid.

To account for all mnp subregions, we form a triple sum

$$\sum_{i}^{m}\sum_{j}^{n}\sum_{k}^{p} f(x_{ijk}, y_{ijk}, z_{ijk})\Delta V.$$

The limit of this sum as m, n, and p increase without bound is the triple integral of f over the solid region D, which we denote

$$\iiint_D f(x, y, z)\, dV.$$

This is contingent upon the existence of the limit. The basic definition of triple integral for a rectangular solid region of integration can be generalized to nonrectangular solid regions. The trick is an obvious extension of the idea we used for double integrals; we do not discuss the details further.

There is a major difference between single and double integrals on the one hand and triple (and higher) integrals on the other hand. In the case of single and double integrals, we were motivated by geometric concepts of area and volume, respectively. The typical term in the sum used to define a triple integral is the product of a function output and a volume. There is no obvious geometric interpretation for this. In the context of specific applications, the product may have an interpretation. For example, if the function output is a mass density, then the typical term represents a mass. Nonetheless, for a generic function the definition of triple integral is a natural generalization of single and double integrals.

We can extend the method of iterated integrals to triple integrals. Consider a function $f : \mathbb{R}^3 \to \mathbb{R}$ continuous over a rectangular volume $D = [a, b] \times [c, d] \times [u, v]$. Analogous to our representation of double integrals, we equate a triple integral of f over D to an iterated integral

$$\int_u^v \int_c^d \int_a^b f(x, y, z)\, dx\, dy\, dz.$$

As before, it is understood that the limits a and b (on the innermost integral sign) are the limits in x (the innermost bookend dx), the limits c and d are the limits on y, and the limits u and v are the limits on z.

The iterative process works just the same way. Write the iterated integral above as

$$\int_u^v \int_c^d \left(\int_a^b f(x, y, z)\, dx\right) dy\, dz.$$

The parentheses indicate that we are to do the (single) definite integral with respect to x and in doing so treat y and z as constants. That definite integral results in a function of y and z, which we call $g(y, z)$. Then the triple integral reduces to a double integral

$$\int_u^v \left(\int_c^d g(y, z)\, dy\right) dz.$$

As we know from our study of double integrals, the definite integral in the parentheses is evaluated in the usual way holding z constant. This produces a function of z, which we call $p(z)$, and leaves the single definite integral

$$\int_u^v p(z)\, dz$$

which can be evaluated using standard techniques. The utility of iterated integrals depends on our ability to evaluate single definite integrals using the second fundamental theorem, which in turn depends on our ability to find antiderivatives.

Example 14.14. Use a triple iterated integral to evaluate the triple definite integral

$$\iiint_D (xy + z)\, dV$$

where $D = [1, 3] \times [-1, 2] \times [0, 3]$.

Solution. The triple integral is equal to an iterated integral given as

$$\iiint_D (xy + z)\, dV = \int_0^3 \int_{-1}^2 \int_1^3 (xy + z)\, dx\, dy\, dz.$$

Using the iterative technique just described, we first evaluate the integral with respect to x, treating y and z as constants:

$$\begin{aligned}
\int_0^3 \int_{-1}^2 \int_1^3 (xy + z)\, dx\, dy\, dz &= \int_0^3 \int_{-1}^2 \left(\frac{1}{2}x^2 y + xz\right)\Bigg|_{x=1}^3 dy\, dz \\
&= \int_0^3 \int_{-1}^2 \left[\frac{1}{2}y(3^2 - 1^2) + z(3-1)\right] dy\, dz \\
&= \int_0^3 \int_{-1}^2 (4y + 2z)\, dy\, dz.
\end{aligned}$$

Next we do the integral with respect to y while holding z constant:

$$\begin{aligned}
\int_0^3 \int_{-1}^2 \int_1^3 (xy + z)\, dx\, dy\, dz &= \int_0^3 \left(4\frac{1}{2}y^2 + 2yz\right)\Bigg|_{y=-1}^2 dz \\
&= \int_0^3 \left[2(2^2 - (-1)^2) + 2z(2 - (-1))\right] dz \\
&= \int_0^3 (6 + 6z)\, dz.
\end{aligned}$$

Finally, we evaluate the remaining definite integral:

$$\int_0^3 \int_{-1}^2 \int_1^3 (xy + z)\, dx\, dy\, dz = \left(6z + 3z^2\right)\Big|_0^3 = 18 + 27 = 45.$$

Thus,

$$\iiint_D (xy + z)\, dV = 45.$$

△

As with double iterated integrals, we can have triple iterated integrals with bounds that depend on the integration variables of the outer integrals. This is illustrated in the next example.

Example 14.15. Evaluate the triple integral

$$\iiint_D (xyz)\, dV$$

where D is the solid region bounded below by the xy-plane, above by the plane $z = 1 - x - y$ and on the sides by the xz- and yz-planes.

Solution. We must first determine bounds on x, y, and z to describe the solid region D. The plane $z = 1 - x - y$ intersects the xy-plane along the line $0 = 1 - x - y$ or $y = 1 - x$. The "base" of the solid is thus a triangular region in the xy-plane described by $0 \le x \le 1$ and $0 \le y \le 1 - x$. For any point in this triangle, the bounds in the vertical direction are given by $0 \le z \le 1 - x - y$. Thus, we have

$$\iiint_D (xyz)\, dV = \int_0^1 \int_0^{1-x} \int_0^{1-x-y} (xyz)\, dz\, dy\, dx.$$

To evaluate the triple iterated integral, we first evaluate the integral with respect to z, treating x and y as constants:

$$\begin{aligned}\int_0^1 \int_0^{1-x} \int_0^{1-x-y} (xyz)\, dz\, dy\, dx &= \int_0^1 \int_0^{1-x} \left(\frac{1}{2}xyz^2\right)\Bigg|_{z=0}^{1-x-y} dy\, dx \\ &= \int_0^1 \int_0^{1-x} \left[\frac{1}{2}xy(1-x-y)^2\right] dy\, dx.\end{aligned}$$

Now we treat the double integral iteratively, first doing the integral with respect to y while holding x constant. The algebra is straightforward, though involved, resulting in

$$\begin{aligned}\int_0^1 \int_0^{1-x} \int_0^{1-x-y} (xyz)\, dz\, dy\, dx &= \frac{1}{2}\int_0^1 \frac{1}{12}(6xy^2 - 12x^2y^2 + 6x^3y^2 - 8xy^3 + 8x^2y^3 + 3xy^4)\Bigg|_{y=0}^{1-x} dx \\ &= \frac{1}{2}\int_0^1 \frac{1}{12}\left[x - 4x^2 + 6x^3 - 4x^4 + x^5\right] dx.\end{aligned}$$

Finally, we evaluate the remaining definite integral:

$$\int_0^1 \int_0^{1-x} \int_0^{1-x-y} (xyz)\, dz\, dy\, dx = \frac{1}{24}\left(\frac{1}{2}x^2 - \frac{4}{3}x^3 + \frac{3}{2}x^4 - \frac{4}{5}x^5 + \frac{1}{6}x^6\right)\Bigg|_0^1 = \frac{1}{720}.$$

Thus, we have

$$\iiint_D (xyz)\, dV = \frac{1}{720}.$$

△

14.3.2 Triple Iterated Integrals in Cylindrical Coordinates

Cylindrical polar coordinates (usually referred to simply as **cylindrical coordinates**) can be used to describe the location of a point in three dimensions. Two of the coordinates in cylindrical coordinates are r and θ, defined exactly as in Equation (14.8) for plane polar coordinates. The third coordinate is z is the same z-coordinate of rectangular coordinates. Therefore, a point in cylindrical coordinates is given by (r, θ, z), with the transformations from rectangular to cylindrical coordinates, and vice versa given by Equation (14.8) along with $z = z$. A generic point in three-dimensional space is shown in Figure 14.13, with the three coordinates r, θ, and z drawn into the picture so that they can be seen in relation to the rectangular coordinates x, y, and z.

The name "cylindrical coordinates" comes from the fact that a right circular cylinder is relatively easy to describe with these coordinates. The curved sides of a cylinder can be part of a surface with $r =$ constant, θ varying over the entire range from 0 to 2π, and z varying between two fixed values. Those two fixed values of z determine the flat end surfaces of the cylinder. Cylindrical coordinates can be particularly useful in problems that involve three-dimensional objects with cylindrical symmetry—that is, symmetry with respect to one axis (as in the right circular cylinder).

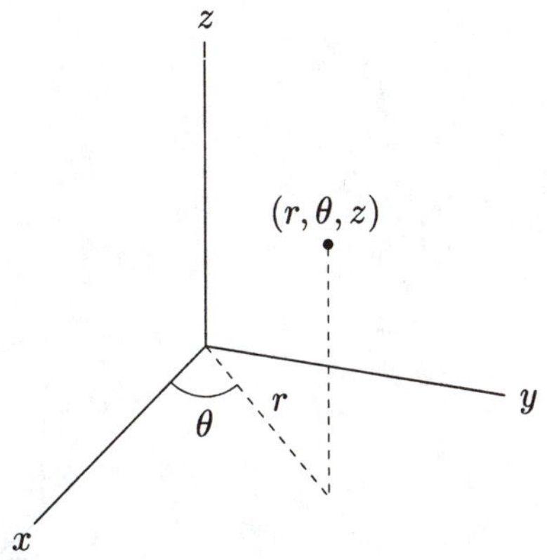

Figure 14.13. Typical point in cylindrical coordinates.

Example 14.16. A 3-cm-thick pie with a 30-cm diameter is cut into eight equal slices. Describe the boundaries of one slice of the pie using cylindrical coordinates.

Solution. We take the bottom surface of the pie to be in the xy-plane, with the $+z$-axis running straight up through the center. We also take one side of the slice to be along the $+x$-axis, so $\theta = 0$ along that side. The other side is one-eighth of the way around the circle, making an angle $(1/8)(2\pi) = \pi/4$ with the $+x$-axis. The outer edge of the pie is at a distance $r = 15$ cm from the z-axis. Therefore, the slice is bounded by $r = 0$ and $r = 15$ cm; $z = 0$ and $z = 3$ cm; $\theta = 0$ and $\theta = \pi/4$. △

Example 14.17. Express the point in rectangular coordinates $(x, y, z) = (-2.3, -1.4, 5.0)$ in cylindrical coordinates.

Solution. We know that the z-coordinate is unaffected, so $z = 5.0$. The transformation of the other two coordinates is given in Equation (14.8). For the radial coordinate,

$$r = \sqrt{x^2 + y^2} = \sqrt{(-2.3)^2 + (-1.4)^2} = 2.7$$

where we have taken the positive root, because by definition r is positive. For the angular coordinate,

$$\theta = \tan^{-1}\frac{y}{x} = \tan^{-1}\left(\frac{-1.4}{-2.3}\right) = \tan^{-1} 0.6087.$$

In evaluating the inverse tangent, we must be careful because the correct answer is not necessarily given by the principal value with $-\pi/2 \leq \theta \leq \pi/2$. In the xy-plane, a point with both x- and y-coordinates negative lies in the third quadrant, with the corresponding polar coordinate θ in the range $(\pi, 3\pi/2)$. Since the angle θ of cylindrical coordinates is defined the same way, it must also be in that range here (since both x- and y-coordinates are negative). The principal value of the inverse tangent is $\tan^{-1}(0.6087) = 0.547$, so the actual value in this case is $\theta = 0.547 + \pi \approx 3.7$. Therefore, the desired cylindrical coordinates are $(r, \theta, z) = (2.7, 3.7, 5.0)$. △

Earlier we considered triple iterated integrals in rectangular coordinates, expressed as

$$\int_u^v \int_c^d \int_a^b f(x, y, z)\, dx\, dy\, dz$$

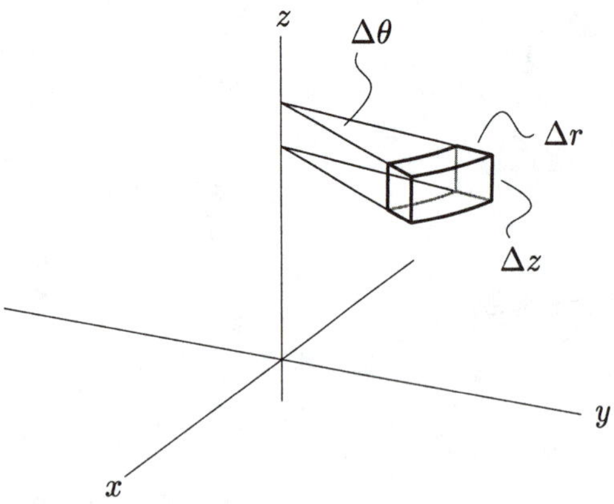

(a) Typical subregion

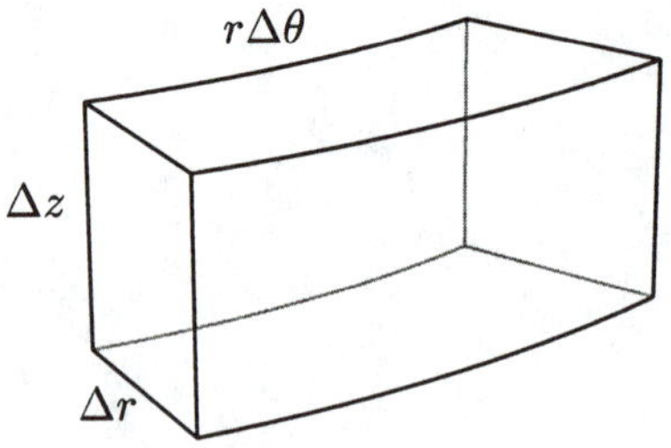

(b) Lengths of "box" sides

Figure 14.14. Typical subregion in cylindrical coordinates of extent Δr by $\Delta\theta$ by Δz.

where we have expressed the volume element dV as $dx\,dy\,dz$ in rectangular coordinates and included the limits of integration.

Based on our experience with double integrals in polar coordinates, we need to express the "volume element" dV in cylindrical coordinates. A typical subregion in cylindrical coordinates is shown in Figure 14.14(a). The subregion is defined by intervals of size Δr, $\Delta\theta$, and Δz. The base of this subregion has the same shape as the polar subregion shown in Figure 14.10. The volume of the cylindrical subregion is the product of the area of the base and the height (see Figure 14.14(b). The base area is approximately $\Delta r(r\Delta\theta)$ and the height of the cylindrical subregion is Δz, so

$$\Delta V \approx (r\Delta r\Delta\theta)(\Delta z)$$

or

$$\Delta V \approx r\Delta r\Delta\theta\Delta z.$$

where the results are approximate because the subregion is *not* a rectangular solid. As before, the result becomes exact in the limit as the volume approaches zero, so the volume element is

$$dV = r\,dr\,d\theta\,dz. \tag{14.10}$$

The real meaning of this is that in cylindrical coordinates we have an iterated integral of the form

$$\iiint_D f(x,y,z)\,dV = \int_c^d \int_\alpha^\beta \int_a^b f(r\cos\theta, r\sin\theta, z)\,r\,dr\,d\theta\,dz \tag{14.11}$$

where a, b, α, β, c, and d are appropriate bounds on the r-, θ-, and z-coordinates to describe the solid region D.

Example 14.18. Provide evidence that the volume element in Equation (14.10) is correct by integrating the function $f(x,y,z) = 1$ over a cylinder of radius R and height h to derive the correct formula for the volume of a cylinder.

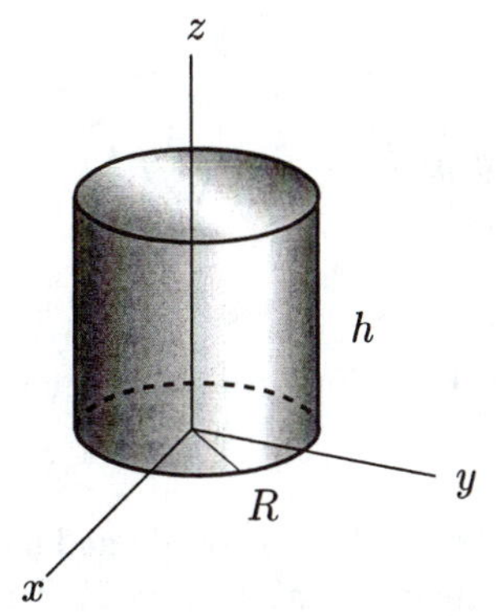

Figure 14.15. Cylinder of Example 14.18.

Solution. Let the base of the cylinder lie in the xy-plane, with the center of the base at the origin (Figure 14.15). The the limits of integration are 0 to R for the variable r, 0 to 2π for the variable θ, and 0 to h for the variable z. The triple integral is

$$\int_0^h \int_0^{2\pi} \int_0^R r\,dr\,d\theta\,dz.$$

We can perform an iterated integral, or noticing that the integrand factors and limits of integration are all constant, we can write

$$\int_0^h \int_0^{2\pi} \int_0^R r\,dr\,d\theta\,dz = \int_0^R r\,dr \int_0^{2\pi} d\theta \int_0^h dz = \left(\frac{1}{2}R^2\right)(2\pi)(h) = \pi R^2 h$$

which is the correct formula for the volume of a cylinder of radius R and height h. △

Example 14.19. Compute the triple integral of the function $f(x, y, z) = x^2 z$ over the volume described in the preceding example.

Solution. First, we must make the change of variable into cylindrical coordinates. Since $x = r\cos\theta$, we have

$$f(r, \theta, z) = (r\cos\theta)^2 z = zr^2\cos^2\theta.$$

The triple integral is

$$\int_0^h \int_0^{2\pi} \int_0^R (zr^2\cos^2\theta)\,r\,dr\,d\theta\,dz = \int_0^h \int_0^{2\pi} \int_0^R zr^3\cos^2\theta\,dr\,d\theta\,dz.$$

Again, the limits of integration are constant, so we write

$$\int_0^h \int_0^{2\pi} \int_0^R zr^3\cos^2\theta\,dr\,d\theta\,dz = \int_0^R r^3\,dr \int_0^{2\pi} \cos^2\theta\,d\theta \int_0^h z\,dz.$$

The integrals with respect to r and z are straightforward. With a table of indefinite integrals, we use the second fundamental theorem of calculus to evaluate the definite integral with respect to θ:

$$\int_0^{2\pi} \cos^2\theta\,d\theta = \left(\frac{\theta}{2} + \frac{1}{4}\sin(2\theta)\right)\Big|_0^{2\pi} = \left(\frac{2\pi}{2} + \frac{1}{4}\sin(4\pi)\right) - (0 + 0) = \pi.$$

Then our triple integral is

$$\int_0^R r^3\,dr \int_0^{2\pi} \cos^2\theta\,d\theta \int_0^h z\,dz = \left(\frac{1}{4}R^4\right)(\pi)\left(\frac{h^2}{2}\right) = \frac{1}{8}\pi R^4 h^2. \qquad \triangle$$

Example 14.20. Construct a triple integral in cylindrical coordinates to find the volume of a cone with height h and base radius R.

Solution. Let's invert the cone, putting the point at the origin and the top at $z = h$, as in Figure 14.16. We find the volume by integrating the the function $f(x, y, z) = 1$. The limits of integration follow from the geometry of the cone. However, in this problem the limits are not constant. If we look at a horizontal slice at height z, the limits on r are 0 to the outer surface of the cone, at $r = z\tan\alpha = zR/h$. Therefore, we must do the integration with respect to r before we integrate with respect to z. The order of the integration with respect to θ is irrelevant, since there is no θ dependence in the integrand and the limits on θ are 0 to 2π. Therefore, we integrate with respect to θ first:

$$\int_0^h \int_0^{zR/h} \int_0^{2\pi} r\,d\theta\,dr\,dz = 2\pi \int_0^h \int_0^{zR/h} r\,dr\,dz.$$

Now integrating with respect to r,

$$\int_0^h \int_0^{zR/h} \int_0^{2\pi} r\,d\theta\,dr\,dz = 2\pi \int_0^h \left(\frac{1}{2}r^2\right)\Bigg|_0^{zR/h} dz = 2\pi \int_0^h \frac{1}{2}\left(\frac{zR}{h}\right)^2 dz = \pi\frac{R^2}{h^2}\int_0^h z^2\,dz$$

$$= \pi\frac{R^2}{h^2}\left(\frac{1}{3}z^3\right)\Bigg|_0^h = \frac{1}{3}\pi R^2 h$$

which we recognize as the correct formula for the volume of a cone. $\triangle$

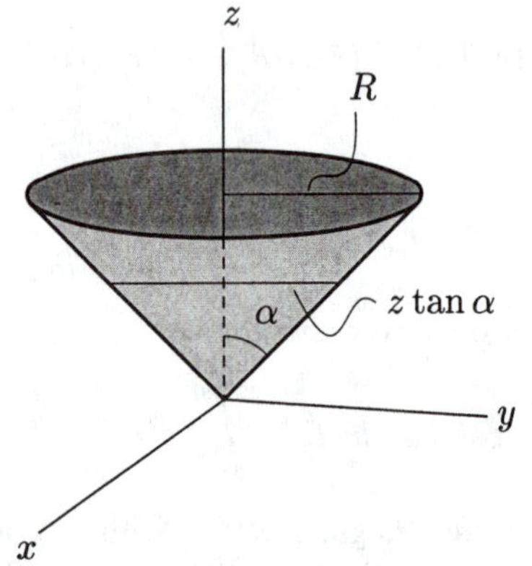

Figure 14.16. Cone of Example 14.20.

Example 14.21. Construct a triple integral in cylindrical coordinates to find the volume of a hemisphere of radius R.

Solution. To find the total volume, we integrate the function $f(x, y, z) = 1$. Let's put the base of the hemisphere in the xy-plane, as shown in Figure 14.17. As in the preceding example, we take a horizontal slice at a height z to establish the limits on r, which go from the z-axis ($r = 0$) to the edge of the hemisphere at $r = \sqrt{R^2 - z^2}$. Once again, we must perform the r integration prior to the z integration. The θ integration can be done any time, so we do it first:

$$\int_0^R \int_0^{\sqrt{R^2-z^2}} \int_0^{2\pi} r\, d\theta\, dr\, dz = 2\pi \int_0^R \int_0^{\sqrt{R^2-z^2}} r\, dr\, dz.$$

Now integrating with respect to r:

$$\int_0^R \int_0^{\sqrt{R^2-z^2}} \int_0^{2\pi} r\, d\theta\, dr\, dz = 2\pi \int_0^R \left(\frac{1}{2}r^2\right)\Bigg|_0^{\sqrt{R^2-z^2}} dz = \pi \int_0^R (R^2 - z^2)\, dz.$$

The integration with respect to z is straightforward:

$$\int_0^R \int_0^{\sqrt{R^2-z^2}} \int_0^{2\pi} r\, d\theta\, dr\, dz = \pi \left(R^2 z - \frac{1}{3}z^3\right)\Bigg|_0^R = \pi \left(R^3 - \frac{1}{3}R^3\right) = \frac{2}{3}\pi R^3$$

which we recognize as the correct volume of a hemisphere (half of $4\pi R^3/3$, the volume of a sphere). △

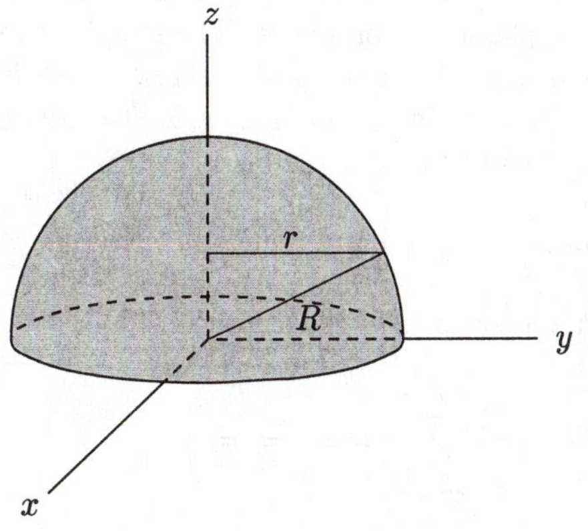

Figure 14.17. Hemisphere of Example 14.21.

14.3.3 Triple Iterated Integrals in Spherical Coordinates

Another three-dimensional coordinate system that has numerous applications is called **spherical polar coordinates**, or simply **spherical coordinates**. The three coordinates (ρ, ϕ, θ) are defined as illustrated in Figure 14.18. For a point P, the coordinate ρ measures the distance from the origin to P, so by the Pythagorean theorem $\rho = \sqrt{x^2 + y^2 + z^2}$ (and as in polar coordinates we take the positive root).

The coordinate ϕ is defined as the angle between a line drawn from the origin to P and the $+z$-axis. We always take ϕ to be the *smaller* of the two angles thus defined, so ϕ is always in the range $[0, \pi]$.

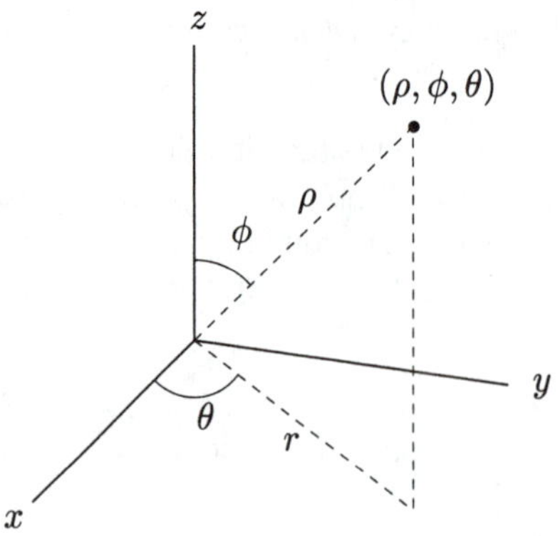

Figure 14.18. Spherical coordinates for a typical point.

From this definition, it can be seen that the transformation from spherical to rectangular coordinates is $z = \rho\cos\phi$. Then the inverse transformation (from rectangular to spherical coordinates) is $\phi = \cos^{-1}(z/\rho) = \cos^{-1}(z/\sqrt{x^2+y^2+z^2})$.

To define the coordinate θ, we look at the projection of the line from the origin to P onto the xy-plane. Then θ is the angle this projection makes with the $+x$-axis, as measured counterclockwise from the $+x$-axis. Notice that this makes θ the same angle as in polar and cylindrical coordinates, so $\tan\theta = (y/x)$. Keep in mind that we must choose the relevant solution if we solve for θ.

We can also use the projection of the line from the origin to P into the xy-plane to find the rectangular coordinates x and y in terms of spherical coordinates. The length of that projection is r as defined in cylindrical coordinates (see Figure 14.18). In terms of the spherical coordinate ρ, notice that $r = \rho\sin\phi$, so $x = r\cos\theta = \rho\sin\phi\cos\theta$ and $y = r\sin\theta = \rho\sin\phi\sin\theta$. To summarize, the transformations from spherical coordinates to rectangular coordinates are

$$x = \rho\sin\phi\cos\theta, \qquad y = \rho\sin\phi\sin\theta, \qquad \text{and} \qquad z = \rho\cos\phi. \tag{14.12}$$

The transformations from rectangular coordinates to spherical coordinates are

$$\rho = \sqrt{x^2+y^2+z^2}, \qquad \phi = \cos^{-1}\left(\frac{z}{\sqrt{x^2+y^2+z^2}}\right), \qquad \text{and} \qquad \theta = \tan^{-1}\left(\frac{y}{x}\right). \tag{14.13}$$

The name "spherical coordinates" comes from the fact that it is straightforward to define a sphere using spherical coordinates. A sphere is a surface defined by $\rho = $ constant, with ϕ and θ covering their full ranges (0 to π and 0 to 2π, respectively). Spherical coordinates are particularly useful in problems that involve objects with spherical symmetry—that is, symmetry with respect to the origin—so that some physical property is a function of ρ alone and independent of ϕ and θ.

Example 14.22. An orange with diameter 8 cm having its center at the origin is cut into 16 equal slices with cuts that all pass along the z-axis. Describe one slice of the orange using spherical coordinates.

Solution. Let's use a slice that has one side touching the x-axis. Then the angle θ in this slice varies between 0 and $2\pi/16 = \pi/8$. The coordinate ρ varies from 0 to the outer radius (4 cm). The coordinate ϕ covers the full range of values from 0 (the top of the orange) to π (the bottom of the orange). △

Example 14.23. Express the point in rectangular coordinates $(x, y, z) = (-2.3, -1.4, 5.0)$ in spherical coordinates.

Solution. We use the transformations in Equation (14.12) to compute

$$\rho = \sqrt{x^2 + y^2 + z^2} = \sqrt{(-2.3)^2 + (-1.4)^2 + (5.0)^2} \approx 5.7$$

and

$$\phi = \cos^{-1}\left(\frac{z}{\sqrt{x^2 + y^2 + z^2}}\right) = \cos^{-1}\left(\frac{5.0}{\sqrt{(-2.3)^2 + (-1.4)^2 + (5.0)^2}}\right) = 0.49.$$

Since ϕ must be between 0 and π, this value of the inverse cosine is the correct one. We know from Example 14.2 that $\theta = 3.7$. Thus, the point is spherical coordinates is $(\rho, \phi, \theta) = (5.7, 0.49, 3.7)$. △

Following the procedure for constructing triple integrals in cylindrical coordinates, we need to construct a volume element dV expressed in terms of the spherical coordinates ρ, ϕ, and θ. An approximation of the volume element is shown in Figure 14.19(a), in which we have taken a typical spherical subregion defined by small changes $\Delta\rho$, $\Delta\phi$, and $\Delta\theta$ in the three coordinates. From the definition of spherical coordinates (and some trigonometry), it follows that three of the edges of this slice have lengths $\Delta\rho$, $\rho\Delta\phi$, and $(\rho\sin\phi)\Delta\theta$ as shown in Figure 14.19(b). Approximating the region just described as a rectangular solid, its approximate volume is

$$\Delta V \approx (\Delta\rho)(\rho\Delta\phi)(\rho\sin\phi\Delta\theta) = \rho^2 \sin\phi\Delta\rho\Delta\phi\Delta\theta \tag{14.14}$$

so the volume element is

$$dV = \rho^2 \sin\phi\, d\rho\, d\phi\, d\theta. \tag{14.15}$$

What this really means is that in spherical coordinates we have an iterated integral of the form

$$\iiint_D f(x, y, z)\, dV = \int_\alpha^\beta \int_\gamma^\delta \int_a^b f(\rho\sin\phi\cos\theta, \rho\sin\phi\sin\theta, \rho\cos\theta)\rho^2 \sin\phi\, d\rho\, d\phi\, d\theta$$

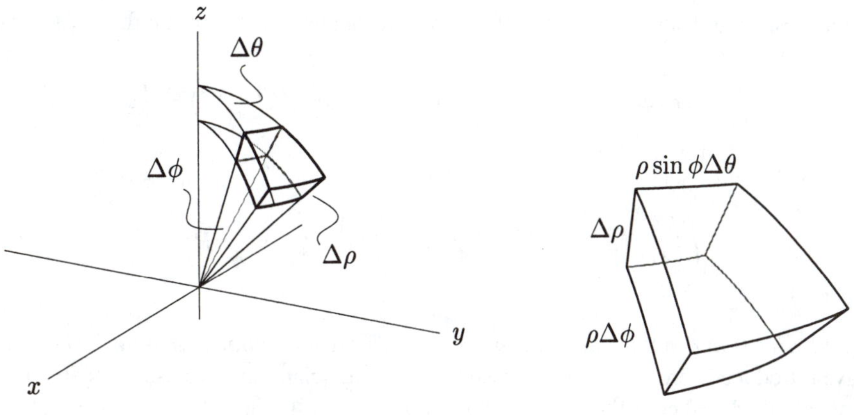

(a) Typical subregion (b) Lengths of "box" sides

Figure 14.19. Typical subregion in spherical coordinates of extent $\Delta\rho$ by $\Delta\phi$ by $\Delta\theta$.

where α and β are appropriate limits on θ, γ and δ are appropriate limits on ϕ and a and b are appropriate limits on ρ. This is best understood through examples.

Example 14.24. Rework Example 14.21 (finding the volume of a hemisphere) by constructing the appropriate triple integral in spherical coordinates.

Solution. If we place the base of the hemisphere on the xy-plane as before (Figure 14.17), we see that the limits of integration in spherical coordinates are as follows: ρ varies from 0 to R; ϕ varies from 0 to $\pi/2$ (just a hemisphere); and θ varies from 0 to 2π. As before, we need only integrate the volume element (with integrand 1) to find the total volume. With the limits all constant, the triple integral can be separated into the product of three single integrals, and the evaluation is straightforward:

$$\int_0^{2\pi}\int_0^{\pi/2}\int_0^{R}\rho^2\sin\phi\,d\rho\,d\phi\,d\theta=\int_0^{2\pi}d\theta\int_0^{\pi/2}\sin\phi\,d\phi\int_0^{R}\rho^2\,d\rho=(2\pi)(1)\left(\frac{1}{3}R^3\right)=\frac{2}{3}\pi R^3.$$

The evaluation in this example is much simpler than when we worked the problem in cylindrical coordinates. By choosing a coordinate system to match the symmetry of the problem, it is usually possible to simplify the evaluation of a triple integral. △

Example 14.25. Find the triple integral of $f(x, y, z) = z$ over the hemisphere described in the previous example.

Solution. Using the transformation from rectangular to spherical coordinates $z = \rho\cos\phi$, the triple integral is

$$\iiint_D f(\rho,\phi,\theta)\,dV=\int_0^{2\pi}\int_0^{\pi/2}\int_0^{R}(\rho\cos\phi)\rho^2\sin\phi\,d\rho\,d\phi\,d\theta=\int_0^{2\pi}\int_0^{\pi/2}\int_0^{R}\rho^3\cos\phi\sin\phi\,d\rho\,d\phi\,d\theta$$

where we have used the limits of integration from the preceding example. With constant limits of integration, the triple integral can be rewritten as the product of three single integrals

$$\iiint_D f(\rho,\phi,\theta)\,dV=\int_0^{2\pi}d\theta\int_0^{\pi/2}\cos\phi\sin\phi\,d\phi\int_0^{R}\rho^3\,d\rho.$$

The integrals with respect to θ and ρ are straightforward. We can use a substitution for the ϕ integration:

$$\int_0^{\pi/2}\cos\phi\,\sin\phi\,d\phi=\left(\frac{1}{2}\sin^2\phi\right)\Big|_0^{\pi/2}=\frac{1}{2}(1^2-0^2)=\frac{1}{2}.$$

Thus,

$$\iiint_D f(\rho,\phi,\theta)\,dV=(2\pi)\left(\frac{1}{2}\right)\left(\frac{1}{4}R^4\right)=\frac{1}{4}\pi R^4.$$

△

A warning: Physicists and mathematicians usually use different notations for spherical coordinates. The notation we have introduced here is commonly used by mathematicians, and we use it because the coordinate θ is the same as in cylindrical coordinates. Physicists normally write the spherical coordinates as (r, θ, ϕ), with r defined the same way we have defined ρ, the angle θ defined the same way we have defined ϕ, and the angle ϕ defined the same way we have defined θ. If you look in any other physics book, you will probably encounter this change in notation.

14.4 Applications of Multiple Integrals in Physics

In this section we introduce some applications of multiple integrals in physics, concentrating on the concepts of center of mass and rotational inertia that were introduced in Chapters 9 and 10. We suggest you review the appropriate sections briefly before continuing. In the remainder of the book, we will use multiple integrals as they are needed in new physics problems.

14.4.1 Center of Mass

Let's begin with a review of what we know about center of mass. In Chapter 9, we defined the center of mass of a collection of n point particles on the x-axis as

$$X = \frac{1}{M}\sum_{i=1}^{n} m_i x_i \tag{9.4}$$

where the ith point particle is at position x_i and has mass m_i. Then we extended the definition to a collection of n point particles in three dimensions. Now the ith point particle lies at the point (x_i, y_i, z_i) and the three components of the center of mass (a position vector $\mathbb{R}$) are

$$X = \frac{1}{M}\sum_{i=1}^{n} m_i x_i, \qquad Y = \frac{1}{M}\sum_{i=1}^{n} m_i y_i, \qquad \text{and} \qquad Z = \frac{1}{M}\sum_{i=1}^{n} m_i z_i. \tag{9.5}$$

The position of the center of mass can be expressed as

$$\vec{R} = X\,\hat{\imath} + Y\,\hat{\jmath} + Z\,\hat{k}$$

or

$$\vec{R} = \frac{1}{M}\sum_{i=1}^{n} m_i \vec{r}_i. \tag{9.6}$$

For a continuous distribution of mass along the x-axis between two points $x = a$ and $x = b$, we found that the total mass of the distribution is

$$M = \int_a^b \lambda(x)\,dx \tag{9.9}$$

and the position of the center of mass is

$$X = \frac{1}{M}\int_a^b \lambda(x)\,x\,dx \tag{9.8}$$

where the linear mass density (the local mass per unit length) is given by the function $\lambda(x)$.

How can we compute the center of mass of a continuous distribution of matter in three dimensions? Notice that when we went from one dimension to three dimensions in our study of point particles, the definition of the x-component of the center of mass remained unchanged [compare Equations (9.4) and (9.5)]. Equation (9.8) for the center of mass of a continuous distribution of matter followed directly from the definition of center of mass for point particles; we used the limit of a Riemann sum to turn the sum into a definite integral. Based on this experience, we should expect the x-component of the center of mass of a continuous distribution of matter in three dimensions is given by an equation that resembles Equation (9.8). However, two important changes need to be made. First, there is the fact that in general the density is a function of all three coordinates x, y, and z. We follow the convention generally used by physicists of using different Greek letters for mass density, depending on whether the mass is distributed over one, two or three dimensions. The one-dimensional (linear) mass density is $\lambda(x)$; the two-dimensional (surface) mass density is $\sigma(x,y)$, and the three-dimensional (volume) mass density is $\rho(x,y,z)$. The second important change in converting Equation (9.8) to three dimensions lies in the fact that the definite integral serves the purpose of summing the contributions in a "weighted average" over the entire distribution. Thus, the definite integral must be a triple integral to include all the matter in the distribution.

With these changes in mind, the x-component of the center of mass of a continuous distribution of matter in three dimensions is

$$X = \frac{1}{M}\iiint_D \rho(x,y,z)\,x\,dV \tag{14.16}$$

where as usual the D is the solid region occupied by the distribution. Similarly, the y- and z-components of the center of mass are given by the analogous expressions

$$Y = \frac{1}{M}\iiint_D \rho(x,y,z)\,y\,dV \quad \text{and} \quad Z = \frac{1}{M}\iiint_D \rho(x,y,z)\,z\,dV. \tag{14.17}$$

As before, the center of mass in three dimensions is a vector $\vec{R}$ with components X, Y, and Z, and can be described in most compact form using Equations (14.16) and (14.17) as

$$\vec{R} = \frac{1}{M}\iiint_D \rho(x,y,z)\,\vec{r}\,dV.$$

Each of these expression for the center of mass involve the total mass M. By analogy with Equation (9.9), we can write in three dimensions

$$M = \iiint_D \rho(x,y,z)\,dV \tag{14.18}$$

which is used to compute the total mass of a distribution.

Example 14.26. A rectangular solid block of aluminum with sides of length 8.0 cm, 4.0 cm, and 3.0 cm has a uniform mass density 2.70 $\times 10^3$ kg/m^3. Use a triple integral to find the total mass of the aluminum and the center of mass of the block.

Solution. Let one corner of the block lie at the origin, with its faces parallel to the three coordinate axes, and let the sides parallel to the x-, y-, and z-axis be 8.0 cm, 4.0 cm, and 3.0 cm, respectively. Then the total mass is given by Equation (14.18)

$$M = \iiint_D \rho(x,y,z)\,dV = \int_0^{0.03}\int_0^{0.04}\int_0^{0.08} \rho\,dx\,dy\,dz.$$

Since ρ is a constant and the limits of integration are constant,

$$M = \rho\int_0^{0.08} dx \int_0^{0.04} dy \int_0^{0.03} dz = (2.70 \times 10^3\ \text{kg/m}^3)(0.08\ \text{m})(0.04\ \text{m})(0.03\ \text{m}) = 0.26\ \text{kg}.$$

For the center of mass we have

$$X = \frac{1}{M} \iiint_D \rho(x,y,z)\, x\, dV = \frac{1}{M} \int_0^{0.03} \int_0^{0.04} \int_0^{0.08} \rho\, x\, dx\, dy\, dz.$$

Once again, both ρ and the limits of integration are constant so

$$X = \frac{\rho}{M} \int_0^{0.08} x\, dx \int_0^{0.04} dy \int_0^{0.03} dz.$$

Notice that the integration over x has changed, but the other integrals are the same as before. Thus,

$$X = \frac{2.70 \times 10^3 \text{ kg/m}^3}{0.26 \text{ kg}} \left[\frac{1}{2}(0.08 \text{ m})^2\right](0.04 \text{ m})(0.03 \text{ m}) = 0.40 \text{ m}.$$

Not surprisingly, the x-component of the center of mass of a uniform rectangular block lies halfway from one end to the other. Similarly,

$$Y = \frac{2.70 \times 10^3 \text{ kg/m}^3}{0.26 \text{ kg}} (0.08 \text{ m}) \left[\frac{1}{2}(0.04 \text{ m})^2\right](0.03 \text{ m}) = 0.20 \text{ m}$$

and

$$Z = \frac{2.70 \times 10^3 \text{ kg/m}^3}{0.26 \text{ kg}} (0.08 \text{ m})(0.04 \text{ m}) \left[\frac{1}{2}(0.03 \text{ m})^2\right] = 0.15 \text{ m}. \qquad \triangle$$

In the previous example, we deliberately looked at a simple case in which calculus was not needed. For a uniform solid, the mass equals the density times the volume, and the center of mass is at the geometric center of the regular solid. The method we developed is perfectly general, however, and can be applied to nonuniform solids, where it *is* necessary to use calculus.

Example 14.27. Consider a rectangular block with the same geometry as in the previous example, but with a density that varies as a function of position according to the function $\rho(x,y,z) = 5.0 \times 10^4 x$ (with outputs in kg/m^3 for inputs in m). Find the total mass of the block and the location of the center of mass.

Solution. It is not easy to find the total mass without doing the required triple integral. For this geometry,

$$M = \iiint_D \rho(x,y,z)\, dV = \int_0^{0.03} \int_0^{0.04} \int_0^{0.08} \rho\, dx\, dy\, dz$$

as before. Inserting the density function,

$$M = \int_0^{0.03} \int_0^{0.04} \int_0^{0.08} (5.0 \times 10^4 x)\, dx\, dy\, dz$$

We can factor out the constant and separate the triple integral into three single integrals, because the limits of integration are constant:

$$M = (5 \times 10^4) \int_0^{0.08} x\, dx \int_0^{0.04} dy \int_0^{0.03} dz = (5 \times 10^4) \left[\frac{1}{2}(0.08)^2\right](0.04)(0.03) = 0.192 \text{ kg}.$$

The x-component of the center of mass is difficult to predict. However, given that the density is proportional to x, we expect that the result will be closer to the $x = 0.08$ m end than to the $x = 0$ end. Because the density is independent of y and z, those components of the center of mass should be the same as in the previous example (at the halfway point along each of these dimensions). Following the prescription for center of mass,

$$X = \frac{1}{M}\iiint\limits_D \rho(x,y,z)\,x\,dV = \frac{1}{M}\int_0^{0.03}\int_0^{0.04}\int_0^{0.08} \rho\,x\,dx\,dy\,dz.$$

Inserting the density function and factoring out constants results in

$$X = \frac{1}{M}\left(5\times 10^4\right)\int_0^{0.08} x^2\,dx\int_0^{0.04} dy\int_0^{0.03} dz = \frac{5\times 10^4}{0.192}\left[\frac{1}{3}(0.08)^3\right](0.04)(0.03) = 0.053 \text{ m}.$$

This is a reasonable result, since it is somewhat greater than $x = 0.04$ m but still well inside the block. For the other components,

$$\begin{aligned} Y = \frac{1}{M}\iiint\limits_D \rho(x,y,z)\,y\,dV &= \frac{1}{M}\int_0^{0.03}\int_0^{0.04}\int_0^{0.08} \rho\,y\,dx\,dy\,dz \\ &= \frac{1}{M}\left(5\times 10^4\right)\int_0^{0.08} x\,dx\int_0^{0.04} y\,dy\int_0^{0.03} dz \\ &= \frac{5\times 10^4}{0.192}\left[\frac{1}{2}(0.08)^2\right]\left[\frac{1}{2}(0.04)^2\right](0.03) = 0.02 \text{ m} \end{aligned}$$

and

$$\begin{aligned} Z = \frac{1}{M}\iiint\limits_D \rho(x,y,z)\,z\,dV &= \frac{1}{M}\int_0^{0.03}\int_0^{0.04}\int_0^{0.08} \rho\,z\,dx\,dy\,dz \\ &= \frac{1}{M}\left(5\times 10^4\right)\int_0^{0.08} x\,dx\int_0^{0.04} dy\int_0^{0.03} z\,dz \\ &= \frac{5\times 10^4}{0.192}\left[\frac{1}{2}(0.08)^2\right](0.04)\left[\frac{1}{2}(0.03)^2\right] = 0.015 \text{ m} \end{aligned}$$

as expected. △

Example 14.28. A uniform thin metal plate is cut into a flat semicircular surface. If the radius of the semicircle is R, find the center of mass of the metal surface.

Solution. This problem calls for a *two-dimensional* center of mass. Let the semicircle lie in the xy-plane with its diameter on the x-axis, with the center of the circle at the origin, as shown in Figure 14.20. By analogy with the three-dimensional results [Equations (14.15) and (14.16)], we have

$$X = \frac{1}{M}\iint\limits_R \sigma(x,y)x\,dA \qquad \text{and} \qquad y = \frac{1}{M}\iint\limits_R \sigma(x,y)y\,dA$$

where $\sigma(x,y)$ is the density (mass per unit area). In this problem, $\sigma(x,y) = \sigma = \text{constant}$. In performing the double integral, we integrate over y first. Looking at a vertical slice of the semicircular region at fixed x

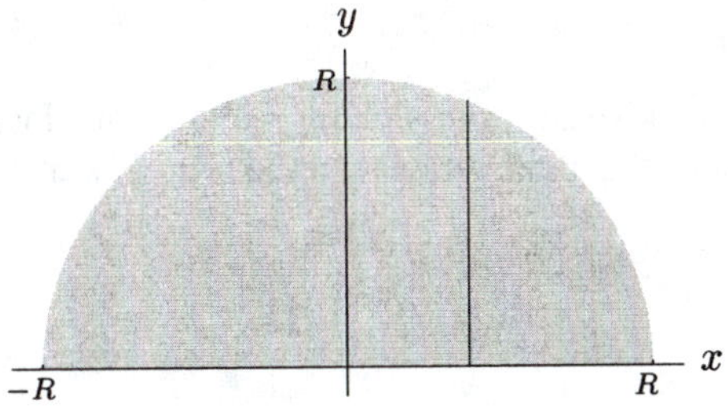

Figure 14.20. The plate in Example 14.28.

(Figure 14.20), we see that the limits on y are 0 to $\sqrt{R^2 - x^2}$. The limits on x are $-R$ to R, and the double integral for X is

$$X = \frac{1}{M} \iint_R \sigma(x,y)\, x\, dA = \frac{\sigma}{M} \int_{-R}^{R} \int_0^{\sqrt{R^2-x^2}} x\, dy\, dx.$$

Integrating with respect to y,

$$X = \frac{\sigma}{M} \int_{-R}^{R} (xy)\Big|_{y=0}^{\sqrt{R^2-x^2}} dx = \frac{\sigma}{M} \int_{-R}^{R} x\sqrt{R^2 - x^2}\, dx.$$

From a table of integrals (or using substitution),

$$X = \frac{\sigma}{M} \left(-\frac{1}{3} (R^2 - x^2)^{3/2} \right) \Big|_{-R}^{R} = 0.$$

This result is expected, given the symmetry of the semicircle. However, we expect $Y > 0$. We can construct the double integral for Y over the same limits:

$$Y = \frac{1}{M} \iint_R \sigma(x,y) y\, dA = \frac{\sigma}{M} \int_{-R}^{R} \int_0^{\sqrt{R^2-x^2}} y\, dy\, dx.$$

Integrating with respect to y,

$$Y = \frac{\sigma}{M} \int_{-R}^{R} \left(\frac{1}{2} y^2 \right) \Big|_{y=0}^{\sqrt{R^2-x^2}} dx = \frac{\sigma}{2M} \int_{-R}^{R} (R^2 - x^2)\, dx.$$

The integration with respect to x is straightforward:

$$Y = \frac{\sigma}{2M} \left(R^2 x - \frac{1}{3} x^3 \right) \Big|_{-R}^{R} = \frac{\sigma}{2M} \left[\left(R^3 - \frac{1}{3} R^3 \right) - \left(-R^3 + \frac{1}{3} R^3 \right) \right] = \frac{\sigma}{2M} \frac{4R^3}{3} = \frac{2\sigma R^3}{3M}.$$

We can simplify the result by noting that by definition of density σ =mass/area= $M/(\pi R^2/2)$ for the semicircle. Therefore,

$$Y = \frac{2R^3}{3M} \frac{2M}{\pi R^2} = \frac{4R}{3\pi}.$$

△

14.4.2 Rotational Inertia

In Chapter 10, we used calculus to derive an expression for the rotational inertia of a continuous distribution of matter along the x-axis. We found that the rotational inertia is approximated by the Riemann sum

$$I = \sum_{i=1}^{n} \lambda(x_i)\, x_i^2 \Delta x \tag{10.9}$$

where the entire interval $x : [a, b]$ over which the matter is distributed has been broken into n subintervals of size Δx. In this expression, $\lambda(x)$ is the function that describes the linear mass density of the distribution. Then in the limit as the number of subintervals becomes infinite, the Riemann sum becomes a definite integral

$$I = \int_a^b \lambda(x)\, x^2\, dx \tag{10.10}$$

which gives the exact rotational inertia.

In extending rotational inertia to two-dimensional objects (say in the xy-plane), we can follow the same procedure. First, we break the two-dimensional region into subregions Δx by Δy, as in our introduction to double integrals earlier in this chapter. In two dimensions, we use the area mass density $\sigma(x, y)$ (the local mass per unit area), and in terms of the mass density, we express the mass of a subregion as

$$m_{ij} = \sigma(x_i, y_j)\Delta x \Delta y. \tag{14.19}$$

The definition of rotational inertia for point particles tells us to multiply the mass of each particle by the square of the particle's distance (r) from the axis of rotation. Using this point-particle definition, we can approximate the rotational inertia of a continuous distribution of matter in two dimensions as

$$I = \sum_{j=1}^{m} \sum_{i=1}^{n} m_{ij} r_{ij}^2$$

or using Equation (14.19)

$$I = \sum_{j=1}^{m} \sum_{i=1}^{n} \sigma(x_i, y_j) r_{ij}^2 \Delta x \Delta y.$$

As before, we have divided the interval $x : [a, b]$ into n subintervals and the interval $y : [c, d]$ into m subintervals. In this notation, r_{ij} is the perpendicular distance of the point (x_i, y_j) from the axis of rotation.

Taking the limit as m and n approach infinity, the double Riemann sum becomes a double integral

$$I = \int_c^d \int_a^b \sigma(x, y) r^2\, dx\, dy \tag{14.20}$$

which is the exact expression for the rotational inertia. By analogy, we can write the rotational inertia of a three-dimensional object in terms of its mass density $\rho(x, y, z)$:

$$I = \int_u^v \int_c^d \int_a^b \rho(x, y, z) r^2\, dx\, dy\, dz. \tag{14.21}$$

Now let's apply these results in some familiar examples.

Example 14.29. Find the rotational inertia of a thin uniform circular disk of mass M and radius R with respect to a perpendicular axis passing through its center.

Solution. The general form of the double integral is given by Equation (14.20). However, the geometry of this problem suggests a change to polar coordinates (naturally we choose to place the disk in the xy-plane with its center at the origin). This means we should replace $dx\, dy$ with the area element in polar coordinates $r\, dr\, d\theta$. Also, in this geometry the general variable r describing the distance from the axis of rotation is exactly the coordinate r of polar coordinates. Therefore, the double integral is

$$I = \int_0^{2\pi} \int_0^R \sigma r^2 \, r \, dr \, d\theta.$$

The limits of integration follow from the geometry of the disk. We factor the constant density σ from the integrals, and separate the double integral with constant limits into a product of single integrals:

$$I = \sigma \int_0^R r^3 \, dr \int_0^{2\pi} d\theta = \sigma \left(\frac{1}{4} R^4 \right) (2\pi) = \frac{\sigma \pi R^4}{2}.$$

By definition of mass density as mass per unit area, $\sigma = M/(\pi R^2)$, and

$$I = \frac{M}{\pi R^2} \frac{\pi R^4}{2} = \frac{1}{2} M R^2.$$

This result is clearly correct dimensionally. △

Example 14.30. Find the rotational inertia of a uniform solid cylinder of mass M, radius R, and height h about the cylinder's axis of symmetry—that is, an axis passing through the cylinder's center and perpendicular to the ends.

Solution. We can consider the cylinder to consist of a large number of thin disks, as described in the previous example. Each disk is equivalent with respect to the axis of rotation, and therefore the cylinder's rotational inertia is

$$I = \frac{1}{2} M R^2.$$

To show this more formally, we construct a triple integral as in Equation (14.21). The geometry of the cylinder suggests a change to cylindrical coordinates, so we replace the volume element with $r\, dr\, d\theta\, dz$. The density ρ is constant, and the general radius r is the same as the coordinate r in cylindrical coordinates. Thus,

$$I = \int_0^h \int_0^{2\pi} \int_0^R \rho r^2 \, r \, dr \, d\theta \, dz = \rho \int_0^h \int_0^{2\pi} \int_0^R r^3 \, dr \, d\theta \, dz.$$

Again, all the limits of integration are constant, so

$$I = \rho \int_0^R r^3 \, dr \int_0^{2\pi} d\theta \int_0^h dz = \rho \left(\frac{1}{4} R^4 \right) (2\pi)(h) = \frac{\rho \pi h R^4}{2}.$$

The density is the mass divided by volume, or $\rho = M/(\pi R^2 h)$, and thus

$$I = \frac{M}{\pi R^2 h} \frac{\rho \pi h R^4}{2} = \frac{1}{2} M R^2$$

as expected. △

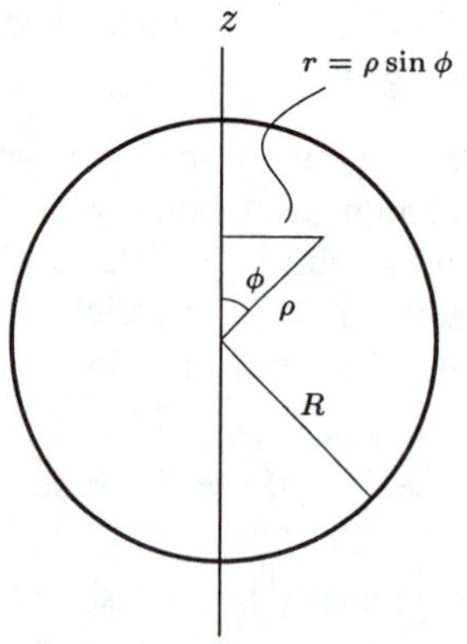

Figure 14.21. Cross section through the axis of rotation for the sphere in Example 14.31.

Example 14.31. Find the rotational inertia of a uniform solid sphere of mass M and radius R about an axis passing through the sphere's center.

Solution. As you might guess, it is wise to choose spherical coordinates for this problem. So as not to confuse the mass density with the spherical coordinate ρ, we use ρ_m for the mass density of the sphere. The general form to follow is Equation (14.21), but now we must replace the volume element with the volume element in spherical coordinates $\rho^2 \sin\phi \, d\rho \, d\phi \, d\theta$. We choose the z-axis as the axis of rotation. Then the perpendicular distance r of any point in the sphere from the axis of $R = \rho \sin\phi$, as shown in Figure 14.21. Inserting these results into Equation (14.21) along with the proper limits of integration (which follow naturally from the geometry of the sphere),

$$I = \int_0^{2\pi} \int_0^{\pi} \int_0^{R} \rho_m \left(\rho \sin\phi\right)^2 \rho^2 \sin\phi \, d\rho \, d\phi \, d\theta = \rho_m \int_0^{2\pi} \int_0^{\pi} \int_0^{R} \rho^4 \sin^3\phi \, d\rho \, d\phi \, d\theta$$

where we have factored out the constant density ρ_m.

With constant limits of integration we can write

$$I = \int_0^R \rho^4 \, d\rho \int_0^{\pi} \sin^3\phi \, d\phi \int_0^{2\pi} d\theta.$$

The integrals with respect to ρ and θ are straightforward. From a table of integrals,

$$\int_0^{\pi} \sin^3\phi \, d\phi = -\frac{1}{3}\left[(\cos\phi)(2 + \sin^2\phi)\right]\Big|_0^{\pi} = -\frac{1}{3}\left[(-1)(2) - (1)(2)\right] = \frac{4}{3}.$$

Thus,

$$I = \rho_m \left(\frac{1}{5}R^5\right)\left(\frac{4}{3}\right)(2\pi) = \frac{8\pi \rho_m R^5}{15}.$$

The constant density (mass/volume) is $\rho_m = M/(4\pi R^3/3)$, so

$$I = \frac{3M}{4\pi R^3}\frac{8\pi R^5}{15} = \frac{2}{5}MR^2.$$

Note this is the result we used in Chapter 10. △

14.5 Problems

14.1 Double Integrals

For Problems 1–8, evaluate the given double integral over the rectangular region given by the limits of integration by performing an iterated integral in the order shown.

1. $\displaystyle\int_0^1 \int_{-1}^3 x^2\, dx\, dy$

2. $\displaystyle\int_1^4 \int_1^3 (x^2 + y^2)\, dx\, dy$

3. $\displaystyle\int_1^5 \int_{-1}^3 \frac{2x}{y}\, dx\, dy$

4. $\displaystyle\int_{-2}^{-1} \int_{-1}^2 ye^x\, dx\, dy$

5. $\displaystyle\int_0^\pi \int_{\pi/2}^{3\pi/4} x \cos y\, dx\, dy$

6. $\displaystyle\int_1^3 \int_0^4 (5x^4 + y)\, dy\, dx$

7. $\displaystyle\int_{-1}^0 \int_1^2 3xe^{4y}\, dy\, dx$

8. $\displaystyle\int_1^3 \int_1^2 \frac{1}{x+y}\, dy\, dx$

For Problems 9–12, compute a lower sum and an upper sum for the given double integral. Use four subintervals in each direction.

9. $\displaystyle\iint_R e^{-(x^2+y^2)}\, dV$ where $R = [0,2] \times [0,2]$

10. $\displaystyle\iint_R \frac{1}{x^4 + y^4}\, dV$ where $R = [1,2] \times [1,2]$

11. $\displaystyle\iint_R \frac{1}{x^4 + y^4}\, dV$ where $R = [-2,-1] \times [3,7]$

12. $\displaystyle\iint_R (x^2 + y^3)\, dV$ where $R = [0,2] \times [0,2]$

14.2 Double Integrals Over Other Regions

For Problems 1–6, evaluate the double integral given.

1. $\displaystyle\int_0^3 \int_1^x 3xy^2\, dy\, dx$

2. $\displaystyle\int_1^3 \int_x^{x^2} 4y^4\, dy\, dx$

3. $\displaystyle\int_0^1 \int_x^{\sqrt{x}} (x^2 + y^2)\, dy\, dx$

4. $\displaystyle\int_{-1}^2 \int_y^8 xy\, dx\, dy$

5. $\displaystyle\int_1^2 \int_0^y \frac{1}{x^2 + y^2}\, dx\, dy$

6. $\displaystyle\int_2^4 \int_y^{y^2} x^2y^3\, dx\, dy$

For Problems 7–15, evaluate the double integral of $f(x,y)$ over the region of the xy-plane described.

7. $f(x,y) = 2x(x^2 + y^3)$, the triangle with vertices at $(1,0)$, $(1,2)$, and $(4,0)$

8. $f(x,y) = e^x$, the region between the curves $y = \sqrt{x}$ and $y = x$

9. $f(x,y) = 2xy$, the triangle with vertices at $(0,3)$, $(3,3)$, and $(3,0)$.

10. $f(x,y) = x^3y$, the region bounded by $y = 1$, $x = 2$, and $y = 4 - x$

11. $f(x,y) = 2xy^2$, the region bounded by $y = 0$, $x = 1$, $x = 4$, and $y = 3x/2$

12. $f(x,y) = y^2$, a semicircle of radius 3 in the lower half-plane, with the arc's center at the origin

13. $f(x, y) = x + 3y$, the region bounded by the hyperbola $x^2 - y^2 = 1$, the line $y = -2$ and the line $y = 1$

14. $f(x, y) = x^2 y$, the region in the first quadrant bounded by the ellipses $\frac{x^2}{4} + y^2 = 1$ and $\frac{x^2}{9} + \frac{y^2}{4} = 1$.

15. $f(x, y) = xy^2$, the region shown in the figure below.

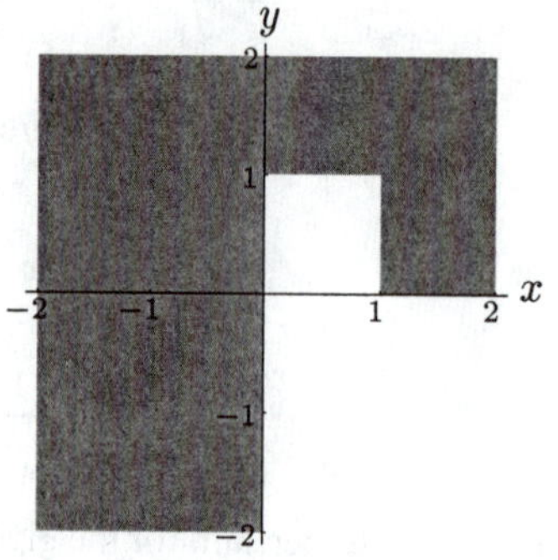

For Problems 16–20, evaluate the double integral of the given function using polar coordinates over the region of the xy-plane described.

16. $f(r, \theta) = 3r$, the region bounded by $r = 5, \theta = \pi$, and $\theta = 4\pi/3$

17. $f(x, y) = x + y$, the region bounded by $r = 2, r = 3, \theta = 0$, and $\theta = \pi/2$

18. $f(r, \theta) = r \sin\theta \cos\theta$, the region bounded by $r = 1, \theta = \pi/4$, and $\theta = \pi/2$

19. $f(r, \theta) = \sin^2\theta$, the region bounded by $r = 1$ and $r = 2$

20. $f(x, y) = x + y$, the region inside the petal of the polar curve $r(\theta) = \sin(4\theta)$ that is in the third quadrant and closest to the y-axis.

For Problems 21–23, find the area of the region described using a double integral.

21. the triangle with vertices at $(1, 0)$, $(1, 2)$, and $(4, 0)$

22. the bounded region between the curves $y = \sqrt{x}$ and $y = x$

23. one "petal" of the polar curve $r(\theta) = \cos(3\theta)$

14.3 Triple Integrals

For Problems 1–6, evaluate the following triple integrals over the rectangular solids defined by the limits of integration. You may perform an iterated integral in any order you choose.

1. $\int_0^1 \int_0^1 \int_0^1 (x^2 + y^2 + z^2)\, dx\, dy\, dz$

2. $\int_{-1}^1 \int_{-1}^1 \int_0^1 4xyz\, dx\, dy\, dz$

3. $\int_1^3 \int_2^4 \int_1^2 \frac{4}{x} yz \, dx \, dy \, dz$

4. $\int_{-1}^5 \int_{-\pi}^{\pi/2} \int_0^{\pi/2} \cos(x+y) \, dx \, dy \, dz$

5. $\int_{-1}^1 \int_1^4 \int_2^7 \ln(xy) \, dx \, dy \, dz$

6. $\int_0^2 \int_{-2}^0 \int_{-1}^1 e^{-x+y+z} \, dx \, dy \, dz$

For Problems 7–10, the given point is in rectangular coordinates. Express each in cylindrical coordinates.

7. $(-1, -3, -3)$

8. $(0, 1.2, 3.0)$

9. $(4.0, -2.7, -2.0)$

10. $(-1.5, 1.9, -8.3)$

For Problems 11–14, the given point is in cylindrical coordinates. Express each in rectangular coordinates.

11. $(3.4, \frac{3\pi}{8}, -1.3)$

12. $(1.2, \frac{7\pi}{8}, 2.2)$

13. $(5.0, \frac{14\pi}{3}, 1.4)$

14. $(14, 6.0, 9.5)$

15. A cylinder of radius R and height h is cut in half by a plane that contains the cylinder's axis. Describe the boundaries of the remaining half-cylinder using cylindrical coordinates.

For Problems 16–19, the given point is in rectangular coordinates. Express each in spherical coordinates.

16. $(0, 1.2, 3.0)$

17. $(-1, -3, -3)$

18. $(2, -1, 3)$

19. $(4.0, -2.7, -2.0)$

For Problems 20–23, the given point is in spherical coordinates. Express each in rectangular coordinates.

20. $(2.0, \pi, 0)$

21. $(3.5, \frac{\pi}{2}, \frac{\pi}{2})$

22. $(4.4, 2.9, 1.7)$

23. $(15, 2.0, 10)$

24. Using spherical coordinates, describe the quarter of Earth that is in *both* the Northern and Western hemispheres.

25. A cone of height h and radius R has its axis of symmetry on the z-axis and is inverted so that just a single point touches the origin. Describe the surface of the cone using cylindrical coordinates and spherical coordinates.

For Problems 26–33, compute the triple integral of the given function over the given limits in cylindrical or spherical coordinates.

26. $f(r, \theta, z) = r \cos\theta \sin\theta, \quad r : [0, 5], \quad \theta : [0, \frac{\pi}{2}], \quad z : [0, 2]$

27. $f(r, \theta, z) = 2r^2, \quad r : [0, 3], \quad \theta : [0, 2\pi], \quad z : [-1, 1]$

28. $f(r, \theta, z) = -rz, \quad r : [0, 9], \quad \theta : [\frac{\pi}{2}, 2\pi], \quad z : [0, 1]$

29. $f(r, \theta, z) = 3z \sin\theta, \quad r : [1, 3], \quad \theta : [\frac{\pi}{2}, \pi], \quad z : [0, 1]$

30. $f(\rho, \phi, \theta) = 3\rho, \quad \rho : [0, 4], \quad \phi : [0, \frac{\pi}{2}], \quad \theta : [0, 2\pi]$

31. $f(\rho,\phi,\theta) = \rho^2 \sin\phi, \quad \rho : [1,2], \quad \phi : [0,\pi], \quad \theta : [0,\pi]$

32. $f(\rho,\phi,\theta) = \cos\phi, \quad \rho : [0,1], \quad \phi : [0,\frac{\pi}{2}], \quad \theta : [0,2\pi]$

33. $f(\rho,\phi,\theta) = e^{-\rho}, \quad \rho : [0,\infty], \quad \phi : [0,\pi], \quad \theta : [0,2\pi]$

34. A solid sphere of radius 20 cm lies centered at the origin. A cylindrical cut of radius 5 cm and centered on the z-axis is made through the sphere. What is the volume of material cut and the remaining volume from the original sphere?

14.4 Applications of Multiple Integrals in Physics

1. A thin uniform triangular plate lies in the xy-plane. The vertices of the triangle lie at $(0,0)$, $(0,12$ cm), (4 cm, 12 cm). Find the center of mass of the plate.

2. A thin metal plate has a uniform area mass density of $\sigma = 0.650$ kg/m^2. The plate lies in the xy-plane and its edges are defined by the x-axis, the y-axis, the line $x = 15$ cm, and the line $y = 5 + x$ (with output in cm for input in cm). Find the plate's total mass and the location of its center of mass.

3. Find the center of mass of a cone of radius R and height h.

4. A cone of radius $R = 3$ cm and height $h = 5$ cm lies with its base in the xy-plane and its axis of symmetry along the $+z$-axis. The mass density is not constant, but varies according to the formula $\rho = (5 - z)(1000)$ (with output in kg/m^3 for input in cm). Find the cone's total mass and center of mass.

5. A 15-cm-long cylinder lies with one end in the xy-plane and the other end in the $z = 15$ cm plane. The cylinder's radius is 2.5 cm. Assuming a constant density of 7600 kg/m^3, find the total mass and the center of mass of the cylinder.

6. Repeat the previous problem if the density of the cylinder is not constant but varies according to the function $\rho = 0.6z$ (with output in g/cm^3 for input in cm).

7. A uniform solid sphere with radius 10 cm has its center at the origin. A solid spherical region of radius 4.0 cm centered at the point $(-5.0$ cm, $0.0)$ is removed. Where is the center of mass of the remaining object?

8. Find the rotational inertia of a solid cylinder of mass M and radius R about an axis tangent to the curved surface and parallel to the main axis.

9. Find the rotational inertia of a solid sphere of mass M and radius R about an axis tangent to the surface.

10. A uniform solid cylindrical shell has a mass M, inner radius R_1, and outer radius R_2. Find the rotational inertia about the main axis. Show that the correct results follow in these limits: (a) $R_1 \to 0$, and (b) $R_1 \to R_2$.

11. A very thin spherical shell has a mass M and radius R. Find the shell's rotational inertia about an axis through its center. Compare with the rotational inertia of a solid sphere of the same mass and radius.

12. Find the rotational inertia of a uniform solid cone of mass M, radius R, and height h about the cone's axis of symmetry.

13. A uniform solid spherical shell has a mass M, inner radius R_1, and outer radius R_2. Find the rotational inertia about an axis through the sphere's center. Show that the correct results follow in these limits: (a) $R_1 \to 0$, and (b) $R_1 \to R_2$ (see Problem 11).

14. A thin rectangular plate of length a and width b has mass M. Find the rotational inertia about a perpendicular axis through the plate's center.

Chapter 15

Gravitation

In this chapter we study gravitation in general (not just near Earth's surface). More specifically, we concentrate on *Newton's law of gravitation* and see how it can be used to explain the motion of planets and satellites. This beautiful work represents the culmination of Newton's work on mechanics, and appropriately we use it as the capstone of our study of classical mechanics.

15.1 Background and History

As we enter the 21st century, it is difficult to imagine the level of scientific understanding when Newton began his work in the middle of the 17th century. One of the foremost scientific questions of the day was: what makes the planets move in their orbits? Let's look briefly at the history behind this question and then see how Newton came to answer the question as he did.

We begin in the ancient world. Ancient astronomers looked up and saw the Sun, Moon, planets, and stars appearing to circle Earth once each day. The stars retain a fixed pattern, but the Sun, Moon, and planets move a little each day against the background of fixed stars. The motion of these "wanderers" is generally west to east. However, occasionally (but predictably) the planets exhibit **retrograde motion** and move east to west for days or weeks at a time.

The ancient Greeks developed an elaborate geometric model that made good predictions of the planets' motions. In this system, Earth was fixed at the center of the universe while all the heavenly bodies turned in circles about Earth. In order to explain retrograde motion, the Greeks placed the planets on small circles attached to larger circles. The combined circular motions lead to periodic retrograde motion, as shown in Figure 15.1. The second century C.E. Greek astronomer Claudius Ptolemy used the long history of planetary observations and painstaking mathematical work to refine the Earth-centered model to the point where its description of planetary motions was extremely accurate.

For many centuries Ptolemy's work was the last word in planetary astronomy. In Europe, it had no significant challenges until the 16th century. This longevity was due in part to the theory's mathematical accuracy. It was also due to the reluctance of medieval European thinkers to question the Earth-centered description of the physical world, which they were able to accommodate quite well to their Christian doctrines. This was important, because scholarship in the medieval and Renaissance periods was centered in religious training and thinking. Things were quite different from today, when we think of science and religion as separate areas of research.

A significant change in thought began in 1543, when the Polish astronomer Nicholas Copernicus (1473–1543) published his theory of planetary motion, in which the Sun remains fixed and Earth and the other

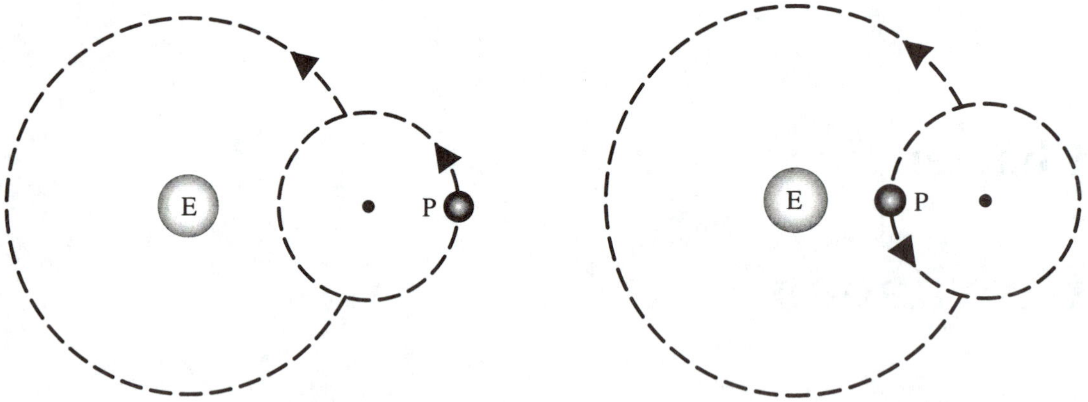

(a) Planet in "normal" motion

(b) Planet in retrograde motion

Figure 15.1. Simple version of a Greek model for the solar system. Planet P in orbit around Earth, E.

planets orbit the Sun. The model developed by Copernicus is similar to Ptolemy's in that it uses circles attached to circles to approximate the motions of the planets. In fact, Copernicus's sun-centered system is mathematically equivalent to Ptolemy's Earth-centered system. But most 16th century thinkers were not compelled to give up their cherished Earth-centered views.

Early in the 17th century Galileo provided a number of significant arguments in support of a Sun-centered solar system. He was the first to use the telescope in astronomy, and he used his telescopic observations of the moons of Jupiter and the phases of Venus in those arguments, along with his physical reasoning that made it more plausible for Earth to be circling around the Sun once a year and rotating on its axis once a day. For his views, Galileo was widely praised in the scientific community but severely censured by the Catholic Church in his native Italy. He was forced to recant his views publicly and was placed under house arrest for the last decade of his life.

Another important contribution was made by Galileo's contemporary, the German astronomer and mathematician Johannes Kepler (1571–1630). Kepler had worked in Denmark with the Danish astronomer Tycho Brahe (1546–1601), who was probably the greatest naked-eye astronomer of all time. In the late 16th century, Brahe made improved measurements of the positions of the planets. Kepler later used that data to show that the planets' orbits are better fit by *ellipses* (with the Sun fixed at one focus of each planet's orbital ellipse) than by the combinations of circles used in the Copernican model. Later in this chapter we look at Kepler's ellipses in some detail.

Thanks to Copernicus, Galileo, Kepler, and others who followed, the view that planets (including Earth) move in ellipses around a fixed Sun had gained wide acceptance by the middle of the 17th century (at least in Protestant countries, including England). But there remained the question with which we started this discussion: what makes the planets move in their orbits? The 17th century scientists prior to Newton were unable to provide a satisfactory answer to this question. Newton succeeded in large part because he had a clear understanding of dynamics (as we developed in Chapters 5) and calculus (as we have developed throughout the text). Newton also made the theoretical breakthrough of applying gravity to the motion of the Moon and planets, and he developed a mathematical model of gravity that successfully predicted the observed elliptical orbits. After studying Kepler's ellipses in Section 15.2, we study Newton's law of gravitation in Section 15.3.

15.2 Kepler's Laws of Planetary Motion

15.2.1 The Geometry of Ellipses

In Kepler's model, the planets travel in elliptical orbits around the Sun. Therefore, we begin with a brief review of the geometry of ellipses.

In Figure 15.2(a), we show an ellipse drawn in the xy-plane with its center at the origin. The equation that defines this ellipse is

$$\frac{x^2}{a^2} + \frac{y^2}{b^2} = 1 \tag{15.1}$$

where a and b are constants. By convention, we set $a \geq b$. In other words, we choose a coordinate system in which the longer axis of the ellipse lies along the x-axis. You can find the points where the ellipse intersects the x-axis by setting $y = 0$ in Equation (15.1) and solving for x. This yields $x = \pm a$. Similarly, the ellipse crosses the y-axis at $y = \pm b$. The ellipse is symmetric with respect to both the x-axis and y-axis. The line segment of length $2a$ from $x = -a$ to $x = a$ is called the **major axis** of the ellipse, and a (the distance from the center to either end of the ellipse along the x-axis) is the length of the **semimajor axis**. Similarly the line segment of length $2b$ from $y = -b$ to $y = b$ is called the **minor axis**, and b is the length of the **semiminor axis**. A circle is a special case of an ellipse, with $a = b$. In this case, $a = b = r$, the radius of the circle.

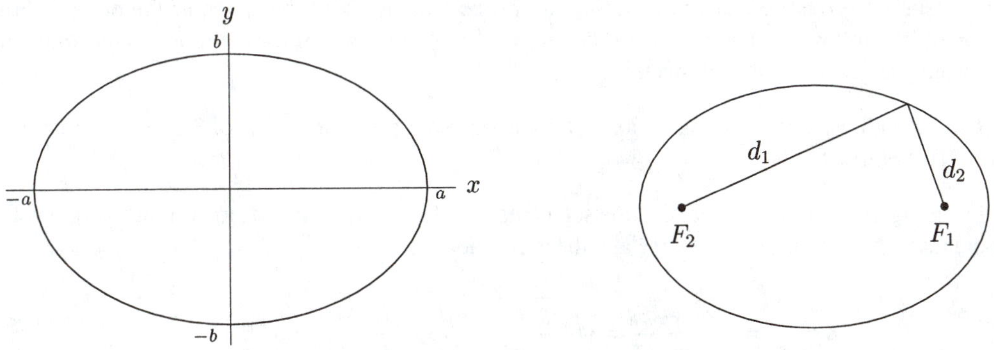

(a) Major and minor axes of an ellipse

(b) The sum $d_1 + d_2$ of the distances from a point to the foci is constant

Figure 15.2. Basic geometry of an ellipse.

An equivalent definition of an ellipse is the locus of all points such that the sum of the distances from two fixed points is a constant. The geometry of this definition is illustrated in Figure 15.2(b). Each of the two fixed points (F_1 and F_2) is called a **focus** of the ellipse. It is straightforward to show that the sum of the distances is $d_1 + d_2 = 2a$.

The general shape of an ellipse is governed by the relative sizes of the parameters a and b. Alternatively, we can use a third parameter c, defined as the distance from the center of the ellipse to either focus (Figure 15.3). From the relation $d_1 + d_2 = 2a$, we see that it is a distance a from the focus to a point on the ellipse. Notice that c can take on values in the interval $[0, a)$. If c is close to a, then the ellipse is long and thin, as shown in Figure 15.3(a), but if c is close to zero, the ellipse is fat, as shown in Figure 15.3(b). The special case $c = 0$ defines a circle.

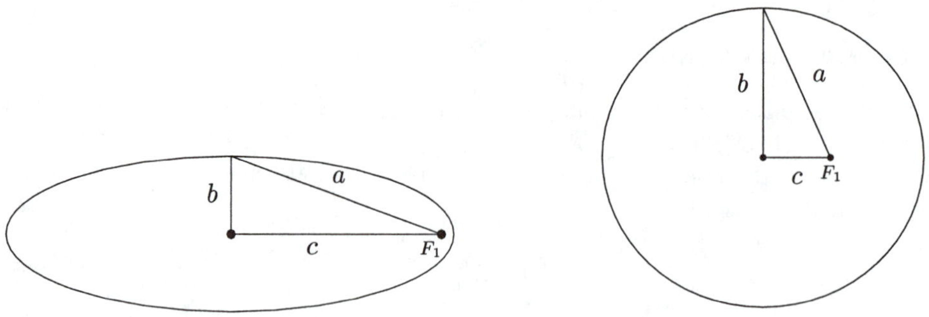

(a) Large eccentricity (b) Small eccentricity

Figure 15.3. Parameters in the eccentricity of an ellipse.

The shape of the ellipse is described by its **eccentricity** e, defined as

$$e = \frac{c}{a}. \tag{15.2}$$

(Don't confuse the parameter e with the base of the natural logarithms.) This single parameter is convenient because it measures the relative sizes of c and a, and hence tells us about the shape of the ellipse. From the definition of c, it follows that e must be in the interval $[0, 1)$, with $e = 0$ for a circle. As the value of e increases, the ellipse becomes more elongated.

Example 15.1. An ellipse has a semimajor axis of 5.4 units and a semiminor axis of 3.5 units. What is the eccentricity of the ellipse?

Solution. From Figure 15.3, we see that the line segments with length a, b, and c form a right triangle with a as the hypotenuse. Therefore, $c = \sqrt{a^2 - b^2}$ and the eccentricity is

$$e = \frac{c}{a} = \frac{\sqrt{a^2 - b^2}}{a} = \sqrt{1 - \frac{b^2}{a^2}} = \sqrt{1 - \frac{(3.5)^2}{(5.4)^2}} = 0.76. \qquad \triangle$$

15.2.2 Kepler's Laws

Tycho Brahe's observations were excellent for their time, but since they were made with the naked eye they are poor by today's standards. Generally, he could measure the positions of the planets to within 1 or 2 minutes of arc (1 minute = $1' = \frac{1}{60}$ degree). The problem of trying to prove that the planetary orbits are ellipses was complicated by the fact that the observations were being made from a planet—Earth—that itself travels on such an ellipse. The eccentricities of the orbits of planets visible to the naked eye are low, and hence the orbits are difficult to distinguish from circles. Nevertheless, after years of hard work, Kepler was able to use some particularly good observations of Mars (which has a relatively high eccentricity of 0.09) to establish his three laws of planetary motion.

In 1609 Kepler formulated the first two laws:

1. The orbit of each planet is an ellipse, with the Sun at one focus.

2. As measured from the Sun, a given planet sweeps out the same area in its orbital plane in a given amount of time anywhere in its orbit.

A generic planetary orbit is shown in Figure 15.4(a). The point in the planet's orbit closest to the Sun is called the **perihelion** (peri- for near and -helion for Sun). The farthest point from the Sun is the **aphelion**. The second law is illustrated in Figure 15.4(b). When a planet is closer to the Sun, it must move faster in order to sweep out the same area (the shaded region in the figure) as it does in the same time interval when it is farther from the Sun.

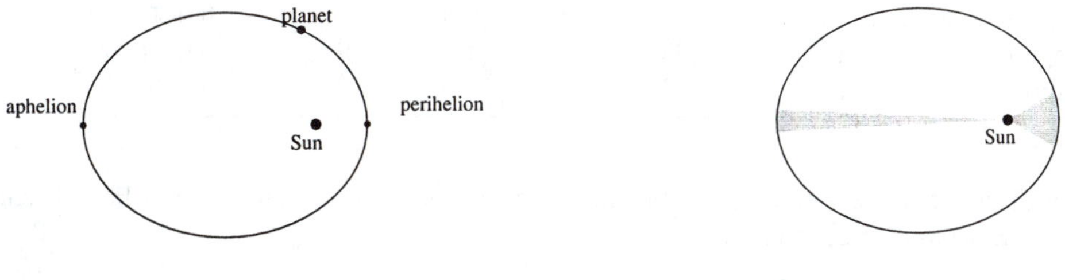

(a) First law (b) Second law

Figure 15.4. Kepler's laws of planetary motion.

In 1619, Kepler presented his third law, sometimes called the *harmonic law*:

3. If T is the period of a planet's orbit around the Sun and a is the semimajor axis of the orbital ellipse, then

$$\frac{a^3}{T^2} = C$$

where C is a constant.

In fact, C is the same for every planet, comet, asteroid, or other object orbiting the Sun. We will see that C depends on the mass of the Sun, so the parameter C has a different value for objects orbiting other objects.

We present some data for the planets and their orbits in Table 15.1. Notice that the eccentricities of most planets' orbits are quite low, meaning that the orbits do not deviate much from circles. Except for Pluto, the planets' orbital planes are quite close to Earth's orbital plane, known as the **ecliptic plane**. However, the planets' major axes do not lie along the same line.

Example 15.2. Use the data for Earth's orbit to find the value of the constant C in Kepler's third law. Be sure to include the correct SI units.

Solution. Plugging in the values from the table:

$$C = \frac{a^3}{T^2} = \frac{(1.496 \times 10^{11}\ \text{m})^3}{[(1.00\ \text{y})(3.156 \times 10^7\ \text{s/y})]^2} = 3.361 \times 10^{18}\ \text{m}^3/\text{s}^2.$$

△

Example 15.3. Using the data for Earth's orbit find the distance from Earth to the Sun at perihelion and aphelion.

Table 15.1. Data on planetary orbits

Planet	Mass (kg)	Orbital period (y)	Semimajor axis (m)	Eccentricity
Mercury	3.20×10^{23}	0.241	5.76×10^{10}	0.2056
Venus	4.87×10^{24}	0.616	1.08×10^{11}	0.0068
Earth	5.976×10^{24}	1.00	1.496×10^{11}	0.0167
Mars	6.42×10^{23}	1.88	2.28×10^{11}	0.0934
Jupiter	1.90×10^{27}	11.9	7.78×10^{11}	0.0484
Saturn	5.69×10^{26}	29.5	1.43×10^{12}	0.0557
Uranus	8.66×10^{25}	84.1	2.87×10^{12}	0.0472
Neptune	1.03×10^{26}	165	4.50×10^{12}	0.0086
Pluto	1.4×10^{22}	249	5.91×10^{12}	0.249

Solution. From Figure 15.3, we see that the perihelion distance is $d_p = a - c$ and the aphelion distance is $d_a = a + c$. From Equation (15.2), we have $c = ea$, so

$$d_p = a - c = a - ea = (1 - e)\,a.$$

Plugging in the numerical values,

$$d_p = (1 - 0.0167)(1.496 \times 10^{11}\ \text{m}) = 1.471 \times 10^{11}\ \text{m}$$

or just over 147 million km.

Similarly, the aphelion distance is

$$d_a = a + c = a + ea = (1 + e)a = (1 + 0.0167)(1.496 \times 10^{11}\ \text{m}) = 1.521 \times 10^{11}\ \text{m}$$

or just over 152 million km. The difference between these two distances is about 5 million km, so Earth is about 5 million km closer to the Sun at perihelion (in early January) than at aphelion (in early July). △

Kepler's three laws hint that there is a universal force at work, or at least one valid throughout the solar system. This is particularly evident from the third law, in which the same constant relates the semimajor axis and period for all the planets. But what is the nature of this force, and how does it produce elliptical orbits? It was left for Isaac Newton to answer this question.

15.3 Newton's Law of Gravitation

15.3.1 Introduction to Newton's Law of Gravitation

The name Newton is often associated with a falling apple. A legend says that Newton saw an apple fall and at that moment realized that the same force—gravity—that makes the apple fall also holds the Moon in its orbit around Earth. Whether true or not, that legend is a useful way to remember Newton's important realization that the attractive force of gravity alone is responsible for the orbital motions of the Moon around Earth and the planets around the Sun. It also illustrates the sudden flash of insight many scientists have had after thinking about a problem for a long time. Alas, for most of us the insights are not as great as Newton's.

Let's pursue the relationship between the apple and the Moon. Approximating the Moon's orbit around Earth to be circular, the Moon's centripetal acceleration a_C is

$$a_C = \frac{v^2}{R} = \frac{\left(\dfrac{2\pi R}{T}\right)^2}{R} = \frac{4\pi^2 R}{T^2}$$

where R is the radius of the Moon's orbit and T its period. Plugging in the (modern) values $R = 3.844 \times 10^8$ m and $T = 27.3$ d$= 2.36 \times 10^6$ s, we have

$$a_C = \frac{4\pi^2\left(3.844 \times 10^8 \text{ m}\right)}{\left(2.36 \times 10^6 \text{ s}\right)^2} = 2.72 \times 10^{-3} \text{ m/s}^2.$$

How does this compare to the acceleration of the falling apple? In round numbers, it is almost exactly equal to $g/3600$ (where $g = 9.80$ m/s^2 is the acceleration of falling bodies near Earth's surface).

This example shows that the magnitude of the gravitational force becomes weaker at greater distances. But how exactly does the magnitude of the force depend on the distance? Newton realized (see Section 15.5) that for spherically symmetric bodies such as the Moon and Earth, the gravitational force can be computed as if the entire mass of the body were a point particle located at the center of the sphere. Therefore, we should compare the distance from the Earth's center to the center of the Moon ($r = 3.84 \times 10^8$ m) with the distance from the Earth's center to the apple, which is just the radius of Earth ($R_E = 6.37 \times 10^6$ m). The ratio of these distances is

$$\frac{R_E}{r} = \frac{6.37 \times 10^6 \text{ m}}{3.84 \times 10^8 \text{ m}} = 1.66 \times 10^{-2}$$

or almost exactly a ratio of 1/60.

To summarize, the gravitational acceleration (proportional to the gravitational force by Newton's second law of motion) of the Moon is 1/3600 that of the apple, while the distance from Earth's center is 60 times greater. With these numbers working out so well, a reasonable hypothesis is that *the magnitude of the gravitational force is proportonal to the inverse-square of the distance.* That is, for two spherically symmetric bodies with a distance r between their centers (or point particles separated by a distance r), the magnitude of the gravitational force goes as

$$F(r) = \frac{\text{constant}}{r^2}.$$

The distance dependence is an important part of Newton's law of gravitation, but it is not all of it. How does the magnitude of the gravitational force depend on the masses of the attracting bodies? Consider two point particles with masses M and m separated by a distance r. From Newton's second law of motion, we expect that the magnitude of the gravitational force on each particle should be proportional to its mass. From Newton's third law of motion, the gravitational forces on the two particles must have the same magnitudes but opposite directions. The only self-consistent way to obey both Newton's second and third laws of motion is for the magnitude of the gravitational force to be proportional to *both* masses—that is, proportional to Mm. Combining this with the inverse-square distance law, the magnitude of the gravitational force is

$$F = (\text{constant})\,\frac{Mm}{r^2}.$$

It is customary to use the symbol G for the constant in the above equation, so

$$F = \frac{GMm}{r^2}. \tag{15.3}$$

Equation (15.3) gives the *magnitude* of the gravitational force. The *direction* of the force on each particle is toward the center of the other particle, as we know by experience (consider the apple or the Moon) and Newton's third law of motion. The numerical value of the **universal gravitation constant** G is approximately

$$G = 6.673 \times 10^{-11} \text{ N·m}^2/\text{s}^2.$$

We can express Newton's law of gravitation in vector form using a unit vector $\hat{r}$ that points from the particle of mass M toward the particle of mass m. Then the force on the particle of mass m is

$$\vec{F} = -\frac{GMm}{r^2}\,\hat{r}\,. \tag{15.4}$$

This is the final form of Newton's law of gravitation.

Let's apply Newton's law of gravitation to the problem of planetary orbits. For the moment, we will assume a circular orbit of a planet of mass m around the Sun, which has mass M. (The mathematics of elliptical orbits is considerably more complicated, and you will encounter it in an advanced mechanics course.) For a circular orbit of radius R and period T, the magnitude of the gravitational force equals the magnitude of the centripetal force, or

$$\frac{GMm}{R^2} = \frac{mv^2}{R}\,.$$

The mass m of the planet cancels, and on rearranging we have

$$v^2 = \frac{GM}{R}\,.$$

Now $v = 2\pi R/T$, so

$$\left(\frac{2\pi R}{T}\right)^2 = \frac{GM}{R}$$

or after rearranging

$$\frac{R^3}{T^2} = \frac{GM}{4\pi^2}\,. \tag{15.5}$$

Recall that a circle is a special case of an ellipse, with $R = a$. Thus, we see that Equation (15.5) is consistent with Kepler's third law of planetary motion. The right side of Equation (15.5) is a constant for the solar system, and the numerical value of the right side is consistent with the value we found in Example 15.2.

Newton showed in his book *Principia* that in general the inverse-square force from a fixed center (e.g., the Sun) leads to orbits that are *conic sections*. You may remember from analytic geometry that the conic sections include the circle, ellipse, parabola, and hyperbola. While planets follow elliptical paths, many asteroids and comets are observed to follow (nonreturning) parabolic or hyperbolic orbits with the Sun at one focus.

Example 15.4. Show that Newton's law of gravitation correctly predicts the gravitational acceleration of bodies dropped near the Earth's surface.

Solution. In this example, we take M to be the mass of Earth and m the mass of the falling body. As in our study of the apple's fall, the distance between the bodies' centers is approximately the radius of Earth R_e. By Newton's second law of motion, we set the magnitude of the gravitational force on the falling body equal to the product of its mass and acceleration:

$$\frac{GMm}{R_E^2} = ma\,.$$

The mass of the body cancels, leaving

$$a = \frac{GM}{R_E^2}\,.$$

As Galileo found, the acceleration of the falling body is independent of its mass. Inserting numerical values we find

$$a = \frac{(6.673 \times 10^{-11}\ \text{N}\cdot\text{m}^2/\text{s}^2)\,(5.976 \times 10^{24}\ \text{kg})}{(6.374 \times 10^6\ \text{m})^2} = 9.82\ \text{m/s}^2.$$

The observed value of g is slightly reduced from this value as we move away from the poles, due to the Earth's rotation (see Chapter 4). △

Example 15.5. Show that the centripetal acceleration of the Moon that was computed previously follows from Newton's law of gravitation.

Solution. Again, using the approximation of a circular orbit, we can equate the magnitude of the gravitational force and the centripetal force to find

$$ma_C = \frac{GMm}{R^2}$$

where a_c is the Moon's centripetal acceleration, m is the Moon's mass, and M is Earth's mass. Then

$$a_C = \frac{GM}{R^2}$$

and inserting numerical values

$$a_C = \frac{(6.673 \times 10^{-11}\ \text{N}\cdot\text{m}^2/\text{s}^2)\,(5.976 \times 10^{24}\ \text{kg})}{(3.84 \times 10^8\ \text{m})^2} = 2.70 \times 10^3\ \text{m/s}^2$$

which is the same result as before, to within rounding errors. △

15.3.2 Newton's Law of Gravitation and Kepler's Second Law

It is possible to show that Newton's law of gravitation predicts Kepler's law of motion for an orbit in the shape of any of the conic sections. We again restrict our analysis to a circular orbit and leave the other orbits for the problems at the end of the chapter.

Our argument is based in the principle of conservation of angular momentum. The gravitational force on an orbiting body is always toward the attracting body (the Sun in the case of planetary motion). Recall that in general $\vec{\tau} = \vec{r} \times \vec{F}$. In this case, the vectors $\vec{r}$ and $\vec{F}$ are in opposite directions, so the cross product is zero and there is zero torque on the orbiting body. Hence, its angular momentum is conserved.

As Figure 15.5 shows, for a circular orbit the radius vector and momentum vector are perpendicular.

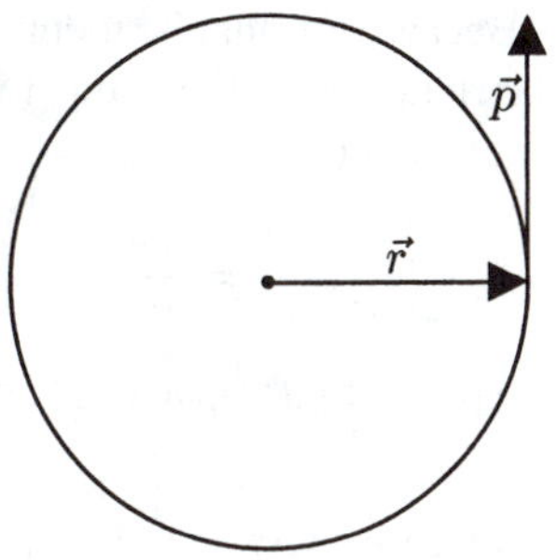

Figure 15.5. Radius vector and momentum vector for an object on a circular orbit.

Thus, the (constant) magnitude of the angular momentum is

$$L = |\vec{r} \times \vec{p}| = rp\sin\frac{\pi}{2} = rp = rmv.$$

We can write the angular momentum in terms of the angular speed of the orbiting planet. In circular motion $v = r\omega$, so

$$L = rm(r\omega) = mr^2\omega. \tag{15.6}$$

Consider now a small angular displacement in the orbit $\Delta\theta$. The region swept out is a circular sector subtending an angle $\Delta\theta$ on a circle of radius r (see Figure 15.6). Therefore, the area is

$$\Delta A = \frac{1}{2}(r)(r\Delta\theta) = \frac{1}{2}r^2\Delta\theta. \tag{15.7}$$

Dividing both sides of Equation (15.7) by the corresponding time interval Δt,

$$\frac{\Delta A}{\Delta t} = \frac{1}{2}r^2\frac{\Delta\theta}{\Delta t}. \tag{15.8}$$

Taking the limit of this expression as Δt approaches zero,

$$\lim_{\Delta t\to 0}\frac{\Delta A}{\Delta t} = \frac{1}{2}r^2\lim_{\Delta t\to 0}\frac{\Delta\theta}{\Delta t} = \frac{1}{2}r^2\omega \tag{15.9}$$

where we have used the definition of angular velocity.

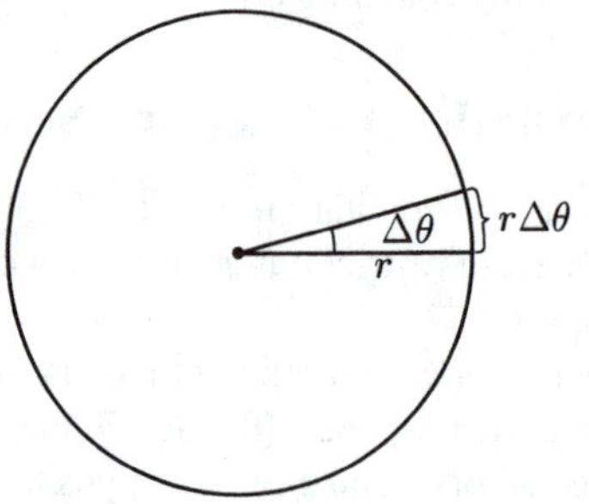

Figure 15.6. Small angular displacement.

Compare Equations (15.6) and (15.9). Because the angular momentum in Equation (15.6) is constant (and the mass of the planet is also constant), the right side of Equation (15.9) is constant. From our knowledge of calculus, we can interpret the left side of Equation (15.9)

$$\lim_{\Delta t\to 0}\frac{\Delta A}{\Delta t} = \frac{dA}{dt}$$

as the instantaneous area per unit time swept out by the planet. Thus, we see that the area per unit time is constant and is equal to

$$\frac{\text{area}}{\text{time}} = \frac{1}{2}r^2\omega = \frac{L}{2m} \tag{15.10}$$

and Kepler's second law is verified in this case.

15.3.3 Newton's Law of Gravitation in Perspective

It is difficult to overestimate the impact of Newton's work on the science of the past 300 years. Of course, to a great extent Newton, by his own admission, stood "on the shoulders of giants" who preceded him: Copernicus, Galileo, and Kepler. And as we have already shown, Newton went beyond his predecessors in several important ways—for example, by showing that orbits in the shapes of conic sections result from in inverse-square force and that a spherically symmetric body acts (gravitationally) like a point particle with its mass concentrated at the sphere's center. In this section we look at two areas of *conceptual* thought strongly affected by Newton's work on gravitation.

In applying the inverse-square force law to planetary motion, we have effectively assumed that the range of the gravitational force is infinite, or at least that it acts over astronomical distances, and that it acts somehow through the nearly perfect vacuum of space. Physicists refer to the action of a force directed from one body to another through a vacuum as **action at a distance**. Most forces in our everyday lives do not appear to be this way, because they involve some direct contact between two bodies: We pull a door open or lift a box from a shelf, a drive shaft supplies torque to a car's wheels, and so on.

Therefore, there is something disquieting about action at a distance, and this fact was not lost on Newton. To his credit, Newton distinguished himself from other philosophers of his day by admitting not to know the answer to this puzzle. The well-quoted phrase on this issue from *Principia* is "Hypothesis non fingo," meaning "I make no hypothesis" on how action at a distance ultimately works. For Newton, it was sufficient to develop the mathematical law and show that the observed phenomena can be derived from the law.

We do see action at a distance in our everyday lives when we deal with electricity and magnetism. In Chapter 17, we will introduce the concept of the field, which provides a way to deal with action at a distance in gravitation as well as electricity and magnetism. The concept of the gravitational field plays a key role in Einstein's theory of general relativity (published in 1915), which has supplanted Newton's theory of gravitation. In recent years, physicists have developed theories of gravitation that involve the exchange of particles (called gravitons) as a way of addressing the action at a distance problem in gravitation, while the exchange of other particles (photons) accounts for electric and magnetic forces.

The concept of a universally valid set of physical laws seems natural enough today but was by no means obvious in the 17th century. The idea that through the power of reason humans can understand fundamental laws governing the universe had a significant impact on the thinking of scientists and philosophers who followed Newton.

Newtonian science played no small role in the Enlightenment, the central intellectual movement in Europe in the 18th century. Many great Enlightenment thinkers studied and admired the work of Newton. This long list included Voltaire in France and Jefferson in America, who saw the power of reason as a basis for their revolutionary democratic views.

15.4 Gravitational Potential Energy

We know from experience that gravity is a conservative force. A projectile maintains the same total mechanical energy throughout its flight (in the absence of air resistance). Planets regularly return to the same place in their orbits with the same speed. Therefore, it is appropriate to consider the potential energy associated with the gravitational force. Earlier (in Chapter 7) we studied the potential energy of bodies near Earth's surface. In this section we apply our previous understanding of potential energy to Newton's law of gravitation to gain a more general understanding of gravitational potential energy. You may want to review the general discussion of potential energy in Chapter 7 before proceeding.

For convenience, let's begin with a one-dimensional version of the problem, as illustrated in Figure 15.7. There are two point particles, one with mass M that remains fixed at the origin, and another with mass m a distance r away. Notice that we have labeled the coordinate axis s rather than x so that our notation will be

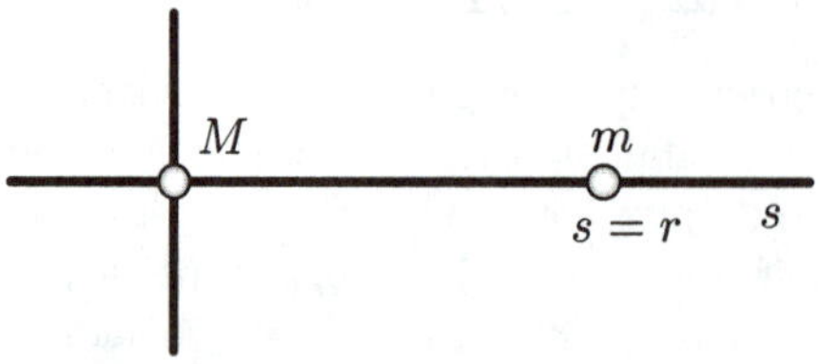

Figure 15.7. Setup for the potential between two point particles.

consistent with that used in our study of potential energy in Section 7.5.5. The definition of potential energy (Definition 7.5) gives us an expression for the *difference in potential energy* of the two particles when the particle with mass m is moved from some point $s = r_a$ to $s = r$. According to Definition 7.5,

$$\Delta U = U(r) - U(r_a) = -\int_{r_a}^{r} F_r(s)\, ds.$$

The integrand $F_r(s)$ is the component of the force on the particle of mass m due to the particle of mass M in the direction of a unit vector that points from M to m. Since the gravitational force is attractive, the force on m is towards M—that is, opposite to that unit vector—so from Newton's law of gravitation we know

$$F_r(s) = -\frac{GMm}{s^2}. \tag{15.11}$$

Using Equation (15.11), we can work out the potential energy difference:

$$U(r) - U(r_a) = -\int_{r_a}^{r} \left(-\frac{GMm}{s^2}\right) ds = GMm \int_{r_a}^{r} \frac{1}{s^2}\, ds$$

where we have factored out the constants G, M, and m. Evaluating the definite integral,

$$U(r) - U(r_a) = GMm \left(-\frac{1}{s}\right)\Bigg|_{r_a}^{r} = GMm \left[-\frac{1}{r} - \left(-\frac{1}{r_a}\right)\right]$$

or

$$U(r) - U(r_a) = GMm \left(\frac{1}{r_a} - \frac{1}{r}\right). \tag{15.12}$$

The result in Equation (15.12) is true but cumbersome to use. We can simplify it by remembering that we are free to define the potential energy to be zero at any place we choose. If we define the gravitational potential energy to be zero at $r_a = \infty$, then Equation (15.12) simplifies to

$$U(r) = -\frac{GMm}{r}. \tag{15.13}$$

Equation (15.13) is a general formula that gives us the potential energy of two masses (M and m) separated by a distance r.

At first, it may seem unusual that the potential energy turns out to be negative for any r (we have defined r to be positive). But remember that it is legal to add an arbitrary constant to the potential energy function (this is equivalent to redefining the place where the potential energy function is zero). The numerical value of the potential energy is not as important as how the potential energy changes with r. Recall from Chapter 7 that the (one-dimensional) force can be derived from the potential energy function [see Equation (7.42)] by taking a derivative:

$$F_s(r) = -\frac{dU(r)}{dr}. \tag{15.14}$$

Taking the derivative as indicated here with the potential energy function in Equation (15.13):

$$F_s(r) = -\frac{d}{dr}\left(-\frac{GMm}{r}\right) = -\frac{GMm}{r^2}$$

which matches the force law we started with, as given by Newton's law of gravitation. This confirms that Equation (15.13) gives a valid potential energy function for gravitation.

In two or three dimensions, the potential energy function is still given by Equation (15.13), because the direction of the s-axis we used in the derivation is arbitrary. The variable r is just the absolute value of the distance between the two point particles.

As another way of seeing this, think again of M as fixed at the origin with m a distance r away. In three dimensions, m can lie anywhere on the surface of a sphere of radius r and have the same potential energy, according to Equation (15.13). This makes sense because the gravitational force on m is always toward M, and thus perpendicular to the surface of that sphere. Recalling that work is computed using the line integral

$$W = \int_{\vec{r}_a}^{\vec{r}_b} \vec{F} \cdot d\vec{r} \tag{7.21}$$

where $\vec{r}_a$ and $\vec{r}_b$ are two points on the sphere of radius r. Because the force (perpendicular to the sphere's surface) is always perpendicular to the displacement (tangent to the sphere's surface), the dot product in Equation (7.21) is zero and hence the work done in moving m from one point to another on the sphere is zero. Then by the work-energy theorem the potential energy on the surface of the sphere is constant. Only changes in position in the *radial* direction change the potential energy.

Example 15.6. In Chapter 7, we found the potential energy change in moving a particle of mass m upward a distance y (near Earth's surface) is $\Delta U = mgy$. Reconcile this result with the general gravitational potential energy function given in Equation (15.13).

Solution. Recall that the gravitational force (and hence the general potential energy) is found assuming Earth is a point particle of mass M at Earth's center. Let's take the starting position of our particle of mass m to be a distance R from the center of Earth, so when it moves up a short distance y, it is then $R+y$ from the center of Earth. Then by Equation (15.13),

$$\Delta U = U(R+y) - U(R) = -\frac{GMm}{R+y} - \left(-\frac{GMm}{R}\right) = GMm\left(\frac{1}{R} - \frac{1}{R+y}\right).$$

Rewriting with a common denominator $R(R+y)$,

$$\Delta U = GMm\left[\frac{R+y}{R(R+y)} - \frac{R}{R(R+y)}\right] = GMm\frac{y}{R(R+y)}.$$

In the denominator, $R(R+y) \cong R^2$, because we have assumed $y << R$. Therefore, to an excellent approximation

$$\Delta U = GMm\frac{y}{R^2}.$$

Near Earth's surface $g = GM/R^2$ (see Example 15.4), so to an excellent approximation

$$\Delta U = mgy$$

as required. △

Example 15.7. Ignoring air resistance, find the speed with which a projectile must be launched straight up from Earth's surface to reach an altitude of 100 km.

Solution. The height is sufficiently large that we should not use the "constant g" version of the potential energy. From Equation (15.13),

$$\Delta U = -\frac{GMm}{r_2} - \left(-\frac{GMm}{r_1}\right)$$

where M is the mass of Earth, m is the mass of the projectile, r_1 is the initial distance from Earth's center, and r_2 is the final distance from Earth's center. Assuming a launch from Earth's surface $r_1 = 6.374 \times 10^6$ m (Earth's radius) and $r_2 = r_1 + 100\text{ km} = 6.474 \times 10^6$ m.

In a conservative system, the total mechanical energy is conserved. If v_1 and v_2 are the speeds corresponding to positions r_1 and r_2, conservation of energy gives

$$\frac{1}{2}mv_1^2 + \left(-\frac{GMm}{r_1}\right) = \frac{1}{2}mv_2^2 + \left(-\frac{GMm}{r_2}\right).$$

At the top of the projectile's flight, $v_2 = 0$, so

$$\frac{1}{2}mv_1^2 = -\frac{GMm}{r_2} - \left(-\frac{GMm}{r_1}\right) = \Delta U.$$

This result makes physical sense in the following way: The kinetic energy *lost* in flight has been *gained* in potential energy in order to conserve the total mechanical energy. Notice that the mass of the projectile cancels, leaving (after rearranging)

$$v_1 = \sqrt{2GM\left(\frac{1}{r_1} - \frac{1}{r_2}\right)}.$$

Inserting numerical values,

$$\begin{aligned} v_1 &= \sqrt{2(6.67 \times 10^{-11}\text{ N·m}^2/\text{kg}^2)(5.976 \times 10^{24}\text{ kg})\left(\frac{1}{6.374 \times 10^6\text{ m}} - \frac{1}{6.474 \times 10^6\text{ m}}\right)} \\ &= 1.39 \times 10^3\text{ m/s}. \end{aligned}$$

△

15.5 The Gravitational Force Due to a Spherically Symmetric Body

In this section we show that the gravitational force of a spherically symmetric body on a particle outside the body is the same as if the entire mass of the spherical body were concentrated at its center. We need actually only show this result for a *uniform spherical shell*, since any spherically symmetric body can be thought of as a collection of such shells.

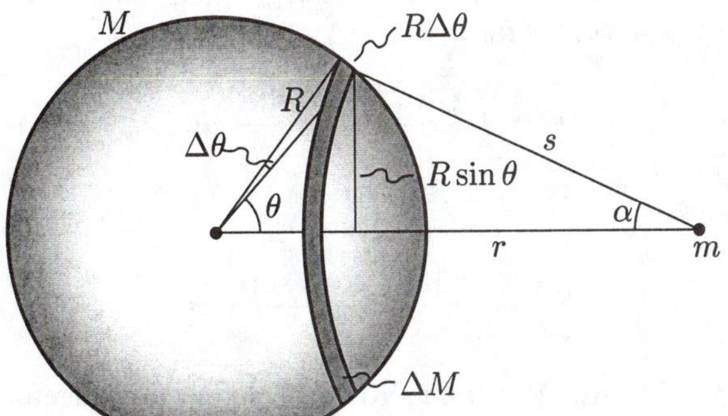

Figure 15.8. Typical spherical shell or radius R and thickness ΔR.

Consider then a very thin shell of radius R, thickness ΔR, and mass M, as shown in Figure 15.8. Look at the thin band that makes up part of the shell. We suppose the band has some mass ΔM equal to the (assumed constant) density ρ multiplied by the volume of the band. For a small band, we can approximate the band as a rectangular solid wrapped around the sphere, with the dimensions of the rectangular solid being the circumference of the band ($2\pi R\sin\theta$) by the width ($R\,\Delta\theta$) by the thickness (ΔR). Thus, we have the approximation $\Delta M = \rho\Delta V$:

$$\Delta M = \rho\Delta V = \rho(2\pi R\sin\theta)(R\Delta\theta)(\Delta R) = 2\pi\rho R^2\sin\theta\,\Delta R\,\Delta\theta. \tag{15.15}$$

Now the density of the shell is the entire mass M divided by the volume, which can be approximated by the surface area $4\pi R^2$ multiplied by the thickness ΔR. Thus,

$$\rho = \frac{M}{4\pi R^2\,\Delta R}$$

and inserting this result into Equation (15.15) gives

$$\Delta M = 2\pi\left(\frac{M}{4\pi R^2\Delta R}\right)R^2\sin\theta\Delta R\,\Delta\theta$$

or

$$\Delta M = \frac{1}{2}M\sin\theta\Delta\theta. \tag{15.16}$$

The magnitude of the force on the external particle of mass m due to the band of mass ΔM is given by Newton's law of gravitation as

$$\Delta F = \frac{Gm\Delta M}{s^2}\cos\alpha = \frac{GmM\cos\alpha\sin\theta}{2s^2}\Delta\theta \tag{15.17}$$

where the factor $\cos\alpha$ comes from the fact that we must take only the component of the force along the axis of symmetry (the other components cancel out over the entire band).

The total force on m due to all such bands over the spherical shell can be found by constructing a Riemann sum over all allowed values of θ, with each term in the sum of the form in Equation (15.17). But in the familiar process in calculus, the Riemann sum becomes a definite integral in the limit as $\Delta\theta$ approaches zero, and so the total force is given by an integral of the form

$$F = \int \frac{GmM \cos\alpha \sin\theta}{2s^2}\, d\theta$$

or, after factoring out constants,

$$F = \frac{GmM}{2} \int \frac{\cos\alpha \sin\theta}{s^2}\, d\theta. \tag{15.18}$$

The integral in Equation (15.18) resulted from a Riemann sum and is therefore a definite integral (over the entire spherical shell), but we have not placed limits on it, because we have to make a change of variable before integrating. This is because the three variables s, θ, and α depend mutually on one another, and it is necessary to express the integral as a function of a single variable before integrating. From the law of cosines, we have

$$s^2 = R^2 + r^2 - 2Rr\cos\theta.$$

Implicit differentiation (with r and R constant) gives

$$2s\, ds = -2Rr(-\sin\theta)\, d\theta = 2Rr \sin\theta\, d\theta.$$

Rearranging,

$$\sin\theta\, d\theta = \frac{s}{Rr}\, ds. \tag{15.19}$$

Applying the law of cosines with the angle α,

$$R^2 = r^2 + s^2 - 2rs\cos\alpha$$

or

$$\cos\alpha = \frac{r^2 + s^2 - R^2}{2rs}. \tag{15.20}$$

Equations (15.19) and (15.20) can be substituted into the integral in Equation (15.18):

$$F = \frac{GmM}{2} \int \frac{1}{s^2} \left(\frac{r^2 + s^2 - R^2}{2rs}\right) \left(\frac{s}{Rr}\right) ds.$$

After some algebra, this reduces to

$$F = \frac{GmM}{4r^2 R} \int_{r-R}^{r+R} \left(1 - \frac{r^2 - R^2}{s^2}\right) ds \tag{15.21}$$

where we have inserted the proper limits of integration with respect to s in order to cover the entire shell. The evaluation of the definite integral in Equation (15.21) is now straightforward but messy enough for us to leave as an exercise for the student. The result is

$$F = \frac{GmM}{r^2}$$

which is the same as if the mass of the shell were concentrated at the shell's center, a distance r from the point particle m.

An interesting sidelight of this calculation (which we also leave for the student to verify) is that the force on a test particle m placed *inside* a spherical shell is *zero*. Therefore, such a test particle placed somewhere inside a uniform solid sphere is affected only by the mass that is "inside of it" (i.e., at a smaller radius). The force on that particle is still toward the center of the sphere due to the symmetry of the sphere, and the magnitude of that force is given by Newton's law of gravitation as

$$F = \frac{GmM'}{r^2}$$

where r is the distance from the test particle to the center of the sphere and M' is the mass of the solid sphere that is inside of it. For a uniform solid sphere with constant density ρ throughout,

$$M' = \rho\left(\frac{4}{3}\pi r^3\right) = \frac{4}{3}\pi\rho r^3.$$

Thus,

$$F = \frac{GmM'}{r^2} = \frac{Gm}{r^2}\left(\frac{4}{3}\pi\rho r^3\right)$$

or

$$F = \frac{4}{3}\pi\rho Gmr. \tag{15.22}$$

Rather than decreasing as the inverse square, the magnitude of the force on the test particle increases and reaches a maximum at the surface of the sphere. The force on the test particle at the center of the sphere is zero, which makes sense given the symmetry of the problem.

Example 15.8. Suppose that a tunnel could be cut straight through Earth along a diameter. If a small object is dropped from rest at one end of the tunnel, show that the resulting motion is simple harmonic motion along the diameter (assume Earth is a solid sphere with constant density ρ). Find the period of the simple harmonic motion.

Solution. If we let the x-axis be along the diameter in the tunnel, we can say from Equation (15.22) that the x-component of the force on the particle is

$$F_x = -\frac{4}{3}\pi\rho Gmx$$

where the negative sign is due to the fact that the force is always toward the Earth's center. From Newton's second law, $F_x = ma_x$, or

$$F_x = ma_x = m\frac{d^2x}{dt^2} = -\frac{4}{3}\pi\rho Gmx.$$

This reduces to

$$\frac{d^2x}{dt^2} = -\frac{4}{3}\pi\rho Gx$$

which we recognize from Chapter 12 [see Equation (12.4)] as the equation of a harmonic oscillator, because the factor $\frac{4}{3}\pi\rho G$ is constant.

From our study of harmonic oscillators [Equation (12.7)], the angular frequency ω of oscillation is

$$\omega = \sqrt{\frac{4}{3}\pi\rho G}$$

and the period is

$$T = \frac{2\pi}{\omega} = \frac{2\pi}{\sqrt{\frac{4}{3}\pi\rho G}} = \sqrt{\frac{3\pi}{\rho G}}.$$

To find a numerical value of T, we need to know the density of Earth. Assuming the density is constant, we have

$$\rho = \frac{M}{\frac{4}{3}\pi R^3}$$

so

$$T = \sqrt{\frac{3\pi}{\left(\frac{3M}{4\pi R^3}\right)G}} = 2\pi\sqrt{\frac{R^3}{MG}}.$$

Evaluating numerically,

$$T = 2\pi\sqrt{\frac{(6.374 \times 10^6 \text{ m})^3}{(5.976 \times 10^{24} \text{ kg})(6.67 \times 10^{-11} \text{ N·m}^2/\text{kg}^2)}} = 5060 \text{ s}$$

or about 84 minutes. We don't recommend this as an engineering project! △

15.6 More on Earth, the Planets, and Satellites

15.6.1 Variation of g Over Earth's Surface

In Chapter 4, we noted how the gravitational acceleration of bodies near Earth's surface varies due to Earth's rotation. Earth's rotation also causes Earth (which is slightly elastic) to bulge at the equator and flatten at the poles. This makes points on the equator farther from the center of Earth than points on the poles, contributing to a lower value of g at the equator. In this section we look at the general problem of how g varies due to altitude for distances small compared with Earth's radius.

Let's assume that Earth is a symmetric sphere of mass M and radius R. Then at the surface of this sphere we know (see Example 15.4) that

$$g = \frac{GM}{R^2}. \tag{15.23}$$

At an altitude h above Earth's surface, we must replace R with $R + h$, the new distance from the center of Earth, and so

$$g = \frac{GM}{(R+h)^2}. \tag{15.24}$$

Factoring the R from the parentheses in the denominator,

$$g = \frac{GM}{R^2\left(1+\frac{h}{R}\right)^2} = \frac{GM}{R^2}\left(1+\frac{h}{R}\right)^{-2}.$$

We have rewritten the expression for g in this way so that we can apply the binomial series (see Table 11.7) to express g as

$$g = \frac{GM}{R^2}\left(1 - 2\frac{h}{R} + \frac{(-2)(-3)}{2!}\left(\frac{h}{R}\right)^2 + \cdots\right). \tag{15.25}$$

At first glance the binomial theorem seems to have complicated matters, turning a relatively straightforward expression into an infinite series. But if $h \ll R$, the terms of order 2 and higher in the series are negligibly small, and therefore to a good approximation

$$g = \frac{GM}{R^2}\left(1 - 2\frac{h}{R}\right) \tag{15.26}$$

Equation (15.26) is an excellent approximation of g as long as $h \ll R$. It has the advantage [compared with the exact expression in Equation (15.24)] of showing *that g decreases linearly with altitude near Earth's surface*.

Example 15.9. Compute g at Earth's surface (sea level) and the top of Mt. Everest ($h = 8848$ m), the latter using both the exact expression and the approximation in Equation (15.26).

Solution. Using the standard values for the mass and radius of Earth, we have at sea level

$$g = \frac{GM}{R^2} = \frac{(6.67\times10^{-11}\ \text{N·m}^2/\text{kg}^2)(5.976\times10^{24}\ \text{kg})}{(6.374\times10^6\ \text{m})^2} = 9.811\ \text{m/s}^2.$$

At the top of Mt. Everest, the exact expression gives (keeping all digits until the end of the calculation)

$$g = \frac{GM}{(R+h)^2} = \frac{(6.67\times10^{-11}\ \text{N·m}^2/\text{kg}^2)(5.976\times10^{24}\ \text{kg})}{(6.374\times10^6\ \text{m} + 8848\ \text{m})^2} = 9.784\ \text{m/s}^2.$$

Using Equation (15.26) as an approximation,

$$g = \frac{GM}{R^2}\left(1-2\frac{h}{R}\right) = \left[\frac{(6.67\times10^{-11}\ \text{N·m}^2/\text{kg}^2)(5.976\times10^{24}\ \text{kg})}{(6.374\times10^6\ \text{m})^2}\right]\left[1-2\left(\frac{8848\ \text{m}}{6.374\times10^6\ \text{m}}\right)\right]$$
$$= 9.784\ \text{m/s}^2.$$

To four significant digits, there is no difference between the two numerical values. Evidently the approximation is an excellent one for these heights. △

Example 15.10. Find the approximate linear rate of change of g per unit altitude near Earth's surface, in units of m/s^2 per km.

Solution. From Equation (15.26),

$$g = \frac{GM}{R^2}\left(1-2\frac{h}{R}\right) = g_0\left(1-2\frac{h}{R}\right)$$

where g_0 is the value of g at sea level. The difference between g at some altitude h and the sea level g_0 is

$$\Delta g = g - g_0 = g_0\left(1 - 2\frac{h}{R}\right) - g_0 = -\frac{2g_0 h}{R}.$$

The change in g per unit altitude is the ratio of Δg to h, or

$$\frac{\Delta g}{h} = -\frac{2g_0}{R} = -\frac{2GM}{R^3}.$$

Inserting numerical values,

$$\frac{\Delta g}{h} = -\frac{2(6.67 \times 10^{-11}\ \mathrm{N{\cdot}m^2/kg^2})(5.976 \times 10^{24}\ \mathrm{kg})}{(6.374 \times 10^6\ \mathrm{m})^3} = -3.078 \times 10^{-6}\mathrm{s}^{-2}$$
$$= -3.078 \times 10^{-3}(\mathrm{m/s^2})/\mathrm{km}.$$

The numerical value of g drops by just over 0.0031 m/s^2 for every kilometer of altitude gained. △

It is important to remember that Earth is neither spherical nor spherically symmetric, so all the models we have presented in this section are good but not exact. There can be fairly significant local variations in g based on the density of the materials in different parts of Earth. Generally, a higher density of materials in Earth creates a stronger downward pull and a higher value of g. Geologists use extremely precise gravity meters to search for oil, which has a much lower density than the surrounding rocks.

15.6.2 Escape Speed

One useful application of the concept of gravitational potential energy is in finding the speed an object needs to completely escape a planet's gravity. Let's consider a rocket escaping Earth's gravity as an example. A rocket of mass m launched from Earth's surface with a speed v_0 has a total mechanical energy (kinetic plus potential)

$$E = K + U = \frac{1}{2}mv_0^2 - \frac{GMm}{R} \tag{15.27}$$

where R is Earth's radius and M is Earth's mass. As the rocket travels away from Earth, its kinetic energy decreases (since the force of gravity is pulling it back), while its potential energy increases. At some distance r from Earth's center, the rocket's total energy is the same as before (gravity is a conservative force), and

$$E = K + U = \frac{1}{2}mv^2 - \frac{GMm}{r} \tag{15.28}$$

with speed $v < v_0$.

If the rocket *just escapes* Earth's gravity, its speed approaches zero at a large distance from earth (approaching infinity). Thus, both the kinetic energy and potential energy approach zero as the distance from Earth approaches infinity. Under these conditions we conclude $E = K + U = 0$. To reach that point, the rocket was launched with **escape speed** v_{esc}, defined as the *minimum* speed needed to escape the planet's gravity. Because $E = 0$, we have

$$0 = \frac{1}{2}mv_{\mathrm{esc}}^2 - \frac{GMm}{R}.$$

Solving for v_{esc}, we find

$$v_{\mathrm{esc}} = \sqrt{\frac{2GM}{R}}. \tag{15.29}$$

Notice that the escape speed is independent of the direction of the velocity vector. We can take Equation (15.29) as a general result for escape from any spherically symmetric object with mass M and radius R.

Example 15.11. Compute the escape speed from (a) the Earth's surface (ignoring air resistance), (b) the surface of the Moon, and (c) the surface of a uniform solid spherical asteroid with density 5000 kg/m^3 and radius 1.00 km.

Solution. (a) Inserting the numerical values into Equation (15.29),

$$v_{\text{esc}} = \sqrt{\frac{2GM}{R}} = \sqrt{\frac{2(6.67 \times 10^{-11}\ \text{N}\cdot\text{m}^2/\text{kg}^2)(5.976 \times 10^{24}\ \text{kg})}{6.374 \times 10^6\ \text{m}}} = 1.12 \times 10^4\ \text{m/s}$$

or about 11.2 km/s.

(b) Using the numerical values for the Moon,

$$v_{\text{esc}} = \sqrt{\frac{2GM}{R}} = \sqrt{\frac{2(6.67 \times 10^{-11}\ \text{N}\cdot\text{m}^2/\text{kg}^2)(7.35 \times 10^{22}\ \text{kg})}{1.74 \times 10^6\ \text{m}}} = 2.37 \times 10^3\ \text{m/s}$$

or 2.37 km/s.

(c) For the asteroid, we can use the fact that for a solid uniform sphere mass equals density time volume, so

$$v_{\text{esc}} = \sqrt{\frac{2G\rho V}{R}} = \sqrt{\frac{2G\rho\left(\frac{4}{3}\pi R^3\right)}{R}} = \sqrt{\frac{8\pi G\rho R^2}{3}}$$

Inserting the numerical values,

$$v_{\text{esc}} = \sqrt{\frac{8\pi(6.67 \times 10^{-11}\ \text{N}\cdot\text{m}^2/\text{kg}^2)\,(5000\ \text{kg/m}^3)\,(1000\ \text{m})^2}{3}} = 1.67\ \text{m/s}.$$

For this escape speed, a rocket wouldn't be needed—a modest jump will do the job! △

15.6.3 Energy in Orbital Motion

Suppose a satellite of mass m is in a circular orbit of radius r around a planet of mass M. The total mechanical energy of the system is

$$E = K + U = \frac{1}{2}mv^2 - \frac{GMm}{r}. \tag{15.30}$$

The speed v of the satellite is constant in a circular orbit in order to maintain a constant total mechanical energy.

We can obtain a simpler expression for the total mechanical energy if we remember that the gravitational force is entirely responsible for the centripetal force. Equating the algebraic expressions for the magnitudes of the gravitational and centripetal forces,

$$\frac{mv^2}{r} = \frac{GMm}{r^2}.$$

Multiplying through by $r/2$,

$$\frac{1}{2}mv^2 = \frac{GMm}{2r}. \tag{15.31}$$

Using Equation (15.31) for the kinetic energy in Equation (15.30), we find

$$E = \frac{GMm}{2r} - \frac{GMm}{r}$$

or

$$E = -\frac{GMm}{2r}. \tag{15.32}$$

Therefore, the total energy of a satellite in circular orbit is *negative* and varies inversely with the radius of the orbit. You should not be shocked to find a negative total mechanical energy; once again it is the result of our choice to set the zero of potential energy at an infinite distance. In fact, for this choice of the zero of potential energy, a negative total energy is characteristic of bound orbits (circles and ellipses for the inverse-square force). In an advanced mechanics class, you will learn that the total energy of a satellite in an elliptical orbit is negative and inversely proportional to the length of the semimajor axis, while the total energy is zero in a parabolic orbit and positive in a hyperbolic orbit.

Example 15.12. A satellite of mass m is originally in a circular orbit of radius r around Earth. It is then moved to a circular orbit of radius $2r$. Find the change in the total energy, kinetic energy, and potential energy between those two configurations.

Solution. We can analyze the total energy simply with Equation (15.32). The original energy is

$$E_i = -\frac{GMm}{2r}$$

(where M is Earth's mass) and the final energy is

$$E_f = -\frac{GMm}{2(2r)} = \frac{E_i}{2}.$$

The total energy is half its original value. The change in total energy is

$$\Delta E = E_f - E_i = -\frac{E_i}{2} = \frac{GMm}{4r}.$$

Therefore, the total energy increases.

The potential energy changes from

$$U_i = -\frac{GMm}{r}$$

to

$$U_f = -\frac{GMm}{2r}$$

so the potential energy is also halved, and the change in potential energy is

$$\Delta U = U_f - U_i = +\frac{GMm}{2r}.$$

The potential energy has *increased* because the orbit is larger.

By conservation of energy, we know that

$$\Delta E = \Delta K + \Delta U.$$

Therefore the change in kinetic energy is

$$\Delta K = \Delta E - \Delta U = -\frac{GMm}{4r}.$$

The kinetic energy must *decrease* by the same amount the total energy *increases*. The satellite must slow down in order to move to an orbit of higher radius! This also makes sense if we think of the gravitational force. Earth's gravitational pull on a satellite in a larger circular orbit is weaker, and therefore the satellite does not have to travel as fast in order to keep up with the "fall" toward Earth. △

15.6.4 Orbital Periods of Satellites

As we saw earlier, when a small body orbits a large body of mass M in a circular orbit, the relationship between the period and radius of the orbit is

$$\frac{R^3}{T^2} = \frac{GM}{4\pi^2} \tag{15.5}$$

Therefore, we can write the period of the orbit as a function of the radius by rearranging Equation (15.5) to solve for T:

$$T = \left(\frac{4\pi^2}{GM}\right)^{1/2} R^{3/2}. \tag{15.33}$$

Let's consider what this means for a satellite in a circular orbit around Earth. For a low Earth orbit, for example at an altitude of 100 km above sea level, the radius of the orbit is

$$R = R_E + 100 \text{ km} = 6.374 \times 10^6 \text{ m} + 1.00 \times 10^5 \text{ m} = 6.474 \times 10^6 \text{ m}.$$

Inserting this numerical value along with the mass of Earth in Equation (15.33) gives

$$T = \left[\frac{4\pi^2}{(6.67 \times 10^{-11} \text{ N·m}^2/\text{kg}^2)(5.976 \times 10^{24} \text{ kg})}\right]^{1/2} (6.474 \times 10^6 \text{m})^{3/2} = 5184 \text{ s}$$

or about 86 minutes. This is about the lowest practical satellite orbit, and so the period we just calculated is the shortest practical period for a satellite.

We can turn around the period/radius relationship to solve for the radius as a function of the period:

$$R = \left(\frac{GM}{4\pi^2}\right)^{1/3} T^{2/3}. \tag{15.34}$$

One practical use of this result is to find the period of a **geosynchronous** satellite, which orbits Earth with a period of 24 hours. Such a satellite will (if orbiting over the equator) remain above the same point on Earth at all times and is therefore ideal for relaying messages (including radio, telephone, and television) from one point on earth to another. The period of a geosynchronous satellite is 24 h = 86,400 s, so by Equation (15.34) the radius of the orbit is

$$R = \left[\frac{(6.67 \times 10^{-11} \text{ N·m}^2/\text{kg}^2)(5.976 \times 10^{24} \text{ kg})}{4\pi^2}\right]^{1/3} (86,400 \text{ s})^{2/3} = 4.22 \times 10^7 \text{ m}.$$

This radius is more than six times Earth's radius, so the geosynchronous orbit is quite high!

We could also rearrange Equation (15.33) or (15.34) to solve for the *mass* of a body as a function of the period and radius of a satellite in circular orbit around the body. This is a convenient way to determine the mass of a planet, as long as we can observe such a satellite (or launch an artificial one ourselves). Of course, this all depends on an accurate knowledge of the universal gravitation constant G. Otherwise, the best we could do is solve for the product GM. As we will see in the next section, an accurate measurement of G was not done in Newton's day, and G is one of the most difficult fundamental constants of nature to measure accurately.

15.7 The Constant *G* and Fundamental Forces

Looking at Newton's inverse-square force law,

$$F = \frac{GMm}{r^2} \tag{15.3}$$

we see that to determine the universal gravitation constant G it is necessary to measure two masses, a distance, and a force. This fact prevents our using planets or moons to determine G, because the masses of such large bodies cannot be determined independently.

The first good experimental measurement of G was made by Henry Cavendish in 1798 (over a century after the publication of *Principia*!). The experiment uses a device shown schematically in Figure 15.9; this device is appropriately called a **Cavendish balance**. In the Cavendish balance, a light rod is used to connect two solid spherical balls of mass m, usually made out of a dense metal such as lead in order to make them as massive as possible. The rod is suspended with a thin fiber so that it is balanced with the rod in a horizontal position. With the system at rest, two other spherical balls (each with mass M) can be placed near the other balls as shown. The gravitational attraction of each ball of mass m for its neighbor of mass M causes a net torque on the fiber, so there is a slight twist that can be measured. The amount of twist can be compared with a known torque to determine the force of attraction between the balls. Then knowing the masses, the center-to-center distances (r in the diagram), and forces, the gravitation constant G can be determined.

Cavendish was able to determine the correct order of magnitude for G. Today, G is known to about five significant digits. That may sound precise for most purposes, but it is really imprecise compared with other fundamental constants of nature. What makes it so difficult to measure G is that the gravitational force is by

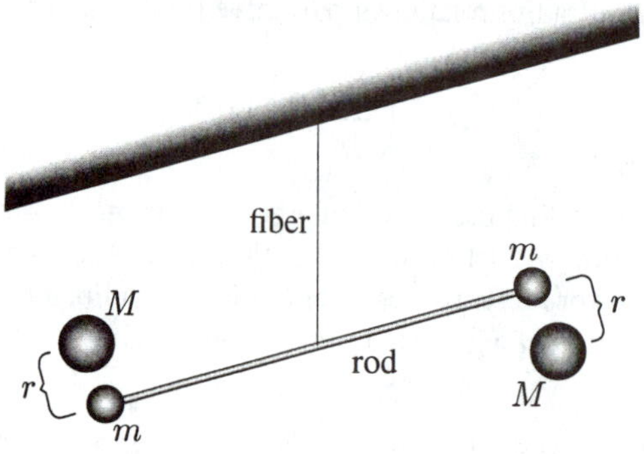

Figure 15.9. Cavendish balance.

far the weakest of the fundamental forces in nature. Our experience tells us that gravity is strong, but we are fooled by living in the vicinity of a planet with mass 6×10^{24} kg!

There are four fundamental forces in nature. One is gravitation, which we have studied in this chapter. Another is the **electromagnetic force**, which we study in the next chapter and which we encounter frequently in our everyday lives. We will see that the mathematical models we use to describe electromagnetism and gravitation are similar. The **nuclear force** (also known as the **strong force**) is effective only within the nucleus and serves to bind together the protons and neutrons in the nucleus. The **weak force** operates in certain types of nuclear decay. While it is difficult to make an exact comparison of the strengths of the four forces (because they do not operate simultaneously), the best order-of-magnitude comparison of these forces with similar particles on similar distance scales is shown in Table 15.2.

Table 15.2. Relative strengths of the fundamental forces.

Force	Relative strength
Nuclear	1
Electromagnetic	10^{-2}
Weak	10^{-10}
Gravitational	10^{-42}

15.8 Problems

15.1 Background and History

There are no problems for this section.

15.2 Kepler's Laws of Planetary Motion

1. Draw ellipses with the following eccentricities: 0.01, 0.1, 0.5, 0.9, 0.99.

2. On the same diagram draw the orbits of earth and Halley's comet, which has a semimajor axis of 2.67×10^{12} m and an eccentricity of 0.967. Draw the orbits in the same plane to show their relative shapes, even though the plane of the orbit of Halley's comet is highly inclined with respect to the ecliptic plane.

3. Using the data in the previous problem determine the period of Halley's comet in its orbit around the Sun.

4. Find the eccentricity of an ellipse in which the major axis is exactly twice the minor axis.

5. For the ellipse described in the preceding problem, find the locations of the two foci.

6. Using the data in Table 15.1, compute the constant $C = a^3/T^2$ for the orbits of Mars, Jupiter, and Saturn.

7. Find the distance of Mars from the Sun at aphelion and perihelion.

8. In the 1920s, the German scientist Walter Hohmann proposed that the most efficient way to travel from earth to an outer planet is an elliptical path (now called a **Hohmann ellipse**), such that Earth is at

the perihelion and the outer planet at the aphelion of the Hohmann ellipse. Find the maximum and minimum values of the semimajor axis and eccentricity of the Hohmann ellipse for a trip from Earth to Mars.

9. The Moon orbits Earth with a period of 27.3 days, and the semimajor axis of the Moon's orbit around Earth is 3.84×10^8 m. Using this data, find the constant $C = a^3/T^2$ for satellites orbiting Earth. Use your computed value of C to find the period of a satellite in a circular orbit 500 km above Earth's surface.

15.3 Newton's Law of Gravitation

1. The Sun has a radius 6.96×10^8 m and a mass 1.99×10^{30} kg. (a) Find the gravitational acceleration of a particle dropped near the surface of the Sun. (b) Find Earth's centripetal acceleration in its orbit around the Sun. (c) Using the results of (a) and (b), show that the gravitational force from the Sun follows an inverse-square distance relationship.

2. The Moon has a radius 1.74×10^6 m and a mass 7.35×10^{22} kg. Find the gravitational acceleration of an object dropped near the surface of the Moon. Compare your result with g near the surface of Earth.

3. Assuming Earth is in a circular orbit around the Sun (with radius 1.495×10^{11} m and period exactly 1 year), compute the mass of the Sun.

4. Use the planetary data in tables to find the gravitational acceleration of a particle near the surface of Mercury, Mars, and Saturn.

5. Show that Kepler's second law holds for a noncircular orbit generated by an inverse-square central force and that the area per unit time is still $L/2m$, where L is the magnitude of the orbiting body's angular momentum and m is its mass.

6. Find the force of gravitational attraction between two protons in a nucleus. Assume that the protons are spherically symmetric and the distance between their centers is 2.1×10^{-15} m.

7. Consider a binary star system, with two identical stars of mass M orbiting about their common center of mass. If the stars are separated by a distance R, find the period of each star's orbit. Evaluate numerically if $M = 2.0 \times 10^{30}$ kg and $R = 1.0 \times 10^{12}$ m.

8. Repeat the previous problem for two stars with different masses: $M = 6.5 \times 10^{30}$ kg and $m = 2.0 \times 10^{30}$ kg with $R = 1.0 \times 10^{12}$ m.

9. A 1.0 kg block is dropped near Earth's surface. Using Newton's third law of motion, determine the magnitude of the gravitation force exerted on the *Earth* by the *block* and find Earth's acceleration toward the block.

10. **Tidal forces** are the differential gravitational forces on an extended body due to another massive body. (a) For example, a drop of water on the side of earth facing the Moon has a slightly larger gravitational acceleration toward the Moon than an identical drop on the opposite side of Earth. Find that difference in gravitational acceleration Δa. (b) Find same difference Δa for an asteroid that has strayed too close to Saturn. Assume the asteroid's radius is 500 km, and it is 7.50×10^7 m from the center of Saturn. In such a case, tidal forces can break apart the asteroid.

11. (a) Use Earth's average distance from the Sun (the semimajor axis a) to estimate the numerical value of Earth's orbital angular momentum. (b) Using your answer to (a) find Earth's orbital speed at aphelion and perihelion. (c) Use your answer to (b) to estimate the area swept out by the earth in its orbit in 1 hour's time at aphelion and perihelion. (d) Show that your results in (c) are in agreement with Kepler's second law.

12. Consider a uniform solid cube with mass M side R. A point particle of mass m is a distance R from the cube's center. Using a computer, perform the following calculation. Break the cube into 1000 pieces, each piece a cube $0.1R$ on a side. Sum the gravitational attraction between the point particle and all 1000 pieces. Due to symmetry, you need only consider the component of the force directed from the point particle toward the center of the cube. Is the total gravitational force between the cube and the point particle greater than, less than, or equal to the attraction between two point particles separated by a distance R? Explain.

13. Suppose the universal law of gravitation were *not* an inverse-square law, but instead had the form $F = GMm/r^n$ where $n \neq 2$. Find the relationship between the radius and period of a circular orbit, and show that this relationship disagrees with Kepler's third law.

14. Compute the numerical value of $GM/4\pi^2$ using the mass of the Sun. Compare your result with the value of a^3/T^2 obtained in Example 15.2.

15.4 Gravitational Potential Energy

1. Compute the potential energy of (a) the Earth-Sun system, and (b) the Earth-Moon system.

2. Ignoring air resistance find the speed of a small object when it strikes Earth if is dropped from rest at a height of (a) 10 km, (b) 1000 km, and (c) 10^{11} m.

3. In Example 15.7, we stated that it is wrong to use the "constant g" version of the gravitational potential energy ($\Delta U = mgy$) to find the launch speed of a projectile intended to reach an altitude of 100 km. Find the numerical value of the error in the launch speed in this case.

4. Ignoring air resistance, find the speed with which a rocket should be launched from earth in order to reach the Moon. (*Hint*: you need to take the Moon's gravitational pull into account.)

5. What is the change in potential energy of the Earth-Sun system when Earth goes from aphelion to perihelion? Use this difference to find the change in Earth's orbital speed from aphelion and perihelion.

6. (a) Find the speed of a satellite in a circular orbit 3000 km above Earth's surface. (b) What additional speed would such a satellite need in order to completely escape Earth starting form that position?

7. Halley's comet reaches a maximum speed of about 54.6 km/s at perihelion. Using the data from Problem 2 of Section 15.2, find the speed of Halley's comet at aphelion.

8. For an object in an elliptical orbit around the Sun with semimajor axis a and semiminor axis b, find the ratio of the speed at aphelion to the speed at perihelion, expressing your answer as a function of a and b.

9. Find the total mechanical energy of a 500-kg satellite in a circular orbit 1500 km above Earth's surface.

15.5 The Gravitational Force Due to a Spherically Symmetric Body

1. Carry out the evaluation of the definite integral in Equation (15.21) and show that the result is

$$F = \frac{GMm}{r^2}$$

2. Repeat the calculation in the text for a particle of mass m inside a uniform spherical shell to show that the gravitational force on the particle is zero.

3. Consider a tunnel through Earth as described in Example 15.8 but cut along a chord (not necessarily a diameter) from one point on Earth's surface to another. Assume Earth has uniform density. Show that a particle dropped from one end of the tunnel undergoes simple harmonic motion and that the period of oscillation is independent of the length of the chord.

15.6 More on Earth, the Planets, and Satellites

1. By how much does g differ between the top and base of a 210-m-tall building?

2. A geologist measures g in an area in which Earth's crust is solid rock with density 3500 kg/m^3. A later measurement of g is made at the same altitude and latitude but directly above a spherical pool of water (density = 1000 kg/m^3) with a diameter of 2.5 km and the top of the spherical pool tangent to the surface of Earth. By how much is the second measurement of g changed from the previous one?

3. Ignoring the effects of Earth's gravity, find the escape speed from the solar system starting from Earth's orbit.

4. An astronaut can jump 0.65 m in a vertical standing jump on earth. How large an asteroid can the astronaut escape by jumping? Assume the asteroid is a uniform solid sphere with a constant density of 3000 kg/m^3.

5. Find the escape speed from the surface of Mars.

6. (a) What is the speed of a geosynchronous satellite? (b) How much more speed does such a satellite have to gain in order to escape Earth from that position?

7. A satellite is launched with exactly half the escape speed from a planet of radius R and enters a circular orbit. Find the radius of the circular orbit as a function of R.

8. Because the Moon has no atmosphere, an extremely low orbit is possible. Find the period of a circular orbit just above the surface of the Moon. The Moon's mass is 7.35×10^{22} kg and its radius is 1740 km.

9. On December 24, 1968, Frank Borman, Jim Lovell, and Bill Anders became the first humans to orbit the Moon. Their orbit was nearly circular, ranging from 59.7 miles to 60.7 miles above the lunar surface (1 mile = 1.609 km). (a) Find the period of their orbit and compare with the result of the previous problem. (b) Estimate the speed of the orbit by assuming a circular orbit 60.2 miles above the surface.

10. Find the total mechanical energy, kinetic energy, and speed of Venus in its (assumed circular) orbit.

11. Two asteroids are each solid spheres with uniform density 4800 kg/m^3. The radii of the asteroids are each 1.50 km. They begin in free space a great distance apart and are attracted directly toward one another by their mutual gravitational force. Find the speeds of the asteroids at the instant they collide.

12. Jupiter's moon Io has an orbit with a semimajor axis 4.22×10^5 km and a period of 1.77 days. Use this data to find the mass of Jupiter.

13. Find the height above Earth of a satellite with a circular orbit of period exactly 2 hours.

14. After it stops burning, a star with a mass of 4.0×10^{30} kg can collapse into a *neutron star* that has the same mass but a radius of only 10 km. (a) Find the escape speed from the surface of such a neutron star. (b) What is the orbital period of a satellite in a circular orbit just above the neutron star's surface?

15. An Earth satellite has an elliptical orbit, with a maximum altitude of 50,000 km above Earth and a minimum altitude of 4000 km. (a) Find the eccentricity of the orbit. (b) The satellite's speed at perigee is 8060 m/s. Find its speed at apogee. (c) Show that the satellite's total mechanical energy in this orbit is the same as in a circular orbit with the same semimajor axis.

15.7 The Constant G and Fundamental Forces

1. In a Cavendish experiment, find the force of attraction between two solid spherical lead balls with masses 0.150 kg and 2.35 kg separated by a center-to-center distance of 6.5 cm.

2. In a hydrogen atom, the proton and electron are held together at a distance of 5.3×10^{-11} m by their electrostatic attraction. Find the force of gravitational attraction between the proton and electron.

Chapter 16

Electric Charges and Forces

In this brief chapter, we introduce the concept of electric charge and describe some if its important properties. We then introduce and use Coulomb's law, which like Newton's law of gravitation (Chapter 15) is an inverse-square force between point particles. Coulomb's law can be used to find the force acting on point particles in the presence of other point particles or a continuous distribution of charge, and we give some examples of each.

16.1 Introduction to Electric Charge

16.1.1 Background and Definitions

Electric charges play important roles in everyday life. Electricity was known to the ancient Greeks. The Greek word $\eta\lambda\epsilon\kappa\tau\rho o\nu$ and the Latin word *electrum* are used to describe we call *amber*, a fossilized resin that is particularly susceptible to electrical charging when rubbed with a cloth or animal fur (this was observed as early as the sixth century B.C.E. by Thales of Miletus). A material charged in this manner is said to exhibit **static electricity**, because once placed on the material, the charges are static.

Despite many centuries of study, human understanding of electricity had not advanced significantly over the Greeks by 1700. Throughout much of the 18th century, there were two competing fundamental theories of static electricity. In the **single-fluid** theory advanced by the American Benjamin Franklin (1706–1790), the observed electrical forces were caused by the excess or deficiency of an electrical fluid that sometimes flows from one material to another (such as when the amber is rubbed). Materials then exhibit an attraction or repulsion based on the fluid's tendency to equalize: Those with an excess (plus) and deficiency (minus) are attracted to one another, while a plus is repelled from another plus (and minus from another minus), so as to avoid a further imbalance of the fluid. This theory is not correct, but we have retained the plus and minus designations.

The competing theory in the 18th century (which eventually proved to be correct) was advanced by the Frenchman Charles Du Fay (1698–1739), among others. This theory holds that there are two kinds of electricity. These are what we now call positive and negative. Du Fay's fundamental rule was "like charges repel, and opposite charges attract."

The Englishman Stephen Gray (1696–1736) made the important observation that materials can be classified into two kinds, based on their ability to conduct electricity. Materials that conduct electricity well are called **conductors**, and those that do not are called **insulators**. Most materials readily fall into one of these two categories. Two other categories have become important recently. **Semiconductors** conduct electricity much better than insulators but not as well as conductors, and unlike conductors, their ability to conduct improves with increasing temperature. Semiconductors are important components in modern electronic devices, especially computers. **Superconductors** usually behave like conductors, but at extremely low temperatures they become perfect conductors.

It is straightforward to demonstrate the existence of two kinds of charge. All materials contain both kinds of charge—positive and negative—normally with an equal amount of each, so that the net charge is zero. Rubbing a plastic rod with fur will (for most kinds of plastic) transfer some negative charge from the fur to the plastic, leaving a net negative charge on the rod. Because plastic is an insulator, the charge will not run off onto your hand. Now if you touch the rod to a light plastic or paper ball hanging from an insulating thread, some of the negative charge leaks off onto the ball. This process is known as **charging by conduction**. At this point, both the rod and the ball have an excess of negative charge. Now if you move the rod near the ball, you will see a repulsive force in action as the ball is bushed away from the rod.

Next try rubbing a glass rod with a piece of silk. This process transfers some negative charge from the glass to the silk, leaving a net positive charge on the glass (in later chapters we will discuss why only the negative charges are mobile). If you move the positively charged glass rod near the ball you previously charged negative, the ball is attracted to the rod.

Various other experiments give consistent results, always with opposites attracting and like charges repelling. For example, you can use the positively charged glass rod to give a positive charge (by conduction) to a previously uncharged plastic or paper ball. Then the positive rod and positive ball repel each other, and the positive ball and negative ball (from the previous experiment) attract each other.

If you observe carefully when charging by conduction as described above, you should notice that the uncharged ball is slightly attracted to the charged rod *before* the charge transfer takes place. The reason for this is a phenomenon known as **charge polarization**. By definition, charges within the uncharged insulating ball are not free to move. However, molecules within the ball may have more positive charge on one end and more negative on the other. If so, they act as **electric dipoles** that are not free to move but can change the orientation of their poles, so that (for example) when a positively charged rod is brought nearby, the negative portion of the dipole is closer to the rod. The net effect of the positive rod on all such dipoles (shown in Figure 16.1) is a weak attraction.

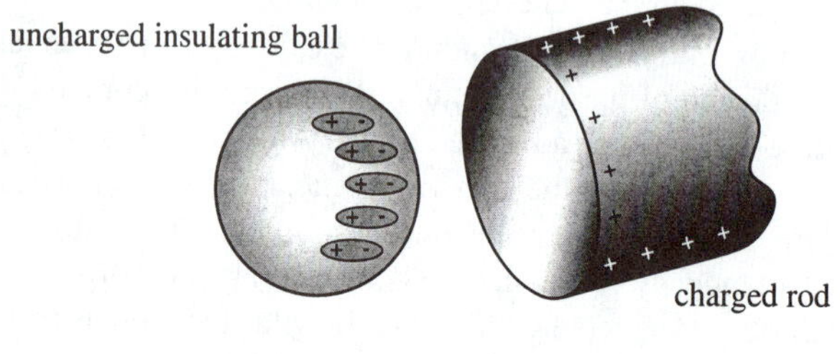

Figure 16.1. The end of a charged rod is brought near an insulating uncharged ball.

16.1.2 Charge Units, Conservation, and Quantization

The SI unit of charge is the **coulomb** (C), named for the French physicist Charles Augustin de Coulomb (1736–1806), who discovered a quantitative force law for electrostatic charges (Section 16.2). The coulomb is not a base unit of the SI system, but rather is defined in terms of the base unit **ampere** (A). The ampere is the unit of electric current (rate of flow of charge), with 1 A = 1 C/s. Therefore 1 C = 1 A·s. We will see how the base unit ampere is defined in Chapter 25.

Physicists believe that charge is a **conserved** quantity. That is, no charge is ever created or destroyed. It may be transferred from one body to another, as in charging by conduction. Beams of charged particles can be sent from one place to another (as in your television) and can even travel vast distances through the universe. But the total amount of charge in any closed system is always the same.

In nuclear reactions, a particle can change into particles with different charges, but the total charge of the system must be conserved. For example, in a process called beta decay, a neutron (n), which has zero charge, decays into three particles: a proton (p) with positive charge, an electron (e) with negative charge, and an antineutrino ($\bar{\nu}$) with zero charge, in the reaction

$$\mathrm{n} \rightarrow \mathrm{p} + \mathrm{e} + \bar{\nu}\,.$$

The charges on the proton and electron are equal in magnitude and but with opposite signs, so the total charge after the reaction is zero, just as before.

The charges on the proton and electron are constants called $+e$ and $-e$, respectively, where the approximate numerical value of e is 1.602×10^{-19} C. The charge on the electron was measured accurately by Robert Millikan in 1911. We will describe that experiment in Chapter 17. (Note: The quantity e, the magnitude of the electronic *charge*, is written with the same symbol we used above for the *particle* itself, except that the charge is italicized. You should not confuse the two, because they will not appear in the same context.)

Physicists observe that charge is **quantized**. That is, free charges always appear in integer multiples of e. Most observed elementary particles have a charge of $+e$, $-e$, or zero (the neutron, the third particle that makes up ordinary matter, has a charge of zero). A few have charges of $+2e$ or $-2e$. Modern elementary particle theory holds that protons and neutrons are composite particles made up of three **quarks**. According to theory, quarks can have charges $\pm e/3$ or $\pm 2e/3$. However, free quarks have not been observed. They are always bound together in threes to form **baryons** (such as protons and neutrons) or pairs to form **mesons**. For example, in a proton (a baryon) there are two "up" quarks with charge $+2e/3$ bound together with a "down" quark with charge $-e/3$ to yield the net observed charge of $+e$. Similarly, a neutron is composed of one up quark and two down quarks. In a positive pi meson, an up quark combines with an "anti-down" quark with charge $+e/3$ to yield the net observed charge of $+e$.

16.2 Coulomb's Law

In 1785, Coulomb used a torsion balance (Figure 16.2) to make quantitative measurements of the forces between different charges. The device is similar to the one Cavendish used to measure G in 1798 (Section 15.7). With fixed point particles with charge q_1 and q_2 (we will always use either lowercase q or uppercase Q to designate charge), the distance between the charged particles can be varied. The torsion balance then indicates the force acting on q_1 as a function of that distance.

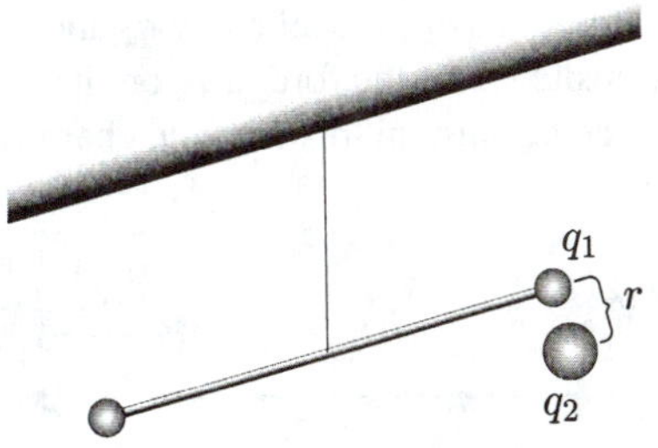

Figure 16.2. A torsion balance.

Physicists often use the shorthand language "point charge" to refer to a point particle that is charged. We will adopt that usage in order to avoid repeating the awkward phrase "point particle with charge q" or similar constructions.

Coulomb found that for point charges the magnitude of the force varies as the inverse-square of the distance and is proportional to the magnitude of each charge. Of course, the direction of the force depends on whether the charges have like or unlike signs. For unlike signs, the force on q_1 is directly toward q_2, and for like signs the force is directly away from q_2. From these results we can express the magnitude of the force (called the **electrostatic force**) as

$$F = \frac{k|q_1||q_2|}{r^2} \tag{16.1}$$

where k is a constant and r is the distance between the two point charges. This is the scalar form of **Coulomb's law**.

The constant k has an experimentally determined value of about

$$k = 8.988 \times 10^9 \text{ N·m}^2/\text{C}^2.$$

Notice that in these units if charge is measured in coulombs and distance in meters, then the force has units of newtons. We often write k in terms of another constant ϵ_0, called the permittivity of free space:

$$k = \frac{1}{4\pi\epsilon_0}$$

where the numerical approximation of ϵ_0 is

$$\epsilon_0 = 8.854 \times 10^{-12} \text{ C}^2/(\text{N·m}^2).$$

Because force is a vector, we should express Coulomb's law in terms of the vector force acting on one of the two charges. We must take care is specifying the direction, however, because by Newton's third law the forces on the two charges are equal in magnitude and opposite in direction. A systematic way of doing so uses the unit vector defined in Figure 16.3. Let $\hat{r}_{12}$ be a unit vector that points directly from the charge q_2 to q_1. With this definition, the vector form of Coulomb's law gives the force $\vec{F}_{12}$ on the charge q_1 due to the charge q_2 as

$$\vec{F}_{12} = \frac{k\, q_1 q_2}{r^2}\, \hat{r}_{12}\,. \tag{16.2}$$

Let's examine the results of Equation (16.2) in various possible cases. When the two charges have the same sign, the force $\vec{F}_{12}$ on the charge q_1 is in the same direction as the unit vector $\hat{r}_{12}$. This is correct, because we know that under these conditions q_1 is repelled by q_2 and experiences a force directly away from q_2. However, if the charges have opposite signs, the force $\vec{F}_{12}$ on the charge q_1 is in the opposite direction as the unit vector $\hat{r}_{12}$. This is also correct, because in this case the charge q_1 is attracted directly toward q_2.

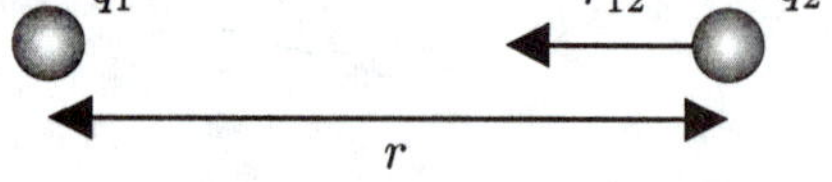

Figure 16.3. The unit vector $\hat{r}_{12}$ points from the charge q_2 to q_1.

Similarly, to express the force $\vec{F}_{21}$ on the charge q_2 due to the charge q_1, we define a unit vector $\hat{r}_{21}$ that points from q_1 to q_2 and, consistent with Equation (16.2), we have

$$\vec{F}_{21} = \frac{k\,q_1 q_2}{r^2}\,\hat{r}_{21}\,. \tag{16.3}$$

We leave it as an exercise for the student to verify that the force vector $\vec{F}_{21}$ has the correct direction for all possible values of q_1 and q_2.

Example 16.1. Given two point charges $q_1 = 5\ \mu$C and $q_2 = 10\ \mu$C, with q_1 at the origin, find the force on q_2 if q_2 is at the point (a) $(1\text{ m}, 0, 0)$; (b) $(1\text{ m}, 1\text{ m}, 0)$; or (c) $(1\text{ m}, 1\text{ m}, 1\text{ m})$.

Solution. (a) The two positive charges repel, so the force on q_2 is in the $+x$-direction. The magnitude of the force is

$$F_{21} = \frac{k|q_1||q_2|}{r^2} = \frac{(8.988 \times 10^9\text{ N·m}^2/\text{C}^2)(5 \times 10^{-6}\text{ C})(10 \times 10^{-6}\text{ C})}{(1\text{ m})^2} = 0.45\text{ N}.$$

Therefore, the force (a vector) is

$$\vec{F}_{21} = 0.45\text{ N}\,\hat{\imath}\,.$$

(b) For this case, the distance between the particles is $r = \sqrt{(1\text{ m})^2 + (1\text{ m})^2} = \sqrt{2}$ m. The magnitude of the force is

$$F_{21} = \frac{(8.988 \times 10^9\text{ N·m}^2/\text{C}^2)(5 \times 10^{-6}\text{ C})(10 \times 10^{-6}\text{ C})}{(\sqrt{2}\text{ m})^2} = 0.225\text{ N}.$$

This force is in the direction of a unit vector $\hat{r}$ that lies at an angle $\pi/2$ above the x-axis. Therefore,

$$\hat{r} = \frac{1}{\sqrt{2}}(\hat{\imath} + \hat{\jmath})$$

and the force is

$$\vec{F}_{21} = 0.225\text{ N}\,\hat{r} = (0.225\text{ N})\left[\frac{1}{\sqrt{2}}(\hat{\imath} + \hat{\jmath})\right] = 0.16\text{ N}\,(\hat{\imath} + \hat{\jmath}).$$

(c) For this case, the distance between the particles is $r = \sqrt{(1\text{ m})^2 + (1\text{ m})^2 + (1\text{ m})^2} = \sqrt{3}$ m. The magnitude of the force is

$$F_{21} = \frac{(8.988 \times 10^9\text{ N·m}^2/\text{C}^2)(5 \times 10^{-6}\text{ C})(10 \times 10^{-6}\text{ C})}{(\sqrt{3}\text{ m})^2} = 0.150\text{ N}.$$

This force is in the direction of a unit vector $\hat{r}$ that makes an equal angle with all three coordinate axes. Therefore,

$$\hat{r} = \frac{1}{\sqrt{3}}(\hat{\imath} + \hat{\jmath} + \hat{k})$$

and the force is

$$\vec{F}_{21} = 0.150\text{ N}\,\hat{r} = (0.150\text{ N})\left[\frac{1}{\sqrt{3}}(\hat{\imath} + \hat{\jmath} + \hat{k})\right] = 0.087\text{ N}\,(\hat{\imath} + \hat{\jmath} + \hat{k}).$$ △

Like Newton's law of gravitation (also an inverse-square force law), Coulomb's law holds for two separate spherically symmetric distributions of charge. The force in Equations (16.1), (16.2), or (16.3) is then computed using the spheres' center-to-center distance, as if all the charge in each sphere were concentrated at its center. This fact follows from the inverse-square nature of the force, which combined with the geometry of the sphere, gives us the shell theorem (Section 15.5). Remember that this works only for spherically symmetric bodies. In other cases, you may *not* assume that the electrical force is the same as if the charge were concentrated at the body's geometric center or center of mass.

The similarity of Coulomb's law to Newton's law of gravitation is striking. Both are inverse-square laws in distance, and both are directly proportional to each relevant quantity (mass or charge). Both forces are essentially infinite in range—that is, acting over vast distances, although the strength does fall off in the inverse-square relationship. One important difference lies in the fact that there is only one kind of mass, and the gravitational force is always attractive, while there are two kinds of charge, making the electrostatic force either attractive or repulsive.

We can also consider the relative difficulty of the experiments done by Cavendish and Coulomb to measure G and k, respectively. Cavendish's experiment is complicated somewhat by the difficulty in maintaining a spherically symmetric charge distribution. In the presence of other charges, an originally symmetric distribution can become distorted, thereby making the verification of the inverse-square law and the measurement of k more difficult. Charges can also leak from the experimental apparatus if the insulation is not good, and induced charges on nearby conducting surfaces can disrupt the measurement. On the other hand, as we discussed in Section 15.7, the electrostatic force is many orders of magnitude stronger than the gravitational force, for similar (charged!) objects at similar distances. This helps overcome the difficulties in Coulomb's experiment.

In the following examples, we will get some practice with Coulomb's law and use it to compare the electrostatic force with the gravitational force.

Example 16.2. In a hydrogen atom, the average distance between the proton and electron is 5.29×10^{-11} m. Compute the magnitudes of both the electrostatic and gravitational forces between the two particles at that distance. Find the ratio of the two force magnitudes.

Solution. The electrostatic force is given by Coulomb's law as

$$F_e = \frac{k|q_1||q_2|}{r^2}.$$

In this case, $|q_1| = |q_2| = e$. Inserting the numerical values of k and r, we find

$$F_e = \frac{ke^2}{r^2} = \frac{(8.988 \times 10^9\ \mathrm{N{\cdot}m^2/C^2})(1.602 \times 10^{-19}\ \mathrm{C})^2}{(5.29 \times 10^{-11}\ \mathrm{m})^2} = 8.24 \times 10^{-8}\ \mathrm{N}.$$

The gravitational force is given by Newton's law of gravitation as

$$F_g = \frac{GMm}{r^2}.$$

From tables, the proton's mass is $M = 1.67 \times 10^{-27}$ kg, and the mass of the electron is 9.11×10^{-31} kg. Thus,

$$F_g = \frac{(6.673 \times 10^{-11}\ \mathrm{N{\cdot}m^2/kg^2})(1.67 \times 10^{-27}\ \mathrm{kg})(9.11 \times 10^{-31}\mathrm{kg})}{(5.29 \times 10^{-11}\ \mathrm{m})^2} = 3.63 \times 10^{-47}\ \mathrm{N}.$$

The ratio of the two force magnitudes is

$$\frac{F(\text{electrostatic})}{F(\text{gravitational})} = \frac{8.24 \times 10^{-8}\ \text{N}}{3.63 \times 10^{-47}\ \text{N}} = 2.27 \times 10^{39}. \qquad \triangle$$

The strengths of the electrostatic and gravitational forces are roughly 40 orders of magnitude different in the hydrogen atom, which is consistent with the relative strengths given in Section 15.7. One conclusion we may draw from this is that the large bodies we see in the solar system (the Sun, planets, moons, comets, and asteroids) are all electrically neutral, or extremely close to it. If not, we would notice an electrostatic force among these bodies. But we do not notice that force, and the motions of planets and moons are driven almost completely by gravity.

Example 16.3. Through friction, a dust particle picks up an excess charge of $q = 3.4 \times 10^{-10}$ C. Assume a spherical dust particle with radius $r = 5.0 \times 10^{-3}$ cm and uniform density $\rho = 3500\ \text{kg/m}^3$. The dust particle is directly above an identical particle with the same charge. For what distance between the two particles does the electrostatic force of repulsion just balance the weight of the dust particle?

Solution. To find the weight of the dust particle, we need its mass, which is density times volume. Equating the magnitudes of the electrostatic and gravitational forces gives

$$\frac{kq^2}{d^2} = mg = (\rho V)g = \rho\left(\frac{4}{3}\pi r^3\right)g$$

where d is the desired distance. Solving for d, we have

$$d^2 = \frac{kq^2}{\frac{4}{3}\pi \rho g r^3}$$

so

$$d = \sqrt{\frac{3kq^2}{4\pi \rho g r^3}}.$$

Inserting numerical values results in

$$d = \sqrt{\frac{3(8.988 \times 10^9\ \text{N}\cdot\text{m}^2/\text{C}^2)(3.4 \times 10^{-10}\ \text{C})^2}{4\pi(3500\ \text{kg/m}^3)(9.80\ \text{m/s}^2)(5.0 \times 10^{-5}\ \text{m})^3}} = 0.24\ \text{m}. \qquad \triangle$$

In this example, a rather small amount of charge (3.4×10^{-10} C) is needed to balance the force of gravity, again demonstrating the relative strength of the electrostatic and gravitational forces. This raises the question: What is a significant amount of charge? As with other physical quantities you have studied, it is a good idea for you to develop an intuitive feel for when the order of magnitude is reasonable. The excess charge you can develop by rubbing a plastic rod or walking across a carpet on a dry day may be on the order of nanocoulombs (10^{-9} C). Electronic devices called *capacitors* (we will study capacitors in Chapter 19) routinely store charges on the order of microcoulombs to millicoulombs in simple electric circuits. Excess charges approaching 1 C are considered quite large. At first, you may feel uncomfortable with "1" being a large number! But remember that the elementary charge is $e = 1.60 \times 10^{-19}$ C, so 1 C contains almost 10^{19} elementary charges.

16.3 Forces Due to Multiple or Extended Charges

16.3.1 Forces Due to Multiple Point Charges

Consider the configuration of three point charges shown in Figure 16.4. This is the same situation as in Example 16.1, but with a third point charge q_3 added. Now what is the force on q_2? To find the total force on q_2, we use Coulomb's law to find separately the force $\vec{F}_{21}$ on q_2 due to q_1 and the force $\vec{F}_{23}$ on q_2 due to q_3. Then, according to the **principle of superposition**, the total force on q_2 is the sum of the forces $\vec{F}_{21} + \vec{F}_{23}$. If we have more charges present, the principle of superposition tells us that the net force on a charge is the sum of the forces due to each of the other charges.

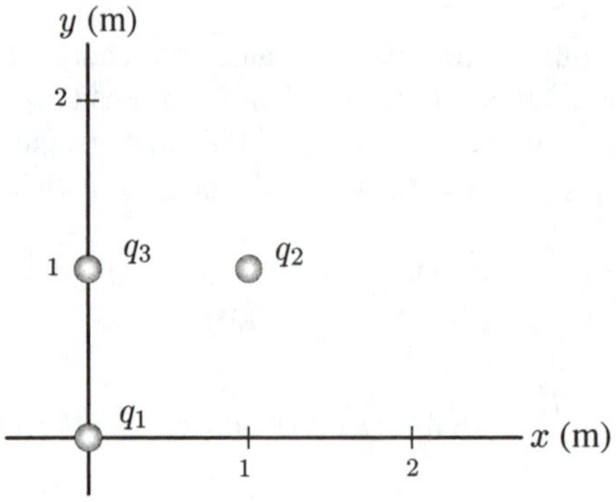

Figure 16.4. Configuration of three charges.

Example 16.4. As in Example 16.1 (and Figure 16.4), a point charge $q_1 = 5\,\mu$C is at the origin, and a charge $q_2 = 10\,\mu$C is at $(1\text{ m}, 1\text{ m}, 0)$. A third charge $q_3 = -7\,\mu$C is placed at $(0, 1\text{ m}, 0)$. Find the total force on q_2.

Solution. From Example 16.1, we know

$$\vec{F}_{21} = 0.16\text{ N}(\hat{\imath} + \hat{\jmath}).$$

It remains to find $\vec{F}_{23}$. We know q_2 is attracted toward q_3, so the force $\vec{F}_{23}$ is in the $-x$-direction. Its magnitude is

$$F_{32} = \frac{k|q_1||q_2|}{r^2} = \frac{(8.988 \times 10^9\text{ N}\cdot\text{m}^2/\text{C}^2)(7 \times 10^{-6}\text{ C})(10 \times 10^{-6}\text{ C})}{(1\text{ m})^2} = 0.63\text{ N}$$

or $\vec{F}_{32} = -0.63\text{ N}\,\hat{\imath}$. Therefore, the net force on q_2 is

$$\vec{F} = \vec{F}_{21} + \vec{F}_{32} = 0.16\text{ N}(\hat{\imath} + \hat{\jmath}) - 0.63\text{ N}\,\hat{\imath} = (-0.47\,\hat{\imath} + 0.16\,\hat{\jmath})\text{N}. \qquad \triangle$$

Example 16.5. Two point charges are fixed on the x-axis, as shown in Figure 16.5, with $q_1 = 8.0\,\mu$C at $x_1 = 0.35$ m and $q_2 = 12.0\,\mu$C at $x_2 = 0.90$ m. (a) Where on the x-axis could a third point charge $q_3 = 5\,\mu$C be placed so that the net force on it is zero? (b) Repeat for the same configuration, but $q_2 = -12.0\,\mu$C.

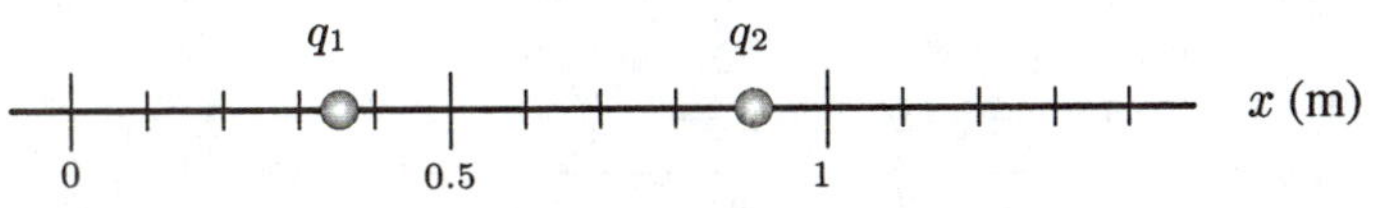

Figure 16.5. Positions of charges in Example 16.5.

Solution. (a) If the third point charge is placed to the left of q_1 or the right of q_2, it will be repelled by both and have a net force not equal to zero. However, if it is between the other two point charges, the repulsive force due to q_1 is to the right, and the repulsive force due to q_2 is to the left. Therefore, these forces can add to zero if their magnitudes are equal. Letting x denote the position of q_3 and setting $F_{31} = F_{32}$, we have

$$\frac{k|q_1||q_3|}{(x - x_1)^2} = \frac{k|q_2||q_3|}{(x_2 - x)^2}.$$

Cancelling common factors and cross multiplying gives

$$|q_1|\,(x_2 - x)^2 = |q_2|\,(x - x_1)^2,$$

which after some algebra leads to

$$x = \frac{x_1 + x_2\sqrt{\frac{q_1}{q_2}}}{1 + \sqrt{\frac{q_1}{q_2}}}$$

where we have dropped the absolute value signs because all charges are positive. Inserting numerical values,

$$x = \frac{0.35 \text{ m} + (0.90 \text{ m})\sqrt{\frac{8.0\,\mu\text{C}}{12.0\,\mu\text{C}}}}{1 + \sqrt{\frac{8.0\,\mu\text{C}}{12.0\,\mu\text{C}}}} = 0.60 \text{ m}.$$

This result is reasonable, because it is dimensionally correct, lies between the two charges (as expected), and in fact lies closer to q_1 than q_2. This makes sense, because q_1 is a smaller charge than q_2, and so q_3 must lie closer to it in order to experience an electrostatic force of the same magnitude.

(b) Now the particle with charge q_3 is repelled by q_1 but attracted toward q_2. If q_3 were placed between those charges, there would be a net force on it to the right. If q_3 were placed at $x > x_2$, then $\vec{F}_{31}$ is to the right, and $\vec{F}_{32}$ is to the left. However, since $q_2 > q_1$ and q_3 is always closer to q_2, those forces can never cancel. This leaves the possibility $x < x_1$. Now $\vec{F}_{31}$ is to the left, and $\vec{F}_{32}$ is to the right, and with q_3 closer to the smaller charge, there is a chance of cancellation. Once again, we equate the magnitudes

$$\frac{k|q_1||q_3|}{(x_1 - x)^2} = \frac{k|q_2||q_3|}{(x_2 - x)^2}.$$

What differs here is that $q_2 < 0$, so $|q_2| = -q_2$. With this change, the algebra leads to

$$x = \frac{x_1 - x_2\sqrt{-\frac{q_1}{q_2}}}{1 + \sqrt{-\frac{q_1}{q_2}}}.$$

Inserting the numerical values,

$$x = \frac{0.35\text{ m} - (0.90\text{ m})\sqrt{-\frac{8.0\,\mu\text{C}}{-12.0\,\mu\text{C}}}}{1 + \sqrt{-\frac{8.0\,\mu\text{C}}{-12.0\,\mu\text{C}}}} = -0.21\text{ m}.$$

This result is reasonable, because it is dimensionally correct and lies to the left of q_1, as expected. △

16.3.2 Forces Due to Continuous Distributions of Charge—One Dimension

What is the force on a point charge due to some collection of charge distributed throughout a volume? By the principle of superposition, we should add the contributions of all the charges in the distribution. We will study a one-dimensional example first, and then extend the procedure to three dimensions.

You may wonder whether a "continuous distribution" of charge is possible. Given that charge is quantized, isn't any collection of charge just a group of point charges $+e$ and $-e$? Technically this is true, but as a practical matter the elementary charge is so small that it is reasonable to assume a continuous distribution. Even a small collection of charge on the order of 1 nC contains on the order of 10^{10} elementary charges. Indeed, the same argument could be made regarding mass, where a piece of matter is ultimately made up of a huge number or protons, neutrons, and electrons. In spite of this, we are able to model matter as essentially continuous distributions of mass to find total mass, center of mass, rotational inertia (Section 14.4), and gravitational force (Section 15.5).

As a one-dimensional example, consider a thin rod of length L along the x-axis with one end at the origin, as shown in Figure 16.6(a). The rod has a net positive charge Q spread uniformly over its surface (as we will see in later chapters, this is quite realistic for a conducting rod). A positive point charge q_0 lies at some point $x = R$ (where $R > L$) to the right of the rod. We wish to find the net electrostatic force acting on q_0 due to all the charge on the rod.

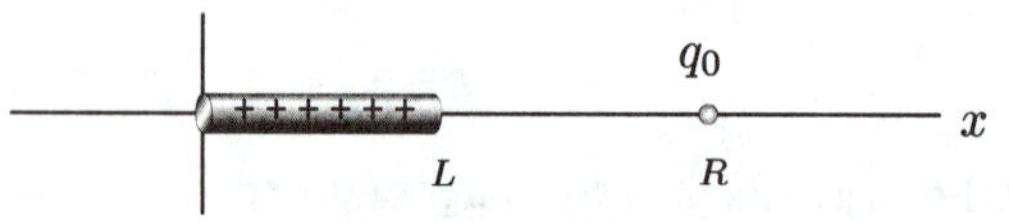

(a) Charged rod near a point charge

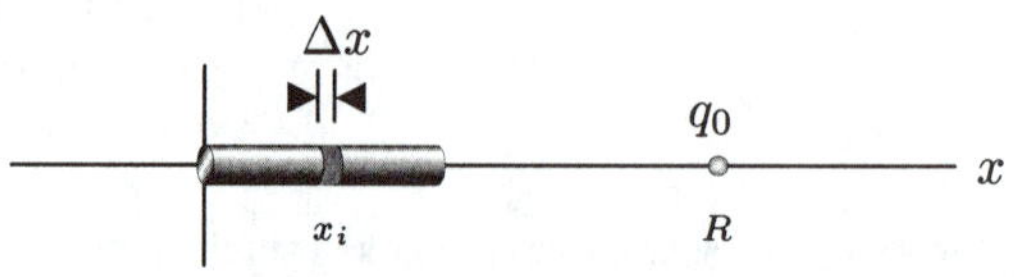

(b) Typical subinterval on the rod

Figure 16.6. Setting up the problem of the charged rod near a point charge.

We assume that there exists a constant **charge density** (defined to be the charge per unit length), which we will call λ. Because we have assumed the charge per unit length to be constant, it is clear that the charge density equals the total charge divided by the total length, or $\lambda = Q/L$.

Following our usual procedure from calculus, we break the interval $[0, L]$ into n equal subintervals of length $\Delta x = L/n$. A generic subinterval [shown in Figure 16.6(b)] at some position $x = x_i$ lies a distance $R - x_i$ from the point charge q_0. How much charge Δq is in the subinterval? From the definition of charge density, it is $\Delta q = \lambda\,\Delta x$. By treating the subinterval as a point charge, we can find the *approximate* magnitude of the force on q_0 due to that subinterval of charge using Coulomb's law:

$$F_i = \frac{kq_0\,\Delta q}{(R - x_i)^2}\,. \tag{16.4}$$

In this problem, all charges are positive, so the direction of this force is to the right. That direction is the same for any subinterval Δx we choose, so the net force on q_0 is to the right and has a magnitude

$$F = \sum_{i=1}^{n} F_i = \sum_{i=1}^{n} \frac{kq_0\,\Delta q}{(R - x_i)^2} \tag{16.5}$$

or, because $\Delta q = \lambda\,\Delta x$,

$$F = \sum_{i=1}^{n} \frac{kq_0\,\lambda\,\Delta x}{(R - x_i)^2}\,. \tag{16.6}$$

From calculus we recognize this sum as a Riemann sum. In the limit as $n \to \infty$ (or $\Delta x \to 0$), the Riemann sum becomes a definite integral, which gives the *exact* magnitude of the force. The definite integral is

$$F = \int_0^L \frac{kq_0\,\lambda}{(R - x)^2}\,dx$$

where we have included the limits of integration 0 to L corresponding to the beginning and end of the interval.

We now turn our attention to evaluating this definite integral. We can factor the constants k, q_0, and L from the definite integral, leaving

$$F = kq_0\,\lambda \int_0^L \frac{1}{(R - x)^2}\,dx\,. \tag{16.7}$$

The evaluation of this definite integral is straightforward using the second fundamental theorem of calculus. Using a table of indefinite integrals (or using the substitution $u = R - x$), we find

$$F = kq_0\,\lambda \left(\frac{1}{R - x}\right)\Bigg|_0^L = kq_0\,\lambda \left(\frac{1}{R - L} - \frac{1}{R}\right).$$

We can express the result in terms of Q by recalling $\lambda = Q/L$. Thus,

$$F = \frac{kq_0 Q}{L}\left(\frac{1}{R - L} - \frac{1}{R}\right). \tag{16.8}$$

This result is correct dimensionally. Also notice that $F > 0$, as required for any $R > L$. Notice that the force is *not* the same as if all the charge were concentrated at the rod's center, because this is not a spherically symmetric distribution of charge.

Example 16.6. For the example of the uniformly charged rod in the text, show that if q_0 is a great distance away ($R >> L$) the result is approximated by two point charges Q and q_0 separated by a distance R.

Solution. Looking at the expression in parentheses in Equation (16.8), we have

$$\frac{1}{R - L} - \frac{1}{R} = \frac{R - (R - L)}{R(R - L)} = \frac{L}{R(R - L)}.$$

Therefore,

$$F = \frac{kq_0 Q}{L}\left[\frac{L}{R(R - L)}\right] = \frac{kq_0 Q}{R(R - L)}.$$

Now when $R \gg L$, the denominator is approximately equal to R^2, so

$$F \approx \frac{kq_0 Q}{R^2},$$

which is the same as for point charges Q and q_0 separated by a distance R. This approximation makes physical sense, because from a great distance away, it is difficult to distinguish the fine structure of the rod from a point charge Q. △

Thus far, we have been dealing with a rather special case. We need to generalize our method so that we can find the force on a point charge due to *any* one-dimensional charge distribution. Without loss of generality, we can restrict the one-dimensional charge distribution to the x-axis. However, the charge density may not be constant, so in general $\lambda(x)$ is not constant, and the charge density *cannot* be factored outside the definite integral as in Equation (16.7).

Also, the point charge q_0 may not lie on the x-axis, a situation we show in Figure 16.7. For a segment of the line charge of length Δx that lies a distance r from the point charge q_0, the direction of the electrostatic force is in the direction of the unit vector $\hat{r}$ as shown, along a line from the segment of length Δx to the point charge. Notice that both r and $\hat{r}$ vary as a function of the position of the Δx segment. With these considerations in mind, we can revise Equation (16.7) to write the vector force on q_0 as

$$\vec{F} = kq_0 \int_{x_1}^{x_2} \frac{\lambda(x)\,\hat{r}}{r^2}\,dx. \tag{16.9}$$

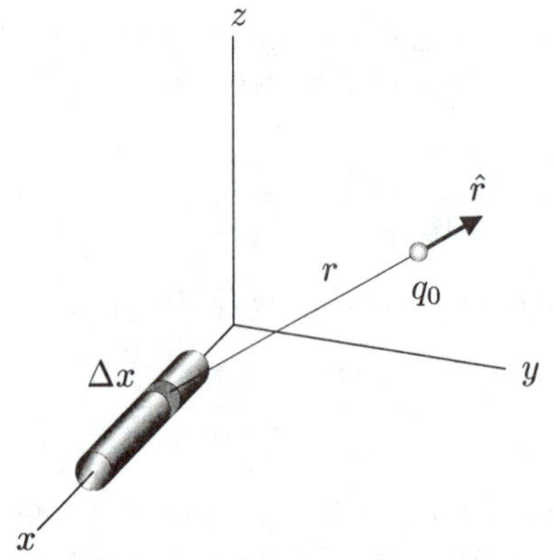

Figure 16.7. Typical subinterval on a line charge.

In solving problems with Equation (16.9), it is important to keep several things in mind. First, the distance r usually depends on the value of x. Therefore, it is necessary to perform some kind of substitution, expressing the entire integrand as a function of x before doing the definite integral. In some cases, it may be appropriate to change the variable of integration, but in any case the integrand must correspond to the variable of integration. Also, the unit vector $\hat{r}$ may vary its direction as x varies over the interval $[x_1, x_2]$. In the following example, we take this into account and use Equation (16.9) to find the force.

Example 16.7. A linear charge distribution from $x = 0$ to $x = L$ on the x-axis has a net charge of $-Q$ distributed uniformly from $x = 0$ to $x = L/2$ and a net charge of $+Q$ distributed uniformly from $x = L/2$ to $x = L$. Find the force on a positive point charge q_0 at $x = R$ where $R > L$.

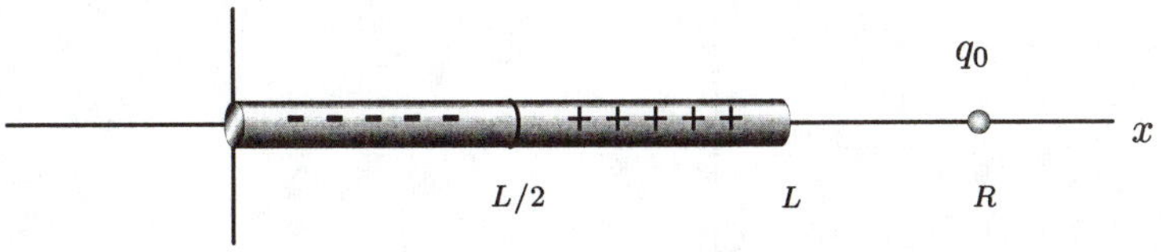

Figure 16.8. Schematic for Example 16.7.

Solution. The situation is shown in Figure 16.8. Even though the net charge on the rod is zero, there is a net force to the right on q_0, because it is closer to the positive charge than the negative charge. In general, the force due to any one-dimensional distribution is given by Equation (16.9).

Let's look at the negative and positive parts of the distribution separately, because the charge density $\lambda(x)$ changes from negative to positive. For the negative portion on the interval $x : [0, L/2]$, we have $\lambda(x) = (-Q)/(L/2) = -2Q/L$. With $\hat{r} = \hat{\imath}$ and $r = R - x$ as before, this means that the force $\vec{F}_-$ on the point charge q_0 due to the negative part of the line charge is

$$\vec{F}_- = kq_0 \int_0^{L/2} \frac{\left(-\frac{2Q}{L}\right)\hat{\imath}}{(R-x)^2}\, dx\,.$$

Factoring out constants,

$$\vec{F}_- = -\frac{2kq_0Q}{L}\,\hat{\imath} \int_0^{L/2} \frac{1}{(R-x)^2}\, dx\,.$$

Similarly, the force $\vec{F}_+$ on q_0 due to the positive distribution of charge from $L/2$ to L is

$$\vec{F}_+ = \frac{2kq_0Q}{L}\,\hat{\imath} \int_{L/2}^{L} \frac{1}{(R-x)^2}\, dx\,.$$

(You should satisfy yourself that this is true, because we have left out some steps.) Then the total force on q_0 is

$$\vec{F} = \vec{F}_+ + \vec{F}_- = \frac{2kq_0Q}{L}\,\hat{\imath}\left[\int_{L/2}^{L} \frac{1}{(R-x)^2}\, dx - \int_0^{L/2} \frac{1}{(R-x)^2}\, dx\right].$$

Evaluating the definite integrals as before,

$$\begin{aligned}\vec{F} &= \frac{2kq_0Q}{L}\,\hat{\imath}\left[\left(\frac{1}{R-x}\right)\Big|_{L/2}^{L} - \left(\frac{1}{R-x}\right)\Big|_0^{L/2}\right] = \frac{2kq_0Q}{L}\hat{\imath}\left[\frac{1}{R-L} - \frac{1}{R-L/2} - \left(\frac{1}{R-L/2} - \frac{1}{R}\right)\right]\\ &= \frac{2kq_0Q}{L}\,\hat{\imath}\left[\frac{1}{R-L} - \frac{2}{R-L/2} + \frac{1}{R}\right].\end{aligned}$$

Expressing the bracketed quantity with a common denominator, we find

$$\vec{F} = \frac{2kq_0Q}{L}\,\hat{\imath}\left[\frac{L^2}{(R-L)(2R-L)R}\right] = \frac{2kq_0QL}{(R-L)(2R-L)R}\,\hat{\imath}.$$

This result is correct dimensionally, and it is clear that for any $R > L$ the force $\vec{F}$ is in the $+x$-direction, as we predicted. △

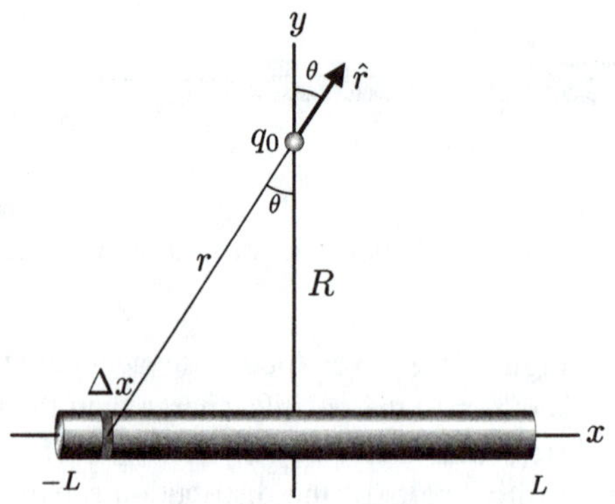

Figure 16.9. Generic segment on the charged rod of Example 16.8.

Example 16.8. A positive uniformly charged rod with total charge Q lies on the x-axis from $x = -L/2$ to $x = +L/2$. Find the force on a positive point charge q_0 at $y = R$ on the $+y$-axis.

Solution. We can express the uniform charge density as $\lambda = Q/L$. The difficulty in this problem lies in the fact that the force on q_0 due to a generic segment of length Δx (see Figure 16.9) is in the direction of $\hat{r}$, which varies with x. However, due to symmetry, we can see that the x-components of the electrostatic force cancel, because for every segment Δx with $x < 0$, there is a corresponding segment with $x > 0$ that will cancel its x-component of force. The y-components of force are positive for all the Δx, so the net force must be in the $+y$-direction. In general, $\hat{r}$ makes an angle θ with the y-axis, so the y-component of $\hat{r}$ is $\cos\theta$. Therefore, by Equation (16.9),

$$\vec{F} = kq_0 \int_{-L/2}^{L/2} \frac{(Q/L)(\cos\theta\,\hat{\jmath})}{r^2}\,dx.$$

We must express the definite integral as a function of a single variable. In this case, we can express r and θ in terms of x. From the diagram,

$$r = \sqrt{x^2 + R^2}$$

and

$$\cos\theta = \frac{R}{r} = \frac{R}{\sqrt{x^2 + R^2}}.$$

Substituting these expressions into the definite integral and factoring out constants, we find

$$\vec{F} = \frac{kq_0QR}{L}\,\hat{\jmath} \int_{-L/2}^{L/2} \frac{1}{(x^2 + R^2)^{3/2}}\,dx\,.$$

From integral tables,

$$\vec{F} = \frac{kq_0QR}{L}\,\hat{\jmath}\left(\frac{1}{R^2}\frac{x}{\sqrt{x^2+R^2}}\right)\Bigg|_{-L/2}^{L/2} = \frac{kq_0Q}{LR}\,\hat{\jmath}\left[\frac{L/2}{\sqrt{(L/2)^2+R^2}} - \frac{-L/2}{\sqrt{(-L/2)^2+R^2}}\right]$$

$$= \frac{kq_0Q}{R\sqrt{(L/2)^2+R^2}}\,\hat{\jmath}\,.$$

Notice that the result is dimensionally correct, and for any $R > 0$ we choose the resultant force is in the $+y$-direction. $\triangle$

Thus far, we have worked examples with a high degree of symmetry. But this will not always be the case. Consider a uniformly charged rod on the x-axis as in the last example, but a point charge q_0 at some point (x_0, y_0) in the xy-plane not necessarily on either the x-axis or the y-axis, as shown in Figure 16.10. What is the force on q_0 due to the charged rod? In principle, the answer is still given by Equation (16.9). It is possible (though more difficult) to write the distance r in terms of the parameters of the problem. However, because $\hat{r}$ varies with x, it is no longer straightforward to find the x- and y-components of the resultant force $\vec{F}$.

In this case, you are better off exploring numerical methods that will allow you to find the x- and y-components of $\vec{F}$ for particular values of the parameters x_0 and y_0. To do this, you need to convert the definite integral in Equation (16.9) back into a Riemann sum. Breaking the interval $[-L/2, L/2]$ into subintervals, you can compute the force on q_0 due to the charge in each subinterval. The sum of the forces from all subintervals is the approximate value of the total force on q_0. You are asked to explore this kind of numerical integration in the problems.

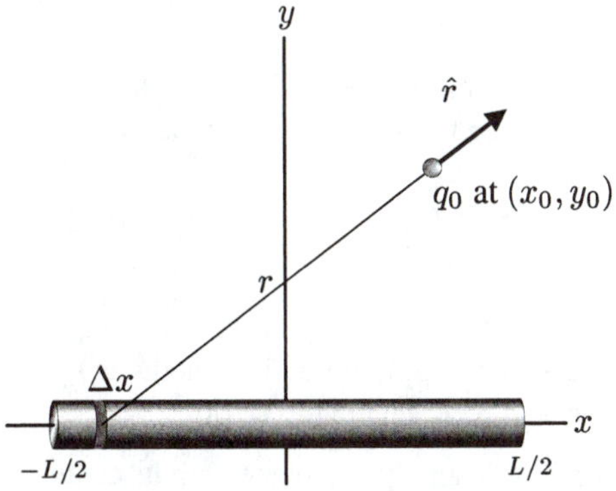

Figure 16.10. Generic segment on a charged rod in relation to a point not on the axis of symmetry.

16.3.3 Forces Due to Continuous Distributions of Charge—More Than One Dimension

We can modify Equation (16.9) to find the force on a point charge due to a continuous distribution of charge in more than one dimension. For example, consider the two-dimensional distribution of charge in the xy-plane, as shown in Figure 16.11(a). Let R be the region in which the charge is confined, and let the charge density (charge per unit area) in the xy-plane be described by the function $\sigma(x, y)$. The force on q_0 due to a small piece of area ΔA within the distribution depends on the distance r and is in the $\hat{r}$ direction. To integrate over the entire charge distribution requires a double integral, and by analogy with Equation (16.9), the total force on q_0 is

$$\vec{F} = kq_0 \iint\limits_R \frac{\sigma(x, y)\,\hat{r}}{r^2}\, dA\,. \tag{16.10}$$

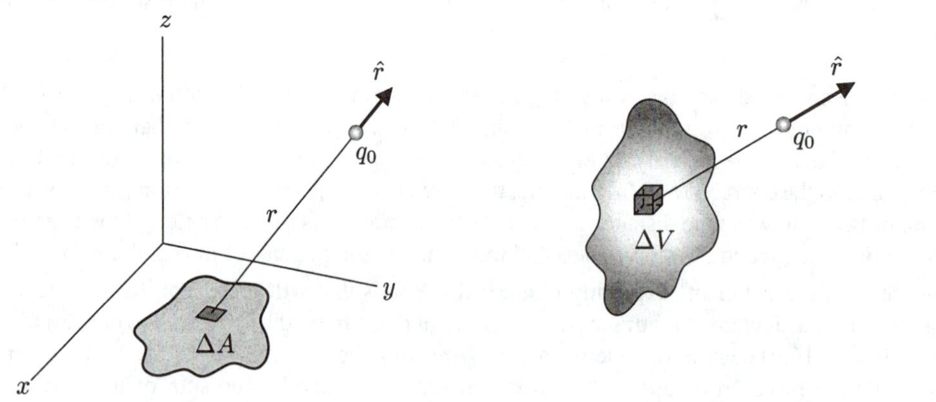

(a) Two-dimensional distribution of charge in the xy-plane (b) Three-dimensional distribution of charge

Figure 16.11. Continuous distributions of charge in two and three dimensions.

Similarly, we express the force on a point charge q_0 due to a three-dimensional charge distribution [Figure 16.11(b)] with charge density $\rho(x, y, z)$ (charge per unit volume) as the triple integral

$$\vec{F} = kq_0 \iiint_D \frac{\rho(x, y, z)\,\hat{r}}{r^2}\,dV. \tag{16.11}$$

In practice, the integrals in Equations (16.10) and (16.11) can be quite difficult to do, especially in cases lacking symmetry. It is often necessary to apply numerical methods in these cases. We will conclude with some high-symmetry examples that can be done in closed form.

Example 16.9. A thin ring of radius R lies in the xy-plane with its center at the origin. The ring is uniformly charged with a net charge $+Q$. Find the force due to the ring on a positive point charge q_0 at the point $z = z_0$ on the z-axis.

Solution. Even though the ring is not along a straight line, it is can be analyzed as a one-dimensional distribution of charge along a circle. The force on q_0 due to a generic segment of length Δs (see Figure 16.12) is in the $\hat{r}$ direction. Upon integration over the whole ring, the x- and y-components of the force cancel, leaving only a z-component. The z-component of $\hat{r}$ is $\cos\theta$. Therefore, the total force on q_0 is [as in Equation (16.9)]

$$\vec{F} = kq_0 \int \frac{\lambda \cos\theta\,\hat{k}}{r^2}\,ds$$

where λ is the constant linear charge density along the ring. The integral is a definite integral, but for now let's omit the limits on s. Notice that for this geometry θ and r are the same for every segment of length Δs. Therefore, we may treat them as constants and factor them (along with the constants $\hat{k}$ and λ) from the definite integral, leaving

$$\vec{F} = \frac{kq_0\lambda\cos\theta}{r^2}\,\hat{k}\int ds.$$

We are left with $\int ds$ over the whole ring, which is just the circumference of the ring $2\pi R$.

$$\vec{F} = \frac{kq_0\lambda\cos\theta}{r^2}\,\hat{k}(2\pi R) = \frac{2\pi kq_0\lambda R\cos\theta}{r^2}\,\hat{k}.$$

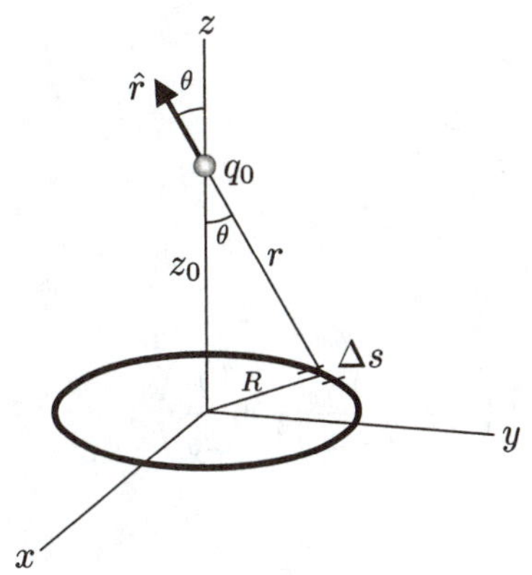

Figure 16.12. Ring of charge in Example 16.9.

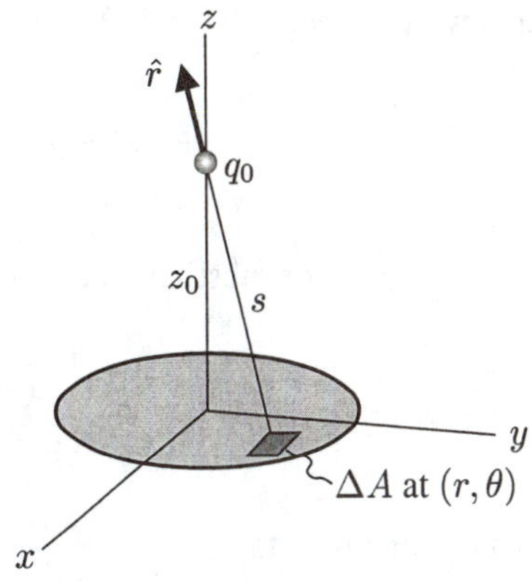

Figure 16.13. Disk of charge in Example 16.10.

To simplify this alphabet soup of English and Greek letters, note that $\lambda = Q/2\pi R$, $\cos\theta = z_0/r$ and $r^2 = R^2 + z_0^2$. Therefore,

$$\vec{F} = \frac{kq_0 Q z_0}{(R^2 + z_0^2)^{3/2}}\,\hat{k}.$$

This result is dimensionally correct. Also, notice that the magnitude of the force approaches zero in the limit as z_0 approaches zero, as symmetry tells us it must. △

Example 16.10. A positively charged disk with uniform charge density σ and radius R lies in the xy-plane with its center at the origin. A point charge q_0 is on the z-axis at $z = z_0$. Find the force on q_0 due to the charged disk.

Solution. In general, we should use Equation (16.10) for a distribution of charge in two dimensions. Because of the radial symmetry of this distribution, we are better off not using $dA = dx\,dy$ from Cartesian coordinates. Instead, we choose polar coordinates, with $dA = r\,dr\,d\theta$. Because we use r in polar coordinates, we define s to be the distance from the point (r, θ) in the disk to q_0, as shown in Figure 16.13. We know from symmetry that the net force on q_0 is in the $+z$-direction, and as in the previous example the z-component of $\hat{r}$ is z_0/s. Thus, by Equation (16.10)

$$\vec{F} = kq_0 \iint\limits_R \frac{\sigma(z_0/s)\,\hat{k}}{s^2} r\,dr\,d\theta$$

where σ is the constant charge density. (Note that we are using R in two different contexts, as the region in which the charge density lies and the radius of the disk.) Factoring out the constants leaves

$$\vec{F} = kq_0\sigma z_0\,\hat{k} \iint\limits_R \frac{1}{s^3} r\,dr\,d\theta\,.$$

From geometry, $s = \sqrt{r^2 + z_0^2}$, and using an iterated double integral with the limits of integration,

$$\vec{F} = kq_0\sigma z_0\,\hat{k} \int_0^{2\pi} \int_0^R \frac{r}{(r^2 + z_0^2)^{3/2}} dr\,d\theta\,.$$

The integration with respect to θ yields 2π, so

$$\vec{F} = 2\pi k q_0 \sigma z_0 \, \hat{k} \int_0^R \frac{r}{(r^2 + z_0^2)^{3/2}} dr \ .$$

Using a table of integrals (or substitution $u = r^2 + z_0^2$),

$$\vec{F} = 2\pi k q_0 \sigma z_0 \, \hat{k} \left(-\frac{1}{\sqrt{r^2 + z_0^2}} \right) \Bigg|_0^R = 2\pi k q_0 \sigma z_0 \, \hat{k} \left(-\frac{1}{\sqrt{R^2 + z_0^2}} + \frac{1}{z_0} \right) .$$

This result can be simplified to

$$\vec{F} = 2\pi k q_0 \sigma \, \hat{k} \left(1 - \frac{z_0}{\sqrt{R^2 + z_0^2}} \right) .$$

You should check to see that this result is dimensionally correct. Notice that in the limit as z_0 approaches zero the force is simply

$$\vec{F} = 2\pi k q_0 \sigma \, \hat{k}$$

or, because $k = 1/4\pi\epsilon_0$,

$$\vec{F} = \frac{q_0 \sigma}{2\epsilon_0} \hat{k} \, .$$

△

16.4 Problems

16.1 Introduction to Electric Charge

1. How many elementary charges are there in exactly 1 C of charge? How many are there in 1 μC?

2. (a) How much charge is contained in 1 kg of electrons? (b) How much charge is contained in 1 kg of protons?

3. (a) Assuming roughly half of Earth's mass is made up of protons, find the total amount of positive charge contained in Earth. (b) If an object containing a net charge of +1 C is grounded (its charge is transferred to Earth), by what fraction does the positive charge in Earth change?

4. Suppose that the charge on the electron and proton are not the same, but instead differ by one part in 10^{10}. (a) What would be the net charge on a 1-kg sample of helium gas? (b) What would be the net charge on Earth (assume half of Earth's mass is protons)?

5. Compute the charge in 1 mole of protons.

16.2 Coulomb's Law

1. Treating them as point particles, find the electrostatic force between two quarks of charge $+2e/3$ separated by a distance of 5.0×10^{-16} m in a proton.

2. Suppose all the electrons could be removed from Earth. Estimate the electrostatic force on a proton just above Earth's surface under these conditions. Compare your answer with the weight of the proton.

3. Suppose all the electrons could be removed from the Moon and Earth. Estimate the electrostatic force on the Moon. Compare your answer with the Moon's gravitational force due to Earth.

4. A free electron and free proton are exactly 1 cm apart. Find the acceleration of the electron (magnitude and direction) and the acceleration of the proton (magnitude and direction).

5. Two identical conducting spheres carry net charges of 1.65 μC and 5.61 μC, respectively. Their centers are separated by 25 cm. (a) What is the force on each sphere? (b) A conducting wire is connected between the two spheres momentarily so that the charge can transfer between the spheres. Find the charge on each sphere and the force between them.

6. Two identical pendulums of length L hang side by side with the bobs (each having mass m) just touching. An equal charge q is placed on each bob. Find an equation relating θ (the angle each pendulum makes with the vertical) to the other parameters.

7. Two ping-pong balls with mass 1.4 g each carry a net charge of 0.350 μC. One ball is held fixed. At what height does the second ball come to rest if it is placed directly above the fixed ball?

8. Suppose Earth carries some excess positive charge spread uniformly over its surface. (a) How large should the magnitude of that charge be in order to balance a proton (against its weight) just above Earth's surface? (b) Under these conditions, find the net downward acceleration (electrostatic plus gravitational) of an electron just above Earth's surface.

9. Find the force on two 0.1-C charges separated by a distance of 10 m.

10. A spring is held fixed at one end, and the other end hangs vertically downward. Attached to the bottom of the (insulating) spring is a ball carrying a net charge of +8.5 μC. The weight of the 0.120-kg ball causes the spring to stretch 24.1 cm beyond its equilibrium position. Another charged ball is placed directly below that ball, and this causes the ball on the spring to move downward another 3.50 cm before coming to rest 11.5 cm above the second ball. What is the charge on the second ball?

16.3 Forces Due to Multiple or Extended Charges

1. Four identical point charges of 5.6 nC are placed at the corners of a square 1.5 cm on a side. Find the force acting on each point charge.

2. Three identical point charges of 0.15 μC are placed at the vertices of an equilateral triangle with edges of length 50 cm. Find the force acting on each point charge.

3. Two point charges q and $3q$ are separated by a distance d. Where could another point charge be placed so that the net force on it is zero? Does it matter whether the third point charge is positive or negative? Explain.

4. Four identical point charges Q are placed at the corners of a square with side d. Find the net force acting on each point charge.

5. Four point charges are placed at the corners of a square of side d. Across one diagonal from each other lie two $+q$ point charges, and across the other diagonal lie two $-q$ charges. (a) Find the net force on each point charge. (b) Find the net force on another point charge Q placed at the center of the square.

6. A point charge $+6.5\,\mu$C is located at the origin, and another point charge $-4.2\,\mu$C is located at (35 cm, 45 cm, 0). Where can a third charge $+1.6\,\mu$C be placed so that the force on it is zero?

7. A thin rod of length L carries a uniform charge density, with a total charge $+Q$. Find the magnitude and direction of the force on point charge $-q$ located along the same line as the rod and a distance L from one end.

8. A thin rod lies along the x-axis from $x = -L/2$ to $x = +L/2$. The charge density of the rod is not uniform, but is given by the function $\lambda(x) = Cx$ (where C is a constant). (a) What are the dimensions of C? (b) Find the force on a positive point charge q_0 that lies at the point $x = L$ on the x-axis.

9. Repeat Problem 8 if the positive point charge q_0 lies at the point $y = L$ on the y-axis.

10. Repeat Problem 8 if the rod's charge density is $\lambda(x) = C\,|x|$, where C is a constant.

11. Assume that the entire x-axis (in both directions) has a uniform positive charge density λ. Find the force acting on a negative charge $-q_0$ at the point $y = d$ on the y-axis.

12. A uniformly charged thin rod (with total charge 10 μC) lies along the x-axis from $x = -50$ cm to $x = +50$ cm. Use numerical techniques to approximate the force on a point charge 10 μC at the point (30 cm, 40 cm, 0). Try breaking the rod into different numbers of subintervals (say 10, 100, and 1000) and see how this affects your results.

13. A circle of radius R lies in the xy-plane with its center at the origin. The semicircle with $y > 0$ has a uniform charge density and total charge $+Q$, while the semicircle with $y < 0$ has a uniform charge density and total charge $-Q$. Find the force on a positive point charge q_0 at the origin.

14. A uniformly charged thin rod with total charge Q lies on the x-axis from $x = 0$ to $x = L$. Find the net force on a point charge q_0 at $y = L$ on the y-axis. [*Hint*: Find the x- and y-components of the force separately.]

15. Another way to work Example 16.10 is to consider the disk to be made up of a number of thin rings of radius r and thickness Δr. Use the result of Example 16.9 for the force on q_0 due to each ring. Then construct a definite integral over all such rings in the disk and show that the total force is the same as we found in Example 16.10.

16. Prove the shell theorem (see Section 15.5) for the electrostatic force.

17. Two thin circular disks of radius R lie parallel to the xy-plane with their centers along the z-axis. Each disk carries a uniform surface charge density. One disk lies in the $z = +d$ plane and carries a total charge $+Q$, while the second disk lies in the $z = -d$ plane and carries a total charge $-Q$. (a) Find the net force on a charge q_0 at the origin. (b) Show that if $R \gg d$, the magnitude of the force on q_0 is approximately $q_0\sigma/\epsilon_0$.

Chapter 17

Vector Fields and Electric Fields

In Chapter 13, we studied functions of more than one variable $f : \mathbb{R}^n \to \mathbb{R}$. In this chapter, we introduce functions of the type $\vec{F} : \mathbb{R}^n \to \mathbb{R}^n$ (with emphasis on $n = 2$ and $n = 3$ as usual). Functions of this type are called *vector fields*. Vector fields are used to model a wide variety of physical phenomena. For these applications, the input generally represents a location in a plane ($n = 2$) or space ($n = 3$), and the output represents the value of the physical quantity at that location. In introducing vector fields, we use the velocity of a fluid as an example, because physical intuition for this is straightforward. We saw examples of vectors fields in Chapter 12, where we studied systems of two first-order differential equations using *tangent vector fields*.

In the second and third sections of the chapter, we study *electric fields*. Electric fields are used to model the interactions of charged objects.

17.1 Introduction to Vector Fields

Our primary goal for this section is to understand the basic symbolic, numeric, and graphic representations of vector fields. In later chapters, we return to consider the calculus of vector fields.

A **vector field** is a function of the type $\vec{F} : \mathbb{R}^n \to \mathbb{R}^n$. That is, each input is a vector with n components and each output is also a vector with n components. We will deal primarily with $n = 2$ and $n = 3$.

In many cases, it is useful to think of the inputs or the outputs in terms of components. For a generic function $\vec{F} : \mathbb{R}^2 \to \mathbb{R}^2$, we denote these with $\vec{F} : (x, y) \mapsto \vec{F}(x, y) = \langle P(x, y), Q(x, y)\rangle$. Note that the two output components are each functions of two variables. We could choose to investigate vector fields by reducing the problem to a study of the two functions P and Q. However, this viewpoint misses the essential vector nature of the outputs. Applications in physics lead us to consider the "vectorness" as important.

As a specific example, consider the vector field defined by

$$\vec{F}(x, y) = \langle x + y, xy\rangle.$$

The outputs have components $P(x, y) = x + y$ and $Q(x, y) = xy$. For the specific input $(x, y) = (2, 5)$, we compute the output $\vec{F}(2, 5) = \langle 7, 10\rangle$.

For a generic function $\vec{F} : \mathbb{R}^3 \to \mathbb{R}^3$, we denote the components with $\vec{F} : (x, y, z) \mapsto \vec{F}(x, y, z) = \langle P(x, y, z), Q(x, y, z), R(x, y, z)\rangle$. For example, the vector field defined by

$$\vec{F}(x, y, z) = \langle yz, xz, xy\rangle$$

has output components $P(x,y,z) = yz$, $Q(x,y,z) = xz$, and $R(x,y,z) = xy$. For the specific input $(x,y,z) = (3,7,2)$, we compute the output $\vec{F}(3,7,2) = \langle 14, 6, 21\rangle$.

In many situations from physics, the output is given in terms of the magnitude and direction of the input. As example, consider the gravitational force an object of mass M exerts on an object of mass m. Choose a coordinate system with origin at the location of the mass M object. Let $\vec{r}$ be the position of the mass m object. The gravitational force is given by Equation (15.4) as

$$\vec{F}(\vec{r}) = -GMm\frac{\hat{r}}{r^2} \tag{15.4}$$

where, as usual, $\hat{r}$ is a unit vector in the direction of $\vec{r}$ and r is the magnitude of $\vec{r}$. We can express this in terms of components. Let $\vec{r} = \langle x, y, z\rangle$. Then $r = \sqrt{x^2+y^2+z^2}$ and

$$\hat{r} = \frac{\vec{r}}{r} = \frac{\langle x, y, z\rangle}{\sqrt{x^2+y^2+z^2}}.$$

Substituting into Equation (15.4) gives

$$\begin{aligned}\vec{F}(x,y,z) &= -GMm\frac{\langle x,y,z\rangle}{(x^2+y^2+z^2)^{3/2}}\\ &= \langle -\frac{GMmx}{(x^2+y^2+z^2)^{3/2}}, -\frac{GMmy}{(x^2+y^2+z^2)^{3/2}}, -\frac{GMmz}{(x^2+y^2+z^2)^{3/2}}\rangle.\end{aligned}$$

The form in Equation (15.4) is compact, but there are situations in which we need to know the components explicitly.

17.1.1 Visualizing Vector Fields

We begin by considering a situation in which data for a function $\vec{F} : \mathbb{R}^2 \to \mathbb{R}^2$ is collected. For example, satellites in orbit around Earth collect data on surface wind velocity over the oceans. In this case, the inputs are locations on a portion of Earth's surface (small enough that the region can be thought of as a region of a plane), and the outputs are surface wind velocities. The surface wind velocity includes only the two components in the (approximate) plane of the region. A small sample of imagined data is given in Table 17.1. Note that the inputs are position coordinates (in units of kilometers with respect to some chosen origin), and the outputs are velocity components (in units of meters per second).

Table 17.1. Surface wind velocity data

Location (km)	Wind velocity (m/s)
$(1.0, 1.0)$	$\langle 0.2, 0.5\rangle$
$(2.0, 1.5)$	$\langle 0.3, -0.5\rangle$
$(-2.0, 2.0)$	$\langle 0.2, -0.9\rangle$
$(-2.0, -1.0)$	$\langle -0.4, 0.5\rangle$
$(1.0, -1.5)$	$\langle -0.1, 1.0\rangle$

How can we visualize this data? Note that we have four dimensions to deal with, two for the inputs and two for the outputs. We handle this by making a plot of the input plane and then "attaching" output planes to points of that input plane. This is illustrated in Figure 17.1(a) for the data in Table 17.1. The output plane attached to a input point is indicated by a set of output coordinate axes with origin at the input point. Each

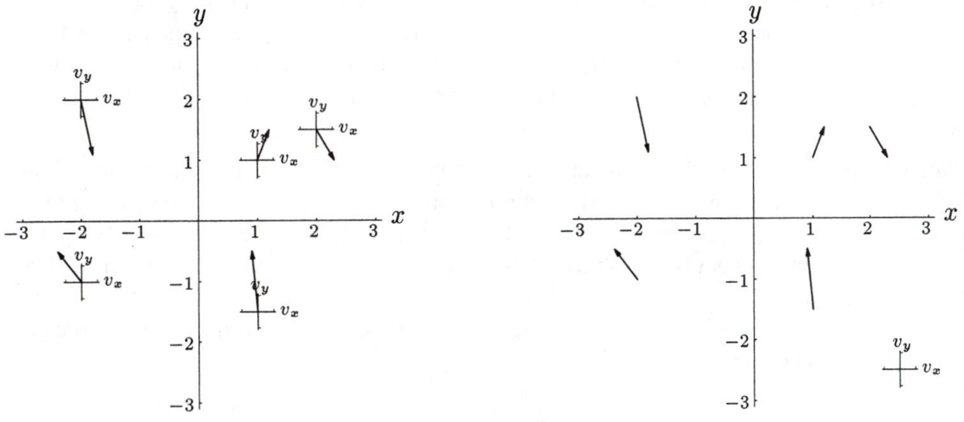

(a) Output coordinate system "attached" to each input

(b) One output coordinate system to set scale

Figure 17.1. Plot of wind velocity vectors given in Table 17.1.

output is represented as an arrow. Note that the scale of the output coordinate system is independent of the scale of the input coordinate system. In fact, for this example (and many applied examples), the input and output quantities have different physical units. We can simplify the picture somewhat by showing only one output coordinate system to set the scale [see Figure 17.1(b)]. The output scale is generally chosen so that the arrows representing the outputs do not overlap. This will depend, of course, on how many outputs we choose to plot.

If we are given a vector field in symbolic form, we can generate a plot by computing outputs for inputs we choose.

Example 17.1. Plot the vector field given by $\vec{F}(x, y) = \langle x - y^2 + 1, x^2 + y^2 + 1 \rangle$.

Solution. To generate the plot, we need to pick inputs and compute corresponding outputs. Generally we pick inputs on a rectangular grid. A small sample of computed outputs is given in Table 17.2. These outputs are plotted in Figure 17.2 along with a larger sample of outputs.

Table 17.2. Outputs for Example 17.1

(x, y)	$\vec{F}(x, y)$
$(0, 0)$	$\langle 1.00, 1.00 \rangle$
$(0.5, 0)$	$\langle 1.50, 1.25 \rangle$
$(0.5, 0.5)$	$\langle 1.25, 1.50 \rangle$
$(0, 0.5)$	$\langle 0.75, 1.25 \rangle$
$(-0.5, 0.5)$	$\langle 0.25, 1.50 \rangle$
$(-0.5, 0)$	$\langle 0.50, 1.25 \rangle$
$(-0.5, -0.5)$	$\langle 0.25, 1.50 \rangle$
$(0, -0.5)$	$\langle 0.75, 1.25 \rangle$
$(0.5, -0.5)$	$\langle 1.25, 1.50 \rangle$

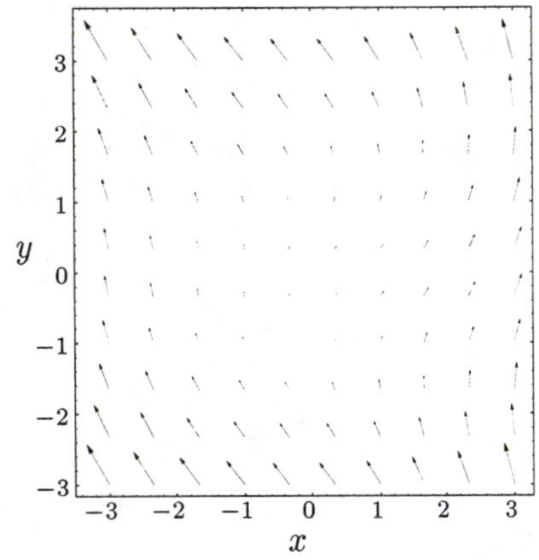

Figure 17.2. The vector field of Example 17.1.

We can also look at the outputs qualitatively. Note that the second component $Q(x,y) = x^2 + y^2 + 1$ is positive for all (x, y). Thus, all outputs have a positive vertical component. The first component $P(x,y) = x - y^2 + 1$ is negative for $y^2 > x + 1$. This is reflected by the fact that the output vectors point to the left at the top and bottom of the plot and point to the right in the portion of the plot near the x-axis (where y^2 is small). △

For vector fields $\vec{F} : \mathbb{R}^3 \to \mathbb{R}^3$, visualization is much more difficult. In principle, we can follow the same plan of plotting outputs as arrows at various input points in space. In practice, this type of plot is difficult to draw on a flat sheet of paper or a computer screen. Using a computer, however, we can view such a plot from various angles or produce an animation showing the plot revolving. We leave it to the reader to explore the options available locally.

For vector fields $\vec{F} : \mathbb{R}^3 \to \mathbb{R}^3$, a verbal description can be useful. As example, consider the gravitational field given by Equation (15.4):

$$\vec{F}(\vec{r}) = -GMm\frac{\hat{r}}{r^2}. \tag{15.4}$$

We describe this verbally in the following way: At each input, the output points radially in to the origin with magnitude that is inversely proportional to the square of the distance from the origin.

17.1.2 Linear Vector Fields

Linear functions play an important role in our understanding of general functions. For functions of one variable $f : \mathbb{R} \to \mathbb{R}$, tangent lines are the graphs of linear functions. Those linear functions have the form $g(x) = f(a) + f'(a)(x - a)$ and are thus characterized by the parameters $f(a)$ and $f'(a)$. The derivative output $f'(a)$ gives the slope of the tangent line. In Chapter 18, we will see that for functions of two variables $f : \mathbb{R}^2 \to \mathbb{R}$, tangent planes are the graphs of linear functions. A linear function of two variables has the form $g(x,y) = A + B(x - a) + C(y - a)$.

A linear vector field with inputs and outputs in $\mathbb{R}^2$ has the form

$$\vec{G}(x,y) = \langle A, B\rangle + \langle px + qy, rx + sy\rangle \tag{17.1}$$

where A, B, p, q, r, and s are constants. The constant term is $\langle A, B\rangle$. Figure 17.3 shows a plot of $\vec{G}(x,y) = \langle 1, -2\rangle$. Such a field with vectors all having the same magnitude and direction is called a constant or **uniform** field.

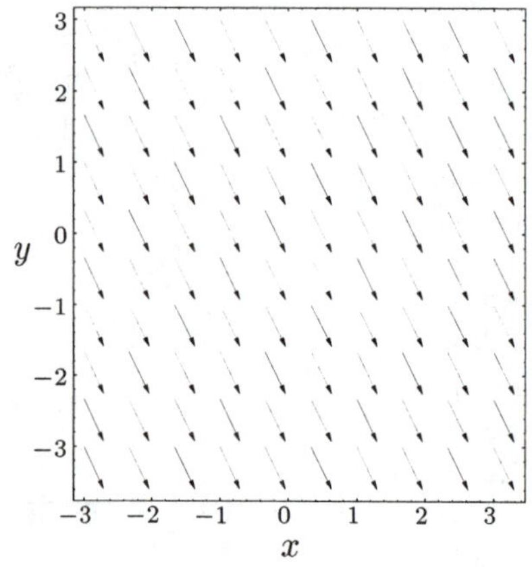

Figure 17.3. Plot of $\langle 1, -2\rangle$ as an example of a constant vector field.

The first-degree terms are characterized by the four parameters p, q, r, and s in Equation (17.1). We explore some of the possibilities in the following examples.

Example 17.2. Plot and characterize the linear vector field $\vec{G}(x, y) = \langle x, y \rangle$.

Solution. At each input point, the output vector points radially out from the origin. The magnitude of each output is equal to the distance of the input from the origin. If the output scale were equal to the input scale, the length of each output vector would be the distance from the origin to the input point. A plot is shown in Figure 17.4(a). In this plot, the output scale is smaller than the input scale.

The linear vector field $\vec{H}(x, y) = \langle -x, -y \rangle$ has outputs opposite to the outputs of $\vec{G}(x, y) = \langle x, y \rangle$. The outputs of $\vec{H}$ point radially in to the origin. △

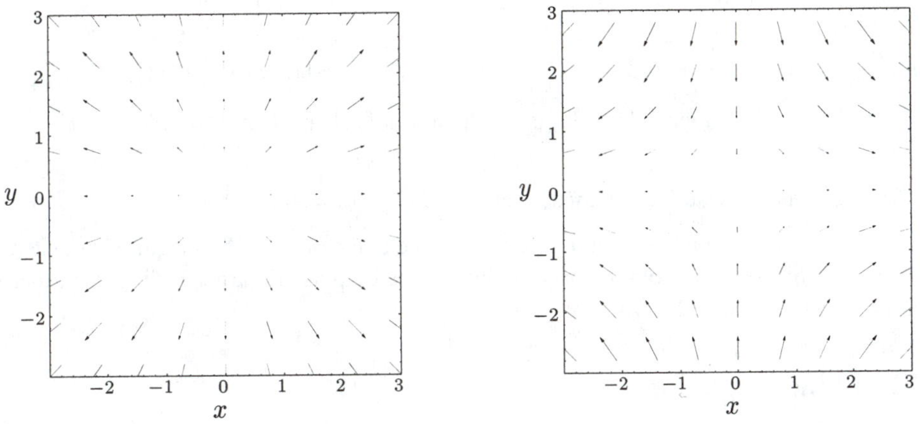

(a) Linear vector field $\langle x, y \rangle$

(b) Linear vector field $\langle x, -y \rangle$

Figure 17.4. Plots for Examples 17.2 and 17.3.

Example 17.3. Plot and characterize the linear vector field $\vec{G}(x, y) = \langle x, -y \rangle$.

Solution. Along the y-axis, the output vectors point toward the origin, while along the x-axis, the output vectors point away from the origin. In the first quadrant of the input plane, the outputs have positive first component and negative second component. These outputs thus point to the right and down when plotted in the xy-plane. Similar observations can be made for the other quadrants. The magnitude of each output is equal to the distance of the input from the origin. That is, the outputs are larger for inputs further from the origin. A plot is shown in Figure 17.4(b).

△

Example 17.4. Plot and characterize the linear vector field $\vec{G}(x, y) = \langle y, x \rangle$.

Solution. Along the positive y-axis, the output vectors point in the positive x-direction, while along the negative y-axis, the output vectors point in the negative x-direction. Similarly, along the positive x-axis, the output vectors point in the positive y-direction, while along the negative x-axis, the output vectors point in the negative y-direction. Everywhere on the x-axis, the output vectors point away from the origin. The magnitude of each output is equal to the distance of the input from the origin. A plot is shown in Figure 17.5(a). △

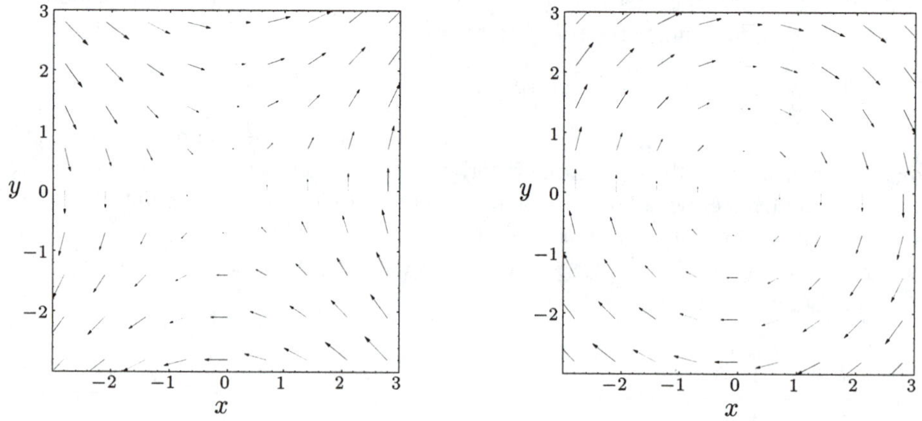

(a) Linear vector field $\langle y, x\rangle$

(b) Linear vector field $\langle y, -x\rangle$

Figure 17.5. Plots for Examples 17.4 and 17.5.

Example 17.5. Plot and characterize the linear vector field $\vec{G}(x, y) = \langle y, -x\rangle$.

Solution. The output vectors point tangent to circles centered at the origin. The magnitude of each output is equal to the distance of the input from the origin. That is, the outputs are larger for inputs farther from the origin. A plot is shown in Figure 17.5(b). △

17.1.3 Field Lines

We can add one more facet to our plot of a vector field $\vec{F} : \mathbb{R}^2 \to \mathbb{R}^2$. To introduce this, let's think of the outputs $\vec{F}(x, y)$ as the velocities of a thin sheet of fluid flowing in a plane. We assume the fluid is in **steady state**, meaning that the velocity field does not change in time. Note, however, that a small test object in the fluid will move in time. A **flowline** is the path of such a test object. To be precise, we assume the test object moves at the fluid velocity at each point in the plane. A flowline is then a path that has the output $\vec{F}(x, y)$ as tangent at each point (x, y). We can sketch a flowline on a plot of a vector field by starting at some input point and following the output vectors. An example is shown in Figure 17.6.

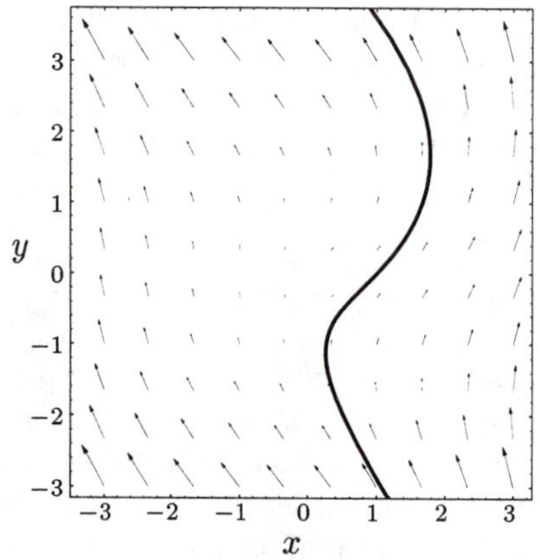

Figure 17.6. An example of a field line.

In the case of a generic vector field, we use the term **field line** for a curve that has output vectors as tangents. (A flowline is thus a field line for a vector field that models the velocity of a planar fluid.) We can easily determine field lines for a few of the linear vector fields from above. Figure 17.7(a) shows some field lines for the vector field $\vec{F}(x, y) = \langle x, y \rangle$. Because the output vectors all point radially outward, each field line is a ray pointing directed radially out from an initial point. A second example is shown in Figure 17.7(b) with the vector field $\vec{F}(x, y) = \langle y, -x \rangle$. In this case, field lines are circles centered at the origin.

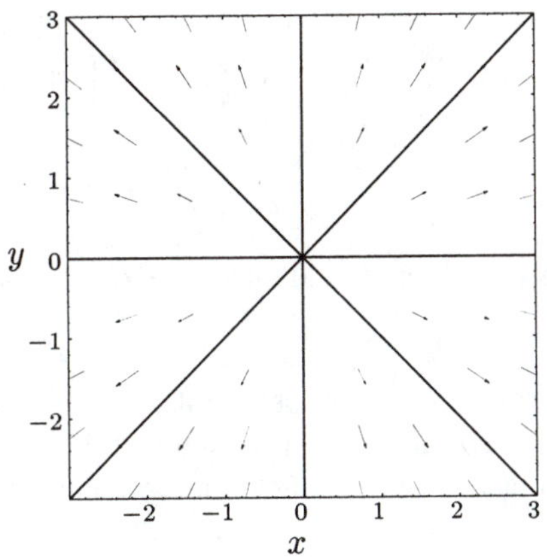

(a) Linear vector field $\langle x, y \rangle$

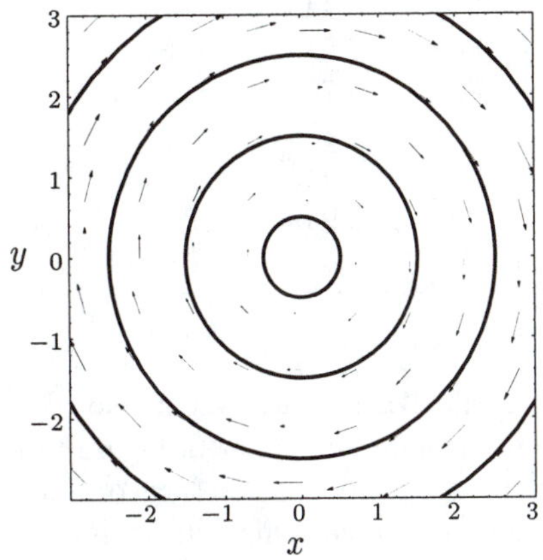

(b) Linear vector field $\langle y, -x \rangle$

Figure 17.7. Field lines for some linear vector fields.

Let's introduce some notation, so we can compute field lines quantitatively for general vector fields. Consider a vector field $\vec{F} : \mathbb{R}^2 \to \mathbb{R}^2$ with $\vec{F} : (x, y) \to \langle P(x, y), Q(x, y) \rangle$. Assume a field line is the output curve of a function $\vec{r} : \mathbb{R} \to \mathbb{R}^2$, with $\vec{r} : u \to \vec{r}(u) = \langle x(u), y(u) \rangle$. On the one hand, the outputs of the derivative $\vec{r}\,'$ are tangent to the output curve. On the other hand, by definition, the outputs of $\vec{F}$ are tangent to the output curve. We can thus equate these to get $\vec{r}\,'(u) = \vec{F}(\vec{r}(u))$. In terms of components, we have

$$\frac{dx}{du} = P(x, y)$$
$$\frac{dy}{du} = Q(x, y).$$

This is a system of two first-order differential equations. We studied such systems in Chapter 12, where we saw how to numerically compute approximations for solutions curves using Euler's method. We illustrate this in the next example.

Example 17.6. Plot representative field lines for the vector field $\vec{F}(x, y) = \langle x - y^2 + 1, x^2 + y^2 + 1 \rangle$.

Solution. The field lines are solution curves for the system

$$\frac{dx}{du} = x - y^2 + 1$$
$$\frac{dy}{du} = x^2 + y^2 + 1.$$

For each field line, we need to choose an initial point. We can then compute an approximate solution using Euler's method. Table 17.3 gives results computed using $\Delta u = 0.1$ for the field line with initial point $(0, 0)$.

Table 17.3. Results for Example 17.6

n	u	$x(u)$	$y(u)$
0	0.0	0.000	0.000
1	0.1	0.100	0.100
2	0.2	0.209	0.202
3	0.3	0.326	0.310
4	0.4	0.449	0.431
5	0.5	0.575	0.569
6	0.6	0.700	0.735
7	0.7	0.816	0.938
8	0.8	0.910	1.193
9	0.9	0.959	1.517
10	1.0	0.924	1.940

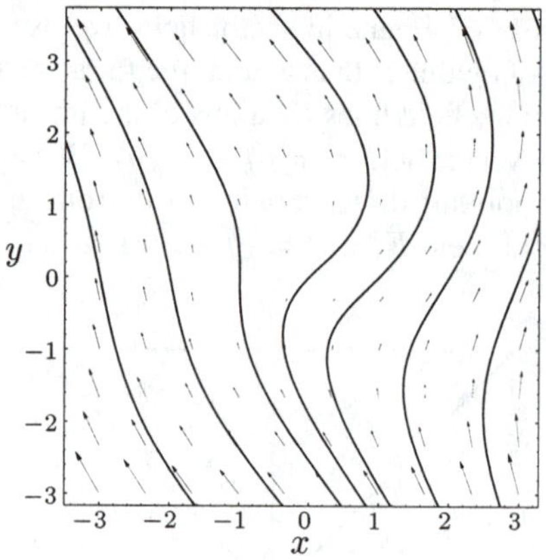

Figure 17.8. Plot of field lines for Example 17.6.

Many software applications are available to compute field lines using techniques similar in spirit to Euler's method but often more sophisticated. Figure 17.8 is a plot produced using such software. If this vector field modeled the fluid velocity of a thin sheet of fluid flowing in a plane, then each field line would be the path of a small test object released at the initial point. △

A caution is in order here: A field line is the path of a test object *only* if the vector field represents *velocities*. For example, if the outputs of a vector field represent forces, a test object will not follow the field lines, because the path of an object is generally not tangent to forces acting on the object.

17.2 Electric Fields

17.2.1 Background and History

Now we can begin to apply the concept of the vector field to our study of electrostatics. We'll take advantage of the similarities between the gravitational and electrostatic forces and continue to use gravity as an example of a field, because gravity is such a significant and obvious factor in our everyday lives and is therefore more familiar to us. Because both of these forces follow inverse-square force laws, certain similarities can be exploited.

One important similarity between the electrostatic and gravitational forces is that they are both long-range forces that operate with "action at a distance" (see Section 15.3). That is, two point particles separated by a great distance in the vacuum of space still interact via the gravitational and electrostatic forces. Remember that this concept bothered Newton, but he offered no explanation of how it could work. In the late 1940s, the American physicist Richard Feynman (1918–1988) and others developed a successful *quantum field theory* that explained the electrostatic force between two charged particles using an exchange of *photons*, particles that travel at the speed of light.

Quantum field theory is beyond the scope of this course. However, the concept of the electric field is an important step toward the modern theory, and by understanding electric fields we will be able to solve a much wider range of problems than we could with Coulomb's law alone. The electric field was developed in the nineteenth century, with significant contributions by the English physicist Michael Faraday (1791–1867). The electric field is a mathematical tool that can be used to describe and model electromagnetic phenomena, but it stops short of addressing the fundamental nature of action at a distance.

17.2.2 Introduction and Definitions

Try to envision the **gravitational field** of the room in which you sit, defined for any position in the room as the gravitational acceleration vector $\vec{g}$ of a point particle at that position. To an excellent approximation, all the $\vec{g}$ vectors in the room are the same; they all point essentially straight down with about the same magnitude (Figure 17.9). Recall from Section 17.1 that a field with vectors all having the same magnitude and direction is called a constant or uniform field.

From your study of gravitation in Chapter 15, you know that the magnitude and direction of $\vec{g}$ varies over Earth's surface and with altitude above Earth's surface. A wider-range view of Earth's gravitational field is shown in Figure 17.9(b). The magnitude of the gravitational acceleration of a point particle falls off as the inverse square of the distance from Earth's center, and the direction of the acceleration is toward Earth's center.

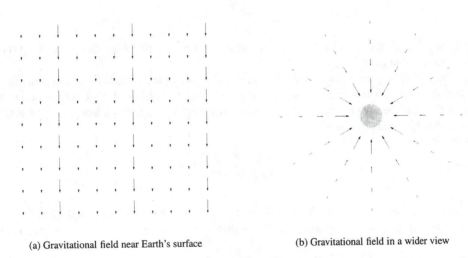

(a) Gravitational field near Earth's surface

(b) Gravitational field in a wider view

Figure 17.9. Gravitational field due to Earth.

In Figure 17.9(b), we still consider $\vec{g}$ to be the acceleration (vector) of a point particle at every position in space. We can write a symbolic representation of $\vec{g}$ based on Newton's law of gravitation, which says that the gravitational force on a point particle of mass m due to another particle of mass M is

$$\vec{F} = -\frac{GMm}{r^2}\hat{r} \tag{15.4}$$

where r is the distance between the two particles and $\hat{r}$ is a unit vector directed from M to m. To find the acceleration $\vec{g}$ of a point particle with mass m due to Earth (mass M), we can use Equation (15.4) along with Newton's second law of motion to write

$$\vec{F} = -\frac{GMm}{r^2}\hat{r} = m\vec{g} \tag{17.2}$$

where r is measured from Earth's center.

Solving for $\vec{g}$ in terms of Earth's mass and the distance r, we have

$$\vec{g} = -\frac{GM}{r^2}\hat{r}. \tag{17.3}$$

This is the symbolic representation of Earth's gravitational field shown in Figure 17.9(b). This representation of $\vec{g}$ fits the concept of vector field we developed earlier in this chapter; it is a vector quantity $\vec{g} : \mathbb{R}^3 \to \mathbb{R}^3$ defined for every point in a region of space (outside Earth).

The gravitational field is a property of Earth, independent of the point particle m (notice that only M and not m appears in our expression for $\vec{g}$). The gravitational field surrounds Earth and "waits" for us to introduce a test particle, which then experiences a force $\vec{F} = m\vec{g}$ when placed in the field. The unit vector $\hat{r}$ points from Earth's center to the point where we wish to know the field. If the gravitational field were due to some distribution of mass that is not spherically symmetric, Equation (17.3) would no longer hold. But there is still a gravitational field due to the combined effects of all the masses in the distribution. We can "map" the field experimentally by taking a test particle (with mass m) around to different locations, measuring the gravitational force $\vec{F}$ on the particle, and then computing

$$\vec{g} = \frac{\vec{F}}{m} \tag{17.4}$$

for each location. The *direction* of the field is significant, because it tells us the direction of the force on a test particle placed at that point.

This brief study of the gravitational field gives us a way to define the electric field. Suppose there exists some distribution of charges in space, as shown in Figure 17.10. In Section 16.3 we found that, in general, the force on a test particle with charge q_0 due to such a distribution is found by integrating over the entire distribution of charge density $\rho(x, y, z)$:

$$\vec{F} = kq_0 \iiint\limits_D \frac{\rho(x, y, z)\,\hat{r}}{r^2}\, dV. \tag{16.11}$$

Clearly the force $\vec{F}$ depends on the location of the test particle relative to the charge distribution. At each point in space, the **electric field** is defined as the force on the test particle divided by its charge, that is,

$$\vec{E} = \frac{\vec{F}}{q_0}. \tag{17.5}$$

Notice that the definition of electric field is precisely analogous to the expression for the gravitational field in Equation (17.4). This should not be surprising, because we saw in our study of Coulomb's law how *charge* plays a role analogous to that of *mass* in gravitation.

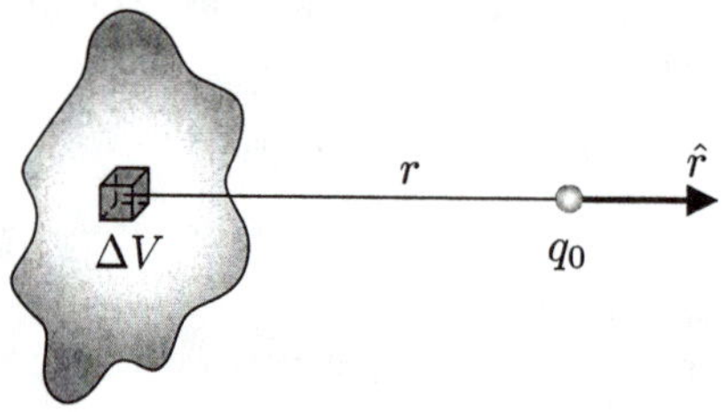

Figure 17.10. Schematic for analyzing the electric field due to a distribution of charge.

17.2.3 The Electric Field of a Point Charge

A symbolic expression for the electric field due to a point charge can be readily found, since the force law is given by Coulomb's law. The force on a test particle of charge q_0 due to a point charge q is given by Equation (16.2) as

$$\vec{F} = \frac{kqq_0}{r^2}\,\hat{r}$$

where r is the distance between the two charges and $\hat{r}$ is a unit vector pointing from q to q_0. From the definition of the electric field [Equation (17.5)], the electric field due to a point particle with charge q is

$$\vec{E} = \frac{\vec{F}}{q_0} = \frac{kq}{r^2}\,\hat{r}. \tag{17.6}$$

Note the similarity between this electric field and the gravitational field due to a point particle, as given in Equation (17.3), with q playing the role in electrostatics that M played in gravitation. However, some extra analysis is called for here, because there are two kinds of charge, whereas there is only one kind of mass. In Figure 17.11(a), we show the electric field due to a positively charged point particle. With $q > 0$ in Equation (17.6), the field vector points radially outward at every point in space, and the magnitude of the field vector falls off as the inverse square of the distance from q. With $q > 0$, Equation (17.6) tells us that $\vec{E}$ points in the same direction as $\hat{r}$.

However, when $q < 0$, the electric field vector points radially inward, as shown in Figure 17.11(b), because Equation (17.6) tells us that $\vec{E}$ points in the *opposite* direction of $\hat{r}$. The magnitude of the field vector still falls off as the inverse square of the distance from q.

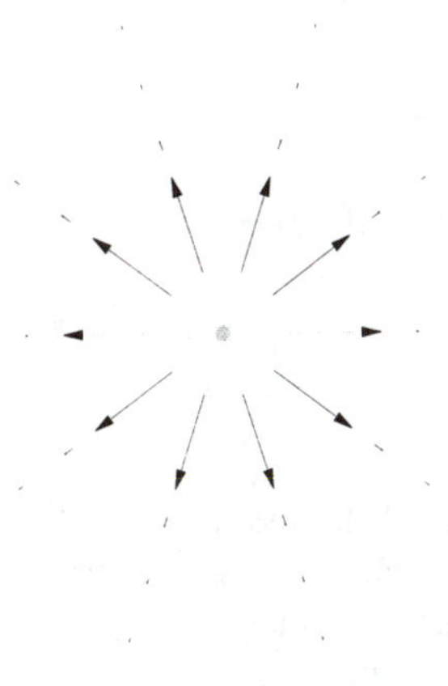

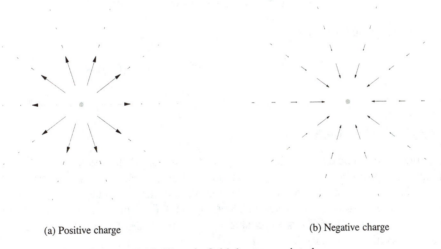

(a) Positive charge

(b) Negative charge

Figure 17.11. Electric field due to a point charge.

Earlier we saw that the direction of the gravitational field vector at a given position indicates the direction of the force on a point particle brought to that position. From the definition of the electric field, we can conclude the following: *The direction of the electric field vector at a given position indicates the direction of the force on a* ***positive*** *point particle brought to that position.* However, the force on a point particle of negative charge located at the same position is *opposite* to the direction of the electric field vector. You should convince yourself that this rule is consistent in all possible cases with the rule that like charges repel and opposites attract. For example, the electric field due to a positive charge is shown in Figure 17.11(a). A positive test charge placed in the field experiences a force in the direction of the field lines, radially outward, consistent with the positive charge being repelled by the other positive charge. A negative charge placed in the same field experiences a force in the direction opposite to the field lines, radially inward, consistent with the negative charge being attracted to the positive charge.

From Equation (17.5), we can see that the dimensions of electric field are force divided by charge. Therefore, in the SI system the electric field can be given in units of N/C. At a distance of 1.0 m from a modest charge of 1 μC, the magnitude of the electric field is

$$E = \frac{kq}{r^2} = \frac{(8.988 \times 10^9 \text{ N·m}^2/\text{C}^2)(1 \times 10^{-6} \text{ C})}{(1 \text{ m})^2} = 9 \times 10^3 \text{ N/C}.$$

You should not be surprised to see electric field magnitudes well up into the thousands or millions of N/C.

Example 17.7. A point charge $Q = +3.50\,\mu$C lies at the origin. Find the electric field at the point (35.0 cm, 46.0 cm) in the xy-plane.

Solution. The electric field due to a point charge is given by Equation (17.6). The magnitude of the electric field at that point is

$$E = \frac{kQ}{r^2} = \frac{(8.988 \times 10^9 \text{ N·m}^2/\text{C}^2)(3.50 \times 10^{-6} \text{ C})}{(0.350 \text{ m})^2 + (0.460 \text{ m})^2} = 9.42 \times 10^4 \text{ N/C}.$$

The unit vector $\hat{r}$ from the origin to the point in question is

$$\hat{r} = \frac{\vec{r}}{r} = \frac{(0.350\,\hat{\imath} + 0.460\,\hat{\jmath}) \text{ m}}{\sqrt{(0.350 \text{ m})^2 + (0.460 \text{ m})^2}} = 0.606\,\hat{\imath} + 0.796\,\hat{\jmath}.$$

Therefore, the electric field (vector) is

$$\vec{E} = \frac{kQ}{r^2}\hat{r} = E\hat{r} = (5.71 \times 10^4\,\hat{\imath} + 7.50 \times 10^4\,\hat{\jmath}) \text{ N/C}.$$

△

17.2.4 The Electric Dipole

In computing the electric field due to multiple charges or continuous distributions of charge, we assume that the principle of superposition holds as it did in our study of electrostatic forces. Restated in the context of electric fields, the principle of superposition states that the electric field due to a collection of charges is the sum of the electric fields of the individual charges. Given the definition of electric field in terms of the electrostatic force [Equation (17.5)], it is reasonable that if the principle of superposition holds true for forces, then it is true for electric fields as well.

As an example, let's find the electric field due to two point charges with the same charge magnitude but opposite signs. Such a configuration is called an **electric dipole**. We are free to place our coordinate system wherever we want with respect to the charges, so for convenience we place the charges on the x-axis, with a charge $+Q$ at $x = a$ and a charge $-Q$ at $x = -a$. The distance between the charges is thus $2a$. The electric dipole is not merely a simple exercise in computing electric fields. Physicists and chemists often use it to model charge distributions in matter, for example, polar molecules.

It is difficult (and not particularly enlightening) to write an expression for the electric field of the dipole at an arbitrary point (x, y, z) in Cartesian coordinates. Instead, we look at a few important special cases. Let's

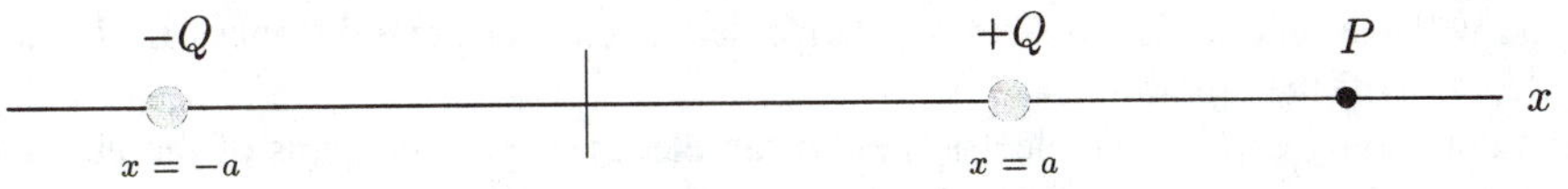

Figure 17.12. An electric dipole.

begin by finding the electric field at a point P on the $+x$-axis, with $x > a$, as shown in Figure 17.12. We consider the electric fields due to each charge separately and then combine them, following the prescription of the principle of superposition. By Equation (17.6), we can write the electric field $\vec{E}_-$ at point P due to the negative charge in the dipole as

$$\vec{E}_- = \frac{k(-Q)}{(x-a)^2}\,\hat{r}.$$

In this case, $\hat{r}$ is a unit vector pointing from the $-Q$ charge to P, so $\hat{r} = \hat{\imath}$. Therefore,

$$\vec{E}_- = -\frac{kQ}{(x+a)^2}\,\hat{\imath}.$$

Similarly, the electric field $\vec{E}_+$ at P due to the positive charge in the dipole is

$$\vec{E}_+ = \frac{kQ}{(x-a)^2}\,\hat{\imath}.$$

The total electric field at P is $\vec{E} = \vec{E}_+ + \vec{E}_-$, according to the principle of superposition. Combining our results and simplifying,

$$\begin{aligned}\vec{E} &= \vec{E}_+ + \vec{E}_- = \left[\frac{kQ}{(x-a)^2} - \frac{kQ}{(x+a)^2}\right]\hat{\imath} = kq\left[\frac{1}{(x-a)^2} - \frac{1}{(x+a)^2}\right]\hat{\imath} \\ &= kQ\left[\frac{(x+a)^2 - (x-a)^2}{(x-a)^2(x+a)^2}\right]\hat{\imath}\end{aligned}$$

or finally

$$\vec{E} = kQ\left[\frac{4xa}{(x^2-a^2)^2}\right]\hat{\imath}. \tag{17.7}$$

You should check to see that this result is dimensionally correct.

It is interesting to find the distance dependence of the electric field at large distances, that is, when $x \gg a$. In that limit, the factor $x^2 - a^2$ in the denominator can be approximated by x^2, leaving

$$\vec{E} \cong k\frac{4aQ}{x^3}\,\hat{\imath}. \tag{17.8}$$

The magnitude of the electric field falls off as the inverse cube of the distance, rather than the inverse-square dependence of the point charge. It makes physical sense that the field's magnitude should fall off faster than the inverse square, because at great distances we "see" a net charge of zero. The specific inverse-cube dependence could not have been predicted, though, without going through the calculation explicitly. In other examples, you will see that the inverse-cube distance dependence at great distances is characteristic of the electric dipole, even off the dipole's axis.

Physicists commonly express the electric field of an electric dipole in terms of the **electric dipole moment**, a vector ($\vec{p}$) with magnitude $2aQ$ (the product of the magnitudes of the charges and the distance between them) and direction from the negative charge toward the positive charge. Thus, with the coordinate system shown in Figure 17.12, $\vec{p} = 2aQ\,\hat{\imath}$. In terms of the electric dipole moment vector, the electric field at $x > a$ on the x-axis is

$$\vec{E} = k\left[\frac{2x\,\vec{p}}{(x^2 - a^2)^2}\right]$$

and the approximate electric field at $x \gg a$ is

$$\vec{E} \approx k\frac{2\,\vec{p}}{x^3}.$$

The next two examples are further explorations of the electric dipole field.

Example 17.8. For the electric dipole shown in Figure 17.12, find the electric field at the origin. Express your answer in terms of Cartesian unit vectors and then in terms of the electric dipole moment $\vec{p}$.

Solution. As in the text, the field at the origin due to the negative charge is

$$\vec{E}_- = -\frac{kQ}{a^2}\,\hat{\imath}.$$

The field due to the positive charge also points to the left, and therefore is

$$\vec{E}_+ = -\frac{kQ}{a^2}\,\hat{\imath}.$$

The net electric field at the origin is

$$\vec{E} = \vec{E}_+ + \vec{E}_- = -\frac{2kQ}{a^2}\,\hat{\imath}.$$

In terms of the electric dipole moment $\vec{p} = 2aQ\,\hat{\imath}$,

$$\vec{E} = -\frac{k}{a^3}\,\vec{p}.$$

△

Example 17.9. For the electric dipole with charge $-Q$ at $x = -a$ and $+Q$ at $x = a$ (both on the x-axis), find the electric field at a point $y > 0$ on the y-axis. Express your answer in terms of Cartesian unit vectors and in terms of the electric dipole moment. Find the distance dependence of the electric field when $y \gg a$.

Solution. This situation is shown in Figure 17.13. Note the directions of $\vec{E}_+$ and $\vec{E}_-$ (the electric fields due to the positive and negative charges, respectively). From symmetry, we can see that the sum $\vec{E} = \vec{E}_+ + \vec{E}_-$ points in the $-x$-direction. Therefore, we need to know only the x-component of each electric field. The x-component of $\vec{E}_+$ is

$$E_{+x} = -E_+\cos\theta = -\left(\frac{kQ}{r^2}\right)\cos\theta.$$

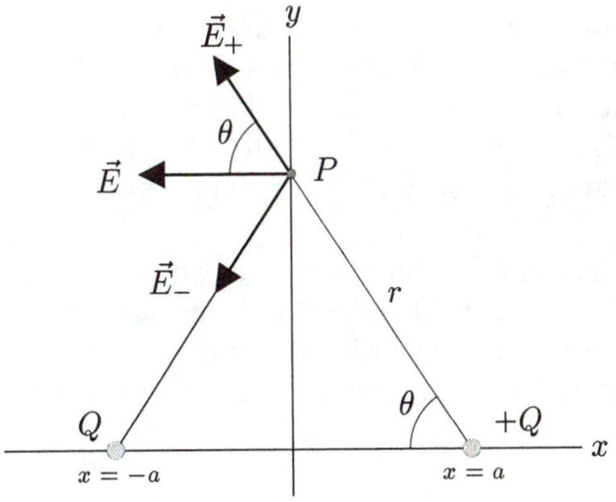

Figure 17.13. Schematic for Example 17.9.

Because $\cos\theta = a/r$, we have

$$E_{+x} = -\left(\frac{kQ}{r^2}\right)\left(\frac{a}{r}\right) = -\frac{kQa}{r^3}.$$

Similarly, we can show that the x-component of $\vec{E}_-$ is

$$E_{-x} = -\frac{kQa}{r^3}$$

and thus the total electric field at P is

$$\vec{E} = (E_{+x} + E_{-x})\,\hat{\imath} = -\frac{2kQa}{r^3}\,\hat{\imath}.$$

From the symmetry of the dipole, each charge is a distance $r = \sqrt{y^2 + a^2}$ from P. Therefore, we may substitute for r and express the electric field in terms of y:

$$\vec{E} = -\frac{2kQa}{(y^2 + a^2)^{3/2}}\,\hat{\imath}.$$

In terms of the electric dipole moment $\vec{p} = 2aQ\,\hat{\imath}$, we have

$$\vec{E} = -\frac{k\vec{p}}{(y^2 + a^2)^{3/2}}.$$

In the limit where $y \gg a$, the denominator is approximately equal to $\left(y^2\right)^{3/2} = y^3$. Therefore, the electric field is approximated by

$$\vec{E} \cong -\frac{2kQa}{y^3}\,\hat{\imath} = -\frac{k\vec{p}}{y^3}.$$

Note that in each example we have seen that the electric field is proportional to $\vec{p}$ and at large distances falls off as the inverse cube of the distance. △

The calculation of the electric field vector (due to an electric dipole) at a position that is neither on the dipole's axis nor the perpendicular bisector becomes more tedious. There are two reasons for the difficulty: The distance from that position to each point charge is different, and the fields $\vec{E}_+$ and $\vec{E}_-$ due to the individual charges do not point in the same direction. Therefore, it is not straightforward to write a symbolic expression for the electric field of the dipole as a function of position (x, y, z), and we will not pursue that calculation here.

However, this is yet another problem for which it is worthwhile to pursue a computer solution. A computer-generated map of the electric field of the electric dipole in the xy-plane is shown in Figure 17.14. Because of the symmetry of the electric dipole with respect to the x-axis, this picture is sufficient to give us a clear idea of the shape of the three-dimensional field. The picture looks exactly the same in any other plane that contains the x-axis.

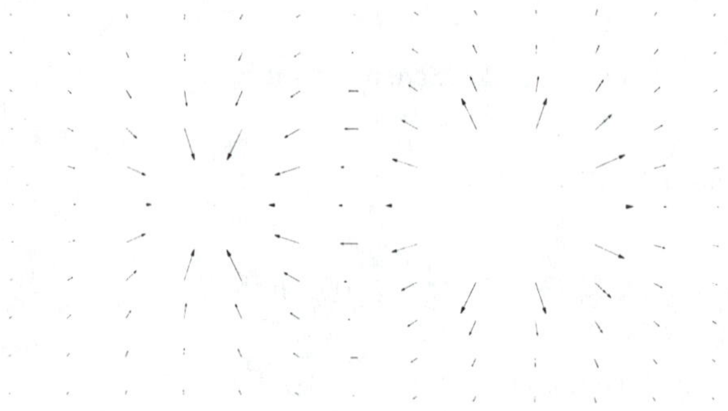

Figure 17.14. Electric field of a dipole.

17.2.5 Electric Field Lines

In Section 17.1.3, we introduced the idea of *field lines*. Physicists often draw electric field lines rather than electric field vectors. An electric field line is a curve that "follows" the electric field vectors in the sense that at each point the electric field vector is tangent to the electric field line. The field lines begin on positive point charges and terminate on negative point charges, because the electric field vectors point away from positive point charges and toward negative point charges. Figure 17.15 shows some electric field lines for the electric field of a dipole.

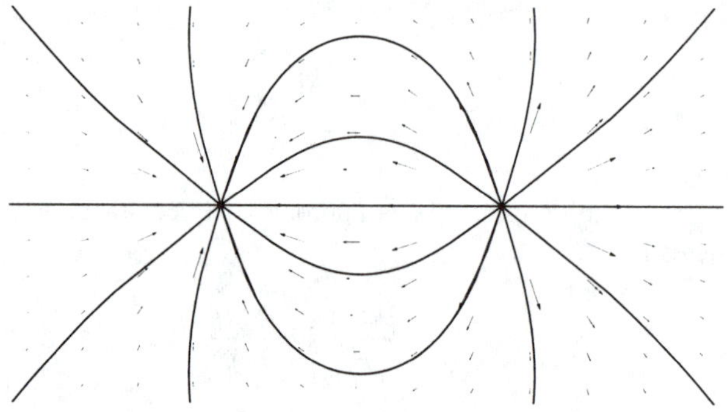

Figure 17.15. Electric field lines for an electric dipole.

In drawing electric field lines, it is conventional to make the number of field lines leaving from a positive charge or terminating on a negative charge proportional to the magnitude of the charge. With this convention, *the density of electric field lines in the diagram is proportional to the magnitude of the electric field at that location.* Why is this so? Consider the electric field of a (positive) point charge. In this case, the electric field lines radiate outward in every direction (Figure 17.16). A sphere of radius r with the point charge at the center has a surface area $4\pi r^2$, and so the density of field lines on the surface of any such sphere is proportional to the inverse square of r. But you know that the electric field magnitude is also proportional to the inverse square of r, so the electric field magnitude is proportional to the density of electric field lines. This result also holds for other charge distributions, due to the principle of superposition.

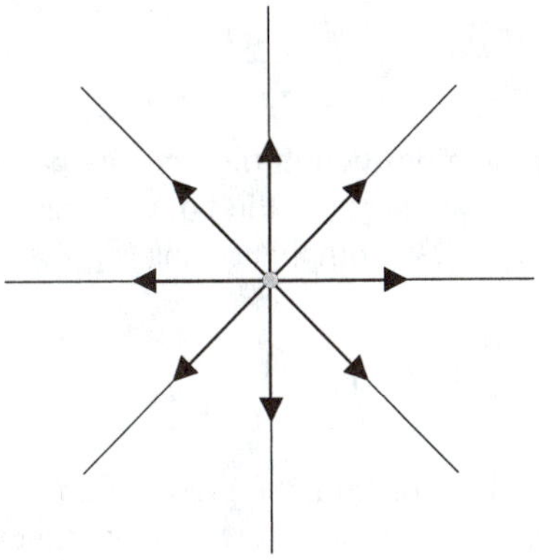

Figure 17.16. Electric field lines for a positive point charge.

Electric field lines may not cross each other. This would imply the existence of more than one value of the electric field at the point of crossing, which is physically meaningless. In situations where you might think electric field lines would cross, as in the electric dipole, you can see (Figure 17.15) that they actually curve so as not to cross, and so that every field line leaving the positive charge terminates on the negative one. This is not the case with two charges of unequal magnitude, however (Figure 17.17), because twice as many lines leave from the $+2Q$ charge as terminate on the $-Q$ charge. In this case, the lines still curve so that they never cross, and the density of field lines is still proportional to the magnitude of $\vec{E}$ at every location.

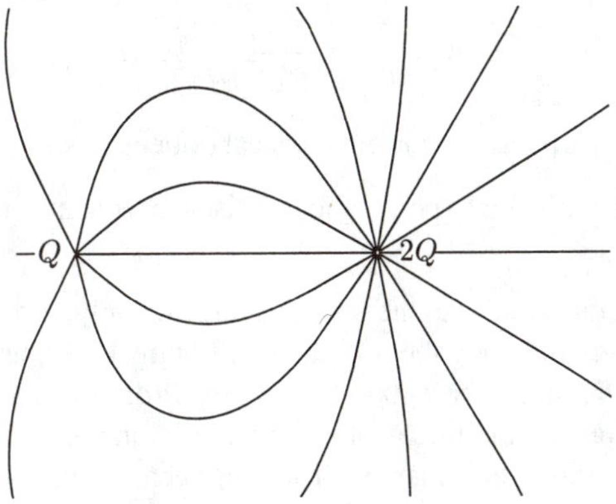

Figure 17.17. Electric field lines for a charge of $2Q$ at $(1, 0)$ and a charge of $-Q$ at $(-1, 0)$, where $Q > 0$.

17.2.6 The Electric Field Due to a Continuous Charge Distribution

In Chapter 16, we learned how to compute the force on a point charge due to a continuous distribution of charge in one, two, or three dimensions. In Equation (17.5), we defined the electric field in terms of the force on a point charge. Therefore, we can use that definition of the electric field to develop strategies for finding electric fields of continuous charge distributions. These will appear quite similar to the strategies we used to find electrostatic forces in Chapter 16.

For example, consider the one-dimensional case, in which some distribution of charge lies along the x-axis with charge density $\lambda(x)$ on the interval $x : [x_1, x_2]$. According to Equation (16.9), the force on a point charge q_0 outside of that distribution is

$$\vec{F} = kq_0 \int_{x_1}^{x_2} \frac{\lambda(x)\,\hat{r}}{r^2}\,dx. \tag{16.9}$$

where r is the (variable) distance from each position in the line charge to the point charge, and $\hat{r}$ is a unit vector directed from that position in the line charge to the point charge. To find the electric field due to a continuous distribution of charge along a line, we simply use this expression for the force in Equation (17.5). The electric field is

$$\vec{E} = k \int_{x_1}^{x_2} \frac{\lambda(x)\,\hat{r}}{r^2}\,dx. \tag{17.9}$$

This expression differs by only a constant factor from the prescription we used for force in Chapter 16; the definite integral is exactly the same. This is why we can use the same strategies we used previously for setting up and evaluating the integral in Equation (17.9).

Example 17.10. A positive uniformly charged rod with total charge Q lies on the x-axis from $x = -L/2$ to $x = +L/2$. Find the electric field at a point $y = R$ on the $+y$-axis.

Solution. In this case, we set up and computed the integral in Equation (17.9) when we found the force a point charge q_0 at the point $y = R$ on the $+y$-axis:

$$\vec{F} = \frac{kq_0 Q}{R\sqrt{(L/2)^2 + R^2}}\,\hat{\jmath}.$$

The desired electric field is that force divided by the point charge q_0:

$$\vec{E} = \frac{\vec{F}}{q_0} = \frac{kQ}{R\sqrt{(L/2)^2 + R^2}}\,\hat{\jmath}.$$

You should verify that this result has the correct physical dimensions. △

Example 17.11. Find the electric field a perpendicular distance R from an infinite line of charge with uniform charge density λ.

Solution. This example is quite similar to the previous one; the only difference is in the limits of integration. Let the infinite line of charge lie along the x-axis as in Figure 17.18, and let the point P (where we wish to know the electric field) lie along the y-axis at $y = R$. From symmetry, we see that the x-components of the electric field cancel (as did the force components in Example 16.8), so the net electric field is in the $+y$-direction (assuming a positive line charge). The unit vector $\hat{r}$ makes an angle θ with the y-axis, so the y-component of $\hat{r}$ is $\cos\theta$, and by Equation (17.9),

$$\vec{E} = k \int_{-\infty}^{\infty} \frac{\lambda(\cos\theta\,\hat{\jmath})}{r^2}\,dx.$$

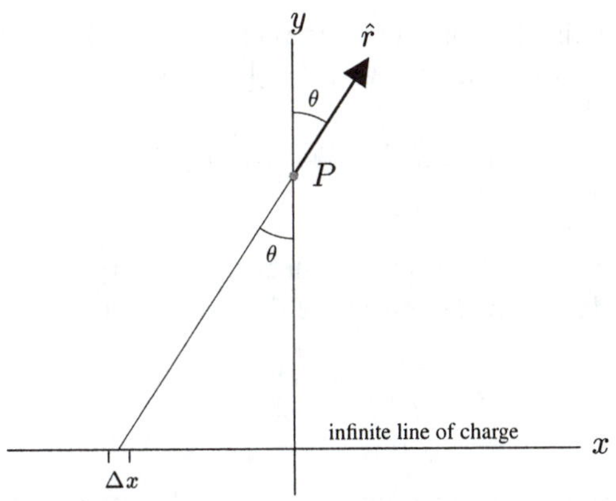

Figure 17.18. The setup of Example 17.11.

We must express the definite integral as a function of a single variable. From the diagram,

$$r = \sqrt{x^2 + R^2}$$

and

$$\cos\theta = \frac{R}{r} = \frac{R}{\sqrt{x^2 + R^2}}.$$

Substituting these expressions into the definite integral and factoring out constants, we find

$$\vec{E} = k\lambda R\,\hat{\jmath} \int_{-\infty}^{\infty} \frac{1}{(x^2 + R^2)^{3/2}}\,dx.$$

From integral tables,

$$\vec{E} = k\lambda R\,\hat{\jmath} \left(\frac{1}{R^2} \frac{x}{\sqrt{x^2 + R^2}} \right)\bigg|_{-\infty}^{\infty}.$$

This is an improper integral, which is easily evaluated because

$$\lim_{x\to\infty} \frac{x}{\sqrt{x^2 + R^2}} = 1.$$

Then

$$\vec{E} = \frac{k\lambda}{R}\,\hat{\jmath} \lim_{x\to\infty} \left(\frac{x}{\sqrt{x^2 + R^2}} \right)\bigg|_{-\infty}^{\infty} = \frac{k\lambda}{R}\,\hat{\jmath}\,[1 - (-1)]$$

or

$$\vec{E} = \frac{2k\lambda}{R}\,\hat{\jmath}.$$

The electric field is finite, even though the total charge in the line is infinite, because of the inverse-square distance dependence of the electric field. △

For a two-dimensional charge distribution with charge density $\sigma(x,y)$, the result in Equation (16.10) can be combined with the definition of electric field to give the electric field due to such a distribution:

$$\vec{E} = k \iint_R \frac{\sigma(x,y)\,\hat{r}}{r^2}\,dA. \tag{17.10}$$

Similarly, for a three-dimensional charge distribution with charge density $\rho(x,y,z)$, Equation (16.11) can be combined with the definition of electric field to give the electric field due to such a distribution:

$$\vec{E} = k \iiint_D \frac{\rho(x,y,z)\,\hat{r}}{r^2}\,dV. \tag{17.11}$$

Example 17.12. A thin disk of radius R and has uniform charge density σ. (a) Find the electric field due to the disk at a point on the disk's axis of symmetry a distance z_0 above the disk. (b) Find the limit of $\vec{E}$ as R approaches infinity, and thereby find the electric field near an infinite plane of charge with uniform charge density σ.

Solution. (a) In Example 16.10, we found the force on a point charge q_0 at this location to be

$$\vec{F} = 2\pi k q_0 \sigma\,\hat{k}\left(1 - \frac{z_0}{\sqrt{R^2 + z_0^2}}\right)$$

where we used the z-axis as the axis of symmetry. Therefore, by the definition of electric field, the electric field at the desired location is

$$\vec{E} = 2\pi k \sigma\,\hat{k}\left(1 - \frac{z_0}{\sqrt{R^2 + z_0^2}}\right).$$

(b) We have

$$\lim_{R\to\infty} \frac{z_0}{\sqrt{R^2 + z_0^2}} = 0$$

and therefore

$$\lim_{R\to\infty} \vec{E} = 2\pi k \sigma\,\hat{k}.$$

This expression can be simplified. Using $k = 1/4\pi\epsilon_0$, we find

$$\lim_{R\to\infty} \vec{E} = \frac{\sigma}{2\epsilon_0}\,\hat{k}.$$

Very close to a thin disk, the extent of the disk appears nearly infinite. Adding more charge by expanding the disk to radii greater than R does not change the electric field at this point significantly. Therefore, the electric field near an infinite plane of charge with uniform charge density σ is

$$\vec{E} = \frac{\sigma}{2\epsilon_0}\,\hat{k}$$

with the unit vector $\hat{k}$ directed away from and perpendicular to the plane of charge (see Figure 17.19). Notice that if the charge were *negative*, the electric field would be directed *toward* the plane of charge (perpendicular to it). △

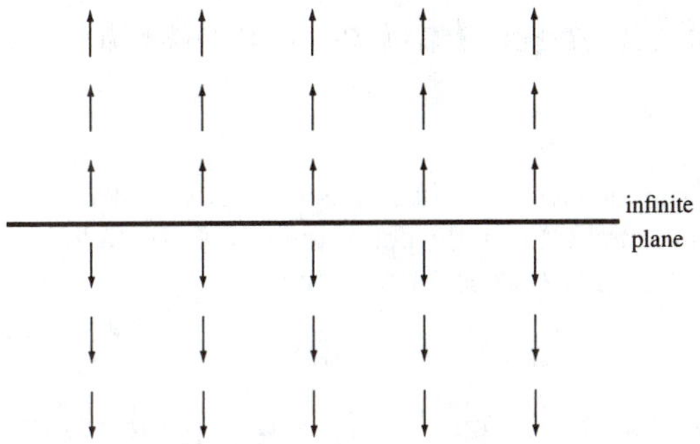

Figure 17.19. Electric field near an unbounded plane charge distribution. This cross-sectional view shows the plane edge-on.

Example 17.13. Use the result of the previous example to find the electric field between two infinite parallel planes with charge densities $+\sigma$ and $-\sigma$ (i.e., equal in magnitude but opposite in sign).

Solution. For convenience, let the $+\sigma$-plane lie in the xy-plane and the $-\sigma$-plane lie at some $z > 0$. The electric field due to the positive plane points straight up and has a magnitude $\sigma/2\epsilon_0$. The electric field due to the negative plane points straight up (toward the negative plane) and has the same magnitude. By the principle of superposition, we simply add these fields to obtain

$$\vec{E} = \left(\frac{\sigma}{2\epsilon_0}\hat{k}\right) + \left(\frac{\sigma}{2\epsilon_0}\hat{k}\right) = \frac{\sigma}{\epsilon_0}\hat{k}.$$

The resultant field has a magnitude σ/ϵ_0 and points straight up—that is, from the positive plane to the negative plane (Figure 17.20). △

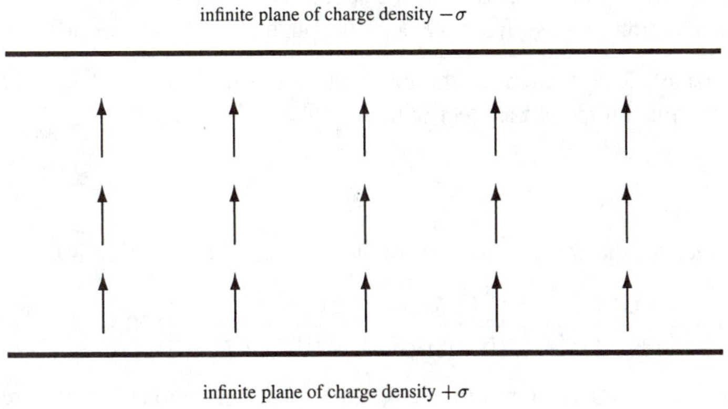

Figure 17.20. Schematic for Example 17.13.

17.3 The Motion of Charged Particles in an Electric Field

17.3.1 Point Charges

One significant advantage of using the electric field concept is that it provides a convenient way to determine the motion of charges placed in the field. From the definition of electric field [see Equation (17.5)], we know that a charge q placed in an electric field $\vec{E}$ experiences a force

$$\vec{F} = q\vec{E}. \tag{17.12}$$

As you know from your study of dynamics, once the force on a particle is known, its motion can then be understood. Using Newton's second law, we can find the acceleration of a particle with charge q and mass m in an electric field. Setting the force in Equation (17.12) equal to $m\vec{a}$, the instantaneous acceleration $\vec{a}$ is

$$\vec{a} = \frac{q\vec{E}}{m}. \tag{17.13}$$

The motion of charged particles in electric fields is important throughout physics and engineering. For example, in a television or computer monitor electrons are directed to specific locations on the screen by electric fields.

As an example, let's consider the motion of a charge in a *uniform* electric field. At the end of the last section, we saw how to create a uniform electric field (or at least an excellent approximation of one) by using parallel planes with charge densities equal in magnitude and opposite in sign. If the parallel planes are made of conducting material, excess charge placed on them tends to spread itself uniformly over the surface. Then as we saw, the electric field in the region between the planes is nearly uniform, with magnitude σ/ϵ_0 (where σ is the magnitude of the charge density on each plane) and direction perpendicular to both planes, pointing from the positively charged plane to the negatively charged one, as in Figure 17.20.

With a uniform electric field, the acceleration $\vec{a}$ in Equation (17.13) is *constant.* We studied of the special case of a particle with constant acceleration extensively when we studied kinematics. Therefore, the motion of a charged particle in a uniform electric field is relatively easy to understand.

Example 17.14. A pair of parallel conducting plates carries uniform charge densities of 4.50×10^{-7} C/m^2 and -4.50×10^{-7} C/m^2. The plates are separated by a distance of 0.360 cm and are large enough so that the electric field between them can be considered uniform. If an electron is released form rest at the negative plate, find (a) the acceleration of the electron, (b) the speed with which it strikes the positive plate, and (c) the time it takes for the electron to travel from the negative plate to the positive one.

Solution. (a) The (constant) acceleration of the electron in the uniform electric field is given by Equation (17.13). Because the magnitude of the electric field is $E = \sigma/\epsilon_0$, we have

$$a = \frac{qE}{m} = \frac{q\sigma}{\epsilon_0 m}.$$

Using the tabulated values for the charge and mass of the electron, the numerical value of the acceleration is

$$a = \frac{(1.60 \times 10^{-19}\ \text{C})(4.50 \times 10^{-7}\ \text{C/m}^2)}{(8.85 \times 10^{-12}\ \text{C}^2/(\text{N}\cdot\text{m}^2))(9.11 \times 10^{-31}\ \text{kg})} = 8.93 \times 10^{15}\ \text{m/s}^2.$$

This may seem like an outrageously high numerical value for acceleration. But it is reasonable for such a light particle (the electron).

(b) From kinematics (with constant acceleration),

$$v^2 = v_0^2 + 2a(x - x_0).$$

In this case, the initial speed is $v_0 = 0$, and the displacement $x - x_0$ is 0.360 cm = 3.60×10^{-3} m. The numerical value of the speed is

$$v = \sqrt{2a(x - x_0)} = \sqrt{2\,(8.93 \times 10^{15}\ \text{m/s}^2)\,(3.60 \times 10^{-3}\ \text{m})} = 8.02 \times 10^6\ \text{m/s}.$$

Once again, this may appear to be a high value, but it is common for an electron.

(c) With such a high acceleration and speed, the time should be short. From kinematics,

$$v = v_0 + at$$

(with constant acceleration), so with $v_0 = 0$

$$t = \frac{v}{a} = \frac{8.02 \times 10^6\ \text{m/s}}{8.93 \times 10^{15}\ \text{m/s}^2} = 8.98 \times 10^{-10}\ \text{s}.$$

△

In the last example, the motion of the charged particle "dropped" from rest is analogous to the dropping of a particle near Earth with approximately constant acceleration $\vec{g}$, except that the acceleration of the electron in the electric field is much greater! The next example extends this comparison to two-dimensional motion.

Example 17.15. In a television tube, an electron is projected horizontally with a speed of 7.55×10^6 m/s. It passes through a 25.0-cm-wide region that has a constant electric field of 2.16×10^3 N/C directed vertically downward, as in Figure 17.21. (a) Find the acceleration of the electron in the region containing the electric field. (b) What is the vertical deflection of the electron? (c) What is the velocity of the electron as it emerges from the electric field?

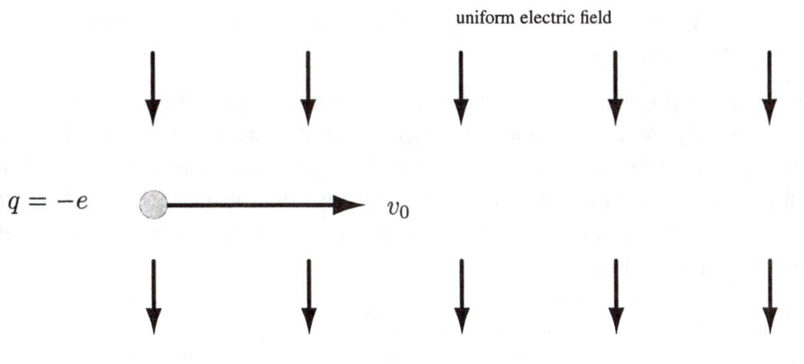

Figure 17.21. Electron entering a region of uniform electric field.

Solution. (a) First, note that because the electron is negatively charged, the force on it is in the opposite direction of the electric field, or vertically *upward.* Therefore, the acceleration is directed vertically upward. Let's take the $+x$-direction to be horizontal, along the electron's initial line of motion, so that the initial velocity is $\vec{v}_0 = v_0\,\hat{\imath}$, with $v_0 = 7.55 \times 10^6$ m/s. Let the $+y$-direction be straight up, so that the electric field is $\vec{E} = -E_0\,\hat{\jmath}$, with $E_0 = 2.16 \times 10^3$ N/C. The electron's acceleration [Equation (17.13)] is

$$\vec{a} = \frac{q\vec{E}}{m} = -\frac{e\vec{E}}{m}$$

where we have used $q = -e$, the charge on the electron. Inserting numerical values,

$$\vec{a} = -\frac{(1.60 \times 10^{-19}\ \text{C})(-2.16 \times 10^{3}\ \hat{\jmath}\ \text{N/C})}{9.11 \times 10^{-31}\ \text{kg}} = 3.79 \times 10^{14}\ \hat{\jmath}\ \text{m/s}^2.$$

(b) Because the y-component of the initial velocity is zero, the vertical deflection $y - y_0$ is given by kinematics as

$$y - y_0 = v_{oy}t + \frac{1}{2}a_y t^2 = \frac{1}{2}a_y t^2$$

where we have used $v_{0y} = 0$. The numerical value of a_y is given by the solution to part (a). What is the time t? Since there is no x-component of acceleration, we know $x - x_0 = v_0 t$, so

$$t = \frac{x - x_0}{v_0} = \frac{0.250\ \text{m}}{7.55 \times 10^{6}\ \text{m/s}} = 3.31 \times 10^{-8}\ \text{s}.$$

Therefore, the vertical displacement is

$$y - y_0 = \frac{1}{2}a_y t^2 = \frac{1}{2}(3.79 \times 10^{14}\ \text{m/s}^2)(3.31 \times 10^{-8}\ \text{s})^2 = 0.208\ \text{m} = 20.8\ \text{cm}.$$

(c) In general, under constant acceleration the velocity is

$$\vec{v} = \vec{v}_0 + \vec{a}t.$$

Inserting the numerical values,

$$\vec{v} = 7.55 \times 10^{6}\ \hat{\imath}\ \text{m/s} + (3.79 \times 10^{14}\ \hat{\jmath}\ \text{m/s}^2)(3.31 \times 10^{-8}\ \text{s}) = (7.55 \times 10^{6}\ \hat{\imath} + 1.25 \times 10^{7}\ \hat{\jmath})\ \text{m/s}.$$

The vertical component of the velocity is actually somewhat greater than the horizontal component when the electron emerges from the electric field. △

The motion of the electron projected horizontally into a vertical electric field is analogous to a horizontally thrown projectile in a vertical gravitational field. The only differences are in the magnitude of the acceleration and the fact that, due to the charge on the electron and the direction of the electric field, the electron is deflected upward rather than downward. (Of course, the electron would be deflected downward if the electric field were straight up instead of straight down.) As in projectile motion, the electron's motion is parabolic, as is evident by substituting the equation

$$t = \frac{x - x_0}{v_0}$$

into the equation

$$y - y_0 = \frac{1}{2}a_y t^2$$

which yields

$$y - y_0 = \frac{1}{2}a_y \left(\frac{x - x_0}{v_0}\right)^2 = \frac{a_y}{2v_0^2}(x - x_0)^2. \tag{17.14}$$

With a_y and v_0 both constant, this is the equation of a parabola.

Example 17.16. Find the magnitude of the vertical electric field needed to balance (a) a proton, and (b) a dust particle with mass 9.0×10^{-16} kg and charge $+e$ against the force of gravity.

Solution. (a) With a positive charge $+e$, the electric field should be directed *upward*, in order to balance the downward gravitational force. Equating the magnitudes of the two forces,

$$eE = mg.$$

Therefore, the magnitude of the electric field needed is

$$E = \frac{mg}{e}.$$

For the proton,

$$E = \frac{(1.67 \times 10^{-27}\ \text{kg})(9.80\ \text{m/s}^2)}{1.60 \times 10^{-19}\ \text{C}} = 1.02 \times 10^{-7}\ \text{N/C}.$$

This is a very small electric field.

(b) For the dust particle,

$$E = \frac{(9.0 \times 10^{-16}\ \text{kg})(9.80\ \text{m/s}^2)}{1.60 \times 10^{-19}\ \text{C}} = 5.51 \times 10^{4}\ \text{N/C}.$$ △

17.3.2 Measurement of the Charge of the Electron

The preceding example suggests that charged objects can be easily suspended against gravity with modest vertical electric fields, which is the basic idea behind the classic experimental determination of the electronic charge. This experiment was first done by the American physicist Robert Millikan in 1911; today we know it as the **Millikan oil-drop experiment** or simply the Millikan experiment.

Suppose a small sphere carries a net charge q (which can be positive or negative). This charged sphere can be balanced against gravity with a uniform vertical electric field. Because the gravitational force is always directed straight down, the direction of the electric field should be straight up of $q > 0$ and straight down if $q < 0$. The situation is shown for a positive charge q in Figure 17.22.

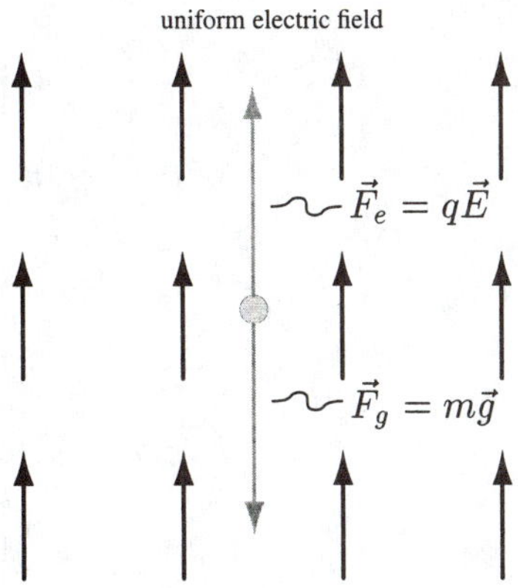

Figure 17.22. Schematic of the Millikan oil-drop experiment.

The sphere is just balanced (with zero net force) if the magnitude of the electric force equals the magnitude of the gravitational force, or $qE = mg$. In a Millikan experiment, a small sphere of known mass is observed through a microscope, while the electric field (usually between parallel conducting plates, so that the field is uniform) is adjusted to achieve this balance. Then the charge a sphere that has been balanced is

$$q = \frac{mg}{E}. \tag{17.15}$$

If the experiment is repeated enough times, and q is found with good precision, a pattern should begin to emerge, because of the quantization of charge in units of $\pm e$. Spheres with charges equal to $\pm e, \pm 2e, \pm 3e$, and so on, will be found, and the average value of e can be computed from all the data. The experimental apparatus is normally designed so that initially uncharged spheres pick up a small charge (no more than a few elementary charges) through frictional forces as they are sprayed into the region containing the electric field. In some cases, radioactive materials that emit electrons or positrons are placed nearby, so that a sphere can be given an extra charge $\pm e$.

In the Millikan experiments used in most undergraduate physics laboratories today, small plastic spheres with known density and uniform radii are used. This way each sphere that is observed has (theoretically) the same mass. In Millikan's original experiment, he used oil droplets of varying (but extremely small) size. When the electric field is off, an oil drop falls and soon reaches terminal speed due to its interaction with the air. Millikan was able to work out the fluid dynamics needed to understand how the terminal speed is related to the radius of the oil drop, and from this he could find the mass of a drop. Then the electric field is turned on and the experiment proceeds as we described, with Equation (17.15) giving the charge on each drop observed.

17.3.3 An Electric Dipole in a Uniform Electric Field

In Figure 17.23, we show a schematic of an electric dipole in an electric field. The axis of the dipole is not necessarily aligned with the field, and in general the dipole's axis makes some angle θ with the field. The electrostatic force on the positive charge $+Q$ is $Q\vec{E}$, and the electrostatic force on the negative charge $-Q$ is $-Q\vec{E}$. Therefore, the net force on the dipole is $Q\vec{E} - Q\vec{E} = 0$.

However, because the forces act at different points, there is a *net torque* on the dipole. We will find the net torque with respect to the geometric center of the dipole. In Figure 17.23, the tendency of both forces is to rotate the dipole counterclockwise as we view it, so the net torque is directed out of the page. Recall that

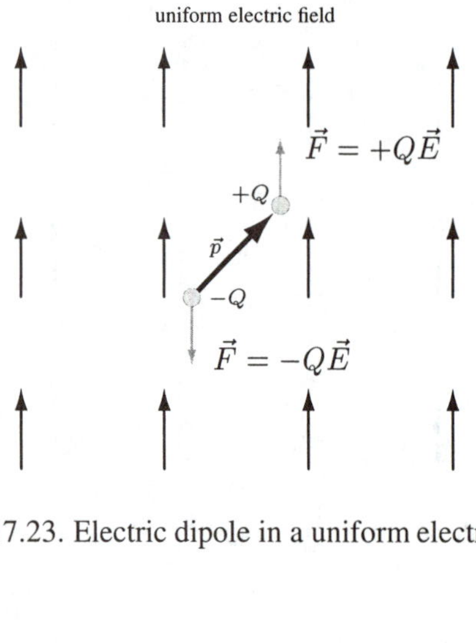

Figure 17.23. Electric dipole in a uniform electric field.

the distance from the center of the dipole to each point charge is a, so the magnitude of the torque is

$$\tau = (a)(QE)\sin\theta + (a)(QE)\sin\theta = 2aQE\sin\theta. \tag{17.16}$$

A more compact expression for the (vector) torque arises from the fact that the electric dipole vector $\vec{p}$ has a magnitude $2aQ$ and points from the negative charge to the positive charge. Therefore, $\vec{p}$ makes an angle θ with the electric field vector $\vec{E}$, and we can write

$$\vec{\tau} = \vec{p} \times \vec{E}. \tag{17.17}$$

Looking at the schematic, it makes sense that the torque is zero when $\theta = 0$ or π (when the dipole's axis is aligned with the electric field) and has a maximum magnitude when $\theta = \pi/2$.

There is some potential energy associated with the dipole's position in the electric field, analogous to the potential energy of a mass in a gravitational field. Because the net force on the electric dipole in a uniform electric field is zero, that potential energy is not associated with the translational motion of the dipole. Rather, with nonzero net torque on the dipole, the potential energy is associated with rotational motion. Recall that torque is the rotational analog of force, and angular position is the rotational analog of linear position. Therefore, by analogy with Definition 7.4 (the definition of potential energy) the change in potential energy associated with rotating the dipole's axis from an angle θ_a to an angle θ with the electric field is

$$U(\theta) - U(\theta_a) = \int_{\theta_a}^{\theta} \tau \, d\theta'$$

where we have used θ' as the variable of integration in order not to confuse it with the limit θ. Also, the minus sign from Definition 7.5 is absent, because in this case we are using the torque applied *by* the field rather than the force of an external agent doing word *against* the field. From Equation (17.16), we have $\tau = 2aQE\sin\theta$, so

$$U(\theta) - U(\theta_a) = \int_{\theta_a}^{\theta} 2aQE\sin\theta' \, d\theta' = 2aQE \int_{\theta_a}^{\theta} \sin\theta' \, d\theta'.$$

This definite integral is easily evaluated using the second fundamental theorem of calculus:

$$U(\theta) - U(\theta_a) = 2aQE(-\cos\theta')\Big|_{\theta_a}^{\theta} = 2aQE\left(-\cos\theta + \cos\theta_a\right).$$

We are free to define the zero point of potential energy anywhere we choose. We choose $U(\theta_a) = 0$ with $\theta_a = \pi/2$. Therefore,

$$U(\theta) = -2aQE\,\cos\theta. \tag{17.18}$$

As before, we can express this result more compactly in terms of the electric dipole moment vector $\vec{p}$:

$$U(\theta) = -\vec{p} \cdot \vec{E}. \tag{17.19}$$

The minimum of potential energy is $U = -pE$ when $\theta = 0$, and the maximum potential energy is $U = +pE$ when $\theta = \pi$. This makes sense, because the position $\theta = 0$ is a point of *stable equilibrium* with respect to rotation, and $\theta = \pi$ is a point of *unstable equilibrium*. In this respect, the rotational motion of the dipole in a uniform electric field is analogous to the motion of a simple pendulum with a rigid rod that is free to swing through all angles between $\theta = 0$ (the bob straight down) and $\theta = \pi$ (the bob straight up).

Example 17.17. A water molecule (H_2O) has a permanent electric dipole moment of about 6.33×10^{-30} C·m. (a) Find the magnitude of the torque on a water molecule that has its electric dipole moment making an angle $\pi/6$ with a uniform electric field of magnitude 5.35×10^4 N/C. (b) Find the potential energy of the same electric dipole moment described in (a). (c) Find the energy (from an external source) needed to change the alignment of one mole of water molecules from an angle $\theta = 0$ to an angle $\theta = \pi/2$ with respect to a uniform electric field of magnitude 5.35×10^4 N/C.

Solution. (a) The magnitude of the torque is

$$\tau = pE \sin\theta.$$

Inserting numerical values,

$$\tau = (6.33 \times 10^{-30} \text{ C·m})(5.35 \times 10^4 \text{ N/C}) \sin\frac{\pi}{6} = 1.69 \times 10^{-25} \text{ N·m}.$$

(b) From the text,

$$U = -pE \cos\theta.$$

Inserting numerical values,

$$U = -(6.33 \times 10^{-30} \text{ C·m})(5.35 \times 10^4 \text{ N/C}) \cos\frac{\pi}{6} = -2.93 \times 10^{-25} \text{ N·m} = -2.93 \times 10^{-25} \text{ J}.$$

Because we are dealing with potential energy rather than torque, it is appropriate to use joules rather than N·m.

(c) For one molecule,

$$U_f - U_i = -pE\cos\theta_f - (-pE\cos\theta_i) = pE(\cos\theta_i - \cos\theta_f).$$

With $\theta_i = 0$ and $\theta_f = \pi/2$,

$$U_f - U_i = (6.33 \times 10^{-30} \text{ C·m})(5.35 \times 10^4 \text{ N/C}) \left[\cos 0 - \cos\frac{\pi}{2}\right] = 3.38 \times 10^{-25} \text{ J}.$$

For one mole, we must supply an amount of energy equal to the change in potential energy that we just found multiplied by the number of molecules (Avogadro's number):

$$N_A(U_f - U_i) = (6.02 \times 10^{23})(3.38 \times 10^{-25} \text{ J}) = 0.203 \text{ J}.$$

As long as we are dealing with a single molecule, the numerical values of the torque and potential energy are small, but for one mole of water, the numerical value is significant. △

17.4 Problems

17.1 Introduction to Vector Fields

For Problems 1–2, find a symbolic form for the vector field $\vec{F} : \mathbb{R}^2 \to \mathbb{R}^2$ described verbally.

1. outputs tangent to radial lines from the origin of the input plane with magnitude inversely proportional to the cube of the distance from the origin

2. outputs tangent to circles centered at the origin of the input plane (in a clockwise orientation) with magnitude proportional to the cube of the distance from the origin

For Problems 3–8, find the output components of the given vector field $\vec{F} : \mathbb{R}^2 \to \mathbb{R}^2$ in terms of the input components (x, y).

3. $\vec{F}(\vec{r}) = \vec{r}$
4. $\vec{F}(\vec{r}) = \hat{r}$
5. $\vec{F}(\vec{r}) = \dfrac{5\hat{r}}{r^2}$
6. $\vec{F}(\vec{r}) = \dfrac{2\hat{r}}{r}$
7. $\vec{F}(\vec{r}) = r^2\hat{r}$
8. $\vec{F}(\vec{r}) = 4r\hat{r}$

For Problems 9–14, find the output components of the given vector field $\vec{F} : \mathbb{R}^3 \to \mathbb{R}^3$ in terms of the input components (x, y, z).

9. $\vec{F}(\vec{r}) = \vec{r}$
10. $\vec{F}(\vec{r}) = \hat{r}$
11. $\vec{F}(\vec{r}) = \dfrac{5\hat{r}}{r^2}$
12. $\vec{F}(\vec{r}) = \dfrac{2\hat{r}}{r}$
13. $\vec{F}(\vec{r}) = r^2\hat{r}$
14. $\vec{F}(\vec{r}) = 4r\hat{r}$

For Problems 15–18, plot, by hand, the given vector field for (x, y) in $[-2, 2] \times [-2, 2]$. Also, sketch some representative field lines.

15. $\vec{F}(x, y) = \langle y, 2\rangle$
16. $\vec{F}(x, y) = \langle x, 2\rangle$
17. $\vec{F}(x, y) = \langle 2, x\rangle$
18. $\vec{F}(x, y) = \langle 2, y\rangle$

For Problems 19–22, use Euler's method to compute an approximate field line for the given vector field and initial point. Compute 10 steps with a step size of 0.1.

19. $\vec{F}(x, y) = \langle x, y\rangle, \quad (1, 1)$
20. $\vec{F}(x, y) = \langle y, -x\rangle, \quad (1, 0)$
21. $\vec{F}(y, 2) = \langle y, 2\rangle, \quad (-1, 0)$
22. $\vec{F}(x, y) = \langle x^2 + 1, xy\rangle, \quad (0, 1)$

For Problems 23–27, make a computer plot of the given vector field for the indicated region. Also, plot some representative field lines.

23. $\vec{F}(x, y) = \langle xy, y\rangle$ for $-2 < x < 2$ and $-2 < y < 2$

24. $\vec{F}(\vec{r}) = \vec{r}/r$ for $-2 < x < 2$ and $-2 < y < 2$

25. $\vec{F}(x, y) = \langle -y, x\rangle$ for $-2 < x < 2$ and $-2 < y < 2$

26. $\vec{F}(x, y) = \langle \sin x, \cos y\rangle$ for $-\pi < x < \pi$ and $-\pi < y < \pi$

27. $\vec{F}(x, y) = \langle \sin y, \cos x\rangle$ for $-\pi < x < \pi$ and $-\pi < y < \pi$

28. Two vector fields are said to be **topologically equivalent** if the field lines of one vector field can be **deformed** (stretched or bent but not broken or joined) to look like the field lines of the other. Develop a classification of linear vector fields of the form $\vec{F}(x, y) = \langle px + qy, rx + sy \rangle$ into classes in which the members are all topologically equivalent. [Hint: One approach is to make vector field plots using a variety of choices for p, q, r, and s.]

17.2 Electric Fields

1. A point charge $+23.4\ \mu$C lies at the origin. Find the electric field at the following points: (a) $(0, 0, 1.35$ m); (b) (28.9 cm, 15.0 cm, 0); and (c) (-1.20 m, -3.49 m, 0.650 m).

2. A point charge $-1.05\ \mu$C lies at the origin. Find the electric field at the following points: (a) (3.45 m, 0, 0); (b) (12.3 m, 4.55 m, 0); and (c) (3.21 m, -2.30 m, 4.10 m).

3. A point charge -12.0 nC lies at the point (-45.0 cm, 13.5 cm, 0). (a) Find the electric field at the point (12.7 cm, -5.30 cm, 0). (b) Find the force acting on a point charge -6.50 nC at that location.

4. A point charge $+1.50\ \mu$C lies at the point (1.10 m, 0.950 m, -1.67 m). (a) Find the electric field at the point (1.00 m, 1.00 m, 1.00 m). (b) Find the force acting on a point charge $-132\ \mu$C at that location.

5. (a) Assuming half of Earth's mass is made up of protons, find the electric field at Earth's surface if all the electrons were removed. (b) What would be the force on a proton at the surface under these conditions? (c) Repeat parts (a) and (b) if only one electron per million were removed. (This illustrates the overall electrical neutrality of Earth.)

6. (a) Find the magnitude of the electric field acting on the electron on a hydrogen atom, assuming the electron is in a circular orbit of radius 5.29×10^{-11} m around the proton. (b) Use your answer in (a) to find the magnitude of the electrostatic force acting on the electron, and compare this with the gravitational force on the electron due to the proton.

7. Find the magnitude of a point charge needed to create a 1000-N/C electric field at a distance of 1 m.

8. An electric dipole lies on the x-axis, with $-Q$ at $x = -a$ and $+Q$ at $x = +a$. Find the electric field at points $x < -a$ on the x-axis.

9. For the electric dipole described in Problem 8, find the electric field at points $x : (-a, a)$ on the x-axis.

10. For the electric dipole described in Problem 8, find the electric field (a) at points $y < 0$ on the y-axis, and (b) at all points in the z-axis.

11. Two $+Q$ point charges lie on the x-axis, with one at $x = -a$ and the other at $x = +a$. Find the electric field at the point $x > a$ and $x < -a$ on the x-axis.

12. For the charge distribution described in Problem 11, find the electric field at all points on the y-axis.

13. An *electric quadrupole* lies in the xy-plane, with point charges $+Q$ at (a, a) and $(-a, -a)$ and point charges $-Q$ at $(a, -a)$ and $(-a, a)$. (a) Find the electric field at all points $x > a$ on the x-axis. *Hint*: Think of the quadrupole as a pair of dipoles and use the known electric field of the dipole. (b) Find the distance dependence of the electric field magnitude for distant points $x >> a$ on the x-axis.

14. Four point charges lie in the xy-plane: $+Q$ at (a, a); $+2Q$ at $(-a, -a)$; $-Q$ at $(-a, a)$; and $-2Q$ at $(a, -a)$. Find the electric field at the origin.

15. Draw the continuous electric field lines in the plane containing (a) two identical $+Q$ point charges, and (b) two identical $-Q$ point charges.

16. Draw the continuous electric field lines in the plane containing the point charges $+Q$ and $-4Q$.

17. For the electric dipole on the x-axis (as described in the text), compute the electric field at ten random points (at least two in each quadrant) in the xy-plane. [*Hint*: Using a computer will make the calculation much easier!] Show that each result is consistent with the electric field of the dipole shown in Figures 17.7 and 17.8.

18. A negative uniformly charged rod with total charge $-Q$ lies on the x-axis from $x = 0$ to $x = L$. Find the electric field at a point $x > L$ on the x-axis.

19. For the charge distribution in Problem 18, find the electric field at the point $(L/2, R)$ in the xy-plane.

20. A positively charged rod lies on the x-axis from $x = 0$ to $x = L$. The total charge is $+Q$, but the charge density is not uniform and is given by $\lambda(x) = Cx$ where C is a constant. (a) Find the constant C in terms of Q and L. (b) Find the electric field at points $R > a$ on the x-axis.

21. A charged rod lies on the x-axis from $x = -L$ to $x = L$. the charge density along the rod is $\lambda(x) = Cx$ where C is a positive constant. Find the electric field at (a) points $R > L$ on the x-axis and (b) points $R < -L$ on the x-axis.

22. For the charge distribution Problem 21, find the electric field at points $y > 0$ on the y-axis.

23. Find the electric field a perpendicular distance R from an infinite line of negative charge with uniform charge density λ.

24. What is the magnitude of the force on a 1.0-μC test particle a distance 1.0 m from an infinite line of charge with uniform charge density 2.5μC/m?

25. A total charge +3.0 μC lies distributed uniformly along the x-axis between the points $x = -2.0$ m and $x = +2.0$ m. A total charge -3.0 μC lies distributed uniformly along the y-axis between $y = -2.0$ m and $y = +2.0$ m. Find the electric field at (a) (3.0 m, 0), and (b) $(0, 3.0$ m).

26. Two point charges lie on the y-axis: +0.25 μC at $y = 25$ cm and +0.16 μC at $y = 75$ cm. Find the place(s) where the electric field is zero.

27. Two point charges lie on the z-axis: +1.65 μC at $z = 1.00$ m and -2.30 μC at $y = 2.50$ m. Find the place(s) where the electric field is zero.

28. A ring of radius R lies in the xy-plane with its center at the origin. The ring carries a uniform charge density and a total charge $+Q$. Find the electric field at all points on the z-axis.

29. In this problem you will examine the uniformity of the electric field between the two parallel disks. Suppose the two disks each have a 25.0-cm radius and are separated by 1.00 cm. They carry uniform charge densities of +2.00 μC/m^2 and -2.00 μC/m^2. (a) Find the magnitude of the electric field between the disks using the approximation $E = \sigma/\epsilon_0$. (b) Use the result of Example 17.12 to find the *exact* value of E exactly halfway between the disks at their center, and compare your result with the approximation in (a). (c) Find the exact value of E between the disks at a point on their mutual axis but only 0.25 cm from one disk. Compare this result with the results from both (a) and (b).

17.3 The Motion of Charged Particles in an Electric Field

1. A proton is in a uniform electric field $\vec{E} = (2.35 \times 10^3\,\hat{\imath} - 7.42 \times 10^3\,\hat{\jmath})$ N/C. Find (a) the force on the proton, and (b) the acceleration of the proton.

2. Two large parallel conducting plates are separated by 0.75 cm and carry uniform charge densities equal in magnitude and opposite in sign. A proton between the plates, and the proton's acceleration is 1.39×10^{11} m/s^2. Find the charge density on the plates.

3. Consider again the electron in the television tube as described in Example 17.15. If the electric field were turned off, how far would the electron fall (during its 25.0-cm horizontal flight) under the influence of gravity? Compare your answer with the vertical deflection due to the electric field.

4. In a particle accelerator, an electric field $\vec{E}_0$ causes protons to have an acceleration $\vec{a}_0$. What electric field will give electrons an acceleration $\vec{a}_0$?

5. A particle with mass m and charge $+Q$ has a velocity $\vec{v} = v_0\,\hat{\imath}$ at the origin. The particle is in a uniform electric field $\vec{E} = E_0\,\hat{k}$. (a) Write an equation for the particle's velocity as a function of time. (b) Write an equation for the particle's position as a function of time. What is the shape of the trajectory?

6. Repeat the preceding problem if the charge is $-Q$.

7. A test particle with mass 1.00×10^{-12} kg and charge $-500e$ has a velocity $(1.25\,\hat{\imath} + 2.35\,\hat{\jmath})$ m/s. The particle enters a uniform electric field $\vec{E} = (2.34 \times 10^3\,\hat{\imath} - 1.98 \times 10^3\,\hat{\jmath} + 1.04 \times 10^3\,\hat{k})$ N/C. Find the particle's velocity and displacement after 5.00 s has elapsed.

8. In the Millikan experiment, an oil droplet (density = 890 kg/m^3) with a radius of 3.50×10^{-7} m is just balanced by an electric field of 3260 N/C. What is the charge on the droplet?

9. In a Millikan experiment, a 4.50×10^{-18} kg plastic sphere with a charge $-4e$ is suspended between two large horizontal parallel plates. If the parallel plates have charge densities equal in magnitude and opposite in sign, what is the magnitude of the charge density? Which plate (upper or lower) is the positive one?

10. A particle with mass 3.55×10^{-4} kg and charge $25.0\,\mu$C is placed in an electric field that varies according to the function $\vec{E} = (2.45 \times 10^4\,x)\,\hat{\imath}$ N/C. (a) If the particle begins at rest at the position $(1.00$ m$, 0, 0)$, show that it executes simple harmonic motion. (b) What are the turning points of the motion? (c) What is the period of simple harmonic motion?

11. An electric dipole consists of two charges $+1.0\,\mu$C and $-1.0\,\mu$C separated by a distance of 25 mm. (a) Find the magnitude of the electric dipole moment. (b) The electric dipole vector makes an angle $3\pi/2$ with a uniform electric field of magnitude 10^5 N/C. Find the torque on the dipole and its potential energy in this configuration.

12. An electric dipole has a dipole moment of magnitude 2.35×10^{-29} C·m and is in a uniform electric field of magnitude 1.35×10^6 N/C. Compute the work an external agent must do in order to change the dipole's orientation with respect to the field from π to 0.

13. An electric dipole with moment $\vec{p}$ is slightly displaced from the equilibrium position $\theta = 0$ with respect to a uniform electric field $\vec{E}$. By making the small-angle approximation $\sin\theta \approx \theta$, show that the dipole makes small (rotational) oscillations about the equilibrium position that approximate simple harmonic motion. Find the approximate expression for the period of simple harmonic motion in terms of the parameters given.

Chapter 18

Partial Derivatives, Gradients, and Electric Potential

In this chapter, we introduce several notions of derivative for functions $f : \mathbb{R}^n \to \mathbb{R}$. These are *partial derivative*, *gradient*, and *directional derivative*. These have applications in a number of areas, including physics, and we see some of these applications in this chapter. Our plan is first to develop an understanding of the basics of partial derivative, gradient, and directional derivative. We then apply these new tools to gain a greater understanding of the physics of charges and electric fields. We use partial derivatives, along with the concepts of work and energy, to develop a new physical concept: *electric potential*. Understanding the electric potential provides an alternative, sometimes easier way of approaching some problems involving charges and electric fields and even allows us to consider problems that could not have been solved using methods previously discussed in this book. In later chapters, electric potential is used in the study of electric circuits.

18.1 Partial Derivatives

18.1.1 Definitions

For a function $f : \mathbb{R} \to \mathbb{R}$ of a single variable, the derivative f' tells us the slope of the tangent line. The derivative is defined as

$$f'(x) = \lim_{\Delta x \to 0} \frac{f(x + \Delta x) - f(x)}{\Delta x}$$

provided the limit exists. Figure 18.1(a) illustrates the standard geometric interpretation. For the input $x = a$, the derivative output $f'(a)$ is the slope of the line tangent to the graph of f at the point $(a, f(a))$. As a simple example, consider the function $f(x) = x^2$ (Figure 18.1(b)). We have $f'(x) = 2x$, so geometrically the slope of the tangent line is equal to $2x$ for any input x.

Now consider a function $f : \mathbb{R}^2 \to \mathbb{R}$ of two variables. For a function of one variable, the definition of derivative is motivated by the geometric concept of a tangent line slope. For a function of two variables, the graph of a function is a surface in space. At a particular input, say (a, b), we can look at the cross sections parallel to the x-axis and the y-axis. Each cross section of the surface is a curve and we examine the tangent lines to these curves.

As an example, consider the quadratic function defined by $f(x, y) = x^2 + y^2$. The output $f(a, b)$ gives the height of the graph at the point (a, b) in the xy-plane. For the input $(1, 2)$, we have $f(1, 2) = 5$. The

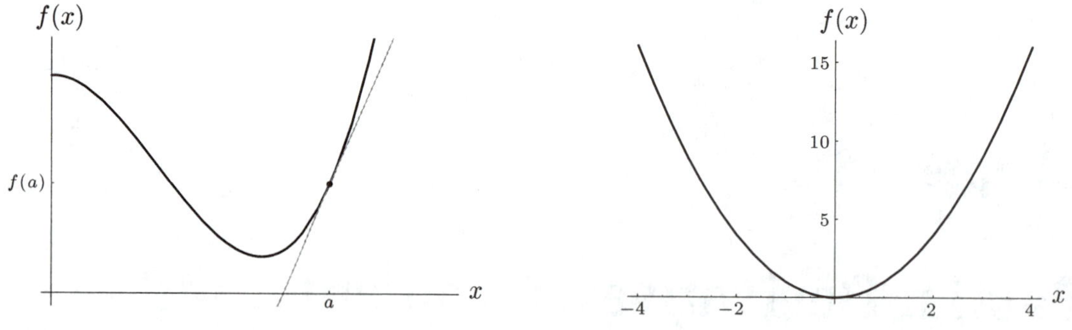

(a) Derivative output $f'(a)$ is the slope of the tangent line at the point $(a, f(a))$ on the curve

(b) Graph of the function $f(x) = x^2$

Figure 18.1. Geometric interpretation of derivative for $f : \mathbb{R} \to \mathbb{R}$.

cross section parallel to the x-axis is shown in Figure 18.2(a) and the cross section parallel to the y-axis is shown in Figure 18.2(b). For each of these, we visualize a line tangent to the cross section graph at the point corresponding to $(1, 2, 5)$. From the graph we can read off approximate slopes for these tangent lines. In the cross section parallel to the x-axis, the tangent line has a slope of 2. In the cross section parallel to the y-axis, the tangent line has a slope of 4.

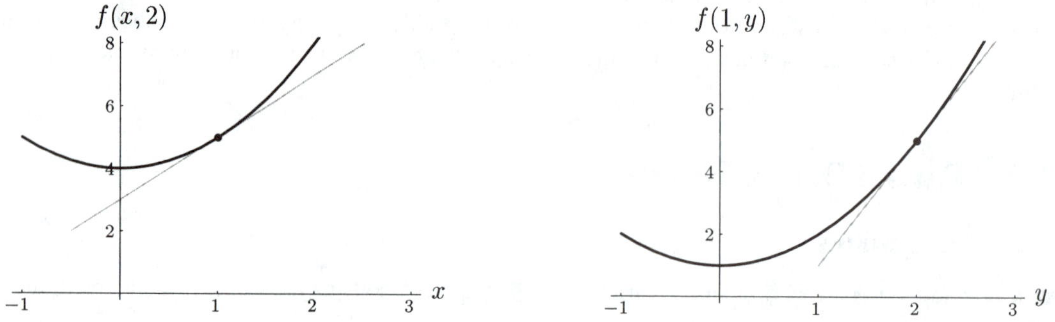

(a) Section parallel to the x-axis

(b) Section parallel to the y-axis

Figure 18.2. Tangent lines in cross sections of $f(x, y) = x^2 + y^2$.

In general, for a function of two variables, we will deal with two slopes for each input, one for the section parallel to the x-axis and one for the section parallel to the y-axis. In order to compute these slopes, we need to develop a relevant notion of *derivative*. Consider the tangent line in the section parallel to the x-axis. To approximate the slope, we compute the slope of a secant line through the points $(x, f(x, y))$ and $(x + \Delta x, f(x + \Delta x, y))$ as

$$\frac{f(x + \Delta x, y) - f(x, y)}{\Delta x}.$$

Note that in this *difference quotient*, the y input is held constant. To define a derivative, we will take a limit as Δx goes to zero. We also have a tangent line in the section parallel to the y-axis. The slope of an approximating secant line is given by

$$\frac{f(x, y+\Delta y)-f(x, y)}{\Delta y}.$$

Here, the x input is held constant. To get a derivative, we will take a limit as Δy goes to zero. These ideas are made complete in the following definition.

Definition 18.1. The **partial derivative** of $f : (x, y) \mapsto f(x, y)$ with respect to x is denoted f_x or $\partial f/\partial x$ and defined by

$$f_x(x, y) = \frac{\partial f}{\partial x}(x, y) \equiv \lim_{\Delta x \to 0} \frac{f(x+\Delta x, y)-f(x, y)}{\Delta x}$$

provided the limit exists.

The partial derivative of $f : (x, y) \mapsto f(x, y)$ with respect to y is denoted f_y or $\partial f/\partial y$ and defined by

$$f_y(x, y) = \frac{\partial f}{\partial y}(x, y) \equiv \lim_{\Delta y \to 0} \frac{f(x, y+\Delta y)-f(x, y)}{\Delta y}$$

provided the limit exists.

The prime (Newtonian) symbol f' is not used for partial derivatives, because there is no indication whether the partial derivative is to be taken with respect to x or y. As a substitute for the Newtonian notation, we use f_x to indicate a partial derivative of f with respect to x and f_y to indicate a partial derivative of f with respect to y. In the Leibniz notation, a different symbol (the curved ∂) is used to denote the partial derivatives $\partial f/\partial x$ and $\partial f/\partial y$, as opposed to the ordinary d in the ordinary derivative df/dx.

With ordinary derivatives, it is important to distinguish between f' (or equivalently df/dx), which is the name of the function, and $f'(x)$ which is the output of the function f' for the input x. The same distinction is important with partial derivatives. However, it is common practice to simply write $\partial f/\partial x$ as a shorthand for $\partial f/\partial x(x, y)$ just as we write df/dx as shorthand for $df/dx(x)$. In context, this shorthand should not be ambiguous. If (a, b) is a specific input, we can denote the corresponding output of the partial derivative with respect to x by

$$f_x(a, b), \qquad \frac{\partial f}{\partial x}(a, b), \qquad \text{or} \qquad \left.\frac{\partial f}{\partial x}\right|_{(a,b)}.$$

We also use the notation $\partial/\partial x$ and $\partial/\partial y$ to denote operators. The operator $\partial/\partial x$ has a function $f(x, y)$ as input and gives the partial derivative of that function with respect to x as output. The operator $\partial/\partial x$ has a function $f(x, y)$ as input and gives the partial derivative of that function with respect to y as output.

For functions $f : \mathbb{R}^n \to \mathbb{R}$ of n variables, there are n partial derivatives, one for each input variable. Each is defined as above, by varying one input variable at a time. For example, for functions of three variables $f : (x, y, z) \mapsto f(x, y, z)$, we include z in the appropriate places in the functions and add to the previous definition a partial derivative with respect to z:

$$\begin{aligned} f_x(x, y, z) &= \frac{\partial f}{\partial x}(x, y, z) \equiv \lim_{\Delta x \to 0} \frac{f(x+\Delta x, y, z)-f(x, y, z)}{\Delta x}, \\ f_y(x, y, z) &= \frac{\partial f}{\partial y}(x, y, z) \equiv \lim_{\Delta y \to 0} \frac{f(x, y+\Delta y, z)-f(x, y, z)}{\Delta y}, \qquad \text{and} \\ f_z(x, y, z) &= \frac{\partial f}{\partial z}(x, y, z) \equiv \lim_{\Delta z \to 0} \frac{f(x, y, z+\Delta z)-f(x, y, z)}{\Delta z}. \end{aligned}$$

18.1.2 Computing Partial Derivatives Symbolically

From the definitions of the partial derivative, we can see that the effect is to differentiate with respect to one variable at a time while holding the other input variables constant in the process. This tells us exactly how to compute partial derivatives in practice: *Follow the rules for taking ordinary derivatives (including the product rule, chain rule, etc.) while treating the other input variables as constants.* That is, in computing the partial derivative with respect to x of a function with outputs $f(x, y)$, we view y as a constant. Likewise, in computing the partial derivative with respect to y of a function with outputs $f(x, y)$, we view x as a constant. The following examples illustrate this idea.

Example 18.1. Compute the partial derivatives for the function $f(x, y) = 4x^2y^3 - 2x^2y$. Evaluate each partial derivative for the input $(2, -1)$.

Solution. To compute f_x, treat y as a constant and differentiate with respect to x to get

$$f_x(x, y) = 8xy^3 - 4xy.$$

Evaluating at the input $(2, -1)$ gives

$$f_x(2, -1) = 8(2)(-1)^3 - 4(2)(-1) = -8.$$

The value -8 is the slope of the tangent line in the x-direction. To compute f_y, treat x as a constant and differentiate with respect to y to get

$$f_y(x, y) = 12x^2y^2 - 2x^2.$$

Evaluating at the input $(2, -1)$ gives

$$f_y(2, -1) = 12(2)^2(-1)^2 - 2(2)^2 = 40.$$

The value 40 is the slope of the tangent line in the y-direction. △

Example 18.2. Compute the partial derivatives for the function $f(x, y) = \sin(xy)$. Evaluate each partial derivative for the input $(1, \pi)$.

Solution. The partial derivative with respect to x is, using the chain rule,

$$\frac{\partial f}{\partial x} = y\cos(xy).$$

Also using the chain rule, we compute the partial derivative with respect to y as

$$\frac{\partial f}{\partial y} = x\cos(xy).$$

For the input $(1, \pi)$ we have

$$\left.\frac{\partial f}{\partial x}\right|_{(1,\pi)} = 1\cos\pi = -1 \quad \text{and} \quad \left.\frac{\partial f}{\partial y}\right|_{1,\pi} = \pi\cos\pi = -\pi.$$ △

Example 18.3. Compute the partial derivatives for the function $f(x, y, z) = xyz$. Evaluate each partial derivative for the input $(1, 2, 3)$.

Solution. To compute the partial derivative with respect to x, treat y and z as constants to get

$$f_x(x, y, z) = yz.$$

In similar fashion, we compute

$$f_y(x, y, z) = xz \qquad \text{and} \qquad f_z(x, y, z) = xy.$$

For the input $(1, 2, 3)$ we have

$$f_x(1,2,3) = (2)(3) = 6, \qquad f_y(1,2,3) = (1)(3) = 3, \qquad \text{and} \qquad f_z(1,2,3) = (1)(2) = 2.$$

A geometric interpretation is difficult for these results. However, in applications partial derivatives are often interpreted as rates of change. △

18.1.3 Graphical and Physical Interpretations of Partial Derivatives

As noted in the beginning of this section, each partial derivative gives the slopes of the tangent line to the graph of the function in two particular directions (parallel to the coordinate axis corresponding to that input variable). Now we explore two physical examples that illustrate some of the graphical and physical interpretations of the partial derivative.

Let the height of a hill above the xy-plane be given by the function

$$f(x, y) = 10\, e^{-(x^2+y^2)/4}$$

with appropriate scaling so that with inputs x and y measured in meters, the output $f(x, y)$ is also in meters. This hill is pictured in Figure 18.3.

The partial derivatives of this function with respect to x and y are

$$f_x(x, y) = -5x\, e^{-(x^2+y^2)/4} \qquad \text{and} \qquad f_y(x, y) = -5y\, e^{-(x^2+y^2)/4}.$$

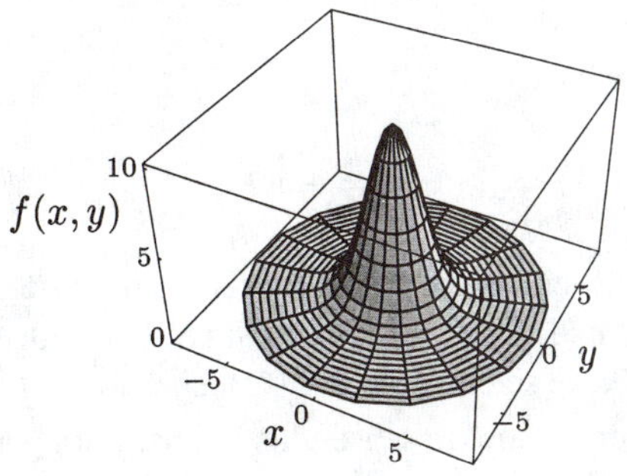

Figure 18.3. Graph of $f(x, y) = 10e^{-(x^2+y^2)/4}$.

Evaluating the partial derivatives at several different inputs (x, y) provides some information. At the input $(0,0)$, the output of both partial derivatives is zero. It should not be hard to convince yourself that the maximum output of f occurs at the input $(0,0)$, because $f(0,0) = 10\,e^0 = 10$ and $f(x,y) < 10$ for any other input, with $e^{-(x^2+y^2)} < 1$. Thus, we see that the input where *both* partial derivatives are zero corresponds to a local maximum in the function output in this example. For the parabolic bowl we studied earlier, both partial derivatives were zero at the input $(0,0)$, but in that case the input was a local *minimum* for outputs of the function. In fact, at any local extremum (maximum or minimum) both partial derivatives will be zero. This is analogous to the fact that for functions of a single variable the derivative is zero at any local extremum. We will study optimization problems for functions of two variables in Chapter 20.

Continuing with the example, we can evaluate the partial derivatives at the input $(0,1)$:

$$f_x(0,1) = -5(0)\,e^{-1/4} = 0, \qquad \text{and} \qquad f_y(0,1) = -5(1)\,e^{-1/4} \approx -3.9.$$

The partial derivative with respect to x is zero at this input, meaning that if we travel a small distance parallel to the x-axis (keeping y fixed) along the function graph we will stay at the same "height" (or output) $f(x,y)$ [Figure 18.4(a)]. In effect, moving "sideways" on the hillside does not change our height. The fact that the partial derivative with respect to y is *negative* means that if we move a small distance parallel to the y-axis (keeping x fixed) we will slide down the hill to a lower height (Figure 18.4(b)). This makes physical sense, because in this direction we are moving directly down the slope.

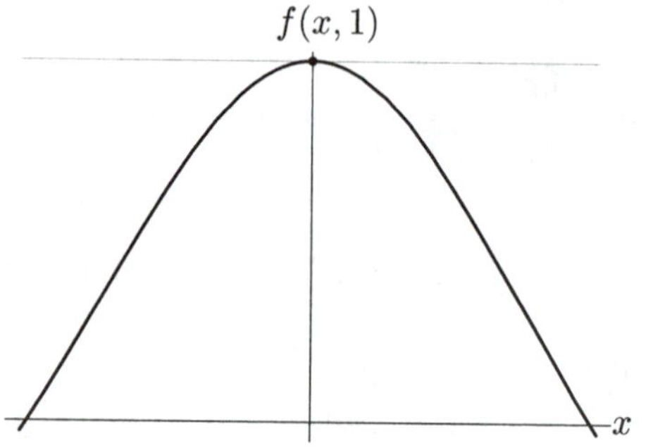

(a) Slope is zero in the x-direction from for the input $(0,1)$

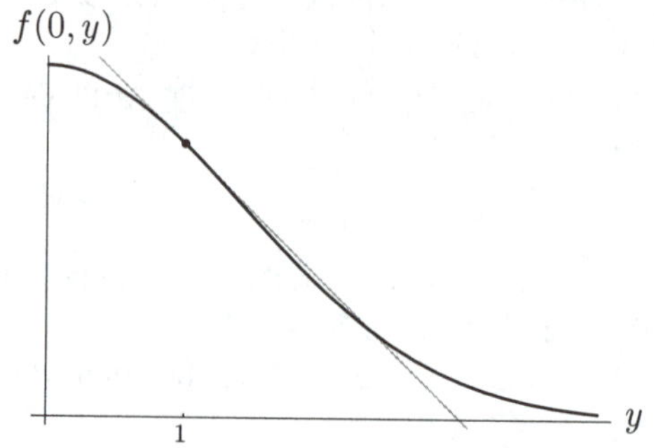

(b) Slope is negative in the y-direction for the input $(0,1)$

Figure 18.4. Slices in the x-direction and the y-direction.

Evaluating at the input $(1,1)$, we have

$$f_x(1,1) = f_y(1,1) = -5(1)\,e^{-1/2} \approx -3.0.$$

Traveling a small distance parallel to either the x- or y-axis results in a slide down the hill. This makes physical sense because above the input $(1,1)$, the hill is "falling off" in both directions. Comparing $f_x(1,1) = f_y(1,1) \approx -3.0$ to $f_y(0,1) \approx -3,9$, we see that the hill is steeper over $(0,1)$ than over $(1,1)$.

Looking back to Figure 18.3, our geometric sense tells us that both partial derivatives will be small for inputs far from the origin. For example, computing the output $f_x(3,3) \approx -0.17$ confirms this intuition. The decaying exponential factor $e^{-(x^2+y^2)/4}$ will dominate the linear factor (x or y) in each partial derivative so outputs of the partial dervivatives approach zero as inputs are taken farther from the origin.

As a second example, let's model the temperature (in kelvins) in the interior of a spherical star of radius R to be given by the function

$$T(x,y,z) = \frac{T_0}{1+\frac{a}{R^2}(x^2+y^2+z^2)}$$

with output in kelvins (K) for inputs x, y, and z in meters. The parameters T_0, a, and R take on specific values for a particular kind of star. We use our Sun as a model, for which $T_0 = 10^6$ K, $a = 180$ (dimensionless), and $R = 6.96 \times 10^8$ m. The function is defined only for inputs with $x^2+y^2+z^2 < R^2$, to limit the domain to the interior of the star. Notice that at the center $(0,0,0)$ the output of the function is $T = T_0$, which for the Sun is $T = 10^6$ K (a typical temperature at the center of a star). At the surface, the inputs satisfy $x^2+y^2+z^2 = R^2$, so the output (using the parameters given above for the Sun) is $T \approx 5500$ K, which is approximately the surface temperature of the Sun.

Computing the partial derivative of our temperature function with respect to one of the input variables gives

$$\frac{\partial T}{\partial x} = -\frac{T_0}{\left[1+\frac{a}{R^2}(x^2+y^2+z^2)\right]^2}\left(\frac{2a}{R^2}x\right) = -\frac{\left(2aT_0/R^2\right)x}{\left[1+\frac{a}{R^2}(x^2+y^2+z^2)\right]^2}.$$

From what we know of partial derivatives, we expect that $\partial T/\partial x$ will tell us how rapidly the output T is changing as we move outward from the origin along the x-axis. We evaluate the partial derivative at several points along the x-axis to determine how rapidly T is changing there. Note that because of the spherical symmetry of the star, our choice of the x-axis is arbitrary; we could obtain similar results by moving along either of the other coordinate axes or along any radial line.

At the input $(0,0,0)$, the partial derivative $\partial T/\partial x$ is equal to zero. This should not be surprising, since it should be clear from the nature of the function T that a local maximum exists in the function's output at the origin.

Halfway out to the surface at $(R/2,0,0)$, we find

$$\left.\frac{\partial T}{\partial x}\right|_{(R/2,0,0)} = -\frac{aT_0/R}{\left[1+\frac{a}{R^2}\left(\frac{R}{2}\right)^2\right]^2} = -\frac{aT_0/R}{\left(1+\frac{a}{4}\right)^2}.$$

This (negative) result tells us that the temperature is decreasing as we proceed outward along the x-axis through the input $(R/2,0,0)$. Notice that the partial derivative $\partial T/\partial x$ has units K/m, because it represents the change in temperature per unit change in distance parallel to the x-axis. Inserting the numerical values for the Sun ($R = 6.96 \times 10^8$ m, $a = 180$, and $T_0 = 10^6$ K), we can compute a numerical output for the partial derivative of T at this input:

$$\left.\frac{\partial T}{\partial x}\right|_{(R/2,0,0)} = -\frac{(180)(10^6\text{K})/(6.96\times 10^8\text{m})}{(1+\frac{180}{4})^2} = -1.2\times 10^{-4}\ \text{K/m}.$$

This may at first seem like a small number, but when we consider that the temperature change from interior to exterior of the star takes place over a vast distance, the result is quite reasonable. Under the assumptions of this model, the local temperature of a star simply does not vary a great deal.

As we reach the surface at $(R,0,0)$, the partial derivative is

$$\left.\frac{\partial T}{\partial x}\right|_{(R,0,0)} = -\frac{2aT_0/R}{(1+a)^2}.$$

Evaluating numerically gives

$$\left.\frac{\partial T}{\partial x}\right|_{(R,0,0)} = -\frac{2(180)(10^6\text{K})/(6.96 \times 10^8\text{m})}{(1+180)^2} = -1.6 \times 10^{-5}\text{K/m}.$$

indicating that at the surface the temperature is still decreasing as a function of increasing distance from the origin, but at a slower rate than when the distance was $R/2$ from the center.

You should notice an important difference between the two examples we studied in this section, related to the fact that the first example used a function of two variables $f(x, y)$ and the second example used a function of three variables $f(x, y, z)$. In the first example, we were able to use the third dimension to represent the output $f(x, y)$ and thereby gain some geometric intuition about the partial derivatives at various locations on the hill. In the second example, there was no simple way for us to represent the outputs of $f : \mathbb{R}^3 \to \mathbb{R}$ geometrically, since this would have required a fourth dimension. In that particular example, we avoided this difficulty by in effect looking only at a one-dimensional slice (along the x-axis) of the function. This was reasonable only because of the high symmetry present in that example, and you should be aware that this is not always so easy to do for functions $f : \mathbb{R}^n \to \mathbb{R}$ with $n > 2$.

18.2 Tangent Planes and Linear Approximations

For a differentiable function $f : \mathbb{R} \to \mathbb{R}$, the derivative output $f'(a)$ gives us the slope of the tangent line at the point $(a, f(a))$. The equation of the tangent line is thus

$$y = f(a) + f'(a)(x - a).$$

Viewing the tangent line as the linear approximation, we write

$$f(x) \approx f(a) + f'(a)(x - a),$$

and our intuition is that the linear approximation is better the smaller we make $x - a$. Our goal in this section is to generalize these ideas to functions $f : \mathbb{R}^n \to \mathbb{R}$.

18.2.1 Functions of Two Variables

To generalize these interpretations to a function $f : \mathbb{R}^2 \to \mathbb{R}$, let's first recall various forms for the equation of a plane. In particular, given a normal vector $\vec{N} = \langle A, B, C\rangle$ and the position $\vec{r}_0 = \langle x_0, y_0, z_0\rangle$ of a point on the plane, we have

$$\vec{N} \cdot (\vec{r} - \vec{r}_0) = 0$$

for the position $\vec{r} = \langle x, y, z\rangle$ of any other point on the plane. Alternatively, we can write the equation as

$$z = z_0 + m_x(x - x_0) + m_y(y - y_0)$$

for the constants x_0, y_0, z_0, m_x, and m_y. Note that m_x and m_y give the slopes of the plane with respect to the x-axis and y-axis, respectively.

The partial derivatives of $f : \mathbb{R}^2 \to \mathbb{R}^2$ give us information about the slopes of tangent lines in sections parallel to the x- and y-axes. In many cases, the two tangent lines at a particular point on the surface define a plane that is tangent to the surface. (We give an example in which this is not the case at the end of this section). A typical case is illustrated in Figure 18.5. The tangent lines are shown in Figure 18.5(a) and the tangent plane is shown in Figure 18.5(b).

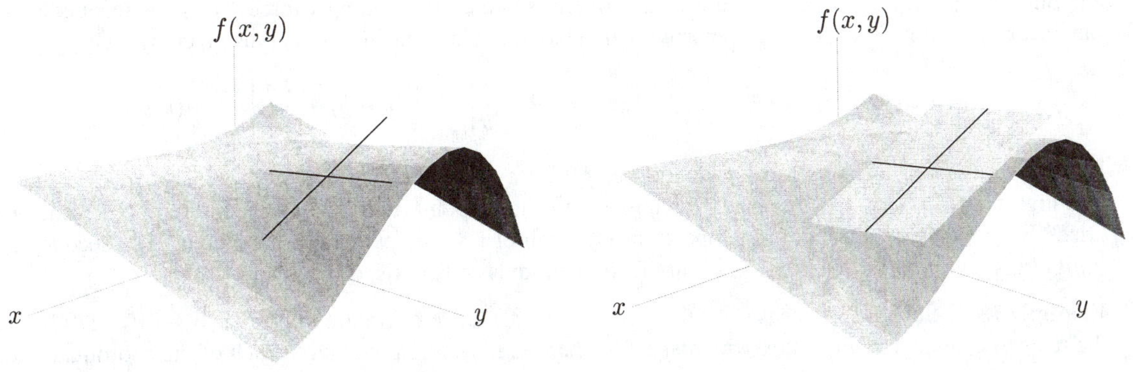

(a) Tangent lines in sections parallel to the coordinate axes

(b) Tangent plane defined by tangent lines

Figure 18.5. A typical tangent plane defined by tangent lines.

Now consider the equation of the tangent plane defined by the tangent lines. From their definitions, it is evident that the partial derivatives give the slopes of the tangent plane. Thus, we have the equation

$$z = f(a,b) + \left.\frac{\partial f}{\partial x}\right|_{(a,b)} (x-a) + \left.\frac{\partial f}{\partial y}\right|_{(a,b)} (y-b)$$

for the tangent plane. Note that we could write this in a different notation as $z = f(a,b) + f_x(a,b)(x-a) + f_y(a,b)(y-b)$. Using the tangent plane equation as the linear approximation, we write

$$f(x,y) \approx f(a,b) + \left.\frac{\partial f}{\partial x}\right|_{(a,b)} (x-a) + \left.\frac{\partial f}{\partial y}\right|_{(a,b)} (y-b). \tag{18.1}$$

Each term on the right side has a geometric interpretation. The first term is the height of the graph above (or below for negative values) the point (a,b) in the xy-plane. The second term gives the rise along the tangent plane for a run of $(x-a)$ in the direction parallel to the x-axis, and the third term gives the rise along the tangent plane for a run of $(y-a)$ in the direction parallel to the y-axis.

Example 18.4. Find the equation of the tangent plane for the function $f(x,y) = x^2 + y^2$ at the input $(1,2)$. Use the linear approximation to estimate $f(1.1, 1.8)$.

Solution. We first compute the output $f(1,2) = 1^2 + 2^2 = 5$. The partial derivatives are $f_x(x,y) = 2x$ and $f_y(x,y) = 2y$. For the input $(1,2)$, we get $f_x(1,2) = 2$ and $f_y(1,2) = 4$. The equation of the tangent plane is thus $z = 5 + 2(x-1) + 4(y-2)$. The linear approximation for $f(1.1, 1.8)$ is

$$f(1.1, 1.8) \approx 5 + 2(1.1-1) + 4(1.8-2) = 5 + 0.2 - 0.8 = 4.6.$$

Of course, having a formula for $f(x,y)$ allows us to compute the exact value of $f(1.1, 1.8) = 4.45$. The utility of the linear approximation comes in other contexts. △

18.2.2 Functions of Three or More Variables

When we consider functions $f : \mathbb{R}^3 \to \mathbb{R}$, a geometric interpretation becomes problematic, but the idea of a linear approximation generalizes with ease from our study of $f : \mathbb{R}^2 \to \mathbb{R}$. For a function with inputs (x, y, z)

and outputs $f(x, y, z)$, there are three partial derivatives and correspondingly three "rises" to account for in the linear approximation. The linear approximation based at the input (a, b, c) is thus given by

$$f(x,y,z) \approx f(a,b,c) + \left.\frac{\partial f}{\partial x}\right|_{(a,b,c)}(x-a) + \left.\frac{\partial f}{\partial y}\right|_{(a,b,c)}(y-b) + \left.\frac{\partial f}{\partial z}\right|_{(a,b,c)}(z-c).$$

It may be easiest to interpret this by thinking of $f(x, y, z)$ as the temperature at a point (x, y, z) in space. The first term $f(a, b, c)$ then represents the temperature at the point (a, b, c). The second term represents the "rise" (which may be negative) in temperature for a run of $(x - a)$ parallel to the x-axis. The second and third terms give similar "rises" for runs parallel to the y-axis and z-axis, respectively.

Example 18.5. Suppose we measure at the point $(2, 5, -6)$ the temperature to be $T(2, 5, -6) = 270$ K and the rates of change in temperature with respect to changes in position parallel to each of the coordinate axes to be

$$\left.\frac{\partial T}{\partial x}\right|_{(2,5,-6)} = 4 \text{ K/m}, \qquad \left.\frac{\partial T}{\partial y}\right|_{(2,5,-6)} = -9 \text{ K/m}, \qquad \text{and} \qquad \left.\frac{\partial T}{\partial z}\right|_{(2,5,-6)} = 2 \text{ K/m}.$$

Find the linear approximation for $T(2.3, 4.9, -5.8)$.

Solution. This involves a straightforward calculation once we get the information into the linear approximation:

$$\begin{aligned} T(2.3,4.9,-5.8) &\approx T(2,5,-6) + \left.\frac{\partial T}{\partial x}\right|_{(2,5,-6)}(2.3-2) + \left.\frac{\partial T}{\partial y}\right|_{(2,5,-6)}(4.9-5) + \left.\frac{\partial T}{\partial z}\right|_{(2,5,-6)}(-5.8+6) \\ &= 270 \text{ K} + (4 \text{ K/m})(0.3 \text{ m}) - (9 \text{ K/m})(-0.1 \text{ m}) + (2 \text{ K/m})(0.2 \text{ m}) \\ &= 272.5 \text{ K}. \end{aligned}$$

△

18.2.3 Differentiability

For a function $f : \mathbb{R} \to \mathbb{R}$ of one variable, we define *differentiability* by the existence of the limit in the defintion of derivative. That is, a function f is differentiable for the input x if

$$\lim_{h \to 0} \frac{f(x+h) - f(x)}{h}$$

exists. In the geometric view, we think of differentiablility as guaranteeing there is a well-defined tangent line. Specifically, if f is differentiable for the input a, then line given by the equation

$$y = f(a) + f'(a)(x - a)$$

is tangent to the graph of f at the point $(a, f(a))$. A second consequence of differentiablility is the guarantee of continuity. That is, if f is differentiable for the input a, then f is continuous for a.

For a function $f : \mathbb{R}^2 \to \mathbb{R}$ of two variables, the definition of differentiablility is not obvious. One possibility is to think in terms of partial derivatives. However, if we want differentiability to guarantee that the plane given by the equation

$$z = f(a,b) + \left.\frac{\partial f}{\partial x}\right|_{(a,b)}(x-a) + \left.\frac{\partial f}{\partial y}\right|_{(a,b)}(y-b)$$

is tangent to the graph of f and to guarantee continuity, it is not enough to require that the partial derivatives of f exist for the input (a, b). We make this explicit in the following example.

Example 18.6. Consider the function f defined by

$$f(x,y)=\begin{cases}\dfrac{2xy}{x^2+y^2} & \text{if } (x,y)\neq(0,0)\\ 0 & \text{if } (x,y)=(0,0).\end{cases}$$

Show that the partial derivatives of f exist for all (x,y) and that f is not continuous for $(0,0)$.

Solution. For $(x,y)\neq(0,0)$, we can compute the partial derivatives using the quotient rule. For the partial derivative with respect to x, we get

$$\frac{\partial f}{\partial x}=\frac{(x^2+y^2)(2y)-2xy(2x)}{(x^2+y^2)^2}=\frac{2y(y^2-x^2)}{(x^2+y^2)^2}.$$

In similar fashion, for the partial derivative with respect to y we get

$$\frac{\partial f}{\partial y}=\frac{(x^2+y^2)(2x)-2xy(2y)}{(x^2+y^2)^2}=\frac{2x(x^2-y^2)}{(x^2+y^2)^2}.$$

For $(x,y)=(0,0)$, we use the definition of partial derivative directly. For the partial derivative with respect to x, we start with

$$\left.\frac{\partial f}{\partial x}\right|_{(0,0)}=\lim_{\Delta x\to 0}\frac{f(\Delta x,0)-f(0,0)}{\Delta x}.$$

From the formula for the function f, we have

$$f(\Delta x,0)=\frac{2(\Delta x)(0)}{(\Delta x)^2+(0)^2}=0 \qquad\text{and}\qquad f(0,0)=0.$$

Thus,

$$\left.\frac{\partial f}{\partial x}\right|_{(0,0)}=\lim_{\Delta x\to 0}\frac{f(\Delta x,0)-f(0,0)}{\Delta x}=\lim_{\Delta x\to 0}\frac{0}{\Delta x}=0.$$

A similar calculation gives us

$$\left.\frac{\partial f}{\partial y}\right|_{(0,0)}=0.$$

To analyze continuity at $(0,0)$, recall the solution of Example 13.24. There we showed that

$$\lim_{(x,y)\to(0,0)}\frac{2xy}{x^2+y^2}$$

does not exist. Thus, f is not continuous for input $(0,0)$. △

A graph of the function f in Example 18.6 is shown in Figure 18.6. From our solution, we know that the both tangent lines in the sections parallel to the coordinate axes have slope zero. The equation of the plane defined by these tangent lines is thus

$$z=f(0,0)+\left.\frac{\partial f}{\partial x}\right|_{(0,0)}(x-0)+\left.\frac{\partial f}{\partial y}\right|_{(0,0)}(y-0)=0.$$

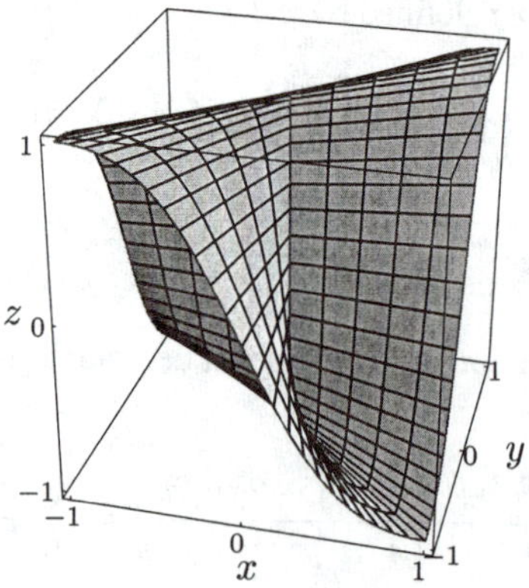

Figure 18.6. Graph of the function in Example 18.6.

The plane is just the xy-plane. From Figure 18.6, it is clear that this is *not* a tangent plane for the point $(0, 0, 0)$. In fact, there is no tangent plane for $(0, 0, 0)$. The point of this example is that merely having partial derivatives exist does not guarantee continuity or a well-defined tangent plane. The existence of partial derivatives alone is not an adequate definition of differentiability for functions of two or more variables.

We do not develop a precise definition of differentiability in this text. In passing, we note that to guarantee continuity and a well-defined tangent plane at an input (a, b), it is sufficient to require that the partial derivatives exist and are continuous in some neighborhood of (a, b). You can assume this to be the case for the functions we encounter in the remainder of the text.

18.3 Gradient and Directional Derivative

The two partial derivatives for a function $f : \mathbb{R} \to \mathbb{R}$ give the two slopes of the tangent plane for each input. In many contexts, it is useful to "package" the partial derivatives as the components of a vector. This vector is called the **gradient vector** of f. We will see how a gradient can be used to compute the directional derivative, which gives the rate of change for function outputs with respect to any direction.

18.3.1 Gradient

To motivate our definition of gradient, we rewrite the linear approximation developed in the previous section. For simplicity, consider functions $f : \mathbb{R}^2 \to \mathbb{R}$ for which the linear approximation is given by

$$f(x, y) \approx f(a, b) + \left.\frac{\partial f}{\partial x}\right|_{(a,b)} (x - a) + \left.\frac{\partial f}{\partial y}\right|_{(a,b)} (y - b).$$

We introduce the vectors

$$\left\langle \left.\frac{\partial f}{\partial x}\right|_{(a,b)}, \left.\frac{\partial f}{\partial y}\right|_{(a,b)} \right\rangle \qquad \text{and} \qquad \langle x - a, y - b\rangle$$

and focus on the last two terms in the linear approximation. We can then rewrite them as a dot product:

$$f(x, y) \approx f(a, b) + \left\langle \left.\frac{\partial f}{\partial x}\right|_{(a,b)}, \left.\frac{\partial f}{\partial y}\right|_{(a,b)} \right\rangle \cdot \langle x - a, y - b\rangle.$$

This formulation separates information about slopes in the first vector from information about "runs" in the second vector. The vector that has partial derivatives as components is called the **gradient vector** of f at (a, b) and is denoted

$$\vec{\nabla} f(a,b) = \left\langle \left.\frac{\partial f}{\partial x}\right|_{(a,b)}, \left.\frac{\partial f}{\partial y}\right|_{(a,b)} \right\rangle.$$

We can extend the notion of a gradient vector to any input (x, y) for which the partial derivatives exist.

Definition 18.2. The **gradient** of a function $f : \mathbb{R}^2 \to \mathbb{R}$ is the vector field $\vec{\nabla} f : \mathbb{R}^2 \to \mathbb{R}^2$ given by

$$\vec{\nabla} f(x,y) = \left\langle \frac{\partial f}{\partial x}, \frac{\partial f}{\partial y} \right\rangle.$$

The symbol $\vec{\nabla}$ is sometimes called "del." We denote the gradient as $\vec{\nabla} f$, because the output is a vector quantity.

Later in this chapter, gradient is used in our study of electric fields in physics. In this section, we will develop some familiarity with the mechanics of working with gradients and some geometric intuition about what a gradient means, as we did previously with partial derivatives.

Example 18.7. Find the gradient of the function $f(x,y) = \sin(\pi xy) + x^2 y$. Evaluate the gradient for the input (2,5).

Solution. The two components of the gradient are the two partial derivatives

$$\frac{\partial f}{\partial x} = \pi y \cos(\pi xy) + 2xy \qquad \text{and} \qquad \frac{\partial f}{\partial y} = \pi x \cos(\pi xy) + x^2.$$

Therefore,

$$\vec{\nabla} f(x,y) = \left\langle \pi y \cos(\pi xy) + 2xy, \pi x \cos(\pi xy) + x^2 \right\rangle.$$

Evaluating the gradient at the input $(2, 5)$ gives

$$\vec{\nabla} f(2,5) = \langle 5\pi \cos(10\pi) + 20, 2\pi \cos(10\pi) + 25 \rangle = \langle 5\pi + 20, 2\pi + 25 \rangle. \qquad \triangle$$

For a function $f : \mathbb{R}^3 \to \mathbb{R}$ of three variables, the definition of gradient involves the three partial derivatives:

$$\vec{\nabla} f(x,y,z) = \left\langle \frac{\partial f}{\partial x}, \frac{\partial f}{\partial y}, \frac{\partial f}{\partial z} \right\rangle.$$

Example 18.8. Find the gradient of the function $f(x,y,z) = e^{xy} + 3y^2 z$. Evaluate the gradient for the input (3,-2,1).

Solution. The three components of the gradient are the three partial derivatives

$$\frac{\partial f}{\partial x} = ye^{xy}, \qquad \frac{\partial f}{\partial y} = xe^{xy} + 6yz, \qquad \text{and} \qquad \frac{\partial f}{\partial z} = 3y^2.$$

Therefore,

$$\vec{\nabla} f = \left\langle ye^{xy}, xe^{xy} + 6yz, 3y^2 \right\rangle.$$

Evaluating the gradient at the input $(3, -2, 1)$ gives

$$\vec{\nabla} f(3, -2, 1) = \langle -2e^{-6}, 3e^{-6} + 6(-2)(1), 3(-2)^2 \rangle \approx \langle -0.0050, -11.99, 12 \rangle . \qquad \triangle$$

It is important to remember that each output of the gradient is a *vector*, so $\vec{\nabla} f$ is a *vector field.* Recall the standard way to visualize a vector field is to think of the output $\vec{\nabla} f(x, y)$ as an arrow with tail at the input point (x, y). For example, consider the function $f(x, y) = -x^2 + y^2$ [a saddle of the type shown in Figure 13.20(b)]. We compute $\vec{\nabla} f(x, y) = \langle -2x, 2y \rangle$. A sketch of this vector field is shown in Figure 18.7.

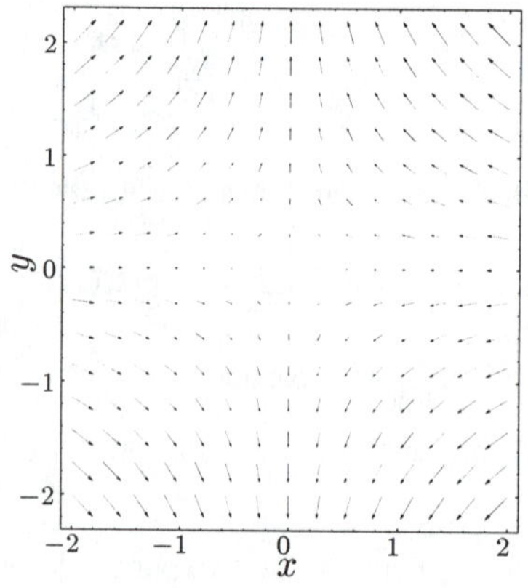

Figure 18.7. The gradient vector field $\vec{\nabla} f = \langle -2x, 2y \rangle$.

Because the components are simply partial derivatives, all of the usual rules of differentiation hold for gradients. These are summarized in the following theorem.

Theorem 18.1. *If* $f : \mathbb{R}^n \to \mathbb{R}$ *and* $g : \mathbb{R}^n \to \mathbb{R}$ *have partial derivatives for an input* (x, y), *then the following hold.*

1. *The constant factor rule:*

$$\vec{\nabla}[cf(x, y)] = c\vec{\nabla} f(x, y)$$

where c is any constant real number.

2. *The sum rule:*

$$\vec{\nabla}[f(x, y) + g(x, y)] = \vec{\nabla} f(x, y) + \vec{\nabla} g(x, y).$$

3. *The product rule:*

$$\vec{\nabla}[f(x, y)g(x, y)] = f(x, y)\vec{\nabla} g(x, y) + g(x, y)\vec{\nabla} f(x, y).$$

4. *The quotient rule:*

$$\vec{\nabla}\left[\frac{f(x, y)}{g(x, y)}\right] = \frac{g(x, y)\vec{\nabla} f(x, y) - f(x, y)\vec{\nabla} g(x, y)}{(g(x, y))^2}.$$

provided there are no issues with division by zero.

Proofs for each conclusion follow from the corresponding results for partial derivatives.

The gradient plays a role in a chain rule we will find useful on occasion. Consider a function $f : \mathbb{R}^2 \to \mathbb{R}$ of two variables and a vector output function $\vec{r} : \mathbb{R} \to \mathbb{R}^2$. The outputs $\vec{r}(t)$ can serve as inputs for f to give $f(\vec{r}(t))$. The chain rule for the derivative of $f(\vec{r}(t))$ with respect to t is given in the following theorem.

Theorem 18.2. *If $f : \mathbb{R}^n \to \mathbb{R}$ and $\vec{r} : \mathbb{R} \to \mathbb{R}^n$ are differentiable, then*

$$\frac{d}{dt}[f(\vec{r}(t))] = \vec{\nabla} f(\vec{r}(t)) \cdot \vec{r}\,'(t).$$

Proof. We sketch the essential idea of a proof. Recall the definition

$$\frac{d}{dt}[f(\vec{r}(t))] = \frac{f(\vec{r}(t+h)) - f(\vec{r}(t))}{h}.$$

To connect this with the gradient of f and the derivative of $\vec{r}$, we use linear approximations. For the function $\vec{r}$, we have

$$\vec{r}(t+h) \approx \vec{r}(t) + \vec{r}\,'(t)h.$$

The term $f(\vec{r}(t+h))$ can thus be approximated as

$$f(\vec{r}(t+h)) \approx f(\vec{r}(t) + \vec{r}\,'(t)h).$$

Now apply a linear approximation to the expression on the right side giving

$$f(\vec{r}(t+h)) \approx f(\vec{r}(t)) + \vec{\nabla} f(\vec{r}(t)) \cdot \vec{r}\,'(t)h.$$

Rearranging this last result gives us

$$\frac{f(\vec{r}(t+h)) - f(\vec{r}(t))}{h} \approx \vec{\nabla} f(\vec{r}(t)) \cdot \vec{r}\,'(t).$$

The expression on the left is the difference quotient in the definition above. Taking a limit on both sides gives us

$$\frac{d}{dt}[f(\vec{r}(t))] = \vec{\nabla} f(\vec{r}(t)) \cdot \vec{r}\,'(t).$$

The detail we omit justifies going from an approximation to equality in the limit. □

18.3.2 Directional Derivatives

We have already seen that one important piece of information carried by the partial derivatives of a function is that they give the rate of change of the function output in directions parallel to each of the coordinate axes. But what if we are interested in how the function output changes as we change inputs in any direction, not necessarily parallel to one of the coordinate axes?

First, fix an input (a, b) at which to explore the rate of change in the function output. To specify information about direction, we introduce a unit vector $\hat{u} = \langle u_1, u_2 \rangle$. To make a "run" of size h in the direction of $\hat{u}$, we can change the input from (a, b) to $(a, b) + h\langle u_1, u_2 \rangle = (a + hu_1, b + hu_2)$. The corresponding "rise" is given by $f(a + hu_1, b + hu_2) - f(a, b)$, and the slope is computed as

$$\frac{f(a + hu_1, b + hu_2) - f(a, b)}{h}.$$

We get a type of derivative by taking the limit as $h \to 0$ of this difference quotient. This is made precise in the following definition.

Definition 18.3. The **directional derivative** of a function $f : \mathbb{R}^2 \to \mathbb{R}$ in the direction of a unit vector $\hat{u}$ at the input (a, b) is

$$D_{\hat{u}} f(a,b) = \lim_{h \to 0} \frac{f(a + hu_1, b + hu_2) - f(a,b)}{h}.$$

For a generic input (x, y), we denote the directional derivative as $D_{\hat{u}} f(x, y)$.

As usual, it is difficult to compute a directional derivative directly from the definition, so we look for an alternative expression. Our strategy for finding an expression that can be used to compute directional derivatives is as follows: First, use the linear approximation to find an approximate change in the output of $f(x, y)$ for a small change $h\hat{u}$ in the input (where h is a small positive real number). Then, by taking a limit as $h \to 0$, we can find an exact expression for the change in the output of $f(x, y)$, and thus from the definition of derivative find the directional derivative.

With a small difference $h\hat{u} = (hu_1, hu_2)$ between the inputs (a, b) and $(a + hu_1, b + hu_2)$, the linear approximation [from Equation (18.1)] gives

$$f(x,y) \approx f(a,b) + \left.\frac{\partial f}{\partial x}\right|_{(a,b)} (hu_1) + \left.\frac{\partial f}{\partial y}\right|_{(a,b)} (hu_2).$$

Substituting into the definition of directional derivative gives

$$D_{\hat{u}} f(a,b) = \lim_{h \to 0} \frac{\left.\frac{\partial f}{\partial x}\right|_{(a,b)} (hu_1) + \left.\frac{\partial f}{\partial y}\right|_{(a,b)} (hu_2)}{h}.$$

The factor h in both numerator and denominator cancels, making the evaluation of the limit trivial:

$$D_{\hat{u}} f(a,b) = \left.\frac{\partial f}{\partial x}\right|_{(a,b)} u_1 + \left.\frac{\partial f}{\partial y}\right|_{(a,b)} u_2.$$

Look carefully at the right side of this equation. The partial derivatives are the two components of $\vec{\nabla} f(a, b)$, while u_1 and u_2 are the components of the vector $\hat{u}$. We recognize this as a dot product, so we have

$$D_{\hat{u}} f(a,b) = \vec{\nabla} f(a,b) \cdot \hat{u}. \tag{18.2}$$

Equation (18.2) gives us a straightforward way to compute the directional derivative of a known function $f(x, y)$ at any input (a, b) in the direction of any unit vector $\hat{u}$ in the xy-plane.

Example 18.9. Compute the directional derivative of $f(x, y) = x^2 + y^2$ at the input $(2, 1)$ in the direction $\hat{u} = \langle 1/\sqrt{2}, 1/\sqrt{2} \rangle$.

Solution. The gradient of f at $(2, 1)$ is

$$\vec{\nabla} f(2,1) = \langle 4, 2 \rangle.$$

Using Equation (18.2), we compute

$$D_{\hat{u}} f(2,1) = \langle 4, 2 \rangle \cdot \left\langle \frac{1}{\sqrt{2}}, \frac{1}{\sqrt{2}} \right\rangle = \frac{6}{\sqrt{2}} = 3\sqrt{2}.$$

This tells us that $3\sqrt{2}$ is the slope of the graph of f for input $(2, 1)$ in the direction $\hat{u} = \langle 1/\sqrt{2}, 1/\sqrt{2} \rangle$. △

The result of Equation (18.2) readily generalizes to $f : \mathbb{R}^3 \to \mathbb{R}$. In this setting, the direction is given by a unit vector $\hat{u} = \langle u_1, u_2, u_3 \rangle$ at an input (a, b, c). The directional derivative is computed using

$$D_{\hat{u}} f(a,b,c) = \vec{\nabla} f(a,b,c) \cdot \hat{u}. \tag{18.3}$$

Example 18.10. Given a function $f(x, y, z) = -2x^2y^3 + 3xz^2$, find the directional derivative at the input $(3, 1, 2)$ in a direction pointing toward the origin.

Solution. First we compute the gradient:

$$\vec{\nabla} f = (-2xy^3 + 3z^2)\,\hat{\imath} - 6x^2y^2\,\hat{\jmath} + 6xz\,\hat{k}.$$

Evaluating at the input $(3, 1, 2)$:

$$\vec{\nabla} f(3, 1, 2) = [-2(3)(1)^3 + 3(2)^2]\,\hat{\imath} - 6(3)^2(1)^2\,\hat{\jmath} + 6(3)(2)\,\hat{k} = 6\,\hat{\imath} - 54\,\hat{\jmath} + 36\,\hat{k}.$$

The vector pointing from $(3, 1, 2)$ to the origin is $\vec{r} = -3\,\hat{\imath} - \hat{\jmath} - 2\,\hat{k}$. Therefore the unit vector $\hat{u}$ in that direction is

$$\hat{u} = \frac{\vec{r}}{r} = \frac{-3\,\hat{\imath} - \hat{\jmath} - 2\,\hat{k}}{\sqrt{3^2 + 1^2 + 2^2}} = \frac{1}{\sqrt{14}}(-3\,\hat{\imath} - \hat{\jmath} - 2\,\hat{k}).$$

Then by Equation (18.3), the directional derivative is

$$D_{\hat{u}} f(3, 1, 2) = \vec{\nabla} f(3, 1, 2) \cdot \hat{u} = \frac{1}{\sqrt{14}}[(6)(-3) + (-54)(-1) + (36)(-2)] = -\frac{36}{\sqrt{14}}.$$

The physical meaning of this directional derivative is that the rate of change of $f(x, y, z)$ in a direction from $(3, 1, 2)$ toward the origin is $-36/\sqrt{14}$ (whatever the units of f are) per unit distance. We will emphasize the physical meaning in an example from physics, in which there are physical units attached to the outputs $f(x, y, z)$. △

Example 18.11. Consider again the model of the temperature field inside a star given in Section 18.1.3. Find the directional derivative of the temperature at the position $(R/2, 0, 0)$ (a) in a direction radially outward from the center, and (b) in any direction *perpendicular* to the radius at that point. For any case in which the directional derivative is nonzero, evaluate it numerically using the radius of the sun $R = 6.96 \times 10^8$ m and the other parameters a and T_0 given previously for the Sun.

Solution. (a) We first need to find the gradient of $T(x, y, z)$ in order to use Equation (18.3) to find the directional derivative. Taking the result of that example for $\partial T/\partial x$, along with similar results for the other partial derivatives,

$$\vec{\nabla} T = -\frac{(2aT_0/R^2)\,x}{\left[1 + \frac{a}{R^2}(x^2 + y^2 + z^2)\right]^2}\,\hat{\imath} - \frac{(2aT_0/R^2)\,y}{\left[1 + \frac{a}{R^2}(x^2 + y^2 + z^2)\right]^2}\,\hat{\jmath} - \frac{(2aT_0/R^2)\,z}{\left[1 + \frac{a}{R^2}(x^2 + y^2 + z^2)\right]^2}\,\hat{k}$$

but at the input $(R/2, 0, 0)$ only the x-component is nonzero, so

$$\vec{\nabla} T(\tfrac{R}{2}, 0, 0) = -\frac{aT_0/R}{\left[1 + \frac{a}{R^2}(\frac{R}{2})^2\right]^2}\,\hat{\imath} \approx -\frac{aT_0/R}{\left(1 + \frac{a}{4}\right)^2}\,\hat{\imath}.$$

Inserting the numerical values of the parameters a, T_0, and R gives

$$\vec{\nabla} T(\tfrac{R}{2}, 0, 0) = -\frac{(180)(10^6\text{ K})/(6.96 \times 10^8\text{m})}{(1 + \frac{180}{4})^2}\,\hat{\imath} = -1.22 \times 10^{-4}\text{ K/m}\,\hat{\imath}.$$

The unit vector pointing radially outward from $(R/2, 0, 0)$ must be $\langle 1, 0, 0\rangle$, so the directional derivative is

$$D_{\hat{u}} f(\tfrac{R}{2}, 0, 0) = \vec{\nabla} T(\frac{R}{2}, 0, 0) \cdot \langle 1, 0, 0\rangle = -1.22 \times 10^{-4}\text{ K/m}.$$

This is essentially the same result we found in the earlier example, which is not surprising, because we happened to pick a direction for the directional derivative that is parallel to one of the coordinate axes. It should be comforting to see that the directional derivative does reduce to the partial derivative in this special case.

(b) Now in a direction perpendicular to the radius $\hat{u} = 0\,\hat{\imath} + u_2\,\hat{\jmath} + u_3\,\hat{k}$ and for any components u_2 and u_3 we choose, the result is

$$D_{\hat{u}} f(\tfrac{R}{2}, 0, 0) = \vec{\nabla} T(\tfrac{R}{2}, 0, 0) \cdot \hat{u} = (-1.22 \times 10^{-4} \text{ K/m}\,\hat{\imath}) \cdot (u_2\,\hat{\jmath} + u_3\,\hat{k}) = 0.$$

The outputs $T(x, y, z)$ do not change for a small displacement perpendicular to the radius. △

18.3.3 Gradients and Level Curves

For a function $f : \mathbb{R}^2 \to \mathbb{R}$, we draw the gradient as a vector field in the input plane. We also draw level curves in the input plane. Is there a geometric relation between gradient vectors and level curves? Figure 18.8 shows two examples of gradients and level curves drawn together. Focus on a particular input through which a level curve passes. The gradient vector at that input is perpendicular to the level curve. Based on these examples, we conjecture that at each input, the gradient vector in perpendicular to the level curve. Justification for this conjecture is given by the following argument.

Consider some portion of a level curve for $f : \mathbb{R}^2 \to \mathbb{R}$. Let $\vec{r}(t)$ with t in some interval (a, b) give a parametrization of this level curve portion. Since the outputs $\vec{r}(t)$ are on a level curve of f, we know that $f(\vec{r}(t))$ is constant for all t in (a, b) and hence

$$\frac{d}{dt}[f(\vec{r}(t))] = 0$$

for all t in (a, b). Using the result in Theorem 18.2, we conclude that $\vec{\nabla} f(\vec{r}(t)) \cdot \vec{r}\,'(t) = 0$ for all t in (a, b). For each t, the vector $\vec{r}\,'(t)$ is tangent to the level curve at the point $\vec{r}(t)$. Since the dot product of the gradient vector and this tangent vector is zero, we conclude that *the gradient vector is perpendicular to the level curve.*

The analogous result is true for a function $f : \mathbb{R}^3 \to \mathbb{R}$. In this case, level curves are replaced by level surfaces. At each input (x, y, z), the gradient vector $\vec{\nabla} f(x, y, z)$ is perpendicular to the level surface through that input.

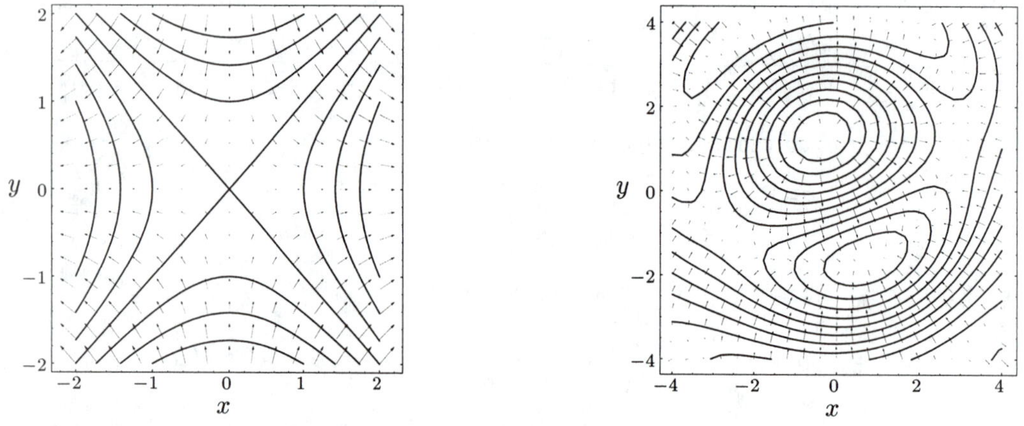

(a) Gradient field and level curves for $f(x, y) = x^2 - y^2$

(b) Gradient field and level curves for a second example

Figure 18.8. Geometric relation between gradient vectors and level curves.

Returning to the case of functions $f : \mathbb{R}^2 \to \mathbb{R}$, we can consider the directional derivative for a particular input (a, b). Using the geometric interpretation of the dot product, we write

$$D_{\hat{u}} f(a, b) = \vec{\nabla} f(a, b) \cdot \hat{u} = \|\vec{\nabla} f(a, b)\| \, \|\hat{u}\| \cos\theta$$

where θ is the angle between $\vec{\nabla} f(a, b)$ and $\hat{u}$. The *maximum* value of this dot product (and hence the maximum rate of increase of the function f) is found for $\cos\theta = 1$. This is at $\theta = 0$, meaning that $\hat{u}$ is in the same direction as the gradient. The directional derivative of f at (a, b) in the direction of the gradient vector is $\|\vec{\nabla} f(a, b)\|$. Let's summarize this result as another general rule: *For each input, the gradient vector points in the direction in which the rate of change of the function f is a maximum. Moreover, the maximum rate of change is the magnitude of the gradient* $\|\vec{\nabla} f(a, b)\|$. We interpret this geometrically by saying each nonzero gradient vector is tangent to the input path corresponding to the steepest path on the graph of the function.

Now think about the sign of the directional derivative $D_{\hat{u}} f(a, b)$ for other directions $\hat{u}$. Draw a line tangent to the level curve through the input (a, b). We know the gradient vector $\vec{\nabla} f(a, b)$ is perpendicular to this tangent line. For either of the two directions parallel to the tangent line, the directional derivative is thus zero. For a direction vector on the same side of the tangent line as the gradient vector, the directional derivative is positive. For a direction vector on the opposite side of the tangent line to the gradient vector, the directional derivative is negative. The directional derivative is maximum in the direction of the gradient vector and is minimum in the direction opposite to the gradient vector. The maximum value of the directional derivative is $\|\vec{\nabla} f(a, b)\|$, and the minimum value of the directional derivative is $-\|\vec{\nabla} f(a, b)\|$.

To generalize these ideas to functions $f : \mathbb{R}^3 \to \mathbb{R}$, we consider a plane tangent to the level surface for a given input (a, b, c). For a direction vector lying in this plane, the directional derivative is zero. For a direction vector on the same side of the tangent plane as the gradient vector, the directional derivative is positive. For a direction vector on the opposite side of the tangent plane to the gradient vector, the directional derivative is negative.

Example 18.12. For the function $f(x, y) = x^2 + 4y^2$, find a vector tangent to the level curve at the input $(-1, 3)$.

Solution. First compute the gradient field $\vec{\nabla} f(x, y) = \langle 2x, 8y \rangle$. For the input $(-1, 3)$, we have $\vec{\nabla} f(-1, 3) = \langle -2, 24 \rangle$. This gradient vector is perpendicular to the level curve. By interchanging the components and taking the negative of one component, we produce the vector $\langle 24, 2 \rangle$ that is perpendicular to the gradient vector. We confirm this by computing the dot product $\langle -2, 24 \rangle \cdot \langle 24, 2 \rangle = -48 + 48 = 0$. Since $\langle 24, 2 \rangle$ is perpendicular to the gradient vector at $(-1, 3)$, it must be tangent to the level curve in question. △

Example 18.13. For the function $f(x, y) = 3x - 2x^2y^4$, find the unit vector $\hat{u}$ pointing in the direction in which the function is increasing most rapidly at the input $(2, 2)$.

Solution. The direction of $\hat{u}$ must be the same as the direction of the gradient, which is given by

$$\vec{\nabla} f(x, y) = \langle 3 - 4xy^4, -8x^2y^3 \rangle.$$

For the input $(2, 2)$, we have

$$\vec{\nabla} f(2, 2) = \langle 3 - 4(2)(2)^4, -8(2)^2(2)^3 \rangle = \langle -125, -256 \rangle.$$

A unit vector $\hat{u}$ pointing in this direction is

$$\hat{u} = \frac{\vec{\nabla} f(2, 2)}{\|\vec{\nabla} f(2, 2)\|} = \frac{\langle -125, -256 \rangle}{\sqrt{125^2 + 256^2}} = \frac{1}{\sqrt{81161}} \langle -125, -256 \rangle.$$

Evaluating numerically results in $\hat{u} \approx \langle -0.439, -0.899 \rangle$. △

18.4 Electric Potential

18.4.1 Review of Work and Energy

The concept of electric potential depends heavily upon some of the concepts developed in earlier chapters, especially work, energy, and the relationship between force and potential energy. Therefore it would be a good idea to review the material in Chapter 7. Here we will discuss some of the highlights of that chapter that pertain to our study of electric potential.

Throughout this chapter, we will assume that frictional forces are absent, and that the electric force between isolated charges, or the force due to an electric field on an isolated charge, is a *conservative* force. (This fact has been confirmed experimentally.) When this is true, the *potential energy* function U becomes a useful tool for solving a variety of physics problems. Here are some basic facts to keep in mind regarding the potential energy:

1. The total mechanical energy $E = K + U$ (where K is the kinetic energy) is a constant. Potential energy can thus be viewed as stored energy that can be recovered and turned into kinetic energy because the total energy E remains constant.

2. The potential energy is solely a function of the position(s) of the particle(s) in the field. For example, in the case of gravitational potential energy, we saw that $U = mgy$ [Equation (7.39)] where y is the height of a mass m above the ground (or some other reference point where $U = 0$).

3. The previous point should remind you that only *differences* in potential energy are physically significant. For example, in the case of gravitational potential energy mentioned above, the more general expression is $\Delta U = mg\Delta y$; that is, the difference in potential energy is proportional to the difference in height. It is often convenient to *arbitrarily* define some position in the field (in this case, $y = 0$) to be the place at which $U \equiv 0$. It is only after such a definition has been made that we may speak of a function U, as opposed to differences ΔU.

4. It is often convenient to compute ΔU using the definition of potential energy (Definition 7.5), which tells us that $\Delta U = U_f - U_i = -$(Work done by the field).

5. The change in the potential energy of a particle when it moves from one position to another in a conservative field is independent of the path taken, from the initial position in the field to the final position.

6. For one-dimensional motion (e.g., using the coordinate x), the force on a particle in the field is given by $F = -dU/dx$ [Equation (7.42)].

18.4.2 Potential Energy and Charges

One dimension Let's begin by considering the two point charges shown in Figure 18.9. The charge q will be held fixed at the origin, and the test charge q_0 can be anywhere on the x-axis with position $x > 0$.

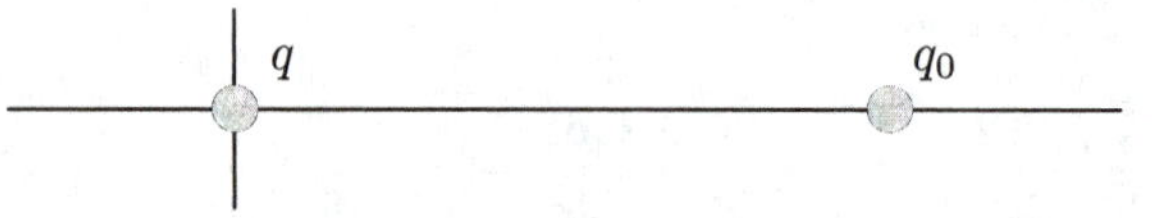

Figure 18.9. Charges q and q_0, with q fixed at the origin and q_0 on the x-axis.

We use the definition of potential energy to compute the change in potential energy when the charge q_0 moves from the position $x = r_i$ to the position $x = r_f$. According to the definition of potential energy, the change in potential energy ΔU is the *negative* of the work done on the moving charge (see Definition 7.5). Thus,

$$\Delta U = -W = -\int_C \vec{F} \cdot d\vec{s} \tag{18.4}$$

where $\vec{F}$ is the force on q_0. In Equation (18.4) we have inserted the subscript C to remind ourselves that the integration takes place over a specific path in space (recall the discussion of line integrals in Chapter 7). Recall that $d\vec{s}$ is part of the shorthand notation introduced in Chapter 7. In the full line-integral notation we would have $\vec{r}\,'(u)\,du$ where u is a variable parametrizing the curve and $\vec{r}\,'(u)$ is a tangent vector. For a path restricted to the x-axis, the line integral ultimately becomes a definite integral, with the variable of integration being x, the position of q_0 along the x-axis. This fact becomes more clear when we note that $d\vec{s} = ds\,\hat{\imath}$, since q_0 is moving along the x-axis, and by Coulomb's law

$$\vec{F} = \frac{kqq_0}{x^2}\,\hat{\imath}.$$

Inserting these expressions into Equation (18.4), we find (remember that $\hat{\imath} \cdot \hat{\imath} = 1$)

$$\Delta U = -\int_{r_i}^{r_f} \frac{kqq_0}{x^2}\,dx.$$

Evaluating the definite integral,

$$\Delta U = -kqq_0 \int_{r_i}^{r_f} \frac{1}{x^2}\,dx = kqq_0 \left.\frac{1}{x}\right|_{r_i}^{r_f}$$

or

$$\Delta U = kqq_0\left(\frac{1}{r_f} - \frac{1}{r_i}\right). \tag{18.5}$$

Before moving on, let's analyze the preliminary result [Equation (18.5)] for its physical significance. Suppose that both q and q_0 are positive. Let's choose (arbitrarily) to make $r_f > r_i$, so that $\Delta U < 0$. This makes physical sense, because the two positive charges repel each other. It takes positive work by an external agent to bring the charges closer together (like compressing a spring), so the potential energy decreases as the distance between the charges increases (like the relaxation of the compressed spring). Also note that the charge q does positive work on the test charge q_0 to move it farther away, and thus by the definition of potential energy, the corresponding change in potential energy must be negative.

What if the charge q were negative (keeping q_0 positive)? Then Equation (18.5) predicts $\Delta U > 0$ for $r_f > r_i$. The situation described in the preceding paragraph is now reversed; using the spring analogy, it is as if we stretched the spring past its equilibrium point, so that further stretching increases the potential energy. Or, from the standpoint of conservation of energy, the negative charge fixed at $x = 0$ does negative work on a positive test charge q_0 as the test charge moves a greater distance away from the origin. Since negative work is done, the change in potential energy is positive. Another useful analogy in this case is gravitation. Because the force between a positive and negative charge is attractive, this situation is similar to that of moving two masses farther apart, which we know from Chapter 15 increases their potential energy. It seems that Equation (18.5) is indeed consistent with our previously developed notions of potential energy.

Two and three dimensions Now we turn our attention to the more general cases of first two and then three dimensions. We will keep the charge q fixed at the origin. Because we are free to place the coordinate axes wherever we want in space, let's define the x-axis by the line containing q and the test charge q_0 in the "initial" configuration. As before, we begin with q_0 a distance r_i from the origin. Then the test particle is moved to some other position in the xy-plane, not necessarily on the x-axis, a distance r_f from the origin, as shown in Figure 18.10(a). We wish to find the change in potential energy ΔU associated with moving the test charge as we have just described.

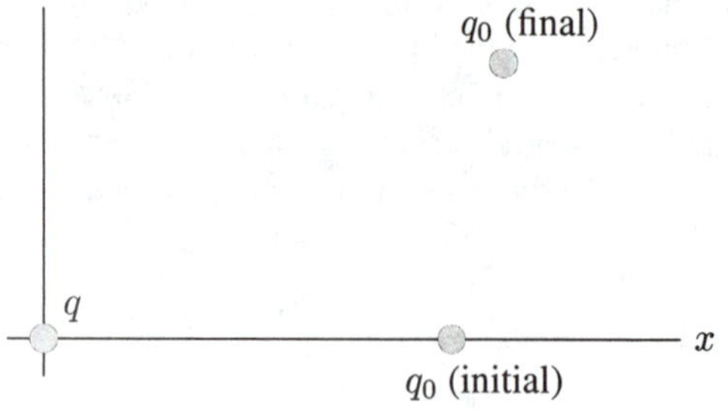

(a) Initial and final positions of the charge

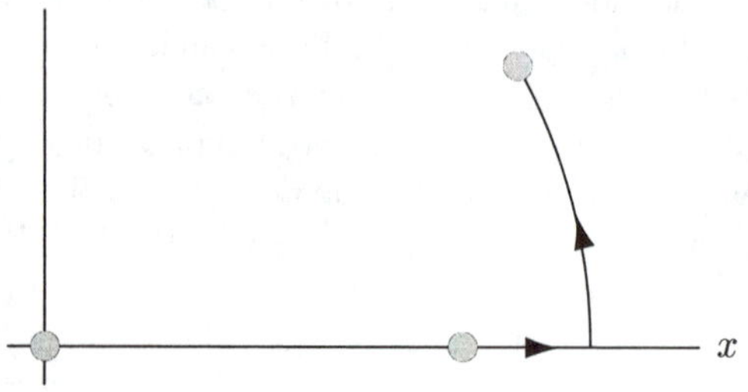

(b) Path split into two parts, one along the x-axis and the other along a circle centered at the origin

Figure 18.10. Charge q_0 is moved from its initial position to another point in the xy-plane.

Recall from Chapter 7 that the work done by a conservative force (and hence the associated change in potential energy) is independent of the path taken. Our calculation is much simpler if we take the path shown in Figure 18.10(b). First, we proceed to take the test charge from its initial point straight out along the x-axis to the position $(r_f, 0)$. The work (or change in potential energy) along this path will be exactly what was found in the one-dimensional case. Now let's complete the path to the final position by traveling along a circular arc of radius r_f centered at the origin. The force acting on the test charge is the Coulomb force, repelling q_0 radially outward from the origin. Therefore, the force vector $\vec{F}$ is perpendicular to the path of the test charge at any position. Since the force vector is perpendicular to the path everywhere along this segment, the dot product $\vec{F} \cdot d\vec{s}$ is zero, and zero work is done in moving along the arc from the position $(r_f, 0)$ on the x-axis to the final position. We conclude that the change in potential energy depends only on the radial distance of the initial and final positions from the origin, with the exact change in potential energy given again by Equation (18.5):

$$\Delta U = kqq_0\left(\frac{1}{r_f} - \frac{1}{r_i}\right). \tag{18.5}$$

In three dimensions, similar arguments can be made based on the path independence of the work and change in potential energy. Instead of a circular arc, the second part of the path (from a position $x = r_f$ on the x-axis to the final position, at any other location a distance r_f from the origin) is along the surface of a sphere of radius r_f centered at the origin. The final result is once again Equation (18.5), which we now recognize as a general result in three dimensions.

The potential energy function $U(r)$ In many cases involving potential energy, it is possible (and useful) to rewrite the *difference* in potential energy ΔU as a *function* U. Just as we did with the gravitational force between two masses in Chapter 15, we will do so here for the electrostatic force between two charges by defining the potential energy to be zero when the charges are an infinite distance apart. Let's use Equation (18.5) to compute the potential energy of two charges (q and q_0) a distance r apart in the following way: Let the initial position correspond to $r_i = r$, and then compute the difference in potential energy in the limit as r_f approaches ∞. Then

$$\Delta U = U_f - U_i = kqq_0\left(\frac{1}{r_f} - \frac{1}{r_i}\right) = \lim_{r_f \to \infty} kqq_0\left(\frac{1}{r_f} - \frac{1}{r}\right) = -kqq_0\frac{1}{r}. \quad (18.6)$$

But $U_f = 0$ by definition, and so by Equation (18.6), our desired potential energy function $U(r)$ for two charges a distance r apart is just U_i, or

$$U(r) = \frac{kqq_0}{r}. \quad (18.7)$$

The function $U(r)$ gives the potential energy of two point charges q and q_0 separated by a distance r. This formula is valid in three dimensions, where r is simply the distance between the charges as given by the distance formula

$$r = \sqrt{(\Delta x)^2 + (\Delta y)^2 + (\Delta z)^2}.$$

Recall that potential energy is a scalar function. In three dimensions, then, it is a function of the type $U : \mathbb{R}^3 \to \mathbb{R}$.

It should not be surprising that Equation (18.7) is quite similar (note the $1/r$ dependence in each case) to the gravitational potential energy function for two point masses separated by a distance r:

$$U(r) = -\frac{GMm}{r}. \quad (15.13)$$

As we know, both the gravitational force between point masses and Coulomb's force between point charges are inverse-square forces. The different constants [kqq_0 in Equation (18.7) and GMm in Equation (15.13)] follow directly from the force laws. Notice also that there is a minus sign in Equation (15.13) that is absent in Equation (18.7). This is due to the fact that the gravitational force is always attractive, and so the potential energy should decrease as the two masses get closer together. With two kinds of charge (positive and negative), the electrostatic force can be either attractive or repulsive. When q and q_0 have the same sign, Equation (18.7) yields a positive output for $U(r)$, because the potential energy should increase as the two mutually repulsive charges get closer together. But when the charges have different signs, the potential energy $U(r)$ becomes negative, as it should for an attractive force, just as for the gravitational attraction of two masses.

Example 18.14. Find the potential energy of a proton and an electron separated by a distance of (a) 1 m, or (b) 5.3×10^{-11} m, as in a hydrogen atom.

Solution. (a) Using Equation (18.7), along with the fact that the charges on the proton and neutron are $+e$ and $-e$, respectively,

$$U(r) = \frac{kqq_0}{r} = \frac{-ke^2}{r} \frac{-(8.99 \times 10^9\ \text{N}\cdot\text{m}^2/\text{C}^2)(1.60 \times 10^{-19}\ \text{C})^2}{1\ \text{m}} = -2.30 \times 10^{-28}\ \text{J}.$$

This is a small amount of energy by any measure.

(b) Using the much smaller radius, the potential energy becomes

$$U = \frac{-(8.99 \times 10^9\ \text{N}\cdot\text{m}^2/\text{C}^2)(1.60 \times 10^{-19}\ \text{C})^2}{5.3 \times 10^{-11}\ \text{m}} = -4.34 \times 10^{-18}\ \text{J}.$$

This may still seem like an incredibly small amount of energy, and by everyday (macroscopic) standards it is. However, the electrostatic potential energy contained in just one mole (Avogadro's number, or 6.02×10^{23} atoms of hydrogen) is $(6.02 \times 10^{23})(-4.34 \times 10^{-18}\text{ J}) = -2.61 \times 10^{6}$ J. Does this seem like an insignificant amount of energy? △

Example 18.15. Repeat the calculation of Example 18.9 for two protons in a nucleus, separated by a distance of about 1.5×10^{-15} m.

Solution. With two $+e$ charges, the potential energy becomes

$$U(r) = \frac{kqq_0}{r} = \frac{ke^2}{r} = \frac{(8.99 \times 10^9\text{ N·m}^2/\text{C}^2)(1.60 \times 10^{-19}\text{ C})^2}{1.5 \times 10^{-15}\text{ m}} = 1.53 \times 10^{-13}\text{ J}.$$

The absolute value of this potential energy is five orders of magnitude larger than that of the electron and proton in the hydrogen atom. This is indicative of the significant difference between chemical reactions (in which the electronic structures of atoms are rearranged, but not the nuclei) and nuclear reactions (in which the protons and neutrons in the nucleus are rearranged). Put another way, this is just about the order of magnitude difference in the energy output of an ordinary chemical explosion (e.g., dynamite) and an atomic bomb, given the same mass input! △

18.4.3 Potential Energy and Electric Potential

Recall from Chapter 17 that we first defined the electric field using the force $\vec{F}$ from Coulomb's law for two point charges, q and q_0. The electric field due to a single, isolated charge q was then defined as $\vec{E} = \vec{F}/q_0$. There exists a similar relationship between the *potential energy* of the two charges q and q_0 [as shown in Figure 18.11(a)] and a quantity known as the **electric potential**. For the two charges shown in Figure 18.11(a), we know from Equation (18.7) that the potential energy is

$$U(r) = \frac{kqq_0}{r}. \tag{18.7}$$

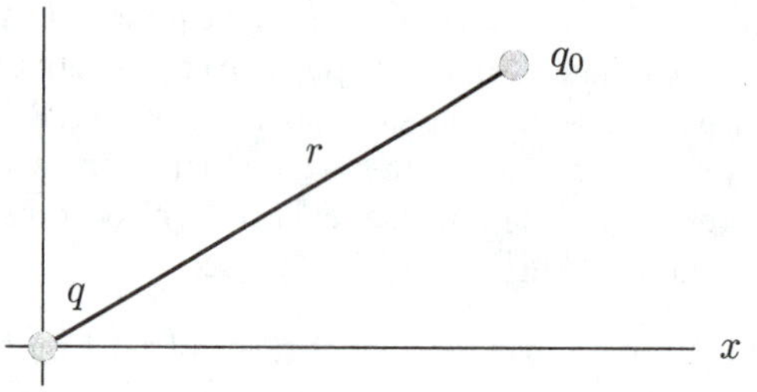

(a) Point charges q and q_0, separated by a distance r

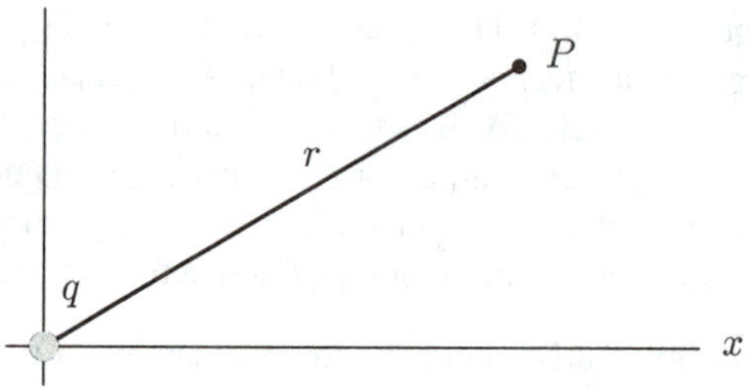

(b) Point P in space is a distance r from the charge q

Figure 18.11. Two point charges in (a) and one isolated charge in (b), used to illustrate the distinction between potential energy and electric potential.

Now consider the situation shown in Figure 18.11(b), with an isolated point charge q at the origin. We define the electric potential V at any position in space to be the potential energy given above divided by the charge q_0 on a test particle. That is,

$$V(r) = \frac{U(r)}{q_0} = \frac{kq}{r}. \tag{18.8}$$

It follows from Equation (18.8) that the electric potential is positive if the point particle's charge is positive, and negative if the charge is negative.

As noted above, in defining $V(r)$ we have followed the same procedure we used to define the electric field. This means that the relationship between V and U is analogous to the relationship between $\vec{E}$ and $\vec{F}$. Should you then think of $V(r)$ as a field? Yes, because it certainly fits the definition of a field; it is a physical quantity defined for all positions in a region of space. But there is one important difference to keep in mind: $\vec{E}$ is a vector field, and V is a scalar field. The output of the function $V(r)$ is a single number (a scalar) for any position in space given as an input, and since it is a scalar there is no direction associated with the electric potential. In the language of mathematics, the vector field $\vec{E}$ is a function of the type $\vec{E} : \mathbb{R}^3 \to \mathbb{R}^3$, while V is a function of the type $V : \mathbb{R}^3 \to \mathbb{R}$. Even though Equation (18.8) appears to be a function of single variable r, it is in fact a function of three variables (in Cartesian space x, y, and z), with r given by the distance formula $r = \sqrt{x^2 + y^2 + z^2}$. In a more general representation, we should write the electric potential explicitly as a function of three variables: $V(x, y, z)$. Or, if cylindrical or spherical coordinates are appropriate for a given geometry, we would write $V(r, \theta, z)$ or $V(\rho, \theta, \phi)$, respectively.

What is the electric potential at a position in space if the point particle with charge q is not located at the origin, but rather at an arbitrary point (x_0, y_0, z_0), as shown in Figure 18.12? The potential still depends only on q and the distance from the charge. We simply reinterpret r in this case to mean the distance from the point particle with charge q, rather than this distance from the origin. With the proper adjustment in the distance formula, Equation (18.8) becomes

$$V(x, y, z) \equiv \frac{kq}{r} = \frac{kq}{\sqrt{(x - x_0)^2 + (y - y_0)^2 + (z - z_0)^2}}.$$

You should realize that the previous expression is not a general result, because it is based upon the function $U(r)$ given by Equation (18.7), where it was assumed that the zero of potential energy was at an infinite distance away. A similar assumption underlies our results to this point regarding electric potential: These results are valid only if the electric potential is taken to be zero at an infinite distance away.

Also, recalling that in general only *differences* in potential energy are physically significant, the same should be true of differences in potential (referred to in physics as **potential differences**). We defined electric potential as the potential energy divided by the charge q_0 on a test particle. Therefore, it makes sense to define potential difference as the difference in potential energy divided by the amount of test charge. Let's find an expression for the potential difference ΔV between two arbitrary positions in space, $\vec{r}_i$ and $\vec{r}_f$. From the

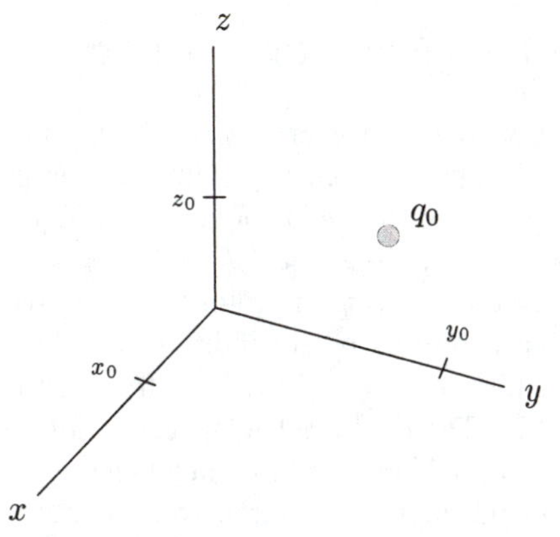

Figure 18.12. Point charge q at the position (x_0, y_0, z_0).

definition of potential difference and the definition of potential energy difference, we find

$$\Delta V = \frac{\Delta U}{q_0} = -\frac{1}{q_0}\int_C \vec{F}\cdot d\vec{s}$$

where as usual C indicates a path from $\vec{r}_i$ to $\vec{r}_f$. But by definition of the electric field, $\vec{E} = \vec{F}/q_0$, so

$$\Delta V = -\int_C \vec{E}\cdot d\vec{s}. \tag{18.9}$$

Another analogy between physical quantities should be apparent from the preceding derivation. The relationship between potential difference and electric field bears a striking resemblance to the relationship between potential energy difference and force. We can effectively compute potential differences from electric fields in the same way (apart from scalar constants) we compute potential energy differences from forces. You may wonder then whether it is possible to exploit this analogy and go in the other direction—that is, compute electric fields from electric potentials, just as you computed forces from potential energy functions (Chapter 7). The answer is that this is not only possible, but an extremely valuable tool in physics, and we study it is some detail in Section 18.5.

Units Using Equation (18.9), we can see that the units for potential difference (or electric potential) are (N/C)·m = N·m/C = J/C. Because electric potential and potential difference are so widely used, a derived unit, the **volt** (V) is defined as

1 volt = 1 joule/coulomb.

The volt is the standard SI unit for electric potential and potential difference. Notice also from Equation (18.9) that defining the volt gives us an alternate set of units for electric field, because 1 N/C = 1 V/m. You may use N/C and V/m interchangeably as the units for electric field.

The definition of the volt also serves as the basis for defining a non-SI unit of energy, the **electron-volt** (eV), which is used widely in atomic and high-energy physics. The potential energy of a test charge $q_0 = e$ (the magnitude of the charge on the electron) at an electric potential V is $U = q_0 V = eV$. When the electric potential is exactly 1 V, the potential energy is defined to be one electron-volt. Then the conversion factor between electron-volts and joules is

$$1\ \text{eV} = (1.602 \times 10^{-19}\ \text{C})(1\ \text{V}) = 1.602\ \times 10^{-19}\ \text{J}. \tag{18.10}$$

An electron-volt is a small amount of energy by everyday standards, but it is useful in atomic systems. For example, in Example 18.9 we found the potential energy of the electron-proton combination in the hydrogen atom to be -4.34×10^{-18} J. Using the conversion factor derived in Equation 18.10, this potential energy is -4.34×10^{-18} J [(1 eV)/1.602 × 10^{-19} J]= -27.1 eV.

In experimental high-energy physics, electrons or protons (or other particles carrying a charge of magnitude e) are often accelerated through significant potential differences. Suppose an electron (with charge $-e$) is accelerated through a potential difference ΔV (with $\Delta V > 0$). Then it experiences a change in potential energy $\Delta U = (-e)(\Delta V) = -e\Delta V$. The total mechanical energy of the electron remains constant in the conservative electric field, so the electron's increase in kinetic energy is $\Delta K = -\Delta U = e\Delta V$. Then it is easy to express the change in kinetic energy in electron-volts. For example, if the potential difference is $\Delta V = 10$ kV, then the change in kinetic energy is 10 keV. Modern particle accelerators give charged particles kinetic energies on the order of MeV to GeV. A few even operate in the TeV range.

Example 18.16. An electron initially at rest in an oscilloscope is accelerated through a potential difference of 42.0 V. Find the electron's kinetic energy (in both eV and J), and find the final speed of the electron.

Solution. As in the text we have $\Delta K = e\Delta V$, so

$$\Delta K = 42.0 \text{ eV} = K - K_0.$$

The initial kinetic energy is $K_0 = 0$, so $K = 42.0$ eV. Converting to joules,

$$K = 42.0 \text{ eV} \frac{1.602 \times 10^{-19} \text{ J}}{1 \text{ eV}} = 6.73 \times 10^{-18} \text{ J}.$$

From the definition of kinetic energy, $K = \frac{1}{2}mv^2$, so solving for v,

$$v = \sqrt{\frac{2K}{m}} = \sqrt{\frac{2\,(6.73 \times 10^{-18} \text{ J})}{9.11 \times 10^{-31} \text{ kg}}} = 3.84 \times 10^6 \text{ m/s}.$$ △

A word of caution: The theory of special relativity tells us that no particle can travel faster than the speed of light (about 3.00×10^8 m/s). The equation $K = \frac{1}{2}mv^2$ is not exact in special relativity. (We do not go through the details of how to revise it in this book.) The result of the previous example is far enough below the speed of light that the inaccuracy is not significant; however, you should not use $K = \frac{1}{2}mv^2$ when speeds are much greater than 1% of the speed of light.

A word on calculating electric potentials What is the electric potential function if there is more than one point charge, or some continuous distribution of charge? We will study methods for computing potentials in some detail in Section 18.6. Briefly stated, however, the general method is the principle of superposition, which we have used for electric fields. (This is reasonable, assuming the superposition principle is valid for electric fields, because we have defined the electric potential in terms of the electric field.)

For example, consider the distribution of point charges shown in Figure 18.13. In order to find the electric potential at some position P, we apply the superposition principle by adding the electric potentials of the individual point charges. Let's call V_1 the contribution to the electric potential at P due to charge q_1, V_2 the contribution due to q_2, and so on, for each of the n total charges. Thus, the potential at P is given by

$$V = V_1 + V_2 + \cdots + V_n = \frac{kq_1}{r_1} + \frac{kq_2}{r_2} + \cdots + \frac{kq_n}{r_n}. \tag{18.11}$$

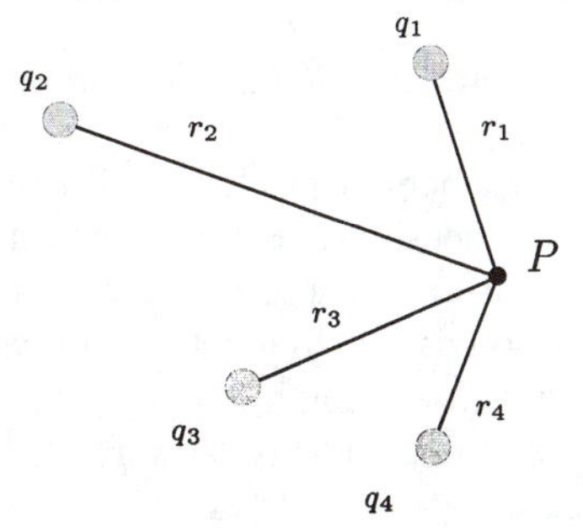

Figure 18.13. Finding the electric potential at a point P due to multiple point charges.

The scalar field $V(x, y, z)$ then consists of the electric potential computed in this manner at every position (x, y, z) in space.

Even though we are using the same superposition principle that applied to electric fields, you should appreciate that the calculation of $V(x, y, z)$ may often be *much* simpler than the calculation of $\vec{E}(x, y, z)$ for the same collection of charges. A scalar is simply one number, not a collection of numbers or a magnitude and direction. With no direction to worry about, it can be easier to compute the scalar electric potential.

18.4.4 Equipotentials

Consider a point charge q isolated in space [Figure 18.14(a)]. As we know from Equation (18.8), the electric potential at any position in space a distance r from the charge is $V = kq/r$. All positions in space that have the same output V therefore have the same r—that is, they lie on a sphere of radius r centered at the charge q. You should recognize from your study of mathematics (in Chapter 13 and earlier sections of this chapter) that each of those spheres is a *level surface* for the function $V(r)$. For the special case of electric potential, we call this level surface an **equipotential** surface, on which all positions share the same electric potential. As shown in Figure 18.14(a), for the isolated point charge the equipotential surfaces are all spheres of different radii; spheres of larger radius correspond to lower numerical values of V (assuming a positive charge q).

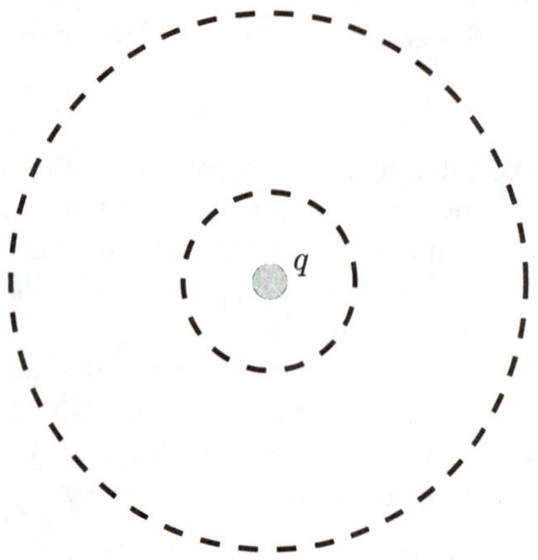

(a) Equipotential surfaces around a single point charge are spheres (shown here in cross section)

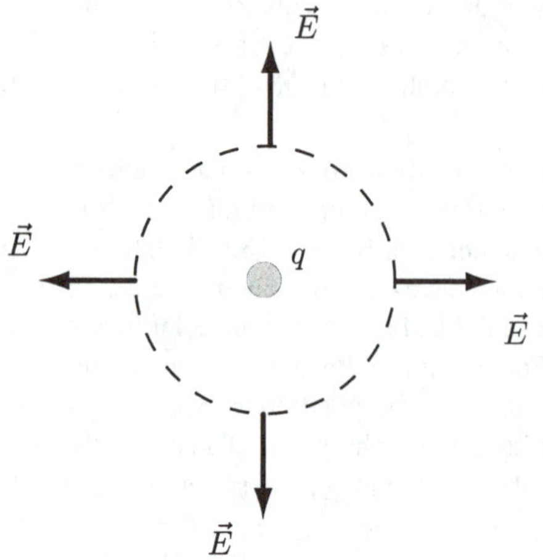

(b) The electric field is normal to the equipotential surfaces

Figure 18.14. Equipotential surfaces and electric field vectors for a point charge.

Graphing equipotential surfaces can be a helpful way to visualize the scalar field $V(x, y, z)$. The fact that we are often interested in two- and three-dimensional fields can make it somewhat more difficult to construct and visualize the equipotential surfaces. This is one area where you can make good use of advanced mathematics software on the computer. It is also possible to graph two-dimensional slices of a three-dimensional field, depending on what you are interested in viewing.

Recall that the electric field vector $\vec{E}$ points radially outward from a positive point charge (or radially inward toward a negative one). This means that the local $\vec{E}$ vector is always perpendicular to the equipotential surface as long as we are considering the $\vec{E}$ and V fields of a point charge [Figure 18.14(b)]. We now show that it is true in general that $\vec{E}$ is always perpendicular to the equipotential surface, regardless of the source of

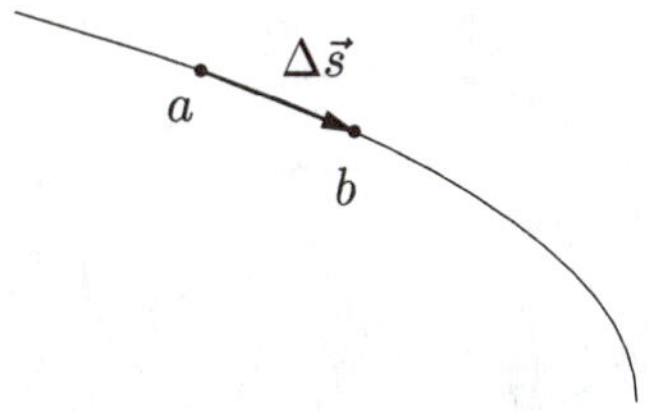

Figure 18.15. Moving from a to b on an equipotential surface along the piece of the path $\Delta\vec{s}$.

the $\vec{E}$ field. Consider the equipotential surface depicted in Figure 18.15. There exists zero potential difference between the positions labeled a and b, separated by the displacement vector $\Delta\vec{s}$, because they lie on the same equipotential. Let the positions of the positions a and b be designated $\vec{r}_a$ and $\vec{r}_b$, respectively. Then by Equation (18.9),

$$\Delta V = -\int_{\vec{r}_a}^{\vec{r}_b} \vec{E} \cdot d\vec{s} = 0.$$

The only way to guarantee that this definite integral is zero for any $\vec{r}_a$ and $\vec{r}_b$ on the equipotential is to make the vectors $\vec{E}$ and $\Delta\vec{s}$ perpendicular to each other. This must be true for any $\Delta\vec{s}$ we choose along the equipotential surface, and therefore $\vec{E}$ is perpendicular to the equipotential surface.

As another example, consider the electric field between two parallel plates, carrying charges of equal magnitudes and opposite signs. In the "infinite plate" approximation (Chapter 17) the electric field lines are straight lines, perpendicular to both plates and uniform in density, indicating a uniform electric field in this region. In order to be perpendicular to the electric field between the plates, the equipotential surfaces must be planes, parallel to each other and to the plates (Figure 18.16).

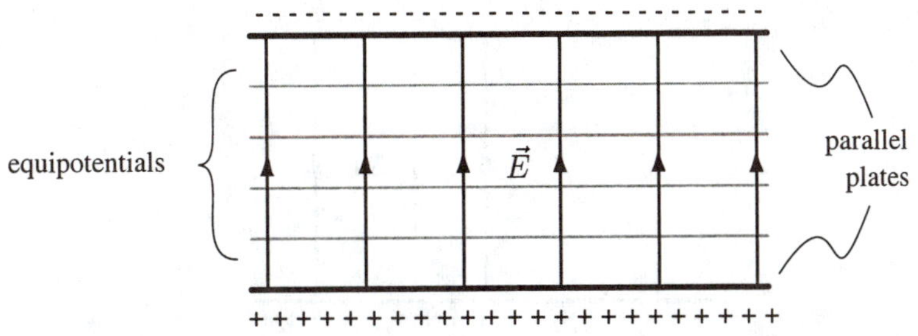

Figure 18.16. The equipotential surfaces and electric field lines for infinite parallel conducting plates carrying charges of equal magnitude and opposite sign. Equipotentials are parallel to the plates and equally spaced. Electric field lines are normal to the equipotentials.

In Figure 18.17, we show a two-dimensional slice of the electric field lines and equipotential surfaces for an electric dipole. Notice how the equipotentials curve in such a way that they are always perpendicular to the electric field lines.

Example 18.17. Consider two parallel plates separated by a distance of 1.00 mm and large enough so that we may approximate them as infinite planes. The plates carry charge densities of opposite sign and equal magnitude, 4.25×10^{-8} C/m^2. (a) Find the electric field between the plates. (b) Use your answer in (a) to find the electric potential as a function of distance from the negative plate. For convenience, let the zero of electric potential ($V = 0$) be at the negatively charged plate.

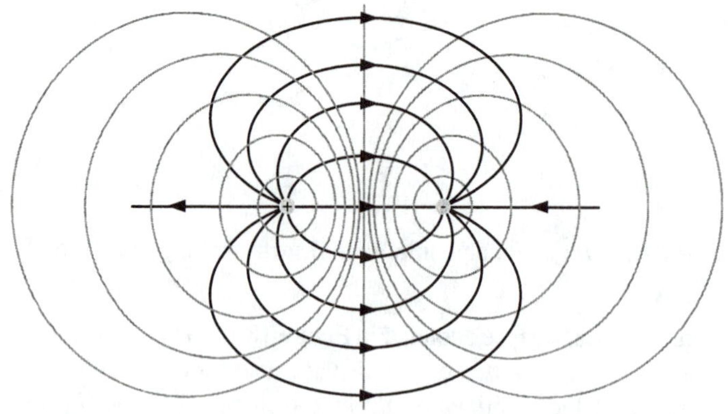

Figure 18.17. Cross sections of the equipotential surfaces (in gray) and electric field lines (in black) for an electric dipole. The electric field lines are normal to the equipotential surfaces.

Solution. (a) From Chapter 17 we know that the electric field in (this approximation) uniform, points from the positive plate to the negative plate (as in Figure 18.18), and has magnitude $E = \sigma/\epsilon_0$. Thus,

$$E = \frac{\sigma}{\epsilon_0} = \frac{4.25 \times 10^{-10}\ \text{C/m}^2}{8.85 \times 10^{-12}\ \text{C}^2{\cdot}\text{N}^{-1}{\cdot}\text{m}^{-2}} = 4.80 \times 10^3\ \text{N/C}.$$

(b) To find the electric potential an arbitrary distance x from the negative plate, we use Equation (18.9), taking our path of integration to be directly from the negative plate to the positive one (Figure 18.18).

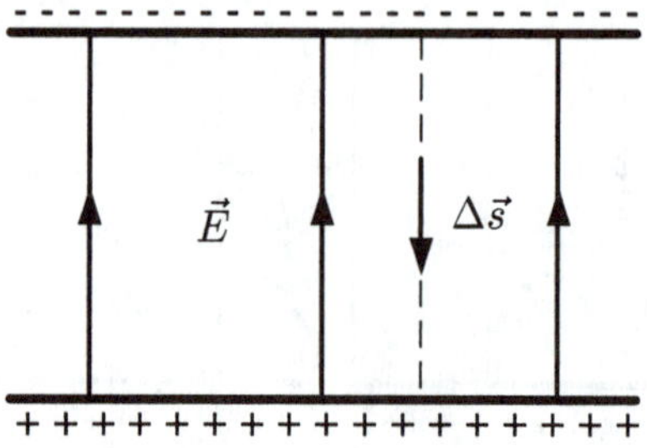

Figure 18.18. The path of integration from one parallel plate to the other. Notice that $\Delta\vec{s}$ is in the same direction as $\vec{E}$.

We evaluate the dot product $\vec{E}\cdot\Delta\vec{s}$ by observing that the vectors $\vec{E}$ and $\Delta\vec{s}$ are in opposite directions, so that the angle between them is π. Thus, $\vec{E}\cdot d\vec{s} = E\,ds\cos\pi = -E\,ds$. Now we evaluate using Equation (18.9), and using the fact that $V(x=0) = 0:$

$$\Delta V = V(x) - V(0) = V(x) = -\int_0^x \vec{E}\cdot d\vec{s} = -\int_0^x -E\,ds = \int_0^x E\,ds.$$

Because E is constant, the definite integral is easily evaluated as

$$V(x) = E\int_0^x ds = Ex$$

where the (constant) value of E was found in part (a).

We can make some observations based on this result. The electric potential varies *linearly* with distance from the negatively charged plate. The equipotential planes are uniformly spaced for equal potential differences ΔV. The electric potential at the positive plate is $V(x = d) = Ed = (4.80 \times 10^3$ N/C)(1 mm) = 4.8 J/C = 4.8 V. △

18.5 Electric Potential and Electric Fields

18.5.1 The One-Dimensional Case

In Chapter 7, we found that in a conservative force field $\vec{F}$ the difference in potential energy ΔU between positions $\vec{r}_a$ and $\vec{r}_b$ joined by a curve C in the field could be calculated using the relation

$$\Delta U = -\int_C \vec{F} \cdot d\vec{s}$$

if the function $\vec{F}$ is known. In many circumstances, we found it useful to use the inverse relation—that is, to find the force if the potential energy function U is known. In one dimension (x), this relation is

$$F_x = -\frac{dU}{dx}. \tag{7.42}$$

Knowing the potential energy function U thus allows us to understand a lot, both qualitatively and quantitatively, about the force on a particle and its subsequent motion.

Because of the similar relationship between electric potential and electric field,

$$\Delta V = -\int_C \vec{E} \cdot d\vec{s}, \tag{18.9}$$

it should be possible to derive the electric field from an electric potential function V in the same way the force can be derived from the potential energy function U. By analogy, we would expect the result for one dimension to be

$$E_x = -\frac{dV}{dx}. \tag{18.12}$$

Let's derive the relation in Equation (18.12) carefully before moving on to two and three dimensions. Consider the electric field $\vec{E} = E_x\,\hat{\imath}$ as shown in Figure 18.19. We evaluate the integral in Equation (18.9) along a path parallel to the x-axis (so that $\vec{E}$ is parallel to $d\vec{s}$ and $\vec{E} \cdot d\vec{s} = E_x\,dx$). Let the starting position ($x = a$) be the place where we define the potential function $V(x)$ to be zero. The endpoint of or integration path is an arbitrary point x. Then by Equation 18.9

$$\Delta V = V(x) - V(a) = V(x) = -\int_C \vec{E} \cdot d\vec{s} = -\int_a^x E_x\,dx$$

or

$$V(x) = -\int_a^x E_x\,dx.$$

You should recognize that the first fundamental theorem of calculus can be applied to this equation to yield

$$E_x = -\frac{dV}{dx} \tag{18.12}$$

as desired.

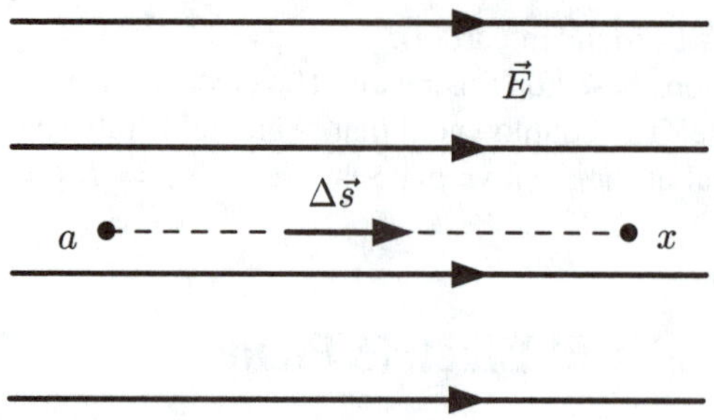

Figure 18.19. Path of integration in the same direction as $\vec{E}$ along a line from a to x.

18.5.2 Two and Three Dimensions

Let's take a path from the position (x, y) on one equipotential surface to the position $(x + \Delta x, y + \Delta y)$ on a neighboring equipotential surface (Figure 18.20). In two dimensions, the path $\Delta\vec{s}$ in Equation (18.9) is $\Delta\vec{s} = \Delta x\,\hat{\imath} + \Delta y\,\hat{\jmath}$. Using the linear approximation from Section 18.2 [Equation (18.1)], the potentials on the two equipotentials are related by

$$V(x + \Delta x, y + \Delta y) \approx V(x, y) + \frac{\partial V}{\partial x}[(x + \Delta x) - x] + \frac{\partial V}{\partial y}[(y + \Delta y) - y].$$

Rearranging and simplifying gives us the potential difference

$$\Delta V = V(x + \Delta x, y + \Delta y) - V(x, y) \approx \frac{\partial V}{\partial x}\Delta x + \frac{\partial V}{\partial y}\Delta y. \tag{18.13}$$

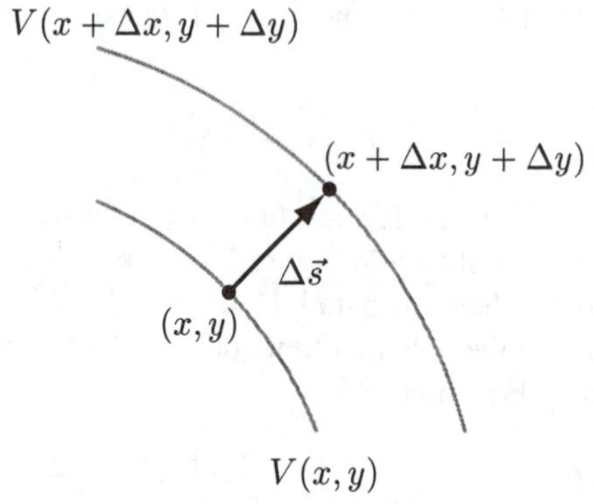

Figure 18.20. Path going from the point (x, y) on one equipotential surface to the point $(x + \Delta x, y + \Delta y)$ on a different equipotential surface.

Now let's evaluate Equation (18.9) over the same path in two dimensions. On the left side, $\Delta V = V(x + \Delta x, y + \Delta y) - V(x, y)$. On the right side, we approximate the integral with one term:

$$-\int_C \vec{E} \cdot d\vec{s} \approx -\vec{E} \cdot \Delta\vec{s} = -(E_x\,\hat{\imath} + E_y\,\hat{\jmath}) \cdot (\Delta x\,\hat{\imath} + \Delta y\,\hat{\jmath}) = -E_x\,\Delta x - E_y\,\Delta y. \tag{18.14}$$

The expressions in Equations (18.13) and (18.14) should both equal the same ΔV. By comparing the Δx and Δy terms in Equations (18.13) and (18.14), it is clear that $E_x = -\partial V/\partial x$ and $E_y = -\partial V/\partial y$. We have confidence in these results, because they reduce to the correct result in the one-dimensional case (where there is no distinction between the total and partial derivative) given by Equation (18.12), and because the physical units are correct (electric field in V/m). This argument can be extended to three dimensions by adding the third dimension to the path $\Delta\vec{s}$, which adds a Δz term to Equations (18.13) and (18.14). Therefore, in three dimensions,

$$E_x = -\frac{\partial V}{\partial x}, \qquad E_y = -\frac{\partial V}{\partial y}, \qquad \text{and} \qquad E_z = -\frac{\partial V}{\partial z}. \tag{18.15}$$

This is the result we have sought. The three components of the electric field are computed by taking the (negative of the) three partial derivatives of the potential function $V(x, y, z)$. This gives us a new strategy to employ for calculating electric fields: *For a given charge distribution, compute the potential $V(x, y, z)$ first and then take the partial derivatives as prescribed by Equations (18.15).* You might argue that since this involves an extra step, why not compute the electric field directly as we did in Chapter 17? The answer to this question relies on the argument we made at the end of Section 18.4: Because the electric potential is a *scalar* function, it may be much easier to compute than the (vector) electric field. We will see some examples in Section 18.6.

Example 18.18. In a given region of space, the electric potential is given by $V(x, y, z) = -3xyz^2 + 4x^3$. (Assume that the units are such that the electric potential is in volts.) Find the electric field in that region.

Solution. According to Equations (18.15) the components of the electric field are

$$E_x = -\frac{\partial V}{\partial x} = -(-3yz^2 + 12x^2) = 3yz^2 - 12x^2,$$

$$E_y = -\frac{\partial V}{\partial y} = -(-3xz^2) = 3xz^2,$$

and

$$E_z = -\frac{\partial V}{\partial z} = -(-6xyz) = 6xyz.$$

You should verify that the correct units for electric field follow from these results. △

From Section 18.3, we know that the three components of the electric field in Equations 18.15 are the negatives of the three components of the *gradient* of the electric potential. Thus, we may collect the three equations into a single gradient and rewrite the electric field as

$$\vec{E} = -\vec{\nabla} V = -\frac{\partial V}{dx}\hat{\imath} - \frac{\partial V}{dy}\hat{\jmath} - \frac{\partial V}{dx}\hat{k}. \tag{18.16}$$

Equation (18.16) is valid for one, two, or three dimensions.

From our previous geometric analysis of the gradient, we know that the gradient vector points "up the hill" from lower to higher outputs of the scalar function (in this case, V). Since the electric field [Equation (18.16)] is the *negative* of the gradient of V, the direction of the electric field vector is "down the hill,"—that is, from locations with *higher* to *lower* electric potential V.

18.6 Calculating Electric Potentials

The formula for the electric potential of a point charge $V = kq/r$ is quite similar in form to the formula for the magnitude of the electric field of a point charge $E = kq/r^2$. Therefore, the methods used to compute potentials (both analytic and numerical) will be similar to those we introduced in Chapter 17 for computing electric fields.

18.6.1 Analytic Methods

Point charges As we pointed out briefly in Section 18.4.3, the principle of superposition may be applied in computing the electric potential of a number of point charges, with the result being

$$V = V_1 + V_2 + \cdots + V_n = \frac{kq_1}{r_1} + \frac{kq_2}{r_2} + \cdots + \frac{kq_n}{r_n} = k\sum_i^n \frac{q_i}{r_i}. \qquad (18.11)$$

This procedure is straightforward, but it can be tedious for a large number of point charges. You should also keep in mind that the goal of computing V in Equation (18.11) is usually not to obtain a *number*—that is, a single output to the function for a given position (or input) (x, y, z)—but rather to find the *function* $V(x, y, z)$, so that the gradient of this function can be used to find the corresponding electric field. This is demonstrated in the following example.

Example 18.19. An electric dipole consists of charges $\pm q$ located at $x = \pm a$ on the x-axis, as shown in Figure 18.21. Find the electric potential $V(x, y)$ in the xy-plane, and use this result to find the electric field $\vec{E}(x, y)$ in the xy-plane. Then find the electric field at $(0, a)$ and $(2a, 0)$ and comment on the results.

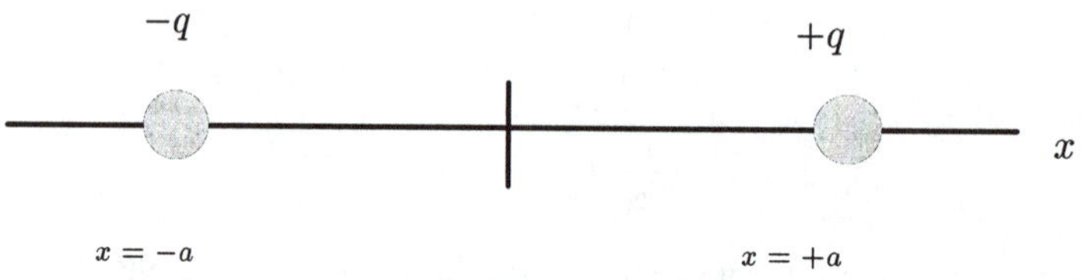

Figure 18.21. Electric dipole.

Solution. From Equation (18.11), the potential is

$$V(x, y) = \frac{kq_1}{r_1} + \frac{kq_2}{r_2} = kq\left(-\frac{1}{r_1} + \frac{1}{r_2}\right).$$

Using the distance formula for r_1 and r_2, we can write V in terms of the Cartesian variables x and y:

$$V(x, y) = kq\left[-\frac{1}{\sqrt{(x+a)^2 + y^2}} + \frac{1}{\sqrt{(x-a)^2 + y^2}}\right].$$

Now the electric field components are given by the appropriate partial derivatives:

$$E_x = -kq\,\frac{\partial}{\partial x}\left[-\frac{1}{\sqrt{[(x+a)^2+y^2]}}+\frac{1}{\sqrt{(x-a)^2+y^2}}\right]$$
$$= kq\left[-\frac{x+a}{[(x+a)^2+y^2]^{3/2}}+\frac{x-a}{[(x-a)^2+y^2]^{3/2}}\right]$$

and

$$E_y = -kq\,\frac{\partial}{\partial y}\left[-\frac{1}{\sqrt{[(x+a)^2+y^2]}}+\frac{1}{\sqrt{(x-a)^2+y^2}}\right]$$
$$= kq\left[-\frac{y}{[(x+a)^2+y^2]^{3/2}}+\frac{y}{[(x-a)^2+y^2]^{3/2}}\right]$$

Evaluating the electric field components at $(0, a)$:

$$E_x = kq\left[-\frac{a}{[a^2+a^2]^{3/2}}+\frac{0-a}{[(-a)^2+a^2]^{3/2}}\right] = -\frac{2kqa}{[2a^2]^{3/2}} = -\frac{kq}{\sqrt{2}a^2}$$

and

$$E_y = -\frac{kqa}{[a^2+a^2]^{3/2}}+\frac{kqa}{[(-a)^2+a^2]^{3/2}} = 0.$$

We conclude that the electric field at a point directly above the midpoint of the dipole points in the $-x$-direction, which is quite consistent with the electric field map of the electric dipole from Chapter 17. You can easily verify that this expression for E_y is also consistent with the result in that chapter.

Evaluating the electric field components at $(2a, 0)$:

$$E_x = kq\left[-\frac{2a+a}{[(2a+a)^2]^{3/2}}+\frac{2a-a}{[(2a-a)^2]^{3/2}}\right] = kq\left[-\frac{3a}{(3a)^3}+\frac{a}{a^3}\right] = kq\left[-\frac{1}{9a^2}+\frac{1}{a^2}\right] = \frac{8kq}{9a^2}$$

and

$$E_y = kq\left[-\frac{0}{[(2a+a)^2]^{3/2}}+\frac{0}{[(2a-a)^2]^{3/2}}\right] = 0.$$

The electric field at the point $(2a, 0)$ points in the $+x$-direction, again consistent with our electric dipole field map and the worked example from Chapter 17. △

This example has demonstrated the effectiveness of using the electric potential to compute the electric field. While it is possible in theory to compute the electric field components directly, it would have been much more difficult then the method we just employed: Compute the scalar electric potential first, and then use the appropriate partial derivatives to find the electric field components.

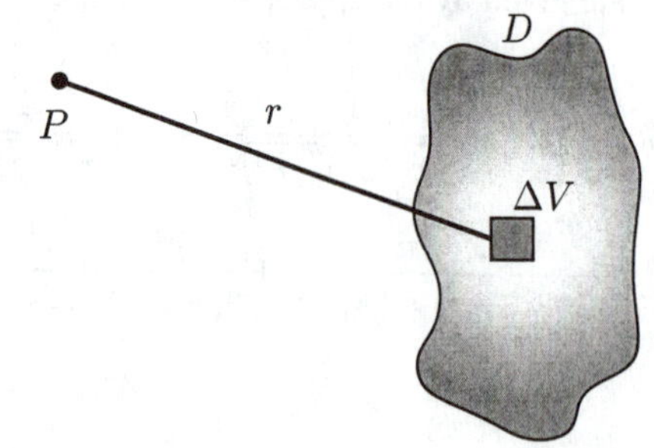

Figure 18.22. Continuous distribution of charge throughout a region D. The charge is broken up into elements Δq, with Δq a distance r away from the point P, where we wish to find the electric potential.

Continuous distributions of charge For a continuous distribution of charge that occupies a region R in space (Figure 18.22), the procedure is in principle the same as for a large number of point charges. Following the procedure we use in calculus, we break the continuous charge distribution into infinitesimal elements dV. Then we treat the sum in Equation (18.11)

$$V = k \sum_i \frac{q_i}{r_i}$$

as a Riemann sum, which becomes a definite integral in the limit in which the discrete charges q_i become charges of infinitesimal size $dq = \rho(x, y, z)\, dV$, where $\rho(x, y, z)$ is the charge density. In that limit the electric potential becomes [see Equation (17.11)]

$$V = k \iiint_D \frac{\rho(x, y, z)}{r}\, dV. \tag{18.17}$$

In Equation (18.17), the D subscript reminds us that this is a triple integral, with the limits of integration depending on the extent of the charge distribution for a given problem. The variable r without a subscript represents the distance from the place at which the potential V is being computed to a given charge element Δq.

Once again, this is very similar to the way in which electric fields were computed directly for a finite charge distribution:

$$\vec{E} = k \iiint_D \frac{\rho(x, y, z)\,\hat{r}}{r^2}\, dV. \tag{17.11}$$

Notice the characteristic difference that the expression for $\vec{E}$ contains the factor $1/r^2$ in the integrand, while the expression for V contains $1/r$. But the significant difference in practical terms is once again that the potential is a scalar quantity, and therefore it is easier to compute in many cases. The analogous expression for two-dimensional charge distributions with charge density $\sigma(x, y)$ over a surface A is

$$V = k \iint_R \frac{\sigma(x, y)}{r}\, dA \tag{18.18}$$

and for one-dimensional charge distributions with charge density $\lambda(x)$ over the interval $[x_a, x_b]$ is

$$V = k \int_{x_a}^{x_b} \frac{\lambda(x)}{r}\, dx. \tag{18.19}$$

Example 18.20. A uniformly charged ring of radius R and total charge Q lies in the xy-plane with its center at the origin. Find the electric potential and electric field everywhere on the z-axis.

Solution. Using Equation (18.19), we need to break the ring into infinitesimal segments of charge ds, as shown in Figure 18.23. For any ds in the ring, the point P is the same distance r away from it; therefore, for this geometry, r is a constant and may be factored outside the definite integral to give

$$V = k \int \frac{\lambda}{r}\, ds = \frac{k\lambda}{r} \int ds.$$

Since the circumference of the ring is $2\pi R$, we know that $\int ds = 2\pi R$. But with a constant charge density $\lambda = Q/(2\pi R)$. Thus,

$$V = \frac{kQ}{r}.$$

Also, since we want to know the potential at any point on the z-axis, we should express our answer as a function of z. (This is important at the next step, when we take the gradient of the electric potential to find the electric field, because the gradient is expressed in terms of the Cartesian coordinates x, y, and z.) From the geometry (Figure 18.23) of the right triangle, $r = \sqrt{R^2 + z^2}$. Combining these results gives

$$V = \frac{kQ}{\sqrt{R^2 + z^2}}.$$

Now to find the electric field everywhere on the z-axis, we note that by symmetry only the z-component of the gradient is nonzero, so

$$\vec{E} = -\vec{\nabla} V = -\frac{\partial V}{\partial z}\,\hat{k} = \frac{kQz}{(R^2 + z^2)^{3/2}}\,\hat{k}.$$

This is the same result as can be obtained for this charge distribution using direct integration as in Chapter 17, but with considerably less effort here. △

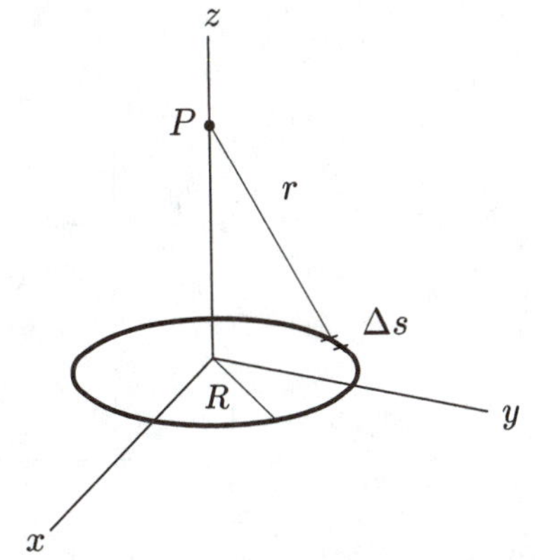

Figure 18.23. Uniformly charged ring lying in the xy-plane, centered at the origin. The point P on the z-axis is the same distance r from any point on the ring.

Example 18.21. Find the electric potential at all positions inside and outside of a solid conducting sphere of radius R that carries a net charge q.

Solution. Because of the spherical symmetry, the electric potential should be a function of r only, where r is the distance from the center of the sphere. We know from the properties of conductors that the charge q is distributed uniformly over the surface of the sphere. In this case, it would be difficult to set up and compute the definite integral in Equation (18.17) (try it!). In such cases, it is wise to take advantage of the fact that we already know the electric field everywhere in space and use Equation (18.9) to find the potential.

For convenience, let's choose a path along the $+x$-axis, with $x = 0$ at the center of the sphere. At an arbitrary position x outside the sphere ($x > R$), the electric field (for a positive charge Q) can be expressed $\vec{E} = (kq/x^2)\hat{r}$, where $\hat{r}$ is now a unit vector pointing radially outward from the origin. Equation (18.9) requires us to integrate over a path to find a potential difference. Let's choose as our starting position $x = \infty$, where the potential is defined to be zero, and the ending position some other point r from the origin, where we wish to know the potential [Figure 18.24(a)]. Then let the path of integration be radially inward, so that the path vector $d\vec{s} = dx\,\hat{r}$. Then $\vec{E} \cdot d\vec{s} = kq/x^2\,dx$, and

$$\Delta V = V(r) - V(\infty) = V(r) = -\int_{\infty}^{r} \frac{kq}{x^2}\,dx$$

so

$$V(r) = -kq\int_{\infty}^{r} \frac{1}{x^2}\,dx = kq\left.\frac{1}{x}\right|_{\infty}^{r} = \frac{kq}{r}.$$

Not surprisingly, the electric potential outside of a spherically symmetric distribution of charge is the same as if all the charge were concentrated at the center.

It is possible to use the same method to find the electric potential inside the sphere. Let's follow the integration path radially inward, this time beginning at $x = R$ and ending at some arbitrary $r < R$. Inside the conducting sphere, the electric field is zero, so $\vec{E} \cdot \Delta\vec{s} = 0$ for any path inside the sphere, and

$$\Delta V = V(r) - V(R) = -\int \vec{E} \cdot d\vec{s} = 0$$

so

$$V(r) = V(R) = \frac{kq}{R} = \text{constant}.$$

A graph of the results for $r > R$ and $r < R$ is shown in Figure 18.24(b). It is left as an exercise to verify that the correct electric fields can be derived from these functions $V(r)$ using $\vec{E} = -\vec{\nabla}V$. △

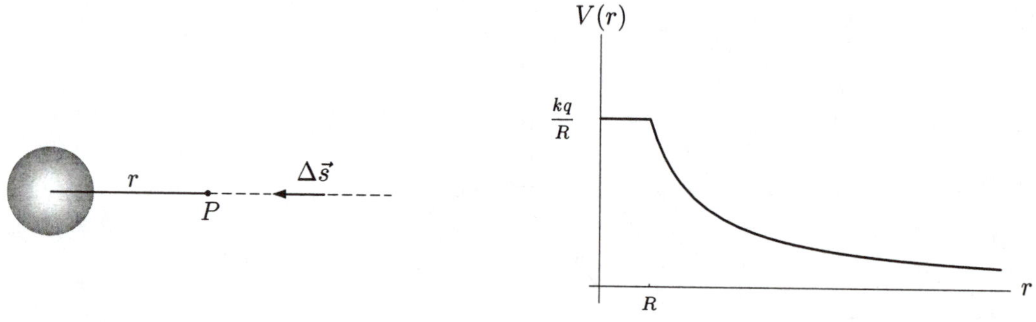

(a) Path to compute V lies on a straight line that goes from infinity to a distance r from the center of sphere

(b) Electric potential V as a function of distance r

Figure 18.24. The electric potential V due to a charged conducting sphere of radius R.

18.6.2 Numerical Methods

For some charge distributions, it is not practical to compute the electric potential analytically for every position in space. For example, consider again the charged ring of Example 18.15. How would we compute the electric potential $V(x, y, z)$ for points not on the z-axis? Without the symmetry that exists along the z-axis, setting up the definite integral in Equation (18.17) would be a formidable task indeed.

One approach to the charged ring problem is to use a calculator or computer to estimate the output of the potential at some desired input (x_0, y_0, z_0). This can be done (Figure 18.25) by conceptually breaking the ring into finite-sized pieces having charge q_i, with each charge q_i a different distance r_i from (x_0, y_0, z_0). It is fairly straightforward to program a computer to compute the distance r_i for each q_i and use the results in the sum

$$V = k \sum_i \frac{q_i}{r_i}$$

in Equation (18.11). The sum becomes a Riemann sum that approximates the exact value of V that would be found by integration, with the accuracy of the approximation dependent on the number of elements q_i in the sum. A different Riemann sum will have to be constructed for each input (x_0, y_0, z_0) of interest; but if this is done for enough inputs, a reasonable map that approximates the function $V(x, y, z)$ can be constructed for any region of interest.

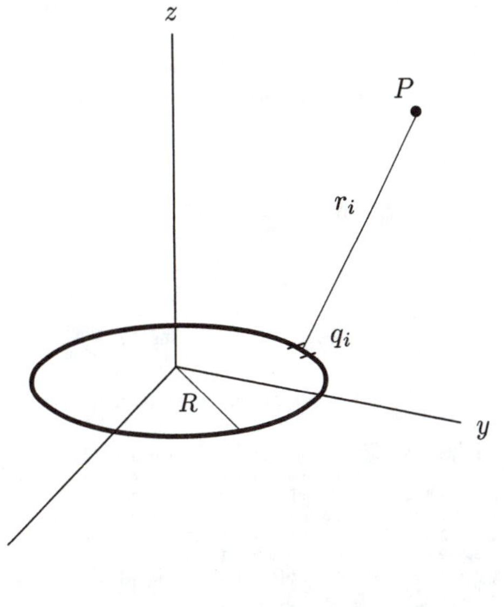

Figure 18.25. Uniformly charged ring lying in the xy-plane, centered at the origin. To find the electric potential at the point P, the ring is broken into finite charge elements q_i.

Another numerical method that does not rely on Riemann sums is illustrated in the following example. Consider two *finite* parallel conducting line segments that lie in the xy-plane as shown in Figure 18.26. Suppose these conducting line segments have equal and opposite charges placed on them. Each conductor then forms an equipotential. If the electric potential on each segment is known, we can use a spreadsheet to construct a map of the electric potentials throughout the xy-plane.

First, we simulate the charged segments by fixing finite segments of two parallel rows of the spreadsheet at the desired potentials (say 0 V and 10 V) as shown in Figure 18.27(a). We also fix the boundary of the

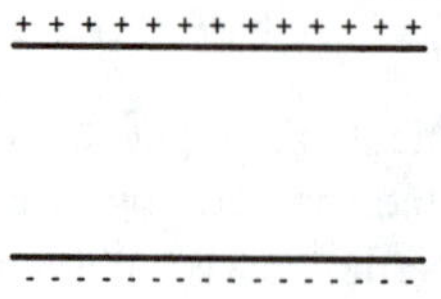

Figure 18.26. Charges of equal magnitudes and opposite signs on finite parallel conducting lines.

region in which we compute with the average of the plate potentials. The remaining spreadsheet cells will be defined as the *average* of the four neighboring cells. This method is based on the mathematical result that the electric potential at any point is the average of the potential on any closed curve around that point. (This type of result is studied in a class on *partial differential equations* and is true for solutions of a partial differential equation called *Laplace's equation*.)

To obtain accurate results, it will be necessary to proceed iteratively—that is, repeat the averaging many times in order to smooth out the variations in the potential map. You can do this on most spreadsheet programs, and it is possible to get a good approximation to the potential in the xy-plane in a short time. A sample of the results is shown in Figure 18.27(b).

5.0	5.0	5.0	5.0	5.0	5.0	5.0	5.0	5.0	5.0	5.0	5.0	5.0	5.0	5.0	5.0	5.0	5.0	5.0	5.0	5.0
5.0																				5.0
5.0																				5.0
5.0																				5.0
5.0																				5.0
5.0																				5.0
5.0					0.0	0.0	0.0	0.0	0.0	0.0	0.0	0.0	0.0	0.0	0.0					5.0
5.0																				5.0
5.0																				5.0
5.0																				5.0
5.0																				5.0
5.0																				5.0
5.0					10.	10.	10.	10.	10.	10.	10.	10.	10.	10.	10.					5.0
5.0																				5.0
5.0																				5.0
5.0																				5.0
5.0																				5.0
5.0																				5.0
5.0	5.0	5.0	5.0	5.0	5.0	5.0	5.0	5.0	5.0	5.0	5.0	5.0	5.0	5.0	5.0	5.0	5.0	5.0	5.0	5.0

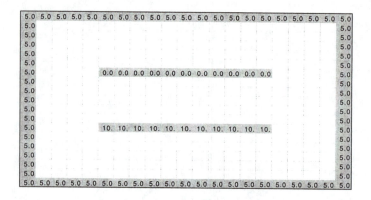

(a) Initial spreadsheet configuration

5.0	5.0	5.0	5.0	5.0	5.0	5.0	5.0	5.0	5.0	5.0	5.0	5.0	5.0	5.0	5.0	5.0	5.0	5.0	5.0	5.0
5.0	4.9	4.8	4.7	4.5	4.4	4.4	4.3	4.3	4.2	4.2	4.2	4.3	4.3	4.4	4.4	4.5	4.7	4.8	4.9	5.0
5.0	4.8	4.6	4.3	4.1	3.9	3.7	3.6	3.5	3.4	3.4	3.4	3.5	3.6	3.7	3.9	4.1	4.3	4.6	4.8	5.0
5.0	4.7	4.3	4.0	3.6	3.2	3.0	2.8	2.7	2.6	2.6	2.6	2.7	2.8	3.0	3.2	3.6	4.0	4.3	4.7	5.0
5.0	4.6	4.2	3.7	3.1	2.5	2.2	2.0	1.8	1.8	1.8	1.8	1.8	2.0	2.2	2.5	3.1	3.7	4.2	4.6	5.0
5.0	4.6	4.1	3.4	2.6	1.6	1.2	1.0	0.9	0.9	0.9	0.9	0.9	1.0	1.2	1.6	2.6	3.4	4.0	4.6	5.0
5.0	4.6	4.1	3.4	2.2	0.0	0.0	0.0	0.0	0.0	0.0	0.0	0.0	0.0	0.0	0.0	2.2	3.4	4.1	4.6	5.0
5.0	4.7	4.3	3.8	3.0	2.1	1.8	1.7	1.7	1.7	1.7	1.7	1.7	1.7	1.8	2.1	3.0	3.8	4.3	4.7	5.0
5.0	4.8	4.6	4.3	4.0	3.6	3.5	3.4	3.4	3.3	3.3	3.3	3.4	3.4	3.5	3.6	4.0	4.3	4.6	4.8	5.0
5.0	5.0	5.0	5.0	5.0	5.0	5.0	5.0	5.0	5.0	5.0	5.0	5.0	5.0	5.0	5.0	5.0	5.0	5.0	5.0	5.0
5.0	5.2	5.4	5.7	6.0	6.3	6.5	6.6	6.6	6.7	6.7	6.7	6.6	6.6	6.5	6.3	6.0	5.7	5.4	5.2	5.0
5.0	5.3	5.7	6.2	7.0	7.9	8.2	8.3	8.3	8.3	8.3	8.3	8.3	8.3	8.2	7.9	7.0	6.2	5.7	5.3	5.0
5.0	5.4	5.9	6.6	7.8	10.	10.	10.	10.	10.	10.	10.	10.	10.	10.	10.	7.8	6.6	5.9	5.4	5.0
5.0	5.4	5.9	6.6	7.4	8.4	8.8	9.0	9.1	9.1	9.1	9.1	9.1	9.0	8.8	8.4	7.4	6.6	5.9	5.4	5.0
5.0	5.4	5.8	6.3	6.9	7.5	7.8	8.0	8.2	8.2	8.2	8.2	8.2	8.0	7.8	7.5	6.9	6.3	5.8	5.4	5.0
5.0	5.3	5.7	6.0	6.4	6.8	7.0	7.2	7.3	7.4	7.4	7.4	7.3	7.2	7.0	6.8	6.4	6.0	5.7	5.3	5.0
5.0	5.2	5.4	5.7	5.9	6.1	6.3	6.4	6.5	6.5	6.6	6.5	6.5	6.4	6.3	6.1	5.9	5.7	5.4	5.2	5.0
5.0	5.1	5.2	5.3	5.5	5.6	5.6	5.7	5.7	5.8	5.8	5.8	5.7	5.7	5.6	5.6	5.5	5.3	5.2	5.1	5.0
5.0	5.0	5.0	5.0	5.0	5.0	5.0	5.0	5.0	5.0	5.0	5.0	5.0	5.0	5.0	5.0	5.0	5.0	5.0	5.0	5.0

(b) Final results after multiple iterations on the spreadsheet

Figure 18.27. Numerical approximation of an electric potential by averaging.

With this type of output, we can visualize (or draw) equipotential lines throughout the plane (Figure 18.28). Because electric field lines must be perpendicular to the equipotentials, it is then possible to fill in the electric field lines. Notice that near the center of the region between the conducting segments, the electric field is approximately uniform, as in the "infinite plate" approximation for parallel plates in three dimensions. Near the edges of the segments, the field lines tend to bow outward, and the field is no longer uniform.

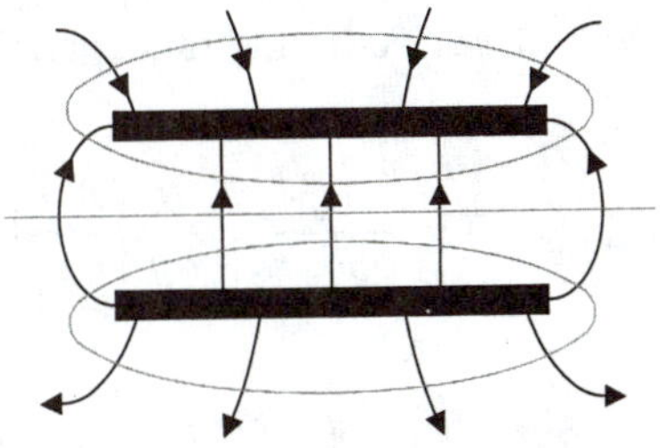

Figure 18.28. Equipotentials (in gray) and electric field lines (in black) drawn using the spreadsheet results. Electric field lines are drawn so that they are perpendicular to the equipotentials.

18.7 Higher-Order Partial Derivatives

With ordinary derivatives, we are accustomed to differentiating successively to find *higher-order derivatives.* For example, the second derivative $f''(x)$ (or d^2f/dx^2) is found by differentiating $f'(x)$ with respect to x, $f'''(x)$ (or d^3f/dx^3) is found by differentiating $f''(x)$ with respect to x, and so on. In this section, we consider higher-order partial derivatives. Since a partial derivative of $f : \mathbb{R}^n \to \mathbb{R}$ may be computed with respect to any of the n independent variables, we must carefully define which variable is being used for each stage of differentiation.

18.7.1 Definitions

Successive partial derivatives with respect to the same variable (e.g., x) are written as

$$f_{xx}(x,y) = \frac{\partial^2 f}{\partial x^2} = \frac{\partial}{\partial x}\left[\frac{\partial f}{\partial x}\right], \qquad f_{xxx}(x,y) = \frac{\partial^3 f}{\partial x^3}\frac{\partial}{\partial x}\left[\frac{\partial^2 f}{\partial x^2}\right], \qquad \ldots .$$

When you calculate these partial derivatives, simply follow the rules for taking partial derivatives at each stage.

What is new with partial derivatives is the possibility of a **mixed partial derivative**, in which differentiation with respect to one variable is followed by differentiation with respect to a different variable. Suppose we take the partial derivative of a function $f(x, y)$ with respect to x, and then take a partial derivative of the result with respect to y. The notation used to describe this process is

$$f_{xy}(x,y) = \frac{\partial^2 f}{\partial y \partial x} = \frac{\partial}{\partial y}\left[\frac{\partial f}{\partial x}\right].$$

Notice that in the notation $\partial^2 f/\partial y \partial x$ the order in which the partial derivatives are taken is indicated by the order of the symbols ∂x and ∂y and proceeds from right to left, just as in the explicit notation containing the square brackets [ ].

For higher-order partial derivatives, the same system is followed. For example, if $f(x, y)$ is successively differentiated with respect to x, y, and x, this partial derivative is written as

$$f_{xyx}(x, y) = \frac{\partial^3 f}{\partial x \partial y \partial x} = \frac{\partial}{\partial x}\left[\frac{\partial}{\partial y}\left[\frac{\partial f}{\partial x}\right]\right].$$

Any two successive partial derivatives with respect to the same variable can be indicated with shorthand notation, as in

$$f_{xxy}(x, y) = \frac{\partial^3 f}{\partial y \partial x^2} = \frac{\partial}{\partial y}\left[\frac{\partial^2 f}{\partial x^2}\right]$$

or

$$f_{yxx}(x, y) = \frac{\partial^3 f}{\partial^2 x \partial y} = \frac{\partial}{\partial x}\left[\frac{\partial^2 f}{\partial x \partial y}\right] = \frac{\partial}{\partial x}\left[\frac{\partial}{\partial x}\left[\frac{\partial f}{\partial y}\right]\right].$$

Example 18.22. Given the function $f(x, y) = 3x^3 y - 4x^2 y^2$, find all possible second-order partial derivatives.

Solution. There are four possible combinations, with two kinds of partial derivatives in pairs:

$$\frac{\partial^2 f}{\partial x^2} = \frac{\partial}{\partial x}\left[9x^2 y - 8xy^2\right] = 18xy - 8y^2,$$

$$\frac{\partial^2 f}{\partial y \partial x} = \frac{\partial}{\partial y}\left[9x^2 y - 8xy^2\right] = 9x^2 - 16xy,$$

$$\frac{\partial^2 f}{\partial x \partial y} = \frac{\partial}{\partial x}\left[3x^3 - 8x^2 y\right] = 9x^2 - 16xy,$$

and

$$\frac{\partial^2 f}{\partial y^2} = \frac{\partial}{\partial y}\left[3x^3 - 8x^2 y\right] = -8x^2.$$ △

Notice in the previous example that $\partial^2 f/\partial y \partial x = \partial^2 f/\partial x \partial y$. This is not a coincidence. Mixed partials containing the same number of derivatives with respect to the same variables but in different order are the same, as long as each derivative is a continuous function. This rule holds no matter how many derivatives and variables there are. So, for example, for derivatives of $f(x, y, z)$ it is true that

$$\frac{\partial^5 f}{\partial x \partial y \partial z \partial y \partial x} = \frac{\partial^5 f}{\partial^2 x \partial^2 y \partial z}$$

as long as the derivatives are continuous.

Example 18.23. For a function $f(x, y, z)$, write all the possible second-order partial derivatives. Assuming the derivatives are continuous, which partials are equal to each other?

Solution. With three types of partial derivatives coming in pairs, the number of possible combinations is $3^2 = 9$. They are given below, with the ones that are equal for continuous functions expressed as equalities:

$$\frac{\partial^2 f}{\partial x^2}, \qquad \frac{\partial^2 f}{\partial y^2}, \qquad \frac{\partial^2 f}{\partial z^2},$$

$$\frac{\partial^2 f}{\partial y \partial x} = \frac{\partial^2 f}{\partial x \partial y}, \qquad \frac{\partial^2 f}{\partial y \partial z} = \frac{\partial^2 f}{\partial z \partial y}, \qquad \text{and} \qquad \frac{\partial^2 f}{\partial z \partial x} = \frac{\partial^2 f}{\partial x \partial z}. \qquad \triangle$$

18.7.2 The Second-Order Taylor Approximation

In Section 18.2.3, we extended the notion of the first-order (linear) Taylor approximation to partial derivatives. Now we consider how to extend the Taylor approximation to include second-order terms with partial derivatives.

Recall that for ordinary derivatives (for functions $f(x)$ of a single variable), the second-order Taylor approximation of $f(x)$ based at the input $x = a$ is

$$f(x) = f(a) + f'(a)(x-a) + \frac{1}{2}f''(a)(x-a)^2.$$

From Equation 18.1 the linear approximation for functions $f(x, y)$ near the input (a, b) is

$$f(x,y) = f(a,b) + \left.\frac{\partial f}{\partial x}\right|_{(a,b)}(x-a) + \left.\frac{\partial f}{\partial y}\right|_{(a,b)}(y-b). \qquad (2.12)$$

How can we extend the Taylor approximation to second-order? Following the construction of Equation (18.1), we should include terms containing all possible second derivatives (as we have seen, there are four of them for functions of two variables). Those four terms should look something like the second derivative terms in the linear approximation above with the appropriate partial derivatives taking the place of $f''(a)$. As we saw in Section 18.2 for the linear approximation, each partial derivative with respect to x is measuring a change in the function's output as we move along parallel to the x-axis, and similarly for y, which is why the partial derivatives are multiplied by $x - a$ or $y - b$. Therefore, the term associated with two successive partial derivatives with respect to x should contain a factor $(x - a)^2$. The term associated with the mixed partial $\partial^2 f/\partial y \partial x$ should contain a factor $(y - b)(x - a)$, and so on. Thus, the second-order Taylor polynomial approximation becomes

$$f(x,y) = f(a,b) + \left.\frac{\partial f}{\partial x}\right|_{(a,b)}(x-a) + \left.\frac{\partial f}{\partial y}\right|_{(a,b)}(y-b) +$$
$$\frac{1}{2}\left[\left.\frac{\partial^2 f}{\partial x^2}\right|_{(a,b)}(x-a)^2 + \left.\frac{\partial^2 f}{\partial x \partial y}\right|_{(a,b)}(x-a)(y-b) + \left.\frac{\partial^2 f}{\partial y \partial x}\right|_{(a,b)}(y-b)(x-a) + \left.\frac{\partial^2 f}{\partial y^2}\right|_{(a,b)}(y-b)^2\right].$$

We may simplify this in the case where the partial derivatives are continuous. Then the mixed partials are equal, so

$$f(x,y) = f(a,b) + \left.\frac{\partial f}{\partial x}\right|_{(a,b)}(x-a) + \left.\frac{\partial f}{\partial y}\right|_{(a,b)}(y-b) +$$
$$\frac{1}{2}\left[\left.\frac{\partial^2 f}{\partial x^2}\right|_{(a,b)}(x-a)^2 + 2\left.\frac{\partial^2 f}{\partial x \partial y}\right|_{(a,b)}(x-a)(y-b) + \left.\frac{\partial^2 f}{\partial y^2}\right|_{(a,b)}(y-b)^2\right]. \qquad (18.20)$$

Example 18.24. Find the second-order Taylor approximation to the function $f(x, y) = x^2 - 2xy$ near the origin.

Solution. We first compute $f(0, 0) = 0$. The partial derivatives are

$$\frac{\partial f}{\partial x} = 2x - 2y \qquad \text{and} \qquad \frac{\partial f}{\partial y} = -2x.$$

Both of these partial derivatives are zero when evaluated at the origin, so neither of the first-order terms survives. The second-order partials are

$$\frac{\partial^2 f}{\partial x^2} = 2, \qquad \frac{\partial^2 f}{\partial x \partial y} = -2, \qquad \text{and} \qquad \frac{\partial^2 f}{\partial y^2} = 0.$$

Combining all of this information, we find the approximation is

$$f(x, y) \approx 0 + 0(x) + 0(y) + \frac{1}{2}\left[2x^2 + 2(-2)xy + 0y^2\right] = x^2 - 2xy.$$

In this case, the second-order approximation has recovered the original function perfectly, because there were no terms of order higher than 2 in $f(x, y)$. △

Example 18.25. Find the second-order Taylor approximation to the function $f(x, y) = x^3 - 2xy$ near the origin.

Solution. As in the preceding example, we compute the first partial derivatives

$$\frac{\partial f}{\partial x} = 3x^2 - 2y \qquad \text{and} \qquad \frac{\partial f}{\partial y} = -2x$$

and the second partial derivatives

$$\frac{\partial^2 f}{\partial x^2} = 6x, \qquad \frac{\partial^2 f}{\partial x \partial y} = -2, \qquad \text{and} \qquad \frac{\partial^2 f}{\partial y^2} = 0.$$

By Equation (18.20), the approximation is

$$f(x, y) = 0 + 0(x) + 0(y) + \frac{1}{2}\left[0x^2 + 2(-2)xy + 0y^2\right] = -2xy.$$

Note that the term x^3 in the original function does not find its way into the approximation. This makes sense, because for coordinates close to zero, x^3 is much smaller than xy. △

The second-order Taylor approximation is especially useful in obtaining polynomial approximations of functions that are not polynomials to begin with.

Example 18.26. Find the second-order Taylor approximation to the function $f(x, y) = ye^x$ near the origin.

Solution. The first partial derivatives are

$$\frac{\partial f}{\partial x} = ye^x \qquad \text{and} \qquad \frac{\partial f}{\partial y} = e^x,$$

and the second partial derivatives are

$$\frac{\partial^2 f}{\partial x^2} = ye^x, \qquad \frac{\partial^2 f}{\partial x \partial y} = e^x, \qquad \text{and} \qquad \frac{\partial^2 f}{\partial y^2} = 0.$$

By Equation (18.20), the approximation is

$$f(x, y) = 0 + (0)x + (1)y + \frac{1}{2}\left[(0)x^2 + 2(1)xy + (0)y^2\right] = y(1 + x).$$

This result makes sense, because we recognize that the original function is the product of y and e^x, while the approximation is the product of y and the *first-order* Taylor polynomial approximation of the function e^x. △

Another major application of ordinary derivatives was in the problem of finding *extrema* of functions, and interpreting those extrema geometrically and physically. In Chapter 20, we will extend the analysis of extrema to partial derivatives.

18.8 Problems

18.1 Partial Derivatives

1. The grid below gives outputs of a function $f : \mathbb{R}^2 \to \mathbb{R}$. Estimate the two partial derivatives of the function at (a) $(0, 0)$, (b) $(0.6, 0.8)$, (c) $(-0.4, 0.8)$, and (d) $(-0.2, -0.6)$.

$x \diagdown y$	−1.0	−0.8	−0.6	−0.4	−0.2	0.0	0.2	0.4	0.6	0.8	1.0
−1.0	5.58	6.02	6.46	6.90	7.34	7.78	8.22	8.66	9.10	9.54	9.98
−0.8	4.33	4.81	5.28	5.75	6.22	6.69	7.17	7.64	8.11	8.58	9.05
−0.6	3.09	3.59	4.10	4.60	5.10	5.61	6.11	6.62	7.12	7.62	8.13
−0.4	1.84	2.38	2.91	3.45	3.99	4.52	5.06	5.59	6.13	6.67	7.20
−0.2	0.60	1.16	1.73	2.30	2.87	3.44	4.00	4.57	5.14	5.71	6.28
0.0	−0.65	−0.05	0.55	1.15	1.75	2.35	2.95	3.55	4.15	4.75	5.35
0.2	−1.90	−1.26	−0.63	0.00	0.63	1.26	1.90	2.53	3.16	3.79	4.42
0.4	−3.14	−2.48	−1.81	−1.15	−0.49	0.18	0.84	1.51	2.17	2.83	3.50
0.6	−4.39	−3.69	−3.00	−2.30	−1.60	−0.91	−0.21	0.48	1.18	1.88	2.57
0.8	−5.63	−4.91	−4.18	−3.45	−2.72	−1.99	−1.27	−0.54	0.19	0.92	1.65
1.0	−6.88	−6.12	−5.36	−4.60	−3.84	−3.08	−2.32	−1.56	−0.80	−0.04	0.72

2. Use the definitions to compute the partial derivatives f_x and f_y for $f(x, y) = xy$.

For Problems 3–10, compute the partial derivatives f_x and f_y of the given function and evaluate each partial derivative at the given input.

3. $f(x, y) = 3x^3 + 4xy^4, \quad (-1, 0)$
4. $f(x, y) = 4e^x - 14xe^y + 6y^2, \quad (2, 2)$
5. $f(x, y) = -4x^y, \quad (2, 1)$
6. $f(x, y) = 6x^2 \ln y, \quad (3, 3)$
7. $f(x, y) = 6xye^y, \quad (2, -1)$
8. $f(x, y) = -2\sin(xy), \quad \left(\sqrt{2\pi}, -\sqrt{\frac{\pi}{2}}\right)$
9. $f(x, y) = \dfrac{1}{\sqrt{x^2 + y^2}}, \quad (-1, 1)$
10. $f(x, y) = \dfrac{-2xy}{\sqrt{x^2 + y^2}}, \quad (1, 2)$

For Problems 11–18, compute the partial derivatives f_x, f_y, and f_z of the given function and evaluate each partial derivative at the given input.

11. $f(x, y, z) = 4xyz + 3x^2 - y^2z^3, \quad (1, 1, 1)$
12. $f(x, y, z) = 3x^2yz - xyz^5, \quad (0, -2, -3)$
13. $f(x, y, z) = \dfrac{-2x}{14y^2 + z}, \quad (5, 4, 3)$
14. $f(x, y, z) = \dfrac{12x}{\sqrt{x^2 + y^2 + z^2}}, \quad (3, 4, 5)$

15. $f(x,y,z) = -3x\ln(yz)$, $(2,2,1)$
16. $f(x,y,z) = xz^3e^{2y^2}$, $(-1,-1,-1)$
17. $f(x,y,z) = e^{xyz}$, $(1,1,4)$
18. $f(x,y,z) = \sin(\pi xyz)$, $(1,1,1)$

For Problems 19–27, compute the indicated partial derivative.

19. $\frac{\partial}{\partial T}[\sigma T^4]$
20. $\frac{\partial}{\partial x}[\sin(kx - \omega t)]$
21. $\frac{\partial}{\partial t}[\sin(kx - \omega t)]$
22. $\frac{\partial}{\partial x}\left[\frac{1}{2}mv^2 + \frac{1}{2}kx^2\right]$
23. $\frac{\partial}{\partial v}\left[\frac{1}{2}mv^2 + \frac{1}{2}kx^2\right]$
24. $\frac{\partial}{\partial t}[N_0e^{-\lambda t}]$
25. $\frac{\partial}{\partial x}[Ae^{ikx} + Be^{-ikx}]$
26. $\frac{\partial}{\partial x}\left[\gamma\left(t - \frac{vx}{c^2}\right)\right]$
27. $\frac{\partial}{\partial t}\left[\gamma\left(t - \frac{vx}{c^2}\right)\right]$
28. Determine the inputs (x,y) for which each of the partial derivatives f_x and f_y exists for the function

$$f(x,y) = \begin{cases} 0 & \text{if } x \neq 0 \text{ and } y \neq 0 \\ 1 & \text{if } x = 0 \text{ or } y = 0. \end{cases}$$

18.2 Tangent Planes and Linear Approximations

1. Compute the linear approximation for $f(4.2, 5.06)$ knowing $f(4,5) = 3.7$, $f_x(4,5) = -2$, and $f_y(4,5) = -3.1$.

2. Compute the linear approximation for $f(-1.1, 18.2, 0.2)$ knowing $f(-1,18,0) = 6$, $f_x(-1,18,0) = -4.3$, $f_y(-1,18,0) = -0.6$, and $f_z(-1,18,0) = 3.2$.

For Problems 3–12, find the equation of the tangent plane for the given function f at the given input.

3. $f(x,y) = 3x^2y - 4y^2$, $(2,3)$
4. $f(x,y) = \frac{x^2+y^2}{2x}$, $(1,1)$
5. $f(x,y) = 3\ln(xy)$, $(2,2)$
6. $f(x,y) = \sin(2x) - \cos(3y)$, $\left(\frac{\pi}{4}, -\frac{\pi}{6}\right)$
7. $f(x,y) = e^{xy}$, $(2,1)$
8. $f(x,y) = e^{-(x^2+y^2)}$, $(-1,1)$
9. $f(x,y) = e^{-(x^2+y^2)}$, $(0,0)$
10. $f(x,y) = x\ln y + y\ln x$, $(2,2)$
11. $f(x,y) = \frac{14}{x-y}$, $(1,3)$
12. $f(x,y) = 4\sqrt{x^2+y^2}$, $(1,2)$

For Problems 13–20, find the linear approximation for the given function f at the given input.

13. $f(x,y,z) = xyz$, $(1,2,3)$
14. $f(x,y,z) = x^2 + y^2 + z^2$, $(3,4,5)$
15. $f(x,y,z) = 3x^2z + y$, $(2,-1,-1)$
16. $f(x,y,z) = \frac{2}{x^2+y^2+z^2}$, $(5,12,13)$
17. $f(x,y,z) = xe^{yz}$, $(1,0,1)$
18. $f(x,y,z) = 3z\ln(xy^2)$, $(2,1,1)$
19. $f(x,y,z) = (x^2+y^2+z^2)^{1/4}$, $(1,2,-1)$
20. $f(x,y,z) = \sin(\pi x)\sin(\pi y)\sin(\pi z)$, $(\frac{1}{2}, \frac{1}{2}, 1)$

21. Find an input for which the plane tangent to the graph of $f(x,y) = x^2 - y^2$ is parallel to the plane with equation $2x + y + 3z - 4 = 0$.

22. Write the linear approximation for a function $f : \mathbb{R}^4 \to \mathbb{R}$.

23. A hiker on a very foggy mountain cannot see past the end of her nose. She knows her altitude is 900 m, and she measures the slopes of the terrain at her location to be 0.24 m/m in the north direction and 0.18 m/m in the east direction. Estimate the hiker's change in altitude if she goes (a) 5 m in the northeast direction; (b) 5 m in the northwest direction; and (c) 5 m in the southwest direction.

24. Consider the function defined in Problem 28 of Section 18.1. Do $f_x(0,0)$ and $f_y(0,0)$ exist? Is f continuous for $(0,0)$? Explain how the answers to these are consistent with the last paragraph of Section 18.2.

18.3 Gradient and Directional Derivative

For Problems 1–4, compute the gradient of the given function $f : \mathbb{R}^2 \to \mathbb{R}$ and plot the vector field.

1. $f(x,y) = xy$
2. $f(x,y) = x^2 + y^2$
3. $f(x,y) = xy(y-x)$
4. $f(x,y) = e^{-(x^2+y^2)}$

For Problems 5–8, compute the gradient of the given function $f : \mathbb{R}^3 \to \mathbb{R}$.

5. $f(x,y,z) = xyz$
6. $f(x,y,z) = x^2 + y^2 + z^2$
7. $f(x,y,z) = \dfrac{1}{\sqrt{x^2+y^2+z^2}}$
8. $f(x,y,z) = e^{-(x^2+y^2+z^2)}$

9. Consider the directional derivative of $f(x,y) = xy$ for the input $(-1,3)$ in the direction $\langle 2,1\rangle/\sqrt{5}$. (a) Compute this directional derivative using the definition. (b) Compute this directional derivative using Equation (18.2).

10. Write an analog of Definition 18.3 for the directional derivative of a function $f : \mathbb{R}^3 \to \mathbb{R}$.

For Problems 11–16, find the directional derivative for the given function f at the input (a,b) in the direction $\vec{u}$ as indicated. Note that the given direction vector $\vec{u}$ is not necessarily a unit vector.

11. $f(x,y) = x^2 + y^2, \quad (2,2), \quad \vec{u} = \hat{\imath} + \hat{\jmath}$
12. $f(x,y) = x^2 + y^2, \quad (2,2), \quad \vec{u} = \hat{\imath} - \hat{\jmath}$
13. $f(x,y) = x^2 + y^2, \quad (4,4), \quad \vec{u} = \hat{\imath} + \hat{\jmath}$
14. $f(x,y) = x^2 - y^2, \quad (2,2), \quad \vec{u} = \hat{\imath} + \hat{\jmath}$
15. $f(x,y) = x^2 - y^2, \quad (2,2), \quad \vec{u} = \hat{\imath} - \hat{\jmath}$
16. $f(x,y) = x^2 - y^2, \quad (2,0), \quad \vec{u} = \hat{\imath} + \hat{\jmath}$

For Problems 17–20, find the directional derivative for the given function f at the input (a,b,c) in the direction $\vec{u}$ as indicated. Note that the given direction vector $\vec{u}$ is not necessarily a unit vector.

17. $f(x,y) = x^2 + y^2 + z^2, \quad (1,2,3), \quad \vec{u} = \hat{\imath} + \hat{\jmath} + \hat{k}$

18. $f(x,y) = x^2 + y^2 + z^2, \quad (1,2,3), \quad \vec{u} = -\hat{\imath} - \hat{\jmath} - \hat{k}$

19. $f(x,y,z) = xyz, \quad (1,4,2), \quad \vec{u} = 2\hat{\imath} + 3\hat{\jmath} - \hat{k}$

20. $f(x,y,z) = \dfrac{1}{\sqrt{x^2+y^2+z^2}}, \quad (-2,-2,-2), \quad \vec{u} = \hat{\imath} + \hat{\jmath} + \hat{k}$

For Problems 21–24, compute the gradient for the given function $f : \mathbb{R}^3 \to \mathbb{R}$ where $\vec{r} = \langle x, y, z \rangle$. Express the result in terms of the unit vector $\hat{r}$ and the magnitude $r = \|\vec{r}\|$.

21. $f(\vec{r}) = r^2$ 22. $f(\vec{r}) = r$ 23. $f(\vec{r}) = r^{-1}$ 24. $f(\vec{r}) = r^{-2}$

25. Consider functions $f : \mathbb{R}^3 \to \mathbb{R}$ of the form $f(\vec{r}) = r^n$ where $\vec{r} = \langle x, y, z \rangle$ and n is an integer. Derive an expression for $\vec{\nabla} f(\vec{r})$ in terms of the unit vector $\hat{r}$ and the magnitude $r = \|\vec{r}\|$.

26. The figure on the left below shows level curves for a function $f : \mathbb{R}^2 \to \mathbb{R}$. (a) Estimate the inputs that give local maxima, minima, and saddle points. (b) Estimate the directional derivative $D_{\hat{u}} f(5, -5)$ where $\vec{u} = \langle -1, 1 \rangle$.

27. The figure on the right below shows the gradient vector field for a function $f : \mathbb{R}^2 \to \mathbb{R}$. (a) Sketch the level curve that passes through the point $(5, -5)$. (b) Sketch the level curve that passes through the point $(8, 5)$. (c) Sketch the path of steepest ascent starting at the point $(2, 6)$. (d) Estimate the input (a, b) for which the graph of f is steepest. (e) Describe the graph of f.

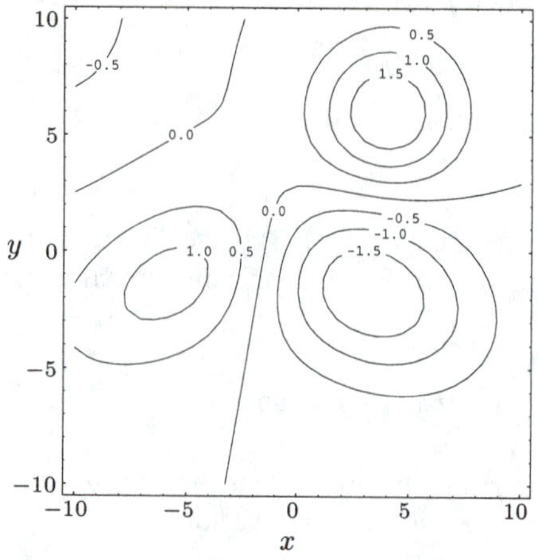

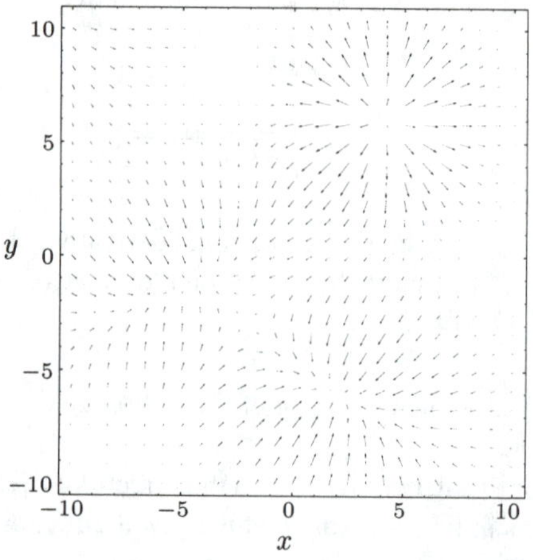

28. The steps in this problem outline a proof of the relation in Equation (18.2). Start by defining the function $g(h) = f(a + hu_1, b + hu_2)$. (a) Use the definition of derivative (for $g : \mathbb{R} \to \mathbb{R}$) to show that $g'(0) = D_{\hat{u}} f(a, b)$. (b) Let $\vec{r}(h) = \langle a + hu_1, b + hu_2 \rangle$ so $g(h) = f(\vec{r}(h))$. Compute $g'(0)$ using the chain rule in Theorem 18.2. (c) Equate the two expressions for $g'(0)$ found in (a) and (b).

18.4 Electric Potential

1. A proton is held fixed at the origin. Calculate the work done by the electric field in moving a second proton from the point $x = 1$ m on the x-axis to the point $x = 2$ m on the x-axis. Compute the change in potential energy for the process just described, and show that it is equal to the negative of the work.

2. Repeat Problem 1 if the second proton begins at the point $(1, 1)$ (measured in meters) in the xy-plane and is moved to the point $(2, 2)$ (meters).

For Problems 3–7, find the total potential energy for the given collection of point charges.

3. $q_1 = 1.9 \times 10^{-9}$ C at (0, 0.25 m), $q_2 = -2.3 \times 10^{-8}$ C at (1.25 m, 0.95 m)

4. $q_1 = -5.5 \times 10^{-10}$ C at (1.0 m, −0.75 m), $q_2 = 4.4 \times 10^{-8}$ C at (1.55 m, −0.05 m)

5. $q_1 = 1.0 \times 10^{-13}$ C at (0.75 m, 0.25 m, 1.00 m), $q_2 = -5.0 \times 10^{-16}$ C at (−0.55 m, 0.75 m, 1.10 m)

6. $q_1 = 5.0 \times 10^{-6}$ C at (−15.0 m, 1.20 m, 2.00 m), $q_2 = 4.0 \times 10^{-7}$ C at (−5.00 m, 16.0 m, 4.70 m)

7. $q_1 = 1.4 \times 10^{-10}$ C at (1.25 m, 0), $q_2 = -1.3 \times 10^{-10}$ C at (1.50 m, 0.90 m), $q_3 = 5.0 \times 10^{-11}$ C at (0, −1.50 m)

8. A 2.5-μC charge with a mass of 1 g is held fixed at the origin. An identical charge that is free to move is released from rest at the point $x = -0.10$ m on the x-axis. (a) Describe qualitatively the subsequent motion of the free charge. (b) Sketch the position vs. time and velocity vs. time for the free charge, assuming it is released at $t = 0$. Your axes need not include numerical scales or units. (c) Find the speed of the free charge when it is 0.20 m from the fixed charge. (d) What is the speed of the free charge when it is a great distance from the fixed charge?

9. A −1.0-nC charge with a mass of 1 g is held fixed at the origin. An identical charge that is free to move is released from rest at the point $z = 0.50$ m on the z-axis. (a) Describe qualitatively the subsequent motion of the free charge. (b) Sketch the position vs. time and velocity vs. time for the free charge, assuming it is released at $t = 0$. Your axes need not include numerical scales or units. (c) Find the speed of the free charge when it is 1.00 m from the fixed charge. (d) What is the speed of the free charge when it is a great distance from the fixed charge?

For Problems 10–12, find the electric potential at the point (a, b, c) for the given collection of charges.

10. $q = 2.5$ pC at the origin; $(a, b, c) = $ (0.50 m, 1.00 m, −0.75 m)

11. $q = -1.60 \times 10^{-19}$ C at (0, 0, 0.25 m); $(a, b, c) = $ (2,00 m, 1.50 m, 2.00 m)

12. $q_1 = 1.50 \times 10^{-10}$ C at the origin, $q_2 = 2.50 \times 10^{-10}$ C at (1.0 m, 1.0 m, 1.0 m); $(a, b, c) = $ (−1.5 m, −1.75 m, −2.0 m)

13. Two charges, $q_1 = 0.50\,\mu$C and $q_2 = 1.0\,\mu$C, are placed on the y-axis at $y = 0.25$ m and $y = 1.25$ m, respectively. Are there any points on the y-axis where the electric potential is zero? Explain, and find such points if they exist.

14. Repeat the preceding problem if q_2 is changed to $-1.0\,\mu$C.

15. Two identical $+q$ charges are a distance d apart. Sketch the equipotential curves in the plane containing the two charges. Is there any point where $V = 0$?

16. Repeat the preceding problem if one of the two charges is changed to $+2q$.

17. Consider two large parallel conducting plates, separated by a distance of 3.0 cm, that have a potential difference of 500 V between their surfaces. (a) What is the electric field in the region between the plates? (b) Sketch the equipotentials between the plates. (c) What is the separation between two equipotentials differing in potential by 100 V? (d) A positively charged particle is released somewhere between the plates. Describe its subsequent motion qualitatively. Will it move toward the higher or lower potential plate? Will its acceleration be constant? Explain. (e) Repeat part (d) for a negatively charged particle. (f) Compute the work done by the field in moving an electron from the higher potential plate to the lower potential plate.

18. Two concentric conducting spheres have radii 5.0 cm and 10.0 cm. The inner and outer spheres carry charges $+1.0\,\mu$C and $-1.0\,\mu$C, respectively. (a) Sketch the equipotential surfaces in the region between the spheres. (b) Compute the potential difference between the spheres. (c) Taking $V = 0$ an infinite distance away from the spheres, what is the electric potential at a point midway between the spheres, that is, 7.5 cm from their common center?

19. An electron moves under the influence of an electric field, and in doing so its speed changes from 2500 m/s to 1500 m/s. What is the potential difference between the two points at which the speed was measured?

20. Find the potential energy of a collection of charge Q spread uniformly over the surface of a sphere. [*Hint:* Compute the work done in assembling the charge in this configuration by bringing infinitesimal bits of charge onto the sphere from an infinite distance. As usual, assume the potential energy an infinite distance away from the final configuration is zero.]

21. In a particular lightning flash, the energy released is 8.0×10^9 J. If the potential difference between the two ends of the flash is 1.2×10^9 V, how much charge is transferred between the two ends of the flash?

22. Two large conducting parallel plates separated by a distance of 1.0 cm carry charge densities of $\pm5.0\ \mu\text{C/m}^2$. (a) What is the electric field between the plates? (b) What is the potential difference between the plates. (c) If an electron is shot from the positive plate directly toward the negative one, with what speed must it be fired so that it just reaches the negative plate before stopping?

18.5 Electric Potential and Electric Fields

For Problems 1–9, find the electric field for the given electric potential function V.

1. $V(x,y) = 2(x^2 - y^2)$
2. $V(x,y,z) = 3(x+y-z)^2$
3. $V(x,y) = \dfrac{2}{x^2+y^2}$
4. $V(x,y) = -\dfrac{1}{\sqrt{x^2+y^2}}$
5. $V(x,y) = 3\sin(2x)\cos(-y)$
6. $V(x,y,z) = 3xy^2 - 9z^3$
7. $V(x,y,z) = \dfrac{1}{\sqrt{x^2+y^2+z^2}}$
8. $V(x,y,z) = \dfrac{2x}{\sqrt{x^2+y^2+z^2}}$
9. $V(x,y,z) = \ln(xyz)$

10. Consider a uniform electric field $\vec{E} = 1500$ N/C $\hat{\imath}$. Find the potential difference $V(\vec{r}_f) - V(\vec{r}_i)$ between the points indicated.

    (a) $\vec{r}_i = \langle 0, 0, 0.50\text{ m}\rangle$, $\vec{r}_f = \langle 1.0\text{ m}, 0, 1.25\text{ m}\rangle$
    (b) $\vec{r}_i = \langle -0.25\text{ m}, 0.45\text{ m}, 0\rangle$, $\vec{r}_f = \langle -0.75\text{ m}, 0.50\text{ m}, 0.25\text{ m}\rangle$
    (c) $\vec{r}_i = \langle 0, 0, 0\rangle$, $\vec{r}_f = \langle 10.0\text{ m}, 10.0\text{ m}, 10.0\text{ m}\rangle$

11. Compute the potential difference between the points indicated in problems using a uniform electric field $\vec{E} = 1500$ N/C $\hat{\imath} + 1000$ N/C $\hat{\jmath} + 750$ N/C $\hat{k}$.

    (a) $\vec{r}_i = \langle 0, 0, 0.50\text{ m}\rangle$, $\vec{r}_f = \langle 1.0\text{ m}, 0, 1.25\text{ m}\rangle$
    (b) $\vec{r}_i = \langle -0.25\text{ m}, 0.45\text{ m}, 0\rangle$, $\vec{r}_f = \langle -0.75\text{ m}, 0.50\text{ m}, 0.25\text{ m}\rangle$
    (c) $\vec{r}_i = \langle 0, 0, 0\rangle$, $\vec{r}_f = \langle 10.0\text{ m}, 10.0\text{ m}, 10.0\text{ m}\rangle$

12. Of the following electric fields, one is the electric field of some potential $V : \mathbb{R}^2 \to \mathbb{R}$ and the other is not (i.e., there is *no* function $V : \mathbb{R}^2 \to \mathbb{R}$ for which the negative of the gradient field is equal to the given vector field). Determine which vector field is an electric field and find the corresponding potential function $V : \mathbb{R}^2 \to \mathbb{R}$. (a) $\vec{E}(x, y) = \langle 3xy, 5y^2 \rangle$, (b) $\vec{E}(x, y) = \langle 3x^2y + 4x, y^3 + x^3 \rangle$.

13. Consider an electric potential given by $V(x, y) = x^4 + y^4 - 4xy + 1$ with output in volts for input in meters. (a) Find the corresponding electric field. (b) Show that the points $(1 \text{ m}, 1 \text{ m})$, $(0, 0)$, and $(-1 \text{ m}, -1 \text{ m})$ are stationary points for V. (c) Determine the shape of the graph of V at each of the stationary points listed in (b)—that is, local maximum, minimum, or saddle point. (d) Describe the motion of a small positive charge placed at rest at $(1 \text{ m}, 1 \text{ m})$. Describe what would happen if the charge were placed near but not at that point.

18.6 Calculating Electric Potentials

1. For the electric dipole described in Example 18.18, find the electric field at the points $(0, -a)$ and $(-2a, 0)$ and comment on the results.

2. For the electric dipole described in Example 18.18, find the electric field at the points (a, a) and $(-a, -a)$ and comment on the results.

3. A uniformly charged disk of radius R and total charge Q lies in the xy-plane with its center at the origin. Find the electric potential at all points on the z-axis. You may use the result of Example 18.19 by considering the disk to consist of a large number of thin rings of different radii.

4. (a) Using the result of the preceding problem, find the electric field due to the charged disk at all points on the z-axis. (b) Find the limiting value of the electric field for the case $0 < z \ll R$. Show that this result is consistent with the electric field of an infinite, uniformly charged plane.

5. A solid nonconducting sphere of radius R carries a uniform charge density throughout its volume, with a total charge of Q. Find the electric potential at all points inside and outside the sphere. Your answers should be functions of R, Q, and r (the distance from the center of the sphere).

6. An infinite line carries a uniform charge density λ per unit length. Find the potential difference between two points at perpendicular distances r_1 and r_2 from the line. Explain why it is *not* possible to set the electric potential to zero an infinite distance away in this case.

7. Using the results of the preceding problem, find the electric field as a function of perpendicular distance from the line. Your result should match that obtained in Chapter 17 for this charge configuration.

8. A thin semicircular ring of radius R lies in the xy-plane with its center of curvature at the origin. It lies in the upper half-plane with $y > 0$. The ring carries a uniform charge density and total charge Q. Find the electric potential and electric field at the origin.

9. A thin rod of length L lies along the y-axis, from $y = -L/2$ to $y = +L/2$. The rod carries a uniform charge density and total charge Q. (a) Find the electric potential at all points on the x-axis. (b) From your result in (a), find the electric field at all points on the x-axis (your result should be consistent with what we found for this problem in Chapter 17. (c) Find and comment on what happens to the resulting electric field in the limit $x \to \infty$.

10. The magnitude of a typical electric field near Earth's surface is about 100 V/m. This field is nearly uniform and directed straight down (i.e., normal to Earth's surface). Use these facts to answer the following questions.

(a) What would the net charge have to be on Earth in order to produce such a field? What is the source of that net charge? What is the charge per unit surface area?

(b) Find the electric potential for points a height h above Earth's surface, assuming $h \ll R$, Earth's radius. For the purposes of this problem it may be convenient to define the zero of potential at Earth's surface.

(c) Consider a test particle of charge q_0 and mass m_0 a height $h \ll R$ above Earth's surface. Write and compare the expressions for the electric potential energy and gravitational potential energy of the test particle. For a test particle of mass 1 g, what must be the charge on the particle so that the electric force on it just balances the gravitational force—that is, so that the net force on the particle is zero?

11. Consider again the arrangement described in Problem 9, except now assume that the rod carries a *nonuniform* charge density $\lambda = \lambda_0 y^2$. (a) Find the electric potential at all points on the x-axis. (b) What can you say about the direction of the electric field along the x-axis?

18.7 Higher-Order Partial Derivatives

For Problems 1–10, compute all the second-order partial derivatives for the given function.

1. $f(x, y) = xy$
2. $f(x, y) = x^3y^2$
3. $f(x, y) = 3\ln(xy)$
4. $f(x, y) = x \sin y$
5. $f(x, y) = \sin(xy)$
6. $f(x, y) = \dfrac{1}{\sqrt{x^2 + y^2}}$
7. $f(x, y, z) = xyz$
8. $f(x, y, z) = x^2 + y^2 + z^2$
9. $f(x, y, z) = 4x^2yz^3 - 5xy^2z^2$
10. $f(x, y, z) = e^{xyz}$

For Problems 11–12, compute all third-order partial derivatives for the given function.

11. $f(x, y) = x^3y^2$
12. $f(x, y, z) = xyz + 4x^2yz^2$

For Problems 13–18, find the second-order Taylor polynomial approximation to the given function based at the indicated point.

13. $f(x, y) = \sin(xy), \quad (0, 0)$
14. $f(x, y) = \sqrt{x^2 + y^2}, \quad (1, 1)$
15. $f(x, y) = x \ln y, \quad (1, 2)$
16. $f(x, y) = ye^x, \quad (0, 0)$
17. $f(x, y) = \sin^2 x \cos y, \quad \left(\frac{\pi}{4}, \frac{\pi}{4}\right)$
18. $f(x, y) = 2\sin x - 3\cos y, \quad \left(\frac{\pi}{2}, -\frac{\pi}{2}\right)$

19. (a) Write a formula for the second-order Taylor approximation of a function $f : \mathbb{R}^3 \to \mathbb{R}$ based at the input (a, b, c). (b) Find the second-order Taylor polynomial for $f(x, y, z) = xyz$ based at $(1, 2, 3)$.

Chapter 19

Capacitors, Dielectrics, and Electric Current

In this chapter, we begin to apply our knowledge of electric charge, fields, and potential in the study of electronic devices. First we consider *capacitors*, devices that store charge in electric circuits. We look at several geometric shapes that can be used for capacitors and understand each based on the concepts of charge and potential. We then examine how to connect capacitors in parallel and series and see how *dielectric materials* affect the operation of capacitors.

Then we turn to the concept of *electric current* and *resistance*. *Ohm's law* gives an important relationship between electric current, resistance, and potential difference, and it helps us understand how energy is used in an electric circuit. For now, we consider only simple, single-loop circuits, but we will apply Ohm's law in more complex circuits in Chapter 21. We conclude this chapter with a comparison of three basic types of materials that conduct electricity: normal *conductors*, *semiconductors*, and *superconductors*.

19.1 Capacitors

19.1.1 Introduction and Definitions

A **capacitor** consists of any pair of conductors, insulated from each other, that carry net charges that are equal in magnitude and opposite in sign ($\pm Q$), as illustrated in Figure 19.1. In electric circuits, capacitors are used to store charge. This is possible because in a capacitor the two conductors are insulated from one another. If the two conductors are connected (say by connecting them with a conducting wire), the excess electrons on the negative conductor flow through the wire to the positive conductor. Thus, zero net charge remains on either conductor, and we say that the capacitor has been **discharged**.

Electric potential is everywhere the same on the surface of a conductor. (This point will be explained in Chapter 23.) Thus, each conductor has a well-defined electric potential, and a potential difference exists between the two conducting surfaces (with the positively charged surface at the higher potential). Let V be the *magnitude* of that potential difference. More properly, we should write ΔV to remind ourselves that it is a potential difference, because V is normally the potential at a specific place. However, it is common to omit the Δ symbol when dealing with capacitors, so that the notation is less cluttered. When dealing with capacitors, we will use V exclusively for the potential difference, and when it is necessary to describe the electric potential at a specific location, we will use some kind of label specifying that place. For example, the electric potential at some position a could be described using a subscript V_a or functional dependence $V(a)$.

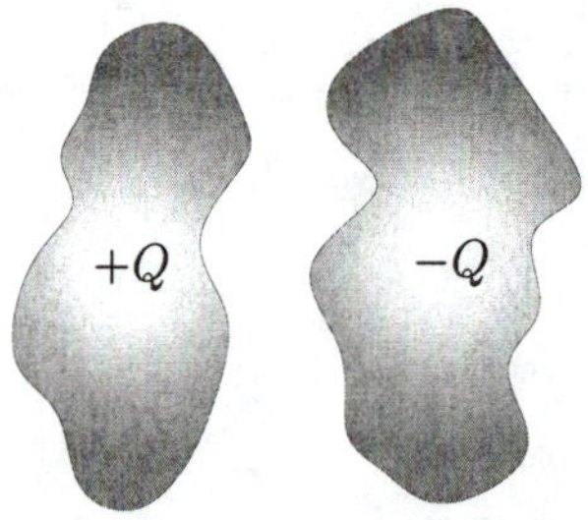

Figure 19.1. Schematic of a capacitor, consisting of two isolated conductors carrying net charges equal in magnitude but opposite signs.

The **capacitance** C of a capacitor is defined as the magnitude of the charge on each conductor divided by the potential difference. Symbolically,

$$C = \frac{Q}{V} \tag{19.1}$$

where Q is the *magnitude* of the charge on each conductor. When charge and potential difference are measured in SI units, the units for capacitance are C/V. For convenience, we use a defined SI unit of capacitance called the **farad** (F), with

$$1\,\text{F} = 1\,\text{C/V}.$$

The farad is named for the English physicist Michael Faraday (1791–1867).

Before going on, we can make several observations about capacitance, based on its definition. First, it is a *scalar* quantity, since it is the quotient of two scalars. It is also clear that capacitance must always be positive, because both Q and V in Equation (19.1) are defined as the magnitudes of their respective quantities. We will see that one farad is an extremely large capacitance. This should not be surprising, because you already know that one coulomb is a large amount of charge, and potential differences on the order of volts are common. Accordingly, it is not unusual to see capacitance measured in μF, nF, or pF.

A final observation follows from the definitions of capacitance and from the fact that the electric potential due to a particular geometric distribution of charge is proportional to the magnitude of the charge (see Chapter 18). That is, for any capacitor in which both conductors have fixed shapes, the ratio Q/V is constant, and hence the capacitance C is *constant*. When we analyze specific capacitor designs in Section 19.1.2, we will see how the capacitance depends only on the geometry of the two conductors and is independent of the charge on the capacitor, because the ratio Q/V cannot vary.

19.1.2 Several Types of Capacitors

In a **parallel-plate capacitor**, two flat conducting plates lie parallel to one another. The cross section may be circular, square, or any other desired shape, provided the cross-sections of the two plates are identical to on another. That is, they should have the same shape and the same surface area A. We will also make the "infinite plane" assumptions: The electric field between the plates is uniform, pointing directly from the positive plate to the negative one (Figure 19.2), and the electric field is zero outside the region between the plates. Remember that this is a good approximation only if the linear surface dimensions of the plates are much larger than the distance d between the plates.

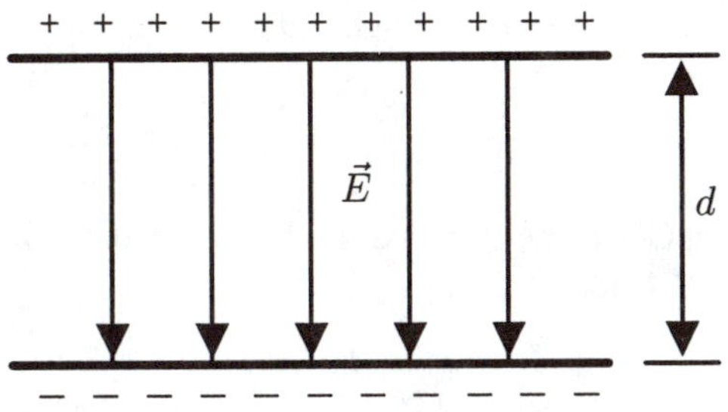

Figure 19.2. Parallel-plate capacitor.

We can compute the capacitance of a parallel-plate capacitor based on our previous study of the electric field and electric potential for parallel conducting plates carrying net charges equal in magnitude and opposite in sign. From the result of Example 18.17, we see that the potential difference between the plates is

$$V = Ed \tag{19.2}$$

where E is the magnitude of the electric field everywhere between the plates in the infinite plate approximation. In Example 17.13, we found that the magnitude of the electric field is

$$E = \frac{\sigma}{\epsilon_0}.$$

The surface charge density is constant for this geometry, so we can say $\sigma = Q/A$ is a constant, where Q is the magnitude of the charge on either plate. Therefore, the magnitude of the charge is

$$Q = \sigma A = \epsilon_0 EA. \tag{19.3}$$

Combining these results with the definition of capacitance [Equation (19.1)],

$$C = \frac{Q}{V} = \frac{\epsilon_0 EA}{Ed}$$

or

$$C = \frac{\epsilon_0 A}{d}. \tag{19.4}$$

Equation (19.4) allows us to find the capacitance of a parallel-plate capacitor if we know the surface area A and plate separation d. This result bears out our statement at the end of Section 19.1.1, that the capacitance of a capacitor depends only on its geometry (in this case, the geometric parameters A and d are significant), and not on the amount of charge contained by the capacitor at a given moment.

Example 19.1. Show that Equation (19.4) is dimensionally correct, and in doing so show that an alternative unit for the constant ϵ_0 is F/m.

Solution. Previously we saw (from Coulomb's law) that the SI unit for the constant ϵ_0 is $\mathrm{C^2/(N{\cdot}m^2)}$. Then the units on the right side of Equation (19.4) reduce to

$$\frac{\left[\mathrm{C^2/(N{\cdot}m^2)}\right](\mathrm{m^2})}{\mathrm{m}} = \frac{\mathrm{C^2}}{\mathrm{N{\cdot}m}}.$$

But 1 N·m = 1 J, and 1 J/C = 1 V, so the units on the right side of Equation (19.4) reduce to

$$\frac{\mathrm{C}^2}{\mathrm{N{\cdot}m}} = \frac{\mathrm{C}^2}{\mathrm{J}} = \frac{\mathrm{C}}{\mathrm{V}}.$$

From the definition of capacitance, C/V is the correct SI unit for capacitance and is equal to a farad.

Now we can solve Equation (19.4) for ϵ_0, giving

$$\epsilon_0 = \frac{Cd}{A}.$$

Because we have verified that this expression gives C in SI units of farads, the units for the right side of this equation are

$$\frac{(\mathrm{F})(\mathrm{m})}{\mathrm{m}^2} = \frac{\mathrm{F}}{\mathrm{m}}.$$

Therefore, it is appropriate to express ϵ_0 in units of F/m, with

$$\epsilon_0 = 8.854 \times 10^{-12}\ \mathrm{F/m} = 8.854\ \mathrm{pF/m}.$$

△

Example 19.2. A parallel-plate capacitor consists of two square plates each 1.50 cm on a side. The plates are separated by a distance of 0.250 mm. Compute (a) the capacitance of this capacitor, and (b) the magnitude of the charge on each plate when the potential difference between them is 12.0 V.

Solution. (a) For the parallel-plate geometry, the capacitance is given by Equation (19.4). The surface area A is simply the square of the side for the square cross section, so

$$C = \frac{\epsilon_0 A}{d} = \frac{(8.854\ \mathrm{pF/m})(0.0150\ \mathrm{m})^2}{2.50 \times 10^{-4}\ \mathrm{m}} = 7.97\ \mathrm{pF}.$$

(b) From the definition of capacitance [Equation (19.1)], the charge is

$$Q = CV = (7.97\ \mathrm{pF})(12.0\ \mathrm{V}) = 95.6\ \mathrm{pC}$$

where we have used the fact that $(1\ \mathrm{F})(1\ \mathrm{V}) = 1\ \mathrm{C}$. △

To build a capacitor with a higher capacitance than we found in the last example, Equation (19.4) dictates that we need a larger surface area A or a smaller separation d. Both of these strategies face practical limitations. A capacitor with a very large surface area could be expensive to build and could clutter an electric circuit. If d is too small, the capacitor can discharge through the gap. For example, if there is air between the capacitor plates, a discharge occurs when the electric field approaches 3 MV/m, equivalent to a potential difference of 3000 V for a 1-mm gap. Later in this chapter, we discuss other methods by which the capacitance of a parallel-plate capacitor can be increased.

Example 19.3. Find the surface area of a 1.0-μF parallel-plate capacitor in which the plates are separated by a distance of 0.10 mm. Assuming a square cross section for the plates, how long is each side?

Solution. From Equation (19.4), the surface area is

$$A = \frac{Cd}{\epsilon_0} = \frac{(1.0 \times 10^{-6}\ \mathrm{F})(1.0 \times 10^{-4}\ \mathrm{m})}{8.854 \times 10^{-12}\ \mathrm{F/m}} = 11.3\ \mathrm{m}^2.$$

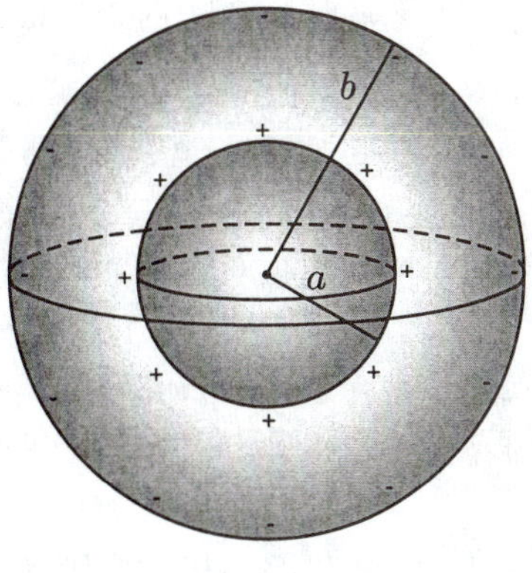

Figure 19.3. Schematic of a spherical capacitor.

The length of a side is

$$s = \sqrt{11.3 \text{ m}^2} = 3.36 \text{ m}.$$

Such dimensions are impractical for electric circuitry! △

We will now examine the operation of a **spherical capacitor**, shown in Figure 19.3. The spherical capacitor consists of two thin concentric conducting spheres, of radii a and b. Let's assume that the inner sphere carries a net charge $+Q$, and the outer sphere carries a net charge $-Q$. (In the problems, you will show that the calculation works just as well with the opposite assumption.) For now, we must assume that the charge is distributed uniformly over the surface of each sphere. (In Chapter 23, we will see why this *must* be so for conducting spheres.)

To calculate the capacitance of a spherical capacitor, we again use Equation (19.1), so we need to know the potential difference between the two spheres. From Chapter 18 (Example 18.21), we know that the electric potential outside a spherically symmetric distribution of charge $+Q$ is

$$V(r) = \frac{kQ}{r}$$

at a distance r from the center of the sphere. In Example 18.21, we also showed that the electric potential *inside* a spherically symmetric charge distribution is constant. Therefore, in the spherical capacitor, the potential difference between the two conducting surfaces is due to the charge on the inner sphere alone. The potential difference between the two surfaces is

$$V = V(a) - V(b) = \frac{kQ}{a} - \frac{kQ}{b}$$

or

$$V = kQ\left(\frac{1}{a} - \frac{1}{b}\right). \tag{19.5}$$

Then the capacitance of the spherical capacitor is

$$C = \frac{Q}{V} = \frac{Q}{kQ\left(\frac{1}{a} - \frac{1}{b}\right)} = \frac{1}{k\left(\frac{1}{a} - \frac{1}{b}\right)}.$$

After some algebra, and using the fact that $1/k = 4\pi\epsilon_0$, this result reduces to

$$C = 4\pi\epsilon_0 \frac{ab}{b-a} . \tag{19.6}$$

You can easily verify that this result has the correct dimensions of capacitance. Notice that the capacitance grows without bound in the limit as the two conducting surfaces approach each other ($b - a = 0$ in this case), just as happened in the parallel-plate capacitor.

Example 19.4. Show that if the separation between the spheres in a spherical capacitor is small compared with the radii of both spheres, the behavior of the capacitor approximates that of a parallel-plate capacitor with plate separation $d \approx b - a$.

Solution. Taking $d = b - a$ as given and $a \approx b \approx r$ for the radii of both spheres, the capacitance in Equation (19.6) becomes approximately

$$C \approx 4\pi\epsilon_0 \frac{r^2}{d} .$$

The surface area of a sphere of radius r is $A = 4\pi r^2$, so this expression reduces to

$$C = \frac{\epsilon_0 A}{d}$$

just as for the parallel-plate capacitor. △

The fact that the radius of the outer sphere did not affect the electric potential inside the spherical capacitor suggests that an **isolated sphere** can act as a capacitor. Strictly speaking, this breaks the rule of having two isolated conductors in a capacitor; however, the situation is no different than if we had an inner sphere of radius a and an outer sphere of much larger radius $b \to \infty$. Taking the appropriate limit of the expression we got for the capacitance of the spherical capacitor, we find for the isolated sphere of radius a

$$C = \lim_{b\to\infty} 4\pi\epsilon_0 \frac{ab}{b-a} = 4\pi\epsilon_0 \lim_{b\to\infty} \frac{ab}{b-a} = 4\pi\epsilon_0 \,(a)$$

or simply

$$C = 4\pi\epsilon_0 a. \tag{19.7}$$

Example 19.5. Treating Earth as an isolated conducting sphere, find its capacitance.

Solution. Using Earth's radius 6.374×10^6 m, we find with Equation (19.7)

$$C = 4\pi\epsilon_0 a = 4\pi(8.854 \times 10^{-12}\ \text{F/m})(6.374 \times 10^6\text{m}) = 7.092 \times 10^{-4}\ \text{F}.$$

This is a fairly significant capacitance, but not as large as many capacitors found in a common physics laboratory! △

19.1.3 Combinations of Capacitors

We now look at how different combinations of capacitors behave in electric circuits. Like other objects, the plates of a capacitor tend to be electrically neutral. There needs to be some external source of the potential difference between the plates and the charges on the plates. We will refer to a source of *constant* potential difference generically as a **battery**, although the chemical cell batteries typically used in flashlights and radios are only one possible source of potential difference. The chemical battery goes back to about 1800, when the Italian physicist Allesandro Volta (1745–1827) found that a stack of alternating copper and zinc plates interleaved with acid-soaked cloth provided a source of nearly constant potential difference under most conditions. The basic principle first observed by Volta is the basis for the chemical cell batteries we still use, although the materials used may differ.

In a **circuit diagram**, the connections between batteries, capacitors, and other electronic devices are shown schematically. In Figure 19.4, we show a circuit diagram in which the two ends of a battery are connected by conducting wires to the two plates of a capacitor. The universal symbol for a capacitor is the two parallel lines of equal length (simulating parallel plates). The battery symbol is two parallel lines of unequal length, to remind you of the potential difference between the two terminals of the battery. The longer line in the battery symbol is the terminal at the higher electrical potential, and the shorter line is at the lower potential. These two terminals are sometimes called "positive" and "negative," respectively, although this can be misleading, because the zero point of electrical potential is arbitrary. Conducting wires are depicted by solid lines in a circuit diagram. In Figure 19.4, you can clearly see the solid lines connecting the two battery terminals to the two ends of the capacitor.

Batteries can be stacked end to end (as in a flashlight) in order to provide a greater potential difference. Suppose we have two batteries, labeled "1" and "2." We connect the negative terminal of battery 1 to the positive terminal of battery 2. Then the potential difference between the open ends (the positive end of 1 and the negative end of 2) is the *sum* of the potential differences of the individual batteries. For example, two 1.5-V batteries connected in this manner produces a 3.0-V source of potential difference. This combination is shown in Figure 19.5.

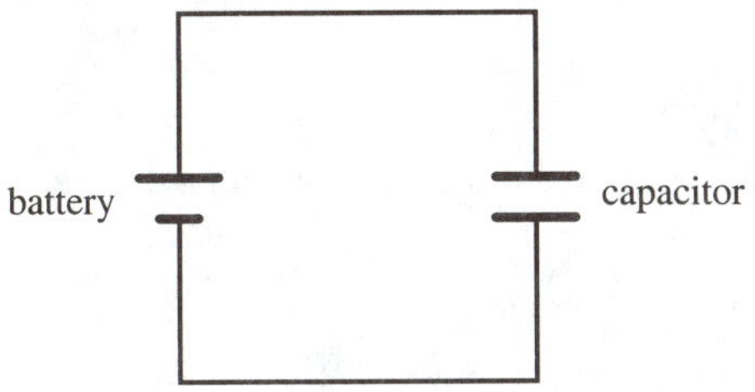

Figure 19.4. Simple circuit containing a battery and a capacitor.

Figure 19.5. Symbol for two batteries.

Now let's analyze the circuit diagram of Figure 19.4. The important thing to remember about a battery is that the potential difference between its terminals is fixed. With the capacitor attached to the battery as shown, the potential difference between the capacitor plates equals the potential difference across the battery. Let V be the magnitude of that potential difference. Then by Equation (19.1) there exist charges of opposite sign and equal magnitude Q on the two capacitor plates, with

$$Q = CV. \tag{19.8}$$

We can easily understand how the charge appears on the capacitor plates when the connection is made. Remember that in a conductor there are free negative charges (electrons). Free electrons are drawn from the (conducting) capacitor plate toward the positive battery terminal. Similarly, electrons are drawn from the negative battery terminal toward the other capacitor plate, so that charges of equal magnitude and opposite sign appear on the capacitor plates. Now if the conducting wires are removed, the capacitor maintains its charge, because the electrons on the negative capacitor plate are still attracted toward the net positive charge on the other plate. In theory, this charge can be maintained indefinitely, but in real capacitors the charge may leak away slowly, because the plates cannot be perfectly insulated from their environment.

Example 19.6. A 0.350-μF capacitor is connected to a 9.0-V battery, as shown in Figure 19.4. Find the magnitude of the charge appearing on each plate of the capacitor.

Solution. The magnitude of the charge is given by Equation (19.8) as

$$Q = CV = (0.350\,\mu\text{F})(9.0\text{ V}) = 3.15\,\mu\text{F}\cdot\text{V}.$$

From the definition of the farad, 1 F·V = 1 C, so

$$Q = 3.15\,\mu\text{C}.$$

△

In Figure 19.6(a), we show a battery (with potential difference of magnitude V) connected to two capacitors C_1 and C_2 **in parallel**. By connecting in parallel, we mean that there is a direct connection (or equivalently, a conducting wire) between the positive battery terminal and one end of each capacitor, and another connection between the negative end of the battery and the other end of each capacitor. We wish to find the **equivalent capacitance** of this combination, meaning that we are to find the capacitance C_{eq} of a single capacitor that would have the same effect as the parallel combination, if it replaced the combination as shown in Figure 19.6(b).

Returning to Figure 19.6(a), we see that there is a potential difference of magnitude V across each capacitor. Therefore, as in Equation (19.8), the magnitude of the charge (plus and minus this amount) on the two capacitors are

$$Q_1 = C_1 V \qquad \text{and} \qquad Q_2 = C_2 V.$$

The total charge stored by the combination is

$$Q = Q_1 + Q_2 = (C_1 + C_2)V.$$

By an equivalent capacitor, we mean one that would store the same net charge if the same potential difference were applied. Therefore,

$$C_{\text{eq}} = \frac{Q}{V} = \frac{(C_1 + C_2)V}{V}$$

or simply

$$C_{\text{eq}} = C_1 + C_2. \tag{19.9}$$

We conclude that *the equivalent capacitance of two capacitors in parallel is the sum of the individual capacitances.*

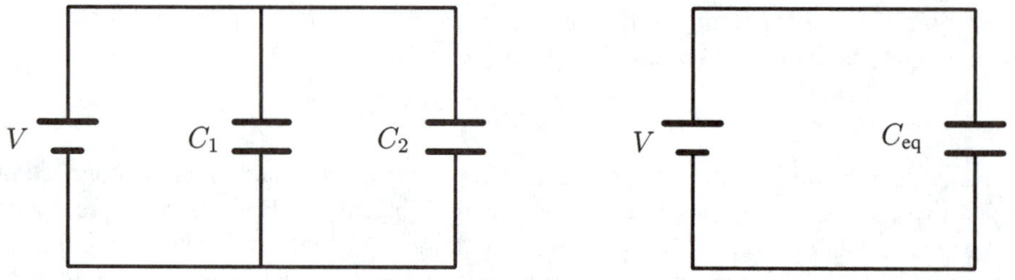

(a) Battery connected to two capacitors in parallel

(b) Two capacitors have been replaced by an equivalent capacitor

Figure 19.6. Defining equivalent capacitance.

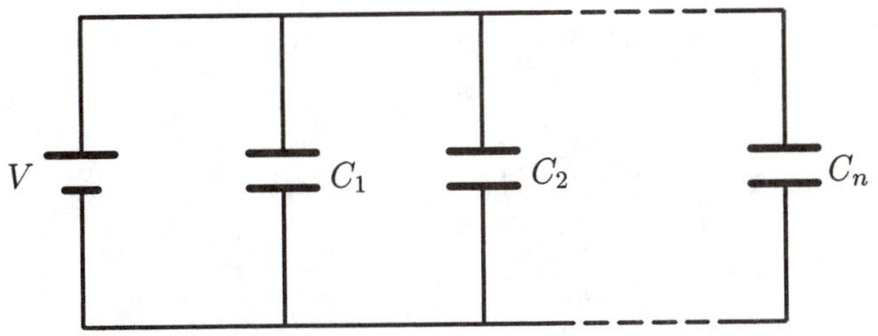

Figure 19.7. Multiple capacitors in parallel.

It is straightforward to show (see the problems) that if more capacitors (say n of them) are connected in parallel as in Figure 19.7, the equivalent capacitance is still the algebraic sum of the capacitances. That is,

$$C_{\text{eq}} = C_1 + C_2 + \cdots + C_n. \tag{19.10}$$

The fundamental reason for this result can be seen in the schematic. If we think of all the capacitors as parallel-plate capacitors with the same plate separation d, the parallel combination of capacitors has an effective surface area equal to the sum of the surface areas of all the capacitors. Because the capacitance of a parallel-plate capacitor is proportional to the area, it makes sense that the capacitance increases as the sum of all the capacitances; the area increases as the sum of the areas.

In Figure 19.8(a), we show a circuit with a battery connected to two capacitors **in series**. By this we mean that the capacitors are joined to each other at one end only, with the free ends then connected to the battery. Once again, we wish to find the equivalent capacitance of this combination—that is, a single capacitor C_{eq} that could replace the series combination with the same effect when connected to the same battery [again refer to Figure 19.6(b)].

The key to solving this problem is to realize that both capacitors must have the same charge magnitude Q. Why is this so? Consider the region of the series circuit surrounded by the dashed line in Figure 19.8(b). Before the battery is connected, there is zero net charge in this region. When electrons in that region move to the plate on C_1 (toward the battery's positive terminal) and deposit a net charge $-Q$ on that plate, conservation of charge dictates that they leave behind a charge $+Q$ on the top plate of C_2. Therefore, the charge magnitude on each capacitor is the same. The total potential difference between the battery terminals is then the sum of

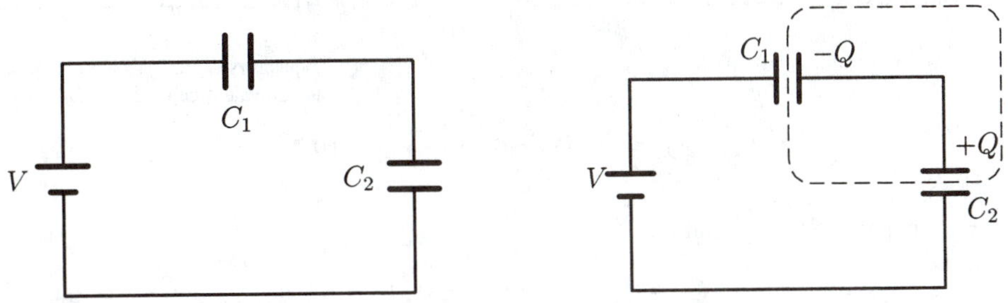

(a) Two capacitors in series

(b) In the region enclosed by the dashed line, the net charge is zero

Figure 19.8. Analyzing equivalent capacitance for capacitors in series.

the potential differences of the capacitor plates, or

$$V = V_1 + V_2.$$

From the definition of capacitance, $V_1 = Q/C_1$ and $V_2 = Q/C_2$, so

$$V = \frac{Q}{C_1} + \frac{Q}{C_2}.$$

Now looking at the equivalent capacitance diagram [Figure 19.6(b)], $V = Q/C_{eq}$, so

$$\frac{Q}{C_{eq}} = \frac{Q}{C_1} + \frac{Q}{C_2}$$

or

$$\frac{1}{C_{eq}} = \frac{1}{C_1} + \frac{1}{C_2}. \tag{19.11}$$

This result can be stated: *The reciprocal of the equivalent capacitance of two capacitors in series is the sum of the reciprocals of the individual capacitances.* Once again, the result is easily extended to n capacitors in series:

$$\frac{1}{C_{eq}} = \frac{1}{C_1} + \frac{1}{C_2} + \cdots + \frac{1}{C_n}. \tag{19.12}$$

Example 19.7. Find the equivalent capacitance of the combination of $C_1 = 3.0\ \mu$F and $C_2 = 6.0\ \mu$F (a) in parallel, as in Figure 19.6(a), and (b) in series, as in Figure 19.8(a). (c) Now a third capacitor $C_3 = 6.0\ \mu$F is combined with the other two capacitors as shown in Figure 19.9(a). What is the equivalent capacitance?

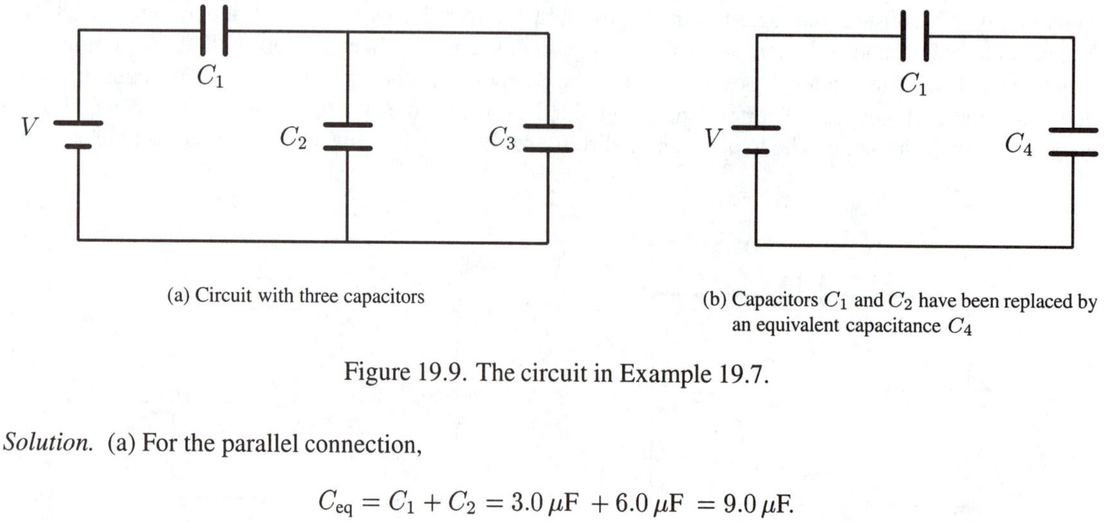

(a) Circuit with three capacitors

(b) Capacitors C_1 and C_2 have been replaced by an equivalent capacitance C_4

Figure 19.9. The circuit in Example 19.7.

Solution. (a) For the parallel connection,

$$C_{eq} = C_1 + C_2 = 3.0\ \mu\text{F} + 6.0\ \mu\text{F} = 9.0\ \mu\text{F}.$$

(b) For the series connection,

$$\frac{1}{C_{eq}} = \frac{1}{C_1} + \frac{1}{C_2} = \frac{1}{3.0\ \mu\text{F}} + \frac{1}{6.0\ \mu\text{F}} = \frac{1}{2.0\ \mu\text{F}}.$$

Taking the reciprocal,

$$C_{\text{eq}} = 2.0\,\mu\text{F}.$$

(c) For a more complex network, we must combine the various parts using the rules established for series and parallel combinations of capacitors. Notice that C_2 and C_3 are in parallel. Therefore, they may be replaced by a single capacitor with capacitance $C_4 = C_2 + C_3 = 12.0\,\mu$F. The remaining network is a *series* containing C_1 and C_4 [Figure 19.9(b)], with equivalent capacitance given by the series rule:

$$\frac{1}{C_{\text{eq}}} = \frac{1}{C_1} + \frac{1}{C_4} = \frac{1}{3.0\,\mu\text{F}} + \frac{1}{12.0\,\mu\text{F}} = \frac{5}{12.0\,\mu\text{F}}\,.$$

Taking the reciprocal,

$$C_{\text{eq}} = \frac{12.0\mu\text{F}}{5} = 2.4\,\mu\text{F}.$$

Notice that we found the equivalent capacitance of the parallel combination first, because it is computed with a simple sum, and we saved the series and its reciprocal sum for last. △

This last example illustrates some important points about series and parallel combinations of capacitors. First, the equivalent capacitance of a parallel combination is greater than the capacitance of either individual capacitor. However, the equivalent capacitance of a series combination is less than the smallest capacitance in the series. Therefore, parallel and series combinations allow you to create larger and smaller capacitances, respectively.

19.1.4 Energy Stored in Capacitors

We can use the principle of conservation of energy to find the potential energy of a charge distribution in a capacitor. The change in potential energy when a capacitor is brought from an uncharged state to a charge Q is equal to the work done by whatever external agent (such as a battery) is moving the charges. This is analogous to the way we found the change in gravitational potential energy of a particle by computing the work needed to lift the particle against gravity.

For convenience, let's imagine charging the capacitor by moving positive charges from one plate of the capacitor to another, putting a positive charge on one plate and leaving a negative charge on the other. (As you know, it is really the other way around, because it is the negative electrons that are free to move in the conductor. However, this assumption is mathematically equivalent and will allow us to avoid dealing with negative signs.) We move the charge incrementally, so that at some time before the capacitor has its full charge Q it has some charge $q < Q$. At that time, the definition of capacitance tells us that the potential difference between the plates is $V = q/C$. From our study of electric potential, we also know that the potential V is the work done per unit charge in moving a test charge from one plate to the other. Expressing this fact as a derivative,

$$V = \frac{dW}{dq}\,. \tag{19.13}$$

Because the potential is the derivative of the work with respect to charge, the net work in taking the capacitor from a charge $q = 0$ to $q = Q$ is the definite integral

$$W = \int_0^Q V\,dq. \tag{19.14}$$

(This is analogous to the kinematics problem in which we begin with $v = dx/dt$ and find the net displacement on a time interval $[a, b]$ with the definite integral $\int_a^b v\,dt$.) With $V = q/C$, we have

$$W = \int_0^Q \frac{q}{C}\,dq = \frac{1}{C}\int_0^Q q\,dq = \frac{1}{C}\frac{q^2}{2}\bigg|_0^Q = \frac{Q^2}{2C}\,.$$

As we noted earlier, the potential energy U of the capacitor is exactly equal to this amount of work, so

$$U = \frac{Q^2}{2C}\,. \tag{19.15}$$

We can use $Q = CV$ to write an equivalent expression for the potential energy of the capacitor as a function of C and V (but not Q):

$$U = \frac{1}{2}CV^2. \tag{19.16}$$

Example 19.8. Find the potential energy of a 10.0-μF capacitor attached directly across a 12.0-V battery.

Solution. Since we are given the potential difference and not the charge, it is convenient to use Equation (19.16) for the potential energy:

$$U = \frac{1}{2}CV^2 = \frac{1}{2}\left(1.0 \times 10^{-5}\text{ F}\right)\left(12.0\text{ V}\right)^2 = 7.2 \times 10^{-4}\text{ J.}$$

Note that in SI units

$$1\text{ F}\cdot\text{V}^2 = 1\frac{\text{C}}{\text{V}}\cdot\text{V}^2 = 1\text{ C}\cdot\text{V} = 1\text{ J.}$$

and so the answer does come out in joules. △

Energy density Physicists often use the quantity **energy density** in a field. For the electric field, the energy density u is defined as the potential energy per unit volume. As an example, let's compute the energy density of a parallel-plate capacitor. We use Equation (19.16) for the potential energy, with Equation (19.4) for the capacitance. In the infinite-plate approximation, the electric field is uniform between the plates, with magnitude $E = V/d$ (where V is the potential difference and d is the distance between the plates). Since we want the potential energy per unit volume, we can use Ad for the volume of the rectangular region between the plates. Combining these results, we find that the energy density between the plates is

$$u = \frac{\text{potential energy}}{\text{volume}} = \frac{\frac{1}{2}CV^2}{Ad} = \frac{V^2}{2Ad}\left(\frac{\epsilon_0 A}{d}\right) = \frac{1}{2}\epsilon_0\left(\frac{V}{d}\right)^2$$

or

$$u = \frac{1}{2}\epsilon_0 E^2. \tag{19.17}$$

It turns out that even though we used the special example of the parallel-plate capacitor to derive Equation (19.17), it is true for any electric field. This is useful to know in cases in which the electric field is not uniform. The definition of energy density tells us that the potential energy associated with an electric field in a region of space R is

$$U = \iiint_D u\,dV \tag{19.18}$$

where u is given by Equation (19.17).

Example 19.9. Use Equation (19.18) to find the potential energy stored in a spherical capacitor with charge Q (i.e., $+Q$ on one sphere and $-Q$ on the other). Show that your result is consistent with the potential energy $U = Q^2/2C$.

Solution. The electric field between the spheres (refer to Figure 19.3) is due to the inner sphere alone and has a magnitude

$$E = \frac{kQ}{r^2}$$

where r is the distance from the center of the sphere. The spherical symmetry of this problem dictates that we use spherical coordinates for the triple integral in Equation (19.18). With $u = \frac{1}{2}\epsilon_0 E^2$, we have

$$U = \iiint\limits_D u\,dV = \frac{1}{2}\epsilon_0 \iiint\limits_D E^2\,dV = \frac{1}{2}\epsilon_0 k^2 Q^2 \iiint\limits_D \frac{1}{r^4}\,dV.$$

In spherical coordinates, $dV = r^2 \sin\phi\,dr\,d\phi\,d\theta$, where we have replaced the ρ of spherical coordinates with the r as defined in this problem. The limits of integration are 0 to 2π for θ, 0 to π for ϕ, and a to b (from the inner sphere to the outer sphere) for r. Carrying out the triple integral,

$$\begin{aligned} U &= \frac{1}{2}\epsilon_0 k^2 Q^2 \int_0^{2\pi}\int_0^{\pi}\int_a^b \frac{1}{r^4} r^2 \sin\phi\,dr\,d\phi\,d\theta \\ &= \frac{1}{2}\epsilon_0 k^2 Q^2 \int_0^{2\pi} d\theta \int_0^{\pi} \sin\phi\,d\phi \int_a^b \frac{1}{r^2}\,dr \\ &= \frac{1}{2}\epsilon_0 k^2 Q^2\,(2\pi)\,(2)\left(\frac{1}{a}-\frac{1}{b}\right) \end{aligned}$$

or after simplifying, with $k = 1/(4\pi\epsilon_0)$:

$$U = \frac{kQ^2}{2}\left(\frac{1}{a}-\frac{1}{b}\right).$$

Now we check this result with $U = Q^2/2C$. From our previous analysis of the spherical capacitor,

$$\frac{1}{2C} = \frac{k}{2}\left(\frac{1}{a}-\frac{1}{b}\right).$$

Therefore,

$$U = \frac{Q^2}{2C} = \frac{kQ^2}{2}\left(\frac{1}{a}-\frac{1}{b}\right),$$

in agreement with our result from integrating the energy density. This example serves to illustrate the general validity of Equation (19.17) for the energy density in an electric field, although the proof of that fact is beyond the scope of this course. △

19.2 Dielectrics

A **dielectric** is any insulating material such as glass, wood, or plastic. We have already seen one way in which dielectric materials respond to electric fields. Recall our discussion of charge polarization at the end

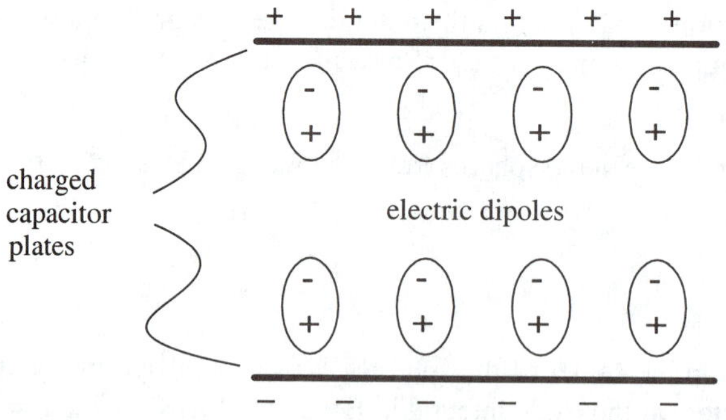

Figure 19.10. Schematic of electric dipoles in a dielectric.

of Section 16.1.1. When a charged rod is brought near an insulator (a dielectric), the electric dipoles within the material can become aligned to some degree, as shown in Figure 16.1. The amount of alignment depends significantly on the dielectric material being studied. We will soon introduce a way to quantify the relative amount of alignment.

In Figure 19.10, we illustrate the alignment of the electric dipoles within a dielectric that has been placed between the charged plates of a capacitor. The negative ends of electric dipoles tend to be attracted toward the positively charged capacitor plate, and the positive ends of electric dipoles tend to be attracted toward the negatively charged capacitor plate. This alignment of dipoles tends to reduce the effective charge on each capacitor plate, thus reducing the electric field between the plates. The factor by which the magnitude of the electric field is reduced is called the **dielectric constant**κ of the dielectric material (κ is therefore a dimensionless number). Since the electric field is reduced for any dielectric, this requires $\kappa > 1$. In Table 19.1, we list the dielectric constants of a few insulators. If there is a vacuum between the capacitor plates, there is no reduction of the electric field, and so $\kappa = 1$ for a vacuum. If we insert a conductor rather than an insulator in the capacitor, enough free charges in the conductor can move toward the capacitor plates (negative charges toward the positive plate, leaving a positive charge facing the positive plate) so that the electric field is completely cancelled and is zero inside the conductor. Then we would say that the dielectric constant of a conductor is infinite.

Table 19.1. Dielectric constants and dielectric strengths of selected materials (measured at 20°C).

Material	Dielectric constant κ	Dielectric Strength E_{max} (V/m)
Vacuum	1.000 (exactly)	none
Air	1.00058	3×10^6
Teflon	2.1	6.0×10^7
Nylon	3.4	1.4×10^7
Paper	3.7	1.6×10^7
Bakelite	4.9	2.4×10^7
Pyrex	5.6	1.4×10^7
Neoprene	6.7	1.2×10^7
Water	80	*
Strontium titanate	256	8.0×10^6

Recall that in the parallel-plate capacitor the potential difference between the plates is

$$V = Ed$$

where E is the magnitude of the electric field and d is the separation between the plates. Because V is proportional to E, we conclude that V is also reduced by a factor κ when a dielectric fills the space between the capacitor plates. In symbols, this is expressed

$$V = \frac{V_0}{\kappa} \tag{19.19}$$

where V_0 is the electric potential under identical conditions with a vacuum between the plates. For a given amount of charge of magnitude Q_0 on the capacitor, the new capacitance is

$$C = \frac{Q_0}{V} = \frac{Q_0}{V_0/\kappa} = \frac{\kappa Q_0}{V_0}$$

or

$$C = \kappa C_0 \tag{19.20}$$

where C_0 is the capacitance of the vacuum-filled capacitor.

Equation (19.20) shows that the capacitance of a dielectric-filled capacitor increases in proportion to the dielectric constant κ. In Section 19.1.2, we showed that it was difficult to build parallel-plate capacitors with high capacitances (say greater than μF) without making the capacitor plates unreasonably large or close together. With dielectrics, we have a way to vary the capacitance significantly and construct capacitors that have extremely high capacitances—on the order of mF or higher.

When dielectrics are used in capacitors, there is a limit to the amount of charge the capacitor can hold. No insulator is a perfect insulator, and if the electric field is strong enough, it can cause electrical conduction inside the insulator. When this happens, the capacitor discharges through the dielectric. The limiting value of the electric field in a dielectric E_{max} is called the **dielectric strength** of the material, and typical values are listed in Table 19.1.

Example 19.10. An air-filled parallel-plate capacitor has square plates measuring 1.10 cm on a side separated by 0.150 mm. (a) Find the capacitance and maximum charge that can be placed on this capacitor. (b) Find the capacitance and maximum charge if a sheet of bakelite replaces the air between the plates.

Solution. (a) Given the number of significant digits in this problem, the effective dielectric constant of air is $\kappa = 1.00$. Using Equation (19.4) for the capacitance of the parallel-plate capacitor,

$$C = \kappa C_0 = \frac{\kappa \epsilon_o A}{d} .$$

Inserting numerical values,

$$C = \frac{(1.00)(8.85 \times 10^{-12}\ \text{F/m})(0.0110\ \text{m})^2}{1.50 \times 10^{-4}\ \text{m}} = 7.14 \times 10^{-12}\ \text{F}$$

or 7.14 pF. To find the maximum charge, recall that $Q = CV$ and $V = Ed$. Therefore,

$$Q_{\text{max}} = CV_{\text{max}} = CE_{\text{max}}d.$$

From Table 19.1, $E_{\text{max}} = 3 \times 10^6$ V/m, so

$$Q_{\text{max}} = (7.14 \times 10^{-12}\ \text{F})(3 \times 10^6\ \text{V/m})(1.50 \times 10^{-4}\ \text{m}) = 3.2 \times 10^{-9}\ \text{F·V}$$

or 3.2 nC.

(b) Repeating the calculations but with $\kappa = 4.9$ and $E_{\text{max}} = 2.4 \times 10^7$ V/m for bakelite,

$$C = \frac{(4.9)(8.85 \times 10^{-12}\ \text{F/m})(0.0110\ \text{m})^2}{1.50 \times 10^{-4}\ \text{m}} = 3.50 \times 10^{-11}\ \text{F} = 35.0\ \text{pF}$$

and

$$Q_{\text{max}} = (3.50 \times 10^{-11}\ \text{F})(2.4 \times 10^7\ \text{V/m})(1.50 \times 10^{-4}\ \text{m}) = 1.26 \times 10^{-7}\ \text{C} = 126\ \text{nC}.$$

△

How does the presence of a dielectric affect energy storage in a capacitor? Suppose a vacuum-filled capacitor with capacitance C_0 is charged to a value Q_0 and disconnected from the charging battery. From Equation (19.15), the energy stored is

$$U_0 = \frac{Q_0^2}{2C_0}\,.$$

When a material with dielectric constant κ is inserted between the plates, the charge in the plates is not affected, but the capacitance is increased by a factor of κ. The new capacitance is $C = \kappa C_0$, and so the new potential energy is

$$U = \frac{Q_0^2}{2C} = \frac{Q_0^2}{2\kappa C_0}$$

or

$$U = \frac{U_0}{\kappa}\,. \tag{19.21}$$

The potential energy is *decreased* by a factor of κ. This can be explained using the concept of conservation of energy. When the dielectric is inserted between the plates, the charged plates attract the partially polarized electric dipole moments in the dielectric. Therefore, we do *negative* work on the dielectric in the process of inserting it between the plates, and the overall potential energy is decreased.

19.3 Electric Current

19.3.1 Introduction and Definitions

In electric circuits, conducting wires are used to carry charge from one place to another. At a given position in the circuit, the **electric current** is defined as the *instantaneous rate* at which charge is flowing past that point. From calculus, we know that the derivative measures the instantaneous rate, so we can write the current I as

$$I = \frac{dq}{dt}\,. \tag{19.22}$$

The SI unit of current is the ampere (A), with

$$1\ \text{A} = 1\ \text{C/s}.$$

As we noted in Chapter 16, the ampere is actually a base unit in the SI system, and the coulomb is defined in terms of it. In Chapter 25, we will see how the ampere is defined.

You know from your study of charge that 1 C is a significant amount of charge. Therefore, you should expect that 1 A is a reasonably high current. Indeed, the current in many electronic devices is on the order of mA, μA, or even nA. Large household appliance such as washing machines and electric clothes dryers may use 10–20 A. Household circuit breakers are often designed to let no more than about 30 A flow through one outlet. It is not unusual to see even higher currents for short periods. For example, a current of up to 200 A may be required from a car battery in order to start a cold engine.

The symbolic definition of electric current in Equation (19.22) shows that current is a *scalar*. However, there is a definite sense of direction in which electric current is flowing in a conducting wire. In this way, the current I is quite similar to the velocity component v_x in one-dimensional motion. In kinematics, $v_x > 0$ if a particle is moving in the $+x$-direction, and $v_x < 0$ if it is moving in the $-x$-direction. Similarly, we will take $I > 0$ when *positive* charges move one direction in a wire, and $I < 0$ when positive charges move in the other direction. Just like the $+x$-direction in one-dimensional motion, the original choice of which direction will give positive current is arbitrary; once that decision is made, we must stick with it throughout the working of a problem in order to maintain consistency.

In problems in which we are concerned about the motion of the individual charge carriers, you should be careful to distinguish between the motion of those charge carriers and the direction of current. In a conductor, it is always the negatively charged electrons that move and carry the charge from one place to another. Thus, their motion is actually *opposite* to that which we have defined as the direction of positive current (Figure 19.11). As long as we are careful to distinguish between the current and the charge carriers, there is no problem, because the same *net* flow of charge is recorded in Equation (19.22), whether negative charges flow one way or positive charges flow the other.

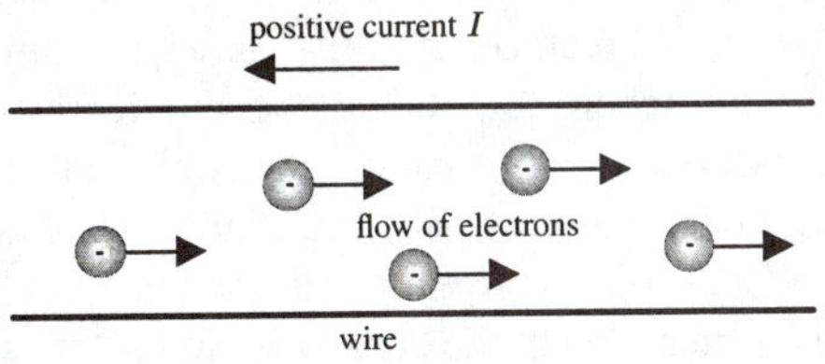

Figure 19.11. Flow of electrons and positive current in a conductor.

In our study of capacitors, we saw that in an electric circuit negative charges flow into the positive terminal of a battery and out of the negative terminal. Therefore, positive current flows out of the positive battery terminal and into the negative terminal. Suppose we connect a conducting wire directly between the two battery terminals. (We do not recommend this; it will discharge the battery rapidly with a high current flow!) Then the flow of positive current in the wire is in the direction shown in Figure 19.12, with the current going from the positive terminal through the wire to the negative terminal.

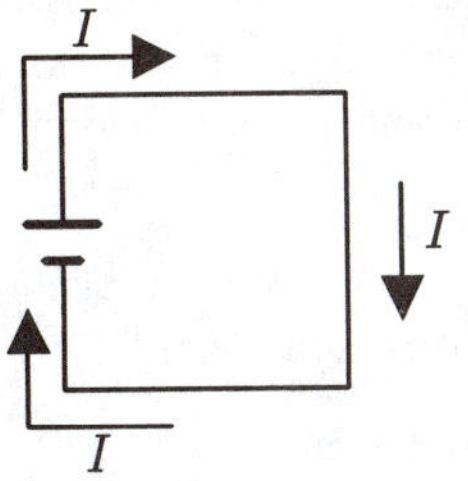

Figure 19.12. Flow of positive current through a wire attached to a battery.

19.3.2 A Microscopic View of Electric Current

We will gain some insight into electric current by considering the motion of individual charge carriers (electron in a conductor). The motion of electrons in a conductor is not at all as regular as in the schematic (Figure 19.11) we used to illustrate current flow. When there is no current flowing in a conductor, the free electrons move randomly due to their thermal energy. This energy is temperature-dependent, but at room temperature free electrons have an average kinetic energy of about 4×10^{-21} J. This may seem like a small amount of energy, but remember that the electron's mass is small. With a kinetic energy $K = 4 \times 10^{-21}\ \mathrm{J} = \frac{1}{2}mv^2$, the speed works out to be nearly 10^5 m/s. In kinetic theory, the speed computed in this manner is called the *root mean square* speed. The speeds of individual electrons vary a good deal, and the directions of the individual electrons' velocities are random.

When a potential difference is introduced between the ends of the conducting wire, an electric field exists in the wire, causing a force on every electron. With an electric field $\vec{E}$, the force on an electron (with charge $-e$) is $\vec{F} = -e\vec{E}$. This creates a current flow, as each free electron is accelerated in a direction *opposite* to the electric field. The direction of positive current is roughly in the direction of the electric field. (We say *roughly* because the direction of $\vec{E}$ inside the wire varies locally and is not absolutely parallel to the wire everywhere.) However, electrons do not accelerate for long before colliding with one of the positive ions in the conductor. Those collisions are the root cause of electrical resistance, which we will study in Section 19.3.3. One result of the electrons' collisions with the positive ions is that the magnitude of an electron's average velocity (in a direction opposite to the applied electric field) is extremely low compared with the root mean square speed due to the thermal energy we described earlier. We call the magnitude of the electron's average velocity the **drift speed** v_d, and we now use dimensional analysis to see how it can be computed.

Consider a section of wire as shown in Figure 19.13, with the drifting electrons passing through a section of wire. In a time Δt, an electron with the average speed v_d passes through a section of length $v_d\Delta t$. The volume of that section of wire is $Av_d\Delta t$, where A is the (assumed uniform) cross-sectional area of the wire. By dimensional analysis, the amount of charge Δq passing through that section of wire equals the charge density of free electrons multiplied by the volume. We will express the charge density of free electrons in terms of the *number density* n (number of free electrons per unit volume) and the charge magnitude e of each electron. Dimensional analysis tells us that the charge density is simply ne, so we find

$$\Delta q = \text{(charge density of free electrons)(volume)} = (ne)(Av_d\Delta t)$$

or

$$\Delta q = neAv_d\Delta t. \tag{19.23}$$

From the definition of current, we have

$$I = \frac{\Delta q}{\Delta t} = neAv_d. \tag{19.24}$$

Equation (19.24) is a significant result, because it relates the macroscopic parameters of the electric current I and the wire's cross-sectional area A to the density of free electrons (n), the electronic charge e, and the electron drift speed v_d. This gives us a way of knowing v_d without having to measure the speeds of individual electrons.

Example 19.11. Suppose a copper wire with circular cross-section and radius 0.150 mm carries a current of 5.42 A. What is the drift speed of the electrons in the wire?

Solution. From Equation (19.24) the drift speed is

$$v_d = \frac{I}{neA}.$$

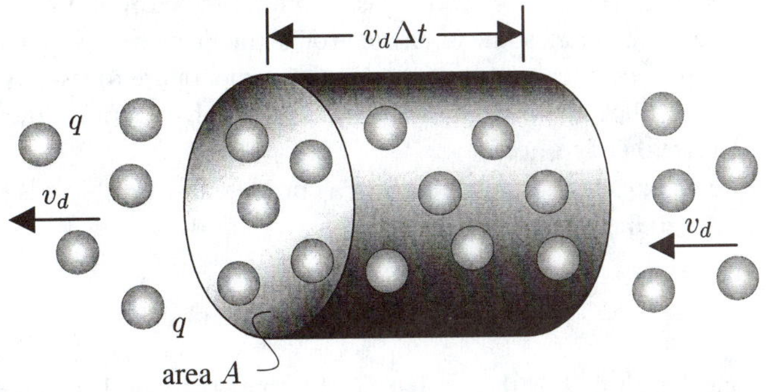

Figure 19.13. Section of a current-carrying wire, illustrating the drift of charges.

Table 19.2. Number density n of free charge carriers in various conductors (at $T = 27°C$)

Conductor	$n(\times 10^{28}\ m^{-3})$	Conductor	$n(\times 10^{28}\ m^{-3})$
Silver	5.86	Aluminum	18.1
Copper	8.47	Tin	14.8
Gold	5.86	Galium	15.4
Iron	17.0	Zinc	13.2
Niobium	5.56	Lead	13.2

In Table 19.2, we give the density of conduction electrons in some common conductors, including copper, which has $n = 8.47 \times 10^{28}\ m^{-3}$. The wire has a circular cross section, with area $A = \pi r^2$. Inserting the numerical values, we find

$$v_d = \frac{5.42\ A}{(8.47 \times 10^{28}\ m^{-3})(1.60 \times 10^{-19}\ C)[(\pi)(1.50 \times 10^{-4}\ m)^2]} = 5.65 \times 10^{-3}\ m/s.$$

This numerical result may seem low, but it is correct. It bears out our statement in the text that the drift speed is normally much lower than the root mean square speed from the thermal energy. In copper, there is one free electron per atom in the conductor. Therefore, the number of conduction electrons in a macroscopic wire is huge, and a relatively low drift speed is needed to produce a significant current. △

Notice that for a given type of conductor, the ratio I/A is directly proportional to the drift speed and is given by

$$J = \frac{I}{A} = nev_d. \tag{19.25}$$

We call J the **current density**. Since the current density is current divided by area, it has SI units A/m^2.

19.3.3 Resistance, Conductivity, and Ohm's Law

In the preceding section, we explained how electrons' drift speeds are limited by their collisions with the positive ions in a conductor. In Equation (19.24), we saw that the electric current I is directly proportional to the drift speed, which means the effect of these collisions is to limit the current flow through the conductor. In each collision between electron and ion, some of an electron's kinetic energy is transferred to the lattice of ions, and this energy shows up as heat, causing the conductor's temperature to rise. It is these two features (limiting of current and heating of heating of the material through which current is flowing) that we associate with the phenomenon of **electrical resistance**.

The physical quantity we call electrical resistance R of a current-carrying object is defined as the ratio of the applied potential difference to the current, or

$$R = \frac{V}{I}. \tag{19.26}$$

A defined SI unit called the **ohm** (Ω), the capital Greek omega) is used to measure resistance, with $1\ \Omega = 1$ V/A. The ohm is named for German physicist Georg Ohm (1787–1854), who made a careful study of the resistances of numerous materials. To a good approximation, many materials follow **Ohm's law**, which states that R is constant for a given piece of material, so that regardless of V and I the ratio $R = V/I$ is the same. We caution that Ohm's law is not really a law in the sense of Newton's law of gravitation or Coulomb's law. It is more like the statement that the coefficient of kinetic friction μ_k is constant for a given material, in that it is an experimental observation that is only approximately true, and the accuracy of the approximation varies greatly with different materials.

People sometime refer (incorrectly) to the equation $V = IR$ as Ohm's law, but as you can see this is really just a rearrangement of the definition of resistance in Equation (19.26). However, for materials that follow Ohm's law (called **ohmic materials**)

$$V = IR \tag{19.27}$$

is a linear relationship, because R is constant. This is illustrated by the graph of V vs. I for an ohmic material (Figure 19.14). The graph is a straight line with a slope R.

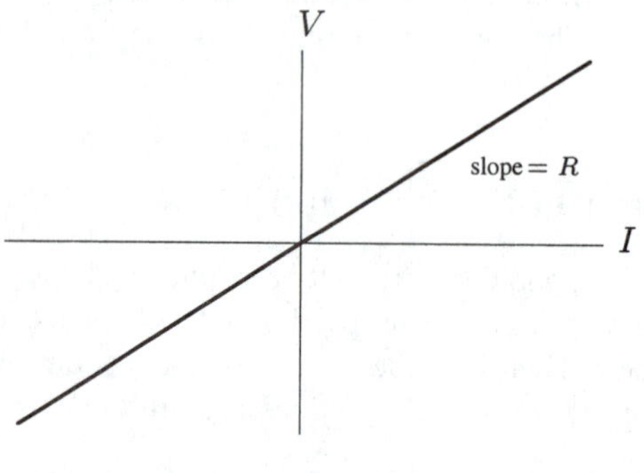

Figure 19.14. Resistance is constant according to Ohm's law.

Ohmic devices constructed specifically for use in electric circuits are called **resistors**. Many different types of resistors are available commercially, with resistances anywhere from a fraction of one ohm to many millions of ohms. Carbon (in the form of graphite) has been a popular resistor material for years, but today resistors are made from a variety of different materials. Often resistors have a cylindrical shape and three colored bands that indicate the numerical value of resistance (see Figure 19.15). There are ten different colors used in these bands, with a number corresponding to each color (see Table 19.3). If the three bands (starting from the one nearer one end) are labeled A, B, and C, then the nominal resistance (in ohms) is

$$R = AB \times 10^C$$

Table 19.3. Resistor color code

Color	Number
Black	0
Brown	1
Red	2
Orange	3
Yellow	4
Green	5
Blue	6
Violet	7
Gray	8
White	9

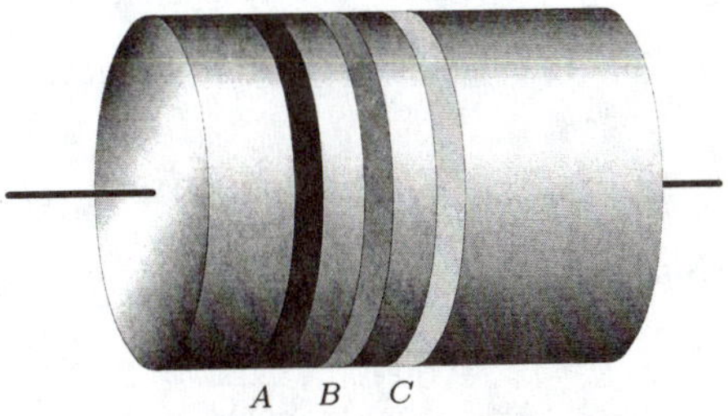

Figure 19.15. Three-bar resistor code.

where AB is *not* A multiplied by B, but rather a two-digit number with A in the ten's place and B in the one's place. For example, if the three colors (in order) are yellow, violet, and red, then the resistance is

$$R = 47 \times 10^2\ \Omega = 4.7 \times 10^3\ \Omega = 4700\ \Omega.$$

Sometimes manufacturers add a fourth colored band to indicate the maximum expected error in the stated value compared with the actual resistance. A gold band indicates an expected error of less than 5%, and a silver band less than 10%.

Resistors are widely used to control the current in electric circuits. Consider the simple circuit shown in Figure 19.16, in which a battery with constant potential difference V is connected to a resistor, which we will assume has constant resistance R (note the symbol for resistors in circuit diagrams, which we introduce in that figure). Positive current flows in the direction shown, from the positive terminal of the battery, through the resistor, and into the negative terminal of the battery. Because no buildup of charge occurs anywhere in the circuit (as in a capacitor), the current I is the same everywhere in the circuit. In this case and in all circuits containing resistors, we will assume that the resistance of the connecting wires is negligible compared with that of the resistor, unless stated otherwise. Therefore, the potential difference across the resistor is V, and the current through the resistor is

$$I = \frac{V}{R}.$$

For a given battery, the amount of current flowing through a resistor varies *inversely* with the resistance. A higher resistance results in less current flow. We will examine current flow through various combinations of resistors (such as series and parallel) in Chapter 21.

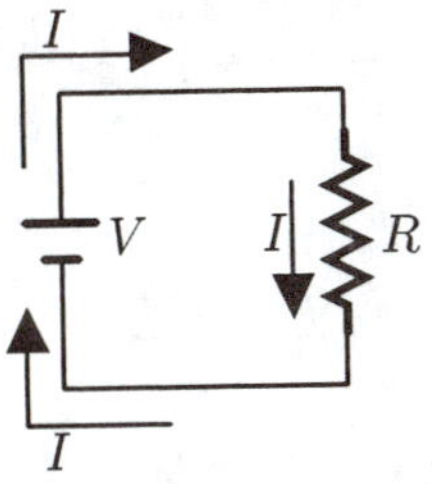

Figure 19.16. Flow of current through a circuit consisting of a resistor attached to a battery.

Example 19.12. A 12-V battery is connected to a single resistor as shown in Figure 19.16. Find the current flowing through the resistor if its resistance is (a) 30 Ω, (b) 30 kΩ, or (c) 30 MΩ.

Solution. (a) From the definition of resistance, $I = V/R$ in each case. With $R = 30\,\Omega$,

$$I = \frac{12\text{ V}}{30\,\Omega} = 0.4\text{ V/}\Omega = 0.4\text{ A}.$$

(b) Similarly,

$$I = \frac{12\text{ V}}{30\text{ k}\Omega} = \frac{12\text{ V}}{30 \times 10^3\,\Omega} = 0.4 \times 10^{-3}\text{ A} = 0.4\text{ mA}.$$

(c) Now

$$I = \frac{12\text{ V}}{30\text{ M}\Omega} = \frac{12\text{ V}}{30 \times 10^6\,\Omega} = 0.4 \times 10^{-6}\text{ A} = 0.4\,\mu\text{A}.$$

△

The numerical value of resistance of a given resistor depends not only on the material in it, but also on the resistor's size and shape. It is observed that for a given material the resistance is proportional to the length L of the resistor (as measured along the direction of current flow) and inversely proportional to the resistor's cross-sectional area A (the surface area perpendicular to the direction of current flow). These facts can be combined in the equation

$$R = \rho\frac{L}{A} \tag{19.28}$$

where the parameter ρ, called the **resistivity**, is a characteristic of the material of which the resistor is made. From Equation (19.28), it is clear that the SI unit for resistivity is Ω·m. The reciprocal of the resistivity is called the **conductivity**, with units $\Omega^{-1}\cdot\text{m}^{-1}$. The symbol σ is used for conductivity, so we can write

$$\sigma = \frac{1}{\rho}\,. \tag{19.29}$$

Please do not confuse these symbols (σ and ρ) with the two- and three-dimensional charge densities we used earlier! They should never appear in the same context.

Some typical resistivities are given in Table 19.4. Notice that different classes of materials have radically different resistivities. Pure metals have the lowest resistivities, followed closely by metal alloys. Another class of material, known as semiconductors (which we will study in Section 19.4), have higher resistivities, and insulators even higher. Notice how many orders of magnitude separate the resistivities of metals and insulators.

With Equation (19.28), we can develop an alternative form of Ohm's law :

$$V = IR = I\left(\rho\frac{L}{A}\right).$$

Rearranging,

$$\frac{I}{A} = \frac{1}{\rho}\frac{V}{L}\,.$$

But recall that $J = I/A$, $\sigma = 1/\rho$, and $E = V/L$. Therefore,

$$J = \sigma E. \tag{19.30}$$

Table 19.4. Resistivities and temperature coefficients of resistivity for selected materials (measured at 20°C)

Material	Resistivity ρ (Ω·m)	Temperature coefficient α (°C^{-1})
Silver	1.59×10^{-8}	3.8×10^{-3}
Copper	1.69×10^{-8}	3.9×10^{-3}
Gold	2.44×10^{-8}	3.4×10^{-3}
Aluminum	2.75×10^{-8}	3.9×10^{-3}
Tungsten	5.61×10^{-8}	3.9×10^{-3}
Platinum	1.06×10^{-7}	3.9×10^{-3}
Lead	2.23×10^{-7}	3.9×10^{-3}
Nichrome	1.50×10^{-6}	4.1×10^{-4}
Carbon	3.52×10^{-5}	-5.0×10^{-4}
Germanium	0.46	-4.8×10^{-2}
Silicon	640	-7.5×10^{-3}
Glass	10^{10} to 10^{14}	
Rubber	10^{13}	
Teflon	10^{14}	

This linear relationship between the current density and applied electric field is analogous to the relationship between current and potential difference. An equivalent expression to Ohm's law is to say that the electrical conductivity σ in Equation (19.30) is constant for any applied electric field.

Of course, Ohm's law is not exact in any material. There is at least one relatively straightforward correction we can make. In conductors, the resistivity tends to increase in nearly linear fashion with increasing temperature, because as the temperature increases, an increase in the random thermal motion of the ions causes more frequent electron-ion collisions. This linear increase is only an approximation, and not a good one at extremely cold (near absolute zero) or extremely high (near the material's melting point) temperatures. We can express this approximation as

$$\rho = \rho_0 \left[1 + \alpha(T - T_0)\right] \tag{19.31}$$

where ρ is the resistivity at temperature T, ρ_0 is the resistivity at some reference temperature T_0, and α is a constant for a given material. The constant α is called the **temperature coefficient of resistivity**. Some typical values of α are given in Table 19.4. Equation (19.31) is reasonably good for most metals near room temperature, as long as the temperature difference $T - T_0$ is not too great. Unfortunately, the coefficient α is not constant at all temperatures (indicating that the relationship between ρ and T is not exactly linear). Notice that the semiconductors in the table have *negative* temperature coefficients, meaning that their resistivity decreases as their temperature increases. This is an important characteristic of semiconductors.

Example 19.13. A thin silver wire of circular cross section has a diameter $d = 25.0\,\mu$m. (a) Find the resistance of a 10.0-m length of this wire. (b) When a high current is fed through the wire, its temperature increases from room temperature (20°C) to 70°C. What is the resistance of the same length of wire?

Solution. From the table of resistivities at room temperature, we have $\rho = 1.59 \times 10^{-8}\,\Omega$·m. The resistance is given by Equation (19.28) as

$$R = \rho\frac{L}{A} = (1.59 \times 10^{-8}\,\Omega\cdot\text{m})\frac{10.0\text{ m}}{(\pi/4)(2.50 \times 10^{-5}\text{ m})^2} = 324\,\Omega.$$

This length of very thin metal wire has a significant resistance.

(b) Taking the previous value of ρ as ρ_0 at room temperature T_0, Equation (19.31) gives a new value ρ at $T = 150°\text{C}$:

$$\rho = \rho_0[1 + \alpha(T - T_0)] = (1.59 \times 10^{-8}\ \Omega\text{·m})[1 + (3.8 \times 10^{-3}\,°\text{C}^{-1})(70°\text{C} - 20°\text{C})]$$
$$= 1.89 \times 10^{-8}\ \Omega\text{·m}.$$

With this new resistivity, the resistance is changed to

$$R = \rho\frac{L}{A} = (1.89 \times 10^{-8}\ \Omega\text{·m})\frac{10.0\ \text{m}}{(\pi/4)(2.50 \times 10^{-5}\ \text{m})^2} = 385\,\Omega.$$

The resistance of this sample of silver has increased by 61 Ω, or nearly 20%. △

19.3.4 Energy and Power in Electric Circuits

When current passes through a resistor, the moving charges lose energy as they collide with the ions in the resistor. This energy shows up in the thermal motion of the ions, and we notice this macroscopically as heat energy (the resistor's temperature increases). In this process, some of the electrical potential energy of the charge carries is lost, because once converted into heat energy it cannot be recovered.

It is not difficult to determine how much electrical energy is converted to heat in a resistor. Consider again the battery and resistor circuit in Figure 19.16. The potential difference between the two ends of the resistor is V. Therefore, a charge Δq undergoes a difference in potential energy $\Delta U = (\Delta q)\,V$ (see Section 18.4) in traveling from one end of the resistor to the other. The *power* dissipated by the resistor is the rate at which energy is lost (Section 7.4), so

$$P = \frac{\Delta U}{\Delta t} = \frac{\Delta q}{\Delta t}V. \tag{19.32}$$

Equation (19.32) gives the average power during a time interval Δt. To find the instantaneous power, we take the limit as $\Delta t \to 0$. Since

$$\lim_{\Delta t \to 0}\frac{\Delta q}{\Delta t} = \frac{dq}{dt} = I$$

we have for the instantaneous power

$$P = IV. \tag{19.33}$$

It is sometimes useful to know the instantaneous power in terms of the resistance R and either the potential difference or the current. Using $V = IR$ for the resistor in Equation (19.33), we can see that equivalent expressions for the instantaneous power are

$$P = I^2R \tag{19.34}$$

and

$$P = \frac{V^2}{R}. \tag{19.35}$$

We leave it as an exercise to verify that the SI unit of power (watts, W) results from the use of the SI units of volts, amperes, and ohms in Equations (19.33), (19.34), and (19.35).

Example 19.14. An electric light bulb has a power output of 100 W when a constant potential difference of 120 V is applied across its terminals. Treating the light bulb as a single resistor R (as in Figure 19.16), find the resistance of the light bulb and the current flowing through it under those conditions.

Solution. We can find the current by rearranging Equation (19.33):

$$I = \frac{P}{V} = \frac{100 \text{ W}}{120 \text{ V}} = 0.833 \text{ A}.$$

We can rearrange Equation (19.35) and compute the resistance:

$$R = \frac{V^2}{P} = \frac{(120 \text{ V})^2}{100 \text{ W}} = 144\,\Omega.$$

As a check, we can put our two results into Equation (19.34):

$$P = IV = (0.833 \text{ A})(144\,\Omega) = 120 \text{ W}$$

which is the correct power. △

19.4 Conductors, Insulators, Semiconductors, and Superconductors

Previously, we distinguished between conductors and insulators. In conductors, free electrons are able to move easily under the influence of an electric field. In an insulator, there are essentially no free electrons. In our study of dielectrics in Section 19.2, we saw how the orientation of the electric dipole moments in the presence of an applied electric field affects the operation of a capacitor.

The reason for the distinct behavior of conductors and insulators follows from the quantum theory of the atom. Atoms consist of a relatively heavy nucleus, made up of protons and neutrons, with a net positive charge due to the charge of the protons. The electrons move around the nucleus and are bound to it by their electrostatic attraction to it, analogous to the way planets are bound in their orbital motion around the Sun. However, classical electromagnetism and dynamics alone are insufficient to explain the motion of electrons. In quantum theory, electrons can exist only in distinct *shells* or *orbitals* of different sizes. As you would expect, electrons in the outermost shells are most weakly bound to the atom. The degree to which these outermost electrons are bound to the atom is responsible for the conducting properties of materials made of those atoms.

In a conductor, the outermost electron is bound very weakly to the atom, and the presence of even a weak applied electric field is sufficient to pull the electron away and make it conduct. But in an insulator, the binding is strong, and applied electric fields cannot pull electrons from the atoms. As the name implies, the behavior of a **semiconductor** is somewhere between a conductor and an insulator. Some evidence of this is in the numerical values of resistivities in Table 19.4. The semiconductors carbon, germanium, and silicon have electrical resistivities covering a wide range of values between the resistivities of conductors and insulators.

The behavior of semiconductors can only be understood using quantum theory. According to quantum theory, there exist only certain allowed energy ranges (called *energy bands*) for electrons in a semiconductor. In Figure 19.17, we illustrate the energy band structure in a semiconductor. Electrons normally exist in the *valence band*. There is a *band gap* between the valence band and the *conduction band*, where electrons have the energy needed to conduct. An applied electric field normally cannot supply enough energy to overcome this gap. However, there can be sufficient thermal energy (due simply to the temperature of the material) to elevate some electrons from the valence band to the conduction band. The higher the temperature, the more electrons are in the conduction band. This is why the resistivity of a semiconductor decreases with increasing

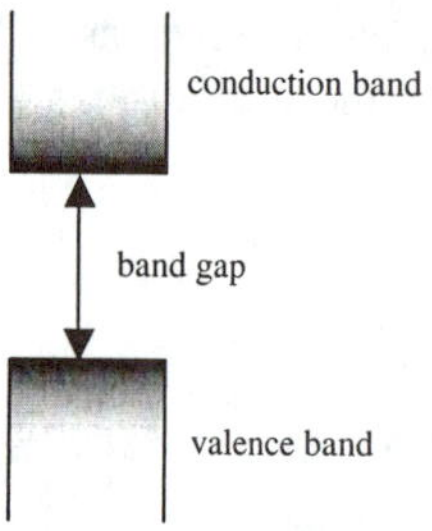

Figure 19.17. Band gap in semiconductors.

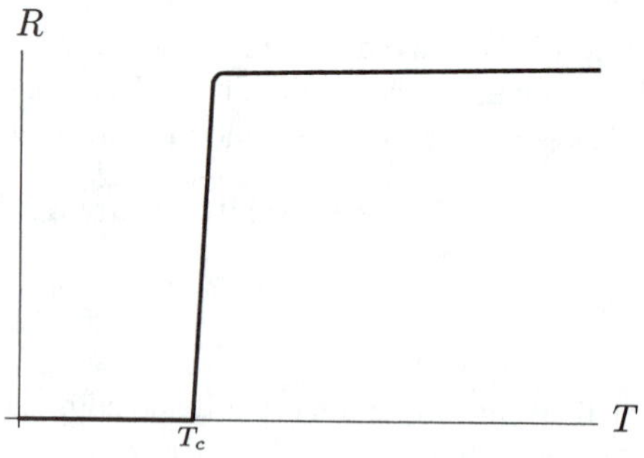

Figure 19.18. Resistance vs. temperature in a superconductor.

temperature. In fact, in most semiconductors the resistivity varies significantly as a function of temperature, especially at lower temperatures. This is why semiconductors are often used as temperature-sensing devices, called *thermistors*.

The behavior of conductors and insulators is also explained by the band theory. In conductors, there is no energy gap between the valence and conduction bands. In insulators, the gap is too great to be overcome by thermal energy. But the conducting mechanism in a **superconductor** is unique. As you see in Figure 19.18, the resistance of a superconductor is zero for all temperatures below the *transition temperature* T_c. Not all materials become superconducting at low temperatures, and the transition temperature varies significantly from one superconducting material to another (Table 19.5). In superconductors, electrons are able to use their mutual interaction with the lattice ions to form pairs that can then travel through the lattice with zero resistance.

Ever since the discovery of superconductors by the Dutch physicist Heike Kamerlingh Onnes (1853–1926) in 1911, the prospect of applications of superconductors has intrigued scientists and engineers. With zero resistance, no power is lost when current travels through a superconductor [see Equation (19.34)], and therefore the use of superconductors in electronic devices and in the transmission of electricity can conceivably result in great savings. Unfortunately, the low temperatures required for superconductivity often make such applications impractical. The 1986 discovery of a new class of ceramic superconductors with T_c above 77 K (the boiling point of liquid nitrogen, an inexpensive, common refrigerant) rekindled interest in superconductivity and led to the development and marketing of new electronic instruments that use superconductivity.

Table 19.5. Superconducting transition temperatures for selected materials

Material	Type	T_c (K)
Mercury	Element	4.2
Lead	Element	7.2
Niobium	Element	9.3
Nb_3Ge	Intermetallic	23
$YBa_2Cu_3O_7$	Ceramic	93
TlBaCaCuO	Ceramic	110–125

Source: C. P. Poole, Jr., *Copper Oxide Superconductors.* (New York: Wiley Interscience, 1988, p. 7.)

19.5 Problems

19.1 Capacitors

1. A parallel-plate capacitor consists of two circular plates separated by a distance of 0.0875 mm. Find the diameter of each plate if the capacitance is 250 pF.

2. Two parallel plates in a capacitor have surface areas of 25.0 cm^2 each. Find the plate separation needed for the capacitance to be (a) 1 nF, (b) 1 μF, (c) 1 mF. Discuss the practicality of building the capacitor in each case.

3. Repeat the derivation of the capacitance of the spherical capacitor if the charge on the inner sphere is $-Q$ and the charge on the outer sphere is Q. Explain why your result is the same as in Equation (19.6).

4. Find the capacitance of a spherical capacitor with radii $a = 1.22$ cm and $b = 1.75$ cm.

5. (a) Repeat Problem 4 with $a = 1.74$ cm and $b = 1.75$ cm. (b) Compare your result with the capacitance of a parallel-plate capacitor with surface area $A = 4\pi a^2$ and plate separation $d = b - a$.

6. The inner sphere of a spherical capacitor has a radius of 0.25 mm, and the outer sphere has a radius of 0.30 mm. (a) Find the capacitance. (b) Find the potential difference between the spheres if the magnitude of the charge on each is 0.15 μC. (c) Find the force on an electron placed 0.275 mm from the spheres' common center.

7. A 120-nF parallel-plate capacitor has a charge of magnitude 0.560 μC on each plate. The plates are separated by a distance of 200 μm. (a) What is the charge density on each plate? (b) An electron is released from the negative plate. How much time does it take to reach the positive plate? (c) A proton is released from the positive plate. How much time does it take to reach the negative plate?

8. Each plate in a parallel-plate capacitor has a surface area of 5.3×10^{-5} m^2. It is found that a potential difference of 55.0 V between the plates results in a charge (positive and negative) of magnitude 14.5 nC appearing on each plate. Find the plate separation.

9. A 10.0-nF capacitor has a potential difference of 10 V applied between the parallel plates, which each have a surface area of 2.67 cm^2. (a) Find the charge on each plate. (b) Find the electric field in the region between the plates.

10. A spherical capacitor has an inner sphere of radius of 12.0 cm and an outer sphere of radius 24.0 cm. The inner sphere carries a net charge of $+0.42\,\mu$C, and the outer sphere carries a net charge of $-0.42\,\mu$C. (a) Find the capacitance. (b) Find the potential difference.

11. For the spherical capacitor described in the previous problem (still carrying the $\pm 0.42\,\mu$C charge), use numerical methods to estimate the time it takes a proton released at the surface of the inner sphere to reach the outer sphere.

12. Two parallel conducting plates are separated by a distance of 0.10 mm. When a potential difference of 950 V is applied between the plates, what is the surface charge density on each plate?

13. How large is an isolated conducting sphere if its capacitance is (a) 1 pF? (b) 1 nF? (c) 1μF?

14. An isolated conducting sphere has a radius of 1.00 m. (a) Find its capacitance. (b) Find the net charge on the sphere if the electric potential at its surface is 240 V. Assume that the electric potential is zero an infinite distance away. (c) Find the magnitude of the electric field just outside the surface of the sphere and show that it is approximately equal to σ/ϵ_0, where σ is the surface charge density.

15. Show that for n capacitors connected in parallel the equivalent capacitance is

$$C_{\text{eq}} = C_1 + C_2 + \cdots + C_n.$$

16. Show that for n capacitors connected in series the equivalent capacitance is given by

$$\frac{1}{C_{\text{eq}}} = \frac{1}{C_1} + \frac{1}{C_2} + \cdots + \frac{1}{C_n}.$$

In Problems 17–22, you have three 3.0-μF capacitors.

17. Find the equivalent capacitance if the three capacitors are connected in parallel.

18. The three capacitors in parallel are connected to a 30-V battery. Find the charge on each capacitor and the potential difference across each capacitor.

19. Find the equivalent capacitance if the three capacitors are connected in series.

20. The three capacitors in series are connected to a 30-V battery. Find the charge on each capacitor and the potential difference across each capacitor.

21. Two of the capacitors are in parallel with each other, and this parallel combination is connected in a series with the third. Find the equivalent capacitance.

22. The two ends of the three-capacitor combination described in the previous problem are connected to a 30-V battery. Find the charge on each capacitor and the potential difference across each capacitor.

23. You have a 230-nF capacitor and wish to combine it with one other capacitor in order to make a net capacitance of 1.50 μF. How large a capacitor do you need, and should it be connected in series or parallel?

24. You have a 950-nF capacitor and wish to combine it with one other capacitor in order to make a net capacitance of 225 nF. How large a capacitor do you need, and should it be connected in series or parallel?

25. You have a 10-μF, a 15-μF, and a 20-μF capacitor. Find the values of all the possible capacitances you can create by making different series and parallel combinations.

26. In the circuit shown, $V = 12$ V and each capacitor has a capacitance of 8.0 μF. Find the potential difference across each capacitor and the charge on each capacitor.

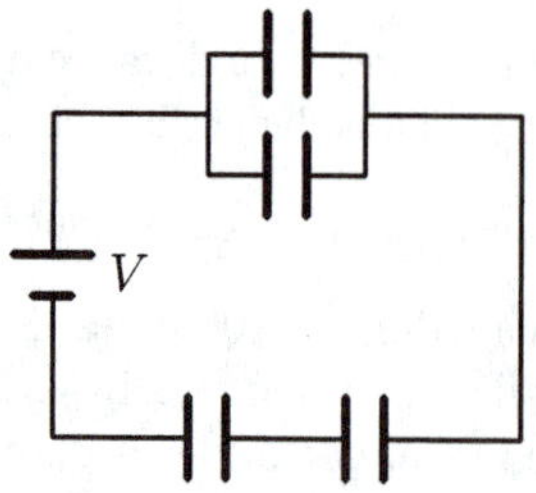

In Problems 27–29, use the following capacitances: $C_1 = 1.0\,\mu$F, $C_2 = 3.0\,\mu$F, $C_3 = 7.0\,\mu$F, $C_4 = 10.0\,\mu$F, $C_5 = 20.0\,\mu$F. In each case, find the equivalent capacitance between the points a and b.

27.

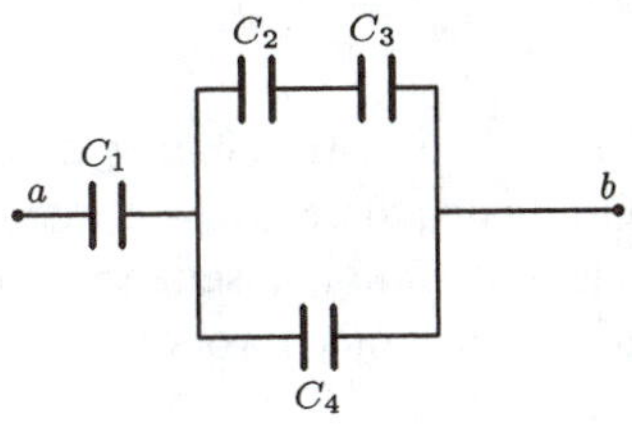

28.

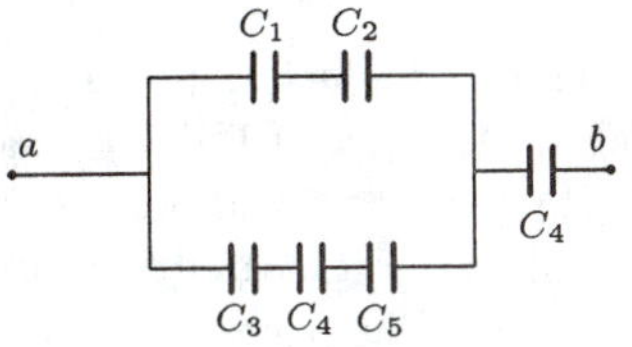

29.

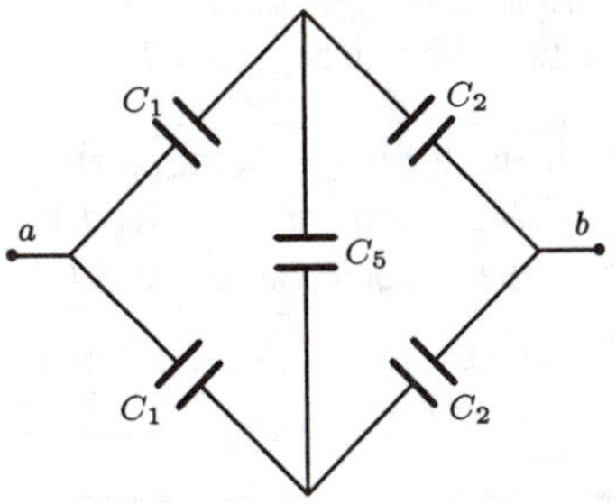

30. A 10-μF parallel-plate capacitor is charged directly with a 6.0-V battery. Find the energy stored in the capacitor.

31. A 150-pF spherical capacitor is used to store an electrical energy of 0.420 J. Find the charge on the capacitor.

32. A spherical capacitor carries charges (positive and negative) of magnitude 210 nC on its two surfaces and stores an energy of 24.5 mJ. If the radius of the outer sphere is 8.5 mm, what is the radius of the inner sphere?

33. A constant potential difference of 54 mV is placed across a 25-pF parallel-plate capacitor. If the plate separation is 0.300 mm, what is the (assumed uniform) energy density between the plates?

34. A spherical capacitor has an inner sphere of radius 20.0 cm and an outer sphere of radius 35.0 cm. With a potential difference of 55.0 V between the spheres, find (a) the charge on each sphere, (b) the total energy stored by the capacitor, and (c) the local energy density just outside the inner sphere and just inside the outer sphere.

35. The energy density between the plates of a parallel-plate capacitor is 0.565×10^{-6} J/m^3. The plates have a surface area of 0.250 cm^2 and are separated by 0.500 mm. Find (a) the electric field between the plates, (b) the charge on each plate, and (c) the potential difference between the plates.

36. A 1.00-m-diameter conducting sphere carries a net charge of 0.0242 C. Find the total energy contained in the electric field surrounding the sphere.

37. The parallel plates of a 10-μF capacitor are separated by 0.142 mm. A potential difference of 35.0 V is placed across the capacitor. (a) Find the energy stored in the capacitor. (b) How much work is done in pulling the charged plates apart to a distance of 0.195 mm?

19.2 Dielectrics

1. A parallel-plate capacitor has plates with surface area 3.25×10^{-5} m^2 separated by 0.12 mm. If the space between the plates is filled with paper, find (a) the capacitance of the capacitor, (b) the charge on each plate when a potential difference of 12.0 V is placed across the plates, and (c) the maximum possible potential difference between the plates.

2. Two parallel conducting plates separated by a distance of 1.00 cm carry charges ± 21.0 pC when a potential difference of 150 V is placed across the plates. Originally, there is air between the plates. (a) What is the electric field between the plates? (b) Now a slab of teflon is inserted between the plates. What is the magnitude of the electric field inside the teflon? (c) How much work was done to insert the teflon between the plates?

3. A bakelite-filled parallel-plate capacitor has capacitance 34.5 nF and plate separation 0.950 mm. What is the maximum charge that can be held by this capacitor?

4. An air-filled parallel-plate capacitor, with plate separation d, has capacitance C_0. A dielectric with dielectric constant κ and thickness $d/2$ is inserted between the plates as shown. What is the new capacitance? Does it matter whether the dielectric is moved closer to one plate? Explain.

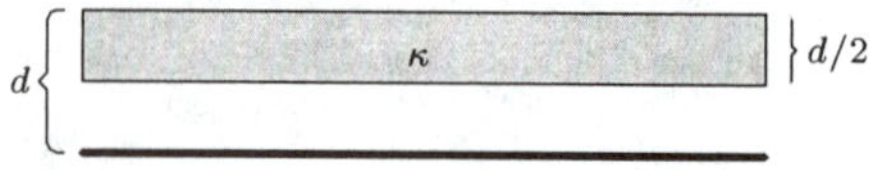

5. An air-filled parallel-plate capacitor, in which the plates have surface area A and the plate separation is d, has capacitance C_0. A dielectric slab of thickness d, surface area $A/2$, and dielectric constant κ is inserted between the plates as shown. What is the new capacitance?

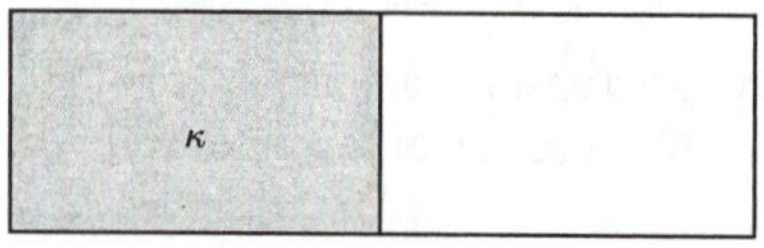

6. An air-filled parallel-plate capacitor, in which the plates have surface area A and the plate separation is d, has capacitance C_0. A dielectric slab of thickness ad, surface area bA (with $a < 1$ and $b < 1$), and dielectric constant κ is inserted between the plates as shown. What is the new capacitance?

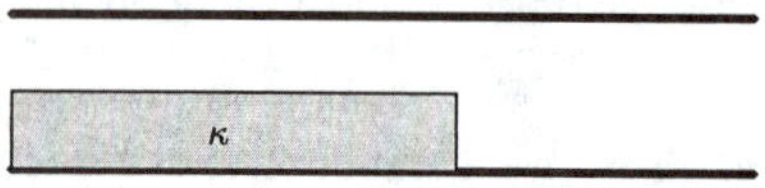

7. A capacitor consists of two concentric conducting spheres, with radii 1.50 cm and 35.0 cm. The region between the spheres is filled with water. What is the capacitance of this spherical capacitor?

19.3 Electric Current

1. The effective current of electrons reaching a computer screen is about 1 mA. How many electrons per second strike the screen?

2. In a lightning strike lasting 45 ms, a total charge of 350 C is transferred to the ground. What is the average current?

3. A potential difference of 25.0 mV exists between the two ends of a 1.25-m-long conducting wire. (a) What is the acceleration of an electron in the wire due to the applied electric field? (b) Suppose the drift speed v_d of electrons in the wire is 5.00 mm/s. Find the time required for an electron accelerated from rest to reach this speed? How far would it travel during that time?

4. A silver wire with a circular cross section carries a current of 500 mA. With this current the electron drift speed is 0.245 mm/s. (a) Find the diameter of the wire. (b) Find the current density.

5. A #20-gauge copper wire with circular cross-section has a diameter of 0.81 mm and is 20.0 m long. (a) Find the resistance of this length of wire. (b) Find the current and the electron drift speed when a potential difference of 60.0 V is applied between the ends of the wire.

6. What should be the diameter of copper wire in order to have a resistance of less than 0.01 Ω per meter of length?

7. Find the resistance between the two ends of a 25.0-cm-long aluminum rod that has a diameter 1.0 cm.

8. (a) Find the resistance (at 20°C) from end to end of a cylindrical carbon resistor that has a radius of 0.0532 mm and a length of 1.54 cm. (b) What is the three-band color code for this resistor? (c) After carrying some current, the temperature of the resistor increases to 40°C. What is its new resistance?

9. Nichrome is an alloy often used in heating elements. Find the rate (in W) at which heat is generated by a nichrome coil that consists of a 5.65 m length of nichrome wire of circular cross section with diameter 0.120 mm when a potential difference of 120 V is applied between the two ends of the wire.

10. What is the potential difference of a battery that can generate 100 W when connected to a 20-Ω resistor? What is the current flowing through the resistor under these conditions?

11. Find the resistances associated with the following color code sequences: (a) red, brown, black; (b) blue, violet, green; (c) brown, black, yellow; (d) gray, green, blue.

12. A platinum wire is originally at 20°C. Find the temperature increase needed to increase the resistance of the wire by (a) 1%, (b) 10%.

13. What is the resistance of the filament of a light bulb if it uses energy at a rate of 75 W with a constant potential difference of 120 V across its terminals?

14. A silver wire has a 0.165 mm diameter. It is 1.45 m long and carries a current of 4.10 A. (a) What is the potential difference between the two ends of the wire? What is the electric field inside the wire? At what rate is energy dissipated in the wire?

15. A 24.0-V battery is connected to a 150-Ω resistor. Find the energy supplied by the battery in exactly one minute.

16. A conducting wire with a cross-sectional area 1.67×10^{-7} m^2 carries a current of 2.53 A. If the electric field in the wire is 0.370 V/m, find the (a) conductivity and (b) resistivity of the conductor. (c) Assuming the conductor is a pure metal, which metal is it?

17. Verify that units of watts are obtained for power when the appropriate SI units are used for resistance, potential difference, and current in Equations (19.33), (19.34), and (19.35).

18. Consider the work done in moving charge through a circuit to be a composite function $W(q(t))$ (where q is charge). Use the chain rule to show that the power dissipated is $P = IV$.

19.4 Conductors, Insulators, Semiconductors, and Superconductors

There are no problems for this section.

Chapter 20

Optimization

Many applications involve determining the minimum or maximum output of a function. The process of finding the relevant minimum or maximum is often referred to as **optimizing**. In this chapter, we study the problem of optimizing a function of two or more variables. All of the analysis parallels the situation for optimizing a function of one variable. We review the one-variable case in the first section to establish terminology and a basic plan. In the next three sections, we develop the details of optimizing for functions of two variables. In the last section, we look at the problem of *constrained optimization*.

20.1 Optimization for Functions of One Variable—Review

We start our review by considering a classic example of optimizing.

Example 20.1. A window is to be made in the shape of a rectangle with a semicircular top as shown in Figure 20.1. Find the dimensions of the window that give the greatest area for a perimeter of 3 m.

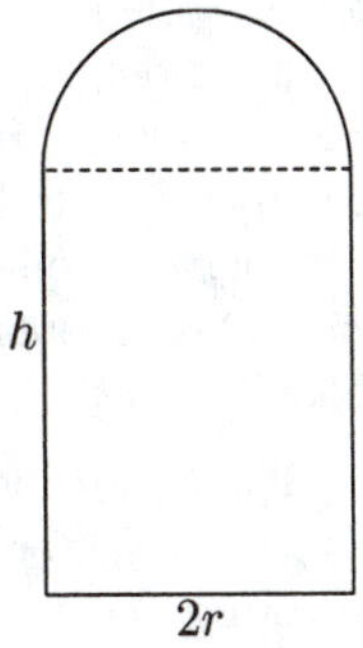

Figure 20.1. Window in the shape of a rectangle with a semicircular cap.

Solution. Let h be the height of the rectangle and r be the radius of the semicircle. The width of the rectangle is thus $2r$, as shown in the figure. The area of the window is given by

$$A = 2rh + \frac{1}{2}\pi r^2, \tag{20.1}$$

and the perimeter is given by

$$P = 2r + 2h + \pi r = 3. \tag{20.2}$$

We express the area as a function of one variable by solving the constraint in Equation (20.2) for h and substituting into the area expression in Equation (20.1). For h, we find

$$h = \frac{3 - (2 + \pi)r}{2}, \tag{20.3}$$

so we get

$$A(r) = 3r - \frac{4 + \pi}{2} r^2. \tag{20.4}$$

We now turn to our knowledge of calculus. If the derivative of A is zero for some input, then the slope of the tangent line for this input is zero. This could correspond to a maximum output. Computing the derivative, we find

$$A'(r) = 3 - (4 + \pi)r.$$

Requiring this derivative to be zero gives us the equation

$$3 - (4 + \pi)r = 0,$$

which is easily solved to give

$$r = \frac{3}{4 + \pi} \approx 0.420. \tag{20.5}$$

The corresponding value of h comes from substituting into Equation (20.3) to get

$$h = \frac{3}{4 + \pi} \approx 0.420 \tag{20.6}$$

△

The preceding example illustrates a situation we often encounter in applications: For a given function $f : \mathbb{R} \to \mathbb{R}$, we want to find an input a so that the output $f(a)$ is greater than other outputs. In some situations, we want the output $f(a)$ to be greater than (or equal to) outputs $f(x)$ for all inputs x "near" the input a, while in other situations we want the output $f(a)$ to be greater than or equal to outputs $f(x)$ for *all possible* inputs x in the relevant domain. Of course, in other applications we are interested in an input a for which the output $f(a)$ is less than (or equal to) other outputs. The following definitions give us some terminology to use in discussing these types of applications.

Definition 20.1. For a function $f : \mathbb{R} \to \mathbb{R}$, an input x_0 is called a **local maximizer** and the corresponding output $f(x_0)$ is called a **local maximum** if $f(x) \leq f(x_0)$ for all x in an open interval containing x_0. An input x_0 is called a **local minimizer**, and the corresponding output $f(x_0)$ is called a **local minimum** if $f(x_0) \leq f(x)$ for all x in an open interval containing x_0.

In Example 20.1, we searched for a local maximizer as a solution of the equation given by setting the first derivative equal to zero. If a function f is differentiable and has a local minimum or a local maximum for input x_0, then the derivative $f'(x_0)$ must equal zero. However, the converse statement is not true. That is, if $f'(x_0) = 0$ for some input x_0, then x_0 is not necessarily a local minimizer or a local maximzer. This is easily seen by considering an example such as $f(x) = x^3$. We have $f'(x) = 3x^2$, so $x_0 = 0$ gives $f'(0) = 0$. However, $f(x) \leq f(0) = 0$ for all $x \leq 0$, so $x_0 = 0$ is not a local minimizer and $f(x) \geq f(0) = 0$ for all $x \geq 0$ so $x_0 = 0$ is not a local maximizer. To assist in our discussion of this idea, we introduce one more piece of terminology.

Definition 20.2. An input x_0 in the domain of a differentiable function f is called a **stationary input** of f if $f'(x_0) = 0$.

In some contexts, we encounter functions that fail to be differentiable at some inputs. An input is called a **critical input** if either $f'(x_0) = 0$ or f is not differentiable at x_0. Most applications involve stationary inputs. In the example shown in Figure 20.2, the inputs x_1, x_2, x_3, and x_4 are stationary inputs. The idea of the previous paragraph can be expressed in this new language: If an input x_0 is a local minimizer or a local maximizer for a differentiable function f, then x_0 is a stationary point. Thus, if we know all stationary inputs, then we know all possible local minimizers and maximizers.

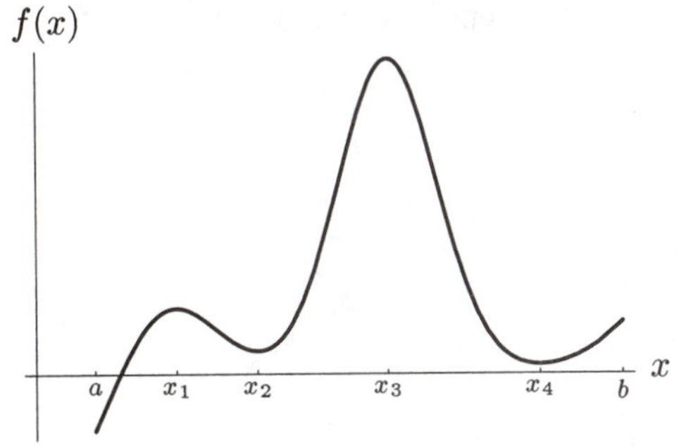

Figure 20.2. Generic example showing stationary inputs.

To check whether a stationary input x_0 is a local minimizer, we can test if the derivative goes from negative for inputs less than x_0 to positive for inputs greater than x_0. If the function is twice differentiable, this is equivalent to $f''(x_0) > 0$ because the first derivative is increasing. In a similar fashion, to check whether a stationary input x_0 is a local maximizer, we can test if the derivative goes from positive for inputs less than x_0 to negative for inputs greater than x_0. If the function is twice differentiable, this is equivalent to $f''(x_0) < 0$ because the first derivative is decreasing. This is the basis for the **second derivative test**.

Example 20.2. Use the second derivative test to show that the stationary point found in Example 20.1 is indeed a local maximizer for the area function.

Solution. We compute the second derivative of $A(r)$ to get

$$A''(r) = -(4 + \pi).$$

The second derivative is negative for all inputs; in particular, it is negative for the stationary point $r = 3/(4 + \pi)$. Thus, this value of r is a local maximizer. △

A given function can have more than one local minimum or local maximum. In some applications, the interest is in the smallest minimum or the largest maximum. To make this precise, we give the following definition.

Definition 20.3. For a function $f : A \to \mathbb{R}$ where A is a subset of $\mathbb{R}$, the input x_0 is called the **global maximizer**, and the corresponding output $f(x_0)$ is called the **global maximum** if $f(x) \leq f(x_0)$ for all x in A. The input x_0 is called the **global minimizer**, and the corresponding output $f(x_0)$ is called the **global minimum** if $f(x_0) \leq f(x)$ for all x in A.

Be careful to note the quantifiers that distinguish *local* from *global*. The term **extremizer** is used to refer to an input that is either a minimizer or a maximizer, and the term **extremum** refers to an output that is either a minimum or a maximum.

A function need not have a global maximum or a global minimum for a given domain A. For example, the function $f(x) = 2x$ with domain $A = \mathbb{R}$ has neither a global minimum or a global maximum, because the outputs are neither bounded below nor bounded above. The function $f(x) = 2x$ also has neither a global minimum nor a global maximum on the domain $A = (3, 5)$. (See Figure 20.3.) It is true that the outputs $f(x)$ are bounded between 6 and 10, but 6 and 10 are *not* outputs of the function for the domain $(3, 5)$. That is, $6 < f(x) < 10$ where it is important to note that the inequalities are strict (i.e., do not include equality). There is no one output of the function that is less than or equal to all other outputs for the domain $(3, 5)$. Likewise, no one output of the function is greater than or equal to all other outputs for the domain $(3, 5)$. The point here is that a lower or upper bound does not qualify as a global minimum or maximum unless it is an output of the function for some input in the chosen domain.

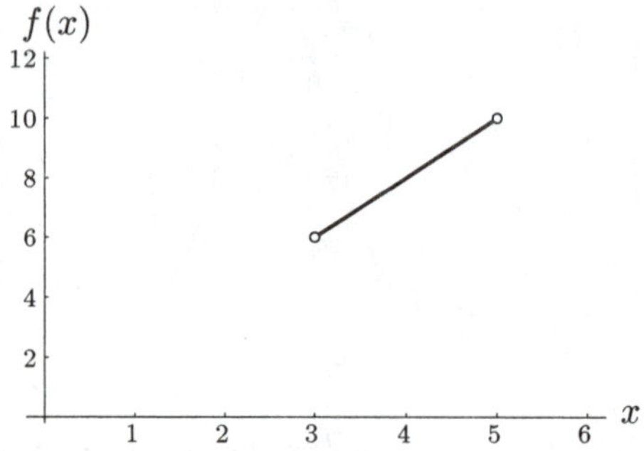

Figure 20.3. Function $f(x) = 2x$ on the domain $(3, 5)$ has neither a global minimum nor a global maximum.

Under what conditions can we guarantee that a given function has a global minimum and a global maximum for a given domain? From the examples in the previous paragraph, it is apparent that we need the domain to be a closed interval $[a, b]$. We also need to require that the function be continuous on this closed interval. This is the content of the following theorem.

Theorem 20.1 (Extreme Value Theorem). *If* $f : [a, b] \to \mathbb{R}$ *is continuous, then* f *has a global minimum and a global maximum on* $[a, b]$.

For an application in which we are interested in finding the global minimum or the global maximum, we must compute the output for each input that is an endpoint on the domain of the function. We then compare these outputs with all of the local minima or maxima to pick the global minimum or maximum. In the example of Figure 20.2, the global minimizer of f for the interval $[a, b]$ is the left endpoint a, and the global maximizer is the local maximizer x_3.

Example 20.3. Show that the local maximum found in Example 20.1 is the global maximum for the relevant domain.

Solution. Given that only positive area makes sense for the context of the application, the relevant domain is the interval $[0, 6/(4 + \pi)]$ for r. The endpoints $r = 0$ and $r = 6/(4 + \pi)$ have outputs $V(0) = 0$ and $V(6/(4 + \pi)) = 0$ (not surprising, given that these conditions were used to determine the endpoints). Comparing these to the local maximum

$$V\left(\frac{3}{4+\pi}\right) = \frac{9}{2(4+\pi)}$$

we see that the global minimum is 0 and the global maximum is $9/[2(4 + \pi)]$.

This analysis is a bit of overkill for the application in Example 20.1, but it does illustrate the ideas in a simple setting. △

Example 20.4. Find all local extremizers and extrema for the function $f(x) = xe^{-x}$ for the interval $[0, 2]$. Also, find the global extremizers and extrema.

Solution. We first find all stationary points. The derivative of f is

$$f'(x) = e^{-x} - xe^{-x} = (1 - x)e^{-x}.$$

Equating to zero gives

$$(1 - x)e^{-x} = 0.$$

Because $e^{-x} > 0$ for all x, the only solution is $x_0 = 1$. For the second derivative test, we compute

$$f''(x) = -2e^{-x} + xe^{-x}.$$

Thus, at the stationary point $x_0 = 1$ we have

$$f''(1) = -2e^{-1} + e^{-1} = -e^{-1} < 0.$$

We conclude that $x_0 = 1$ is a local maximizer with local maximum $f(1) = e^{-1} \approx 0.368$.

To determine the global extrema, we need to compute outputs for the endpoints $a = 0$ and $b = 2$. We find $f(0) = 0$ and $f(2) = 2e^{-2} \approx 0.271$. Comparing the three relevant outputs, we conclude that the global minimizer is $a = 0$ with global minimum $f(0) = 0$, and the global maximizer is $x_0 = 1$ with global maximum $f(1) = e^{-1} \approx 0.368$. △

20.2 Local Extrema for Functions of Several Variables

The theory of local extremizers and local extrema is easily generalized from $f : \mathbb{R} \to \mathbb{R}$ to $f : \mathbb{R}^2 \to \mathbb{R}$. In defining local extremizer and local extrema, we need to take into consideration the fact that inputs are points in $\mathbb{R}^2$. For functions of one variable, the definition of local extrema refers to an open interval containing the extremizer. For functions of two variables, we generalize this to an *open disk* centered at the input in question.

Definition 20.4. The **open disk of radius δ centered at (x_0, y_0)** is the set of all points (x, y) with distance from (x_0, y_0) less than δ. We denote the open disk as $D_\delta(x_0, y_0)$. Thus, we have

$$D_\delta(x_0, y_0) = \left\{(x, y) | \sqrt{(x - x_0)^2 + (y - y_0)^2} < \delta\right\}.$$

The adjective "open" refers to the fact that the inequality is strict, so that the points on the circle of radius δ centered at (x_0, y_0) are *not* in the open disk.

Definition 20.5. For a function $f : \mathbb{R}^2 \to \mathbb{R}$, an input (x_0, y_0) is called a **local maximizer**, and the corresponding output $f(x_0, y_0)$ is called a **local maximum** if $f(x, y) \leq f(x_0, y_0)$ for all (x, y) in an open disk centered at (x_0, y_0). An input (x_0, y_0) is called a **local minimizer**, and the corresponding output $f(x_0, y_0)$ is called a **local minimum** if $f(x_0, y_0) \leq f(x, y)$ for all (x, y) in an open disk centered at (x_0, y_0).

To motivate the definition of *stationary input*, we note the result in the following theorem.

Theorem 20.2. *If (x_0, y_0) is a local extremizer for a differentiable function $f : \mathbb{R}^2 \to \mathbb{R}$, then $\vec{\nabla} f(x_0, y_0) = \langle 0, 0 \rangle$. That is,*

$$\left.\frac{\partial f}{\partial x}\right|_{(x_0,y_0)} = 0 \qquad \textit{and} \qquad \left.\frac{\partial f}{\partial y}\right|_{(x_0,y_0)} = 0.$$

Proof. If (x_0, y_0) is a local extremizer, then the function $f(x, y_0)$ (viewed as a function of one variable x) has a local extreme for the input x_0. Hence, the partial derivative with respect to x must be zero. Likewise, the function $f(x_0, y)$ (viewed as a function of one variable y) has a local extreme for the input y_0, and hence the partial derivative with respect to y must be zero. □

An input (x_0, y_0) is called a **stationary input** if $\vec{\nabla} f(x_0, y_0) = \langle 0, 0 \rangle$. The theorem gives us a strategy for finding local extremizers: Find all stationary inputs, and then test each to determine if it is a local minimizer, a local maximizer, or neither. This is more complex than the analogous situation for functions of one variable, because the condition for a stationary input involves a system of two equations in two unknowns and because it generally takes more work to distinguish among local minimizers, local maximizers, or neither. We cannot simply test the derivative on each side of the stationary input. The following examples illustrate this idea.

Example 20.5. Find and classify all stationary inputs of $f(x, y) = x^2 + y^2$.

Solution. It is straightforward to compute the partial derivatives as

$$\frac{\partial f}{\partial x} = 2x \qquad \text{and} \qquad \frac{\partial f}{\partial y} = 2y.$$

Equating each partial derivative to zero gives the system of equations

$$\begin{aligned} 2x &= 0 \\ 2y &= 0. \end{aligned}$$

The only solution is clearly $(x_0, y_0) = (0, 0)$. Thus, $f(x, y) = x^2 + y^2$ has one stationary input.

To determine the nature of this stationary input, we note that $f(x_0, y_0) = f(0, 0) = 0$. For any other input (x, y), we have $f(x, y) = x^2 + y^2 \geq 0$, because each term is nonnegative. Thus, $f(0, 0) \leq f(x, y)$ for all (x, y), so $(0, 0)$ is a local minimizer and $f(0, 0) = 0$ is a local minimum. In fact, $f(0, 0) = 0$ is the *global minimum* for this example. △

Example 20.6. Find and classify all stationary inputs of $f(x, y) = x^2 - y^2$.

Solution. It is straightforward to compute the partial derivatives as

$$\frac{\partial f}{\partial x} = 2x \qquad \text{and} \qquad \frac{\partial f}{\partial y} = -2y.$$

Equating each partial derivative to zero gives the system of equations

$$\begin{aligned} 2x &= 0 \\ -2y &= 0. \end{aligned}$$

The only solution is clearly $(x_0, y_0) = (0, 0)$. Thus, $f(x, y) = x^2 - y^2$ has one stationary input.

To determine the nature of this stationary input, we first note that $f(x_0, y_0) = f(0, 0) = 0$. Now $f(x, 0) = x^2 \geq 0$ for all x. Therefore, in *any* open disk centered at $(0, 0)$, there is an input with an output greater than $f(0, 0)$, so $(0, 0)$ is not a local maximizer. Likewise, $f(0, y) = -y^2 \leq 0$ for all y, so *any* open disk centered at $(0,0)$ has an input with an output less than $f(0,0)$. Hence, $(0,0)$ is not a local minimizer. This stationary point falls into the "neither" category. △

The last example illustrates what generally occurs for a stationary point that is neither a local minimizer nor a local maximizer. We know from Chapter 13 that the graph of $f(x,y) = x^2 - y^2$ is a "saddle." The graph is locally flat at the stationary input, meaning the tangent plane is horizontal, because both partial derivatives are zero. In one cross section, the graph is concave up, while in a different cross section the graph is concave down. We will refer to this situation as a **saddle point**. For $f(x,y) = x^2 - y^2$, these are the $y = 0$ and $x = 0$ cross sections, respectively. For other cases, the relevant cross sections may not be as easy to find. In the next section, we develop a *second derivative test* to help distinguish saddle points from local extremes.

Example 20.7. Find and classify all stationary inputs of $f(x,y) = x^3y - xy^3$.

Solution. It is straightforward to compute the partial derivatives as

$$\frac{\partial f}{\partial x} = 3x^2y - y^3 \qquad \text{and} \qquad \frac{\partial f}{\partial y} = x^3 - 3xy^2.$$

Equating each partial derivative to zero gives the system of equations

$$\begin{aligned} 3x^2y - y^3 &= y(3x^2 - y^2) = 0 \\ x^3 - 3xy^2 &= x(x^2 - 3y^2) = 0. \end{aligned}$$

From the first equation, we have the conditions $y = 0$ or $3x^2 - y^2 = 0$, and from the second equation we get $x = 0$ or $y^2 - 3x^2 = 0$. There are thus *four* possibilities to solve both equations simultaneously:

$$\begin{array}{cccc} y = 0 & y = 0 & 3x^2 - y^2 = 0 & 3x^2 - y^2 = 0 \\ x = 0 & y^2 - 3x^2 = 0 & x = 0 & x^2 - 3y^2 = 0 \end{array}$$

For each case, the only solution is $(0,0)$. Thus, $(0,0)$ is the only stationary input.

To determine the nature of this stationary input, begin by noting that $f(0,0) = 0$. To understand outputs for inputs near $(0,0)$, it is useful to express the outputs in a factored form as $f(x,y) = xy(x^2 - y^2)$. We can get negative outputs by choosing inputs with x and y of the same sign and $x^2 - y^2 \le 0$. Inputs of the form $(x, 2x)$ give outputs $f(x,2x) = x(2x)(x^2 - 4x^2) = -6x^4 \le 0$ for all x. Similarly, for inputs of the form $(2x, x)$, the outputs are $f(2x,x) = (2x)(x)(4x^2 - x^2) = 6x^4 \ge 0$ for all x. In *any* open disk (no matter how small) centered at the stationary input $(0,0)$, there are inputs of the form $(x, 2x)$ and inputs of the form $(2x, x)$. Hence, $(0,0)$ is neither a local minimizer nor a local maximizer. In fact, the graph is concave down in the cross section defined by $y = 2x$ and concave up in the cross section defined by $2x = y$, so $(0,0)$ corresponds to a saddle point. △

Example 20.8. Find and classify all stationary inputs of $f(x,y) = 2(x^2 + y^2)e^{-(x^2+y^2)}$.

Solution. For this example, the partial derivatives are a bit more difficult to compute. Using the product rule, we find

$$\frac{\partial f}{\partial x} = 4x[1 - (x^2 + y^2)]e^{-(x^2+y^2)} \qquad \text{and} \qquad \frac{\partial f}{\partial y} = 4y[1 - (x^2 + y^2)]e^{-(x^2+y^2)}.$$

Equating each partial derivative to zero gives the system of equations

$$\begin{aligned} 4x[1 - (x^2 + y^2)]e^{-(x^2+y^2)} &= 0 \\ 4y[1 - (x^2 + y^2)]e^{-(x^2+y^2)} &= 0. \end{aligned}$$

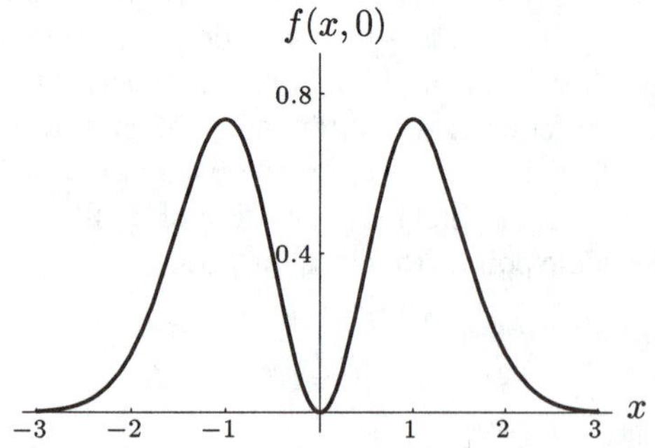

(a) $y = 0$ cross section

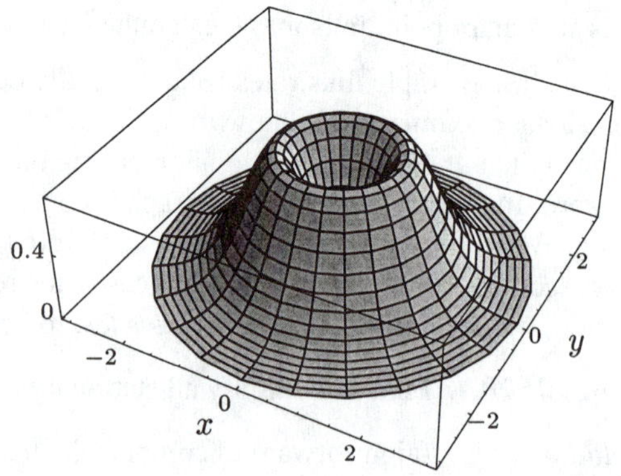

(b) Graph of $f(x, y)$

Figure 20.4. Two plots of the function in Example 20.8.

In each equation, the exponential factor is strictly positive. From the first equation, we have the conditions $x = 0$ or $1 - (x^2 + y^2) = 0$, and from the second equation we get $y = 0$ or $1 - (x^2 + y^2) = 0$. Every solution of $1 - (x^2 + y^2) = 0$ satisfies both equations, so every point on the unit circle centered at the origin is a stationary input. The origin $(0, 0)$ is also a stationary input, since $x = 0$ satisfies the first equation and $y = 0$ satisfies the second equation.

The stationary input $(0, 0)$ is a local minimizer, because $f(0, 0) = 0$ and $f(x, y) \geq 0$ for all (x, y). To see this, note that the factor $x^2 + y^2$ is nonnegative, and the exponential factor is strictly positive.

To understand the nature of the stationary points on the unit circle, we take advantage of the radial symmetry in the function. Because $f(x, y)$ depends only on $x^2 + y^2$ (i.e., the square of the distance from the origin), it is enough to look at one cross section. In the $y = 0$ cross section, we have

$$f(x, 0) = 2x^2 e^{-x^2}.$$

This is shown in Figure 20.4(a). Viewing this as a function of the one variable x, the input $x_0 = 1$ is a stationary input. The derivative with respect to x is

$$\left.\frac{\partial f}{\partial x}\right|_{(x,0)} = 4x(1 - x^2)e^{-x^2}.$$

For an input $x = 1 - h$ with $h > 0$ (so $1 - h$ is just less than $x_0 = 1$), the derivative is positive, because $1 - (1 - h)^2$ is positive. For an input $1 + h$ with $h > 0$ (so $1 + h$ is just more than $x_0 = 1$), the derivative is negative, because $1 - (1 + h)^2$ is negative. Hence, $x_0 = 1$ is a local maximizer in this cross section. From the radial symmetry, we can conclude that each input (x_0, y_0) on the unit circle is a local maximizer for $f(x, y)$. Figure 20.4(b) shows the graph of the function. The local minimum corresponds to the bottom of the central bowl, and the local maximums correspond to the points on the rim. △

20.3 A Second Derivative Test

We turn our attention now to developing a test to distinguish among local minima, local maxima, and saddle points for stationary inputs. The test involves the second partial derivatives of the function $f : \mathbb{R}^2 \to \mathbb{R}$. We assume that f has continuous second partial derivatives, so that the mixed partials are equal.

Let (x_0, y_0) be a stationary input for f. By definition, the first partial derivatives of f are zero for input (x_0, y_0). Using this fact in Equation (18.20) for the second-order Taylor polynomial of f based at (x_0, y_0), we have

$$f(x,y) \approx f(x_0,y_0) + \frac{1}{2}\left[\left.\frac{\partial^2 f}{\partial x^2}\right|_{(x_0,y_0)}(x-x_0)^2 + 2\left.\frac{\partial^2 f}{\partial x \partial y}\right|_{(x_0,y_0)}(x-x_0)(y-y_0) + \left.\frac{\partial^2 f}{\partial y^2}\right|_{(x_0,y_0)}(y-y_0)^2\right].$$

In general, we can restrict our attention to a small enough open disk centered at (x_0, y_0) so that the function f has the same nature (local min, local max, or saddle) as this second order approximation. Thus, we can study the shape of the graph of the Taylor polynomial. Because we are interested in the shape of the graph, we can further simplify matters. The constant term $f(x_0, y_0)$ determines the overall height of the graph but does not affect its shape. We can thus study the related function in which we set this term equal to zero. Removing the factor of $\frac{1}{2}$ is equivalent to changing the vertical scale but will not change the nature of the stationary point. We will thus examine the expression

$$\left.\frac{\partial^2 f}{\partial x^2}\right|_{(x_0,y_0)}(x-x_0)^2 + 2\left.\frac{\partial^2 f}{\partial x \partial y}\right|_{(x_0,y_0)}(x-x_0)(y-y_0) + \left.\frac{\partial^2 f}{\partial y^2}\right|_{(x_0,y_0)}(y-y_0)^2. \tag{20.7}$$

To simplify this expression, we introduce the notation

$$A = \left.\frac{\partial^2 f}{\partial x^2}\right|_{(x_0,y_0)}, \qquad B = \left.\frac{\partial^2 f}{\partial x \partial y}\right|_{(x_0,y_0)}, \text{ and } \qquad C = \left.\frac{\partial^2 f}{\partial y^2}\right|_{(x_0,y_0)}.$$

These are constants for a given function and a given stationary input. We also let

$$u = x - x_0 \qquad \text{and} \qquad v = y - y_0.$$

We focus on $(u, v) = (0, 0)$, because this input in u and v corresponds to the stationary input $(x, y) = (x_0, y_0)$. With this notation, the expression in Equation (20.7) gives the outputs of a function

$$g(u, v) = Au^2 + 2Buv + Cv^2.$$

This is a quadratic function in the input variables u and v. The graphs of such functions are included among the quadric surfaces we studied in Section 13.3.

To understand the shape of the graph of g near $(0, 0)$, we can examine slices with vertical planes containing the $z = g(u, v)$ axis. These planes are defined by lines throught the origin in the uv-plane. Along the line given by $v = mu$, the outputs of g are

$$g(u, mu) = Au^2 + 2Bu(mu) + C(mu)^2 = (A + 2Bm + Cm^2)u^2.$$

From this last expression, we deduce that the concavity of the curve in the section defined by $v = mu$ is determined by the sign of

$$A + 2Bm + Cm^2.$$

The sign clearly depends on the value of m, that is, on which section we have. If the concavity is positive for every value of m, then the stationary input at $(u, v) = (0, 0)$ must be a local minimizer. If the concavity is negative for every value of m, then the stationary input must be a local maximizer. If the concavity is negative in some sections and positive in others, then the stationary input corresponds to a saddle.

Note that the line $u = 0$ is not included among the lines $v = mu$. The line $u = 0$ is the v cross section. The concavity in this section is given by C. Note this is consistent with the fact that the $v = 0$ section has concavity given by A.

To distinguish among the possible cases, we look at the condition under which the concavity is zero. Let m_0 be a value of m for which

$$A + 2Bm_0 + Cm_0^2 = 0.$$

The case $C = 0$ is a special situation, which you are asked to examine in Problem 9. Assuming that C is not equal to zero, we use the quadratic formula to solve for m_0 giving

$$m_0 = \frac{-2B \pm \sqrt{4B^2 - 4AC}}{2C} = \frac{-B \pm \sqrt{B^2 - AC}}{C}. \qquad (20.8)$$

There are two possibilities, depending on the sign of the quantity $B^2 - AC$. If $B^2 - AC > 0$, then there are two solutions for m_0, corresponding to two sections in which the concavity is zero. These lines divide the uv-plane into four "wedges," and the concavity alternates from positive to negative in these wedges (see Figure 20.5). The stationary input $(0, 0)$ thus corresponds to a saddle. If $B^2 - AC > 0$, then there are no (real-valued) solutions. Hence, the concavity is positive in all sections or negative in all sections. Note that in order for this case to occur, A and C must have the same sign. If the signs are both positive, then all sections are concave up, meaning $(0, 0)$ is a local minimizer. If the signs are both negative, then all sections are concave down, meaning $(0, 0)$ is a local maximizer.

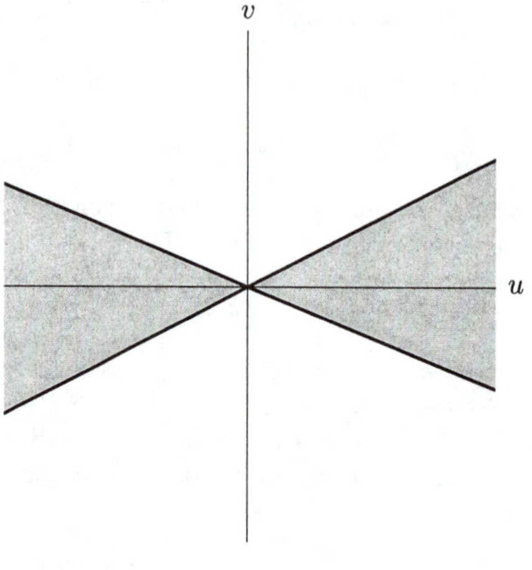

Figure 20.5. For a quadratic saddle there are two sections in which the concavity is zero. These correspond to the two lines through the origin in the uv-plane. The concavity is the same sign for all sections in the gray "wedges" and of opposite sign for all sections in the white "wedges."

We summarize the conclusions of the preceding argument in the following theorem. The argument constitutes a proof of the theorem. For historical reasons, the convention is to define the quantity $AC - B^2$ as the discriminant. We denote this as D.

Theorem 20.3. *Suppose* (x_0, y_0) *is a stationary input for a function* $f : \mathbb{R}^2 \to \mathbb{R}$. *Also, suppose* f *has* *continuous second partial derivatives in some open disk centered at* (x_0, y_0). *Let*

$$D = AC - B^2 = \left.\frac{\partial^2 f}{\partial x^2}\right|_{(x_0,y_0)} \left.\frac{\partial^2 f}{\partial y^2}\right|_{(x_0,y_0)} - \left(\left.\frac{\partial^2 f}{\partial x \partial y}\right|_{(x_0,y_0)}\right)^2.$$

If $D < 0$, *then* (x_0, y_0) *corresponds to a saddle point. If* $D > 0$, *then* (x_0, y_0) *is a local extremizer. In this case, if* $A > 0$ *and* $C > 0$ *then* (x_0, y_0) *is a local minimizer. If* $A < 0$ *and* $C < 0$, *then* (x_0, y_0) *is a local maximizer.*

Note that the theorem is silent in the case $D = AC - B^2 = 0$. We illustrate the use of this second derivative test in the following examples, starting with some simple cases.

Example 20.9. Use the second derivative test to show that $f(x, y) = x^2 + y^2$ has a local minimum at the stationary input $(0, 0)$.

Solution. It is straightforward to compute the second partial derivatives and find $A = 2$, $B = 0$, and $C = 2$. Thus, $D = (2)(2) - 0^2 = 4 > 0$. From the second derivative test, we conclude that $(0, 0)$ is a local minimizer. △

Example 20.10. Use the second derivative test to show that $f(x, y) = x^2 - y^2$ has a saddle corresponding to the stationary input $(0, 0)$.

Solution. It is straightforward to compute the second partial derivatives and find $A = 2$, $B = 0$, and $C = -2$. Thus, $D = (2)(-2) - 0^2 = -4 < 0$. From the second derivative test, we conclude that $(0, 0)$ corresponds to a saddle. △

Example 20.11. Find and classify all stationary inputs for $f(x, y) = x^3 + 6xy - xy^2 + 4$.

Solution. We compute derivatives to get

$$\frac{\partial f}{\partial x} = 3x^2 + 6y - y^2 \quad \text{and} \quad \frac{\partial f}{\partial y} = 6x - 2xy.$$

Equating each partial derivative to zero gives the system of equations

$$\begin{aligned} 3x^2 + 6y - y^2 &= 0 \\ 2x(3 - y) &= 0. \end{aligned}$$

From the second equation, we get two cases: $x = 0$ and $y = 3$. For $x = 0$, the first equation reduces to $6y - y^2 = y(6 - y) = 0$, giving $y = 0$ and $y = 6$. Thus, we have $(0, 0)$ and $(0, 6)$ as stationary inputs. For $y = 3$, the first equation gives $3x^2 + 18 - 9 = 3x^2 + 9 = 0$. This has no (real-valued) solutions, and hence the $y = 3$ case gives no additional stationary inputs.

We now compute second derivatives to get

$$\frac{\partial^2 f}{\partial x^2} = 6x, \qquad \frac{\partial^2 f}{\partial x \partial y} = 6 - 2y\ , \text{and} \qquad \frac{\partial^2 f}{\partial y^2} = -2x.$$

For the stationary input $(0, 0)$, we have $A = 0$, $B = 6$, and $C = 0$. Thus, $D = AC - B^2 = (0)(0) - 6^2 = -36 < 0$. Hence, $(0, 0)$ corresponds to a saddle.

For the stationary input $(0, 6)$, we have $A = 0$, $B = -6$, and $C = 0$. Thus, $D = AC - B^2 = (0)(0) - (-6)^2 = -36 < 0$. Hence, $(0, 6)$ also corresponds to a saddle. △

20.4 Global Extrema

The definitions of global extremizer and global extreme from Section 20.1 can be readily generalized to functions $f : \mathbb{R}^2 \to \mathbb{R}$. We can then consider the conditions under which a given function on a given domain is guaranteed to have a global minimum and a global maximum.

Definition 20.6. For a function $f : A \to \mathbb{R}$, where A is a subset of $\mathbb{R}^2$, the input (x_0, y_0) is called the **global maximizer**, and the corresponding output $f(x_0, y_0)$ is called the **global maximum**, if $f(x, y) \leq f(x_0, y_0)$ for all (x, y) in A. The input (x_0, y_0) is called the **global minimizer**, and the corresponding output $f(x_0, y_0)$ is called the **global minimum**, if $f(x_0, y_0) \leq f(x, y)$ for all (x, y) in A.

The extreme value theorem for functions $f : \mathbb{R} \to \mathbb{R}$ requires that f be continuous on a subset of $\mathbb{R}$ in the form of a closed interval $[a, b]$. To generalize this theorem, we need to introduce some ideas about subsets of $\mathbb{R}^2$. In particular, we need a precise description of those subsets for which a version of the extreme value theorem holds. An obvious extension of closed interval is a closed rectangle defined by

$$[a, b] \times [c, d] = \{(x, y) | a \leq x \leq b, c \leq y \leq d\}.$$

This extension is *not* general enough for our purposes. In particular, we must allow for more general regions as, for example, shown in Figure 20.6. The two notions we want to capture are

- that the region contain all of its *boundary points* (the equivalent of a closed interval containing both endpoints); and

- that the region be of finite size in some sense.

To address the first of these, we need some terminology. Given a subset A in $\mathbb{R}^2$, we define three types of points. A point (x, y) is an **interior point of A** if there is an open disk centered at (x, y) that contains only points in A. A point (x, y) is an **exterior point of A** if there is an open disk centered at (x, y) that contains only points *not* in A. A point (x, y) is a **boundary point of A** if *every* open disk centered at A contains at least one point in A and at least one point not in A.

With this terminology, we can distinguish two special classes of subsets in the plane. A set A is **open** if A contains *no* boundary points of A. A set A is **closed** if A contains *all* boundary points of A. In drawing plane regions, the convention is to use a dashed curve for boundary points that are not included in the set and a solid curve for boundary points that are included in the set. Figure 20.6(a) shows a typical closed set in the plane, and Figure 20.6(b) shows a typical open set in the plane. Note that a set can be neither open nor closed, for example, as shown in Figure 20.6(c).

Now let's turn to the second issue. To give some sense of "finite in size," we can require that the set A be inside some circle of finite radius. Specifically, we say a set A is **bounded** if A is a subset of some disk of finite radius centered at the origin. For example, the upper half plane is not bounded, while all three of the regions shown in Figure 20.6 are bounded.

Note the need for care with this terminology: The terms *bounded* and *boundary point* have common root but independent meanings. Determining whether a set is bounded is not related to determining whether a set contains none, some, or all of its boundary points.

With this language we can give a careful statement of the extreme value theorem for functions of two variables.

Theorem 20.4. *If A is a subset of $\mathbb{R}^2$ that is closed and bounded, and $f : A \to \mathbb{R}$ is continuous, then f has a global minimum and a global maximium.*

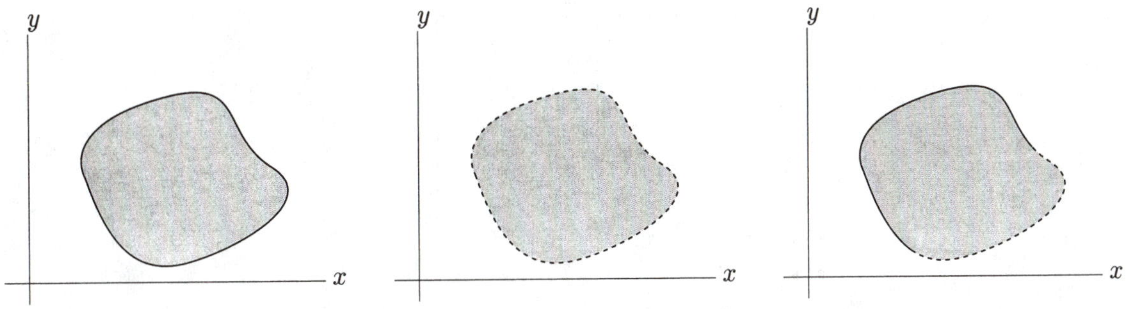

(a) Typical closed set in the plane

(b) Typical open set in the plane

(c) Set that is neither open nor closed

Figure 20.6. Subsets of the plane. The convention is to use a dashed curve for boundary points that are not included in the set and a solid curve for boundary points that are included in the set.

As with a function $f : [a, b] \to \mathbb{R}$, the process for determining the global extremes involves finding all local extrema on the interior of the region and then comparing these to the outputs on the boundary of the region. For a function of one variable with a closed interval $[a, b]$, there are only two endpoints to examine as inputs. For a function $f : A \to \mathbb{R}$, where A is a closed and bounded region, we must examine the entire boundary. We illustrate this for some simple examples here. In the next section, we develop a powerful technique for locating *constrained extrema.*

Example 20.12. Find the global extremes of $f(x, y) = x^2 + y^2$ for the disk of radius 3 centered at the origin.

Solution. It is straightforward to determine that the only stationary input is $(0, 0)$. The corresponding output is $f(0, 0) = 0$. The boundary of the disk is the circle of radius 3 given by the equation $x^2 + y^2 = 3^2$. The outputs of the function for points on this circle are $f(x, y) = 3^2 = 9$. Thus, the global minimum is 0 and the global maximum is 9. △

Example 20.13. Find the global extremes of $f(x, y) = xy - x + 2y$ for the rectangle $[-3, 3] \times [-2, 2]$.

Solution. We compute the first partial derivatives to get

$$\frac{\partial f}{\partial x} = y - 1 \qquad \text{and} \qquad \frac{\partial f}{\partial y} = x + 2.$$

Equating each partial derivative to zero gives the system of equations

$$\begin{aligned} y - 2 &= 0 \\ x + 1 &= 0. \end{aligned}$$

The only solution is $(-2, 1)$. For this stationary input, the corresponding output is $f(-2, 1) = -2$. Note that we need not determine whether this is a local minimum, a local maximum, or a saddle point, because our objective is to find the global extremes.

We must now examine outputs along the boundary. This consists of four line segments that we have labeled A through D in Figure 20.7. Let's look at the function with inputs constrained to each segment, one at a time.

Start with segment A, which has $y = -2$ with x in the interval $[-3, 3]$. For inputs on this segment, the function has outputs $f(x, -2) = -2x - x + 4 = -3x + 4$. This is linear in x with a negative slope, so the

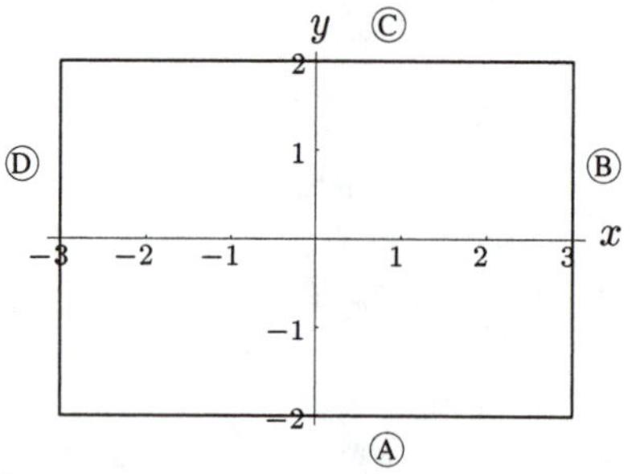

Figure 20.7. Boundary of the region in Example 20.13.

minimum is clearly $f(3,-2) = -5$, and the maximum is clearly $f(-3,-2) = 13$. For segment B, the inputs have $x = 3$ with y in the interval $[-2, 2]$. The outputs are $f(3, y) = 3y - 3 - 2y = y - 3$. The minimum is $f(3,-2) = -5$, and the maximum is $f(3, 2) = -1$. Continuing in a similar fashion, for segment C we have outputs $f(x, 2) = x - 4$. The minimum is $f(-3, 2) = -7$, and the maximum is $f(3, 2) = -1$. Finally, for segment D we have $f(-3, y) = -5y + 3$, with minimum $f(-3, 2) = -7$ and maximum $f(-3, -2) = 13$.

Comparing all of the candidates, we find the global minimum is $f(-3, 2) = -7$, and the global maximum is $f(-3, -2) = 13$. △

20.5 Constrained Optimization

In the previous section, we considered the problem of finding global extremes for a function $f : \mathbb{R}^2 \to \mathbb{R}$. In many cases, this involves locating **constrained extrema**. That is, we need to find the extreme outputs of a function for inputs that are constrained to a curve in the input plane.

As a simple first example, consider finding the extremes of the function $f(x, y) = xy$, with inputs constrained to the circle given by $x^2 + y^2 = 2$. In Figure 20.8(a), we show the constraint curve (the circle of radius $\sqrt{2}$ in this case) and some level curves for the function. We argue that the constrained extremes should correspond to the inputs for which the constraint curve is tangent to a level curve. The informal argument is that these inputs correspond to points at which the constraint curve reaches the extreme level curves.

In a generic case, we can look at this situation by "zooming" in on a region in which a constrained extremizer lives. With sufficient zoom, the graph of f is well approximated by a tangent plane so that the level curves of the function f are approximately parallel lines, as shown in Figure 20.8(b). (This is the generic situation for an input that is not a stationary input.) For each input, there is a level curve of some particular output. On one side of this level curve, outputs are larger, and on the other side, outputs are smaller. This figure also shows a constraint curve. Now consider a point on the constraint curve that is not tangent to the level curve through that point. The curve has points on both sides of the level curve, and so some inputs on the curve have outputs smaller than the output on the level curve, and some inputs on the curve have outputs greater than the output on the level curve. Only at an input point for which the constraint curve is tangent to the level curve is it possible for all nearby inputs on the curve to have outputs always less than or always greater than the output at that input.

To make a more rigorous argument, we consider a symbolic approach. Suppose the constraint curve is parametrized by the function $\vec{r} : \mathbb{R} \to \mathbb{R}^2$. The outputs of the function f for inputs along the curve are given by the composition

$$h(t) = f(\vec{r}(t)) = f(x(t), y(t)).$$

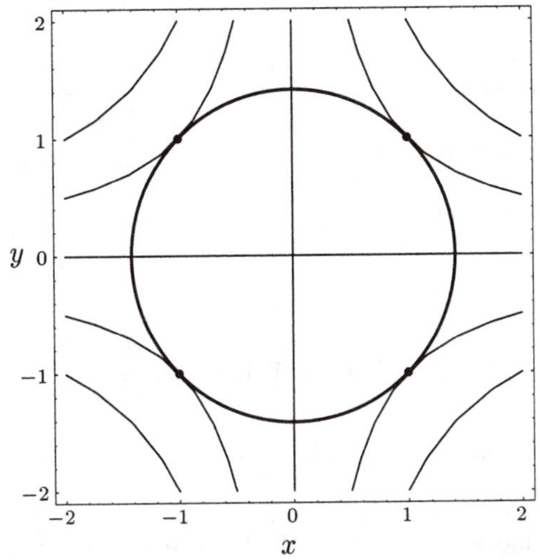

(a) Constraint curve $x^2 + y^2 = 2$ shown with some level curves for the function $f(x, y) = xy$

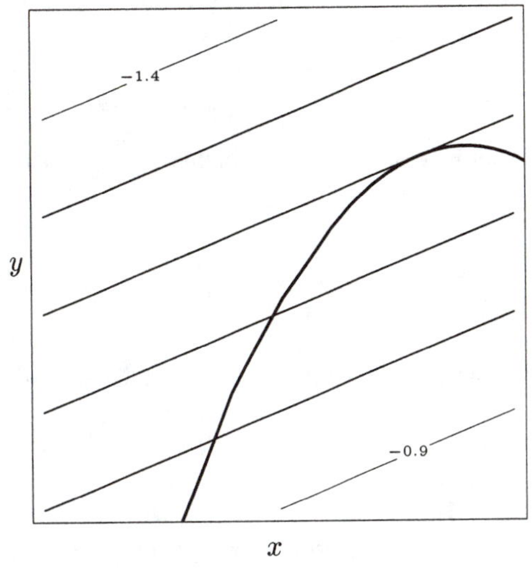

(b) Generic set of level curves viewed locally with a constraint curve

Figure 20.8. Illustrating the relation between a constraint curve and level curves of the function to be optimized.

Note that h is a function of one variable. Stationary inputs for h are determined by equating the first derivative to zero. If t_0 is a stationary input, then

$$h'(t_0) = \frac{d}{dt}\left[f(\vec{r}(t))\right]\Big|_{t_0} = 0.$$

Using a chain rule for the derivative of $f(\vec{r}(t))$, we can express this condition as

$$\vec{\nabla} f\left(\vec{r}(t_0)\right) \cdot \vec{r}\,'(t_0) = 0.$$

We can interpret this as saying the gradient of f at $\vec{r}(t_0)$ is perpendicular to the tangent vector $\vec{r}\,'(t_0)$. The gradient vector is perpendicular to the level curve at $\vec{r}(t_0)$, so we conclude that the constraint curve is tangent to the level curve.

We use this result to our advantage in the following way. Think of the constraint curve as the level curve of some new function $g : \mathbb{R}^2 \to \mathbb{R}$. In the example we started above, this function would be $g(x, y) = x^2 + y^2$, and the constraint curve is the $c = 2$ level curve. At a constrained extremizer, this particular level curve of g must be tangent to some level curve of f. This implies that the gradient of g is parallel to the gradient of f at a constrained extremizer. We can express this symbolically by saying there is some scale factor λ, so that if (x_0, y_0) is a constrained extremizer then

$$\vec{\nabla} f(x_0, y_0) = \lambda \vec{\nabla} g(x_0, y_0). \tag{20.9}$$

Note that we have three unknowns, the two coordinates x_0 and y_0 along with the scale factor λ. The vector equation in Equation (20.9) gives two scalar equations for these three unknowns. The constraint equation gives a third equation. We use these to complete the example we have started.

Example 20.14. Use Equation (20.9) along with the constraint equation $g(x, y) = c$ to find all constrained extremizers for $f(x, y) = xy$ with inputs constrained to the curve $x^2 + y^2 = 2$.

Solution. For this case, the gradient of f is $\vec{\nabla} f(x,y) = \langle y, x\rangle$. The gradient of g is $\vec{\nabla} g(x,y) = \langle 2x, 2y\rangle$. With these, Equation (20.9) is

$$\langle y, x\rangle = \lambda\langle 2x, 2y\rangle = \langle 2\lambda x, 2\lambda y\rangle.$$

Equating components and including the constraint equation, we have the system

$$\begin{aligned} y &= 2\lambda x \\ x &= 2\lambda y \\ x^2 + y^2 &= 2. \end{aligned}$$

We can substitute for y in the second and third equations using the first equation. This gives us the system

$$\begin{aligned} x &= 4\lambda^2 x \\ x^2 + 4\lambda^2 x^2 &= 2. \end{aligned}$$

From the first equation of this system, we get two cases, either $x = 0$ or $4\lambda^2 = 1$. The case $x = 0$ leads to $0 = 2$ in the third equation, so there are no solutions for this. The case $4\lambda^2 = 1$ gives $x^2 + x^2 = 2$ in the third equation, for which the solutions are $x = -1$ and $x = 1$. Note that $\lambda = -1/2$ or $\lambda = +1/2$ in this case.

We now return to the top equation in the original system to determine relevant values of y. There are four possibilities. If $x = -1$ and $\lambda = -1/2$, then $y = 1$. If $x = -1$ and $\lambda = +1/2$, then $y = -1$. If $x = 1$ and $\lambda = -1/2$, then $y = -1$. Finally, if $x = 1$ and $\lambda = 1/2$, then $y = 1$. There are thus a total of four solutions for the system of three equations. As ordered triples (x, y, λ), these are $(-1, -1, 1/2)$, $(-1, 1, -1/2)$, $(1, -1, -1/2)$, and $(1, 1, 1/2)$.

The constrained extremizers are given by the (x, y) pairs alone. The corresponding outputs are easily computed as $f(-1,-1) = 1$, $f(-1,1) = -1$, $f(1,-1) = -1$ and $f(1,1) = 1$. Thus, the constrained minimum is -1, and the constrained maximum is 1. △

The scalar factor λ in Equation (20.9) is called a **Lagrange multiplier**, in honor of the French mathematician Joseph Louis Lagrange (1736–1813). We will refer to the function g as the **constraining function**. This process of locating constrained extremizers is called the **method of Lagrange multipliers**. Note that the numerical value of the Lagrange multipliers is generally not directly relevant in the final result.

Example 20.15. Find the point on the line $6x - 4y = 5$ that is closest to the origin.

Solution. For a point (x, y), the distance to the origin is given by $\sqrt{x^2 + y^2}$. We can simplify matters by considering the square of the distance, that is, $x^2 + y^2$. The point on the line that minimizes $x^2 + y^2$ is the same point that minimizes $\sqrt{x^2 + y^2}$. Thus, we seek the constrained minimizer of the function $f(x,y) = x^2 + y^2$ for inputs on the line $6x - 4y = 5$. We can take $g(x,y) = 6x - 4y$ as the constraining function.

Computing the relevant gradients and writing out the component equations together with the constraint equation gives us the system

$$\begin{aligned} 2x &= 6\lambda \\ 2y &= -4\lambda \\ 6x - 4y &= 5. \end{aligned}$$

We can solve the first equation to get $x = 3\lambda$ and the second equation to get $y = -2\lambda$. Substituting into the third equation gives

$$18\lambda + 8\lambda = 26\lambda = 5.$$

Thus, $\lambda = 5/26$ is the only solution. With this value of λ, we get $x = 15/26$ and $y = -5/13$. From the geometry of the problem, we conclude that the point $(15/26, -5/13)$ is the point on the given line that is closest to the origin. △

In the next example, we return to the problem with which we opened the chapter. In that problem, we found the dimensions of a window that gave the maximum area for a fixed perimeter. In solving the problem, we used the perimeter constraint to eliminate one variable in the area expression, leaving an optimization problem in one variable. We now resolve the problem using the method of Lagrange multipliers.

Example 20.16. A window is to be made in the shape of a rectangle with a semicircular top as shown in Figure 20.1. Find the dimensions of the window that give the greatest area for a perimeter of 3 m.

Solution. From our previous work, we know that the area of the window is given by

$$A(r,h) = 2rh + \frac{1}{2}\pi r^2.$$

The constraining function is

$$P(r,h) = 2r + 2h + \pi r = (2+\pi)r + h$$

with $c = 3$ as the specific level curve to which inputs are constrained.

Computing the relevant gradients and combining Equation (20.9) with the constraint equation gives the system

$$\begin{aligned} 2h + \pi r &= \lambda(2+\pi) \\ 2r &= 2\lambda \\ (2+\pi)r + 2h &= 3. \end{aligned}$$

From the second equation, we have $r = \lambda$. Substituting this into the first equation and solving gives $h = \lambda$. With these results in the third equation, we get

$$(2+\pi)\lambda + 2\lambda = 3$$

so

$$\lambda = \frac{3}{4+\pi}.$$

Thus,

$$r = h = \frac{3}{4+\pi}$$

gives the maximum area for a perimeter of 3 m. △

20.6 Problems

20.1 Optimization for Functions of One Variable

1. Give an example of a domain and function that has a global maximum but not a global minimum.

2. Give an example of a domain and function that has a global minimum but not a global maximum.

3. Give an example of a function that is bounded above and below but does not have a global minimum or a global maximum.

4. Give an example of a function defined on the closed interval $[0, 5]$ that does not have a global minimum or a global maximum.

For Problems 5–8, find all local extremizers and extrema for the given function on the given domain. Also find the global extremizers and extrema if these exist.

5. $f(x) = x^2 e^{-x}, \quad [0, 2]$

6. $g(t) = t \sin t\,, \quad [-\pi, \pi]$

7. $U(r) = \dfrac{1}{r^5} - \dfrac{1}{r^2}\,, \quad (0, 5]$

8. $g(x) = x^3 - 6x^2 + 2x - 1\,, \quad [-1, 6]$

9. A lifeguard must reach a swimmer who is in the water 10 m off the beach from a point that is 20 m down the beach from the lifeguard station. The lifeguard can run on the beach at a speed of 3.0 m/s and swim at a speed of 1.4 m/s. How far down the beach should the lifeguard run in order to reach the swimmer in the minimum time?

10. A beaker is being designed in shape of a right circular cylinder with a base but no top. Find the dimensions of the beaker that has the greatest volume for a fixed surface area S.

11. Consider the potential energy function

$$U(r) = \frac{a}{r^5} - \frac{b}{r^2}$$

where r is the distance from the origin. Find the absolute minimizer of U in terms of a and b.

20.2 Local Extrema for Functions of Several Variables

For Problems 1–6, find and classify all stationary points of the given function.

1. $f(x, y) = xy$

2. $f(x, y) = x^2 + y^2 - xy$

3. $f(x, y) = 2x^2 + y^2 - xy - 7y$

4. $f(x, y) = y^2 - xy + 2x + y + 1$

5. $f(x, y) = x^2 + 5xy - y^3$

6. $f(x, y) = x \sin y$

7. (a) Write a definition analogous to Definition 20.4 for $\mathbb{R}^3$. (b) Write a definition analogous to Definition 20.5 for functions $f : \mathbb{R}^3 \to \mathbb{R}$.

8. Formulate and prove a theorem analogous to Theorem 20.2 for functions $f : \mathbb{R}^3 \to \mathbb{R}$.

For Problems 9–12, find and classify all stationary points of the given function.

9. $f(x, y, z) = x^2 + y^2 + z^2$

10. $f(x, y, z) = x^2 + y^2 - z^2$

11. $f(x, y, z) = x^2 - y^2 + z^2$

12. $f(x, y, z) = xyz$

13. Describe the gradient vector field near a local minimizer for a function $f : \mathbb{R}^2 \to \mathbb{R}$.

14. Describe the gradient vector field near a local maximizer for a function $f : \mathbb{R}^2 \to \mathbb{R}$.

15. Describe the gradient vector field near an input that gives a saddle point for a function $f : \mathbb{R}^2 \to \mathbb{R}$.

16. Describe the level curves near a local minimizer for a function $f : \mathbb{R}^2 \to \mathbb{R}$.

17. Describe the level curves near a local maximizer for a function $f : \mathbb{R}^2 \to \mathbb{R}$.

18. Describe the level curves near an input corresponding to a saddle point for a function $f : \mathbb{R}^2 \to \mathbb{R}$.

20.3 A Second Derivative Test

For Problems 1–6, find and classify all stationary points of the given function.

1. $f(x, y) = xy$
2. $f(x, y) = x^2 + y^2 - xy$
3. $f(x, y) = 2x^2 + y^2 - xy - 7y$
4. $f(x, y) = y^2 - xy + 2x + y + 1$
5. $f(x, y) = x^2 + 5xy - y^3$
6. $f(x, y) = x \sin y$
7. For the function $f(x, y) = x^2 + 2xy + y^2$, show that $(0, 0)$ is a stationary input with $D = 0$. Use the fact that $f(x, y) = (x + y)^2$ to argue that $(0, 0)$ is a local minimizer.
8. Find and classify all stationary inputs of $f(x, y) = 4 + 4x + x^2 + 4y + 2xy + y^2$.
9. In the argument for the second derivative test, we assumed $C \neq 0$ to get Equation (20.8). Examine the case $C = 0$.

20.4 Global Extrema

For Problems 1–8, specify the interior points, the exterior points, and the boundary points of the given set. Determine whether the set is open, closed, or neither. Determine whether the set is bounded or unbounded.

1. $A = \{(x, y) | y > 0\}$
2. $A = \{(x, y) | y \geq 0\}$
3. $A = \left\{ (x, y) \middle| \frac{x^2}{5^2} + \frac{y^2}{3^2} \leq 1 \right\}$
4. $A = \left\{ (x, y) \middle| \frac{x^2}{5^2} + \frac{y^2}{3^2} \geq 1 \right\}$
5. $A = \{(x, y) | x = 0\}$
6. $A = \{(0, 0)\}$
7. $A = \{(x, y) | y \leq 0 \text{ if } x > 0;\ y < 0 \text{ if } x \leq 0\}$
8. $A = \{(x, y) | x \text{ and } y \text{ are rational numbers }\}$

For Problems 9–14, find the global extremes of the given function.

9. $f(x, y) = xy$ for the rectangle $[0, 1] \times [0, 2]$.
10. $f(x, y) = x^2 + y^2$ for the rectangle $[-5, -3] \times [0, 2]$.
11. $f(x, y) = x^3 - xy^2 + 13$ for the rectangle $[-2, 2] \times [-2, 2]$.
12. $f(x, y) = \sin(x^2 + y^2)$ for the rectangle $[-1, 1] \times [-1, 1]$.
13. $f(x, y, z) = x + 3y - 2z$ for the rectangular box $[-1, 1] \times [0, 4] \times [-3, 1]$.
14. $f(x, y, z) = x^2 + y^2 + z^2$ for the rectangular box $[0, 1] \times [0, 1] \times [0, 1]$.

20.5 Constrained Optimization

1. Find the extremes of $f(x, y) = xy$ for inputs constrained to the ellipse $(x/3)^2 + (y/4)^2 = 1$.
2. Find the global extremes of $f(x, y) = xy$ for the domain $A = \{(x, y)|(x/3)^2 + (y/4)^2 \leq 1\}$.
3. Find the extremes of $f(x, y) = x^2 + 5y^2$ for inputs constrained to the ellipse with equation $(x/3)^2 + (y/4)^2 = 1$.
4. Find the global extremes of $f(x, y) = x^2 + 5y^2$ for the domain $A = \{(x, y)|(x/3)^2 + (y/4)^2 \leq 1\}$.
5. Find the point on the line $-2x + 5y = 2$ that is closest to the point $(3, -5)$.
6. Find the point or points on the parabola $y = x^2 + 3x - 4$ that are closest to the origin.
7. Find the point on the plane $6x - y + 2z = 3$ that is closest to the origin.
8. A cylindrical can is being designed to hold a volume of 400 cm^3. The top and bottom of the can must be twice as thick as the sides, and so cost twice as much per area. Find the dimensions of the can that has the lowest cost for material.
9. A rectangular window is being designed to have an area of A m^2. The trim costs s euros per meter for the sill at the bottom and t euros per meter for the sides and top. Find the dimensions for the window that minimize the cost of the trim material.

Chapter 21

dc Circuits

This chapter is devoted to the study of electric circuits powered by a constant source of potential difference (a battery). These circuits are called **direct-current circuits**, or simply **dc circuits**, because the direction of current flow is determined by the polarity of the battery or batteries and does not change. We begin with a brief look at batteries, where we will introduce the concepts of internal resistance and electromotive force. Then we establish the rules for determining the equivalent resistance of various combinations of resistors: series, parallel, and combinations of series and parallel connections. More generally, batteries and resistors can be combined in multiloop circuits. Kirchoff's loop and junction rules allow us to determine the current in any part of a multiloop circuit, provided that the resistances and the potential differences across the batteries are known.

After studying multiloop circuits, we study how various measuring instruments (especially ammeters and voltmeters) are used in the experimental analysis of electric circuits. Finally, capacitors can be combined with resistors to form an RC circuit, and our knowledge of differential equations is used to analyze the charging and discharging of a capacitor in an RC circuit.

21.1 Batteries: Real and Ideal

In Chapter 19, we described a battery as a source of constant potential difference. It is common to refer to a battery as a source of **electromotive force**, or simply **emf**. This term is misleading, because it is not a force. Rather, the physical quantity we call emf is just the magnitude of the potential difference supplied by a battery. To distinguish the emf of a battery from other potential differences in a circuit, we use a special symbol $\mathcal{E}$ for emf. The SI units for $\mathcal{E}$ are volts, since it represents a potential difference. Formally, the emf is the work done per unit charge emitted as current. That is, $\mathcal{E} = dW/dq$. Thus, emf has dimensions of work divided by charge, which is equivalent to dimensions of electric potential.

A battery is only one example of a source of emf, although we will use it throughout this chapter as a paradigm. In a battery, chemical energy is converted into electrical energy. Other important sources of emf include photovoltaic devices, which convert light to electrical energy, and electric generators, which convert mechanical energy into electrical energy. Regardless of the source of emf, it is important to think of it not as a force but as a source of electrical energy. As you saw in Section 19.3.4, a battery (or other source of emf) does work in moving charges through an electric circuit. Some of this work always ends up as heat energy (unless the circuit is entirely superconducting). Since it does work, the battery expends energy in making current flow through a circuit. Sometimes part of the electrical energy can be converted back into mechanical energy, as in an electric motor.

Whether or not it is a battery, we refer to a circuit element that supplies energy as a **source**, and the elements such as resistors and capacitors connected to them as the **load**. The emf of an ideal battery or other source is constant, independent of the load. But no source is ideal, and the principal reason is that the battery (or other source) has some resistance between its two terminals. This resistance is called the **internal resistance** of the battery. We will use the lowercase r to refer to internal resistance and continue (as in Chapter 19) to use the uppercase R for any load resistors.

Consider a circuit in which a battery is connected to single load resistor R, as in Figure 21.1(a). A model that reasonably accounts for the battery's internal resistance is shown in Figure 21.1(b), where the battery consists of a series combination of an ideal emf $\mathcal{E}$ and an internal resistance r. We analyze this circuit by considering the potential differences across each of the three elements ($\mathcal{E}$, r, and R) in the circuit. We assume in this example and throughout this chapter that there is no potential difference in any of the connecting wires. In doing so, we are effectively assuming that the wires have no resistance. As you know, this is not true; however, if the wires are made of a good conducting material such as copper, their resistance can often be ignored, because it is low compared with the resistances of the other circuit elements.

At this point, we introduce a systematic method to analyze the potential differences, which will be used here and in other examples throughout the chapter. First, pick an arbitrary point in the circuit, such as the point labeled P in Figure 21.1(b). Then proceed around the circuit and write the potential difference as you travel across each element. Upon returning to P, the sum of all the potential changes is zero. This is because the electric field is conservative, and you must therefore return to the same electric potential when returning to the same point in a field (or in this case to the same point in a circuit). Because if its importance in circuit analysis, we emphasize this statement: *The sum of the potential changes around a closed loop is zero.* The statement is true for any closed loop in an electric circuit and is known as **Kirchoff's loop rule**, named for the German physicist Gustav Kirchoff (1824–1887).

Following this method, the first element encountered (traveling upward along the left branch in the diagram) is the ideal emf source $\mathcal{E}$, so by definition that potential difference is $\mathcal{E}$. A careful look reveals that we have traveled through a potential *gain* $+\mathcal{E}$, because we are going from the negative terminal to the positive terminal of the battery. Next, traveling through the resistance r, we experience a potential *drop* in moving away from the positive terminal toward the negative terminal. That drop is of magnitude Ir, so the change

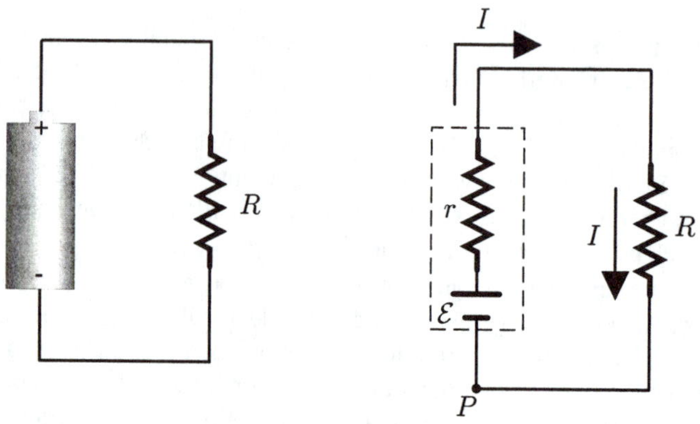

(a) Schematic of the battery

(b) Flow of current in a circuit with a battery connected to a resistor

Figure 21.1. Model for a battery as series combination of an ideal emf $\mathcal{E}$ and an internal resistance r.

in electric potential in traveling across r in this direction is $-Ir$. Similarly, the change in electric potential in traveling across the load resistor R is $-IR$. As you learned in Chapter 19, the current is the same everywhere in this single-loop circuit, because there is no observed buildup of charge anywhere. Summing all the potential changes and setting the sum equal to zero (Kirchoff's loop rule), we have

$$\mathcal{E} - Ir - IR = 0. \tag{21.1}$$

We can analyze this result in a number of ways. First, we can rearrange to solve for the current in the circuit as a function of the emf and resistances:

$$I = \frac{\mathcal{E}}{r + R}. \tag{21.2}$$

This result tells us that the current in the circuit is less than the current $I = \mathcal{E}/R$ in a circuit consisting of an ideal emf with no internal resistance and a load resistor R.

Another result that we can see from Equation (21.1) is that the actual potential difference between the battery terminals is not $\mathcal{E}$, but rather $\mathcal{E} - Ir$. Since the current I also depends on the load resistance R, this explains why the real potential difference across the battery terminals is load-dependent.

How can the internal resistance r of a battery or other source be measured experimentally? One way is to attach a connecting wire with negligible resistance directly between the two battery terminals. The load resistance R is effectively zero, and in this case Equation (21.1) reduces to $\mathcal{E} - Ir = 0$, or $r = \mathcal{E}/I$. Unfortunately, a direct connection between the battery terminals (called a **short circuit**) results in an extremely high current flow. This can rapidly drain the battery and can cause the temperature of the conducting wire to increase significantly, thus increasing its resistance. The other problem with this method is that if we are to compute the internal resistance using $r = \mathcal{E}/I$, this assumes we can make an independent measurement of the battery's emf. As you saw before, this is not a good assumption.

A better way of determining a battery's internal resistance is to make measurements of the current that flows when different loads are put in place of the load resistor R in Figure 21.1. Suppose we measure the current I_1 that flows with a load R_1 in place and the current I_2 that flows with a load R_2. In both cases, Equation (21.1) must be satisfied. Thus, we have two linear equations,

$$\mathcal{E} - I_1 r - I_1 R_1 = 0$$

and

$$\mathcal{E} - I_2 r - I_2 R_2 = 0$$

that can be solved simultaneously for r and $\mathcal{E}$. For example, subtracting the second equation from the first gives

$$(I_2 - I_1)\, r - I_1 R_1 + I_2 R_2 = 0.$$

This equation can be rearranged to solve for r:

$$r = \frac{I_1 R_1 - I_2 R_2}{I_2 - I_1}. \tag{21.3}$$

The resulting value for r can be substituted into either of the original equations to find $\mathcal{E}$. Therefore, this method gives us a way of determining not only the internal resistance, but also the emf $\mathcal{E}$, rather than the potential difference across the whole battery. Numerous measurements of the current that flows with different loads will produce a better experimental determination of both r and $\mathcal{E}$.

Example 21.1. An automobile battery is designed to be a source of extremely high current with a relatively low potential difference (typically about 12 V). When a particular battery is connected to the automobile starter (which has a resistance of 0.058 Ω), a current of 161 A flows. When the same battery is connected to a 50-Ω load in a test, the current is 239.8 mA. (a) Find $\mathcal{E}$ and the internal resistance for this battery. (b) Find the potential difference across the battery terminals when it is operating the starter.

Solution. (a) Following the method outlined in the text, we can use Equation (21.3) to solve for the internal resistance:

$$r = \frac{I_1R_1 - I_2R_2}{I_2 - I_1} = \frac{(161\text{ A})(0.058\,\Omega) - (0.2398\text{ A})(50\,\Omega)}{0.2398\text{ A} - 161\text{ A}} = 0.0165\,\Omega.$$

Now to find the battery's emf, we have from the text

$$\mathcal{E} - I_1r - I_1R_1 = 0.$$

Solving for $\mathcal{E}$,

$$\mathcal{E} = I_1r + I_1R_1 = (161\text{ A})(0.0165\,\Omega) + (161\text{ A})(0.058\,\Omega) = 12.0\text{ V}.$$

(b) The magnitude of the potential difference V across the battery terminals equals the magnitude of the potential difference across the load, or

$$V = I_1R_1 = (161\text{ A})(0.058\,\Omega) = 9.34\text{ V}.$$

This numerical result illustrates that the potential difference across the battery terminals is dependent on the load.

△

Example 21.2. A 1.51-V flashlight battery has an internal resistance of 0.210 Ω. Find the current that flows and the rate of energy dissipation (in watts) (a) when the battery terminals are connected with a wire of negligible resistance, and (b) when a 2.30-kΩ resistor is connected between the terminals.

Solution. (a) From our analysis in the text the current is given by Equation (21.2), with $R = 0$ in this case.

$$I = \frac{\mathcal{E}}{r + R} = \frac{1.51\text{ V}}{0.210\,\Omega + 0} = 7.19\text{ A}.$$

From Chapter 19 [Equation (19.34)], we know that the rate of energy dissipation (the power) in a resistor r is

$$P = I^2r = (7.19\text{ A})^2(0.210\,\Omega) = 10.9\text{ W}.$$

(b) With the added load resistance, the current is much lower:

$$I = \frac{\mathcal{E}}{r + R} = \frac{1.51\text{ V}}{0.210\,\Omega + 2.30 \times 10^3\,\Omega} = 6.56 \times 10^{-4}\text{ A} = 0.656\text{ mA}.$$

The rate of energy consumption is also much lower. We must account for the energy consumed by both resistors (r and R):

$$P = I^2r + I^2R = I^2(r + R) = (6.56 \times 10^{-4}\text{ A})^2(0.210\,\Omega + 2.30 \times 10^3\,\Omega) = 9.90 \times 10^{-4}\text{ W}.$$

A comparison of the numerical results for (a) and (b) shows why a battery will run down more quickly when short-circuited.

△

In the last example, we computed the rate at which energy was used in the resistors. Since the battery is the source of that energy, we could just as well have found the rate at which energy was supplied by the battery. Equation (19.33) suggests that the instantaneous power supplied by a source of emf $\mathcal{E}$ is

$$P = I\mathcal{E}. \tag{21.4}$$

You can easily verify that Equation (21.4) gives the same numerical results as we found in the example.

21.2 Series and Parallel Combinations of Resistors

21.2.1 Resistors in Series

Consider the circuit shown in Figure 21.2(a), with a battery or other source of emf $\mathcal{E}$ connected to two resistors R_1 and R_2 in series. For convenience, we assume that the internal resistance of the battery is so small compared with R_1 and R_2 that it can be neglected. We will continue to make this assumption throughout the reminder of this chapter. Whenever the internal resistance is *not* negligible, you can account for it by replacing the battery with an emf source $\mathcal{E}$ plus an internal resistance r in series, as in Figure 21.1.

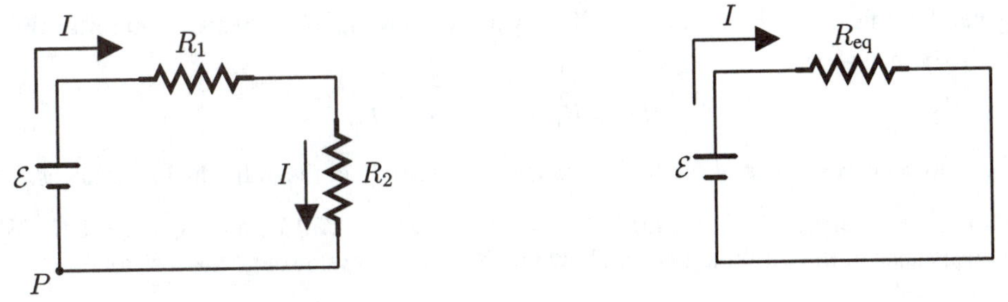

(a) Two resistors in series

(b) Equivalent resistance of the two resistors in series

Figure 21.2. A circuit with two resistors in series.

We wish to consider the following question: With what equivalent resistance R_{eq} could we replace the series combination of R_1 and R_2, so that the current flowing in the circuit is the same? The replacement of the series by R_{eq} is shown in Figure 21.2(b). Remember that the current flowing through a series of resistors is the same everywhere, since no buildup of charge is observed. We can find the current just as we did in Section 21.1, by choosing an arbitrary starting point P, adding all the potential differences encountered as we travel around the loop, and applying Kirchoff's loop rule. Starting at P in Figure 21.2(a) and going around the loop in a clockwise direction, we encounter a potential *gain* $+\mathcal{E}$ going through the battery, and a potential change (actually a *drop)* $-IR_1$ in the first resistor and $-IR_2$ in the second resistor. Then Kirchoff's loop rule says

$$\mathcal{E} - IR_1 - IR_2 = 0.$$

Rearranging to solve for I, we have

$$I = \frac{\mathcal{E}}{R_1 + R_2}. \tag{21.5}$$

Now consider the comparable circuit in Figure 21.2(b). The potential difference across the equivalent resistor is simply $\mathcal{E}$, so

$$I = \frac{\mathcal{E}}{R_{eq}}. \tag{21.6}$$

It was our premise that R_{eq} will produce the same current as the series combination of R_1 and R_2, so the currents in Equations (21.5) and (21.6) are equal. Therefore,

$$R_{eq} = R_1 + R_2. \tag{21.7}$$

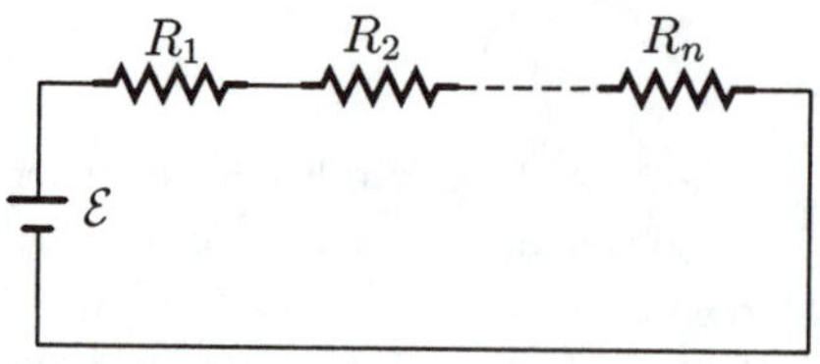

Figure 21.3. A circuit with n resistors in series.

We can generalize this result. By a similar method, you can show that for n resistors in series (Figure 21.3), the equivalent resistance is

$$R_{\text{eq}} = R_1 + R_2 + \cdots + R_n. \tag{21.8}$$

That is, the equivalent resistance of resistors in series is the sum of all the individual resistances.

Example 21.3. In Figure 21.4, $\mathcal{E} = 24.0$ V, $R_1 = 300\ \Omega$, $R_2 = 750\ \Omega$, and $R_3 = 175\ \Omega$. What is the equivalent resistance of the combination, and how much current flows through the resistors?

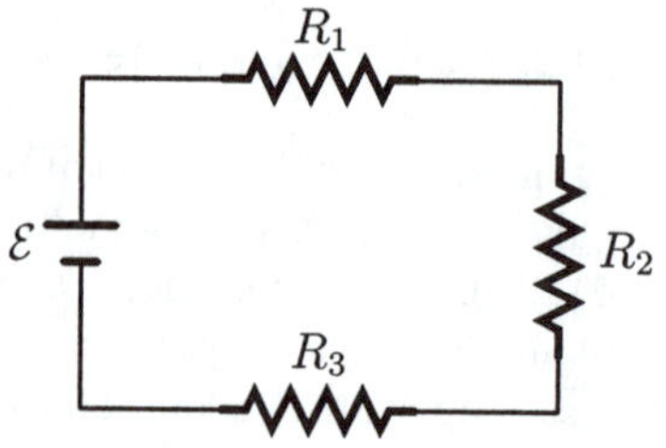

Figure 21.4. Circuit for Example 21.3.

Solution. The equivalent resistance of a combination of resistors in series is the sum of the individual resistances:

$$R_{\text{eq}} = R_1 + R_2 + R_3 = 300\ \Omega + 450\ \Omega + 165\ \Omega = 915\ \Omega.$$

Then the current is

$$I = \frac{\mathcal{E}}{R_{\text{eq}}} = \frac{24.0\text{ V}}{915\ \Omega} = 0.0262\text{ A } = 26.2\text{ mA}.$$ △

21.2.2 Other Single-Loop Circuits

Consider the single-loop circuit shown in Figure 21.5(a), which has three batteries and two resistors. The emfs and resistance values are given. We would like to know how much current will flow in this circuit, and

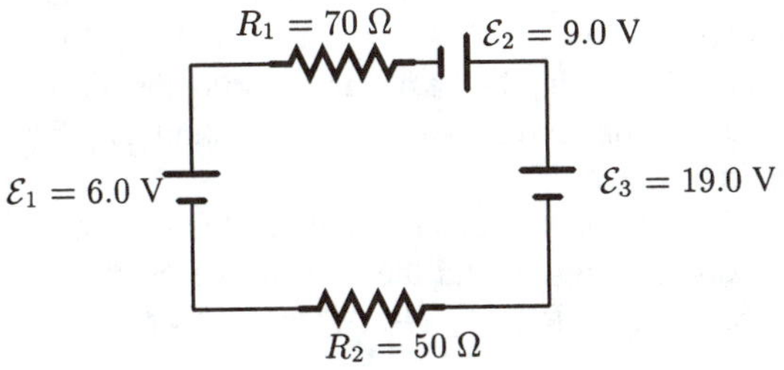

(a) Circuit elements

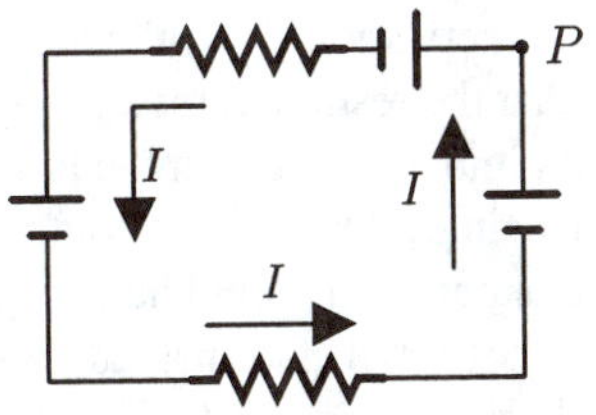

(b) Current flow

Figure 21.5. Single-loop circuit with more than one battery.

in which direction. Once again, Kirchoff's loop rule enables us to solve the problem. Notice, however, that the batteries are not all aligned with their $-$ and $+$ terminals in the same sequence as we proceed around the loop. Thus, it is not immediately obvious in which direction the current will flow.

The procedure in this case is to guess at which direction the current will flow, and proceed using Kirchoff's loop rule. Then, if our guess is wrong, the numerical value of current will be negative, indicating a flow in the opposite direction to that initially chosen.

One other difficulty arises due to the placement of the batteries. When proceeding from one end of a resistor to another, it is not clear whether we are moving from a lower potential to a higher potential, or vice versa. But recall that positive current travels from higher to lower potential. Therefore, if we are proceeding around the loop in the same direction we have chosen for the current, we will encounter a potential *drop* (equal to current times resistance) each time we pass through a resistor. But if we are proceeding in the opposite direction as the current, we will encounter a potential *gain* of the same magnitude. As long as we compute the drops and gains in accordance with the direction we have assumed for the current flow, we will calculate the correct current.

With these rules in mind, let's apply Kirchoff's loop rule to the circuit shown in Figure 21.5(a). We arbitrarily choose the point P to be at the upper right and the current flow to be counterclockwise around the loop, as shown in Figure 21.5(b). Remember that since it is still a single-loop circuit, the current I is the same everywhere. If we proceed counterclockwise from P, we encounter the following potential changes: $-\mathcal{E}_2$ at the battery and $-IR_1$ at the battery and resistor on the top branch, $-\mathcal{E}_1$ at the battery on the left branch, $-IR_2$ at the resistor on the bottom branch, and $+\mathcal{E}_3$ at the battery on the right branch. Kirchoff's loop rule tells us to add these changes and set the sum equal to zero:

$$-\mathcal{E}_2 - IR_1 - \mathcal{E}_1 - IR_2 + \mathcal{E}_3 = 0. \tag{21.9}$$

Solving for I,

$$-\mathcal{E}_2 - \mathcal{E}_1 + \mathcal{E}_3 - I\,(R_1 + R_2) = 0$$

or

$$I = \frac{-\mathcal{E}_2 - \mathcal{E}_1 + \mathcal{E}_3}{R_1 + R_2}\,. \tag{21.10}$$

Inserting numerical values, we find

$$I = \frac{-9.0\ \mathrm{V}\ - 6.0\ \mathrm{V} + 19.0\ \mathrm{V}}{70\,\Omega + 50\,\Omega} = 0.0333\ \mathrm{A}\ = 33.3\ \mathrm{mA}.$$

Because the numerical result is positive, we conclude that we chose the correct direction (counterclockwise) for the current flow. This makes sense, because looking at the polarity of the batteries we see that $\mathcal{E}_3$ is opposed to $\mathcal{E}_1$ and $\mathcal{E}_2$. However, the combined effect of $\mathcal{E}_1$ and $\mathcal{E}_2$ is 6.0 V + 9.0 V = 15.0 V. It is logical that one 19.0-V battery should "win" when opposed to a 15.0-V combination.

Notice that the result expressed symbolically in Equation (21.10) is simply the net emf (taking the different polarities into account with the proper signs) divided by the sum of the resistances. You will find that this is true in any single-loop circuit containing only batteries and resistors.

Let's consider what would have happened had we chosen the current direction (counterclockwise) shown in Figure 21.5(b), but had instead summed the potential changes going around the loop in the clockwise direction. Then (starting from P) we have potential changes: $-\mathcal{E}_3$, $+IR_2$, $+\mathcal{E}_1$, $+IR_1$, and $+\mathcal{E}_2$. The net change is

$$-\mathcal{E}_3 + IR_2 + \mathcal{E}_1 + IR_1 + \mathcal{E}_2 = 0. \tag{21.11}$$

Notice that Equations (21.9) and (21.11) are equivalent, because one is simply equal to the other multiplied by a constant (-1) on both sides. This shows that our choice of the direction in which we sum the potential changes is arbitrary, and the same result I follows in either case.

Similarly, what if we had guessed that the current flowed in the clockwise direction? The proceeding clockwise from P the potential changes are $-\mathcal{E}_3$, $-IR_2$, $+\mathcal{E}_1$, $-IR_1$, and $+\mathcal{E}_2$. The net potential change is

$$-\mathcal{E}_3 - IR_2 + \mathcal{E}_1 - IR_1 + \mathcal{E}_2 = 0,$$

which can be solved for I to yield

$$I = \frac{\mathcal{E}_2 + \mathcal{E}_1 - \mathcal{E}_3}{R_1 + R_2}. \tag{21.12}$$

This is just the negative of the result in Equation (21.10). The numerical value of I from Equation (21.12) has the same magnitude as the current obtained from Equation (21.10), but it is negative. This indicates that our guess that the current flows clockwise is incorrect, and it really flows counterclockwise. This agrees with our previous result, showing that our original guess of the current's direction does not affect the final answer.

21.2.3 Resistors in Parallel

In Figure 21.6(a), we show a battery connected to two resistors in parallel with each other. We wish to find the equivalent resistance of this pair of resistors. That is, what single resistor R_{eq} can we use to replace the parallel combination [as in Figure 21.6(b)] so that the same current flows from the battery?

The key to solving this problem lies in the fact that the resistors share connections at both terminals of the battery. Therefore, the same potential difference $\mathcal{E}$ exists across each resistor, and this enables us to find the current flowing through each resistor:

$$I_1 = \frac{\mathcal{E}}{R_1} \qquad \text{through } R_1$$

and

$$I_2 = \frac{\mathcal{E}}{R_2} \qquad \text{through } R_2.$$

But how much current I flows from the battery? To answer this question, we need to look at what happens at the point labeled P [Figure 21.6(a)], known as a **junction**, because more than one conducting wire meets

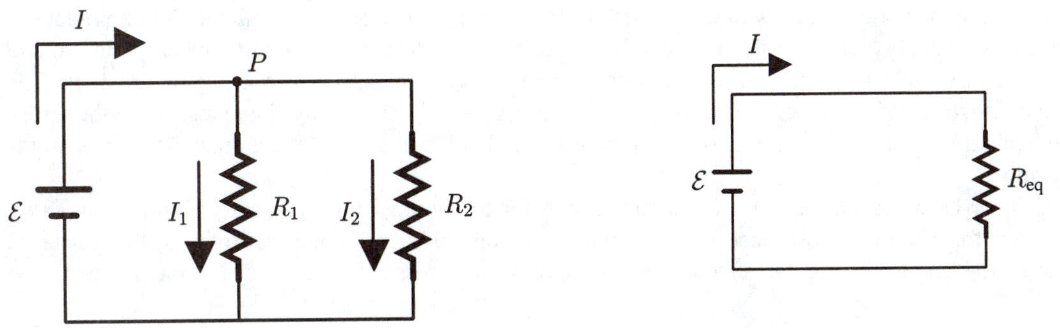

(a) Two resistors in parallel

(b) Equivalent resistance of the two resistors in parallel

Figure 21.6. A circuit with two resistors in parallel.

there. From the polarity of the battery, we know that I flows in the direction shown, from the positive terminal of the battery. Current reaching the junction at P must flow either through R_1 or R_2. But as we have seen, some current flows through each resistor. Therefore, the current I flowing into P must divide between the two resistors. Conservation of charge dictates that

$$I = I_1 + I_2 \tag{21.13}$$

because any charge flowing into P must flow out through R_1 or R_2.

Equation (21.13) is a specific example of a more general rule, called **Kirchoff's junction rule**: *The net current entering a junction equals the net current leaving the junction.* Kirchoff's junction rule follows from a generalization of our statement above, regarding conservation of charge.

Returning to the problem at hand, we see that the current I flowing from the battery is the current that flows through the equivalent resistance R_{eq} in Figure 21.6(b). Therefore,

$$I = \frac{\mathcal{E}}{R_{eq}} \qquad \text{through } R_{eq}.$$

Making the appropriate substitutions in Equation (21.13),

$$\frac{\mathcal{E}}{R_{eq}} = \frac{\mathcal{E}}{R_1} + \frac{\mathcal{E}}{R_2}.$$

Dividing through by the common factor $\mathcal{E}$,

$$\frac{1}{R_{eq}} = \frac{1}{R_1} + \frac{1}{R_2}. \tag{21.14}$$

The reciprocal of the equivalent resistance R_{eq} is the sum of the reciprocals of the individual resistances.

You can show (see the problems) that for n resistors in parallel, the reciprocal of the equivalent resistance R_{eq} is the sum of the reciprocals of the individual resistances. Stated symbolically,

$$\frac{1}{R_{eq}} = \frac{1}{R_1} + \frac{1}{R_2} + \cdots + \frac{1}{R_n}. \tag{21.15}$$

Now that we have established the rules for finding the equivalent resistance of resistors in series or parallel, we can make some comparisons with our previous study (Chapter 19) of series and parallel combinations of capacitors. Simply put, the rules are just the opposite. Resistors in series and capacitors in parallel add directly. Resistors in parallel and capacitors in series add reciprocally [as in Equation (21.15)]. The equivalent resistance of a number of resistors in series is *greater* than that of any single resistor in the series. The equivalent resistance of a number of resistors in parallel is *less* than that of the smallest resistor in the combination.

In Chapter 19, we saw a geometric basis for the rules for combining capacitors, and a similar basis exists for the rules for combining resistances. Recall that for a material with a given resistivity ρ, the resistance depends on the length L (along the direction of current flow) and cross-sectional area A of the resistor:

$$R = \rho\frac{L}{A}. \tag{19.28}$$

Geometrically, adding resistors in series is equivalent to increasing the length of a single resistor. For example, suppose two identical resistors are combined in series. The combination has the same A but a length of $2L$, so the net resistance is twice the original resistance. As Equation (19.28) shows, the resistance increases in proportion to the length, and therefore it makes sense that resistances in series add directly. Similarly, adding resistors in parallel is geometrically equivalent to increasing the area A of a single resistor. If two identical resistors are combined in parallel, the combination has the same L but an area $2A$, so the net resistance is half the original resistance. With A in the denominator of Equation (19.28), it makes sense that resistances in series add reciprocally.

Example 21.4. Three resistors, $R_1 = 8\,\Omega$, $R_2 = 12\,\Omega$, and $R_3 = 6\,\Omega$ are connected in parallel. (a) Find the equivalent resistance of the combination. (b) The parallel combination is connected to a 6.0-V battery. Find the current flowing from the battery and the current in each resistor.

Solution. (a) Following the rule for adding resistors in parallel,

$$\frac{1}{R_{\text{eq}}} = \frac{1}{R_1} + \frac{1}{R_2} + \frac{1}{R_3} = \frac{1}{8\,\Omega} + \frac{1}{12\,\Omega} + \frac{1}{6\,\Omega} = \frac{3+2+4}{24\,\Omega} = \frac{9}{24\,\Omega}$$

Therefore,

$$R_{\text{eq}} = \frac{24}{9}\,\Omega.$$

(b) The circuit is shown in Figure 21.7. The current leaving the battery is determined by the equivalent resistance, with

$$I = \frac{\mathcal{E}}{R_{\text{eq}}} = \frac{6.0\text{ V}}{(24/9)\,\Omega} = 2.25\text{ A}.$$

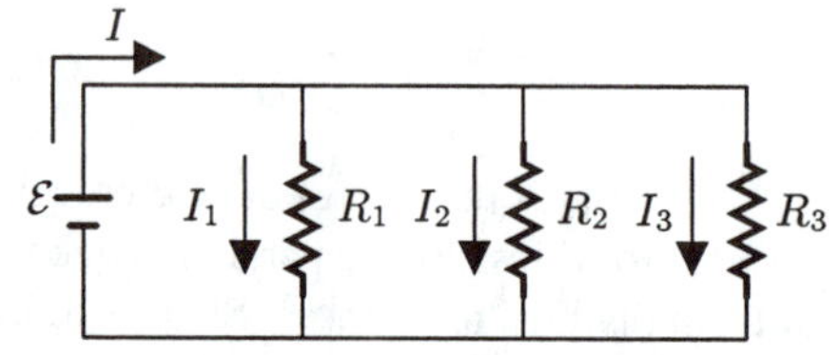

Figure 21.7. Circuit for Example 21.4, with three resistors in parallel.

The same potential difference $\mathcal{E}$ exists across each resistor in the parallel combination, so

$$I_1 = \frac{\mathcal{E}}{R_1} = \frac{6.0\ \text{V}}{8\,\Omega} = 0.75\ \text{A}$$

$$I_2 = \frac{\mathcal{E}}{R_2} = \frac{6.0\ \text{V}}{12\,\Omega} = 0.5\ \text{A}$$

$$I_3 = \frac{\mathcal{E}}{R_3} = \frac{6.0\ \text{V}}{6\,\Omega} = 1.0\ \text{A}.$$

We can check these results, because we know that by Kirchoff's junction rule the current divides at the junction where the three resistors meet, with $I = I_1 + I_2 + I_3$. Using the numerical values just computed,

$$I_1 + I_2 + I_3 = 0.75\ \text{A} + 0.5\ \text{A} + 1.0\ \text{A} = 2.25\ \text{A} = I.$$

Our results are in agreement with Kirchoff's junction rule. △

21.2.4 Combinations of Series and Parallel Connections

Armed with the rules for finding the equivalent resistances of series and parallel combinations of resistors, we can find the equivalent resistance of more complex combinations of resistors in series and parallel. For example, consider the network of four resistors shown in Figure 21.8(a). The resistors R_1 and R_2 are in series with each other and with the parallel combination of R_3 and R_4. We wish to find the equivalent resistance between the points labeled a and b. It is not necessary to attach a battery to the combination and imagine current flowing through the circuit in order to find the equivalent resistance; we need only follow the rules for adding resistors in series and parallel.

In a problem like this, it is generally a good idea to proceed systematically by replacing parallel combinations with equivalent resistances first, since adding reciprocally is more difficult. We will replace the parallel combination R_3 and R_4 with an equivalent resistance R_5, as shown in Figure 21.8(b). The rule for adding resistors in parallel tells us

$$\frac{1}{R_5} = \frac{1}{R_3} + \frac{1}{R_4}$$

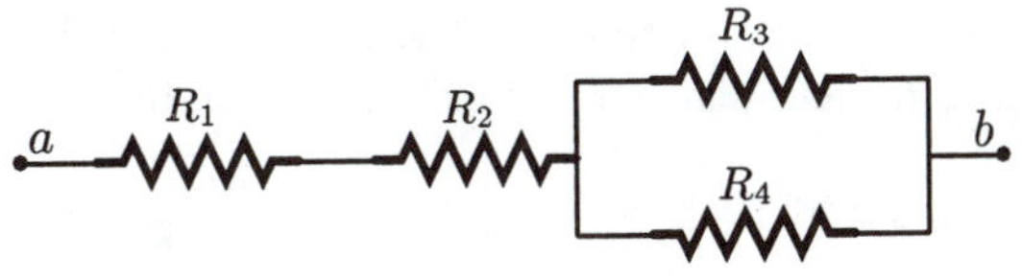

(a) Original combination or resistors

(b) An equivalent resistance replacing the parallel combination

Figure 21.8. Resistors in a combination of series and parallel connections.

or

$$R_5 = \frac{1}{\frac{1}{R_3} + \frac{1}{R_4}} = \frac{1}{\frac{R_4}{R_3 R_4} + \frac{R_3}{R_3 R_4}} = \frac{R_3 R_4}{R_3 + R_4}.$$

Now we have three resistors (R_1, R_2, and R_5) in series, so the equivalent resistance between a and b is given by the series rule:

$$R_{\text{eq}} = R_1 + R_2 + R_5$$

or

$$R_{\text{eq}} = R_1 + R_2 + \frac{R_3 R_4}{R_3 + R_4}.$$

Example 21.5. Consider again the combination of four resistors in Figure 21.8(a). Let $R_1 = 5\,\Omega$, $R_2 = 3\,\Omega$, $R_3 = 8\,\Omega$, and $R_4 = 2\,\Omega$. (a) Find a numerical value of the equivalent resistance between a and b. (b) Now a 9.6-V battery is connected with its two terminals attached to a and b (Figure 21.9). What is the current flowing through each resistor?

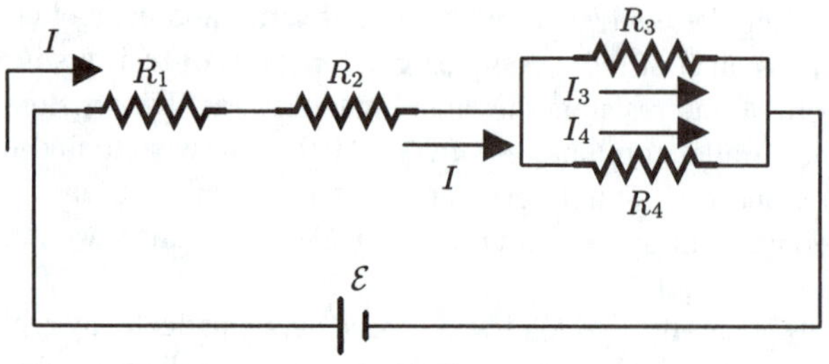

Figure 21.9. Circuit for Example 21.5, with a combination of resistors attached to a battery.

Solution. (a) Inserting the numerical values in the result derived in the text,

$$R_{\text{eq}} = R_1 + R_2 + \frac{R_3 R_4}{R_3 + R_4} = 5\,\Omega + 3\,\Omega + \frac{(8\,\Omega)(2\,\Omega)}{8\,\Omega + 2\,\Omega} = 9.6\,\Omega.$$

(b) For convenience, we have redrawn the circuit with the battery in place ($\mathcal{E} = 9.6$ V) in Figure 21.9. The resistance R_{eq} effectively replaces the entire network of resistors. Therefore, the current leaving the battery is

$$I = \frac{\mathcal{E}}{R_{\text{eq}}} = \frac{9.6\text{ V}}{9.6\,\Omega} = 1.0\text{ A}.$$

The current in a series is the same everywhere, so we conclude that the current flowing through both R_1 and R_2 is also 1.0 A. However, this current divides at the junction where R_3 and R_4 meet. We can find the current through each of those resistors by finding the potential difference across them. Since those resistors are in parallel, the same potential difference V_{34} is across each of them. We apply Kirchoff's loop rule, starting with the battery and proceeding clockwise around the loop:

$$\mathcal{E} - IR_1 - IR_2 - V_{34} = 0.$$

We thus complete the loop by going through R_1 and R_2, followed by *either* R_3 or R_4 (it doesn't matter whether we use R_3 or R_4, since the potential difference V_{34} is the same over each). Therefore,

$$V_{34} = \mathcal{E} - IR_1 - IR_2 = \mathcal{E} - I(R_1 + R_2) = 9.6\text{ V} - (1.0\text{ A})(5\,\Omega + 3\,\Omega) = 1.6\text{ V}.$$

Then the remaining currents are

$$I_3 = \frac{V_{34}}{R_3} = \frac{1.6\text{ V}}{8\,\Omega} = 0.2\text{ A}$$

and

$$I_4 = \frac{V_{34}}{R_4} = \frac{1.6\text{ V}}{2\,\Omega} = 0.8\text{ A}.$$

These results make sense, because by Kirchoff's junction rule we know $I = I_3 + I_4$. Indeed that is the case, with $1.0\text{ A} = 0.2\text{ A} + 0.8\text{ A}$. △

An observation is in order based on the preceding example. When the 1.0-A current reached the junction between R_3 and R_4, it divided so that 0.8 A passed through the 2-Ω resistor and 0.2 A passed through the 8-Ω resistor. This illustrates the fact that in a parallel network of resistors, the current divides so that more of it passes through the lower-valued resistor, with the current inversely proportional to the resistance. This fact has given rise to the commonly heard statement: "Current takes the path of least resistance." You may use that statement, but be careful not to misinterpret it to mean "*All* the current flows through the lower-valued resistor." It is better to remember that the product of current and resistance is the same for each resistor in a parallel network (because the potential difference V is the same across each). Thus, the current $I_n = V/R_n$ in the nth parallel resistor is inversely proportional to R_n, but it is never zero.

21.3 Multiloop Circuits

We now analyze some more complex circuits, such as the one shown in Figure 21.10. What makes this problem more difficult than previous ones we have studied is that there are two closed loops, and because of the placement of the batteries, it is impossible to find an equivalent resistance for the whole network.

Fortunately, we can analyze this two-loop circuit using Kirchoff's two rules (loop and junction), with which you are already familiar. If the resistance and emf values are given as in Figure 21.10, applying Kirchoff's laws allows you to find the current flowing in every part of the circuit. Notice that there are three

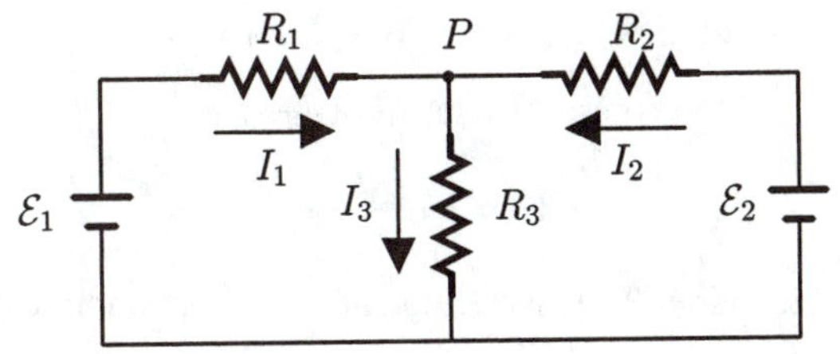

Figure 21.10. Circuit with two loops.

different current values in this circuit: One in the left loop, one in the right loop, and one in the middle branch. In Figure 21.10, we have identified a junction point P where these three currents meet. We will follow the established practice of guessing at the directions of the current flow in each branch; a correct guess results in a positive numerical value for the current, and an incorrect guess in a negative numerical value, as in previous examples.

We start by applying Kirchoff's loop rule to the closed loop on the left, starting at the negative terminal of the battery, proceeding clockwise through R_1, down through R_3, and back to the battery. Summing the potential changes around that closed loop and setting the sum equal to zero,

$$\mathcal{E}_1 - I_1 R_1 - I_3 R_3 = 0. \tag{21.16}$$

Next, we apply Kirchoff's loop rule to the closed loop that starts at the negative terminal of the battery on the right, proceeding counterclockwise through the battery and R_2, down through R_3, and back to the battery. Summing the potential changes around this closed loop and setting the sum equal to zero,

$$\mathcal{E}_2 - I_2 R_2 - I_3 R_3 = 0. \tag{21.17}$$

Why not use the loop that goes all the way around the outside of the circuit, say from the battery on the left through R_1 and R_2 on the top, down through the battery on the right, and back through the bottom wire to the battery on the left? Applying Kirchoff's loop rule around that outer loop gives

$$\mathcal{E}_1 - I_1 R_1 + I_2 R_2 - \mathcal{E}_2 = 0. \tag{21.18}$$

Notice that the sign on the $I_2 R_2$ term is positive, since we are now going against the current flow in that branch, and the sign on $\mathcal{E}_2$ is negative, since along this path we go from the positive terminal to the negative one. It turns out that this equation is of no value in our effort to solve for the currents I_1, I_2, and I_3, because mathematically it provides no new information. We could just as well have obtained Equation (21.18) by subtracting Equation (21.17) from Equation (21.16). The mathematical term for this condition is to say that Equation 21.18 is not *linearly independent* from the other two equations. We will not pursue the topic of linear independence in this course; you should encounter it again in a course in linear algebra, which normally follows calculus.

More information is obtained from Kirchoff's junction rule. Applying the junction rule at the point P gives

$$I_1 + I_2 = I_3. \tag{21.19}$$

At this point, there is sufficient information to solve the problem of finding the three currents, because we have three linearly independent equations [Equations (21.16), (21.17), and (21.19)] in three unknowns (I_1, I_2, and I_3).

The currents can now be found be the algebraic method of elimination by substitution. For example, from Equation (21.19) we have $I_1 = I_3 - I_2$, which can be substituted into Equation (21.16) to yield

$$\mathcal{E}_1 - (I_3 - I_2) R_1 - I_3 R_3 = 0. \tag{21.20}$$

We rearrange Equation (21.17) to express I_2 in terms of I_3:

$$I_2 = \frac{\mathcal{E}_2 - I_3 R_3}{R_2} \tag{21.21}$$

which can be substituted into Equation (21.19) to eliminate I_2. After simplification, the resulting expression for I_3 is

$$I_3 = \frac{\mathcal{E}_1 + \mathcal{E}_2 \frac{R_1}{R_2}}{R_1 + R_2 + \frac{R_1 R_3}{R_2}}. \tag{21.22}$$

With the given emfs and resistances, we use Equation (21.22) to find a numerical value of $I_3 = 0.233$ A. Substituting this numerical value into Equation (21.21) gives $I_2 = 0.025$ A, and with these values of I_2 and I_3, Equation (21.19) gives $I_1 = 0.208$ A. Notice that all the numerical current values have turned out to be positive, indicating we have chosen the correct directions of current flow.

In the two-loop problem we just solved, it was not difficult to use Kirchoff's rules to write the equations relating the various emfs, resistances, and currents. Solving a set of three linear equations with three unknowns can be rather tedious, however. The process grows even more laborious when more loops are added. For example, in three-loop circuit there are three loop equations and three junction equations, for a total of five linear equations that must be solved simultaneously. We urge you to pursue more efficient means, such as matrix methods or advanced software packages, to solve larger sets of equations.

Example 21.6. Considering the parallel resistor network in Figure 21.6(a) as a two-loop circuit, write the equations that follow from Kirchoff's loop and junction rules. Then taking $R_{eq} = \mathcal{E}/I$ (where I is the total current flowing from the battery), derive the rule from finding the equivalent resistance in terms of the resistors R_1 and R_2.

Solution. Going clockwise around the left loop (including the middle branch),

$$\mathcal{E} - I_1 R_1 = 0. \tag{21.23}$$

Going clockwise around the right loop (including the middle branch),

$$I_2 R_2 - I_1 R_1 = 0. \tag{21.24}$$

Applying the junction rule at P,

$$I = I_1 + I_2. \tag{21.25}$$

Using the fact that $\mathcal{E} = I R_{eq}$, Equation (21.23) tells us

$$\mathcal{E} = I_1 R_1 = I R_{eq}.$$

Therefore,

$$R_{eq} = \frac{I_1 R_1}{I}.$$

But from Equation (21.25), $I = I_1 + I_2$, so

$$R_{eq} = \frac{I_1 R_1}{I_1 + I_2}.$$

From Equation (21.24), $I_2 = I_1 R_1 / R_2$, so

$$R_{eq} = \frac{I_1 R_1}{I_1 + I_1 \frac{R_1}{R_2}} = \frac{R_1}{1 + \frac{R_1}{R_2}}.$$

Multiplying by R_2/R_2,

$$R_{eq} = \frac{R_1 R_2}{R_2 + R_1}.$$

Taking the reciprocal,

$$\frac{1}{R_{eq}} = \frac{R_2 + R_1}{R_1 R_2} = \frac{1}{R_1} + \frac{1}{R_2}$$

which is the rule for adding resistors in parallel. △

21.4 Measuring Instruments

Throughout this chapter, it has been implicit that the important physical quantities in a dc circuit—potential difference, current, and resistance—can all be measured. In this section, we briefly describe how to use the instruments that measure those quantities. We generally omit details on the design and construction of those measuring instruments, because today such instruments are made with complex circuitry, often including microprocessors. Science and engineering students who go on to take a course in electronics or electrical engineering will have the opportunity to study those details later.

Before we discuss specific instruments, keep in mind one thing: When any measuring instrument is introduced into an electric circuit, it invariably changes the operation of the circuit. Therefore, electronic instruments should be designed and used in such a way that they disturb the original circuit as little as possible. We will see specific examples of this general principle as we consider different measuring instruments.

A device designed to measure current is called an **ammeter**. In order to measure the current in a particular part of a circuit, all the current must pass through the ammeter. This means that an ammeter must be placed in series with the portion of the circuit in question. For example, suppose we wish to measure the current in the circuit shown in Figure 21.2(a), consisting of a battery connected to two resistors in series. The circuit must be broken temporarily, so that the two terminals of the ammeter can be connected. In a series circuit, it makes no difference where the ammeter is placed, because the current is the same everywhere. In Figure 21.11, we show different possible placements for the ammeter (A) in the circuit in question; any of these placements gives the same reading on the ammeter.

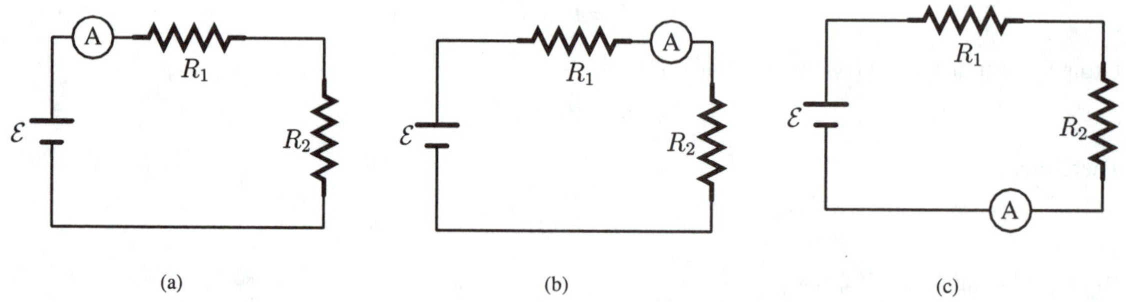

Figure 21.11. Three different positions for an ammeter in a circuit.

Any ammeter has two terminals, usually labeled + and – or + and "common." Ammeters are designed so that positive current should enter the + terminal and leave the other terminal. In modern electronic ammeters, this connection results in a digital output reading with a + sign (e.g., +5.42 mA), and the opposite connection results in a reading with the same magnitude but a negative sign (e.g., −5.42 mA).

For an ammeter to have the least possible effect on the circuit in which it is placed, the ammeter should have the least possible internal resistance. This is because the ammeter is placed in series with other circuit elements, so its internal resistance adds directly to the resistance of that part of the circuit. Typically, ammeters have internal resistances of less than 1 Ω. Therefore, the internal resistance is usually negligible as long as the other resistances is the circuit (such as the one in Figure 21.11) are much larger, say on the order of kΩ or higher. Otherwise, you may need to be concerned with the internal resistance of the ammeter.

Example 21.7. In the circuit of Figure 21.11, suppose $\mathcal{E} = 4.50$ V, $R_1 = 3.00\ \Omega$, and $R_2 = 6.00\ \Omega$. Find the percentage change in the current that flows with or without an ammeter with internal resistance 0.241 Ω.

Solution. Without the ammeter, the current is

$$I = \frac{\mathcal{E}}{R_{\text{eq}}} = \frac{\mathcal{E}}{R_1 + R_2} = \frac{4.50\ \text{V}}{3.00\ \Omega + 6.00\ \Omega} = 0.500\ \text{A}.$$

With the ammeter having resistance R_a in place,

$$I = \frac{\mathcal{E}}{R_{\text{eq}}} = \frac{\mathcal{E}}{R_1 + R_2 + R_a} = \frac{4.50\ \text{V}}{3.00\ \Omega + 6.00\ \Omega + 0.241\ \Omega} = 0.487\ \text{A}.$$

The percentage difference from the original circuit with no ammeter is

$$\frac{0.500\ \text{A}\ - 0.487\ \text{A}}{0.500\ \text{A}} \times 100\% = 2.60\%.$$

This error could be significant in some applications. △

A **voltmeter** is used to measure potential difference. The voltmeter also has two terminals, which are placed at any two points in an electric circuit, in order to measure the potential difference between those two points. For example, if we wish to know the potential difference between the two ends of resistor R_1 in Figure 21.2(a), the voltmeter V is placed as shown in Figure 21.12. Notice that the voltmeter is in parallel with the resistor in that diagram. An old rule of thumb "ammeters in series, voltmeters in parallel" is a good one to remember.

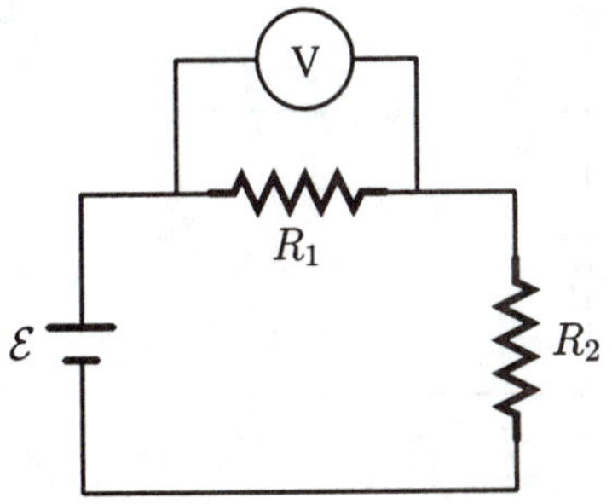

Figure 21.12. Voltmeter placed to measure the potential difference between two ends of a resistor.

The internal machinery of a voltmeter really detects current and then translates that current reading to a potential difference (normally in volts) that appears on the face of the instrument. Let's see how this works. Remember that any measuring instrument should disturb the original circuit as little as possible. Since the voltmeter is connected in parallel, it should have the highest possible internal resistance, so that it draws the smallest possible current that it can still measure accurately. Suppose that I is the current leaving the battery in Figure 21.12. At the junction where the resistor R_1 meets the voltmeter, the current divides between the resistor and voltmeter. If a current I_V flows through the voltmeter, the desired potential difference is simply

$$V = I_V R_V \tag{21.26}$$

where R_V is the internal resistance of the voltmeter. Therefore, if the voltmeter measures I_V, the voltmeter's display is the product of I_V and the known internal resistance R_V.

The **galvanometer** [named for the Italian physiologist Luigi Galvani (1737–1798), who performed experiments on the electrochemistry of living organisms] is the part of the ammeter or voltmeter that actually measures current flow. An ammeter consists of a galvanometer wired in parallel with a low-valued resistor, to produce a device with a net low resistance. A voltmeter consists of a galvanometer wired in series with a high resistance, to produce a device with a high net resistance.

An **ohmmeter** is a device used to measure the resistance of a resistor or combination of resistors. There are many different ohmmeter designs, but they all require some internal source of emf $\mathcal{E}$. For example, the source of emf can be placed in series with an ammeter A and a known resistor R_0, as shown in Figure 21.13 (these elements are all inside the ohmmeter). Then this series combination can be connected to an unknown resistor R. The ammeter measures the current I that flows in the series. If the internal resistance of the ammeter and battery are negligible,

$$I = \frac{\mathcal{E}}{R_{\text{eq}}} = \frac{\mathcal{E}}{R + R_0}.$$

The ohmmeter is then designed to register

$$R = \frac{\mathcal{E}}{I} - R_0. \tag{21.27}$$

Because the ohmmeter contains its own source of emf, it is important to disconnect the resistors you wish to measure from any other sources of emf before using an ohmmeter. Otherwise, the ohmmeter cannot account for the additional emf and the current produced by that emf, and the ohmmeter's reading will be inaccurate.

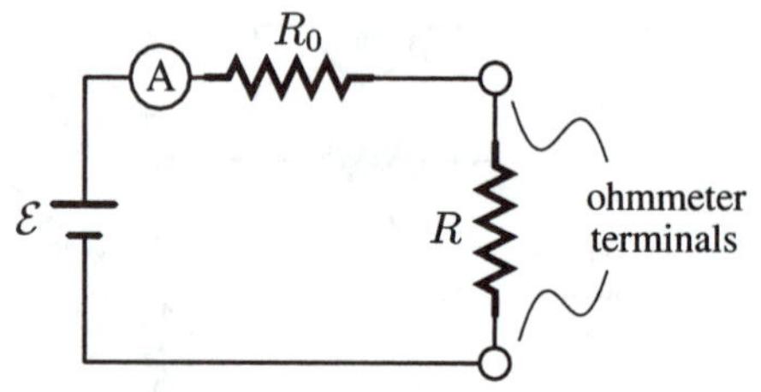

Figure 21.13. Ohmmeter connected across a resistor.

In the spirit of disturbing the system as little as possible, the ohmmeter should be designed to generate the smallest possible current needed to measure a particular resistance accurately. This avoids heating the resistor (or other parts of the circuit), which is important because resistances are temperature-dependent (Chapter 19).

Often one instrument will contain an ammeter, voltmeter, and ohmmeter. Such a device is often called a **VOM** (short for voltmeter-ohmmeter-milliammeter). A modern VOM allows the user to choose a number of different ranges, corresponding to different values of the appropriate series or parallel resistor. Many VOMs are now auto-ranging, meaning that the internal electronics selects the appropriate range based on the load to which it is attached.

21.5 *RC* Circuits

So far in this chapter, we have studied circuits containing only batteries and resistors (and, possibly, measuring instruments). We have seen that the current is constant in these circuits. In Chapter 19, we studied various connections of capacitors and batteries. In those circuits, charge flows from a battery to capacitors almost instantaneously, after which the fully charged capacitors retain their charge, and no further current flows.

In this section, we examine what happens in an ***RC* circuit**, in which a resistor and capacitor are combined in series, and this combination is then attached to a battery. Using Kirchoff's rules, we will find that both the current through the series and the charge on the capacitor are time-dependent.

A simple RC circuit is shown in Figure 21.14. Notice that the series contains a **switch** S that can be opened to break the circuit. When the switch is open, no current flows, and any charge on the capacitor remains there. But when the switch is closed, it forms a conducting path that can carry current. Let's suppose that the capacitor initially carries no charge and the switch is open. At time $t = 0$, we close the switch, allowing current to flow as shown in Figure 21.14(b).

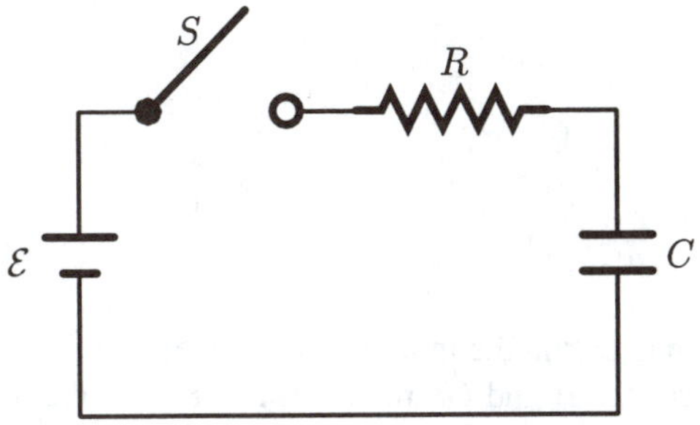

(a) The switch is open, and no current flows

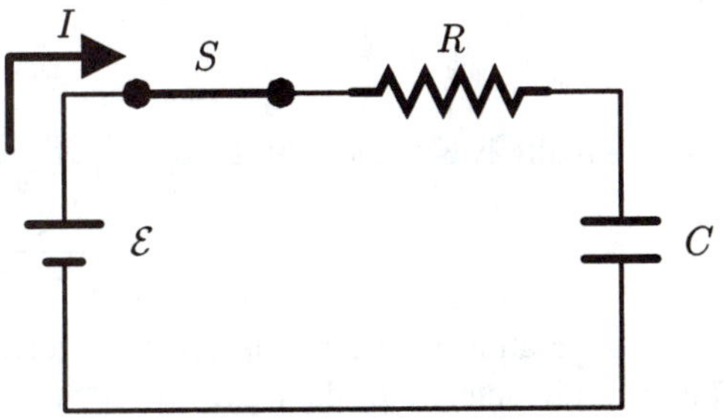

(b) The switch is closed, initiating current flow

Figure 21.14. An RC circuit containing a switch.

We apply Kirchoff's loop rule to the circuit in Figure 21.14(b). Starting at the negative terminal of the battery and proceeding clockwise around the loop,

$$\mathcal{E} - IR - \frac{Q}{C} = 0. \tag{21.28}$$

As usual, there is a potential difference $-IR$ across the resistor (since we are proceeding around the loop in the same direction as the current flow). From the definition of capacitance, we know the magnitude of the potential difference across the capacitor is Q/C. The negative sign in that term arises because we are proceeding from the higher-potential to the lower-potential plate of the capacitor.

Recall that the capacitor is initially uncharged. But as positive current flows in the direction shown in the diagram, positive charge builds up on the upper plate. No charge flows from one capacitor plate to another. However, in this series circuit the same positive current that leaves the positive battery terminal flows into the negative battery terminal. That positive current must come from the lower plate of the capacitor, leaving behind a negative charge on that plate. The two capacitor plates carry charges that are equal in magnitude but opposite in sign at any given time. The variable Q in Equation (21.28) is the *magnitude* of the charge on each plate. As current continues to flow in the circuit, Q increases. We want to know how Q increases in time; in other words, we seek the function $Q(t)$.

Because the charge building up on the capacitor is entirely due to the current in the circuit, the charge and current are related by

$$I = \frac{dQ}{dt}.$$

Therefore, we can rewrite Equation (21.28) as

$$\mathcal{E} - R\frac{dQ}{dt} - \frac{Q}{C} = 0. \tag{21.29}$$

From our study of differential equations in Chapter 6 you should recognize this as a linear, first-order differential equation for the function $Q(t)$ (we assume that the parameters $\mathcal{E}$, R, and C are fixed and known). The solution of the differential equation can be found symbolically using the technique of separation of variables (Chapter 6). First, we rearrange Equation (21.29)

$$\frac{dQ}{dt} = \frac{\mathcal{E}}{R} - \frac{Q}{RC}$$

and then separate to get

$$\frac{1}{\frac{Q}{RC} - \frac{\mathcal{E}}{R}}\,dQ = -\,dt.$$

Next, we multiply both sides by $1/RC$ to get

$$\frac{1}{Q - \mathcal{E}C}\,dQ = -\frac{1}{RC}\,dt. \tag{21.30}$$

The separation of variables is now complete. The next step is to find the indefinite integral of both sides. The indefinite integral of the right side is straightforward, because R and C are constants. But on the left side, we need to employ a substitution. Letting $u = Q - \mathcal{E}C$, we have $dQ = du$, so

$$\int \frac{1}{Q - \mathcal{E}C}\,dQ = \int \frac{1}{u}\,du = \ln u = \ln(Q - \mathcal{E}C).$$

We will omit the constant of integration for the moment, but we must remember to include a constant when we complete the integration of both sides of Equation (21.30). The result of that integration is

$$\ln(Q - \mathcal{E}C) = -\frac{t}{RC} + \text{constant}. \tag{21.31}$$

Obviously we cannot use C for the constant of integration, because C is the capacitance. Anticipating that we will take the exponential of both sides of Equation (21.31) to solve for Q, we let the constant be $\ln A$ (where A is a constant). Then

$$\ln(Q - \mathcal{E}C) = -\frac{t}{RC} + \ln A.$$

Taking the exponential of both sides,

$$Q - \mathcal{E}C = \exp\left(-\frac{t}{RC} + \ln A\right).$$

Using the rules of exponentials,

$$Q - \mathcal{E}C = \exp\left(-\frac{t}{RC}\right)\exp(\ln A) = A\exp\left(-\frac{t}{RC}\right).$$

Solving for Q,

$$Q = \mathcal{E}C + A\exp\left(-\frac{t}{RC}\right). \tag{21.32}$$

Recall from our study of differential equations in Chapter 6 that Equation (21.32), with an expression for $Q(t)$ that contains an arbitrary constant A, is a *general solution* to the differential Equation (21.29). The constant A can be determined from the *initial condition* $Q(0) = 0$, and by inserting that value of A into the general solution we produce a *specific solution*. Substituting $Q = 0$ and $t = 0$ into Equation (21.32), we find

$$0 = \mathcal{E}C + A\exp\left(-\frac{0}{RC}\right) = \mathcal{E}C + A.$$

Therefore, $A = -\mathcal{E}C$, so the specific solution is

$$Q = \mathcal{E}C + (-\mathcal{E}C)\exp\left(-\frac{t}{RC}\right)$$

or

$$Q = \mathcal{E}C\left[1 - \exp\left(-\frac{t}{RC}\right)\right]. \qquad (21.33)$$

Let's examine this result. A graph of the function $Q(t)$ from Equation (21.33) is shown in Figure 21.15(a). Notice that the factor RC in the denominator of the exponential has dimensions of time. It is convenient to mark time in multiples of RC, and we do so on the graph. The **time constant** of an RC circuit is commonly defined as $\tau = RC$. After a long time ($t \gg \tau$), the term $\exp(-t/RC)$ asymptotically approaches zero, so the charge on the capacitor asymptotically approaches $\mathcal{E}C$. This is the same charge that would appear virtually instantaneously on the capacitor without the resistor in the circuit. The resistor simply delays the charging of the capacitor, but it does not change the final charge on the capacitor after a long time has elapsed. After just one time constant ($t = \tau = RC$), the charge on the capacitor is

$$Q = \mathcal{E}C\left[1 - \exp\left(-\frac{RC}{RC}\right)\right] = \mathcal{E}C[1 - e^{-1}] \approx 0.632\,\mathcal{E}C.$$

The current in the circuit in Figure 21.14 can be found using Equation (21.33) with $I = dQ/dt$:

$$I = \frac{dQ}{dt} = \frac{d}{dt}\left\{\mathcal{E}C\left[1 - \exp\left(-\frac{t}{RC}\right)\right]\right\} = \mathcal{E}C\left[-\exp\left(-\frac{t}{RC}\right)\right]\left(-\frac{1}{RC}\right)$$

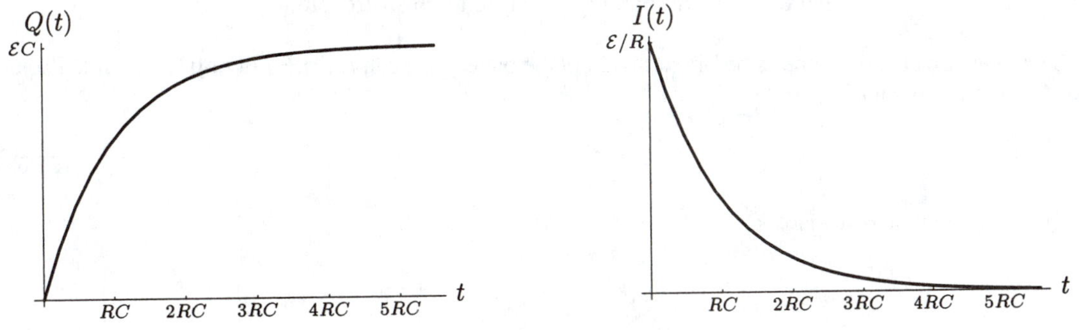

(a) Charge vs. time

(b) Current vs. time

Figure 21.15. Charge and currrent for an RC circuit.

where we have applied the chain rule for taking the derivative of a composite function. Simplifying,

$$I = \frac{\mathcal{E}}{R} \exp\left(-\frac{t}{RC}\right). \tag{21.34}$$

A graph of current vs. time is shown in Figure 21.15(b). At $t = 0$, the current is $\mathcal{E}/R$, which is the same as it would be with no capacitor in the series. After a long time ($t >> \tau$), the current asymptotically approaches zero, which is consistent with our previous observation that the capacitor asymptotically approaches its full charge after a long time. At $t = \tau$, the current is

$$I = \frac{\mathcal{E}}{R} \exp\left(-\frac{RC}{RC}\right) = \frac{\mathcal{E}}{R} e^{-1} \approx 0.378 \frac{\mathcal{E}}{R}.$$

The RC circuit we have been studying (Figure 21.14) is called a **charging** circuit, because charge is supplied from a battery and built up on the capacitor. We can also construct a **discharging** circuit, in which a charged capacitor is discharged through a resistor. The discharging circuit is shown in Figure 21.16(a). It is essentially the same circuit as before, but with the battery removed. Suppose the capacitor initially carries a charge Q_0, and the switch is closed at time $t = 0$. As shown in Figure 21.16(b), positive current flows counterclockwise around the loop, from the higher potential side of the capacitor to the lower potential side. The charge on the capacitor therefore decreases in time. As before, we wish to know how the charge Q on the capacitor changes with time; that is, we seek the function $Q(t)$.

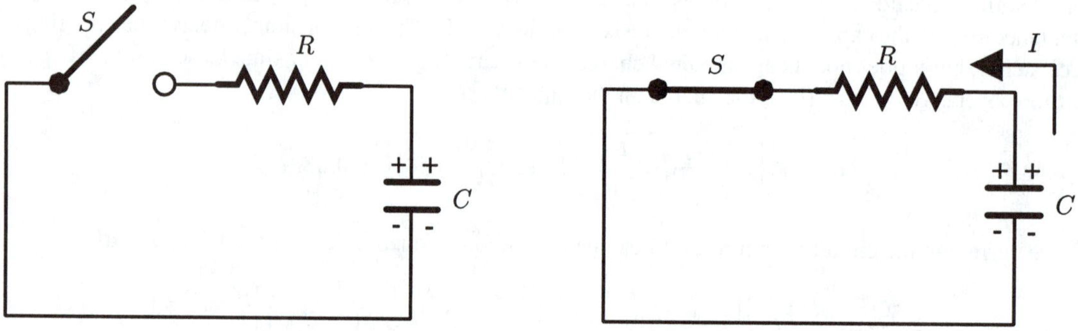

(a) The switch is open, so no current flows

(b) The switch is closed, and current can flow as shown

Figure 21.16. Discharging the capacitor in an RC circuit.

Since the circuit is the same as before but without the battery, we can rewrite Kirchoff's loop rule [Equation (21.29)] without the emf $\mathcal{E}$:

$$-R\frac{dQ}{dt} - \frac{Q}{C} = 0. \tag{21.35}$$

Employing separation of variables,

$$\frac{1}{Q}\, dQ = -\frac{1}{RC}\, dt.$$

This time the integration of both sides is straightforward, resulting in

$$\ln Q = -\frac{t}{RC} + \ln A$$

where as before we have written the constant of integration as $\ln A$, where A is a constant. Taking the exponential of both sides to solve for Q,

$$Q = \exp\left(-\frac{t}{RC} + \ln A\right) = \exp\left(-\frac{t}{RC}\right)\exp(\ln A)$$

or

$$Q = A\exp\left(-\frac{t}{RC}\right). \tag{21.36}$$

Equation (21.36) gives the general solution for $Q(t)$. The specific solution can be found using the fact that $Q = Q_0$ at $t = 0$. Making those substitutions,

$$Q_0 = A\exp\left(-\frac{0}{RC}\right) = Ae^0 = A.$$

Thus, $A = Q_0$, and the specific solution is

$$Q = Q_0\exp\left(-\frac{t}{RC}\right). \tag{21.37}$$

The current follows directly from this result:

$$I = \frac{dQ}{dt} = \frac{d}{dt}\left[Q_0\exp\left(-\frac{t}{RC}\right)\right] = Q_0\left[\exp\left(-\frac{t}{RC}\right)\right]\left(-\frac{1}{RC}\right).$$

Simplifying,

$$I = -\frac{Q_0}{RC}\exp\left(-\frac{t}{RC}\right). \tag{21.38}$$

The negative sign in Equation (21.38) is in agreement with our previous observation that the current flows in the reverse direction, compared with the current flow in the charging circuit.

The results for $Q(t)$ and $I(t)$ are shown in Figures 21.17(a) and 21.17(b), respectively. Again, it is convenient to consider time in multiples of RC. At $t = RC$, both the charge and the current have dropped to

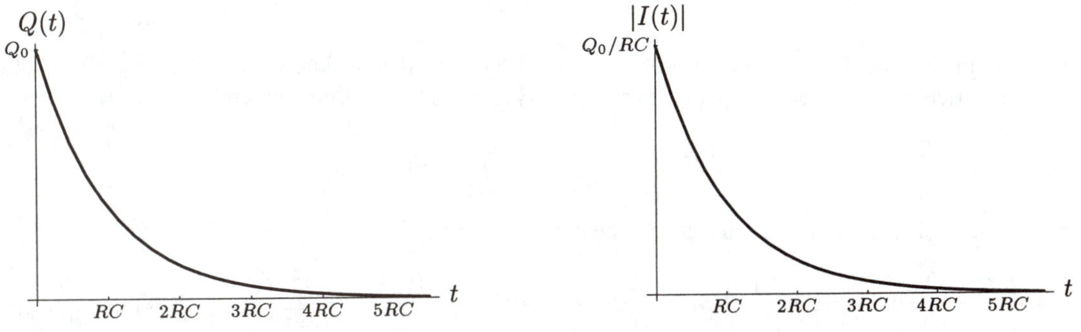

(a) Charge vs. time

(b) Current vs. time

Figure 21.17. Charge and currrent for a discharging RC circuit.

$e^{-1} \approx 0.368$ of their initial values. After a long time, the charge on the capacitor asymptotically approaches zero. The current also approaches zero asymptotically in that limit. This makes sense, because if very little charge remains on the capacitor, there is very little potential difference between the capacitor plates. Since that potential difference is responsible for the flow of current, it makes sense that the current should drop to zero along with the potential difference.

Notice that at any time after the switch is closed,

$$I = -\frac{Q_0}{RC}\exp\left(-\frac{t}{RC}\right) = -\frac{Q}{RC}.$$

But $Q/C = V$ (the potential difference between the capacitor plates), so

$$I = -\frac{V}{R}$$

as we expect in a dc circuit containing a potential difference V across a resistor R.

Finally, let's examine the energy in the discharging RC circuit. Just before the switch is closed, the capacitor stores an amount of energy [see Equation (19.15)]

$$U_0 = \frac{Q_0^2}{2C}.$$

At some later time, the capacitor carries less charge, so it stores energy

$$U_c = \frac{Q^2}{2C}.$$

Using our expression for Q in Equation (21.37),

$$U_c = \frac{\left[Q_0 \exp\left(-\frac{t}{RC}\right)\right]^2}{2C} = \frac{Q_0^2 \exp\left(-\frac{2t}{RC}\right)}{2C}$$

or

$$U_c = U_0 \exp\left(-\frac{2t}{RC}\right). \tag{21.39}$$

Now consider the energy used as current flows through the resistor. From Section 19.3.4, we know that a resistor with a current I flowing through it consumes energy at a rate I^2R. Expressing the rate as a derivative,

$$\frac{dU}{dt} = I^2R$$

where U is the electrical energy present in the circuit. From calculus we know that the change ΔU in the circuit's electrical energy U during a time interval $[t_a, t_b]$ is given by a definite integral

$$\Delta U = \int_{t_a}^{t_b} \frac{dU}{dt}\, dt = \int_{t_a}^{t_b} I^2R\, dt.$$

Using Equation (21.35) for I, we can evaluate the definite integral:

$$\Delta U = \int_{t_a}^{t_b} \left[-\frac{Q_0}{RC}\exp\left(-\frac{t}{RC}\right)\right]^2 R\, dt = \frac{Q_0^2}{RC^2}\int_{t_a}^{t_b} \exp\left(-\frac{2t}{RC}\right) dt = \frac{Q_0^2}{RC^2}\left(-\frac{RC}{2}\right)\exp\left(-\frac{2t}{RC}\right)\Bigg|_{t_a}^{t_b}$$

$$= -\frac{Q_0^2}{2C}\left[\exp\left(-\frac{2t_b}{RC}\right) - \exp\left(-\frac{2t_a}{RC}\right)\right] = -U_0\left[\exp\left(-\frac{2t_b}{RC}\right) - \exp\left(-\frac{2t_a}{RC}\right)\right].$$

To find the total energy dissipated in the resistor in a time interval starting at $t = 0$, let $t_a = 0$ and $t_b = t$ (some arbitrary time), so

$$\Delta U = -U_0 \left[\exp\left(-\frac{2t}{RC}\right) - \exp\left(-\frac{2(0)}{RC}\right)\right]$$

or

$$\Delta U = U_0 \left[1 - \exp\left(-\frac{2t}{RC}\right)\right]. \tag{21.40}$$

Now compare Equations (21.39) and (21.40). At any given time t, the sum of the energy stored in the capacitor and the energy dissipated by the resistor is

$$U_c + \Delta U = U_0 \exp\left(-\frac{2t}{RC}\right) + U_0 \left[1 - \exp\left(-\frac{2t}{RC}\right)\right] = U_0.$$

Thus, all the original energy in the circuit (U_0) is accounted for. After a long time has elapsed ($t >> RC$), very little energy is left in the capacitor, and almost all of it has been dissipated by the resistor.

Example 21.8. A capacitor with an initial charge Q_0 is placed in the discharging circuit of Figure 21.16, in which $R = 120$ kΩ and $C = 2.35\ \mu$F. (a) Find the time constant for this circuit. (b) How much time after the switch is closed does it take for the charge on the capacitor to reach half its initial value?

Solution. (a) From the definition of the time constant,

$$\tau = RC = \left(1.20 \times 10^5\ \Omega\right)\left(2.35 \times 10^{-6}\ \text{F}\right) = 0.282\ \Omega\cdot\text{F} = 0.282\ \text{s}.$$

We leave it for the student (in the problems) to show that $1\ \Omega\cdot\text{F} = 1$ s.

(b) Using $Q = Q_0/2$ in Equation (21.37) for the charge on a discharging capacitor as a function of time,

$$Q = Q_0 \exp\left(-\frac{t}{RC}\right) = \frac{Q_0}{2}.$$

Therefore,

$$\exp\left(-\frac{t}{RC}\right) = \frac{1}{2}.$$

To solve for the time t, take the logarithm of both sides:

$$\ln\left[\exp\left(-\frac{t}{RC}\right)\right] = -\frac{t}{RC} = \ln\left(\frac{1}{2}\right).$$

Rearranging,

$$t = -RC \ln\left(\frac{1}{2}\right).$$

Evaluating numerically,

$$t = -(1.20 \times 10^5\ \Omega)(2.35 \times 10^{-6}\ \text{F})(-0.693) = 0.195\ \text{s}.$$

This is just a little less than RC, which seems reasonable, because we know that at $t = RC$ the charge has dropped to e^{-1} (or about 0.368) of its initial value. △

21.6 Problems

21.1 Batteries: Ideal and Real

1. A battery produces 15.5 mA when it is connected to a 230-Ω load and 22.2 mA when it is connected to a 160-Ω load. What is the internal resistance and emf of the battery?

2. A 12.0-V battery has an internal resistance of 0.560 Ω. The battery is connected to a 25.0-Ω load. Find the current produced and the potential difference across the load.

3. A 48.0-V battery with an internal resistance of 0.255 Ω is connected to a 100-Ω load. (a) Find the current produced. (b) Find the rate at which energy is dissipated by the load resistor and the internal resistance. (c) Show that the net rate of energy dissipation (in both resistors) equals the rate $I\mathcal{E}$ at which energy is generated by the battery.

4. For the automobile battery described in Example 21.1, find the potential difference across the battery's terminals when it is connected to a 50-Ω load. Compare your result with the emf of the battery.

5. Consider again the battery described in Example 21.2. In both cases (short circuit and 2.30-kΩ load) verify that the total rate of energy dissipation in all resistors equals the rate $I\mathcal{E}$ at which energy is generated by the battery.

6. A battery connected to a constant load produces a steady current of 420 mA for 15 minutes. During this time the battery produces 1320 J of energy. What is the battery's emf?

7. How much time must a 12.0-V battery be connected to a 1.50-kΩ load in order to move 1 C of charge through the circuit? What is the energy produced by the battery during that time?

21.2 Series and Parallel Combinations of Resistors

1. Verify Equation (21.8).

2. Verify Equation (21.15).

3. Four resistors, valued at 12 Ω, 15 Ω, 20 Ω, and 35 Ω, are connected in series. (a) Find the equivalent resistance of the combination. (b) The series is connected to a 6.0-V battery. Find the current that flows and the potential difference across each individual resistor. Verify that the sum of those potential differences equals the emf of the battery.

4. The four resistors described in the previous problem are connected in parallel with each other. (a) Find the equivalent resistance of the combination. (b) The parallel combination is connected to a 6.0-V battery. Find the current that flows from the battery and the current in each resistor.

5. You are given a box full of 10-Ω resistors. Describe how to connect the resistors in order to produce the following equivalent resistances: (a) 2 Ω, (b) 35 Ω, (c) 7 Ω, (d) 19 Ω.

In Problems 6–11, assume you have a box full of identical batteries and identical light bulbs. The filament of a light bulb has a significant resistance, so that when current passes through it the filament heats enough to glow. Let the single bulb connected to a single battery (as shown in the diagram) serve as a reference.

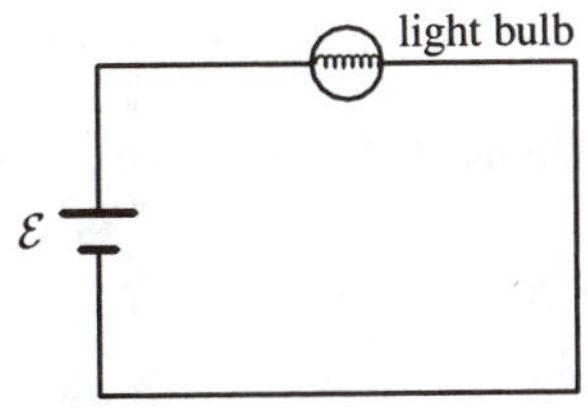

6. Three bulbs are connected in series and attached to a single battery. How does the brightness of each bulb compare to each other and to the reference bulb? What happens to the relative brightness if one of the three bulbs is removed from the series?

7. Three bulbs are connected in parallel and attached to a single battery. How does the brightness of each bulb compare to each other and to the reference bulb? What happens to the relative brightness if one of the three bulbs is removed?

8. Three bulbs are connected as shown, with a parallel connection of two bulbs in series with a third. How does the brightness of each bulb compare to the others and to the reference bulb? What happens to the relative brightness if B_1 is removed? What happens to the relative brightness if B_2 is removed?

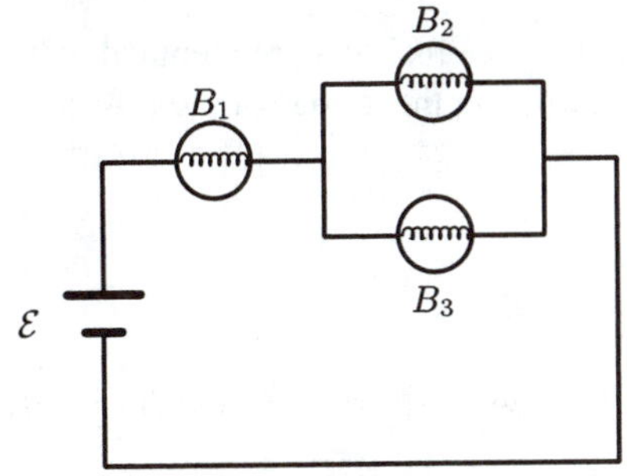

9. Repeat Problem 6 if the single battery is replaced with two batteries in series.

10. Repeat Problem 7 if the single battery is replaced with two batteries in series.

11. Repeat Problem 8 if the single battery is replaced with two batteries in series.

For Problems 12–15, refer to the following circuit diagram.

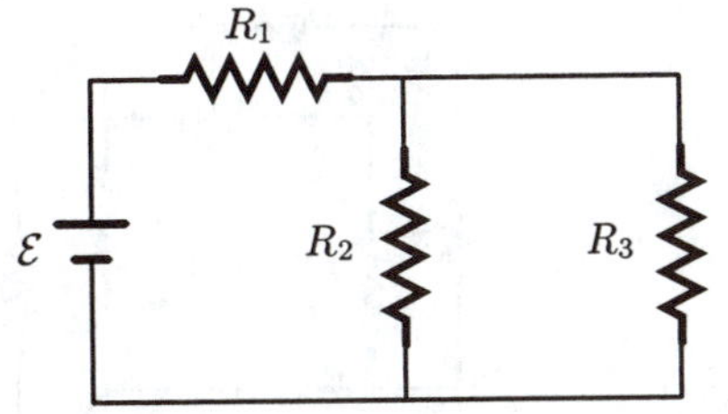

12. If $\mathcal{E} = 9.0$ V, $R_1 = 150\ \Omega$, $R_2 = 250\ \Omega$, and $R_3 = 1000\ \Omega$, find the current in each resistor.

13. Assume all three resistors are identical. If the battery's emf is 12 V and it produces a steady current of 200 mA, what is the value of each resistor?

14. Suppose $\mathcal{E} = 10.0$ V, $R_1 = 200\ \Omega$, and $R_2 = 300\ \Omega$. What value of R_3 should you select to make that battery produce a current of 25.0 mA?

15. The three resistors each have a resistance of 50 kΩ. What should be the battery's emf if it is to produce 1.0 mA of current?

For Problems 16–18, refer to the following circuit diagram.

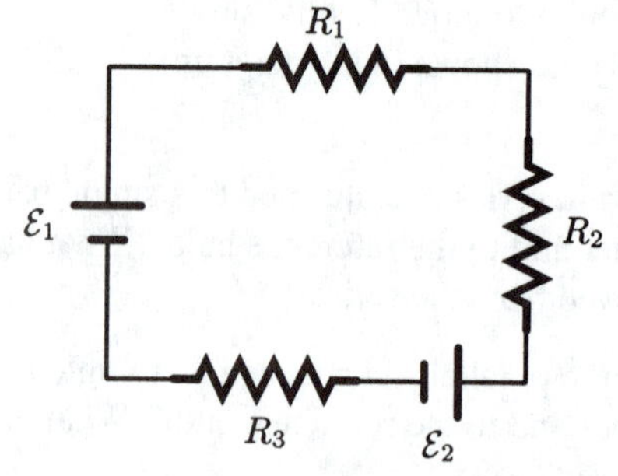

16. Find the current in the loop if $\mathcal{E}_1 = 6.5$ V, $\mathcal{E}_2 = 9.2$ V, $R_1 = 1.5$ kΩ, $R_2 = 0.90$ kΩ, $R_3 = 2.1$ kΩ.

17. Suppose $\mathcal{E}_1 = 1.75$ V, and the three resistors are identical, with the resistance of each equal to 200 Ω. Find two different possible values of $\mathcal{E}_2$ so that a current of 2.15 mA flows in the loop.

18. Suppose $\mathcal{E}_1 = 12.5$ V, and the three resistors are identical, with the resistance of each equal to 600 Ω. Find the value of $\mathcal{E}_2$ if the current in the loop is (a) 20 mA, or (b) zero.

21.3 Multiloop Circuits

Refer to Figure 21.10 for Problems 1–3.

1. Letting $\mathcal{E}_1 = \mathcal{E}_2 = 12$ V, $R_1 = 24\,\Omega$, $R_2 = 30\,\Omega$, and $R_3 = 40\,\Omega$, find the current in each resistor.

2. Repeat Problem 1 if the polarity of the battery on the right is reversed but all numerical values are kept the same.

3. Suppose all three resistors are 100 Ω. If $I_1 = 50$ mA and $I_2 = 150$ mA, find $\mathcal{E}_1, \mathcal{E}_2$, and I_3.

In Problems 4–6 refer to the following figure. There are 12 identical resistors (with resistance R_0) in the shape of a cube, with a resistor on each edge.

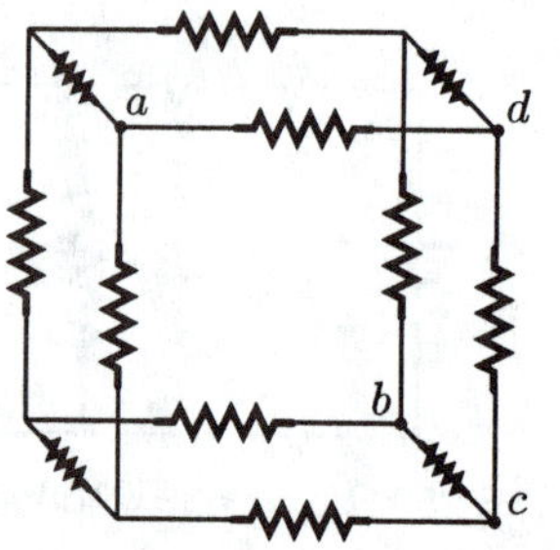

4. Find the equivalent resistance between two points (a and b) on opposite corners of the cube.

5. Find the equivalent resistance between two points (a and c) on opposite corners of one face of the cube.

6. Find the equivalent resistance between two points (a and d) on opposite sides of one edge of the cube.

7. Suppose you have a three-loop circuit containing batteries and resistors of known value. Show that there are (up to) five different values of current in the circuit, and the Kirchoff's rules lead to five equations that can be solved for the five values of current.

8. A device known as a *Wheatstone Bridge* (see diagram) can be used to measure an unknown resistance. Suppose R_x is the resistor we wish to measure. In a Wheatstone bridge, R_2 and R_3 are fixed, and R_1 is adjusted until the ammeter reads zero. Find R_x as a function of the other resistance values when the ammeter reads zero.

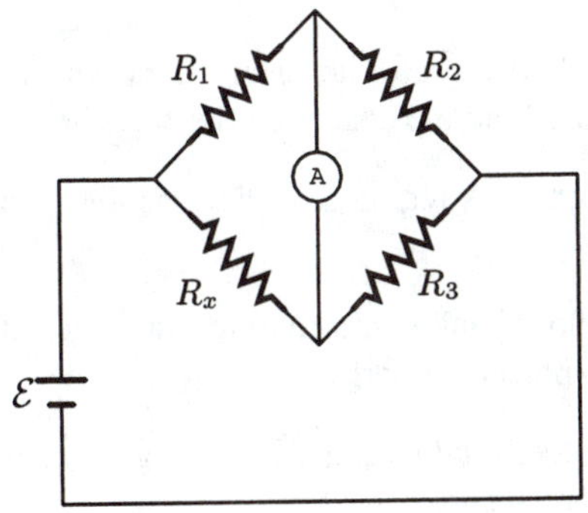

21.4 Measuring Instruments

1. In the circuit of Figure 21.11, let $\mathcal{E} = 6.50$ V, $R_1 = 11.0\,\Omega$, and $R_2 = 16.5\,\Omega$. Find the maximum internal resistance of the ammeter if its presence is to change the current in the circuit by no more than 0.1%.

2. In a series circuit, $\mathcal{E} = 48.0$ V, $R_1 = 24.5$ kΩ, $R_2 = 50.6$ kΩ, and $R_3 = 33.4$ kΩ. (a) Find the current in each resistor. (b) Find the current in each resistor if a voltmeter with internal resistance 1.05 MΩ is connected in parallel with R_1. (c) Find the current in each resistor if the same voltmeter is connected in parallel with R_2 and R_3.

In Problems 3–6, you have a galvanometer with an internal resistance of 34.6 Ω that has a full-scale deflection when a current of 0.100 mA passes through it.

3. With what value of resistor should the galvanometer be combined in parallel to produce an ammeter with an internal resistance of 0.500 Ω? Draw the circuit and show the proper connections. How much current passes through the ammeter when the galvanometer deflects full scale?

4. With what value of resistor should the galvanometer be combined in series to produce a voltmeter with an internal resistance of 500 kΩ? Draw the circuit and show the proper connections. What is the potential difference across the voltmeter when the galvanometer registers full scale?

5. With what value of resistor should the galvanometer be combined in parallel to produce an ammeter with a full-scale reading of 10.0 mA? What is the internal resistance of the resulting ammeter?

6. With what value of resistor should the galvanometer be combined in series to produce a voltmeter with a full-scale reading of 250 V? What is the internal resistance of the resulting voltmeter?

7. An ohmmeter is powered by a 1.5-V battery and has an ammeter that reads 5.0 μA full-scale. What known resistor R_0 should the ohmmeter contain if resistances if the full-scale ammeter reading is to correspond to a resistance of 100 kΩ?

21.5 *RC* Circuits

1. Using the SI units for resistance and capacitance, show that the units for the product RC are seconds.

2. A 25-nF capacitor is connected in series with a 1.45-MΩ resistor. The capacitor is initially uncharged when the series is connected to a 24.0-V battery. (a) What is the maximum charge the capacitor can attain (after a long time)? (b) How much time does it take for the capacitor to reach 50% of the charge you found in (a)? (c) How much time does it take for the capacitor to reach 90% of the charge you found in (a)?

3. A 300-pF capacitor is uncharged and placed in series with a 175-MΩ resistor. The series is then connected to a 6.0-V battery. Find the charge on the capacitor after (a) 1 ms; (b) 10 ms; (c) 100 ms.

4. Find the number of time constants required for a capacitor in an RC charging circuit to reach 99% of its maximum charge.

5. How many time constants does it take for a charging capacitor to reach (a) half of its maximum charge, or (b) half of its maximum potential energy?

6. How many time constants does it take for a discharging capacitor to reach (a) half of its initial charge, or (b) half of its initial potential energy?

7. A capacitor with an initial charge Q_0 is placed in a discharging RC circuit. Show that the time needed to reach a charge of aQ_0 (where $a < 1$ is a dimensionless parameter) is

$$t = RC\ln\left(\frac{1}{a}\right).$$

8. A 0.25-μF capacitor with an initial charge of 250 μC is discharged through a 100-kΩ resistor. (a) How much energy was stored in the capacitor prior to its discharge? (b) How much energy remained in the capacitor at that point in the discharge when its charge was 100 μC? (c) Show that the difference between (a) and (b) is accounted for by the energy dissipated in the resistor.

Chapter 22

Line Integrals

In Chapter 7, we introduced the concept of *line integral* as a tool in defining the physical quantity of *work*. In this chapter, we take a deeper look at line integrals. With the notion of *gradient* in hand, we can give the *fundamental theorem of calculus for line integrals*. We also discuss the connections among this fundamental theorem, *conservative vector fields*, and *potential functions*.

The type of line integral we saw in Chapter 7 represents a sum involving the tangential component of a vector field along a curve. We can also define a type of line integral that represents a sum involving the normal component of a vector field along a curve. We call this a *normal line integral*. Normal line integrals are the topic of the last section and will be a useful prelude to our study of surface integrals in Chapter 23.

22.1 Curves

We begin with a review of curves. There are at least three ways in which we can specify a curve. Let's restrict our attention to curves in the plane for now. The first way to specify a curve is to give a *geometric* description. For example, we can specify one curve as the set of all points in the plane at distance of one unit from a particular fixed point in plane. This curve is, of course, a circle of radius 1 centered at the fixed point. Note that this description involves only the geometric concepts of point and distance. It does not involve a coordinate system, as shown in Figure 22.1(a).

If we introduce a coordinate system, say Cartesian coordinates (x, y), we can give an *analytic* description of a curve. We do this in a number of ways. Take the circle of radius 1 as an example. Let's choose the coordinate system to have its origin at the center of the circle. That is, we assign $(0, 0)$ to the fixed point. Any other point on the circle has coordinates (x, y) that satisfy the equation

$$x^2 + y^2 = 1. \tag{22.1}$$

This equation constitutes an analytic description of the circle [see Figure 22.1(b)]. We can also think of the circle as composed of graphs of functions. We need at least two functions to describe the circle, so that each graph passes the "vertical line test." Solving for y in Equation (22.1), we get the two functions

$$f_+(x) = \sqrt{1 - x^2} \qquad \text{and} \qquad f_-(x) = -\sqrt{1 - x^2}. \tag{22.2}$$

The graph of the first function gives the upper half-circle, and the graph of the second function gives the lower half-circle (with two points in common at $(1, 0)$ and $(-1, 0)$.) In some contexts, it is convenient to interpret an equation such as Equation (22.1) as specifying a level curve for some function $g : \mathbb{R}^2 \to \mathbb{R}$. For

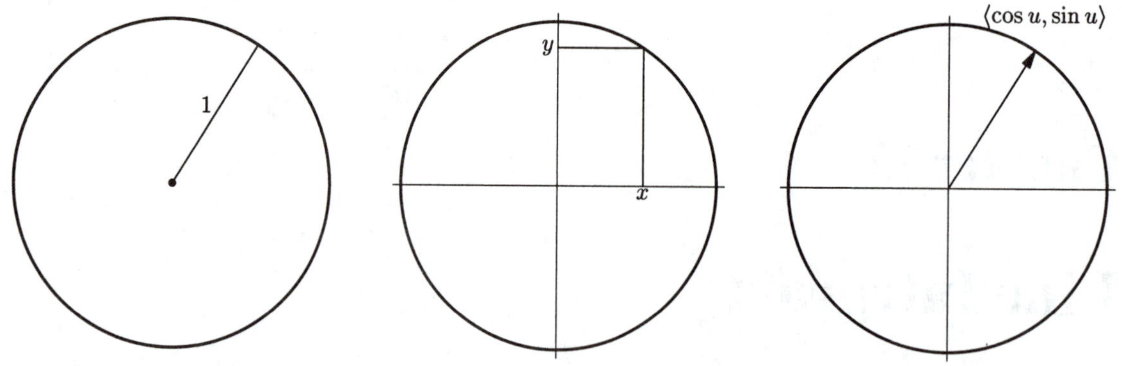

(a) Geometric description (b) Analytic description (c) Parametric description

Figure 22.1. Three ways to describe the unit circle.

the equation of the circle, we could choose $g(x, y) = x^2 + y^2$, and thus view the circle as the 1-level curve of this function. We used this strategy, for example, in the Lagrange multiplier method.

Our third way to specify a curve is to give a *parametric* description. For a curve in the plane, this means finding a vector-output function $\vec{r} : \mathbb{R} \to \mathbb{R}^2$. We let u be the input variable, with u taking values from the interval $[a, b]$. The outputs $\vec{r}(u)$ give position vectors for points on the curve. For a circle of radius one centered at the origin, we can use the parametric description

$$\vec{r}(u) = \langle \cos u, \sin u \rangle \tag{22.3}$$

with u in the interval $[0, 2\pi)$ [see Figure 22.1(c)]. Note that as u goes from 0 to 2π, the outputs $\vec{r}(u)$ trace out the circle exactly once in the counterclockwise direction starting from the point $(1, 0)$.

We use the parametric description of curves to define line integrals. The direction in which the outputs trace out the circle are important. We also require that the function $\vec{r}$ be differentiable for all u in $[a, b]$. It is also important to have a parametrization that does not "double back" on the curve. That is, the outputs should trace through each point on the curve just once as the parameter goes from a to b. We can guarantee this by imposing the condition that the derivative $\vec{r}\,'(u)$ be nonzero for all u in the interval $[a, b]$. We say a curve is **smooth** if it can be parametrized in this way.

The following examples give some strategies for parametrizing a curve described geometrically or analytically.

Example 22.1. Find a parametrization for the unit circle centered at the origin that traces out the circle clockwise from $(1, 0)$.

Solution. We can modify the parametrization in Equation (22.3) to get a clockwise orientation by changing the second component to $-\sin u$. That is, we use the parametrization

$$\vec{r}(u) = \langle \cos u, -\sin u \rangle$$

with u in the interval $[0, 2\pi)$. For $u = 0$, we have $\vec{r}(0) = \langle 1, 0 \rangle$. As u increases from 0, the second component $-\sin u$ is negative, corresponding to tracing out the circle in clockwise direction. △

Example 22.2. Find a parametrization for the line segment from $(1, 2)$ to $(-3, 5)$.

Solution. From our experience with position functions, we know that a straight path is described by a linear function. Thus, we look for a parametrization in the form

$$\vec{r}(u) = \langle a + bu, c + du \rangle$$

where a, b, c, and d are constants that we need to determine for the specific line segment. We can get a specific choice by picking the interval $[0, 1]$ for the parameter u. From the given starting point, we need to have

$$\vec{r}(0) = \langle a, c \rangle = \langle 1, 2 \rangle.$$

We clearly have $a = 1$ and $c = 2$. Using these with $u = 1$ and the given ending point gives the condition

$$\vec{r}(1) = \langle 1 + b, 2 + d \rangle = \langle -3, 5 \rangle.$$

Solving gives $b = -4$ and $d = 3$. A parametrization for the line segment is thus

$$\vec{r}(u) = \langle 1 - 4u, 2 + 3u \rangle$$

for u in $[0, 1]$.

We could clearly find many other parametrizations for this line segment. For example, we could keep the linear form but require the parameter u to range in the interval $[-4, 4]$. This would result in some other choice of the constants a, b, c, and d. △

In some situations, we need to parametrize a curve that is the graph of a given function f for outputs x in a given interval $[a, b]$. One way to parametrize the graph of f is to let $x(u) = u$ and $y(u) = f(u)$ with the parameter u in the same interval $[a, b]$ as the domain of the function. With this choice, the outputs $\vec{r}(u) = \langle u, f(u) \rangle$ automatically fall on the graph of the function f.

Example 22.3. Find a parametrization of the graph of $f(x) = \sin x$ for x in the domain $[0, \pi]$.

Solution. This is easy to accomplish using the strategy given above. We let $x(u) = u$ and $y(u) = \sin u$ to get the parametrization

$$\vec{r}(u) = \langle u, \sin u \rangle$$

with u in the interval $[0, \pi]$. △

Consider now the curve C shown in Figure 22.2, with starting point at $(6, 1)$ and ending point at $(2, 3)$. This curve has a "corner" at the point $(6, 3)$. For this point, there is no well-defined tangent vector, so we cannot expect to find a parametrization that is differentiable at all points.

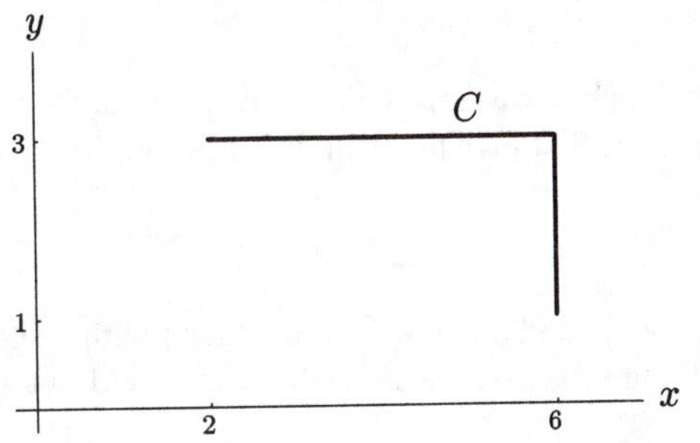

Figure 22.2. Curve with a corner.

To handle this situation, we relax the differentiability requirement but only for the corner point. One way to parametrize the curve is to trace out the vertical segment for u in $[0,1]$ and the horizontal segment for u in $[1,2]$. For the vertical segment, we can use $\vec{r}(u) = \langle 6, 2u+1\rangle$ with u in $[0,1]$. For the horizontal segment we can use $\vec{r}(u) = \langle -4u+10, 3\rangle$ with u in $[1,2]$. To emphasize the fact that this is a single parametrizing function, we write this as

$$\vec{r}(u) = \begin{cases} \langle 6, 2u+1\rangle & \text{for } u \text{ in } [0,1] \\ \langle -4u+10, 3\rangle & \text{for } u \text{ in } [1,2]. \end{cases} \tag{22.4}$$

Of course, this is only one way in which the curve C can be parametrized.

As an alternative to the single parametrized function given in Equation (22.4), we can view the curve C as a **sum of curves**. Let C_1 denote the vertical segment, traced out from $(6,1)$ to $(6,3)$, and let C_2 denote the horizontal segment, traced out from $(6,3)$ to $(2,3)$. We write $C = C_1 + C_2$. The curves C_1 and C_2 can be parametrized independently. One choice is $\vec{r}_1(u) = \langle 6, 2u+1\rangle$ for u in $[0,1]$ for C_1 and $\vec{r}_2(u) = \langle -4u+6, 3\rangle$ for u in $[0,1]$. We are free to use the same interval, $[0,1]$ in this case, as domain for both functions. Think of this as starting the clock over at the point where the curves are joined.

The curve C in Figure 22.2 is an example of a **piecewise smooth curve**. This means that the curve can be split into a finite number of pieces, each of which is smooth.

Finally, we turn our attention to curves in space $\mathbb{R}^3$. Curves in space are often most conveniently described parametrically using a function $\vec{r} : \mathbb{R} \to \mathbb{R}^3$ with input u in some interval $[a,b]$. For example, a helix wrapping three times around the z-axis can be parametrized as

$$\vec{r}(u) = \langle \cos u, \sin u, u\rangle$$

for u in the interval $[0, 6\pi]$. This specific interval gives a helix that starts at the point $(1,0,0)$ and ends at the point $(1,0,6\pi)$. We use the same language of *smooth* if the curve can be parametrized with $\vec{r}\,'(u) \neq \vec{0}$ for all u in the relevant interval, and *piecewise smooth* if the curve can be split into a finite number of pieces, each of which is smooth.

Example 22.4. Find a parametrization for the circle of constant latitude midway between the equator and the north pole on a sphere of radius 1 centered at the origin.

Solution. We can use spherical coordinates for this with $\rho = 1$ to have points on a sphere of radius 1. All of the points on the circle in question have $\phi = \pi/4$. We trace out the circle as the spherical coordinate θ ranges from 0 to 2π. We thus use θ as the parameter. From the transformations given in Equation (14.12), we have $\vec{r}(\theta) = \langle (\sqrt{2}/2)\cos\theta, (\sqrt{2}/2)\sin\theta, \sqrt{2}/2\rangle$ for θ in $[0, 2\pi]$. △

22.2 Line Integrals

In Chapter 7, we defined the work done by a force field along a path in the plane or a path in space. Here we abstract the mathematical details. We first give the definition relevant for a vector field in the plane.

22.2.1 Definition

Consider a vector field $\vec{F} : \mathbb{R}^2 \to \mathbb{R}^2$ and a curve C in the plane with a specified orientation. An example is shown in Figure 22.3(a). In Figure 22.3(b), we show the vector field outputs for some inputs on the curve itself.

In the precise definition below, the line integral of $\vec{F}$ on the curve C is defined in terms of a definite integral. To gain some intuition, consider construction of that definite integral from first principles. This

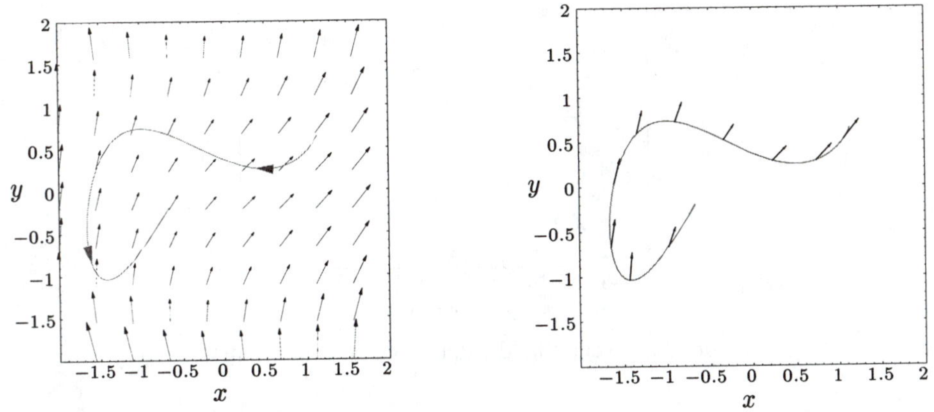

(a) Outputs for $\vec{F}$ throughout the plot region (b) Outputs for $\vec{F}$ along the curve

Figure 22.3. A vector field $\vec{F}$ and a curve C in the plane.

means constructing a sum. Each term in the sum is the product of the tangential component of $\vec{F}$ at a point in the relevant piece and the length of the piece. Let's first set up some notation.

Assume we have parametrized the curve by a vector-output function $\vec{r} : \mathbb{R} \to \mathbb{R}^2$, with the inputs u in some interval $[a, b]$. To construct a definite integral, we split the interval $[a, b]$ into n subintervals of size $\Delta u = (b - a)/n$. (More generally, the subintervals need not be of equal size, but we've chosen this for simplicity here.) This splitting of the interval $[a, b]$ induces a splitting of the curve C, as illustrated in Figure 22.4(a). Note that the curve pieces will generally not be of equal length, even if the subintervals of $[a, b]$ are of equal size. Let u_i be an input in the ith subinterval. In Figure 22.4(b), we show a tangent vector $\vec{r}'(u)$ for each curve piece. The quantity $\|\vec{r}'(u_i)\|$ gives a scale factor that relates the subinterval size Δu to the length of the corresponding curve piece. This is simple to understand, if we think of u as time so that $\|\vec{r}'(u_i)\|$ gives a speed. The product $\|\vec{r}'(u_i)\|\Delta u$ is then the approximate distance traveled during the ith interval of length Δu. This is equal to the approximate length of the ith curve piece, because we have assumed that $\vec{r}'(u) \neq \vec{0}$.

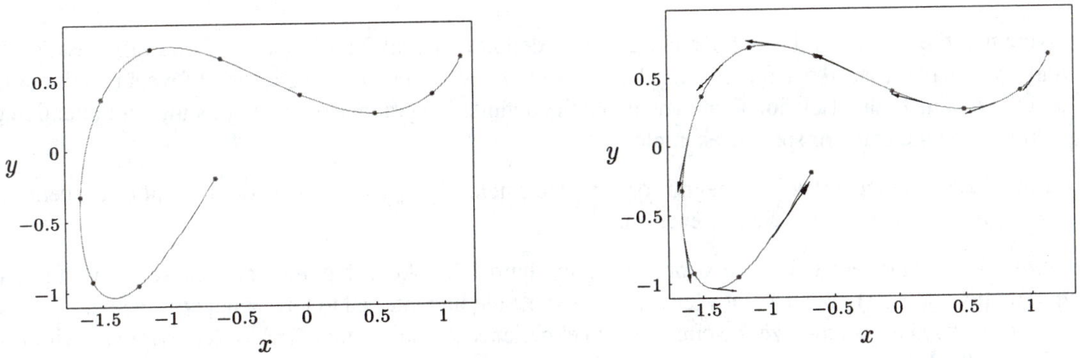

(a) Endpoints of the curve pieces (b) Tangent vector for each curve piece

Figure 22.4. Splitting the curve into pieces.

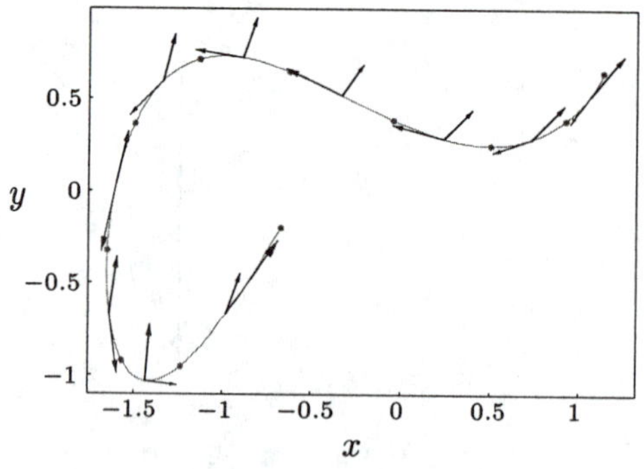

Figure 22.5. Vector field outputs and tangent vectors.

To specify an output of the vector field at a point on the curve, we use $\vec{r}(u_i)$ as the input for $\vec{F}$, giving $\vec{F}(\vec{r}(u_i))$. To get a tangential component, we can dot this with a vector that is tangent to the curve. This is conveniently given by the derivative $\vec{r}\,'(u_i)$. An example with these vectors is shown in Figure 22.5. Now construct a sum, in which the ith term is

$$\vec{F}(\vec{r}(u_i)) \cdot \vec{r}\,'(u_i)\Delta u.$$

To interpret this quantity, use the geometric view of dot product to write

$$\vec{F}(\vec{r}(u_i)) \cdot \vec{r}\,'(u_i)\Delta u = \|\vec{F}(\vec{r}(u_i))\|\|\vec{r}\,'(u_i)\| \sin\theta_i \Delta u = \left(\|\vec{F}(\vec{r}(u_i))\| \sin\theta_i\right) \left(\|\vec{r}\,'(u_i)\|\Delta u\right).$$

In the last expression, we've arranged the expression into two factors. The first factor is the component of $\vec{F}(\vec{r}(u_i))$ tangent to the curve. The second factor is the approximate length of the ith curve piece.

To get a definite integral, we form a sum with one term for each curve piece, and then take a limit as the number of curve pieces goes to infinity. The resulting definite integral appears in the following definition.

Definition 22.1. The **line integral of a vector field $\vec{F} : \mathbb{R}^2 \to \mathbb{R}^2$ on the curve** C parametrized by $\vec{r} : \mathbb{R} \to \mathbb{R}^2$ with input u in the interval $[a, b]$ is defined as

$$\int_C \vec{F} \cdot d\vec{s} = \int_a^b \vec{F}(\vec{r}(u)) \cdot \vec{r}\,'(u)\, du. \tag{22.5}$$

Note that the line integral is defined in terms of a definite integral for a function of one variable. For the definition to make sense, the definite integral must exist. We won't pursue the question of what conditions on $\vec{F}$ and the parametrizing function $\vec{r}$ guarantee that the definite integral exists. Now let's turn our attention to computing line integrals for specific examples.

Example 22.5. Compute the line integral for the vector field $\vec{F}(x, y) = \langle y, -x\rangle$ on the unit circle centered at the origin with counterclockwise orientation.

Solution. The vector field and the curve are shown in Figure 22.6. At each point on the curve, the vector field output points opposite to the tangent vector, so we expect the line integral to have a negative value.

First, we need to parametrize the circle. From experience, we know that $\vec{r}(u) = \langle \cos u, \sin u\rangle$ with u in the interval $[0, 2\pi]$ parametrizes the unit circle with a counterclockwise orientation. Next, we differentiate to get $\vec{r}\,'(u) = \langle -\sin u, \cos u\rangle$. We also need the outputs $\vec{F}(\vec{r}(u))$; these are given by

$$\vec{F}(\vec{r}(u)) = \vec{F}(\cos u, \sin u) = \langle \sin u, -\cos u\rangle.$$

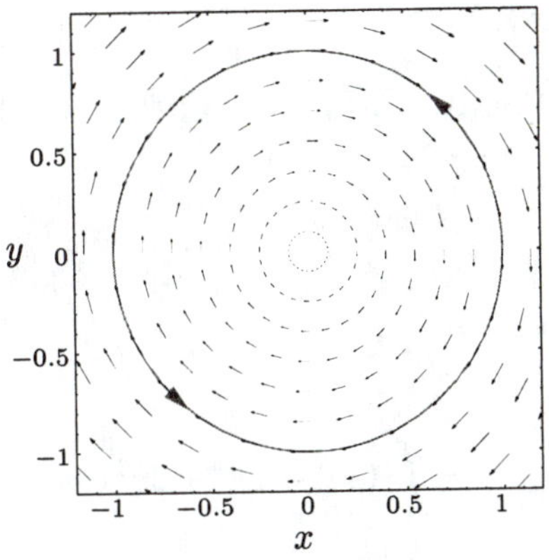

Figure 22.6. Vector field and curve for Example 22.5.

The integrand of the defining definite integral is the dot product

$$\vec{F}(\vec{r}(u)) \cdot \vec{r}'(u) = \langle \sin u, -\cos u\rangle \cdot \langle -\sin u, \cos u\rangle = -\sin^2 u - \cos^2 u = -1.$$

In this case, the integrand is quite simple. Putting these results into the definition and evaluating the resulting definite integral, we get

$$\int_C \vec{F} \cdot d\vec{s} = \int_0^{2\pi} (-1)\, du = -2\pi.$$

The value of the line integral is thus -2π. △

Example 22.6. Compute the line integral for the vector field $\vec{F}(x, y) = \langle xy, x+y\rangle$ on the curve C parametrized by $\vec{r}(u) = \langle u^3, u^2\rangle$ for u in the interval $[0, 2]$.

Solution. The vector field and the curve are shown in Figure 22.7. Along the curve, we see the vector field outputs pointing more in the direction of the tangent vectors than opposite to the tangent vectors. We expect the line integral to have a positive value.

We are given a parametrization of the curve, so we need only differentiate to get

$$\vec{r}'(u) = \langle 3u^2, 2u\rangle.$$

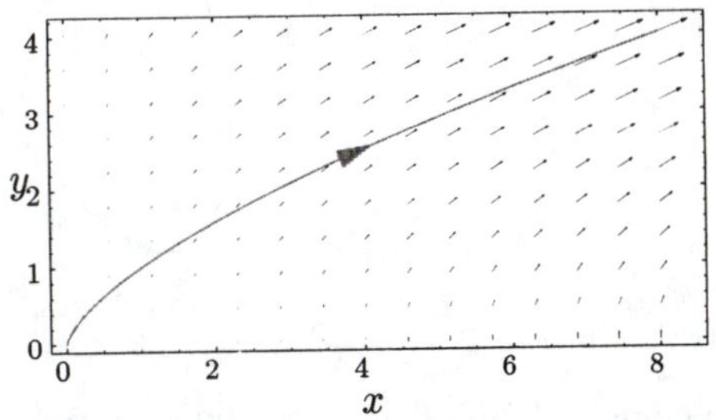

Figure 22.7. Vector field and curve for Example 22.6.

Outputs of $\vec{F}$ along the curve are given by

$$\vec{F}(\vec{r}(u)) = \vec{F}(u^3, u^2) = \langle u^5, u^3 + u^2 \rangle.$$

The relevant dot product is

$$\vec{F}(\vec{r}(u)) \cdot \vec{r}\,'(u) = \langle u^5, u^3 + u^2 \rangle \cdot \langle 3u^2, 2u \rangle = 3u^7 + 2u^4 + 2u^3.$$

From the definition, we have

$$\int_C \vec{F} \cdot d\vec{s} = \int_0^2 (3u^7 + 2u^4 + 2u^3)\, du = \frac{584}{5}.$$

The line integral has a positive value, in agreement with our geometric intuition. △

In some situations, it is useful to view a given curve C as composed of two or more pieces. Suppose C starts at the point P and ends at the point Q, as shown in Figure 22.8(a). We can pick a point R on the curve and define two curves: The curve C_1 starts at the point P and ends at the point R following the curve C, while the curve C_2 starts at the point R and ends at the point Q following the remainder of the curve C. Let's denote this as $C = C_1 + C_2$. If $\vec{r}(u)$ parametrizes C for u in the interval $[a, b]$, there will be some value $u = c$ for which $\vec{r}(c)$ is the point R. Thus, $\vec{r}(u)$ for u in $[a, c]$ parametrizes C_1, and $\vec{r}(u)$ for u in $[c, b]$ parametrizes C_2. Now from the definition of line integral and properties of definite integrals, we get

$$\begin{aligned}\int_C \vec{F} \cdot d\vec{s} &= \int_a^b \vec{F}(\vec{r}(u)) \cdot \vec{r}\,'(u)\, du \\ &= \int_a^c \vec{F}(\vec{r}(u)) \cdot \vec{r}\,'(u)\, du + \int_c^b \vec{F}(\vec{r}(u)) \cdot \vec{r}\,'(u)\, du \\ &= \int_{C_1} \vec{F} \cdot d\vec{s} + \int_{C_2} \vec{F} \cdot d\vec{s}.\end{aligned}$$

It is straightforward to generalize this result to a situation in which the curve C is composed of more than two pieces, with the endpoint of one serving as the start point of the next.

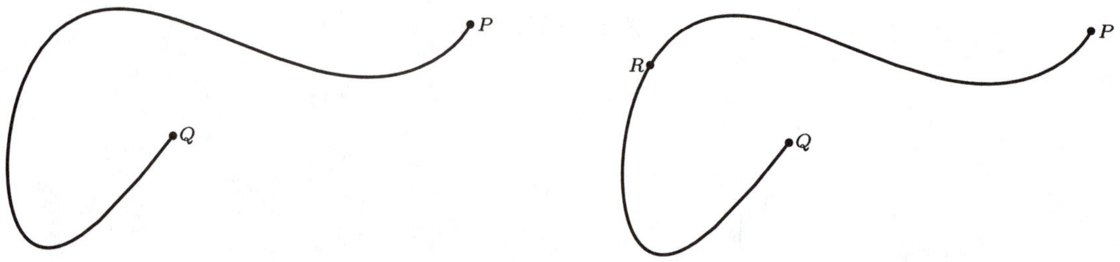

(a) Curve C starts at the point P and ends at the point Q

(b) C splits into two curves at the point R

Figure 22.8. Splitting a curve into two pieces.

22.3 The Fundamental Theorem of Calculus for Line Integrals

In this section, we develop a fundamental theorem of calculus for line integrals. This theorem is a natural extension of the second fundamental theorem of calculus for definite integrals. We restate that theorem for reference. We use g to denote the function to avoid a possibly confusing conflict of notation. Recall that G is an antiderivative of g if $G'(x) = g(x)$ for all x in the relevant domain.

Theorem 22.1 (Second Fundamental Theorem of Calculus). *If the function $g : [a, b] \to \mathbb{R}$ is continuous and the function $G : [a, b] \to \mathbb{R}$ is an antiderivative of g, then*

$$\int_a^b g(x)\, dx = G(b) - G(a).$$

To state an analog of the second fundamental theorem, we need some notion of antiderivative for a vector field $\vec{F} : \mathbb{R}^n \to \mathbb{R}^n$. That is, we need a function that has some type of derivative equal to $\vec{F}$. Since the outputs of $\vec{F}$ are vectors, it is reasonable to think of the *gradient* as the relevant notion of derivative. This implies that the relevant type of function is of the type $V : \mathbb{R}^n \to \mathbb{R}$. That is, we take $V : \mathbb{R}^n \to \mathbb{R}$ as our "antiderivative" of the vector field $\vec{F} : \mathbb{R}^n \to \mathbb{R}^n$ if

$$\vec{\nabla} V(x, y) = \vec{F}(x, y)$$

for all inputs (x, y) in the relevant domain. For historical reasons, the function V is not called an antiderivative for $\vec{F}$. Instead, we say that V is a **potential function** for the vector field $\vec{F}$.

Theorem 22.2 (Fundamental Theorem of Calculus for Line Integrals). *If $V : \mathbb{R}^n \to \mathbb{R}$ is a potential function for the vector field $\vec{F} : \mathbb{R}^n \to \mathbb{R}^n$ on a domain that includes the curve C, then*

$$\int_C \vec{F} \cdot d\vec{s} = V(\vec{r}_2) - V(\vec{r}_1)$$

where $\vec{r}_1$ is the starting point of C and $\vec{r}_2$ is the ending point of C.

Proof. To prove this theorem, we put ourselves into a position in which we can apply the second fundamental theorem for a definite integral. Let $\vec{r}(u)$ for u in $[a, b]$ be a parametrization for the curve C. Since $\vec{r}_1$ is the starting point of the curve, we have $\vec{r}(a) = \vec{r}_1$. Likewise, $\vec{r}(b) = \vec{r}_2$. From the relevant chain rule, we get

$$\frac{d}{du}\left[V(\vec{r}(u))\right] = \vec{\nabla} V(\vec{r}(u)) \cdot \vec{r}\,'(u).$$

Since V is a potential function for $\vec{F}$, we get

$$\frac{d}{du}\left[V(\vec{r}(u))\right] = \vec{F}(\vec{r}(u)) \cdot \vec{r}\,'(u).$$

Using this result in the definition of line integral, we get

$$\int_C \vec{F} \cdot d\vec{s} = \int_a^b \vec{F}(\vec{r}(u)) \cdot \vec{r}\,'\, du = \int_a^b \frac{d}{du}\left[V(\vec{r}(u))\right] du.$$

In the last expression, the integrand is the derivative of the function $V(\vec{r}(u))$, so $V(\vec{r}(u))$ is clearly an antiderivative. Thus, by the second fundamental theorem for definite integrals, we have

$$\int_C \vec{F} \cdot d\vec{s} = \int_a^b \frac{d}{du}\left[V(\vec{r}(u))\right] du = V(\vec{r}(b)) - V(\vec{r}(a)) = V(\vec{r}_2) - V(\vec{r}_1).$$

This completes the proof. □

From the fundamental theorem for line integrals, we get a result about *curve-independence* for the line integral of a vector field that has a potential function.

Corollary. *If C_1 and C_2 are two curves in $\mathbb{R}^n$ that both start at the point $\vec{r}_1$ and end at the point $\vec{r}_2$, and the vector field $\vec{F} : \mathbb{R}^n \to \mathbb{R}^n$ has a potential function, then*

$$\int_{C_1} \vec{F} \cdot d\vec{s} = \int_{C_2} \vec{F} \cdot d\vec{s}.$$

That is, the value of the line integral for $\vec{F}$ depends on the endpoints $\vec{r}_1$ and $\vec{r}_2$ but not on the curve that connects these two points.

Proof. Let V be a potential function for $\vec{F}$. Then by the fundamental theorem for line integrals, we have

$$\int_{C_1} \vec{F} \cdot d\vec{s} = V(\vec{r}_2) - V(\vec{r}_2) = \int_{C_2} \vec{F} \cdot d\vec{s}.$$

This completes the proof. □

Recall from Chapter 7 that a force is defined to be *conservative* if the work done by the force is path-independent. We use this same language for any vector field. That is, a vector field $\vec{F} : \mathbb{R}^n \to \mathbb{R}^n$ is **conservative** if the value of any line integral depends only the endpoints and not on the curve connecting those endpoints. With this language, we can restate the corollary as saying that if a vector field has a potential function, then the vector field is conservative.

Example 22.7. Find a potential function, if one exists, for the vector field $\vec{F}(x, y) = \langle 2x + y, x \rangle$.

Solution. If there is a potential V, then it must satisfy $\vec{\nabla} V(x, y) = \vec{F}(x, y)$ on some domain. In terms of components $\vec{F} = \langle P, Q \rangle$, this is

$$\frac{\partial V}{\partial x} = P(x, y) = 2x + y \tag{22.6}$$

$$\frac{\partial V}{\partial y} = Q(x, y) = x. \tag{22.7}$$

Antidifferentiating the top equation with respect to x gives

$$V(x, y) = x^2 + xy + \phi(y) \tag{22.8}$$

where ϕ can be any function. We include this term to have the most general expression consistent with Equation (22.6). In similar fashion, we antidifferentiate with respect to y on both sides of Equation (22.7) to get

$$V(x, y) = xy + \psi(x) \tag{22.9}$$

where ψ is a second arbitrary function.

The expressions in Equations (22.8) and (22.9) must be consistent if a potential function exists. We can choose $\phi(y) = 0$ and $\psi(x) = x^2$ to get the desired consistency. Thus, the function

$$V(x, y) = x^2 + xy$$

is a potential function for $\vec{F}(x, y) = \langle 2xy + y, x \rangle$. We can get other potential functions by adding a constant term, just as we can generate additional antiderivatives from any one antiderivative. △

Example 22.8. Find a potential function, if one exists, for the vector field $\vec{F}(x, y) = \langle 2x + y, 2x \rangle$.

Solution. If there is a potential V, then it must satisfy $\vec{\nabla}V(x, y) = \vec{F}(x, y)$ on some domain. In terms of components $\vec{F} = \langle P, Q \rangle$, this is

$$\frac{\partial V}{\partial x} = P(x, y) = 2x + y \tag{22.10}$$

$$\frac{\partial V}{\partial y} = Q(x, y) = 2x. \tag{22.11}$$

Antidifferentiating the top equation with respect to x gives

$$V(x, y) = x^2 + xy + \phi(y) \tag{22.12}$$

where ϕ can be any function. We include this term to have the most general expression consistent with Equation (22.10). In similar fashion, we antidifferentiate with respect to y on both sides of Equation (22.11) to get

$$V(x, y) = 2xy + \psi(x) \tag{22.13}$$

where ψ is a second arbitrary function.

The expressions in Equations (22.12) and (22.13) must be consistent if a potential function exists. We can choose $\psi(x) = x^2$ to get x^2 in both expressions, but we are stuck when it comes to the terms xy and $2xy$. No choice for $\phi(y)$ can give us consistency. There is no potential function for the vector field $\vec{F}(x, y) = \langle 2x + y, 2x \rangle$. △

By thinking about the equality of mixed second partial derivatives, we can find a necessary condition for a vector field $\vec{F}(x, y) = \langle P(x, y), Q(x, y) \rangle$ to have a potential function. For this condition, we need to assume the components P and Q have continuous partial derivatives in the entire plane. Now if V is a potential function, then

$$\frac{\partial V}{\partial x} = P(x, y)$$

$$\frac{\partial V}{\partial y} = Q(x, y).$$

Differentiate both sides of the top equation with respect to y to get

$$\frac{\partial^2 V}{\partial y \partial x} = \frac{\partial P}{\partial y}.$$

Likewise, differentiate both sides of the bottom equation with respect to x to get

$$\frac{\partial^2 V}{\partial x \partial y} = \frac{\partial Q}{\partial x}.$$

By equality of mixed partials, we must have

$$\frac{\partial P}{\partial y} = \frac{\partial Q}{\partial x}.$$

We have proved the following theorem.

Theorem 22.3. *If $\vec{F}(x,y) = \langle P(x,y), Q(x,y)\rangle$ has components with continuous partial derivatives and there exists a potential function for $\vec{F}$, then*

$$\frac{\partial P}{\partial y} = \frac{\partial Q}{\partial x}.$$

The utility of this theorem comes in the contrapositive statement: If

$$\frac{\partial P}{\partial y} \neq \frac{\partial Q}{\partial x}$$

for some (x,y), then either there is no potential function for $\vec{F}$ or the components do not have continuous partial derivatives.

Example 22.9. Check the condition given by Theorem 22.3 for the vector fields in Examples 22.7 and 22.8.

Solution. For Example 22.7, the vector field is $\vec{F}(x,y) = \langle 2x+y, x\rangle$. We calculate

$$\frac{\partial P}{\partial y} = 1 \qquad \text{and} \qquad \frac{\partial Q}{\partial x} = 1$$

so

$$\frac{\partial P}{\partial y} = \frac{\partial Q}{\partial x}$$

for all (x,y). The necessary condition is satisfied, and it is possible that a potential function exists. Of course, we know that one does exist, having found one in the solution of Example 22.7.

For Example 22.8, the vector field is $\vec{F}(x,y) = \langle 2x+y, 2x\rangle$. We calculate

$$\frac{\partial P}{\partial y} = 1 \qquad \text{and} \qquad \frac{\partial Q}{\partial x} = 2$$

so

$$\frac{\partial P}{\partial y} \neq \frac{\partial Q}{\partial x}$$

for all (x,y). The necessary condition is not satisfied. The component functions have continuous partial derivatives, so we conclude that no potential function exists for this vector field. This is consistent with our findings in the solution of Example 22.8. △

The generalization of Theorem 22.3 to vector fields $\vec{F} : \mathbb{R}^3 \to \mathbb{R}^3$ requires an idea that we will introduce in Chapter 25. Also, in Chapter 27 we will see that the converse of Theorem 22.3 is true . That is, if $\partial P/\partial x = \partial Q/\partial y$ on a reasonable domain, then a potential function exists for $\vec{F}(x,y) = \langle P(x,y), Q(x,y)\rangle$.

22.4 The Normal Line Integral

Now let's turn our attention to a new concept, the *normal line integral*. This type of line integral involves the *normal* component of a vector field along a curve. In contrast, the line integrals of the previous two sections involve the *tangential* component of a vector field along a curve. It would be fair to call the line integrals we have studied so far *tangential* line integrals, though by convention we just call them line integrals.

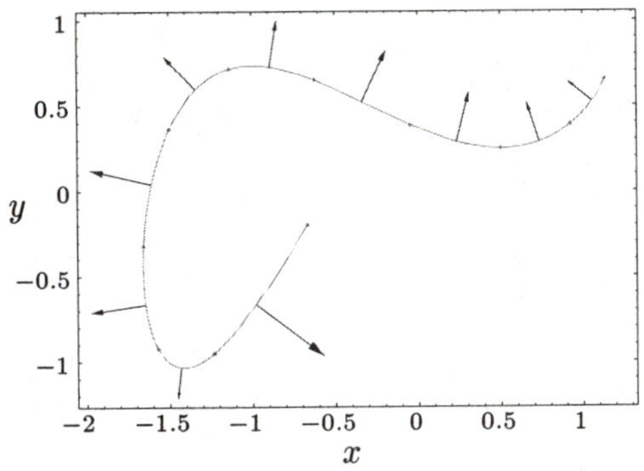

Figure 22.9. Normal vectors along a curve.

It is important to note that the definition of normal line integral requires that we work in the plane. That is, what follows is relevant to curves in the plane and vector fields $\vec{F} : \mathbb{R}^2 \to \mathbb{R}^2$. The idea of normal line integral does not apply to curves in space and vector fields $\vec{F} : \mathbb{R}^3 \to \mathbb{R}^3$. However, in the next chapter, we will generalize the normal line integral in the concept of a *surface integral*.

Assume we have a given vector field $\vec{F} : \mathbb{R}^2 \to \mathbb{R}^2$ and a given curve C in the plane. Further assume we have parametrized the curve by a vector-output function $\vec{r} : \mathbb{R} \to \mathbb{R}^2$ with the inputs u in some interval $[a, b]$. Let $\hat{n}(u)$ be a unit vector normal to the curve at the point $\vec{r}(u)$ corresponding to input u. For a plane curve, there are two normal directions at each point. For the definition of normal line integral, we want to pick the unit normal vector in a consistent manner. The parametrization given by $\vec{r}$ traces out the curve in a particular direction. We choose the unit normal vectors to be always on the left side of the curve or always on the right side of the curve as we traverse the curve from start to end. Note that this is not the choice of unit normal given by the principal unit normal vector defined in Chapter 4. At each point on a curve, the principal unit normal vector points in to the center of the osculating circle. This is to the left at some points and to the right at other points.

Given a specific curve C and a parametrization $\vec{r}(u)$, we can find a unit normal $\hat{n}(u)$ with the following insight. For any vector $\langle a, b\rangle$ in the plane, the vectors $\langle b, -a\rangle$ and $\langle -b, a\rangle$ are perpendicular to $\langle a, b\rangle$. This is easy to see by computing the dot products between $\langle a, b\rangle$ and each of $\langle b, -a\rangle$ and $\langle -b, a\rangle$. We use this insight to get a unit normal vector $\hat{n}(u)$ from the derivative vector $\vec{r}'(u) = \langle x'(u), y'(u)\rangle$. We can choose either $\vec{n}(u) = \langle y'(u), -x'(u)\rangle$ or $\vec{n}(u) = \langle -y'(u), x'(u)\rangle$. To be consistent, we must make the same choice at each point on the curve, as illustrated in Figure 22.9. To get a unit normal vector $\hat{n}(u)$, we must divide by $||\vec{r}'(u)||$.

Example 22.10. Find the unit normals $\hat{n}(u)$ that point to the left on the upper half circle traced out from $(1, 0)$ to $(-1, 0)$.

Solution. We use the parametrization $\vec{r}(u) = \langle \cos u, \sin u\rangle$ with u in the interval $[0, \pi]$. The derivative is

$$\vec{r}'(u) = \langle -\sin u, \cos u\rangle.$$

For the normal vector, we must choose between $\langle \cos u, \sin u\rangle$ and $\langle -\cos u, -\sin u\rangle$. To make this choice, consider the input $u = 0$. The first option gives $\langle 1, 0\rangle$, and the second option gives $\langle -1, 0\rangle$. The second option points to the left for the orientation of this parametrization. We thus choose $\vec{n}(u) = \langle -\cos u, -\sin u\rangle$. Since $||\vec{r}'(u)|| = 1$, we have the unit normal vector $\hat{n}(u) = \langle -\cos u, -\sin u\rangle$. △

Intuitively, the normal line integral of $\vec{F}$ on the curve C is defined by splitting the curve into pieces and then building a sum with one term for each piece. Each term in the sum is the product of the normal

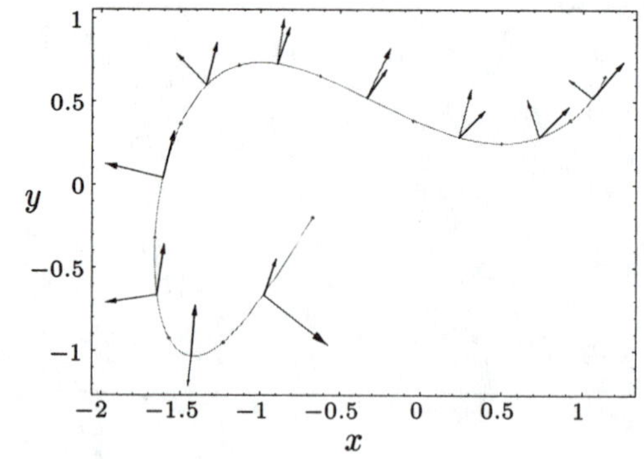

Figure 22.10. Summing the normal component of a vector field along a curve.

component of $\vec{F}$ at a point in the relevant piece and the length of the piece. We will denote the normal line integral as

$$\int_C \vec{F} \cdot \hat{n}\, ds.$$

To specify an output of the vector field at a point on the curve, we use $\vec{r}(u)$ as the input for $\vec{F}$, giving $\vec{F}(\vec{r}(u))$. To get a normal component, we dot this with the chosen unit normal vector $\hat{n}(u)$. To get the correct length for the piece of the curve, we must include a factor of $||\vec{r}\,'(u)||$. These are the elements (illustrated in Figure 22.10) in the following definition.

Definition 22.2. The **normal line integral of a vector field $\vec{F} : \mathbb{R}^2 \to \mathbb{R}^2$ on the curve C** parametrized by $\vec{r} : \mathbb{R} \to \mathbb{R}^2$ with input u in the interval $[a, b]$ is defined as

$$\int_C \vec{F} \cdot \hat{n}\, ds = \int_a^b \vec{F}(\vec{r}(u)) \cdot \hat{n}(u) \|\vec{r}\,'(u)\|\, du \tag{22.14}$$

where the unit normal vectors $\hat{n}(u)$ are chosen to all lie on the same side of the curve.

Note that if we choose the normal vectors as described above, the factor of $||\vec{r}\,'(u)||$ is included in $\vec{n}(u)$, since the components are $\pm x(u)$ and $\pm y(u)$. Thus, we could write the definition as

$$\int_C \vec{F} \cdot \hat{n}\, ds = \int_a^b \vec{F}(\vec{r}(u)) \cdot \vec{n}(u)\, du. \tag{22.15}$$

In many cases of interest, the curve C is closed, meaning that the curve begins and ends at the same point. In that case, the conventional choice is to pick the unit normal vectors that point outward. This is illustrated in the following example.

Example 22.11. Compute the normal line integral of $\vec{F}(x, y) = \langle x, y \rangle$ on the unit circle centered at the origin with outward pointing unit normal vectors.

Solution. We use the parametrization $\vec{r}(u) = \langle \cos u, \sin u \rangle$ with u in $[0, 2\pi]$. The derivative is $\vec{r}\,'(u) = \langle -\sin u, \cos u \rangle$. To get the outward pointing normal vectors, we choose $\vec{n}(u) = \langle \cos u, \sin u \rangle$. Since $||\vec{r}\,'(u)|| = 1$, we have $\hat{n}(u) = \langle \cos u, \sin u \rangle$. Along the curve, we have outputs

$$\vec{F}(\vec{r}(u)) = \vec{F}(\cos u, \sin u) = \langle \cos u, \sin u \rangle.$$

The relevant dot product is

$$\vec{F}(\vec{r}(u)) \cdot \hat{n}(u) = \langle \cos u, \sin u \rangle \cdot \langle \cos u, \sin u \rangle = \cos^2 u + \sin^2 u = 1.$$

From the definition, we get

$$\int_C \vec{F} \cdot \hat{n}\, ds = \int_0^{2\pi} (1)(1)\, du = 2\pi.$$

The value of this normal line integral is thus 2π. △

Example 22.12. Compute the normal line integral of $\vec{F}(x, y) = \langle xy, x + y \rangle$ on the curve parametrized by $\vec{r}(u) = \langle u^3, u^2 \rangle$ for u in $[0, 2]$ with normal vectors pointing to the left.

Solution. We are given a parametrization of the curve, so we need only differentiate to get

$$\vec{r}\,'(u) = \langle 3u^2, 2u \rangle.$$

To get normal vectors pointing to the left, we use the choice $\vec{n}(u) = \langle -2u, 3u^2 \rangle$. Outputs of $\vec{F}$ along the curve are given by

$$\vec{F}(\vec{r}(u)) = \vec{F}(u^3, u^2) = \langle u^5, u^3 + u^2 \rangle.$$

The relevant dot product is

$$\vec{F}(\vec{r}(u)) \cdot \vec{n}(u) = \langle u^5, u^3 + u^2 \rangle \cdot \langle -2u, 3u^2 \rangle = -2u^6 + 3u^5 + 3u^4.$$

From the definition in Equation (22.15), we have

$$\int_C \vec{F} \cdot \hat{n}\, ds = \int_0^2 (-2u^6 + 3u^5 + 3u^4)\, du = \frac{512}{35}.$$

△

22.5 Problems

22.1 Curves

For Problems 1–10, parametrize the given curve.

1. the closed curve consisting of the line segment from $(-2, 0)$ to $(2, 0)$ and the upper half of the circle of radius 2 centered at the origin
2. the graph of the tangent function branch that goes through the origin
3. the branch of the hyperbola given by $x^2 - y^2 = 1$ that lies in the half-plane with $x > 0$
4. the rectangle with corners at $(-2, 3)$, $(4, 3)$, $(4, 5)$, and $(-2, 5)$
5. the curve that goes from $(-3, 1)$ to $(8, -2)$ first along a line segment parallel to the x-axis and then along a line segment parallel to the y-axis

6. the spiral given by the polar relation $r(\theta) = \theta$
7. the curve in $\mathbb{R}^3$ that goes from $(1, 0, 0)$ to $(0, 0, 1)$ along the arc of a circle centered at the origin
8. the semicircle of constant longitude that passes through the point $(2, 2, 0)$ on the sphere of radius 4 centered at the origin
9. the helix that starts at $(3, 0, 0)$, wraps around the z-axis five times, and ends at $(3, 0, 10)$
10. a "conical helix" that starts at the origin and wraps four times around the z-axis on a cone that has vertex at the origin, the positive z-axis as central axis, and a radius of 1 at height 1

22.2 Line Integrals

For Problems 1–6, compute the line integral $\int_C \vec{F} \cdot d\vec{s}$ of the given vector field $\vec{F} : \mathbb{R}^2 \to \mathbb{R}^2$ for the given curve C.

1. $\vec{F}(x, y) = \langle -y, x \rangle$ over the circle of radius 1 centered at the origin and oriented counterclockwise
2. $\vec{F}(x, y) = \langle y, -x \rangle$ over the circle of radius 1 centered at the origin and oriented clockwise
3. $\vec{F}(x, y) = \langle x, y \rangle$ over the circle of radius 1 centered at the origin and oriented counterclockwise
4. $\vec{F}(x, y) = \langle x^2, y \rangle$ over the upper half of the circle of radius 1 centered at the origin and oriented clockwise
5. $\vec{F}(x, y) = \langle y, -x \rangle$ over the square centered at the origin with sides of length 2 parallel to the coordinate axes and oriented counterclockwise
6. $\vec{F}(x, y) = \langle y, -x \rangle$ over the curve parametrized by $\vec{r}(u) = \langle u^2 + 1, u^3 - u \rangle$ for u in $[-1, 1]$

For Problems 7–10, compute the line integral $\int_C \vec{F} \cdot d\vec{s}$ of the given vector field $\vec{F} : \mathbb{R}^3 \to \mathbb{R}^3$ for the given curve C.

7. $\vec{F}(x, y, z) = \langle x, y, z \rangle$ over line segment from $(1, 0, 0)$ to $(1, 0, 2)$
8. $\vec{F}(x, y, z) = \langle x, y, z \rangle$ over the helix thats starts at $(1, 0, 0)$, wraps around the z-axis once, and ends at $(1, 0, 2\pi)$
9. $\vec{F}(x, y, z) = \langle y, z, x \rangle$ over the helix thats starts at $(1, 0, 0)$, wraps around the z-axis once, and ends at $(1, 0, 2\pi)$
10. $\vec{F}(x, y, z) = \langle yz, xz, xy \rangle$ over the line segment from $(0, 0, 0)$ to $(1, 1, 1)$
11. (a) The circle of radius 1 centered at the origin with counterclockwise orientation can be parametrized by the function $\vec{r}(t) = \langle \cos t, \sin t \rangle$ for t in $[0, 2\pi]$. Use this parametrization to compute $\int_C \vec{F} \cdot d\vec{s}$ for $\vec{F}(x, y) = \langle x + y, y^2 \rangle$.
    (b) The same circle also be parametrized by the function $\vec{p}(u) = \langle \cos(2\pi u), \sin(2\pi u) \rangle$ for u in $[0, 1]$. Use this parametrization to compute $\int_C \vec{F} \cdot d\vec{s}$ for $\vec{F}(x, y) = \langle x + y, y^2 \rangle$.
    (c) Compare the results of (a) and (b).
12. Consider two vector output functions $\vec{r} : [a, b] \to \mathbb{R}^n$ and $\vec{p} : [c, d] \to \mathbb{R}^n$ that parametrize the same curve with the same orientation. Assume $\vec{r}(t)$ is related to $\vec{p}(u)$ by a differentiable function $f : [c, d] \to [a, b]$ such that $\vec{r}(f(u)) = \vec{p}(u)$ for all u in $[c, d]$. Also assume that $f(c) = a$ and $f(d) = b$. Show that the line integral $\int_C \vec{F} \cdot d\vec{s}$ computed using $\vec{r}$ is equal to the line integral computed using $\vec{p}$.

22.3 The Fundamental Theorem of Calculus for Line Integrals

For Problems 1–8, find a potential function or show that no potential function exists for the given vector field.

1. $\vec{F}(x,y) = \langle x+y, x-y\rangle$
2. $\vec{F}(x,y) = \langle 3x^2y, x^3+2y\rangle$
3. $\vec{F}(x,y) = \langle 3x^2y, 2x^3+2y\rangle$
4. $\vec{F}(x,y) = \langle \cos xy, \sin xy\rangle$
5. $\vec{F}(x,y) = \langle y+\cos x, x\rangle$
6. $\vec{F}(x,y) = \langle 2xy, x-y\rangle$
7. $\vec{F}(x,y) = \langle y\cos xy, x\cos xy\rangle$
8. $\vec{F}(x,y) = \langle 2xe^{x^2+y^2}, 2ye^{x^2+y^2}\rangle$
9. Consider the vector field $\vec{F}(x,y) = \langle 2xy^2+2x, 2x^2y+1\rangle$.
   (a) Compute the line integral $\int_{C_1}\vec{F}\cdot d\vec{s}$ where C_1 is the line segment from $(0,0)$ to $(1,1)$.
   (b) Compute the line integral $\int_{C_2}\vec{F}\cdot d\vec{s}$ where C_2 is the segment of the curve $y=x^2$ from $(0,0)$ to $(1,1)$.
   (c) Compute the line integral $\int_{C}\vec{F}\cdot d\vec{s}$ where C is any curve from $(0,0)$ to $(1,1)$ by using the fundamental theorem for line integrals.
   (d) Compare the results of (a)–(c).
10. Consider the vector field $\vec{F}(x,y,z) = \langle 2xz+y, x, x^2\rangle$.
    (a) Compute the line integral $\int_{C_1}\vec{F}\cdot d\vec{s}$ where C_1 is the line segment from $(1,0,0)$ to $(1,0,2\pi)$.
    (b) Compute the line integral $\int_{C_2}\vec{F}\cdot d\vec{s}$ where C_2 is the helix that starts at $(1,0,0)$, wraps once around the z-axis, and ends at $(1,0,2\pi)$.
    (c) Show that $V(x,y,z) = x^2z+xy$ is a potential function for $\vec{F}$. Compute the line integral $\int_{C}\vec{F}\cdot d\vec{s}$ where C is any curve from $(1,0,0)$ to $(1,0,2\pi)$ by using the fundamental theorem for line integrals.
    (d) Compare the results of (a)–(c).

For Problems 11–14, use the fundamental theorem for line integrals to compute the line integral $\int_C \vec{F}\cdot d\vec{s}$ for the given vector field and any curve C that starts at $\vec{r}_1$ and ends at $\vec{r}_2$ as given.

11. $\vec{F}(x,y) = \langle y, x\rangle$, $\vec{r}_1 = \langle 1,2\rangle$, $\vec{r}_2 = \langle 4,6\rangle$
12. $\vec{F}(x,y) = \langle y^2, 2xy+1\rangle$, $\vec{r}_1 = \langle 0,1\rangle$, $\vec{r}_2 = \langle -1,5\rangle$
13. $\vec{F}(x,y) = \langle \sin(xy)+xy\cos(xy), x^2\cos(xy)\rangle$, $\vec{r}_1 = \langle 1,2\rangle$, $\vec{r}_2 = \langle 0,0\rangle$
14. $\vec{F}(x,y) = \langle -2x\exp(-x^2-y^2), -2y\exp(-x^2-y^2)\rangle$ $\vec{r}_1 = \langle 0,0\rangle$, $\vec{r}_2 = \langle 2,-3\rangle$

22.4 Normal Line Integrals

For Problems 1–8, compute the normal line integral $\int_C \vec{F}\cdot\hat{n}\,ds$ of the given vector field $\vec{F}:\mathbb{R}^2\to\mathbb{R}^2$ for the given curve C.

1. $\vec{F}(x,y) = \langle y, -x\rangle$ for the circle of radius 1 centered at the origin with outward pointing normals
2. $\vec{F}(x,y) = \langle y, -x\rangle$ for the circle of radius 1 centered at the origin with inward pointing normals

3. $\vec{F}(x,y) = \langle y, -x\rangle$ for the line segment from $(0,0)$ to $(1,1)$ with normals pointing to the right
4. $\vec{F}(x,y) = \langle y, -x\rangle$ for the segment of the curve $y = x^2$ from $(0,0)$ to $(1,1)$ with normals pointing to the right
5. $\vec{F}(x,y) = \langle x, y\rangle$ for the segment of the curve $y = x^2$ from $(0,0)$ to $(1,1)$ with normals pointing to the right
6. $\vec{F}(x,y) = \langle x^2, y\rangle$ for the circle of radius 1 centered at the origin with outward pointing normals
7. $\vec{F}(x,y) = \langle x^2, y\rangle$ for the circle of radius 1 centered at the origin with inward pointing normals
8. $\vec{F}(x,y) = \langle xy, y\rangle$ for the ellipse $x^2/4 + y^2/9 = 1$ with outward pointing normals
9. Consider a vector field $\vec{F} : \mathbb{R}^2 \to \mathbb{R}^2$ and a curve C in $\mathbb{R}^2$. There are two options for the choice of unit normals $\hat{n}$: pointing to the left as the curve is traced out from start to end or pointing to the right as the curve is traced out from start to end. Compare the normal line integral defined using one choice to the normal line integral defined using the other choice.
10. Consider attempting to apply the definition of normal line integral to a vector field $\vec{F} : \mathbb{R}^3 \to \mathbb{R}^3$ and a curve C in $\mathbb{R}^3$. What aspect of the definition fails to be sensible?

Chapter 23

Gauss's Law and Surface Integrals

In this chapter, we study *Gauss's law*. We have already studied Coulomb's law, which can be used to relate electric fields to the charges that create those fields. Gauss's law gives us another way of looking at the relationship. By studying Gauss's law, we gain still more insight into electric fields and acquire another tool to compute electric fields for some charge distributions, particularly those with a high degree of symmetry.

Gauss's law involves the use of a *surface integral*, specifically the surface integral of a vector field over a surface. For the first section, we use an informal construction of a surface integral. In the remainder of the chapter, we study the mathematics of surface integrals in more depth. As part of this study, we develop the idea of *parametrizing a surface*.

23.1 Gauss's Law

Gauss's law is named for the German mathematician and physicist Karl Friedrich Gauss (1777–1855), who was a pioneer in vector calculus. In order to understand Gauss's law, we must first understand the concept of the *flux* of a vector field. As usual, we start with simple cases and then begin to generalize. Gauss's law can be understood as a recasting of Coulomb's law in a way that uses the concept of flux of the electric field. Finally, we consider applications of Gauss's law and implications for the behavior of conductors.

23.1.1 Flux

We begin by considering a uniform vector field, for example, a flowing stream of water in which the velocity $\vec{v}$ of the water is the same at each point in the stream. For convenience, take the vector $\vec{v}$ to be horizontal. Such a uniform vector field (in this case, a velocity field) is pictured in Figure 23.1(a). Now imagine a rectangular loop of cross-sectional area $A = ab$ placed in the stream, so that it is perpendicular to the vector field. Think of the loop as the boundary of a surface though which water can flow. For this special case (uniform velocity field perpendicular to a surface), we define the **flux** of the field through the surface to be simply the product of v (the magnitude of vector $\vec{v}$) and A. In equation form,

$$\Phi_v = vA \tag{23.1}$$

where we have used the Greek letter Φ (uppercase phi) for flux, and the subscript v indicates the flux of a velocity field.

Before moving on, let's consider what this particular flux means physically. Because it is defined to be the product of the velocity magnitude and area, Φ_v is in some sense a measure of the rate at which water

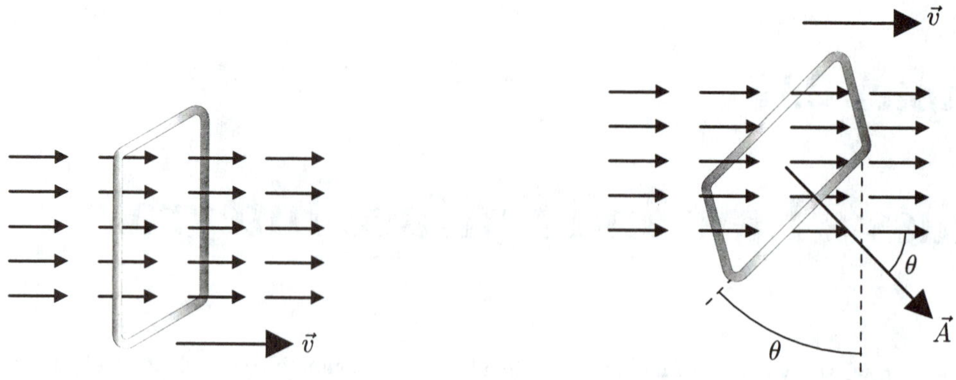

(a) Loop oriented perpendicular to the flow direction

(b) Loop in a general orientation

Figure 23.1. Water flowing through a loop.

flows through the surface bounded by the rectangular loop. We know this, because the flux increases linearly with both v and A, and increasing either factor must increase the amount of water passing through the loop. We can be more exact. Notice that the units of Φ_v are (m/s)(m^2) = m^3/s. This is a volume per unit time. In fact, a careful dimensional analysis reveals just that—that is, for the geometry we are considering, $\Phi_v = vA$ gives *exactly* the volume of water per unit time flowing through the loop of area A. The details are as follows. During a time interval Δt, the volume of water that flows through the loop of area A just fills a rectangluar solid of cross-sectional area A and length $v\Delta t$. Therefore, the solid has volume $Av\Delta t$. The volume per unit time is this volume divided by Δt, or Av.

To make the situation slightly less special, we now consider tipping the rectangular loop as shown in Figure 23.1(b), so that the plane of the loop makes an angle θ with respect to the vertical. How does this affect the water's volume flow rate through the loop, which we have associated with flux? Notice that the projection of the side of length b onto an axis perpendicular to the vector field $\vec{v}$ is just $b\cos\theta$. Thus, the effective surface area is reduced from $A = ab$ to $ab\cos\theta$. The net effect has been to reduce the flux by a factor of $\cos\theta$, so the flux is now

$$\Phi_v = vA\cos\theta. \tag{23.2}$$

Equation (23.2) might remind you of our geometric expression for a dot product, because it is the product of two scalars multiplied by the cosine of an angle. Remember that v is just the magnitude of a vector $\vec{v}$, which is consistent with thinking about the dot product. However, until this point we have not thought of surface area as a vector. We can do so in a way that will allow us to think of flux as a dot product, if we define a *surface area vector $\vec{A}$ for a flat surface* (such as the one we have been considering) *to be a vector with magnitude A* (the ordinary surface area for a flat surface) *and direction perpendicular to that surface.* The vector $\vec{A}$ for our rectangular surface in the stream is shown in Figure 23.1(b). With this definition of the area vector, we can use the geometric expression for the dot product to rewrite Equation (23.2) as

$$\Phi_v = \vec{v}\cdot\vec{A}. \tag{23.3}$$

Equation (23.3) can be considered a general expression for the flux of a uniform velocity field $\vec{v}$ through a flat surface described by area vector $\vec{A}$. There is in fact no need for that surface to be a rectangle in order for this rule to work; the surface need only be flat, so that the direction of $\vec{A}$ can be defined as normal to that surface.

You may have noticed some ambiguity in the way the area vector $\vec{A}$ is defined, because we have only said that it is normal to the surface. In the example we have been considering, the vector $\vec{A}$ shown in Figure 23.1(b) is indeed normal to the surface outlined by the rectangular loop. However, we could just as well have picked a vector $\vec{A}$ that is normal to the surface but in the opposite direction, as pictured in Figure 23.2. Notice that this choice of $\vec{A}$ changes the angle between $\vec{v}$ and $\vec{A}$ from θ to $\pi - \theta$. Now if we use the dot product in Equation (23.3) to compute the flux of this velocity field, we get the *negative* of the previous value, because by identity $\cos(\pi - \theta) = -\cos(\theta)$. To summarize, the flux computed in this way is one of two values that differ by a factor of -1.

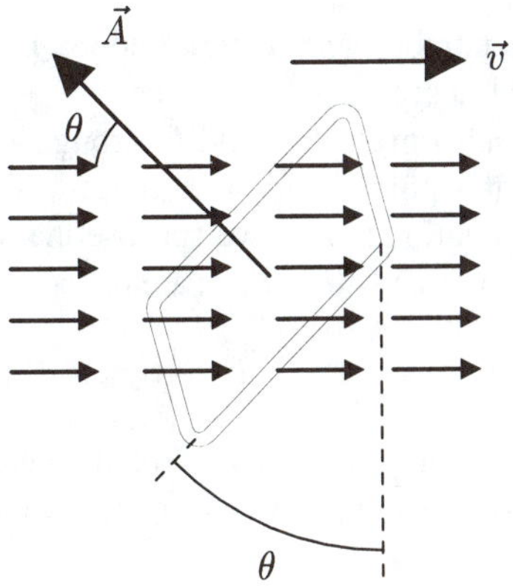

Figure 23.2. Alternative choice for the area vector direction.

How can we deal with this ambiguity? The answer to this question has two parts. First, when thinking about a simple flat surface such as the one we have been studying, we can think of the flux through the surface as the absolute value of the quantity Φ_v given in Equation (23.3). The second part of this answer is that when studying Gauss's law, we will use surfaces that are *closed*, meaning that the surface itself defines the boundary between an inside and outside. Then we will be able to give an unambiguous definition of the direction of the area vector $\vec{A}$, namely that it is normal to the surface and points toward the outside. Then our computation of flux produces a definite value that can be either positive or negative. We defer further discussion of this issue until we are ready to do such computations.

The definition of flux in Equation (23.3), which uses the dot product, is a useful definition of flux for *any* uniform vector field. The nature of the dot product is such that this definition always tells us, in a sense, the rate at which the field $\vec{v}$ passes through the surface with surface area $\vec{A}$. Consistent with this idea is the fact that for vectors $\vec{v}$ and $\vec{A}$ of given magnitude, the flux is maximized when $\vec{v}$ and $\vec{A}$ are in the same direction [so that $\cos\theta = 1$, as in Figure 23.3(a)], and the flux is zero when $\vec{v}$ is perpendicular to $\vec{A}$ [so that $\cos\theta = 0$, as in Figure 23.3(b)]. In this latter case, the field lines are parallel to the surface, and with no field lines passing through the surface, the flux is zero as it should be.

Because flux is defined in terms of a dot product, it is a scalar quantity. This may seem strange at first, because flux measures the amount of a *vector field* passing through a surface. However, for a given vector

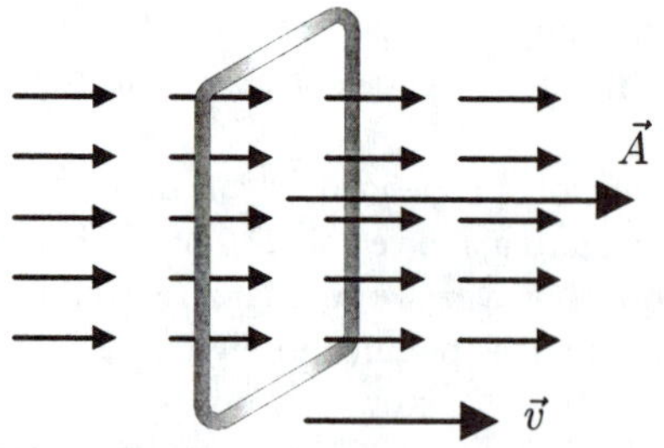

(a) Orientation for maximum flux

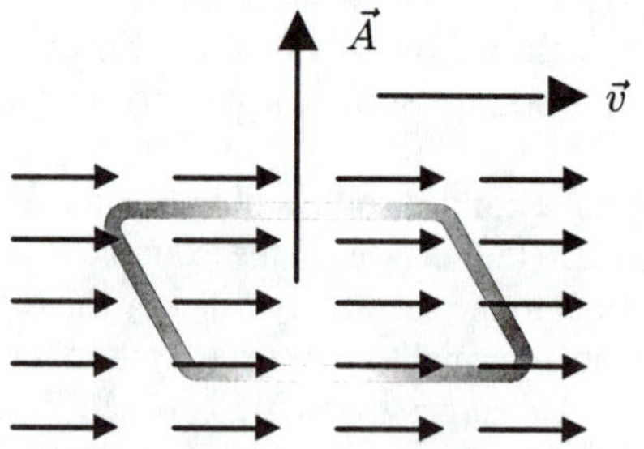

(b) Orientation for minimum flux

Figure 23.3. Maximum and minimum flux of a field through a flat loop.

field and surface, the flux is a single number, and as you know a physical quantity specified by a single number (rather than a set of numbers) is a scalar.

Now we can consider the flux of a uniform electric field $\vec{E}$ through a flat surface. The picture (Figure 23.4) is the same we have been studying to this point, but now with vector field $\vec{E}$ instead of vector field $\vec{v}$. From the generalized definition of flux in Equation (23.3), we surmise that the correct definition of the flux of a uniform electric field $\vec{E}$ through a flat surface with surface area $\vec{A}$ is

$$\Phi_E = \vec{E} \cdot \vec{A} \tag{23.4}$$

where now the subscript E on the symbol Φ_E denotes the flux of an electric field. The units for electric flux are (N/C)(m^2) = N·m^2/C. In the following example, we compute the electric flux associated with an electric field found in nature.

Example 23.1. Near Earth's surface, a uniform electric field of magnitude 100 N/C points straight downward. Compute the flux of this electric field through (a) a square surface 1 m on a side, and (b) a square surface 1 hectare on a side, each oriented in a horizontal plane.

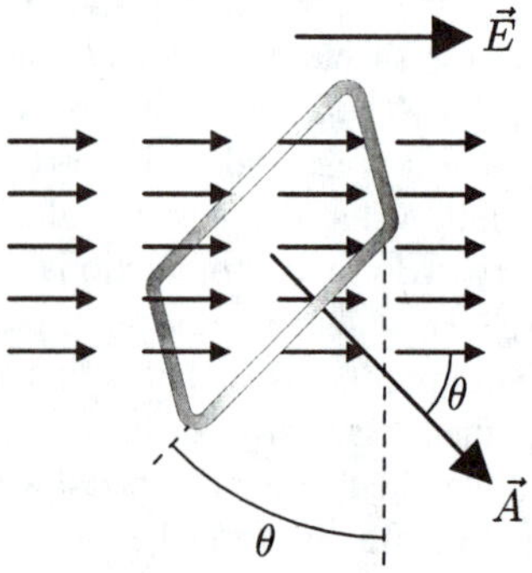

Figure 23.4. Flux of an electric field through a flat surface.

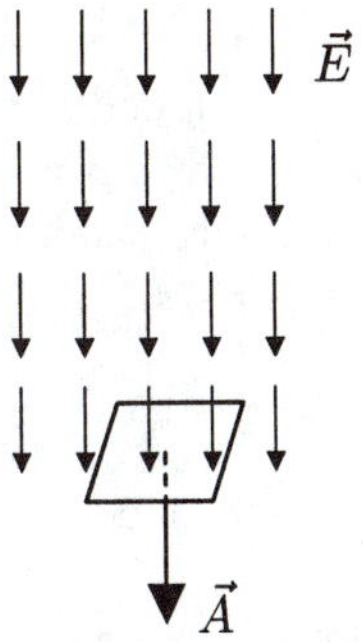

Figure 23.5. Electric flux near the earth's surface.

Solution. (a) The situation is shown in Figure 23.5. The area vector $\vec{A}$ must be oriented normal to the horizontal surface. We choose to make $\vec{A}$ point straight down (rather than straight up) so that $\vec{E}$ and $\vec{A}$ will be in the same direction, and the dot product in Equation (23.4) gives a positive number. With $\vec{E}$ and $\vec{A}$ in the same direction, the angle between them is zero, and we find

$$\Phi_E = \vec{E} \cdot \vec{A} = EA\cos(0) = (100 \text{ N/C})(1 \text{ m}^2)(1) = 100 \text{ N·m}^2/\text{C}.$$

(b) The computation is the same as in (a), except for the magnitude of the surface area. One hectare equals 100 m × 100 m = 10^4 m^2, so we find

$$\Phi_E = (100 \text{ N/C})(10^4 \text{ m}^2)(1) = 1.00 \times 10^6 \text{ N·m}^2/\text{C}.$$ △

23.1.2 A More General Approach to Flux

So far we have defined electric flux only for a uniform electric field and a flat surface [Equation (23.4)]. For the concept of flux to have much practical benefit, we must be able to handle cases in which the electric field is not necessarily uniform and the surface is not necessarily flat. The generic situation is shown in Figure 23.6(a).

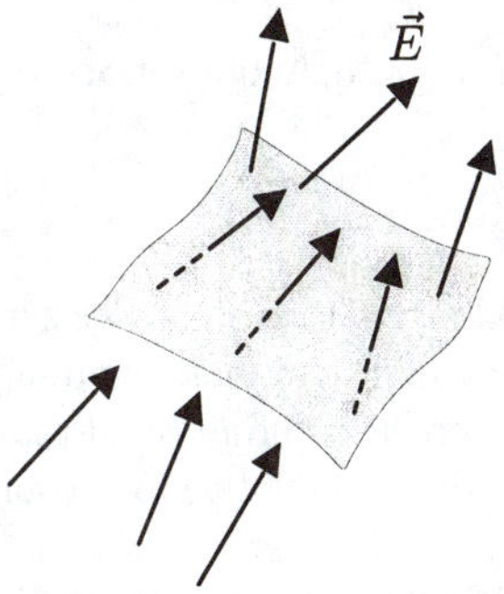

(a) Schematic of the field and surface

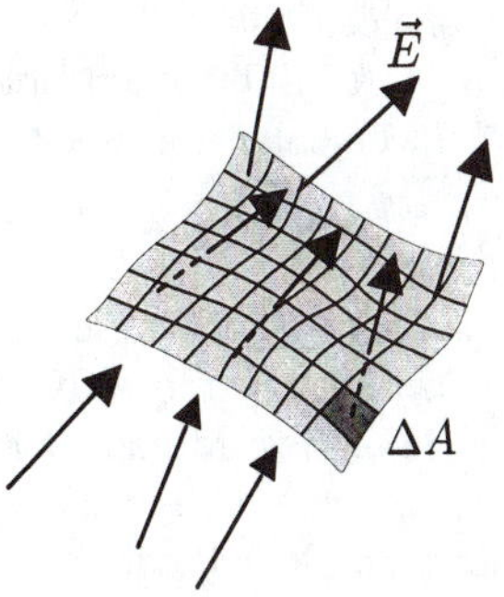

(b) Surface broken into patches of size ΔA

Figure 23.6. Electric flux for a generic field and surface.

Our approach will be a familiar one from calculus. First, we break the surface S into small "patches" [with a typical patch of surface area ΔA shown in Figure 23.6(b)]. If the patches are small enough, the electric field is approximately uniform there, and the patch is approximately flat. Thus, our previous rule for the electric flux due to a uniform field and a flat surface can be used to approximate the electric flux $\Delta\Phi_E$ through one patch. By Equation (23.4),

$$\Delta\Phi_E \approx \vec{E} \cdot \hat{n}\,\Delta A \tag{23.5}$$

where $\vec{E}$ is the approximate electric field in that patch, and $\hat{n}$ is a unit vector that is normal to the surface at some point in the patch ΔA.

The usual construction in calculus then involves summing the $\Delta\Phi_E$ given by Equation (23.5) over every patch on the surface, and then taking a limit as the number of patches approaches infinity. The result is a type of definite integral. This is called a **surface integral** because the domain of integration is a surface. We work here with an informal construction of a surface integral. In Sections 23.2 and 23.3, we develop a precise definition, but first let's outline the steps in our informal construction.

Let the subscript i stand for a generic patch of the surface over which we wish to compute the electric flux. If there are a total of n patches, then the net flux through the surface is approximated by

$$\Phi_E \approx \sum_{i=1}^{n} \vec{E}_i \cdot \hat{n}\,\Delta A_i\,.$$

The result becomes exact in the limit as the number of patches becomes infinite:

$$\Phi_E = \lim_{n\to\infty} \sum_{i=1}^{n} \vec{E}_i \cdot \hat{n}\,\Delta A_i\,.$$

This limit of the sum over the entire surface is our informal definition of surface integral, which can be written symbolically as

$$\Phi_E = \iint_S \vec{E} \cdot \hat{n}\,dA = \iint_S \vec{E} \cdot d\vec{A}. \tag{23.6}$$

From this construction, it is not clear how we actually evaluate a surface integral as written in Equation (23.6). We will defer a general treatment of this kind of integral to Section 23.2. However, it is possible to evaluate this integral in a less formal way in some cases that will turn out to be quite useful. We illustrate how this can be done with a couple of examples before moving on to present Gauss's law using Equation (23.6) for electric flux.

While working through these examples, it is important to keep in mind how the surface integral above was constructed. Specifically, note that the quantity

$$\vec{E}_i \cdot \hat{n}$$

in the sum is a scalar quantity, because it is a dot product. As you consider how to evaluate the surface integral in Equation (23.6), you should therefore think of doing the dot product first, and then summing, just as the symbols indicate. In our less formal approach, we generally use the geometric definition of the dot product. For that reason, it is important to focus first on the direction of the electric field relative to the surface, so that the angle between each $\vec{E}_i$ and each $\hat{n}$ is known.

In the examples that follow, we consider **closed surfaces**, such that the surface defines a boundary between an inside and outside region of space. By convention, we choose a normal vector $\hat{n}$ for a patch of a closed surface to point *outward* from the surface. This removes the ambiguity we encountered previously with surfaces that are not closed.

Example 23.2. Find the net electric flux through the cylinder shown in Figure 23.7(a), with a uniform electric field $\vec{E}$ parallel to the axis of the cylinder.

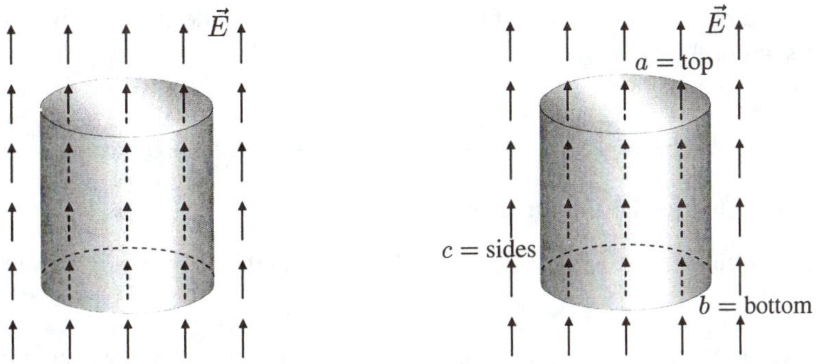

(a) Electric field and cylindrical surface

(b) Considering the surface as three pieces: top, bottom, and sides

Figure 23.7. Computing the flux of a uniform electric field through a cylinder.

Solution. The best approach to this kind of geometry is to think of the cylinder as made up of three separate pieces: the top, bottom, and one continuous curved side. These pieces are labeled a, b, and c, respectively in Figure 23.7(b). Flux is a scalar, and the net flux through the cylinder is then the sum of the fluxes through the three sides. Stated symbolically,

$$\Phi_E = \Phi_{Ea} + \Phi_{Eb} + \Phi_{Ec}$$

where Φ_{Ea} is the electric flux through piece a, and so on. We consider the three pieces one at a time and then put the results together.

Consider first the top of the cylinder (part a). This is a simple flat surface, with area vector $\vec{A}_a$ pointing outward, normal to the surface. For this surface, we can use Equation (23.4) for a uniform electric field and a flat surface. The angle between $\vec{E}$ and $\vec{A}_a$ is zero. Letting A be the magnitude of vector $\vec{A}_a$, we have

$$\Phi_{Ea} = \vec{E} \cdot \vec{A}_a = EA\cos(0) = EA$$

where as usual E is the magnitude of vector $\vec{E}$.

On the bottom of the cylinder, the situation is the same as the top, except that the angle between $\vec{E}$ and $\vec{A}_b$ is π (180°). Thus,

$$\Phi_{Eb} = \vec{E} \cdot \vec{A}_b = EA\cos(\pi) = -EA.$$

What happens along the sides of the cylinder (part c)? Since these sides are curved, we might consider breaking this surface into small pieces ΔA_i and constructing a surface integral as described in the text. Notice, however, that for this electric field, no matter how the ΔA_i are chosen, the corresponding normal vectors $\hat{n}$ will all be perpendicular to the electric field. Therefore, for all such pieces,

$$\vec{E}_i \cdot \hat{n} = 0$$

and the flux through the sides is zero. Stated symbolically,

$$\Phi_{Ec} = 0.$$

We can obtain the same answer by thinking about the vector field geometrically. Notice that the electric field lines depicted in Figures 23.7(a) and 23.7(b) lie parallel to the sides and never "pierce" through them. Because it is this piercing we associate with flux, we can rightly conclude that there is zero flux of this electric field through the sides of the cylinder.

Combining our results,

$$\Phi_E = \Phi_{Ea} + \Phi_{Eb} + \Phi_{Ec} = EA + (-EA) + 0 = 0.$$

The net flux of this electric field through the cylinder is zero. △

Example 23.3. A positive point charge q lies at the origin. Find the electric flux through a sphere of radius R centered at the origin.

Solution. The situation is shown in Figure 23.8(a). We know from our previous study of electric fields that the electric field points radially outward from a positive point charge, and at a distance R the electric field's magnitude is

$$E = \frac{kq}{R^2}.$$

With the electric field directed radially outward, it is perpendicular to the surface of the sphere at all points on the sphere. Thus, for the generic patch of surface area ΔA_i shown in Figure 23.8(a), the vector $\vec{E}$ and the normal unit normal vector $\hat{n}$ point in the same direction. Then the dot product of the electric field and area vectors for any ΔA_i is

$$\vec{E}_i \cdot \hat{n}\, \Delta A_i = (E)(\Delta A_i)\cos(0) = E\Delta A_i.$$

Now to find the electric flux, we use the surface integral in Equation (23.6) to find

$$\Phi_E = \iint \vec{E} \cdot \hat{n}\, dA.$$

Because we have already seen that the dot product in the integrand reduces to $E\Delta A_i$, the surface integral can be expressed as

$$\Phi_E = \iint E\, dA$$

where the symbol dA, as we have seen before in surface integrals, tells us to integrate over a surface. In this problem, E is a constant over the entire surface of integration (the sphere), and thus we can factor E from the integral, leaving

$$\Phi_E = E \iint dA.$$

The remaining integral simply gives the total surface area of the sphere, which we know is $4\pi R^2$ for a sphere of radius R. Combining this with the magnitude of the electric field, we find

$$\Phi_E = (E)(4\pi R^2) = \frac{q}{4\pi\epsilon_0 R^2} \cdot 4\pi R^2$$

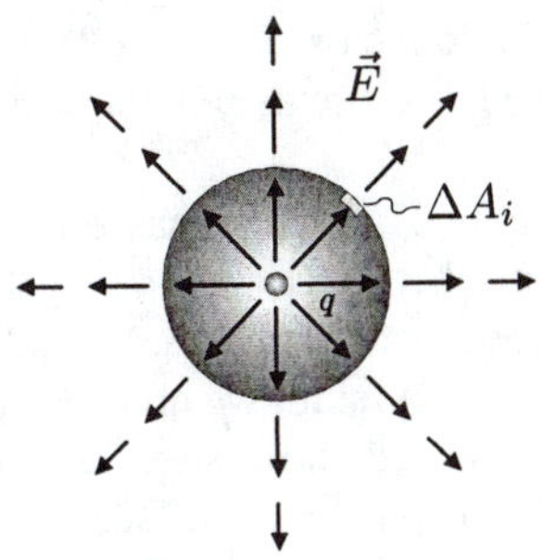

(a) Flux through a sphere of the electric field due to a positive point charge

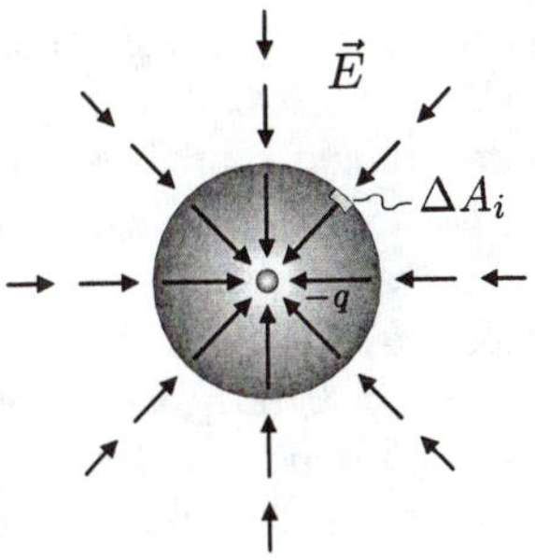

(b) Flux through a sphere of the electric field due to a negative point charge

Figure 23.8. Schematics for Examples 23.3 and 23.4.

or

$$\Phi_E = \frac{q}{\epsilon_0}.$$

In this surprisingly simple expression, it makes sense that the flux of the electric field through the sphere is positive and proportional to the magnitude of the charge at the origin. △

Example 23.4. Repeat the preceding example for a negative charge $-q$ at the origin.

Solution. Everything is the same as the preceding example, except that the electric field, though having the same magnitude, is directed radially inward [Figure 23.8(b)]. This makes the angle between the electric field and area vectors equal to π (180°) everywhere on the surface of the sphere. Where before we had $\cos(0) = 1$, we now have $\cos(\pi) = -1$ in the dot product

$$\vec{E}_i \cdot \hat{n}\Delta A_i.$$

That is,

$$\vec{E}_i \cdot \hat{n}\Delta A_i = (E)\,(\Delta A_i)\cos(\pi) = -E\Delta A_i.$$

This negative sign carries though the rest of the computation, making the net flux through the sphere

$$\Phi_E = \frac{-q}{\epsilon_0}.$$

Now the flux of the electric field through the sphere is negative, but still proportional to the magnitude of the charge at the origin. △

Comparing the preceding three examples illustrates an important point about the flux of an electric field. In Example 23.3, the electric field lines emanating from the positive point charge pierce outward through the sphere, and this corresponds to a net positive flux through the surface. (This is really the effect of our choice to define the area vector as pointing outward from a closed surface.) Similarly, in Example 23.4, the electric field lines point radially inward toward the negative point charge. This situation produces a net negative flux through the surface. The general rule illustrated by these examples is the following: A closed surface that surrounds a net positive charge has a positive flux through it, while a closed surface that surrounds a net negative charge has a negative flux through it. We have not proved this rule in general, and we shall not do so, but it is am important part of Gauss's law.

As a further illustration, consider again the cylinder in Example 23.2, which has zero net flux through its surface. In that example, we did not discuss the location of any charge. However, you should remember that the uniform electric field used in that example can be produced by (for example) an infinite plane with uniform charge density. Clearly, such a plane does not lie within the cylinder we consider in that example, so the net charge within the cylinder is zero. Our general rule presented above states that a closed surface that surrounds a net positive charge has a positive flux through it, while a closed surface that surrounds a net negative charge has a negative flux through it. It is certainly consistent with this rule that a surface enclosing zero net charge has zero net electric flux through it.

23.1.3 Electric Flux and Gauss's Law

Example 23.3 serves as the basis for our presentation of Gauss's law. We state without proof that the result of that Example 23.3 is perfectly true in general for the flux of the electric field through *any* closed surface enclosing a net charge q_{enc}. At this point, we introduce a new notation, a surface integral sign with a circle through it, to be used whenever the surface integral is over a closed surface. With this notation, we state Gauss's law symbolically:

$$\Phi_E = \oiint \vec{E} \cdot \hat{n}\, dA = \frac{q_{\text{enc}}}{\epsilon_0} . \tag{23.7}$$

In words, Gauss's law says that the flux of the electric field through a closed surface is equal to the net charge enclosed by that surface divided by the constant ϵ_0.

Certainly the results of Examples 23.2, 23.3, and 23.4 are consistent with Gauss's law. As noted above, we will not prove that it is true in general. However, one geometric argument supporting Gauss's law is as follows. Think again of the sphere enclosing the positive charge in Example 23.3. What would happen if the sphere were deformed, so that its shape were nonspherical, but so that it still enclosed the same point charge? The electric field lines, which emanate outward from the point charge, must still pass through that deformed surface somewhere. Our geometric concept of flux—how much of a field passes through a surface—tells us that the electric flux due to that point charge is the same, regardless of the shape of the enclosing surface. This lends support to the general validity of Gauss's law.

The closed surfaces we have considered in our examples to this point are often called **Gaussian surfaces**, because they are used with Gauss's law. This language reminds us that the closed surface used in Gauss's law is a geometric construct, whose sole purpose is to compute electric flux. There is not (necessarily) a physical object with that shape placed around a charge distribution.

As another illustration of Gauss's law, consider the electric dipole shown in Figure 23.9. In this figure, we have drawn four possible Gaussian surfaces, labeled S_1, S_2, S_3, and S_4. For this qualitative discussion, it is not necessary that these closed surfaces have any particular shape, because we know Gauss's law works for any closed surface. First, consider S_1, which encloses the positive charge in the dipole. Just like the positive point charge we studied in Example 23.3, all the electric field lines pass outward through the surface. We have seen that this is consistent with a positive flux. Gauss's law predicts a positive flux through S_1, because that surface encloses a net positive charge. Similarly, surface S_2 encloses a net negative charge and has a negative flux though it, because all the electric field lines pierce it pointing inward.

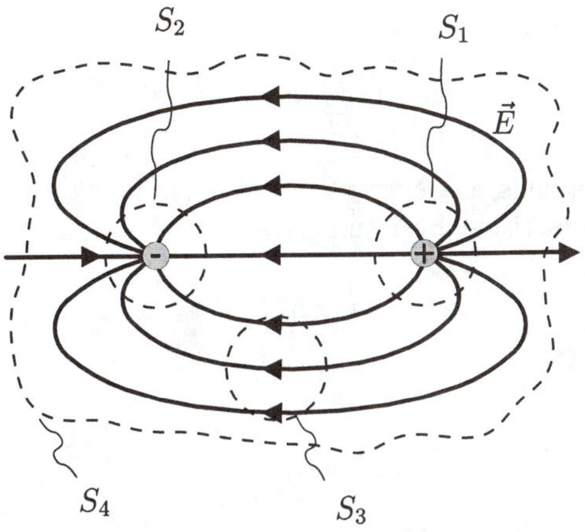

Figure 23.9. Several Gaussian surfaces for an electric dipole.

Surface S_3 in Figure 23.9 encloses no charge. Therefore, Gauss's law says that the flux through that surface is zero. Notice that in the picture just as many electric field lines leave the surface S_3 (positive flux) as enter it (negative flux). The picture makes it plausible that the net flux is zero. Now consider surface S_4. If you count carefully, you will see that just as many electric field lines leave that surface as enter it. This implies a net flux of zero through S_4 This is again consistent with Gauss's law, because the entire dipole (with net charge $+q - q = 0$) is enclosed by S_4.

23.1.4 Application of Gauss's Law

Thus far, it may not be clear why we should have gone through the trouble to develop Gauss's law. Gauss's law is based on Coulomb's law and can be considered a restatement of Coulomb's law, becaus it expresses fundamentally the same information about the electric fields of point charges, but in different language (the language of electric flux). In this section, we present some examples showing how Gauss's law can be used to compute electric fields and will see that in some cases it can be a much simpler way to do that computation.

As a first example, we show how to use Gauss's law to compute the electric field of a point charge. Even though you already know the answer to this problem, going through the exercise helps illustrate how to apply Gauss's law. *The key to applying Gauss's law is to choose a Gaussian surface (over which to integrate) that makes the dot product* $\vec{E} \cdot \hat{n}$ *in Equation* (23.7) *easy to evaluate.* This often means choosing a Gaussian surface that is either perpendicular or parallel to the electric field at each point on the surface. For example, since we know the electric field of the (positive) point charge is directed radially outward, a *spherical* Gaussian surface centered at the point charge is normal to the electric field at all points on the sphere. [Refer again to Figure 23.8(a).] Therefore, for the entire sphere,

$$\vec{E} \cdot \hat{n} = (E)(1)\cos(0) = E\,.$$

Therefore, Gauss's law gives us

$$\oiint \vec{E} \cdot \hat{n}\, dA = \oiint E\, dA = \frac{q_{\text{enc}}}{\epsilon_0} = \frac{q}{\epsilon_0}$$

where we use the fact that the enclosed charge is a single point charge q. By symmetry, the magnitude of the electric field (E) is the same at all points on the sphere, so we can factor the E out of the surface integral, leaving

$$E \oiint dA = \frac{q}{\epsilon_0}.$$

The remaining surface integral is, as we have seen before, just the total surface area of the sphere, which is $4\pi R^2$ for a sphere of radius R. Using this result gives us

$$E\left(4\pi R^2\right) = \frac{q}{\epsilon_o},$$

or after rearranging,

$$E = \frac{q}{4\pi\epsilon_0 R^2},$$

which is the familiar result for the magnitude of the electric field a distance R from a point charge q.

The procedure of picking the proper Gaussian surface so that the surface integral in Gauss's law can be evaluated easily, and then using the resulting equation to solve for E, is one that can be used in a number of cases. We will see how in the following examples.

Example 23.5. Use Gauss's law to find the electric field a perpendicular distance r from a line of charge with uniform, positive charge density λ (charge per unit length).

Solution. The situation is shown in Figure 23.10(a). We know from the symmetry of the infinite line charge that at the desired point the electric field points directly away from the line charge. Our task is to pick a Gaussian surface such that for this electric field the dot product $\vec{E} \cdot \hat{n}$ is easy to evaluate everywhere on that surface. The perfect choice for this turns out to be a cylinder (of radius r and arbitrary height H), as shown in Figure 23.10(b).

Why is this a good choice for a Gaussian surface? First, notice that along the top and bottom of the cylinder, the electric field lines are parallel to the surface and do not pierce the Gaussian surface. From previous examples, you know this means there is zero electric flux along those surfaces. Or, stated symbolically, since $\hat{n}$ is normal to the surface (straight up from the top, straight down from the bottom), $\vec{E}$ is perpendicular to $\hat{n}$ along those surfaces, and $\vec{E} \cdot \hat{n} = 0$ there.

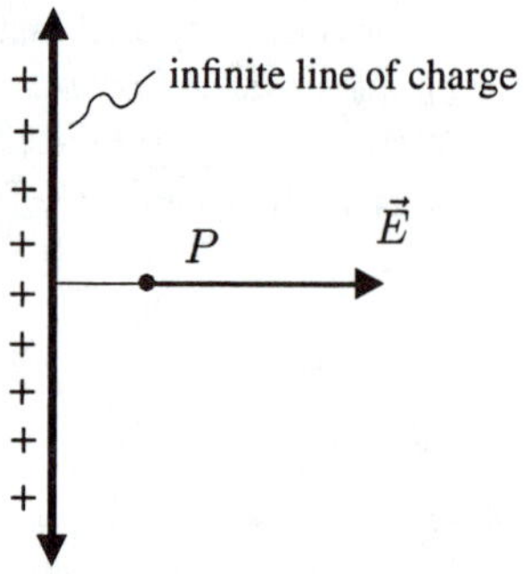

(a) Infinite line of charge

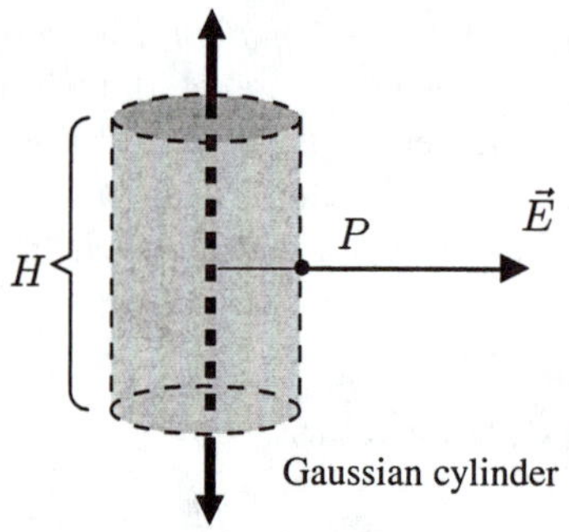

(b) Choosing a Gaussian surface

Figure 23.10. Schematic for Example 23.5.

This leaves only the sides of the cylinder to deal with. The electric field vector points straight away from the line charge everywhere in space, so it must be normal to the sides of the cylinder. Along the sides,

$$\vec{E} \cdot \hat{n} = (E)(1)\cos(0) = E\,.$$

Using this result in Gauss's law,

$$\oiint \vec{E} \cdot \hat{n}\, dA = \iint_{\text{sides}} E\, dA = \frac{q_{\text{enc}}}{\epsilon_0}$$

where only the integral along the sides of the cylinder remains, because, as we argued above, there is zero flux though the top and bottom. By symmetry, E is the same everywhere along the sides of the cylinder. This makes the integrand constant, so it can be factored out of the surface integral. On the right-hand side of the equation, note that by definition of λ (charge per unit length), we have $q_{\text{enc}} = \lambda H$. Combining these results,

$$E \iint_{\text{sides}} dA = \frac{\lambda H}{\epsilon_0}\,.$$

The remaining surface integral is just the surface area of the sides of the cylinder, which from geometry is $2\pi r H$. Hence,

$$E\,(2\pi r H) = \frac{\lambda H}{\epsilon_0}$$

and

$$E = \frac{\lambda}{2\pi\epsilon_0 r}\,.$$

This result is exactly the same as we found in Example 17.11. It is worth noting that using Gauss's law required much less computational effort. We also note that the height H of the Gaussian cylinder does not appear in the result. In fact, the value of H is irrelevant, provided it is small compared with the length of the line charge. △

Example 23.6. Use Gauss's law to find the electric field a perpendicular distance D from a plane of charge with uniform, positive charge density σ (charge per unit area).

Solution. Again, we can use our knowledge of electric fields and symmetry to reason that at the point in question, the electric field must point directly away from the plane. This is important, because it dictates the choice of Gaussian surface for the problem. For this problem, we choose another Gaussian cylinder, of height $2D$, oriented as shown in Figure 23.11.

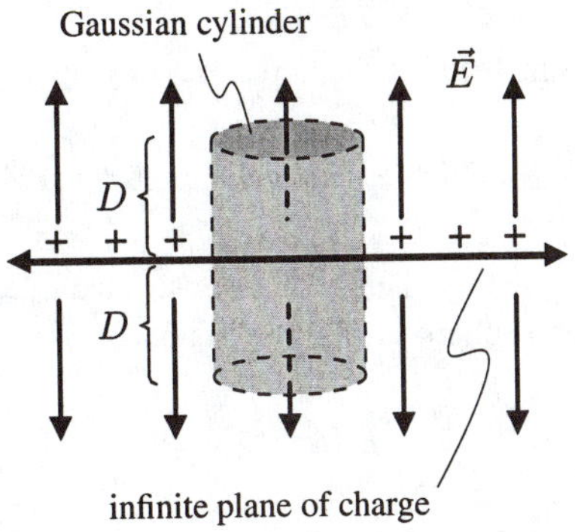

Figure 23.11. To compute the electric field of an infinite plane of charge, we choose a cylinder as the Gaussian surface.

Along the sides of the cylinder, the electric field is parallel to the sides, so there is zero electric flux on that part of the Gaussian surface (again we could say formally that $\vec{E}$ is perpendicular to $\hat{n}$ and $\vec{E}\cdot\hat{n} = 0$). Clearly, there is a net flux through the top and bottom of the cylinder. Along the top and bottom of the cylinder, the electric field is uniform and perpendicular to the flat surface, so the flux through (say) the top is just

$$\Phi_{\text{top}} = \vec{E} \cdot \vec{A} = EA$$

where A is the surface area of the top of the cylinder. By symmetry, the flux through the bottom is the same, so the net flux through the cylinder is $2EA$. Gauss's law then says

$$\oiint \vec{E} \cdot \hat{n}\, dA = 2EA = \frac{q_{\text{enc}}}{\epsilon_0}\,.$$

With a uniform charge density, we can see that $q_{\text{enc}} = \sigma A$, which we can use in the above equation to get

$$2EA = \frac{\sigma A}{\epsilon_0}.$$

Solving for E gives

$$E = \frac{\sigma}{2\epsilon_0}\,.$$

Again, we have found a familiar result with much less computational effort, demonstrating the power and elegance of Gauss's law. △

Example 23.7. Consider a spherical shell of radius R, which has a net positive charge Q spread uniformly over its surface. Use Gauss's law to find the electric field inside and outside the shell.

Solution. By symmetry, the electric field vector points in the radial direction from any point in space—that is, radially outward from the center of the sphere. This makes it sensible to choose as our Gaussian surface a sphere that is concentric with the charged sphere. We need to consider two separate cases: outside the charged sphere and inside the charged sphere.

Case 1: outside the charged sphere. Everywhere on the Gaussian surface [Figure 23.12(a)], the electric field is normal to the surface, and

$$\vec{E} \cdot \hat{n} = (E)(1)\cos(0) = E\,.$$

By symmetry, E is constant along the Gaussian sphere, so applying Gauss's law, we find

$$\oiint \vec{E} \cdot \hat{n}\, dA = \oiint E\, dA = E \oiint dA = \frac{Q}{\epsilon_0}$$

where we have used the fact that $q_{\text{enc}} = Q$. The remaining surface integral is equal to the surface area of the entire Gaussian sphere, or $4\pi r^2$ for a sphere of radius r. Thus,

$$E\left(4\pi r^2\right) = \frac{Q}{\epsilon_0}$$

or

$$E = \frac{Q}{4\pi\epsilon_0 r^2}$$

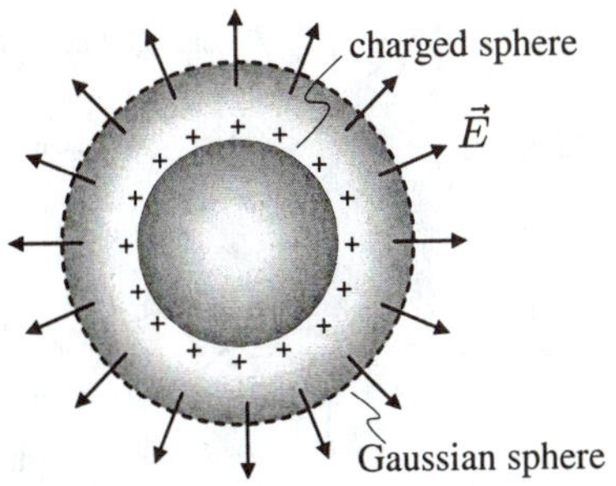

(a) Gaussian sphere outside a charged spherical shell

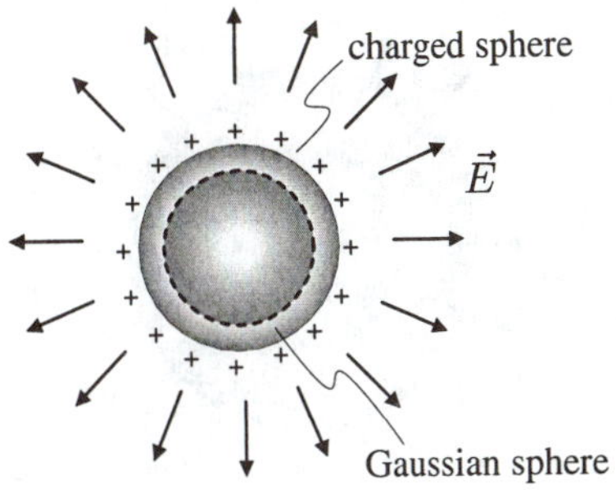

(b) Gaussian sphere inside a charged spherical shell

Figure 23.12. Gaussian surfaces for the two cases considered in Example 23.7.

which is the familiar result, because outside of a uniform sphere of charge, the electric field should act like a point charge Q located at the center of the sphere.

Case 2: inside the charged sphere. The left side of Gauss's law (including evaluation of the surface integral) is unchanged from case 1. However, the charge enclosed by the Gaussian sphere is now zero [Figure 23.12(b)]. Thus,

$$E\left(4\pi r^2\right) = \frac{0}{\epsilon_0}$$

and $E = 0$. The electric field is zero at all points inside the uniformly charged spherical shell. △

Example 23.8. Consider a solid sphere of radius R, which has a net positive charge Q spread uniformly throughout its volume. Use Gauss's law to find the electric field inside and outside the sphere.

Solution. As in the previous example, we use a sphere as our Gaussian surface [Figure 23.13(a)].

Case 1: outside the charged sphere. The situation is quite similar to the first case of the preceding example (outside the sphere). By symmetry, the electric field points radially outward from the center of the sphere, which makes evaluation of the surface integral straightforward. The entire charged sphere (and charge Q) is enclosed in a Gaussian sphere of radius r. Applying Gauss's law with these facts,

$$\oint\!\!\oint \vec{E}\cdot\hat{n}\,dA = \oint\!\!\oint E\,dA = E\oint\!\!\oint dA = \frac{Q}{\epsilon_0}.$$

The remaining surface integral gives the surface area of the Gaussian sphere $4\pi r^2$. Thus,

$$E\left(4\pi r^2\right) = \frac{Q}{\epsilon_0}$$

and

$$E = \frac{Q}{4\pi\epsilon_0 r^2}.$$

Outside the sphere, the electric field is the same as if all the charge were located at the center of the sphere.

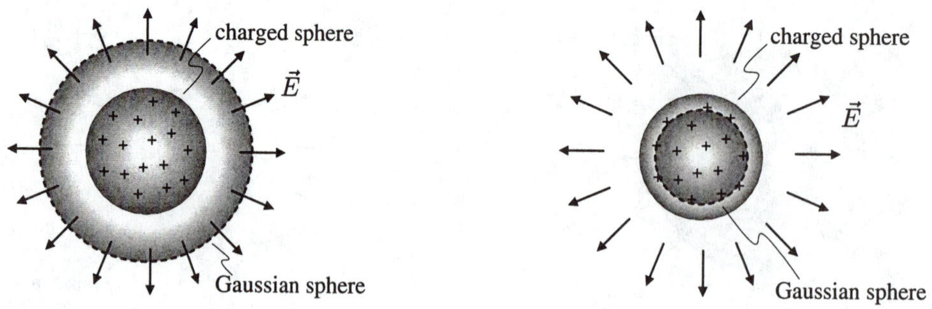

(a) Gaussian sphere outside a uniformly charged solid sphere (b) Gaussian sphere inside a uniformly charged solid sphere

Figure 23.13. Gaussian surfaces for the two cases in Example 23.8.

Case 2: inside the charged sphere. By symmetry, the electric field still points radially outward. The left side of Gauss's law (containing the surface integral) is the same as in case 1. Applying Gauss's law,

$$\oiint \vec{E} \cdot \hat{n}\, dA = \oiint E\, dA = E \oiint dA = \frac{q_{\text{enc}}}{\epsilon_0}.$$

Thus

$$E\left(4\pi r^2\right) = \frac{q_{\text{enc}}}{\epsilon_0}$$

or

$$E = \frac{q_{\text{enc}}}{4\pi\epsilon_0 r^2}.$$

But now [see Figure 23.13(b)], only a fraction of the total charge Q is enclosed by the Gaussian sphere of radius r (which is less than R). With uniform charge density, the ratio of the enclosed charge to the total charge equals the ratio of the enclosed volume to the total volume. That is,

$$\frac{q_{\text{enc}}}{Q} = \frac{\frac{4}{3}\pi r^3}{\frac{4}{3}\pi R^3} = \frac{r^3}{R^3}.$$

Solving for q_{enc},

$$q_{\text{enc}} = Q\frac{r^3}{R^3}.$$

Using this result in our equation for the electric field,

$$E = \frac{Q\,r}{4\pi\epsilon_0 R^3}.$$

Notice that the magnitude of the electric field increases linearly with increasing radius r. This result indicates that the electric field is zero at the center of the sphere, as it must be by symmetry. When $r = R$ (i.e., on the

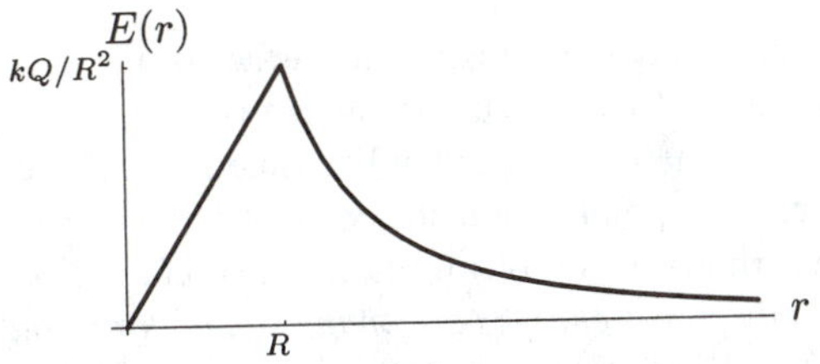

Figure 23.14. Electric field magnitude vs. radius.

surface of the sphere) the results from case 1 and case 2 reduce to the same value, namely

$$E = \frac{Q}{4\pi\epsilon_0 R^2} . \qquad (r = R)$$

A graph of the electric field for all r is shown in Figure 23.14. △

This last example in particular shows how Gauss's law can be used to compute electric fields in cases of high symmetry, with relatively few computational steps. Consider how difficult the last example would have been if you had tried to compute the electric field by direct integration (i.e., summing the individual fields due to all the point charges in the sphere)!

23.1.5 Implications for Properties of Conductors

Gauss's law can be used in a more qualitative fashion to make some observations about charge distributions in conductors. In this section, we make some arguments based on the properties of conductors and Gauss's law. These arguments are not rigorous, but they can be made more rigorous if Gauss's law is covered in more detail, as you may do in an advanced course in electromagnetic theory.

To begin, let's consider what happens if an excess positive charge is put onto a solid conducting sphere. (We could just as well use a negative charge, and all the arguments we will make would work as well, with the direction of the electric field reversed.) In a conductor, charge is free to move, and it is plausible the mutual repulsion will force all the excess positive charge to the outside of the sphere, spread symmetrically. The situation is shown in Figure 23.15(a).

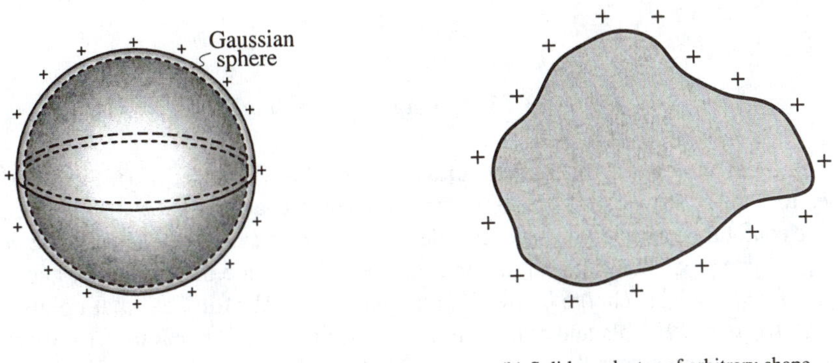

(a) Solid conducting sphere

(b) Solid conductor of arbitrary shape

Figure 23.15. Charge distribution on a conductor.

Now consider a Gaussian sphere drawn just inside the surface of the conducting sphere [also shown in Figure 23.15(a)]. The charge enclosed by the Gaussian sphere is zero, and in fact the physical situation is the same as for spherical conducting shell we studied in Example 23.7. Therefore, we conclude that for this geometry, the electric field is zero at all points inside the conducting sphere.

In fact, experimental evidence shows the following, more general fact: *The electric field inside any closed conductor is zero, provided the only excess charges are on the conductor or outside of it.* This is true regardless of the shape of the conductor. While the statement seems reasonable enough for a conducting sphere, it is less obvious for a closed conductor of arbitrary shape, shown schematically in Figure 23.15(b). In this case, we can still argue that if the excess charges move in such a way as to achieve equilibrium (and they do, as observed experimentally by measuring the electric field at points outside the conducting surface), then the electric field in the conductor should be zero. For a hypothetical test charge q_0 placed inside the conductor to measure the electric field, $\vec{E} = \vec{F}/q_0 = 0$, because in equilibrium $\vec{F} = 0$. This does not depend on anything but the excess charge being on the surface, so it works as well for a solid conductor or a hollowed out conducting shell.

A corollary to the phenomenon just discussed is the fact that if excess charges are held *outside* (not in contact with) a closed conductor, the free charges inside the conductor then rearrange themselves in such a way as to keep the electric field zero inside the conductor. This is true whether or not there is any excess positive or negative charge placed on the conductor. Suppose, for example, a positive charge is held outside a conducting sphere that has zero net charge. Then some of the free negative charges in the conductor are attracted toward the outside charge, as shown in Figure 23.16, leaving excess positive charges on the opposite side of the sphere. The situation quickly reaches an equilibrium, with zero electric field inside the conductor. Experimenters who are making sensitive electronic measurements often do so inside a metal box or enclosed by metal screen, in order to negate the effects of stray electric fields from outside. This setup is sometimes called a "Faraday cage."

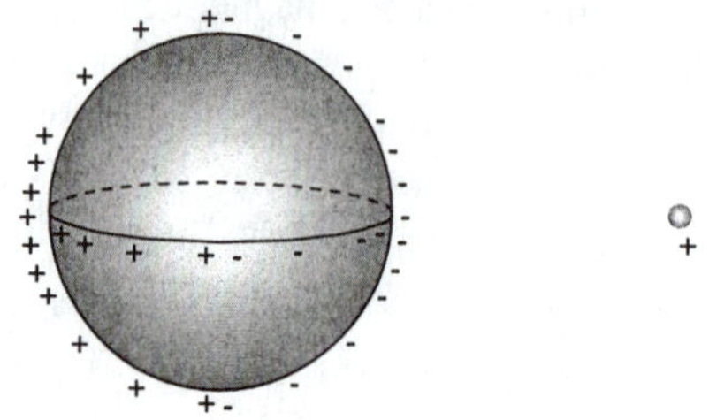

Figure 23.16. A charge outside a conductor.

Next, we consider the closed conducting surface depicted in Figure 23.17(a), in which one part of the surface comes to a fairly sharp point. If some excess positive charge is placed on this conductor, what is the final charge configuration that will make the electric field zero everywhere inside the conductor? To see what happens, consider specifically the situation at the point P, inside the conductor and near the tip of the point. The net effect of all the charge below P is to generate an electric field that points generally upward in the diagram. To counteract this and make the net electric field at P equal to zero, there must be a fairly large concentration of charge near the tip of the point above P. Experimental evidence shows that this is exactly what happens, and that the charge concentration at the sharp point on a conductor tends to be higher then in smooth regions. This is the physics behind the technological instrument called a lightning rod. In an

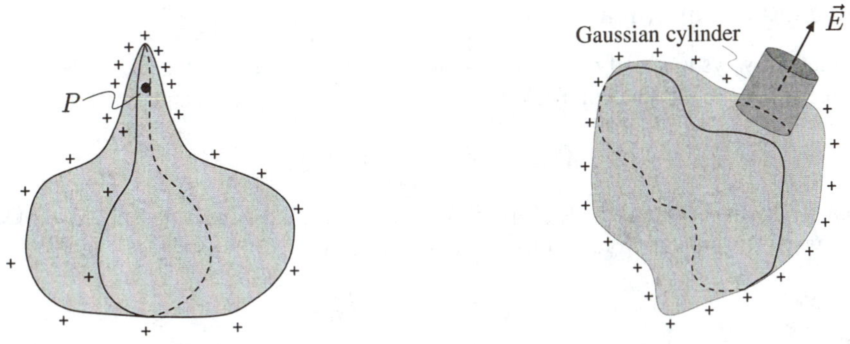

(a) Conducting surface with a sharp point

(b) Using a Gaussian cylinder to compute the electric field near a surface of a conductor

Figure 23.17. Excess charge on a conducting surface.

electrical storm, electrical discharges tend to be drawn to these sharp metal points, which are then connected safely to ground outside of the building on which the lightning rod sits, thus saving the building itself from damage.

Finally, we will use Gauss's law to compute the electric field just outside the surface of a conductor on which some excess (assumed positive) charge has been placed. When the charges are in equilibrium and not moving along the surface of the conductor (again, the experimental fact is that this happens quickly after the excess charge is put on the conductor), the electric field at the surface must be normal to the surface, and it points outward because the excess charge is positive. Now draw a small Gaussian cylinder, as shown in Figure 23.17(b). The surface integral in Gauss's law is easy to evaluate. With zero electric field inside the conductor, there is zero flux through that part of the cylinder. There is also zero flux through the sides of the cylinder that stick out from the conductor's surface, because $\vec{E}$ is normal to the surface and does not pierce the sides. Through the other end of the cylinder with surface area A, the flux is simply $\Phi_E = EA$. Then Gauss's law says

$$\Phi_E = EA = \frac{q_{\text{enc}}}{\epsilon_0}$$

and the desired magnitude of the electric field is

$$E = \frac{q_{\text{enc}}}{\epsilon_0 A}.$$

But q_{enc}/A is just the surface charge density σ at the part of the conductor we are considering, so we find that

$$E = \frac{\sigma}{\epsilon_0}.$$

To summarize, the electric field just outside a conductor points normal to the surface and has a magnitude equal to the local surface charge density divided by the constant ϵ_0. This result is consistent with experimental measurements.

23.2 Surfaces

Our goal in the remaining sections of this chapter is to make precise the concept of *surface integral*. As the first step, we must develop tools for working with *surfaces*. We do so in the current section.

23.2.1 Parameterizing Surfaces

There are a number of ways to specify a surface. In Section 13.2 we looked at surfaces defined by quadratic equations in three variables. As an example, we can describe the unit sphere centered at the origin as the set

$$\{(x, y, z)|x^2 + y^2 + z^2 = 1\}$$

or, more compactly, by the equation $x^2 + y^2 + z^2 = 1$ alone. Graphs of functions $f : \mathbb{R}^2 \to \mathbb{R}$ are surfaces specified in this way. The graph of a function f is the set of points

$$\{(x, y, z)|z = f(x, y)\}.$$

Note that, for example, the unit sphere is *not* the graph of a function. We can, however, break the sphere into pieces each of which is the graph of a function. The upper hemisphere is the graph of

$$f_1(x, y) = \sqrt{1 - (x^2 + y^2)}$$

while the lower hemisphere is the graph of

$$f_2(x, y) = -\sqrt{1 - (x^2 + y^2)}.$$

A more general way to specify surfaces is through a **parametric** description. The analog here is to specifying a plane curve by giving a parametrized description in the form of a function $\vec{r} : \mathbb{R} \to \mathbb{R}^2$. As a simple example, the function $\vec{r}(\theta) = \langle \cos\theta, \sin\theta \rangle$ for θ in the interval $[0, 2\pi]$ parametrizes the unit circle centered at the origin.

To give a parametrized description of a surface, we need a function with outputs in $\mathbb{R}^3$ because the surface "lives" in $\mathbb{R}^3$. A function of the type $\vec{r} : \mathbb{R} \to \mathbb{R}^3$ parametrizes a space curve, so this will not do. To get a surface, we need a function with two-component inputs. That is, we will work with functions of the type $\vec{r} : \mathbb{R}^2 \to \mathbb{R}^3$. As the first example, we find a parametrization of the unit sphere.

Example 23.9. Use spherical coordinates to find a parametrization of the unit sphere centered at the origin.

Solution. In spherical coordinates, ρ represents the distance from origin to the point in question. For all points on the unit sphere, we have $\rho = 1$. To cover all points on the sphere, we allow the coordinate ϕ to range in the interval $[0, \pi]$ and the coordinate θ to range in the interval $[0, 2\pi]$.

From the equations relating Cartesian coordinates (x, y, z) to spherical coordinates (ρ, ϕ, θ), we have

$$\begin{aligned} x &= \rho \sin\phi \cos\theta = \sin\phi\cos\theta \\ y &= \rho \sin\phi \sin\theta = \sin\phi\sin\theta \\ z &= \rho\cos\phi \qquad\;\; = \cos\phi \end{aligned}$$

We can thus use ϕ and θ as the inputs for a function $\vec{r} : \mathbb{R}^2 \to \mathbb{R}^3$. This function is

$$\vec{r}(\phi, \theta) = \langle \sin\phi\cos\theta, \sin\phi\sin\theta, \cos\phi \rangle.$$

We can think of this in terms of a globe. For a fixed θ, allowing ϕ to range from 0 to π traces out a semicircle of constant longitude from the north pole to the south pole. For a fixed ϕ, allowing θ to range from 0 to 2π traces out a circle of constant latitude. With $\phi = \pi/2$, this circle is the equator. △

Example 23.10. Describe the surface given parametrically by $\vec{r}(u, v) = \langle u + v, 2u - 3v, -5u + 2v \rangle$ for u in $\mathbb{R}$ and v in $\mathbb{R}$.

Solution. Note that each component in the output is linear in the inputs u and v. This should lead us to think that the surface is a plane. If we write out the components explicitly, we have

$$x = u + v, \qquad y = 2u - 3v, \qquad \text{and} \qquad z = -5u + 2v.$$

We can "deparametrize" this by solving for u and v in terms of x and y in the first two equations. Doing so gives $u = (-3x + y)/5$ and $v = (2x - y)/5$. We can then substitute these into the last equation to get

$$z = -5\frac{-3x + y}{5} + 2\frac{2x - y}{5}.$$

This can be rewritten as $11x - 3y + 5z = 0$. This is a *linear* equation in three variables and thus describes a plane. This particular plane goes through the origin and has $\langle 11, -3, 5\rangle$ as a normal vector. △

Example 23.11. Describe the surface given parametrically by

$$\vec{r}(u, v) = \langle \cos u(3 + \cos v), \sin u(3 + \cos v), \sin v\rangle$$

for u in $[0, 2\pi]$ and v in $[0, 2\pi]$.

Solution. We can get some feel for the surface by fixing one input and varying the other. For $v = 0$, we have $\vec{r}(u, 0) = \langle 4\cos u, 4\sin u, 0\rangle$. As u ranges from 0 to 2π, this traces out a circle of radius 4 in the xy-plane centered on the origin. For $v = \pi/2$, we have $\vec{r}(u, \pi/2) = \langle 3\cos u, 3\sin u, 1\rangle$. As u ranges from 0 to 2π, this traces out a circle of radius 3 centered on the z-axis at a height of one unit above the xy-plane. For $v = \pi$, we have $\vec{r}(u, \pi) = \langle 2\cos u, 2\sin u, 0\rangle$. As u ranges from 0 to 2π, this traces out a circle of radius 2 in the xy-plane centered on the origin. For $v = 3\pi/2$, we have $\vec{r}(u, \pi/2) = \langle 3\cos u, 3\sin u, -1\rangle$. As u ranges from 0 to 2π, this traces out a circle of radius 3 centered on the z-axis at a height of one unit below the xy-plane.

Now consider some fixed values of u. For $u = 0$, we have $\vec{r} = \langle 3 + \cos v, 0, \sin v\rangle$. As v ranges from 0 to 2π, this traces out a circle in the xz-plane centered at the point $(3, 0, 0)$ on the x-axis. For $u = \pi/2$, we have $\vec{r} = \langle 0, 3 + \cos v, \sin v\rangle$. As v ranges from 0 to 2π, this traces out a circle in the yz-plane centered at the point $(0, 3, 0)$ on the y-axis. For $u = \pi$, we have $\vec{r} = \langle -3 - \cos v, 0, \sin v\rangle$. As v ranges from 0 to 2π, this traces out a circle in the xz-plane centered at the point $(-3, 0, 0)$ on the x-axis. For $u = 3\pi/2$, we have $\vec{r} = \langle 0, -3 - \cos v, \sin v, \sin v\rangle$. As v ranges from 0 to 2π, this traces out a circle in the yz-plane centered at the point $(0, -3, 0)$ on the x-axis.

Note that the circles traced out for constant v fit together with the circles traced out for constant u. To form the full surface, we put all of these circles together and interpolate for values of u and v we did not explicitly examine. We see the surface shown in Figure 23.18. The surface has the same shape as the surface of a donut. This type of surface is called a **torus**. △

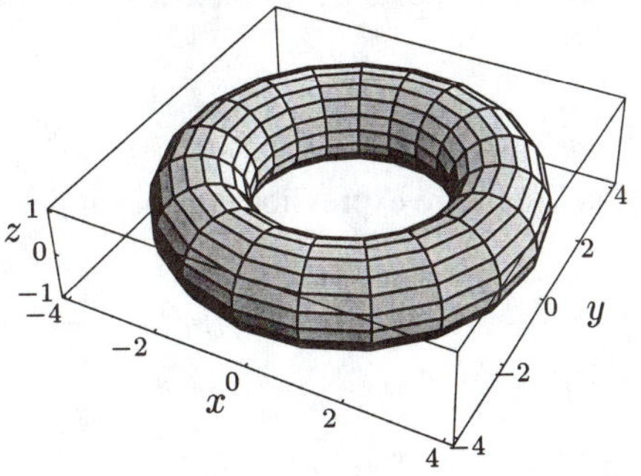

Figure 23.18. Torus parametrized by $\vec{r}(u, v) = \langle \cos u(3 + \cos v), \sin u(3 + \cos v), \sin v\rangle$ for u in $[0, 2\pi]$ and v in $[0, 2\pi]$.

23.2.2 An Area Scale Factor

Consider a generic parametization $\vec{r} : \mathbb{R}^2 \to \mathbb{R}^3$, with input u in $[a, b]$ and input v in $[c, d]$. We can think of the function $\vec{r}$ as *mapping* the rectangle $[a, b] \times [c, d]$ in the uv-plane onto the surface in the xyz-space. Figure 23.19(a) shows a generic example of such a rectangle. For the torus example, the relevant rectangle is $[0, 2\pi] \times [0, 2\pi]$ as shown in Figure 23.19(b). Our definition of surface integral for a surface given parametrically will be in terms of a double integral in the variables u and v. As part of this, we need to understand how a subregion of the rectangle in the uv-plane relates to a piece of the surface in the xyz-space.

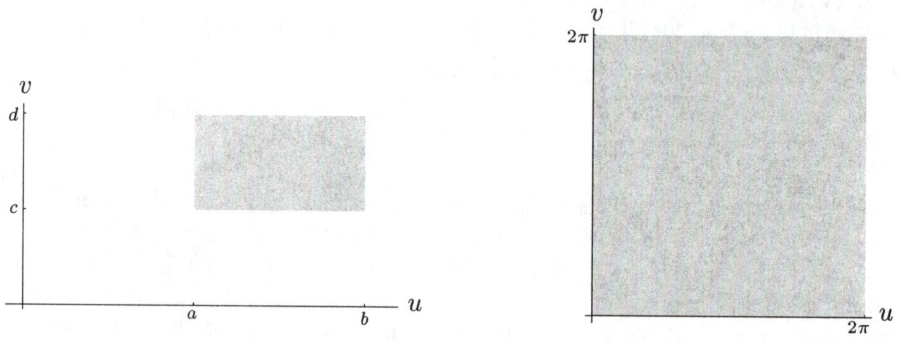

(a) Generic rectangular domain in the uv-plane (b) Domain for the torus example

Figure 23.19. The domain of a parametrized surface is a region in the uv-plane.

At a point (u_0, v_0) in the uv-domain, we can consider a line segment defined by $v = v_0$. The points on this line are mapped onto a space curve that lies on the surface in the xyz-space and goes through the point with position vector $\vec{r}(u_0, v_0)$. This curve is parametrized by $\vec{r}(u, v_0)$, thought of as a function of u. The derivative of $\vec{r}$ with respect to u gives a vector output for input u_0 that is *tangent* to this curve at the position $\vec{r}(u_0, v_0)$, and hence tangent to the surface at this point. Because $\vec{r}$ has a two-component input, we denote this derivative using partial derivative notation as

$$\left. \frac{\partial \vec{r}}{\partial u} \right|_{(u_0, v_0)}.$$

In a similar fashion, the line segment defined by $u = u_0$ maps onto another curve on the surface through the same point with position vector $\vec{r}(u_0, v_0)$. The tangent vector from this curve is denoted

$$\left. \frac{\partial \vec{r}}{\partial v} \right|_{(u_0, v_0)}.$$

We introduce some notation to simplify future expressions. For an input (u, v), let

$$\vec{T}_u(u, v) = \frac{\partial \vec{r}}{\partial u}$$

and let

$$\vec{T}_v(u, v) = \frac{\partial \vec{r}}{\partial v}.$$

From these definitions, $\vec{T}_u(u, v)$ and $\vec{T}_v(u, v)$ are vectors tangent to the surface at the point with position vector $\vec{r}(u, v)$.

Example 23.12. Compute $\vec{T}_u(u,v)$ and $\vec{T}_v(u,v)$ for the parametrization of the unit sphere given by $\vec{r}(u,v) = \langle \sin u \cos v, \sin u \sin v, \cos u \rangle$ at the point corresponding to $(u,v) = (\pi/2, 0)$.

Solution. We must compute the relevant derivatives. These are

$$\vec{T}_u(u,v) = \frac{\partial \vec{r}}{\partial u} = \langle \cos u \cos v, \cos u \sin v, -\sin u \rangle$$

and

$$\vec{T}_v(u,v) = \frac{\partial \vec{r}}{\partial v} = \langle -\sin u \sin v, \sin u \cos v, 0 \rangle.$$

Inputting $(u,v) = (\pi/2, 0)$ gives

$$\vec{T}_u(\tfrac{\pi}{2}, 0) = \langle 0, 0, -1 \rangle$$

and

$$\vec{T}_v(\tfrac{\pi}{2}, 0) = \langle 0, 1, 0 \rangle.$$

These results make sense in light of the fact that the point $\vec{r}(\pi/2, 0) = \langle 1, 0, 0 \rangle$ is on the equator. The curve defined by $v = 0$ is a longitude curve, and the curve defined by $u = 0$ is the equator itself. △

For a given point on the surface, we define the **tangent plane** to be the plane containing the vectors $\vec{T}_u$ and $\vec{T}_v$ for that point. With the two tangent vectors $\vec{T}_u$ and $\vec{T}_v$ in hand, we can easily get a vector *normal* to the tangent plane at a particular point by computing the cross product. That is, at the point $\vec{r}(u,v)$ on the surface, the vector

$$\vec{n}(u,v) = \vec{T}_u(u,v) \times \vec{T}_v(u,v) \tag{23.8}$$

is a normal vector for the tangent plane.

Example 23.13. Compute the normal vector field $\vec{n}(u,v)$ for the sphere as parametrized in the previous example. Compute the specific normal vector for $(u,v) = (\pi/2, 0)$.

Solution. This is a straightforward calculation. Using the results of the previous example, we have

$$\begin{aligned} \vec{n}(u,v) = \vec{T}_u(u,v) \times \vec{T}_v(u,v) &= \langle \cos u \cos v, \cos u \sin v, -\sin u \rangle \times \langle -\sin u \sin v, \sin u \cos v, 0 \rangle \\ &= \langle \sin^2 u \cos v, \sin^2 u \sin v, \cos u \sin u \rangle. \end{aligned} \tag{23.9}$$

For the input $(u,v) = (\pi/2, 0)$, we have

$$\vec{n}(\tfrac{\pi}{2}, 0) = \langle 1, 0, 0 \rangle.$$

This result is consistent with our visualization of the sphere. The input $(\pi/2, 0)$ gives the point on the equator directly above the x-axis, so the normal vector $\vec{n}(\pi/2, 0) = \langle 1, 0, 0 \rangle$ is correct. △

There is a very nice bonus in the normal vector field defined in Equation (23.8). The magnitude

$$\|\vec{n}(u,v)\| = \left\| \vec{T}_u(u,v) \times \vec{T}_v(u,v) \right\|$$

gives a scale factor relating the area of a small rectangle in the uv-plane to the surface area of the corresponding "patch" on the surface in the xyz-space. We turn now to an argument justifying this claim.

Consider a rectangle in the uv-plane with corners at (u, v), $(u+\Delta u, v)$, $(u, v+\Delta v)$ and $(u+\Delta u, v+\Delta v)$. The sides are of length Δu and Δv, so the area is $\Delta u \Delta v$. These four corners are mapped to the four points $\vec{r}(u, v)$, $\vec{r}(u + \Delta u, v)$, $\vec{r}(u, v + \Delta v)$, and $\vec{r}(u + \Delta u, v + \Delta v)$. We now make an approximation by thinking of the rectangle as being mapped onto a parallelogram defined by the three points $\vec{r}(u, v)$, $\vec{r}(u + \Delta u, v)$, and $\vec{r}(u, v + \Delta v)$. Two adjacent edges of this parallelogram are given by the difference vectors $\vec{r}(u + \Delta u, v) - \vec{r}(u, v)$ and $\vec{r}(u, v + \Delta v) - \vec{r}(u, v)$. Using linear approximations, we have

$$\vec{r}(u + \Delta u, v) - \vec{r}(u, v) \approx \left.\frac{\partial \vec{r}}{\partial u}\right|_{(u,v)} \Delta u = \vec{T}_u(u, v)\Delta u$$

and

$$\vec{r}(u, v + \Delta v) - \vec{r}(u, v) \approx \left.\frac{\partial \vec{r}}{\partial v}\right|_{(u,v)} \Delta v = \vec{T}_v(u, v)\Delta v.$$

Thus, the vectors $\vec{T}_u(u, v)\Delta u$ and $\vec{T}_v(u, v)\Delta v$ approximate the edges of the approximating parallelogram. Note that both of these approximations improve if the Δu and Δv are made smaller.

Now recall that the cross product of two vectors has a magnitude equal to the area of the parallelogram defined by the vectors. Thus, our parallelogram has an area given approximately by

$$\left\|\vec{T}_u(u, v) \times \vec{T}_v(u, v)\right\| \Delta u \Delta v$$

where we have factored the scalars Δu and Δv out of the cross product. Since $\Delta u \Delta v$ gives the area of the rectangle, the quantity

$$\left\|\vec{T}_u(u, v) \times \vec{T}_v(u, v)\right\|$$

gives us the scale factor relating area in the uv-plane to surface area on the surface in the xyz-space. Note that in general this scale factor is different for each point in the uv-plane.

Example 23.14. Compute the area scale factor for sphere parametrized as in the previous two examples. Compute the area scale factor specifically for $(u, v) = (\pi/4, 0)$ and $(\pi/2, 0)$.

Solution. With the result in Equation (23.9), we compute the area scale factor as

$$\begin{aligned} \|\vec{n}(u, v)\| &= \sqrt{\sin^4 u \cos^2 v + \sin^4 u \sin^2 v + \cos^2 u \sin^2 v} = \sqrt{\sin^4 u + \cos^2 u \sin^2 u} \\ &= \sqrt{\sin^2 u(\sin^2 u + \cos^2 u)} = |\sin u|. \end{aligned}$$

Since $\sin u$ is not negative for u in the chosen interval $[0, \pi]$, the absolute value has no effect. We thus have

$$\|\vec{n}(u, v)\| = \sin u \tag{23.10}$$

as the area scale factor for the parametrization of the sphere we have been using.

For the input $(u, v) = (\pi/4, 0)$, the area scale factor is

$$\left\|\vec{n}\left(\tfrac{\pi}{4}, 0\right)\right\| = \sin\frac{\pi}{4} = \frac{\sqrt{2}}{2}.$$

A region of area $\Delta u \Delta v$ centered at $(\pi/4, 0)$ in the uv-plane is thus mapped to a patch of surface area approximately $(\sqrt{2}/2)\Delta u \Delta v$ of the sphere.

For the input $(u, v) = (\pi/2, 0)$, the area scale factor is

$$\left\|\vec{n}\left(\frac{\pi}{2}, 0\right)\right\| = \sin\frac{\pi}{2} = 1.$$

An region of area $\Delta u \Delta v$ centered at $(\pi/2, 0)$ in the uv-plane is thus mapped to a patch of surface area approximately $\Delta u \Delta v$ of the sphere.

For this parametrization of the sphere, the area scale factor goes from 0 to 1 and back to 0 as u goes from 0 to $\pi/2$ to π. There is no dependence on v. You can see this on any globe. The area bounded by a pair of adjacent latitudes and a pair of adjacent longitudes is largest at the equator and smallest at the poles. $\triangle$

23.3 Surface Integrals

23.3.1 Surface Area

As a prelude to defining surface integral, consider the question of constructing a double integral that gives the total surface area for a surface described parametrically. Suppose the surface S is given parametrically by a function $\vec{r}(u, v)$, with (u, v) in the rectangle $[a, b] \times [c, d]$. Dividing this rectangle into subrectangles, each of size Δu by Δv, induces a division of the surface into small surface patches. From the arguments of the preceding section, the area of a generic patch is approximated by

$$\Delta A \approx \left\|\vec{T}_u(u, v) \times \vec{T}_v(u, v)\right\| \Delta u \Delta v.$$

We can approximate the total surface area with a double sum

$$A \approx \sum_i \sum_j \left\|\vec{T}_u(u, v) \times \vec{T}_v(u, v)\right\| \Delta u \Delta v.$$

In the limit as $\Delta u \to 0$ and $\Delta v \to 0$, we get a double integral defining the exact surface area as

$$A = \iint_S \left\|\vec{T}_u(u, v) \times \vec{T}_v(u, v)\right\| dA = \int_S \|\vec{n}(u, v)\| \, dA. \tag{23.11}$$

Note that this is a double integral of the type we studied in Chapter 14. The region of integration is a rectangle in the uv-plane, and the integrand is a function with (u, v) as input. For any specific case, we can evaluate the double integral using iterated integrals. We illustrate in the following example.

Example 23.15. Compute the surface area for a sphere of radius R.

Solution. To use Equation (23.11), we must parametrize the sphere. We can use a minor modification of the parametrization we found in Example 23.9. To account for the fact that the sphere here has radius R, we use

$$\vec{r}(u, v) = R\langle \sin u \cos v, \sin u \sin v, \cos u \rangle.$$

with (u, v) in $[0, \pi] \times [0, 2\pi]$. We can use the results of Example 23.12 by including a factor of R in both tangents, giving

$$\vec{T}_u(u, v) = \frac{\partial \vec{r}}{\partial u} = R\langle \cos u \cos v, \cos u \sin v, -\sin u \rangle$$

and

$$\vec{T}_v(u,v) = \frac{\partial \vec{r}}{\partial v} = R\langle -\sin u \sin v, \sin u \cos v, 0\rangle.$$

The cross product of these is

$$\vec{n}(u,v) = \vec{T}_u(u,v) \times \vec{T}_v(u,v) = R^2\langle \sin^2 u \cos v, \sin^2 u \sin v, \cos u \sin u\rangle.$$

The area scale factor is thus

$$\|\vec{n}(u,v)\| = \left\|\vec{T}_u(u,v) \times \vec{T}_v(u,v)\right\| = R^2 \sin u.$$

This is the integrand for the double integral in Equation (23.11).

With the relevant limits of integration, the double integral is expressed as

$$A = \iint_S \left\|\vec{T}_u(u,v) \times \vec{T}_v(u,v)\right\| \, dA = \int_0^{2\pi} \int_0^{\pi} R^2 \sin u \, du dv.$$

The iterated integral is easily evaluated giving

$$A = R^2 \int_0^{2\pi} dv \int_0^{\pi} \sin u \, du = R^2(2\pi)(2) = 4\pi R^2.$$

This result should be familiar to you. △

23.3.2 A Definition of Surface Integral

We are now prepared to put together a definition of surface integral that makes precise the informal notions introduced in Section 23.1. Recall that our motivation there was to define *flux* for a given electric field (not necessarily uniform) and a given surface (not necessarily flat). In this section, we will think generically of a vector field $\vec{F} : \mathbb{R}^3 \to \mathbb{R}^3$ and a surface S parametrized by a function $\vec{r} : \mathbb{R}^2 \to \mathbb{R}^3$ with inputs (u,v) in a rectangle $[a,b] \times [c,d]$. The surface integral is defined by first considering this rectangle to be split into subrectangles with sides of length Δu and Δv. For each subrectangle, we have a corresponding patch on the surface. For a generic patch, the vector $\vec{T}_u \times \vec{T}_v$ is normal to the surface. Now consider the quantity

$$\vec{F} \cdot (\vec{T}_u \times \vec{T}_v)\Delta u \Delta v. \tag{23.12}$$

We can sum quantities of this type for each subrectangle in a double sum

$$\sum_i \sum_j \vec{F} \cdot (\vec{T}_u \times \vec{T}_v)\Delta u \Delta v.$$

We define the **surface integral of the vector field $\vec{F}$ over the surface** S to be the double integral that results in the limit of this double sum as $\Delta u \to 0$ and $\Delta v \to 0$. That is,

$$\iint_S \vec{F} \cdot d\vec{A} = \iint_{[a,b]\times[c,d]} \vec{F}(\vec{r}(u,v)) \cdot (\vec{T}_u(u,v) \times \vec{T}_v(u,v)) \, dA \tag{23.13}$$

Using the notation $\vec{n}(u, v)$ for the normal vector field, we can write this as

$$\iint_S \vec{F} \cdot d\vec{A} = \iint_{[a,b]\times[c,d]} \vec{F}(\vec{r}(u,v)) \cdot \vec{n}(u,v)\, dA$$

Note that we have defined a new object (the surface integral) in terms of a familiar object (a double integral). Now we illustrate how to evaluate a surface integral from the definition in the following example.

Example 23.16. Compute the surface integral of the constant vector field $\vec{F}(x, y, z) = \langle 1, 0, 1\rangle$ for the upper hemisphere of radius 1 centered at the origin.

Solution. We can use some of our results from previous examples. In Example 23.9, we found a parametrization for the sphere. It allows us to parametrize the hemisphere here by restricting the inputs to an appropriate rectangle. We will use

$$\vec{r}(u, v) = \langle \sin u \cos v, \sin u \sin v, \cos u\rangle$$

with (u, v) in $[0, \pi/2] \times [0, 2\pi]$. Note that u ranges only up to $\pi/2$.

In Example 23.15, we computed the cross product

$$\vec{T}_u(u, v) \times \vec{T}_v(u, v) = \langle \sin^2 u \cos v, \sin^2 u \sin v, \cos u \sin u\rangle.$$

This is correct here, because we are using the same parametrization except for the domain. The integrand in Equation (23.13) is the dot product

$$\begin{aligned}\vec{F}(\vec{r}(u,v)) \cdot (\vec{T}_u(u,v) \times \vec{T}_v(u,v)) &= \langle 1, 0, 1\rangle \cdot \langle \sin^2 u \cos v, \sin^2 u \sin v, \cos u \sin u\rangle \\ &= \sin^2 u \cos v + \cos u \sin u.\end{aligned}$$

We thus have

$$\begin{aligned}\iint_S \vec{F} \cdot d\vec{A} &= \iint_{[0,\pi/2]\times[0,2\pi]} \vec{F}(\vec{r}(u,v)) \cdot (\vec{T}_u(u,v) \times \vec{T}_v(u,v))\, dA = \iint_{[0,\pi/2]\times[0,2\pi]} \left(\sin^2 u \cos v + \cos u \sin v\right) dA \\ &= \int_0^{2\pi}\int_0^{\pi/2} \left(\sin^2 u \cos v + \cos u \sin u\right) du\, dv.\end{aligned}$$

The remaining iterated integral can be evaluated in the usual way. The final result is π. That is,

$$\iint_S \vec{F} \cdot d\vec{A} = \pi$$

for the given vector field and surface. △

To relate this definition to the less formal construction in Equations (23.5) through (23.6), let $\hat{n}$ be a unit vector in the direction of $\vec{T}_u \times \vec{T}_v$. We can thus write

$$\vec{T}_u \times \vec{T}_v = \hat{n} \left\|\vec{T}_u \times \vec{T}_v\right\|.$$

Substituting this expression into that of Equation (23.12) gives

$$\left(\vec{F} \cdot \hat{n}\right) \left\|\vec{T}_u \times \vec{T}_v\right\| \Delta u \Delta v. \tag{23.14}$$

From our discussion in Section 23.2, recall that the expression

$$\left\|\vec{T}_u \times \vec{T}_v\right\| \Delta u \Delta v$$

approximates the area of the surface patch corresponding to a rectangle of area $\Delta u \Delta v$. This quantity thus plays the same role as ΔA in Equation (23.5).

Using the unit vector $\hat{n}$ and the expression in Equation (23.14), we can rewrite the left side of Equation (23.13) to get

$$\iint\limits_S \vec{F} \cdot d\vec{A} = \iint\limits_{[a,b]\times[c,d]} \left(\vec{F}(\vec{r}(u,v)) \cdot \hat{n}(\vec{r}(u,v))\right) \left\|\vec{T}_u(u,v) \times \vec{T}_v(u,v)\right\| \, dA \tag{23.15}$$

In applications involving Gauss's Law, we are often able to pick the surface in such a way that the quantity $\vec{F}(\vec{r}(u,v)) \cdot \hat{n}(\vec{r}(u,v))$ is constant for all points on the surface. This quantity can thus be factored out of the double integral, leaving

$$\iint\limits_S \vec{F} \cdot d\vec{A} = (\vec{F} \cdot \hat{n}) \iint\limits_{[a,b]\times[c,d]} \left\|\vec{T}_u(u,v) \times \vec{T}_v(u,v)\right\| \, dA.$$

This remaining double integral is precisely the definition of surface area for the surface S. Under these circumstances, we thus have

$$\iint\limits_S \vec{F} \cdot d\vec{A} = (\vec{F} \cdot \hat{n})(\text{surface area of } S).$$

This justifies the technique used for several examples in Section 23.1.

23.3.3 Orientability and Orientation

In our definition and discussion of surface integrals, we were very casual about the issue of **orientability**. In particular, we worked with the implicit assumption that the surface S has two well-defined "sides" and that computed normal vectors all point to the same side of the surface. For example, a sphere has, in an intuitive sense, two sides: the side facing in and the side facing out. As another example, the graph of any continuous function $f : \mathbb{R}^2 \to \mathbb{R}$ has two sides, one facing down and one facing up. A surface is **orientable** if the surface has two well-defined sides. We are relying on an intuitive sense of what this means. Some surfaces are **not orientable**. The Möbius strip is the classic example of a surface that is not orientable. The concept of flux is not meaningful for a surface that is not orientable.

For an orientable surface, we must also address the issue of **orientation**. This is the question of what side the normal vectors point toward. For example, if we use the parametrization of the sphere from Example 23.9, then the computed normal vectors $\vec{T}_u \times \vec{T}_v$ point outward at each point on the sphere. However, it is possible to parametrize the sphere so that the computed normal vectors point inward at each point. Furthermore, it is possible to parametrize the sphere so that the computed normal vectors point inward at some points and point outward at other points. Our interpretation of surface integral as flux requires that we use a parametrization for which the computed normals all point to the same side of the surface. For a closed surface such as a sphere, the convention is to use a parametrization that results in outward pointing normal vectors.

Finally, we have the question of whether the value of a surface integral depends on what parametrization we use for the surface. The answer is that the value of a surface integral for a given surface S and a given vector field $\vec{F}$ is the same for any two parametrizations that have the same orientation.

23.4 Problems

23.1 Gauss's Law

1. Find the electric flux through a flat rectangular surface 30 cm by 20 cm, lying in a uniform electric field of magnitude 2.0×10^4 N/C, if the electric field vector makes and angle of (a) 30° or (b) 60° with the surface.

2. A river is on average 900 m wide and 6.0 m deep. Find the flux of water (in units of m^3/s) if the river is flowing at a rate of 2.5 m/s.

3. (a) In the preceding problem, find the energy available per kilometer length of the river. (b) In practice, some of this energy can be changed to electrical energy in a hydroelectric power plant. If the power plant is 10% efficient, find its power output in watts.

4. A biophysicist measures the rate of blood flow through an artery to be 0.106 cm^3/s. If the opening in the (circular) artery at that point is 1.5 mm in diameter, what is the speed of the blood flow?

5. A large nonconducting plane carries a uniform surface charge density 15 $\mu C/m^2$. Find the electric flux through a 10 cm by 10 cm square parallel to and just above the charged plane.

6. Repeat the preceding problem if the square is tilted at a 30° angle to the charged plane.

7. A 150-nC point charge is located at the geometric center of a cube. (a) Find the electric flux through each face of the cube. [*Hint*: Use Gauss's law and symmetry arguments rather than trying to compute the flux directly.] (b) Explain why the result is the same regardless of the size of the cube.

8. Repeat the preceding problem if the point charge is located at one corner of the cube.

9. Assume Gauss's law to be true, and use it to derive Coulomb's law.

10. A cube 20 cm on a side lies completely in the first octant and has one corner at the origin. Find the electric flux through each face of the cube if there is an electric field $\vec{E} = 300x\,\hat{\imath}$ N/C.

11. A cube 20 cm on a side lies with its center at the origin and its edges parallel to the Cartesian coordinate axes. The electric field present is the superposition of a uniform electric field $-3.0 \times 10^5\,\hat{k}$ N/C and the field of a 1.2-μC point charge at the origin. Find the electric flux through each face of the cube.

12. Find the flux of a *nonuniform* electric field $4.5 \times 10^4 x\,\hat{k}$ N/C through a square that lies in the xy-plane with its corners at (0,0), (50 cm, 0), (50 cm, 50 cm), and (0, 50 cm).

13. Derive Gauss's law for *gravitation*. [*Hint*: Consider a point particle of mass m and find the gravitational field $\vec{g}$ in the space around the particle. Then compute the flux of the gravitational field

$$\Phi_g = \iint \vec{g} \cdot \hat{n}\, dA$$

through a Gaussian sphere, and argue that your result is true in general for a Gaussian surface enclosing mass.]

14. Using the results of the preceding problem, find the gravitational field inside and outside a solid sphere of uniform mass density, total mass M, and radius R.

15. Consider a large conducting plane with charge density σ on its surface, and a similar but nonconducting plane with charge density σ on its surface. Use Gauss's law to explain why the electric field just outside the conducting plane has magnitude σ/ϵ_0, but the field just outside the nonconductor is $\sigma/2\epsilon_0$.

16. On a large flat section of ground, Earth's electric field is 100 N/C pointing straight down. (a) Find the surface charge density on that part of Earth, assuming Earth is a conductor. (b) If the surface charge density is the same everywhere on Earth, find the total charge on Earth's surface.

17. A long, straight, nonconducting cylindrical wire of radius a carries a uniform charge density ρ per unit volume. Use Gauss's law to find the electric field (a) at points outside the wire and (b) at points inside the wire. (c) Show that your answers to (a) and (b) agree at the surface of the wire.

In Problems 18–19, consider a solid conducting sphere of radius a that is surrounded by a concentric conducting spherical shell with inner radius b and outer radius c ($a < b < c$). An excess positive charge Q has been placed on the inner sphere. The net charge on the conducting shell is zero.

18. Determine the electric field at the following points in space (r is measured from the center of the sphere): (a) $r < a$; (b) $a < r < b$; (c) $b < r < c$; and (d) $r > c$.

19. What is the net charge on each surface (the surface of the solid sphere and the inner and outer surfaces of the shell)?

In Problems 20–21, consider a solid conducting sphere of radius a that is surrounded by a concentric conducting spherical shell with inner radius b and outer radius c ($a < b < c$). An excess positive charge Q has been placed on the inner sphere, and an excessive negative charge $-3Q$ has been placed on the outer sphere.

20. Use Gauss's law to determine the electric field at the following points in space (r is measured from the center of the sphere): (a) $r < a$; (b) $a < r < b$; (c) $b < r < c$; and (d) $r > c$.

21. What is the net charge on each surface (the surface of the solid sphere and the inner and outer surfaces of the shell)?

22. A solid nonconducting sphere of radius a is surrounded by a concentric nonconducting spherical shell with inner radius b and outer radius c ($a < b < c$). The inner sphere carries a charge Q spread uniformly throughout its volume, and the outer shell carries a charge $-Q$ spread uniformly throughout its volume. Find the electric field at all points in space by considering each region of space separately: (a) $r < a$; (b) $a < r < b$; (c) $b < r < c$; and (d) $r > c$.

23. A very long nonconducting cylindrical shell has inner radius a and outer radius b. The material carries a uniform charge density ρ per unit volume. Find the electric field at points r (as measured from the central axis of the shell) given by (a) $r < a$; (b) $a < r < b$; and (c) $r > b$.

Problems 24–26 deal with a nonconducting sphere of radius R that carries a nonuniform charge density given by the function $\rho = \rho_0 r$, where ρ_0 is a constant and r is the distance from the center of the sphere.

24. Find the total charge in the sphere. (Your answer should be a function of ρ_0 and R.)

25. Find the electric field at points outside the sphere ($r > R$).

26. Find the electric field at points inside the sphere ($r > R$).

27. In a simplified model of an atom, the nucleus is a point charge $+Ze$, and the electronic charge $-Ze$ is uniformly distributed throughout a sphere of radius a that is centered at the nucleus. Find the electric field everywhere, both inside the atom (as a function of r, the distance from the nucleus) and outside the atom.

23.2 Surfaces

For Problems 1–10, give a parametrization of the described surface.

1. the upper hemisphere of radius 2 centered at the origin
2. the lower hemisphere of radius 2 centered at the origin
3. the hemisphere of radius 2 centered at the origin with positive y-coordinates
4. the hemisphere of radius 2 centered at the origin with negative y-coordinates
5. the graph of the function $f(x, y) = x^2 + y^2$
6. the graph of the function $f(x, y) = \sin(xy)$
7. the right circular cylinder of radius 2 centered on the z-axis
8. the right circular cylinder of radius 5 centered on the y-axis
9. the right circular cone centered on the z-axis with vertex at the origin and radius 2 for $z = 1$
10. the torus formed by rotating the circle $(x - 5)^2 + z^2 = 4$ around the z-axis
11. Using the parametrization for the unit sphere given in Example 23.9, compute the area scale factor $||\vec{T}_u(u, v) \times \vec{T}_v(u, v)||$ for some point on the equator and for some point midway between the south pole and the equator.
12. Using the parametrization for the unit sphere given in Example 23.9, compute the area scale factor $||\vec{T}_u(u, v) \times \vec{T}_v(u, v)||$ for the north pole.
13. Using the parametrization for the torus in Example 23.11, compute the area scale factor $||\vec{n}(u, v)||$ for two different points.
14. Consider the parametrization $\vec{r}(u, v) = \langle \cos u, \sin u, v\rangle$ for u in $[0, 2\pi]$ and v in $\mathbb{R}$. (a) Describe the surface given by this parametrization. (b) Compute the area scale factor for this parametrization.
15. Consider the parametrization $\vec{r}(u, v) = \langle v \cos u, v \sin u, v\rangle$ for u in $[0, 2\pi]$ and v in $\mathbb{R}$. (a) Describe the surface given by this parametrization. (b) Compute the area scale factor for this parametrization.

23.3 Surface Integrals

For Problems 1–4, compute the surface area of the given surface.

1. a right circular cylinder (no top or base) of radius R and height H
2. the torus in Example 23.11.
3. a cone (no base) of radius R and height H.
4. the graph of $f(x, y) = x^2 + y^2$ for the domain consisting of the unit circle centered on the origin of the xy-plane.

For Problems 5–11, compute the surface integral for the given vector field and surface.

5. $\vec{F}(x, y, z) = \langle 0, 0, x\rangle$ for the upper hemisphere of radius 1 centered at the origin

6. $\vec{F}(x, y, z) = \langle 0, 0, x \rangle$ for the portion of the sphere of radius 1 centered at the origin that is in the first octant

7. $\vec{F}(\vec{r}) = \vec{r}/r^3$ for the unit sphere centered at the origin.

8. $\vec{F}(x, y, z) = \langle x, y, z \rangle$ for the right circular cylinder of radius 1 centered on the z-axis and extending from $z = -1$ to $z = 1$

9. $\vec{F}(x, y, z) = \langle yz, xz, xy \rangle$ for the square at height $z = 1$ with (x, y) in $[0, 1] \times [0, 1]$

10. $\vec{F}(x, y, z) = \langle yz, xz, xy \rangle$ for the square at height $z = 3$ with (x, y) in $[0, 1] \times [0, 1]$

11. $\vec{F}(x, y, z) = \langle x, y, xyz \rangle$ for the right circular cone centered on the z-axis, vertex at the origin, radius 1 for height $z = 1$, extending from $z = 0$ to $z = 2$

12. Cut two strips of paper, each about one inch wide, from the long edge of a sheet of notebook paper. With one, join the short edges together (with tape or a staple) to form a cylinder. With the other, do a half twist along the length of the strip and then join the short edges together. This forms a Möbius strip. Convince yourself that the cylinder has two sides and that the Möbius strip does not have two sides.

Chapter 24

Magnetic Fields I

Throughout much of the remaining chapters, we will study various aspects of magnetism. Magnetism is common in nature and is responsible for many observed phenomena, on scales as large as stars and planets and as small as subatomic particles. It is also a rich source of applications in science, engineering, and technology. These applications include motors, transformers, magnetic deflection of particle beams, electromagnets for lifting metal objects, and the storage of information on magnetic tape.

In this chapter, we begin our study of magnetic fields. After looking briefly at the history of magnetism and defining some basic concepts, we examine how charged particles behave in magnetic fields and in combined electric and magnetic fields. The study of moving charges in magnetic fields can also be applied to current-carrying conductor in magnetic fields, which in turn can be applied to closed loops of current and magnetic dipoles that are placed in magnetic fields. The Hall effect is one specific example of how a current-carrying conductor behaves in a magnetic field, and we consider some ways the Hall effect is used in research. As we study magnetism in this chapter and the chapters that follow, we begin to establish the connections that exist between electric and magnetic fields, culminating with Maxwell's equations in Chapter 27.

24.1 Background and History

Magnetism was observed by the ancient Greeks, although the exact origins of its discovery remain unknown. Near Magnesia (in Asia Minor) the Greeks found lodestone, an iron ore that is now usually called magnetite (Fe_3O_4), and they observed that lodestone exerted forces (at a distance) on small pieces of iron. The existence of two distinct *poles* on a permanent magnet (such as iron) was known by the thirteenth century. In 1600, the English physician William Gilbert (1544–1603) published his book *De Magnete*, in which he reported the results of numerous systematic experiments designed to explain more systematically the behavior of magnetic materials. Among other things, Gilbert used a magnetic "terrella"—a spherical stone—to model Earth, in order to show that the entire Earth behaves like a magnet.

At the beginning of the nineteenth century, the nature of magnetic forces remained a mystery. Later, we will see how the work of nineteenth century physicists including Oersted, Ampere, Faraday, and Maxwell, greatly improved human understanding of magnetism and established the intimate linkage between electricity and magnetism.

You may already be familiar with some of the properties of a bar magnet [shown in Figure 24.1(a)]. The magnet has a "north pole" (labeled N) and a "south pole" (labeled S). If a second magnet is brought near the first, the rules for attraction and repulsion are the same as for positive and negative charges: Like poles repel, and unlike poles attract. Historically, the labels north pole and south pole follow from the use of a small

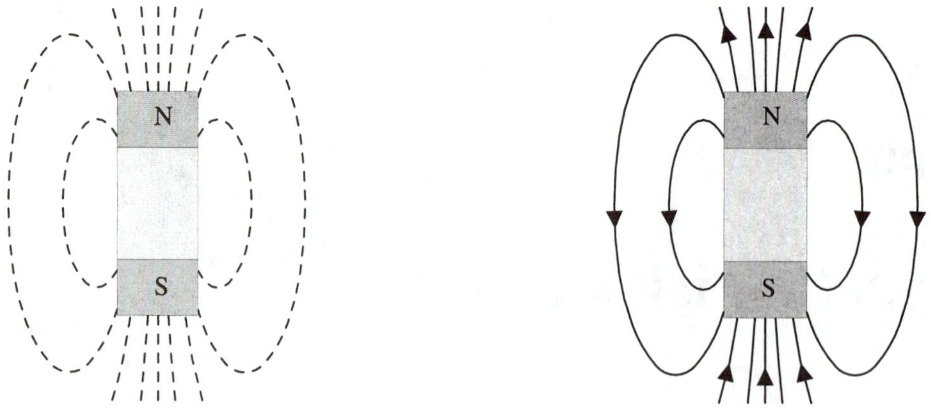

(a) Schematic of a bar magnet

(b) Magnetic field lines for a bar magnet

Figure 24.1. A bar magnet.

magnet as a compass needle. The north pole of a compass needle points generally toward the north (and the south pole of the compass generally to the south). We say *generally* because Earth's north and south magnetic poles drift over time and do not necessarily coincide with the geographic poles. The north magnetic pole is currently in the area of Greenland. Based on the rules of attraction and repulsion, you can deduce that the "north magnetic pole" is actually the *south* pole of a magnet, because it attracts the north poles of compasses.

You have probably seen or performed an experiment in which small iron filings are sprinkled on a piece of paper near a bar magnet. The filings tend to arrange themselves roughly in the pattern shown by the dashed lines in Figure 24.1(a). Each filing is itself a magnet that is becoming aligned with the **magnetic field** at that particular location. The resulting pattern should remind you of the electric field of an electric dipole—just think of replacing the north and south poles with positive and negative charges. This is one of the similarities between electricity and magnetism: The shape of the dipole field is roughly the same. Just as we did for electric fields in Chapter 17 (see Figure 17.15), we can draw continuous magnetic field lines for the magnetic dipole [Figure 24.1(b)]. As with electric fields, the magnetic field at any location is tangent to the continuous line, and the density of field lines is proportional to the magnitude of the magnetic field in that region (we discuss dimensions and units for magnetic fields in Section 24.2). By convention, the field lines leave the north pole and terminate on the south pole. The "opposites attract" rule dictates that the south pole of an iron filing in the field of a bar magnet lies on the part of the line that is closer to the bar magnet's north pole, and vice versa. The magnetic field is a *vector field*, because there is both a magnitude and direction associated with the field at each point in space. The symbol $\vec{B}$ is used to represent the magnetic field.

There is one important difference between electric and magnetic fields. In *De Magnete*, Gilbert reported that if a bar magnet is cut into two pieces (shown in Figure 24.2, the result is not a distinct north pole and south pole, but rather two smaller magnets, each with a north and south pole. We summarize this result by saying that there are *no magnetic monopoles*. While electric charges can be isolated into elementary positive and negative charges, there is no comparable magnetic "charge" (i.e., an isolated north or south pole) as far as we know, although the possible existence of such an entity has been the subject of considerable theoretical speculation over the years. We return to this important point in this and later chapters, when we consider the fundamental cause of magnetism and the connections between electricity and magnetism.

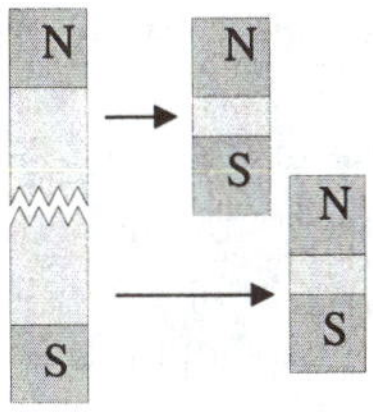

Figure 24.2. A bar magnet sliced into two pieces, each of which is a smaller bar magnet.

24.2 The Magnetic Force on an Electric Charge

24.2.1 The Magnetic Force and the Lorentz Force

It is observed experimentally that a point particle with charge q moving through a magnetic field experiences a force according to the relation

$$\vec{F} = q\vec{v} \times \vec{B} \tag{24.1}$$

where $\vec{v}$ is the particle's velocity and $\vec{B}$ is the magnetic field. Equation (24.1) is the basis for understanding the motion of a charged particle in a magnetic field.

Recall that the result of the cross product of two vectors is another vector that is perpendicular to both vectors in the product. Because the two vectors in the cross product define a plane, there are two possible directions for a vector perpendicular to both. The correct direction of the resultant vector is given by the right-hand rule: Starting with the fingers of your right hand in the direction of the first vector in the product, curl your fingers (through the angle less than π) into the direction of the second vector in the product, and then your thumb points in the direction of the resultant vector.

In Equation (24.1), it is important to notice that the cross product $\vec{v} \times \vec{B}$ is multiplied by the charge q. If q is positive, then multiplying $\vec{v} \times \vec{B}$ by q gives a force $\vec{F}$ that points in the same direction as the vector $\vec{v} \times \vec{B}$. However, if q is negative, multiplying $\vec{v} \times \vec{B}$ by q gives a force $\vec{F}$ that points in the *opposite* direction as the vector $\vec{v} \times \vec{B}$. We illustrate this fact with an example. In Figure 24.3(a), we show a positive charge with its velocity in the $+x$-direction. The charge is passing through a region in which the magnetic field is uniform and in the $+y$-direction. [Later we address the problem of how to make a uniform, or approximately uniform, magnetic field, because it is not obvious how to do so using the dipole field of Figure 24.1(b).]

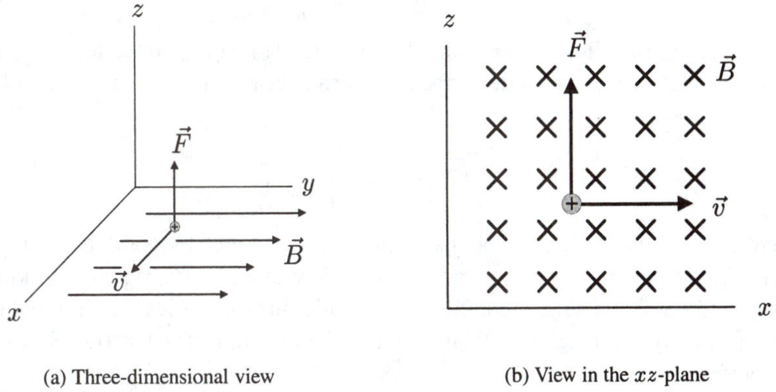

(a) Three-dimensional view (b) View in the xz-plane

Figure 24.3. Force on a positive point charge with velocity $\vec{v}$ in the $+x$-direction.

What is the direction of the force on this particle? By the right-hand rule, $\vec{v} \times \vec{B}$ is in the $+z$-direction. Because the charge is positive, the force $\vec{F} = q\vec{v} \times \vec{B}$ is also in the $+z$-direction. In Figure 24.3(b), we offer a different view of the same situation. In this view, we have made the xz-plane the plane of the paper and used the convention that the $\times$ indicates a vector going straight into the page (in this convention, a dot symbol $\cdot$ indicates a vector coming straight out of the page). Sometimes it is easier to visualize the spatial relationships between the vectors in this representation, and you should feel free to use either the representation in Figure 24.3(a) or the one in Figure 24.3(b).

Now consider a *negatively* charged particle with the same velocity (in the $+x$-direction) in the same uniform magnetic field (in the $+y$-direction), as shown in Figure 24.4(a). Because $\vec{v}$ and $\vec{B}$ are the same as for the positively charge particle we just considered, $\vec{v} \times \vec{B}$ is still in the $+z$-direction. However, multiplying this vector by the negative scalar q produces a vector in the $-z$-direction. Therefore, the force on the negative particle is in the $-z$-direction, as shown in Figure 24.4(a). In Figure 24.4(b), we show the same situation with the xz-plane in the plane of the page.

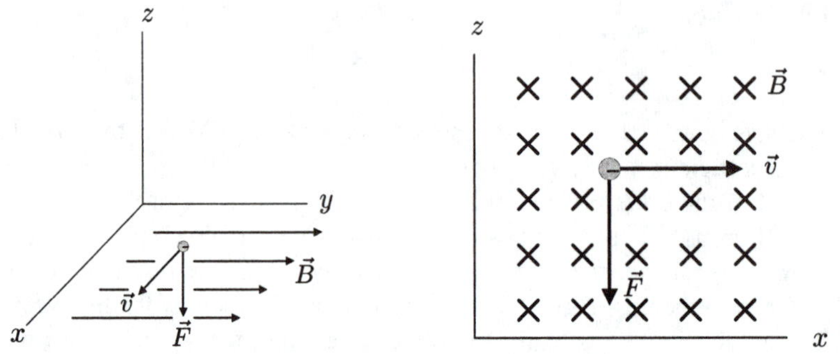

(a) Three-dimensional view

(b) View in the xz-plane

Figure 24.4. Force on a negative point charge with velocity $\vec{v}$ in the $+x$-direction.

Dimensions and Units The dimensions of the magnetic field can be deduced from Equation (24.1). The dimensions are force divided by (product of charge and velocity). Expressed in SI units, this is

$$\frac{\text{N}}{\text{C·m/s}} = \frac{\text{N}}{\text{A·m}}.$$

For convenience, these units are collected in a defined SI unit, the **tesla** (T), named for the Croatian-American physicist Nicola Tesla (1856–1943), who developed numerous practical applications of electromagnetism. That is,

$$1\text{ T} = 1\,\frac{\text{N}}{\text{A·m}}.$$

A magnetic field of 1 T is rather large. The magnetic fields produced by good laboratory electromagnets are on the order of 1 T. Common bar magnets produce much smaller fields, typically on the order of 10^{-3} or 10^{-2} T. The magnitude of Earth's magnetic field (at Earth's surface) varies, but in most places it is about 5×10^{-5} T. We occasionally see magnetic fields given in the cgs unit of gauss (G). The conversion factor between tesla and gauss is

$$1\text{ T} = 10^4\text{ G}.$$

Thus, the magnitude of Earth's magnetic field is typically about 0.5 G.

Example 24.1. An electron is traveling at a speed of 5.4×10^5 m/s in the $+y$-direction when it enters a uniform magnetic field $\vec{B} = 0.25\ \text{T}\ \hat{k}$. Find (a) the force on the electron, and (b) the electron's acceleration the moment it enters the magnetic field.

Solution. (a) The magnetic force is given by $\vec{F} = q\vec{v} \times \vec{B}$. Using the numerical value of the charge of the electron, and $\vec{v} = 5.4 \times 10^5$ m/s $\hat{\jmath}$, we have

$$\vec{F} = (-1.60 \times 10^{-19}\ \text{C})(5.4 \times 10^5\ \text{m/s}\ \hat{\jmath}) \times (0.25\ \text{T}\ \hat{k}) = -2.14 \times 10^{-14}\ \text{C·m·T/s}\ \hat{\jmath} \times \hat{k}.$$

Using the definition of the tesla, we see that the units reduce to newtons, as you should expect, because we have used SI units consistently. From the rules of cross products, $\hat{\jmath} \times \hat{k} = \hat{\imath}$. Therefore, the force on the electron is

$$\vec{F} = -2.14 \times 10^{-14}\ \text{N}\ \hat{\imath}.$$

You should verify that applying the right-hand rule gives a force in the $-x$-direction, consistent with this result.

(b) Because $\vec{F} = m\vec{a}$, we have

$$\vec{a} = \frac{\vec{F}}{m} = \frac{-2.14 \times 10^{-14}\ \text{N}\ \hat{\imath}}{9.11 \times 10^{-31}\ \text{kg}} = -2.35 \times 10^{16}\ \text{m/s}^2\ \hat{\imath}.$$

The magnitude of the acceleration is large, but this is reasonable for such a light particle. △

The magnetic force and the Lorentz force In many applications, electric and magnetic fields coexist in the same space. In computing the force on a charged particle in this situation, it is observed experimentally that the principle of superposition holds. That is, to find the net force on a charged particle in a region containing both an electric field and a magnetic field, we simply add the electric and magnetic forces:

$$\vec{F} = \vec{F}(\text{electric}) + \vec{F}(\text{magnetic}) = q\vec{E} + q\,\vec{v} \times \vec{B}.$$

Simplifying,

$$\vec{F} = q\left[\vec{E} + \left(\vec{v} \times \vec{B}\right)\right]. \tag{24.2}$$

This result is known as the **Lorentz force law**, named for the Dutch physicist Hendrik A. Lorentz (1853–1928).

Example 24.2. A proton with velocity $\vec{v} = (2.4 \times 10^4\,\hat{\imath} - 1.9 \times 10^4\,\hat{\jmath})$ m/s enters a region containing a uniform electric field $\vec{E} = (240\,\hat{\imath} + 360\,\hat{k})$ N/C and a magnetic field $\vec{B} = (-2.6 \times 10^{-3}\,\hat{\imath} - 3.9 \times 10^{-3}\,\hat{k})$ T. Find (a) the force on the proton, and (b) the acceleration of the proton.

Solution. (a) Recall that, in general, we can write the cross product in terms of its components [see Equation (10.21)]:

$$\vec{v} \times \vec{B} = (v_y B_z - v_z B_y)\,\hat{\imath} + (v_z B_x - v_x B_z)\,\hat{\jmath} + (v_x B_y - v_y B_x)\,\hat{k}.$$

In this problem, $v_z = 0$ and $B_y = 0$, so the cross product simplifies to

$$\vec{v} \times \vec{B} = (v_y B_z)\,\hat{\imath} + (-v_x B_z)\,\hat{\jmath} + (-v_y B_x)\,\hat{k}.$$

We can also write the electric field in terms of its vector components, noting that in this case $E_y = 0$. Thus, the Lorentz force becomes

$$\vec{F} = q\left[\vec{E} + \left(\vec{v} \times \vec{B}\right)\right] = q\left[\left(E_x + v_y B_z\right)\hat{\imath} + \left(-v_x B_z\right)\hat{\jmath} + \left(E_z - v_y B_x\right)\hat{k}\right].$$

Now we will insert the numerical values. We leave it as an exercise (see the problems) to show that the SI units for the product of velocity and magnetic field are identical to those for electric field (N/C).

$$\begin{aligned}\vec{F} &= (1.60 \times 10^{-19}\ \mathrm{C})\big[(240\ \mathrm{N/C} + (-1.9 \times 10^{4}\ \mathrm{m/s})(-3.9 \times 10^{-3}\ \mathrm{T}))\,\hat{\imath} \\ &\quad + (-2.4 \times 10^{4}\ \mathrm{m/s})(-3.9 \times 10^{-3}\ \mathrm{T})\,\hat{\jmath} + \left(360\ \mathrm{N/C} - (-1.9 \times 10^{4}\ \mathrm{m/s})(-2.6 \times 10^{-3}\ \mathrm{T})\right)\hat{k}\big] \\ &= (1.60 \times 10^{-19}\ \mathrm{C})\big[(314.1\,\hat{\imath} + 93.6\,\hat{\jmath} + 310.6\,\hat{k})\ \mathrm{N/C}\big] \\ &= (5.03 \times 10^{-17}\,\hat{\imath} + 1.50 \times 10^{-17}\,\hat{\jmath} + 4.97 \times 10^{-17}\,\hat{k})\ \mathrm{N}.\end{aligned}$$

(b) The acceleration of the proton is

$$\begin{aligned}\vec{a} = \frac{\vec{F}}{m} &= \frac{(5.03 \times 10^{-17}\,\hat{\imath} + 1.50 \times 10^{-17}\,\hat{\jmath} + 4.97 \times 10^{-17}\,\hat{k})\ \mathrm{N}}{1.67 \times 10^{-27}\ \mathrm{kg}} \\ &= (3.01 \times 10^{10}\,\hat{\imath} + 8.98 \times 10^{9}\,\hat{\jmath} + 2.98 \times 10^{10}\,\hat{k})\ \mathrm{m/s^2}.\end{aligned}$$

This is a reasonable result for a proton. △

24.2.2 J.J. Thomson's Experiment

In 1897, the English physicist Joseph John Thomson (1856–1940) (widely known as "J.J.") was studying the properties of what were called "cathode rays." This unknown radiation was so called because it was emitted from a cathode (a metal plate or filament at low electric potential) when the cathode was heated. This radiation could be accelerated toward an anode (another piece of metal at a higher electric potential). At the time, physicists disagreed about whether cathode rays were a wavelike phenomenon or a beam of tiny particles. It was this question that Thomson set about to answer. By deflecting cathode rays in electric and magnetic fields, Thomson showed conclusively that cathode rays act as charged particles. In fact, they are electrons, and the quantitative results of Thomson's experiments gave the first reasonable determination of the ratio of the electron's charge to its mass. Recall (see Section 17.3) that the electronic charge was measured independently by Millikan some years later.

A schematic of Thomson's experiment is shown in Figure 24.5. Parallel conducting plates are used to create a nearly uniform electric field $\vec{E}$ by putting a potential difference across the two plates. Let's suppose that the top plate in the diagram is at the higher potential, so that the (assumed uniform) electric field points straight down as shown. If an electron with speed v_0 is projected between the plates as shown (in a direction parallel to the plates and thus perpendicular to the electric field), it experiences a force due to the electric field. In general, the electric force is $\vec{F} = q\vec{E}$. In this case, the charge q is negative, so the force due to the electric field is straight up in the diagram, opposite to the electric field. For a constant electric field, this force is constant, and the electron will be deflected upward along a parabolic path, as we saw in Section 17.3.

Now consider the effect of a uniform magnetic field $\vec{B}$, directed straight into the page, as shown in the diagram (Figure 24.5). Using the right-hand rule, you should convince yourself that the force $\vec{F} = q\vec{v} \times \vec{B}$ due to the magnetic field is straight down in the diagram. This is the opposite direction of the force due to the electric field, as we discussed above. Therefore, it is plausible that a proper adjustment of the parameters $\vec{E}$, $\vec{v}$, and $\vec{B}$ will result in a net force of zero on the electron. Under these conditions, it can be observed by noticing that a beam of electrons passes through the region of interest with no deflection whatsoever.

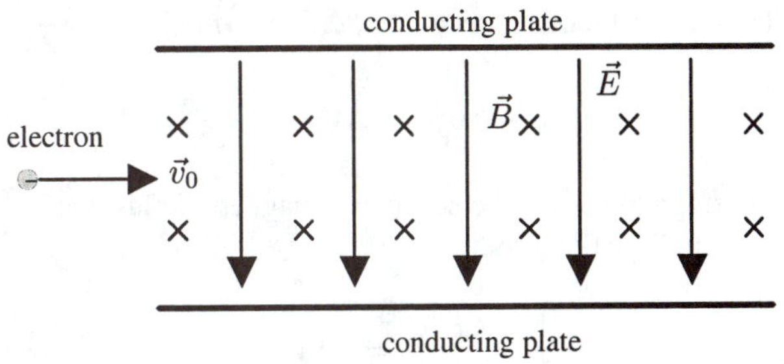

Figure 24.5. Electron entering a region between parallel conducting plates with crossed electric and magnetic fields.

The condition for no deflection is that the total (Lorentz) force is zero— that is,

$$\vec{F} = q\left[\vec{E} + \left(\vec{v} \times \vec{B}\right)\right] = 0.$$

We divide both sides by the scalar q, leaving $\vec{E} + (\vec{v} \times \vec{B}) = 0$, or

$$\vec{E} = -\vec{v} \times \vec{B}.$$

Taking the magnitude of both sides,

$$|\vec{E}| = |-\vec{v} \times \vec{B}|. \tag{24.3}$$

The left side of Equation (24.3) is simply E, the magnitude of the electric field between the plates. The magnitude of the cross product on the right side of the equation is

$$|-\vec{v} \times \vec{B}| = |\vec{v} \times \vec{B}| = vB \sin\theta,$$

where θ is the angle between the vectors $\vec{v}$ and $\vec{B}$. Because the electron experience no net force and no deflection, $v = v_0 =$ constant, and $\theta = \pi/2$, so $\sin\theta = 1$. Therefore,

$$|-\vec{v} \times \vec{B}| = v_0 B$$

and Equation (24.3) becomes

$$E = v_0 B. \tag{24.4}$$

This is the condition that must be satisfied for electrons to pass through the crossed electric and magnetic fields without deflection.

How can the speed v_0 of an electron be determined? Electrons are too small to see, so a time-of-flight measurement was out of the question in Thomson's day (it can now be done electronically). However, recall that electrons are accelerated from the cathode to the anode through a potential difference ΔV. The change in the kinetic energy ΔK of the electrons passing through that potential difference equals the negative of the change in the potential energy, or

$$\Delta K = -\Delta U = -q\Delta V$$

where we have used $\Delta U = q\Delta V$ from Section 18.4. The charge on the electron is $q = -e$, so

$$\Delta K = e\Delta V.$$

If we assume the electrons leave the cathode nearly at rest, $\Delta K = \frac{1}{2}mv_0^2 - 0 = \frac{1}{2}mv_0^2$, so

$$\frac{1}{2}mv_0^2 = e\Delta V. \tag{24.5}$$

For electrons that then pass through the crossed electric and magnetic fields, we have from Equation (24.4) $v_0 = E/B$. Inserting this for v_0 in Equation (24.5),

$$\frac{1}{2}m\left(\frac{E}{B}\right)^2 = e\Delta V.$$

Upon rearranging, we find

$$\frac{e}{m} = \frac{E^2}{2B^2\Delta V}. \tag{24.6}$$

This is the fundamental result of Thomson's experiment. Experimentally determined values of the parameters E, B, and ΔV allow us to compute the ratio e/m for the electron. Thomson reported a value of e/m that is on the correct order of magnitude but about 35% lower than today's accepted value of

$$\frac{e}{m} = 1.759 \times 10^{11} \text{ C/kg}.$$

Thomson's result was astonishing to many, because the numerical value of e/m is so much higher than the ratio of the electronic charge to the mass of the smallest atom, hydrogen. This value, which can be obtained by electrolysis experiments, is

$$\frac{e}{M_H} = 9.58 \times 10^8 \text{ C/kg}.$$

The difference in the ratios is due to the fact that the mass of the electron is nearly 2000 times smaller than the mass of the hydrogen atom. Thomson's experiment was an important step toward understanding the structure of the atom, which was unknown in 1897. In Chapter 28, we will see how Niels Bohr developed a successful model of the hydrogen atom in 1913, armed with the knowledge of the electronic charge and mass.

Example 24.3. In a replication of Thomson's experiment, electrons are accelerated from a cathode to an anode through a potential difference of 35.0 V. It is found that these electrons pass undeflected though crossed electric and magnetic fields (as in Figure 24.5) when the magnitudes of the electric and magnetic fields are $E = 1450$ N/C and $B = 4.19 \times 10^{-4}$ T. (a) What is the ratio e/m as determined from this data? (b) What was the approximate speed of the electrons?

Solution. (a) Using the numerical values obtained for E, B, and ΔV, we find

$$\frac{e}{m} = \frac{E^2}{2B^2\Delta V} = \frac{(1450 \text{ N/C})^2}{2(4.19 \times 10^{-4} \text{ T})^2(35.0 \text{ V})} = 1.71 \times 10^{11} \text{ C/kg},$$

which is rather close to the accepted value.

(b) The condition for no deflection of the electrons is [Equation (24.4)] $E = v_0 B$. Therefore,

$$v_0 = \frac{E}{B} = \frac{1450 \text{ N/C}}{4.19 \times 10^{-4} \text{ T}} = 3.46 \times 10^6 \text{ m/s}.$$

△

24.2.3 Other Applications

Suppose a particle with positive charge q is moving as shown in Figure 24.6(a), with its velocity $\vec{v}$ perpendicular to a uniform magnetic field $\vec{B}$. Assuming there are no other forces acting on the particle, what is its subsequent motion through the field? At the instant shown in the diagram, we know from the right-hand rule that the direction of the force is straight down in the drawing. Over a parameterized path, the work done on the particle by the magnetic field is

$$W = \int \vec{F} \cdot d\vec{s} = \int \vec{F} \cdot \vec{v}\, dt.$$

But because $\vec{F}$ is perpendicular to $\vec{v}$, the work done is zero. By the work-energy theorem, the speed does not change, even though the particle is deflected downward (in the diagram). With $\vec{v}$ and F in the plane of the page, the particle remains in that plane. As the particle follows its curved path in the plane, the force remains constant in magnitude and perpendicular to the velocity. You should recognize (see Chapter 4) the conditions just given as those for *uniform circular motion.* The charged particle travels in a circle, moving clockwise in the plane, as shown in Figure 24.6(b). If a *negatively charged* particle is sent into the field with the same initial velocity, the magnetic force on it is in the opposite direction, causing it to travel counterclockwise in a circle in the same plane.

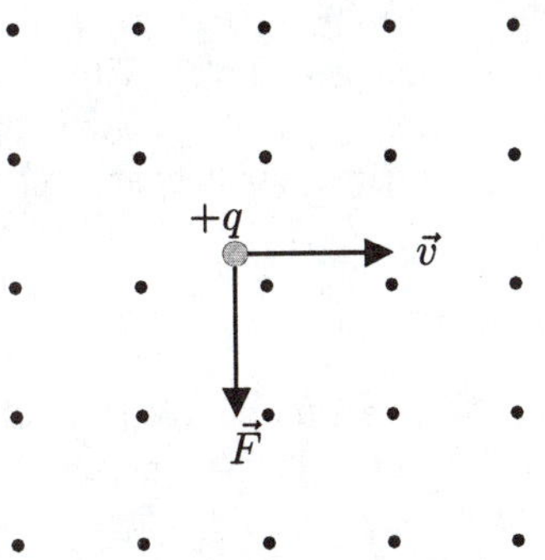

(a) Relation among velocity, field, and force at one instant of time

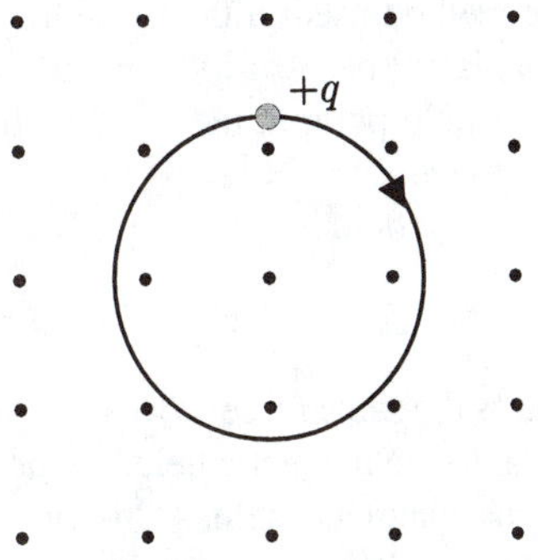

(b) Trajectory of the particle

Figure 24.6. Particle with positive charge moving in a uniform magnetic field.

The magnitude of the force on the charged particle is

$$F = |q\vec{v} \times \vec{B}| = q|\vec{v} \times \vec{B}| = qvB \sin\frac{\pi}{2} = qvB$$

where we have used the fact that velocity $\vec{v}$ makes an angle $\pi/2$ with the magnetic field $\vec{B}$. As we saw above, the speed is constant, so this force has the same magnitude throughout the circle. In fact, it is the *centripetal force* responsible for uniform circular motion. Recalling that the centripetal force on a particle of mass m traveling around a circle of radius R is $F = mv^2/R$, we have

$$\frac{mv^2}{R} = qvB.$$

Therefore, the radius of the circle can be expressed in terms of the other parameters of the problem as

$$R = \frac{mv}{qB}. \tag{24.7}$$

Example 24.4. A particle with a charge q is moving in the xy-plane, so its velocity can be expressed

$$\vec{v} = v_x\,\hat{\imath} + v_y\,\hat{\jmath}.$$

There exists a uniform magnetic field

$$\vec{B} = B_0\,\hat{k}$$

perpendicular to the particle's velocity. Find an expression for the magnetic force on the charged particle, and use that expression to show that the force is perpendicular to the velocity, regardless of the values of v_x and v_y.

Solution. In terms of components, the cross product is

$$\vec{F} = q[(v_y B_0)\hat{\imath} - (v_x B_0)\,\hat{\jmath}].$$

To show that one vector is perpendicular to another, compute the dot product:

$$\vec{F}\cdot\vec{v} = q[(v_y B_0)\,\hat{\imath} - (v_x B_0)\,\hat{\jmath}]\cdot[v_x\,\hat{\imath} + v_y\,\hat{\jmath}] = q[v_y B_0 v_x - v_x B_0 v_y] = 0.$$

Because the dot product of $\vec{F}$ and $\vec{v}$ is zero, the two vectors are perpendicular. △

Following the method used in the preceding example, we can study the trajectory of a charged particle with a velocity that is not necessarily perpendicular to a uniform magnetic field. Without loss of generality, we let the magnetic field point in the $+z$-direction, so $\vec{B} = B_0\,\hat{k}$. In general, the velocity vector may have three nonzero components, so $\vec{v} = v_x\,\hat{\imath}+v_y\,\hat{\jmath}+v_z\,\hat{k}$. The force on a particle with charge q having this velocity in the uniform magnetic field is

$$\vec{F} = q\vec{v}\times\vec{B} = q[(v_y B_0)\,\hat{\imath} - (v_x B_0)\,\hat{\jmath}]. \tag{24.8}$$

Notice that this is the same force as we found in Example 24.4, where the charged particle was moving in a plane perpendicular to the magnetic field. Evidently, the velocity component (in this case, the z-component) is unaffected by the magnetic field. This makes sense, because the cross product of two vectors in the same direction is zero. It is easy to show (see the problems) that this force is still perpendicular to the velocity vector, so no work is done on the charged particle and its speed remains constant. We can think of the trajectory as a superposition of uniform circular motion in the xy-plane with constant motion in the z-direction. The resulting spiral path is known as a **helix** (more precisely, a right-circular helix).

One application of the motion of a charged particle in a uniform magnetic field is in the **bubble chamber**, a device used by experimental high-energy physicists to study the properties of charged particles ejected from various collisions and nuclear reactions. A bubble chamber is normally filled with liquid hydrogen. When a fast-moving charged particle travels through the liquid, it vaporizes small amounts of the liquid and leaves a visible trail, which can be photographed. Normally, the liquid container is placed between the poles of a large electromagnet (electromagnetism will be discussed in Chapter 25), which creates a nearly uniform magnetic field in the region of interest. As we have seen, a charged particle follows a curved trajectory in a magnetic field, with the trajectory being a circle in a precisely uniform field. From the photograph, the radius of curvature of the charged particle can be measured. As we have also seen, the direction of the charged particle's curve reveals whether it is positively or negatively charged.

For example, suppose a particle of known charge q is observed to have a radius of curvature R when moving perpendicular to a uniform magnetic field of magnitude B. By Equation (24.7), the momentum of that particle is

$$p = mv = qBR. \tag{24.9}$$

If the velocity of the particle can be measured independently (e.g., by a time of flight measurement in a region with no magnetic field), the mass of the particle can then be found. If the charge is unknown, then the charge to mass ratio of the particle can still be found:

$$\frac{q}{m} = \frac{v}{BR}. \tag{24.10}$$

One word of caution: If the speed of the charged particle is a significant fraction of the speed of light, the dynamics are governed by the rules of special relativity, and the momentum of a particle is no longer simply its mass times its velocity. However, the relation $p = qBR$ still holds in special relativity and can be used along with the correct relationship between mass, velocity, and momentum to relate these quantities to the charge, magnetic field, and radius of curvature.

Example 24.5. In a bubble chamber with a uniform magnetic field of magnitude 1.45 T, a positive pi-meson traveling with a speed 9.49×10^5 m/s perpendicular to the field is observed to have a radius of curvature 1.02 mm. (a) Find the charge to mass ratio of the pi-meson. (b) Assuming the change on the meson is e, what is its mass? (c) Describe the path of electron with the same speed in the same magnetic field.

Solution. (a) The charge to mass ration is given by Equation (24.10):

$$\frac{q}{m} = \frac{v}{BR} = \frac{9.49 \times 10^5 \text{ m/s}}{(1.45 \text{ T})(1.02 \times 10^{-3} \text{ m})} = 6.42 \times 10^8 \text{ T}^{-1}\text{s}^{-1} = 6.42 \times 10^8 \text{ C/kg}.$$

(b) With $q = e$, we have

$$m = \frac{e}{q/m} = \frac{1.602 \times 10^{-19} \text{ C}}{6.42 \times 10^8 \text{ C/kg}} = 2.50 \times 10^{-28} \text{ kg}.$$

This is significantly greater than the mass of an electron and less than the mass of a proton. In fact, mesons tend to be intermediate in mass between the lightest particles, leptons (which include electrons, muons, and neutrinos) and the heavy particles known as baryons (which include protons and neutrons).

(c) First, because an electron has a negative charge, its curvature is opposite to that of the positive pi-meson. Its radius of curvature is

$$R = \frac{mv}{qB} = \frac{(9.11 \times 10^{-31} \text{ kg})(9.49 \times 10^5 \text{ m/s})}{(1.602 \times 10^{-19} \text{ C})(1.45 \text{ T})} = 3.72 \times 10^{-6} \text{ kg·m·s}^{-1}\text{C}^{-1}\text{T}^{-1} = 3.72\ \mu\text{m}.$$

An electron has less momentum than a pi-meson with the same speed, and is therefore bent into a smaller circular path by the magnetic field. △

Consider again the circular motion of a charged particle moving perpendicular to a uniform magnetic field. The speed of the particle is the circumference of the circle divided by the period T, or $v = 2\pi R/T$. But from Equation (24.7), the speed is also given by $v = qBR/m$. Equating these two expressions for the speed, we find

$$\frac{2\pi R}{T} = \frac{qBR}{m}.$$

Solving for the period of revolution,

$$T = \frac{2\pi m}{qB}. \tag{24.11}$$

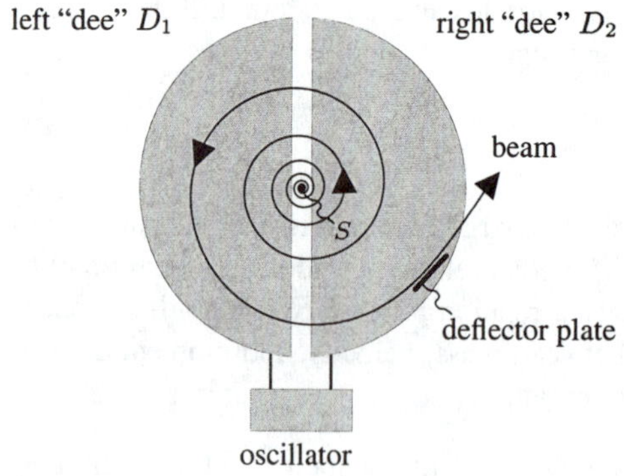

Figure 24.7. A cyclotron.

Notice that the period is independent of the radius, for a particle with a given charge to mass ratio q/m. Using this principle, Ernest O. Lawrence (1901–1958) developed the **cyclotron** in the early 1930s. A schematic of a cyclotron is shown in Figure 24.7. The circular region containing a uniform magnetic field is separated into two halves, called "dees," because of their shape. A particle injected from a source S near the center of the cyclotron with a low speed has a circular trajectory with a small radius. A potential difference ΔV is placed between the two dees, so that when the particle moves from one dee (D_1 in the diagram) to the other (D_2), it gets a boost of kinetic energy $\Delta K = q\Delta V$. When the particle is in D_2, the sign of the potential difference is reversed, so the particle gets another kinetic energy boost when it returns to D_1. The process can be repeated at regular intervals, because the period is independent of the radius. The frequency of circular motion, known as the **cyclotron frequency**, is the reciprocal of the period:

$$\nu_c = \frac{1}{T} = \frac{qB}{2\pi m}. \tag{24.12}$$

The potential difference on the dees must be switched at a frequency of $2\nu_c$, because a particle crosses from one dee to another twice in each orbit. As you can see from Equation (24.9), the radius of the particle's orbit grows in proportion to its speed. Thus, the highest possible speed is limited by the size of the cyclotron.

During World War II, Lawrence designed cyclotrons that were built at Oak Ridge, Tennessee, for the purpose of accelerating uranium-containing ions, so that they could be separated according to their isotope mass (see the discussion of the mass spectrograph, below). Cyclotrons have been used by physicists to accelerate charged particles that can then be shot into fixed targets, in order to initiate nuclear reactions.

In a **synchrotron**, charged particles are accelerated around a circular ring of fixed radius. With a fixed radius, the period of circular motion is not independent of the particle's speed, and therefore careful design is needed to (1) adjust the magnetic fields to keep the particles in the ring, and (2) give particles accelerating boosts of energy at the proper times, to accelerate them to the highest possible energies. The largest particle accelerators in the world, such as those at Fermilab in Illinois and CERN on the Switzerland/France border, are large synchrotrons that can accelerate particles to kinetic energies on the order of 1 TeV.

In Figure 24.8, we show a schematic of a **mass spectrometer**. A particle with charge q can be accelerated from rest through a known potential difference ΔV to achieve a kinetic energy $K = q\Delta V$. If the particle is then injected perpendicular to a uniform magnetic field (with magnitude B), it follows a circular trajectory. For nonrelativistic speeds

$$K = \frac{1}{2}mv^2 = q\Delta V,$$

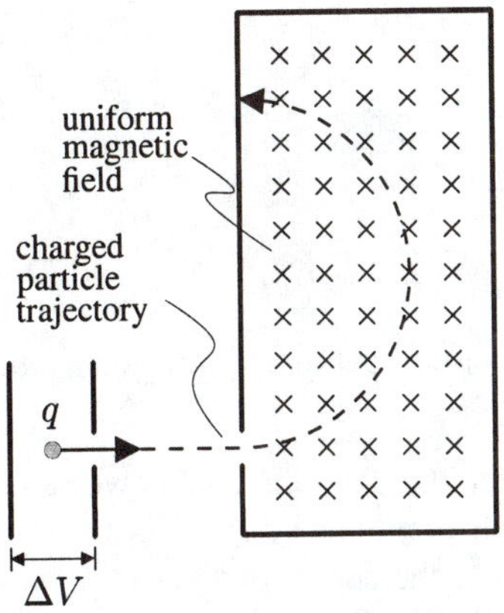

Figure 24.8. A mass spectrometer.

and thus the speed is

$$v = \sqrt{\frac{2q\Delta V}{m}}.$$

Then by Equation (24.7), the speed of the charged particle is related to the radius of curvature R by

$$v = \frac{qBR}{m}.$$

Equating these two expressions for the speed,

$$\sqrt{\frac{2q\Delta V}{m}} = \frac{qBR}{m}.$$

Rearranging,

$$\frac{q}{m} = \frac{2\Delta V}{B^2R^2}. \tag{24.13}$$

Therefore, if the magnetic field is known and the radius of curvature can be measured, the particle's charge to mass ratio can be found. In many cases, the particle's charge is known (for ions it is some integer multiple of e), in which case the particle's mass can be deduced. Particles with the same charge will be separated according to their masses, with larger masses having larger radii of curvature. The mass spectrometer was developed by Francis Aston in 1919 and has been a useful tool for physicists and chemists interested in separating atoms or molecules by mass.

24.3 The Force on a Current-Carrying Conductor

Consider a conducting wire, as represented in Figure 24.9(a), which carries some current I. What happens if this current-carrying wire is in a magnetic field? A charge q traveling with a velocity $\vec{v}$ in the wire experiences a force $\vec{F} = q\vec{v} \times \vec{B}$ in the magnetic field $\vec{B}$. What we want to know is the net force on the wire, due to the forces on all the moving charges.

(a) Current-carrying wire (b) One segment of the wire

Figure 24.9. Setup for computing the force on a current-carrying wire.

We do not show the magnetic field in the diagram, so that we can leave open the questions of the field's direction with respect to the wire and whether the field is uniform. Notice also that the wire is not necessarily straight. Therefore, both the velocity $\vec{v}$ of the charge carriers in the wire and the value of the magnetic field $\vec{B}$ can be different at each point in the wire. To find the net force on the wire, it is necessary to sum the forces on all the individual charge carriers. This should remind you of a Riemann sum from calculus. Specifically, the construction we need is that of a line integral. Think of the wire as a curve C. Following our normal procedure in calculus, let's break the curve that runs from point P to point Q into n smaller segments $\Delta\vec{s}$ [Figure 24.9(b)], with the direction of $\Delta\vec{s}$ tangent to the wire and in the direction of positive current flow. From the definition of current, the amount of charge Δq_i passing through the ith segment in a time Δt is

$$\Delta q_i = I\Delta t. \tag{24.14}$$

The velocity of a charge carries passing through the ith segment $\Delta\vec{s}_i$ is approximately

$$\vec{v}_i = \frac{\Delta\vec{s}_i}{\Delta t}. \tag{24.15}$$

We can write the force $\vec{F}_i$ on the ith segment of the wire as

$$\vec{F}_i = \Delta q_i \, \vec{v}_i \times \vec{B}_i \tag{24.16}$$

where $\vec{B}_i$ is the value of the magnetic field at that segment. Using the result of Equations (24.14) and (24.15) in Equation (24.16),

$$\vec{F}_i = (I\Delta t)\left(\frac{\Delta\vec{s}_i}{\Delta t} \times \vec{B}_i\right)$$

or

$$\vec{F}_i = I\left(\Delta\vec{s}_i \times \vec{B}_i\right).$$

The net force on the entire length of wire (from P to Q) is

$$\vec{F} = \sum_{i=1}^{n} \vec{F}_i = \sum_{i=1}^{n} I\left(\Delta\vec{s}_i \times \vec{B}_i\right). \tag{24.17}$$

Equation (24.17) expresses the force as a Riemann sum. It is only an approximation, because of the assumptions we have made [Equation (24.15), as well as assuming that the magnetic field is constant over each of the small segments]. The result becomes exact in the limit as n approaches infinity, or equivalently, as $\Delta\vec{s}_i$ approaches zero. Then as you know from calculus, the Riemann sum

$$\vec{F} = \lim_{n\to\infty} \sum_{i=1}^{n} I\left(\Delta\vec{s}_i \times \vec{B}_i\right)$$

becomes a type of line integral, which we denote as

$$\vec{F} = \int_C I\, d\vec{s} \times \vec{B}. \tag{24.18}$$

To be precise about the definition, we must think of the curve C as parametrized by $\vec{r}(u)$ for u in some interval $[a, b]$. We define

$$\int_C I\, d\vec{s} \times \vec{B} = \int_a^b I\vec{r}\,'(u) \times \vec{B}(\vec{r}\,'(u))\, du.$$

In practice, we rarely use this definition to compute the force on a current-carrying wire. Instead, we use geometric arguments to evaluate the line integral.

Equation (24.18) is the general expression for the force on a wire carrying a current I through a magnetic field $\vec{B}$ from point P to point Q. The symbol $d\vec{s}$ represents $\Delta\vec{s}$ in the Riemann sum in the limit as $\Delta\vec{s}$ approaches zero. Think of $d\vec{s}$ as a very short segment of wire pointing in a direction tangent to the wire and in the direction of positive current flow. Therefore, you need to exercise caution in computing the cross product inside the integral, because in general $d\vec{s}$ can point in different directions along different parts of the wire. This is similar to the problem of finding the electric field of a continuous charge distribution using Equation (17.11), in which the unit vector $\hat{r}$ points in different directions for different parts of the charge distribution. We will see how to take this feature into account as we look at applications of Equation (24.18).

As our first application, consider a straight segment of wire of length L carrying a current I. The wire makes an angle θ with a uniform magnetic field, as shown in Figure 24.10, and we wish to know the force on the wire. It is just this sort of problem for which the general prescription in Equation (24.18) works well. First, we will analyze the cross product in the integral. For a straight wire, the direction of $d\vec{s}$ is the same for every part of the wire; it is along the wire in the direction of the positive current flow. Therefore, the direction of the vector $d\vec{s} \times \vec{B}$ (as given by the right-hand rule) is straight into the page, perpendicular to both the wire and the magnetic field. Every $d\vec{s}$ in the wire experiences a force in this direction, and so the net force $\vec{F}$ is in the same direction.

The magnitude of $d\vec{s} \times \vec{B}$ is

$$\left\|d\vec{s} \times \vec{B}\right\| = \left\|d\vec{s}\right\| \left\|\vec{B}\right\| \sin\theta.$$

Taking $ds = \left\|d\vec{s}\right\|$ and $B = \left\|\vec{B}\right\|$,

$$\left\|d\vec{s} \times \vec{B}\right\| = (ds)(B)\sin\theta. \tag{24.19}$$

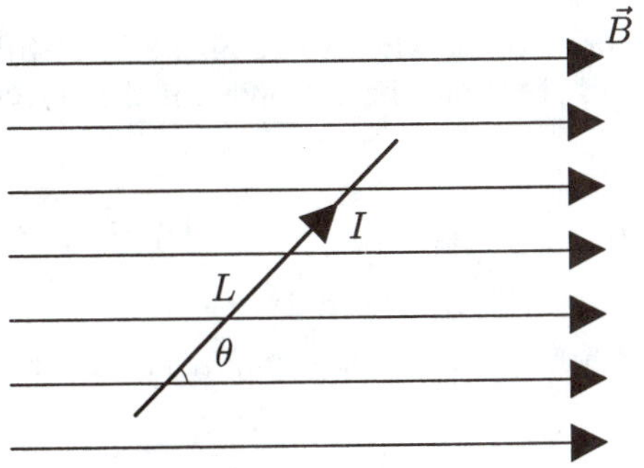

Figure 24.10. Straight segment of wire in a uniform magnetic field.

The magnitude of the force on the wire is given by Equation (24.18) as

$$F = \|\vec{F}\| = \left\| \int_C I\, d\vec{s} \times \vec{B} \right\|$$

where C is the curve in the shape of the wire. Because I is constant and $d\vec{s}$ and $\vec{B}$ are constant for the length of wire,

$$F = I \int_C \|d\vec{s} \times \vec{B}\|.$$

Using Equation (24.19) and factoring constants out of the integral,

$$F = I \int_C (ds)(B) \sin\theta = IB \sin\theta \int_C ds.$$

The remaining line integral is simply the length L of the wire, so

$$F = ILB \sin\theta. \tag{24.20}$$

Equation (24.20) gives the magnitude of the force on length L of straight wire carrying a current I, when the wire makes an angle θ with a uniform magnetic field of magnitude B. Recall that we determined earlier that the direction of the force on the wire in Figure 24.10 is into the page. It is possible to combine the magnitude and direction in a single expression, written as a cross product. Define $\vec{L}$ as a vector with magnitude L (the length of the straight wire segment) and direction in the direction of positive current flow. Then the formula

$$\vec{F} = I\vec{L} \times \vec{B} \tag{24.21}$$

gives the correct magnitude and direction of the force. You can use this result to find the magnitude and direction of the force on a straight wire in a uniform magnetic field, and we will do so in the next section.

Example 24.6. A semicircular loop of wire lies in the xy-plane, with the center of curvature at the origin and the endpoints on the x-axis, as shown in Figure 24.11(a). The radius of curvature is R, and a constant current I flows in the counterclockwise direction. A uniform magnetic field of magnitude B points in the $+y$-direction. What is the force on the wire?

Solution. For any infinitesimal segment of the wire $d\vec{s}$ [see Figure 24.11(b)], the right-hand rule tells us that $d\vec{s} \times \vec{B}$ points straight into the page. Therefore, the net force is in that direction. The magnitude of the force is

$$F = \left\| \int_C I\, d\vec{s} \times \vec{B} \right\| = \int_C I(ds)(B) \sin\theta$$

where C is a curve in the shape of the wire. In this problem, I and B are constant and can be factored out of the integral:

$$F = IB \int_C ds\, \sin\theta.$$

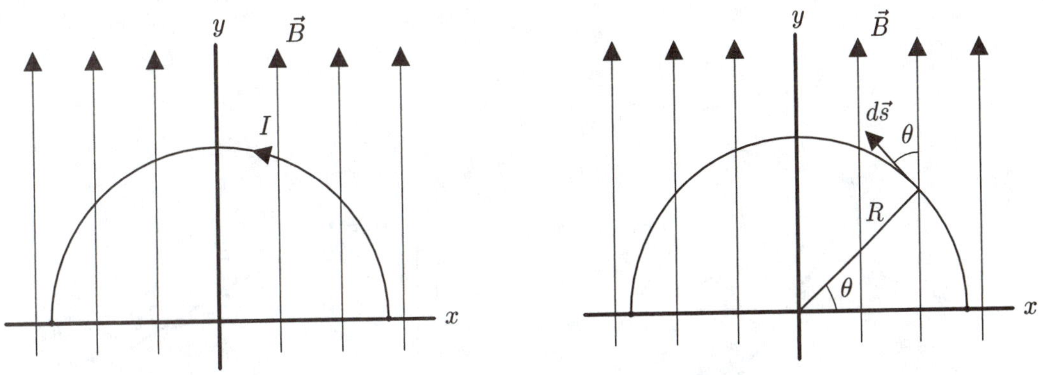

(a) Semicircular current-carrying wire in a magnetic field

(b) Analyzing the force on a generic segment

Figure 24.11. Schematic for Example 24.6.

However, $\sin\theta$ is not constant. But notice that the angle θ between $d\vec{s}$ and $\vec{B}$ is the same angle that the radius makes with the $+x$-axis—that is, it is the angular coordinate θ of polar coordinates. We can express the integral in polar coordinates. The arc length is $ds = R\,d\theta$, and the endpoints of the wire are at $\theta = 0$ and $\theta = \pi$. Thus,

$$F = IB\int_0^{\pi}(R\,d\theta)\sin\theta = IBR\int_0^{\pi}\sin\theta\,d\theta.$$

Evaluating the definite integral is straightforward:

$$F = IBR(-\cos\theta)\Big|_0^{\pi} = -IBR[-1-(1)] = 2IBR.$$

Notice that the result is dimensionally correct. △

24.4 Current Loops and Magnetic Dipoles in Magnetic Fields

24.4.1 Torque and Potential Energy

In this section, we extend the study of forces on current-carrying wires in magnetic fields to closed loops. This is a realistic and sensible thing to try, because you know from your study of electric circuits that a complete circuit is needed for current to flow.

Consider the closed *rectangular* current loop of wire shown in Figure 24.12, with sides of length a and b. The current I is everywhere the same in the wire, and the loop sits in a uniform magnetic field of magnitude B pointing in the $+z$-direction. The sides of the rectangle labeled 1 and 3 are parallel to the x-axis, but the sides labeled 2 and 4 make an angle θ with the y-axis, so that the plane of the loop makes an angle θ with the xy-plane. In Figure 24.12(b), we show a view looking down a line parallel to the x-axis passing through the center of the loop, so that side 2 is closest to us, side 1 is at the upper left, and side 3 at the lower right. Side 4 is hidden from view behind side 2.

We can use Equation (24.21) from the preceding section to determine the force on each of the four sides of the loop. Using the right-hand rule, we see that the force on side 1 is in the $-y$-direction [as shown in

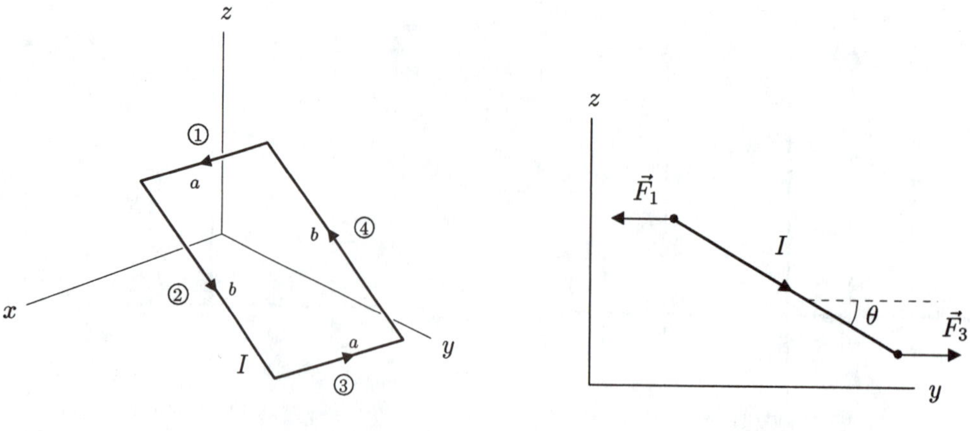

(a) View in three dimensions (b) View in the yz-plane

Figure 24.12. Closed rectangular current loop in a uniform magnetic field.

Figure 24.12(b)]. From Equation (24.21), the magnitude of the force on side 1 is

$$F_1 = IaB \sin \frac{\pi}{2} = IaB.$$

Similarly, the force on side 3 is in the $+y$-direction [also shown in Figure 24.12(b)] and has a magnitude

$$F_3 = IaB \sin \frac{\pi}{2} = IaB.$$

Notice that $F_1 = F_3$. Turning to the other two sides, the right-hand rule tells us that the force on side 2 is in the $+x$-direction and has a magnitude

$$F_2 = IbB \sin\left(\frac{\pi}{2} + \theta\right)$$

and the force on side 4 is in the $-x$-direction and has a magnitude

$$F_4 = IbB \sin\left(\frac{\pi}{2} - \theta\right).$$

From trigonometry we know

$$\sin\left(\frac{\pi}{2} + \theta\right) = \sin\left(\frac{\pi}{2} - \theta\right)$$

and therefore $F_2 = F_4$.

Looking at the forces in pairs, we see that $\vec{F}_1$ and $\vec{F}_3$ have equal magnitudes and opposite directions, and therefore their (vector) sum is zero. The same is true for the pair $\vec{F}_2$ and $\vec{F}_4$. The net force on the rectangular wire loop is

$$\vec{F}_1 + \vec{F}_2 + \vec{F}_3 + \vec{F}_4 = 0.$$

However, there is a net *torque* on the loop, due to the combined effect of $\vec{F}_1$ and $\vec{F}_3$. Recall that the torque on a rigid body about some point P due to a force $\vec{F}$ is $\vec{\tau} = \vec{r} \times \vec{F}$, where $\vec{r}$ is a displacement vector from P

to the point where the force is applied. We compute the torque about the center of the loop. According to the right-hand rule, both $\vec{F}_1$ and $\vec{F}_3$ produce a torque in the $+x$-direction with magnitude $rF\sin\theta$, or

$$\tau_1 = \tau_3 = \left(\frac{b}{2}\right)(IaB)\sin\theta.$$

The magnitude of the net torque is

$$\tau = \tau_1 + \tau_3 = 2\tau_1$$

or

$$\tau = IabB\sin\theta. \tag{24.22}$$

We now develop a shorthand notation that allows us to generalize from the previous example and express the torque as a vector quantity. For the geometry of the example we have been considering (the rectangular wire carrying a current I), the **magnetic dipole moment** $\vec{\mu}$ is defined as a vector that has a *magnitude* equal to the product of the current and the area enclosed by the loop and a *direction* perpendicular to the plane of the loop. There are two directions (opposite to one another) perpendicular to the plane of the loop, so we use another right-hand rule to define the proper one. If you curl the fingers of your right hand around the loop in the direction of positive current flow, your right thumb points in the direction of the magnetic dipole moment vector. In Figure 24.13, we show the rectangular current-carrying loop from the same view as in Figure 24.12(b), but with the magnetic dipole moment vector drawn in. For the rectangular loop, the magnitude of the magnetic dipole moment is

$$\mu = (\text{current})(\text{area}) = (I)(ab) = Iab.$$

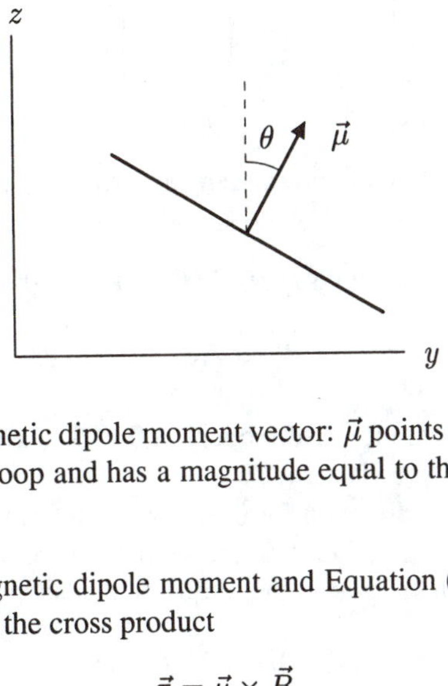

Figure 24.13. Defining the magnetic dipole moment vector: $\vec{\mu}$ points perpendicular to the plane of the current-carrying loop and has a magnitude equal to the product of the current and the area of the loop.

From this definition of the magnetic dipole moment and Equation (24.22), you can see that the vector torque can be expressed in terms of the cross product

$$\vec{\tau} = \vec{\mu} \times \vec{B}. \tag{24.23}$$

You should verify that under the conditions described previously for the rectangular loop, Equation (24.23) gives the correct magnitude and direction of the torque. It turns out that Equation (24.23) is valid for *any* planar loop, although the proof of this fact is beyond the scope of this course.

One important application using a current-carrying loop in a magnetic field is an electric motor. Suppose the rectangular current-carrying loop in Figure 24.12 is fixed about an axis parallel to the x-axis that passes through the center of the loop. We have already shown that there is a torque about this axis, in the $+x$-direction when the loop is oriented as shown in that diagram. The torque is in the $+x$-direction as long as the angle θ shown in Figure 24.12(b) (or Figure 24.13) is between 0 and π. However, when that angle is between π and 2π, the torque is in the $-x$-direction. In order to produce an efficient motor, an engineer would like the torque to be in the same direction no matter what the orientation of the loop, so that the loop is always being turned in the same direction. One way to accomplish this is to reverse the direction of the current whenever the angle θ is between π and 2π.

Example 24.7. The circular loop shown in Figure 24.14 lies in the xy-plane. The loop carries a constant current I as shown, and there exists a uniform magnetic field of magnitude B in the $+y$-direction. By adding the contribution of each segment ds on the loop (i.e., computing a definite integral), (a) show that the net force on the loop is zero, and (b) show that the net torque on the loop about the x-axis agrees with that given by Equation (24.23).

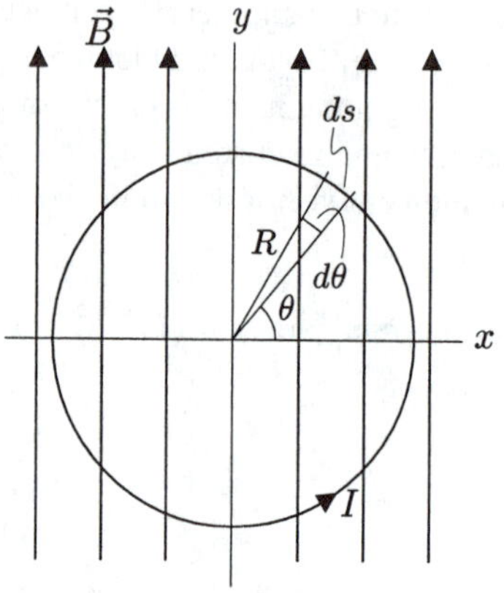

Figure 24.14. Circular loop of current-carrying wire in a uniform magnetic field.

Solution. (a) The magnitude of the force on the segment ds is given by Equation (24.18) as

$$\left\| I\, d\vec{s} \times \vec{B} \right\| = I(ds)(B)\sin\theta.$$

However, the *direction* of the force is in the $-z$-direction for $0 < \theta < \pi$ and in the $+z$-direction for $\pi < \theta < 2\pi$. Therefore, by symmetry, the net force is zero.

(b) Because the force is in the $-z$-direction for $0 < \theta < \pi$ and in the $+z$-direction for $\pi < \theta < 2\pi$, a net torque exists in the $-x$-direction. Consider first the semicircle from $\theta = 0$ to $\theta = \pi$. Noticing that the segment ds is a distance $R\sin\theta$ from the x-axis, the torque on that segment about the x-axis is

$$(R\sin\theta)[I(ds)(B)\sin\theta],$$

and therefore the torque on the upper semicircle is

$$\tau_1 = \int_0^{\pi} R\sin\theta)[I(ds)(B)\sin\theta].$$

From the geometry of the circle, $ds = R\,d\theta$. Inserting this expression and factoring out constants from the definite integral, we have

$$\tau_1 = IR^2 B \int_0^{\pi} \sin^2\theta\, d\theta.$$

Evaluating the definite integral using a standard table, trig identity, or computer package, we have

$$\int_0^{\pi} \sin^2\theta\, d\theta = \frac{\pi}{2}.$$

Therefore,

$$\tau_1 = \frac{IR^2 B\pi}{2}.$$

By symmetry, we know the torque on the lower semicircle (from $\theta = \pi$ to $\theta = 2\pi$) is the same. Thus, the net torque on the circular loop has a magnitude

$$\tau = 2\tau_1 = IR^2 B\pi.$$

Because the area of the loop is πR^2, we have for the magnitude of the magnetic dipole moment $\mu = I(\pi R^2)$, and so $\tau = \mu B$. It is straightforward to show that this agrees with Equation (24.23). Starting with the equation $\vec{\tau} = \vec{\mu} \times \vec{B}$, we see that in this case $\vec{\mu}$ points in the $+z$-direction, so $\vec{\tau} = \vec{\mu} \times \vec{B}$ points in the $-x$-direction as required. With the angle between $\vec{\mu}$ and $\vec{B}$ equal to $\pi/2$, we have $\tau = \|\vec{\mu} \times \vec{B}\| = \mu B \sin(\pi/2) = \mu B$, in agreement with the result we found from computing the definite integral. △

Another application of the study of a current loop in a magnetic field occurs in atomic physics. Electrons and other subatomic particles possess an intrinsic angular momentum, analogous to a classical object such as a billiard ball spinning about an axis. Because electrons are charged, this gives them an intrinsic magnetic dipole moment, which is subject to a torque in the presence of an applied magnetic field. However, according to quantum theory, the angular momentum of the electron is *quantized*—that is, it can take on only certain specific values. Thus, the torque on the electron does not necessarily change its angular momentum. It turns out that for subatomic particles, a more important property than torque is the potential energy of a magnetic dipole in a magnetic field, and it is to this subject that we turn our attention now.

In linear motion, the change in potential energy associated with a conservative force is the negative of the work done by that force, when an object is moved against the field by an external force. Similarly, in rotational motion, the change in potential energy associated with a torque is the negative of the work done by that torque, when an object is moved against the field by an external torque. Thus, in turning a rigid body from an angle θ_a to θ_b the change is potential energy is

$$\Delta U = -\text{ Work done by torque} = \int_{\theta_a}^{\theta_b} \tau\, d\theta.$$

Notice that this situation is analogous to the torque of an electric dipole in an electric field, which we studied in Section 17.3. As in that problem, the minus sign has vanished, because we are considering the torque applied *by* the field, rather than the force of an external agent doing work *against* the field. In the case we are now considering, the rigid body has a magnetic dipole moment $\vec{\mu}$, and is turned from one angle to another by the torque due to a magnetic field. For a uniform magnetic field $\vec{B}$,

$$\tau = \left\|\vec{\mu} \times \vec{B}\right\| = \mu B \sin\theta$$

where θ is the angle between $\vec{\mu}$ and $\vec{B}$. The change in potential energy is

$$\Delta U = \int_{\theta_a}^{\theta_b} \tau \, d\theta = \int_{\theta_a}^{\theta_b} \mu B \sin\theta \, d\theta = \mu B \int_{\theta_a}^{\theta_b} \sin\theta \, d\theta$$

where we have factored the constant magnitudes μ and B out of the integral. Now the definite integral is straightforward:

$$\Delta U = \mu B(-\cos\theta)\Big|_{\theta_a}^{\theta_b}$$

or

$$\Delta U = -\mu B\left(\cos\theta_b - \cos\theta_a\right). \tag{24.24}$$

As usual, we can turn the potential energy *difference* into a potential energy *function* by assigning a potential energy of zero to a specific location. Because we have a cosine function, it is convenient to define $U = 0$ at a fixed reference angle $\theta_a = \pi/2$. Then at any other angle $\theta = \theta_b$, the potential energy is

$$\Delta U = U(\theta) - U(\theta_a) = U(\theta) - 0 = -\mu B\left(\cos\theta - \cos\frac{\pi}{2}\right) = -\mu B\cos\theta.$$

Thus, the desired potential energy function is

$$U(\theta) = -\mu B\cos\theta, \tag{24.25}$$

which clearly satisfies the requirement $U(\theta) = 0$. Because θ is the angle between $\vec{\mu}$ and $\vec{B}$, we can express the potential energy function more compactly in terms of the dot product

$$U(\theta) = -\vec{\mu}\cdot\vec{B}. \tag{24.26}$$

The results expressed in Equations (24.26) and (24.26) make sense if we look at how the potential energy varies as a function of θ, the angle between $\vec{\mu}$ and $\vec{B}$. The potential energy varies from a minimum of $U = -\mu B$ at $\theta = 0$ to $U = +\mu B$ at $\theta = \pi$. The torque on the magnetic dipole in the magnetic field tends to turn the dipole toward alignment with the field at $\theta = 0$. In doing so, the field does work on the dipole, lowering its potential energy whenever the angle θ is reduced (from $\theta = \pi$ toward $\theta = 0$). This is analogous to Earth's gravitational field doing work by moving a particle from some altitude toward the ground.

Example 24.8. Find the maximum difference in potential energy for two different orientations of these magnetic dipole moments in a uniform 0.25-T magnetic field: (a) a square coil of wire 12 cm on a side carrying a current of 10 A, and (b) a small grain of iron with a permanent magnetic dipole moment of magnitude $\mu = 4.5 \times 10^{-14}$ J/T.

Solution. In general, the greatest difference in potential energy occurs between the orientations $\theta = 0$ and $\theta = \pi$ (where θ is the angle between $\vec{\mu}$ and $\vec{B}$). Then

$$\Delta U = -\mu B\cos\pi - (-\mu B\cos 0) = -\mu B(-1-1) = 2\mu B.$$

(a) In this case, the magnitude of the magnetic dipole moment is

$$\mu = IA = (10\text{ A})(0.12\text{ m})^2 = 0.144\text{ A·m}^2\text{·T} = 0.144\text{ J/T}.$$

Then

$$\Delta U = 2\mu B = 2(0.144\text{ J/T})(0.25\text{ T}) = 0.072\text{ J}.$$

(b) Similarly,

$$\Delta U = 2\mu B = 2(4.5\times10^{-14}\text{ J/T})(0.25\text{ T}) = 2.3\times10^{-14}\text{ J}.$$

△

24.4.2 Comparison with Electric Dipoles in Electric Fields

Our discussion of torque and potential energy of magnetic dipoles in magnetic fields should remind you of our previous study (Section 17.3). There we found that the torque on an electric dipole moment $\vec{p}$ in a uniform electric field $\vec{E}$ is

$$\vec{\tau} = \vec{p} \times \vec{E}, \tag{17.17}$$

which looks quite similar to the analogous expression in magnetism,

$$\vec{\tau} = \vec{\mu} \times \vec{B}. \tag{24.23}$$

The potential energy function of the electric dipole in the uniform electric field is

$$U(\theta) = -\vec{p} \cdot \vec{E}, \tag{17.19}$$

which is again quite similar to the analogous expression in magnetism,

$$U(\theta) = -\vec{\mu} \cdot \vec{B}. \tag{24.26}$$

Evidently, there is something quite similar about electricity and magnetism! This is but one of many intriguing similarities and connections between electricity and magnetism. This particular set of equations, relating the behavior of electric and magnetic dipoles, is rather striking when we recall that the electric dipole consists of a pair of isolated electric charges—monopoles—while magnetic monopoles do not exist, and in fact, our model for the magnetic dipole was a closed current loop. To make the connection between electricity and magnetism more complete, recall that fundamentally a "current loop" is a circulating *electric* charge that creates a *magnetic* dipole. This connection will be explained in Chapter 25, when we study electromagnetism, and a more complete connection will be established with Maxwell's equations in Chapter 27.

24.5 The Hall Effect

Another application of moving charges in a magnetic field is in the **Hall effect**, named for the physicist Edwin Hall, who discovered it in 1879. In Figure 24.15, we show a schematic of the experimental setup designed to observe the Hall effect. A thin strip of metal is attached to a battery has shown in Figure 24.15(a), so that a current I flows through the metal. For simplicity, let's assume the metal strip has a rectangular cross section (perpendicular to the current flow), with the width d as shown in the diagram.

Positive current flows from the positive terminal to the negative terminal of the battery. But in a metal, the charge carriers are electrons, which flow in the opposite direction with an average velocity $\vec{v}$ as shown. The magnitude of this average velocity is the drift speed we discussed in Section 19.3.2. To show the Hall effect, a (normally uniform) magnetic field is applied perpendicular to the current-carrying strip; in this case, the magnetic field is directed straight into the page as shown. The force on a moving charge is $\vec{F}_B = q\vec{v} \times \vec{B}$, so by the right-hand rule the magnetic field causes a force toward the left on the moving electrons. As a result, there is a surplus of negative charge on the left side of the strip (as we view it in the diagram), and a surplus of equal magnitude of positive charge on the right side of the strip, because the net charge on the strip is zero. The buildup of charges causes a potential difference, with the right (positive) side at a higher electric potential than the left (negative) side. This is essentially what the Hall effect is about, and the transverse potential difference is called the **Hall potential difference**. The Hall potential difference can be measured directly by attaching a voltmeter across the two sides of the strip, as shown in Figure 24.15(b).

The charge buildup we have just described does not continue indefinitely. As you know from Chapter 18, the existence of a potential difference between the two sides of the strip means that there is an electric field

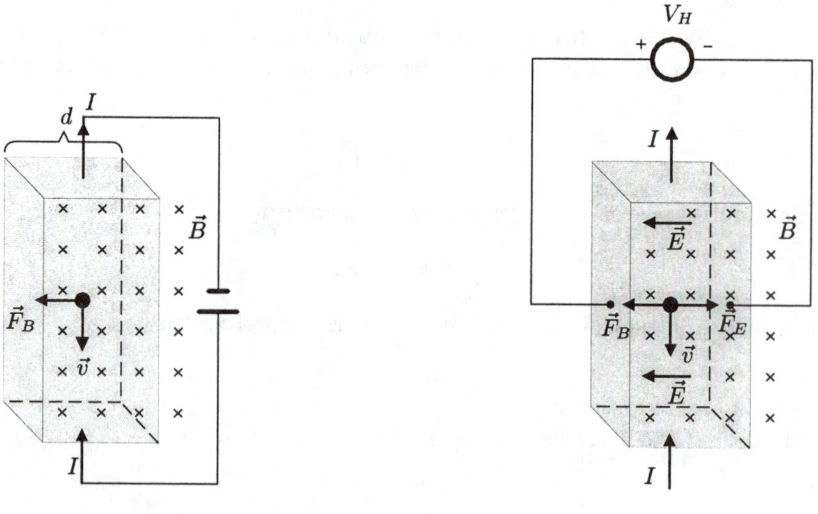

(a) Battery attached to the metal strip

(b) Voltmeter placed to measure the Hall voltage V_H

Figure 24.15. Setup for studying the Hall effect.

in the metal, pointing from right to left as shown. This electric field causes a force $\vec{F}_E = q\vec{E}$ to the *right* on the electrons ($q = -e$). The charge buildup continues until the electric field is sufficiently high so that $\|\vec{F}_E\| = \|\vec{F}_B\|$. At this point, because $\vec{F}_E$ and $\vec{F}_B$ point in opposite directions, we have $\vec{F}_E + \vec{F}_B = 0$, and the electrons passing through the metal are no longer deflected sideways. This is a similar physical situation as we encountered in the Thomson experiment: crossed electric and magnetic fields producing no net force on electrons passing though the crossed fields. Normally, this equilibrium occurs quite quickly.

When the magnitudes of the electric and magnetic fields are equal,

$$\|\vec{F}_E\| = \|\vec{F}_B\|,$$

so

$$eE = evB$$

or

$$v = \frac{E}{B}. \tag{24.27}$$

Recall from Section 19.3.2 [see Equation (19.24)] that the electron drift speed in a metal is

$$v = \frac{I}{neA},$$

where A is the cross-sectional area of the wire and n is the density (number per unit volume) of conduction electrons. Also, we know from Chapter 18 (see Example 18.17) that the magnitude of the electric field (E) due to the charge built up on the sides of the strip is $E = V_H/d$, where V_H is the Hall potential difference measured by the voltmeter in Figure 24.15(b). Inserting these relationships for v and E in Equation (24.27),

$$\frac{I}{neA} = \frac{V_H/d}{B}.$$

Solving for n,

$$n = \frac{IBd}{eAV_H}. \tag{24.28}$$

In Equation (24.28), we have chosen to solve for n to demonstrate a point: The Hall effect allows us to make an experimental measurement of n, the density of conduction electrons in a metal. All the quantities on the right side of the equation are easily measured, and their numerical values can be inserted into the equation to obtain a numerical result for n. We illustrate this fact with an example.

Example 24.9. In a Hall effect measurement, a copper strip is made with rectangular cross section, as shown in Figure 24.15. The strip is made with a width $d = 1.05$ cm and a thickness of 2.50 μm. A student does an experiment in which the current passing through the strip is 7.25 A and the uniform magnetic field perpendicular to the strip is 0.750 T. If the student measures a Hall potential difference of 0.165 mV, what is the measured value of the conduction electron density in copper? Compare this result with the accepted value of 8.47×10^{28} m^{-3}.

Solution. The number density of conduction electrons is given by Equation (23.28). With a rectangular cross section, the cross-sectional area A is simply the width times the thickness. Inserting numerical values,

$$n = \frac{(7.25\ \text{A})(0.750\ \text{T})(0.0105\ \text{m})}{(1.60 \times 10^{-19}\ \text{C})(0.0105\ \text{m})(2.50 \times 10^{-6}\ \text{m})(0.165 \times 10^{-3}\ \text{V})} = 8.24 \times 10^{28}\ \text{A·T·m}^{-1}\text{C}^{-1}\text{V}^{-1}.$$

The SI units reduce to m^{-3}, so

$$n = 8.24 \times 10^{28}\ \text{m}^{-3}.$$

This is quite close to the accepted value, with a percentage error of

$$\frac{8.47 \times 10^{28}\ \text{m}^{-3} - 8.24 \times 10^{28}\ \text{m}^{-3}}{8.47 \times 10^{28}\ \text{m}^{-3}} \times 100\% = 2.72\%.$$

The fact that the numerical result is low may be a random error in the experiment, or it may be due to the fact that impurities in this copper sample make its electrical conductivity lower than that of pure copper. If the student varies the current and magnetic field, measuring the Hall potential difference under a variety of conditions, she can determine the extent of the random errors in the experiment and get a better experimental value for n. △

You have just seen how the Hall effect can be used to measure the density of conduction electrons in a metal. It can also be used to determine the nature of the charge carriers in a semiconductor. In some semiconductors, there exist regions deficient in electrons called *holes* (although the material is still electrically neutral overall). These holes can act as positive charges that conduct by traveling in the direction we assign to positive current. Some semiconductors have a majority of holes, and others a majority of electrons.

Referring again to Figure 24.15(a), if the material is a semiconductor with hole conduction, the positive hole would travel from the bottom to the top of the diagram. The magnetic force $\vec{F} = q\vec{v} \times \vec{B}$ on a positive charge carrier moving that way is to the left in the diagram, causing a buildup of positive charge on the left side of the strip and leaving a buildup of negative charge on the left side. This is the opposite of the situation we encountered with electron conduction in metals. Therefore, the *sign* of the Hall potential difference tells us whether the majority of charge carriers are electrons or holes.

24.6 Problems

24.1 Background and History

There are no problems for this section.

24.2 The Magnetic Force on an Electric Charge

In Problems 1–7, find the force on the given point charge q moving with the given velocity in a *uniform* magnetic field.

1. $q = 1.50\,\mu$C, moving in the $-x$-direction at 250 m/s in a 0.25-T magnetic field pointing in the $+z$-direction

2. $q = 4.95 \times 10^{-12}$ C, moving in the $-z$-direction at 1340 m/s in a 6.25-T magnetic field pointing in the $-y$-direction

3. $q = -0.035$ C, moving in the $-x$-direction at 55 m/s in a 1.75-T magnetic field pointing in the $+x$-direction

4. $q = -0.035$ C, moving in the $-x$-direction at 55 m/s in a 1.75-T magnetic field pointing in the $+y$-direction

5. $q = -4.50$ mC moving with a velocity $(25.0\,\hat{\imath} + 30.0\,\hat{\jmath}) \times 10^3$ m/s in a magnetic field $\vec{B} = 35.0\,\hat{k}$ mT

6. $q = -e$ moving with a velocity $6.65 \times 10^6\,\hat{\jmath}$ m/s in a magnetic field $\vec{B} = 0.36(\hat{\imath} + \hat{\jmath} + \hat{k})$T

7. $q = 2e$ moving with a velocity $(980\,\hat{\imath} + 317\,\hat{\jmath} - 875\,\hat{k}) \times 10^3$ m/s in a magnetic field $\vec{B} = (1.45\hat{\imath} - 0.350\hat{\jmath} - 1.05\hat{k})$ T

8. At a certain location on Earth's surface, Earth's magnetic field is $\vec{B} = (1.45\,\hat{\imath} + 4.43\,\hat{\jmath} - 2.65\,\hat{k}) \times 10^{-5}$ T, where the $+x$-axis is east, the $+y$-axis is north, and the $+z$-axis is straight up. Find the force on a proton traveling with a speed of 5.00×10^5 m/s (a) due west, and (b) straight down vertically.

9. Show that the SI units of speed multiplied by the SI units of magnetic field produce the SI units of electric field.

10. Find the Lorentz force on a proton traveling at 4.55×10^5 m/s in the $+x$-direction in a region containing a uniform electric field $\vec{E} = 23,400\,\hat{\jmath}$ V/m and a uniform magnetic field $\vec{B} = 156\,\hat{k}$ mT.

11. The *solar wind* contains charged particles that travel outward from the Sun. Some of these particles tend to collect near the north and south poles of Earth, and as they are accelerated in Earth's magnetic field they radiate light, causing the displays known as *auroras*. Explain why charged particles should tend to collect this way, and describe a typical trajectory for such a particle.

12. Show that if SI units are used on the right side of Equation (24.6), those units reduce to C/kg.

13. (a) Using the data for the Thomson experiment described in Example 24.3, write an equation for the trajectory of the electron if the magnetic field is turned off. (b) Suppose the region containing the fields is 1.0 cm long (in the direction of the electron's undeflected path). By how much is the electron deflected in a direction perpendicular to its path by the electric field when the magnetic field is turned off?

14. A proton with a velocity 4500 $\hat{\jmath}$ m/s enters a region in which there exists a uniform magnetic field of 450 $\hat{k}$ mT. What electric field should be applied in this region so that the proton experiences no net force?

15. Using the data of Example 24.3, and assuming the electron travels horizontally a distance of 10.0 cm through the crossed electric and magnetic fields. (a) Over that distance how far does the electron fall due to its weight? (b) By how much would the electric field have to be changed in order to compensate for that fall? Your results should show that the gravitational force on the electron is insignificant in this experiment.

16. For a uniform electric field of magnitude 4.2×10^5 V/m, what is the magnitude of a uniform magnetic field needed to pass an electron of speed 1.5×10^6 m/s with no deflection? Draw the orientation of the vectors $\vec{v}$, $\vec{E}$, and $\vec{B}$ needed for this to occur.

17. Show that the force in Equation (24.8) is perpendicular to the velocity.

18. A particle with mass m, charge q, and velocity $\vec{v}_0 = v_{0x}\,\hat{\imath} + v_{0z}\,\hat{k}$ is injected into a uniform magnetic field $\vec{B}$=$B_0\,\hat{k}$ at time $t = 0$. Write an equation for the particle's (a) velocity, and (b) position as functions of time.

19. A proton traveling at 6.50×10^5 m/s travels in a circular path perpendicular to a uniform magnetic field of magnitude 2.00 T. (a) What is the radius of the circle? Show the magnetic field and the direction the proton moves around the circular path in a diagram. (b) Repeat for an electron with the same speed in the same field.

20. A particle with mass m and charge q moves in a circular trajectory in a plane perpendicular to a uniform magnetic field of magnitude B. (a) Find an expression for the period of one revolution in terms of the variables given. (b) If a particle with mass 5.34×10^{-11} kg moves in a circle perpendicular to a uniform magnetic filed of magnitude 430 mT, and the period of one revolution is 1.50 ms, what is the charge on the particle?

21. Find the radius of curvature of a positron (the antiparticle of an electron, with the same mass as the electron and charge $+e$) in a bubble chamber in which the (uniform) magnetic field has a magnitude 4.00 T. Assume the positron moves perpendicular to the field and was accelerated from rest through a potential difference of 25.0 V before entering the field. Draw the trajectories of an electron and positron that enter the bubble chamber along the same line.

22. You have at your disposal a fairly uniform magnetic field of exactly 1.00 T. How large a bubble chamber would you need to detect 1 keV (kinetic energy) (a) electrons, (b) alpha particles (helium nuclei), (c) pi-mesons (see Example 24.5)?

23. A cyclotron operates with a frequency of 400 kHz. The diameter of the cyclotron is 15.0 cm. (a) With what speed are alpha particles (helium nuclei) emitted from this cyclotron? (b) How large a magnetic field is needed in this case?

24. A proton orbits around a synchrotron with a diameter of 14.5 m with a period of 95.0 μs. How large a magnetic field is required?

25. Describe the design of a synchrotron that will accelerate alpha particles (helium nuclei) to a kinetic energy of 5 MeV. Be sure to choose magnets with realistic values of the magnetic field.

26. You wish to make a mass spectrometer in a vacuum chamber 50 cm on a side. How large a magnetic field is needed to detect singly ionized helium atoms with a kinetic energy of 250 eV?

27. In a mass spectrometer, singly ionized iron atoms are accelerated through a 120-V potential difference and then injected into a region with a transverse magnetic field of magnitude 0.42 T. Describe the path followed by these ions in the magnetic field.

28. One early use of mass spectrometers in World War II was to separate the uranium isotopes ^{235}U and ^{238}U. Suppose a singly ionized atom of each isotope is accelerated to a kinetic energy of 150 keV in a cyclotron and then injected into the mass spectrometer with a uniform magnetic field of 3.0 T. What is the radius of curvature of each species? Discuss whether you think this would be an effective way of separating the isotopes.

29. In the preceding problem, find the frequency of the cyclotron.

Problems 30–32 refer to a device called a *velocity selector*, in which (similar to the Thomson experiment) particles with a certain speed are "selected" because they pass through crossed electric and magnetic fields with no deflection.

30. Explain why for fixed values of E and B all charged particles with a particular speed will pass through undeflected, regardless of their charge or mass.

31. An uncharged particle with any speed will pass through the velocity selector without deflection. How could you tell if you had a beam of uncharged particles?

32. You wish to use a parallel-plate capacitor as part of a velocity selector, with a uniform 1.25-T magnetic field parallel to the plates, which are separated by 0.45 mm. What potential difference should be placed across the plates in order to select charged particles moving at 15,000 m/s?

24.3 The Force on a Current-Carrying Conductor

1. A conducting wire is bent into the shape of a right triangle has sides of length $a, a,$ and hypotenuse $\sqrt{2}a$. The wire forms a closed conducting loop that carries a constant current I. The plane of the loop is perpendicular to a uniform magnetic field of magnitude B. (a) Find the force (magnitude and direction) on each side of the triangular loop. (b) Find the net force on the loop.

2. Repeat the preceding problem if the uniform magnetic field is directed parallel to triangle's hypotenuse.

3. A 7.70-cm length of straight wire lies in a horizontal plane, running along a north/south line. The wire carries a 3.45-A current from south to north through a uniform 1.25-T magnetic field directed straight up. Find the force on the wire.

4. A cylindrical copper wire has a diameter of 0.150 mm and is 10.0 cm long. The wire is in a horizontal plane and carried a current of 2.15 A. Find the magnitude and direction of a uniform magnetic field needed so that the wire is suspended at rest in Earth's gravitational field. [*Note*: The density of copper is 8960 kg/m^3.]

5. A wire forming three-fourths of a circular loop of radius R lies in the xy-plane, with the center of curvature at the origin. The two ends of the wire just touch the $+x$-axis and the $-y$-axis. The wire carries a current I in the counterclockwise direction around the loop, and there exists a uniform magnetic field $\vec{B} = B\,\hat{k}$. Find the net force on the loop of wire.

6. A square loop of wire lies in the xy-plane, with its corners at the origin, $(0, a)$, (a, a), and $(a, 0)$. A current I flows clockwise around the loop. Find the net force on the loop when there is a uniform magnetic field of magnitude B (a) in the $+z$-direction, (b) in the $+x$-direction, (c) parallel to the diagonal of the square directed form the origin to (a, a).

7. A straight conducting wire carries a 7.5-A current in the $+y$-direction. Find the magnetic field that causes a force per unit length on the wire of 0.14 N/m in the $-z$-direction.

8. A straight wire carries a 4.35-A current along a line making an equal angle with the $+x$, $+y$, and $+z$-axes (assume the current starts at the origin and is directed into the first octant). Find the force per unit length acting on this wire in a uniform magnetic field $\vec{B} = 500\,\hat{\imath}$ mT.

24.4 Current Loops and Magnetic Dipoles in Magnetic Fields

1. Show that if SI units are used for $\vec{\mu}$ and $\vec{B}$ in the expression for torque in Equation (24.23), the correct SI units for torque are obtained.

2. Consider again the circular loop in Example 24.7 (Figure 24.14), carrying a constant current I, with a uniform magnetic field in the $+y$-direction. Let the loop still be free to rotate about the x-axis, but suppose now the loop has rotated so that it makes and angle α with the xy-plane. Compute directly the net torque on the loop about the x-axis and show that your result agrees with Equation (24.23).

3. A circular loop of wire has its center at the origin and a diameter of 5.5 cm. It is fixed so that it can turn about the y-axis. Find the torque about the y-axis when the loop carries a 350 mA current in the counterclockwise direction (as viewed from above the xy-plane) and there exists a uniform 1.40-T magnetic field (a) in the $+x$-direction, (b) in the $+y$-direction, (c) in the $+z$-direction.

In Problems 4–7, consider an approximately planar coil of wire consisting of n loops in the same geometric shape (circle, square, etc.), with current flowing the same direction around each loop.

4. Explain why the magnetic dipole moment has a direction perpendicular to the plane of the coil, as given by the right-hand rule for a single loop, and a magnitude given by $\mu = nIA$.

5. A coil consists of 150 loops of wire in the form of a rectangle, with sides 2.5 cm and 3.5 cm. The coil lies parallel to the xy-plane and carries a 500-mA current in the clockwise direction (viewed from above). What is the magnetic dipole moment of this coil?

6. A circular coil of wire has 40 loops and a 9.0-cm diameter. The coil rests in the xz-plane and carries a current of 1.30 A. (a) Find the direction of the uniform magnetic field that will produce the greatest magnitude torque on the coil about its diameter parallel to the z-axis. (b) Find the torque on the coil if a 950-mT uniform magnetic field is applied in the direction you found in (a).

7. You have a 1.50-m length of conducting wire that carries a current of 650 mA. You wish to make a circular coil out of the wire that has a magnetic dipole moment of 9.70×10^{-3} A·m^2. Find the radius of the coil and the number of loops of wire in the coil.

In Problems 8–10, consider a square loop 12.0 cm on a side. The loop is free to turn about an axis passing through its center and parallel to two of the sides. The loop carries a 750-mA current.

8. Describe the orientations of the loop relative to a uniform magnetic field that will produce the lowest potential energy and the highest possible potential energy. If the magnetic field's magnitude is 2.42 T, compute the lowest and highest possible potential energy.

9. The total mass of the loop is 24.0 g. What is its rotational inertia about the fixed axis of rotation?

10. If the loop is released from rest near the orientation with the highest possible potential energy, what is its angular velocity when it reaches its lowest potential energy?

11. A current loop with a magnetic dipole moment $\vec{\mu}$ is in a uniform magnetic field $\vec{B}$. Initially, $\vec{\mu}$ and $\vec{B}$ are in the same direction. (a) What is the torque on the dipole in this orientation? (b) Now suppose the dipole is displaced slightly from this orientation, so that $\vec{\mu}$ makes a small angle with $\vec{B}$. Show that the dipole executes simple harmonic (angular) motion. Find the period of simple harmonic motion as a function of μ, B, and the rotational inertia of the dipole about its rotation axis.

24.5 The Hall Effect

1. In the Hall effect, a metal strip has a length (in the direction of current flow) L, width d, and thickness h. Find the density of conduction electrons as a function of these parameters, the current in the strip, the magnetic field, and the Hall potential difference.

2. In Example 24.9, show that the units $\mathrm{A \cdot T \cdot m^{-1} C^{-1} V^{-1}}$ reduce to $\mathrm{m^{-3}}$.

3. A Hall-effect experiment done with a silver strip that is 2.05 cm wide and 3.50 μm thick. When the strip carries a current of 1.42 A, a 1.55-T transverse magnetic field causes a Hall potential difference of 67.0 μV. (a) What is the density of conduction electrons in silver? (b) How many conduction electrons are there per atom of silver? [*Note*: The density of silver is 10.49 g/cm^3.]

4. Using the experimental parameter in the Hall-effect experiment done with copper in Example 24.9, how would the thickness of the strip have to be changed in order to produce a Hall potential difference of exactly 1.00 mV?

5. A Hall-effect probe is often used to measure magnetic fields. If a Hall-effect probe is made with a copper strip of thickness 2.00 μm and the current flowing through this strip of 500 mA, find the transverse magnetic field present when the Hall potential difference is (a) 1.0 μV, (b) 0.1 mV.

Chapter 25

Magnetic Fields II; Divergence and Curl

We continue our study of magnetic fields by looking at how magnetic fields are generated. Because of the many similarities (some of which you have already seen) between electricity and magnetism, let's begin by thinking briefly about what causes electric fields and how we compute them. Static electric fields are caused by charges, and we have several different strategies for using a known arrangement of charges to compute the resulting electric field. First, the electric field is the force on a charged test particle divided by that charge. Second, the electric field can be computed by breaking a continuous charge distribution with charge density ρ into infinitesimal volumes dV and then integrating over the entire charge distribution:

$$\vec{E} = k \iiint\limits_{D} \frac{\rho \hat{r}}{r^2} dV. \tag{17.11}$$

When the vector integral in Equation (17.11) is too difficult, it may be easier to integrate over the charge distribution to find the (scalar) electric potential V. Then the electric filed is given by

$$\vec{E} = -\vec{\nabla} V. \tag{18.16}$$

Finally, for cases in which the charge distribution is highly symmetrical, Gauss's law provides a convenient way to compute the electric field.

Just as static electric fields are due to charges, *static magnetic fields are due to currents*. The Biot-Savart law is analogous to Equation (17.11), for it allows us to compute the magnetic field generated by a specific distribution of current. Ampère's law is analogous to Gauss's law, in that it allows us to find the magnetic field generated by the current for cases in which the current distribution is highly symmetrical. The Biot-Savart law and Ampère's law are the two methods we will study for computing static magnetic fields.

In magnetism, no method is comparable to finding an electric field by taking the force on a charged test particle divided by the particle's charge, because there are no magnetic monopoles. There is, however, a method comparable to using the electric potential in Equation 18.16. For magnetic fields, a *vector potential* function can used to compute the magnetic field. In fact, the magnetic field is the *curl* of the vector potential, and we will study the curl later in this chapter. However, we will not pursue the vector potential; it is customarily left to an advanced course in electromagnetic theory.

Later in this chapter, we use our understanding of magnetic fields to examine the forces that parallel current-carrying wires exert on one another. We then introduce Maxwell's displacement current—a correction to Ampère's law—and briefly examine magnetism in matter. Finally, we return to more pure mathematics and study two more operators that involve partial derivatives: the divergence and the curl.

25.1 The Biot-Savart Law

In 1820, the Danish physicist Christian Ørsted (1777–1851) discovered that an electric current caused a nearby magnetic compass needle to deflect. This provided experimental evidence that electric current is responsible for static magnetic fields and also an important link between electricity and magnetism. Soon after Ørsted's discovery, the French physicists Jean Baptiste Biot and Felix Savart developed a mathematical formula for computing the magnetic field produced by a particular distribution of current. We now call this formula the **Biot-Savart law**.

We use a pair of examples to illustrate the Biot-Savart law and its similarity with the law for finding electric fields from a distribution of charge. In Example 17.11, we found the electric field a perpendicular distance R from an infinitely long line carrying a uniform charge density λ per unit length. Let's briefly review the highlights of that example (you should review the entire example). The situation is illustrated in Figure 17.18, with the charge lying along the x-axis, and the point P where we wish to find the electric field on the $+y$-axis. The general expression for the electric field due to a line of charge on the x-axis is

$$\vec{E} = k\int_{x_1}^{x_2} \frac{\lambda(x)\,\hat{r}}{r^2}dx, \qquad (17.9)$$

where the charge lies in the interval $[x_1, x_2]$ and the unit vector $\hat{r}$ points from each infinitesimal segment dx in the charge distribution to the point P where we wish to know the electric field. From symmetry, we argued that the electric field at P points in the $+y$-direction in this case, and therefore in computing the definite integral we need only keep the y-component of $\hat{r}$, which is $\cos\theta$. Therefore, for an infinite line with uniform charge density λ,

$$\vec{E} = k\int_{-\infty}^{\infty} \frac{\lambda(\cos\theta\,\hat{\jmath})}{r^2}\,dx.$$

After using trigonometry to express r in terms of x, the definite integral was evaluated to yield

$$\vec{E} = \frac{2k\lambda}{R}\,\hat{\jmath}.$$

Now let's consider the comparable example in magnetism, illustrated in Figure 25.1(a). An infinite wire on the x-axis carries a constant current I in the $+x$-direction, and we wish to know the magnetic field at P, which lies on the $+y$-axis a perpendicular distance R from the wire. Just as when we found the electric field, we break the line into segments Δx, with a representative segment Δx shown in Figure 25.1(a). As before, let r be the distance from Δx to P, and let the unit vector $\hat{r}$ point from Δx toward P.

Experiments show that the magnetic field in this example is given by a definite integral similar to the one in Equation (17.9). In magnetism, we replace the electrostatic constant $k\ (= 1/4\pi\epsilon_0)$ by a different constant $\mu_0/4\pi$. In the SI system of units, the constant μ_0 has the value

$$\mu_0 = 4\pi \times 10^{-7}\ \text{T·m/A}$$

and is called the **permeability of free space**. The magnetic field at P due to the segment Δx is *not* in the direction of $\hat{r}$, but rather is in the direction of $\Delta\vec{x} \times \hat{r}$, where $\Delta\vec{x}$ is a displacement vector in the direction of positive current flow. It turns out (as confirmed by experiment) that this cross product also gives the correct magnitude of the magnetic field at P. The charge density λ that produced the electric field is replaced in magnetism by the current I, which produces the magnetic field. Finally, we sum the contributions of all the Δx and take a limit as Δx approaches zero to obtain a definite integral. With these changes, the magnetic field is given by the formula

$$\vec{B} = \frac{\mu_0}{4\pi}\int_{x_1}^{x_2} \frac{I\,d\vec{x} \times \hat{r}}{r^2}$$

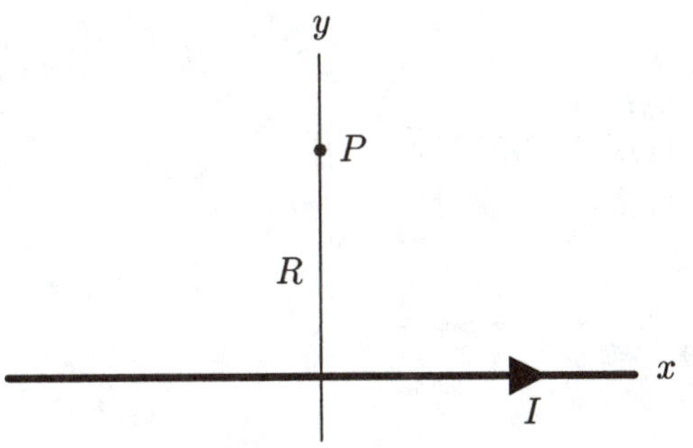

(a) A long straight wire lies along the x-axis, and the point P lies on the y-axis

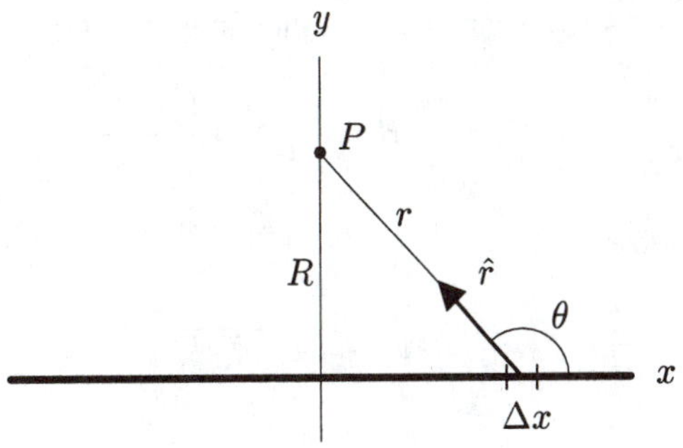

(b) The segment Δx lies a distance r from point P

Figure 25.1. Determining the magnetic field due to a current-carrying wire.

or, because we are considering an infinite wire in this example,

$$\vec{B} = \frac{\mu_0}{4\pi} \int_{-\infty}^{\infty} \frac{I\, d\vec{x} \times \hat{r}}{r^2}. \tag{25.1}$$

Equation (25.1) is actually a special case of the Biot-Savart law. Before generalizing the law, we proceed to compute the definite integral in Equation (25.1) and evaluate the magnetic field at P. In the problem we are considering, the current I is constant and can be factored outside the integral. Also, notice that for any dx we choose, the direction of the cross product $d\vec{x} \times \hat{r}$ is straight up from the page—that is, in the $+z$-direction. With an angle θ between the vectors $d\vec{x}$ and $\hat{r}$ [see Figure 25.1(b)], the magnitude of the cross product is

$$\|d\vec{x} \times \hat{r}\| = (dx)(1)\sin\theta$$

and the cross product is

$$d\vec{x} \times \hat{r} = \hat{k}\, \sin\theta\, dx.$$

Inserting this result in Equation (25.1),

$$\vec{B} = \frac{\mu_0 I}{4\pi}\hat{k} \int_{-\infty}^{\infty} \frac{1}{r^2} \sin\theta\, dx. \tag{25.2}$$

In Equation (25.2), the variables r, θ, and x are all interdependent, and so we have to rewrite the definite integral in terms of one variable before evaluating the integral. With $\sin\theta = \sin(\pi - \theta) = R/r$ and $r^2 = R^2 + x^2$, we have

$$\vec{B} = \frac{\mu_0 I}{4\pi}\hat{k} \int_{-\infty}^{\infty} \frac{1}{r^2}\frac{R}{r}\, dx,$$

or

$$\vec{B} = \frac{\mu_0 I R}{4\pi}\hat{k} \int_{-\infty}^{\infty} \frac{1}{(R^2 + x^2)^{3/2}}\, dx.$$

This is an improper integral, but one that is easily evaluated by taking the appropriate limits ($x \to \infty$ and $x \to -\infty$). From tables of indefinite integrals,

$$\vec{B} = \frac{\mu_0 I R}{4\pi}\hat{k}\left(\frac{x}{R^2\sqrt{R^2+x^2}}\right)\bigg|_{-\infty}^{\infty} = \frac{\mu_0 I}{4\pi R}\hat{k}\left(\frac{x}{\sqrt{R^2+x^2}}\right)\bigg|_{-\infty}^{\infty}$$

With

$$\lim_{x\to\infty}\frac{x}{\sqrt{R^2+x^2}} = 1 \qquad \text{and} \qquad \lim_{x\to-\infty}\frac{x}{\sqrt{R^2+x^2}} = -1,$$

we have

$$\vec{B} = \frac{\mu_0 I}{4\pi R}\,\hat{k}\,[1-(-1)]\,,$$

or

$$\vec{B} = \frac{\mu_0 I}{2\pi R}\,\hat{k}. \tag{25.3}$$

Notice that the functional form of the magnetic field in Equation (25.3) is quite similar to that of the electric field of a long straight wire with a uniform static charge. Both share a $1/R$ dependence and are proportional to the physical quantity (λ or I) responsible for generating the field. However, the directions of the field vectors $\vec{E}$ and $\vec{B}$ are quite different.

We can use the result in Equation (25.3) to describe more generally the magnetic field of an infinitely long straight wire carrying a constant current I. Let's assume that the wire still lies along the x-axis, with current flowing in the $+x$-direction. What is the magnetic field at some other point in the yz-plane, a perpendicular distance R from the wire? From the geometry of Figures 25.1(a) and 25.1(b), we know the magnetic field is perpendicular to a plane containing the wire and the unit vector $\hat{r}$. In Figure 25.2(a), we show the resulting magnetic field at various points in the yz-plane at the same distance R from the wire on the x-axis. From Equation (25.3), we know the magnitude B is the same for all points at the same distance R.

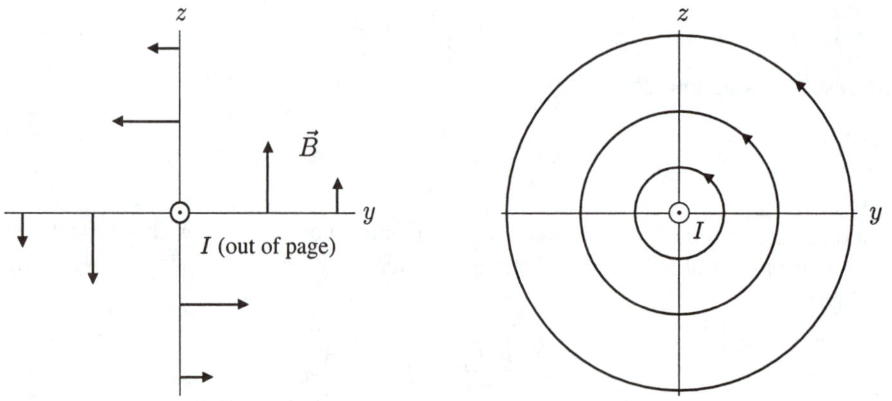

(a) Magnetic field vectors dure to a long straight current-carrying wire

(b) Magnetic field lines in the plane for the same wire

Figure 25.2. The magnetic field around a current-carrying wire.

At different distances from the wire, the field magnitude drops as $1/R$. We can illustrate this, and the whole magnetic field of the straight wire, using continuous magnetic field lines. In Figure 25.2(b), we show a view looking back along the $+x$-axis toward the origin. The current in the straight wire is coming straight out of the page in this view, and the continuous field lines are concentric circles around the wire. The circles grow farther apart at increasing distances from the wire, illustrating the diminishing field strength there.

From this view, we can derive another right-hand rule, this one not for cross products, but rather specifically for the magnetic field generated by a long straight wire. Point your right thumb in the direction in which current flows in the straight wire. Then if you curl your fingers, they follow the circular magnetic field lines in the proper direction.

Equation (25.1) is not the most general form of the Biot-Savart law. In general, a current-carrying wire need not lie along the x-axis or even be straight. We assume that the current is entirely along a curve C, which can be broken into small line segments Δs (Figure 25.3). We can then define a vector $\Delta\vec{s}$ to lie tangent to the wire and pointing in the direction of positive current flow. Now we sum the contributions due to each $\Delta\vec{s}$ and take the limit as $\Delta\vec{s}$ approaches zero to obtain a line integral. The integral in Equation (25.1) can then be expressed as a line integral along the curve C:

$$\vec{B} = \frac{\mu_0}{4\pi}\int_C \frac{I\,d\vec{s}\times\hat{r}}{r^2}. \tag{25.4}$$

This is the general form of the Biot-Savart law. It gives the magnetic field at some point P outside the wire, due to the current from one end of the wire (the curve C) to the other. As before, the unit vector $\hat{r}$ points from each segment Δs to the point P (see Figure 25.3).

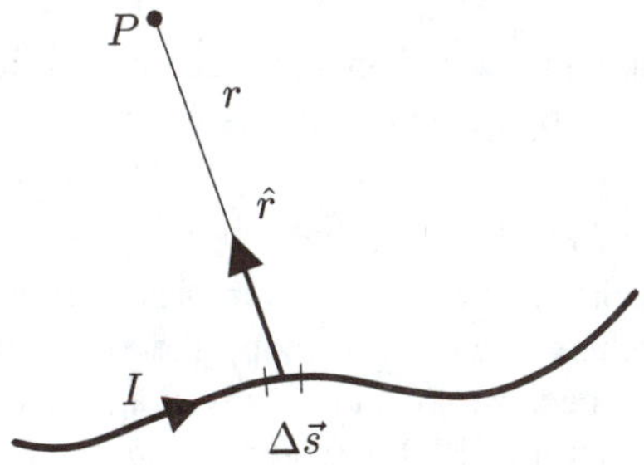

Figure 25.3. Analyzing the magnetic field due to a general current-carrying wire.

Example 25.1. A conducting wire in the shape of a circle of radius R carries a constant current I. Use the Biot-Savart law to find the magnetic field at the center of the loop.

Solution. Let's place the loop in the xy-plane with its center at the origin, and let the current flow in the counterclockwise direction as viewed from above (Figure 25.4). For an arbitrary segment ds as shown, the unit vector $\hat{r}$ points radially inward. Because the vector $d\vec{s}$ is tangent to the wire, in this case it is tangent to the circle, and therefore $d\vec{s}$ and $\hat{r}$ are perpendicular. Thus,

$$\|d\vec{s}\times\hat{r}\| = (ds)(1) = ds.$$

By the right-hand rule, the vector $d\vec{s}\times\hat{r}$ points in the $+z$-direction for each ds on the loop. Therefore, from the Biot-Savart law [Equation (25.4)],

$$\vec{B} = \frac{\mu_0}{4\pi}\hat{k}\int_C \frac{I\,ds}{r^2}.$$

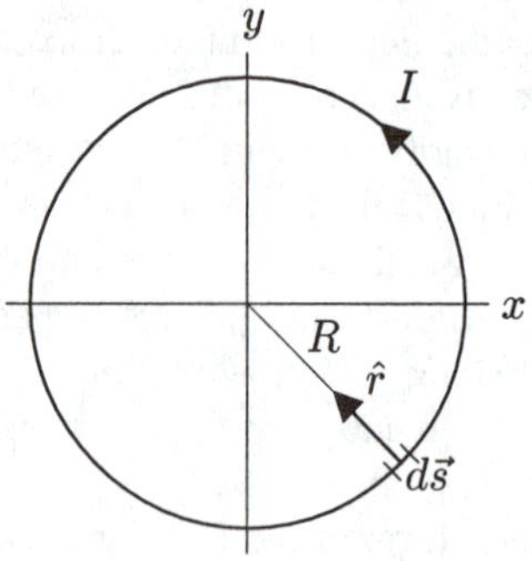

Figure 25.4. A current-carrying loop, considered in Example 24.1.

The current I is constant, and the distance $r = R$ is constant over the entire curve. Factoring these constants from the integral leaves

$$\vec{B} = \frac{\mu_0 I}{4\pi R^2}\,\hat{k}\int_C ds.$$

The remaining line integral is simply $\int_C ds = 2\pi R$, the circumference of the circle. Therefore,

$$\vec{B} = \frac{\mu_0 I}{4\pi R^2}\,\hat{k}\,(2\pi R) = \frac{\mu_0 I}{2R}\,\hat{k}.$$

We can generalize this result (taking it away from the specific coordinate system we have used) with yet another right-hand rule. For a circular loop carrying a constant current, curl your (right-hand) fingers around the loop in the direction of the current. Then your thumb points in the direction of the magnetic field at the center of the loop, perpendicular to the plane of the loop. △

When is it useful to use the Biot-Savart law? In our examples (the magnetic field near an infinite straight wire and the magnetic field at the center of a circular loop), the evaluation of the definite integral in Equation (25.4) was not extremely tedious, because we had the advantage of symmetry. This is similar to what we found using the definite integral in Equation (17.9) to compute electric fields directly.

When such symmetry is not present, it may again be advisable to approximate the definite integral using numerical methods. For example, suppose you want to know the electric field everywhere in space in the vicinity of the circular ring of current in Example 25.1. You should try setting up the definite integral in Equation (25.4) for some random point (x, y, z) in order to convince yourself that this would not be an easy problem to do analytically.

A representation of the magnetic field in the vicinity of the circular ring (using continuous field lines) is shown in Figure 25.5. It is no accident that this field bears some resemblance to that of the bar magnet (Figure 24.1). At the beginning of this section, we remarked that static magnetic fields are due to currents. But what about permanent bar magnets, such as the bar magnet shown in Figure 24.1? Not long after Ørsted discovered electromagnetism, the French physicist Andrè Marie Ampère (1775–1836), for whom the SI current unit is named, made the brilliant observation that there are not two distinct kinds of magnetism (permanent magnets and electromagnets), but only one. Ampère wrote in his *Recueil d' observations électrodynamiques* (1822): "... a magnet should be considered an assemblage of electric currents, which flow in planes perpendicular to its axis ... ." This should remind you of the current loop we used to model the magnetic dipole moment in Section 24.4. It is remarkable that Ampère was able to make this statement a century before the modern theory of atomic angular momentum was formulated. It is the intrinsic angular momentum of the electrons in atoms that is responsible for magnetism in matter.

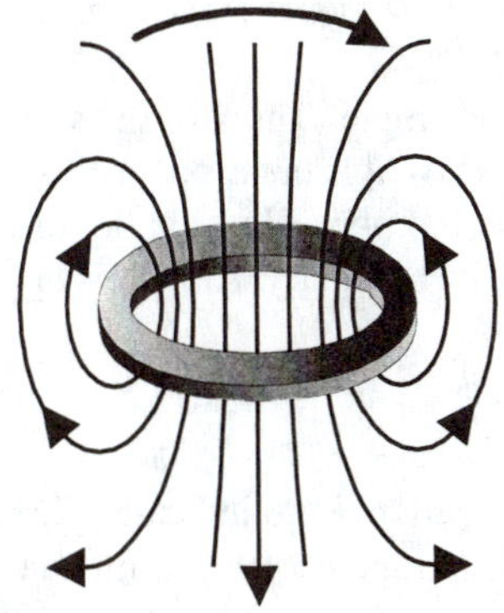

Figure 25.5. Magentic field lines in the vicinity of a circular ring carrying current I.

25.2 Ampère's Law

Ampère's law provides another convenient way of determining the magnetic field due to a distribution of current in certain high-symmetry situations. It is awkward to try to develop Ampère's law from first principles, so we will state it without proof. By way of introducing Ampère's law, we will demonstrate its use in an example with which you are already familiar: the infinitely long straight wire.

Refer again to Figure 25.2(b), which shows the continuous magnetic field lines generated by a long straight wire carrying a constant current I. The magnetic field lines form concentric circles around the wire, with the magnitude of the field given by Equation (25.3) as $B = \mu_0 I/2\pi R$ at a perpendicular distance R from the wire. We evaluate the line integral

$$\int_C \vec{B} \cdot d\vec{s}$$

along a curve C consisting of one of those concentric circles, at a radius R. We follow the path in the same direction as the magnetic field lines, so that the vectors $\vec{B}$ and $d\vec{s}$ are in the same direction at every point on the path. Therefore, by the rules of dot products

$$\vec{B} \cdot d\vec{s} = (B)(ds)\cos(0) = B\,ds$$

and

$$\int_C \vec{B} \cdot d\vec{s} = \int_C B\,ds = \int_C \left(\frac{\mu_0 I}{2\pi R}\right) ds, \tag{25.5}$$

where we have inserted the magnitude B from above. The current I is constant, and along our chosen path R is constant, so these can be factored from the integral to give

$$\int_C \vec{B} \cdot d\vec{s} = \frac{\mu_0 I}{2\pi R} \int_C ds.$$

The remaining line integral is simply the circumference of the circle, or $2\pi R$, so our result is

$$\int_C \vec{B} \cdot d\vec{s} = \frac{\mu_0 I}{2\pi R}(2\pi R),$$

or simply

$$\int_C \vec{B} \cdot d\vec{s} = \mu_0 I. \tag{25.6}$$

Ampère's law is a generalization of this example. Think of the surface (in this case, a flat disk) that has the curve C as its boundary. To use Ampère's law, it is therefore necessary that the curve C be closed. Notice that the current I passed through that surface. **Ampère's law** applies to *any* closed curve C that has a current I_C passing through the surface that has C as its boundary. Ampère's law says that

$$\oint_C \vec{B} \cdot d\vec{s} = \mu_0 I_C\,. \tag{25.7}$$

In Equation (25.7), we have introduced the notation $\oint_C$ to indicate a line integral over a closed curve C.

The procedure for using Equation (25.7) (Ampère's law) to find an *unknown* magnetic field is straightforward in principle. First, it is necessary to define a curve C such that you know the direction of the unknown magnetic field along the curve. This allows you to rewrite the dot product in the line integral as $\vec{B} \cdot d\vec{s} = (B)(ds)\cos\theta$, where θ is the angle between the vectors $\vec{B}$ and $d\vec{s}$ at each point along C. Normally this (and the subsequent evaluation of the line integral) is only practical if there is a high degree of symmetry. But if the line integral can be evaluated, then you can solve for the magnitude B.

For example, suppose we did *not* know the magnitude of B for the infinite straight wire but wanted to find it using Ampère's law. From our previous study of the infinite straight wire, we know that the field lines are concentric circles, as shown in Figure 25.2(b). Then as before, $\vec{B} \cdot d\vec{s} = (B)(ds)\cos\theta = B\,ds$ at all points along the curve. Thus, the line integral is

$$\oint_C \vec{B} \cdot d\vec{s} = \oint_C B\,ds = B \oint_C ds.$$

In the last step, we have reasoned that B is constant along the path because of the symmetry of the situation, so as a constant it has been factored from the integral. Then we know $\int_C ds = 2\pi R$, and applying this result in Ampère's law [Equation (25.7)]:

$$\oint_C \vec{B} \cdot d\vec{s} = B(2\pi R) = \mu_0 I$$

where $I_C = I$, the current in the single wire. Solving for B,

$$B = \frac{\mu_0 I}{2\pi R},$$

which is the correct result.

From the preceding paragraph, you should see that the mathematical details of using Ampère's law can be *much* simpler than attacking the same problem with the Biot-Savart law. This is the great advantage in using Ampère's law. The disadvantage is that, although Ampère's law is perfectly general, it is only practical to use in situations when there is a good deal of symmetry. It is most convenient when you can choose a path so that the vectors $\vec{B}$ and $d\vec{s}$ are either parallel or perpendicular, thus making the evaluation of the dot product straightforward. These facts about the use of Ampère's law should remind you of using Gauss's law to find static electric fields, where it is also necessary to have a high degree of symmetry and some knowledge of the direction of the electric field on the Gaussian surface, and where it is useful to have the vectors $\vec{E}$ and $\hat{n}$ be parallel or perpendicular.

To clarify our discussion of Ampère's law, we should point out that the current I_C is the ***net current*** passing through the surface having C as its boundary. Even though electric current is not a vector quantity, it has a definite sense of direction. Just as when we studied dc circuits, you should think of current flowing in one direction as positive and the other direction as negative. Then the net current I_C to be used in Ampère's law is the *difference* between the currents flowing in opposite directions. To illustrate this, consider again the long straight wire along the x-axis shown in Figure 25.2(b). Now suppose a second infinite wire also lies along the x-axis and carries current in the opposite direction. (Physically it is impossible for the two wires to occupy the same space, but in practice they can be very thin and close together.) The magnetic field of one wire is as shown in Figure 25.2(b). For the other wire (not shown, with current going into the page) the magnetic field is also concentric circles, but with the field lines *clockwise* in the yz-plane. By the principle of superposition, the net magnetic field is the sum of these two fields. In applying Ampère's law, the net current I_C passing through a circular disk of radius R in the yz-plane is the difference between the currents flowing in the two opposite directions. With this definition of I_C, applying Ampère's law as before leads to

$$B = \frac{\mu_0 I_C}{2\pi R}.$$

For example, if the two currents are equal in magnitude, then $I_C = 0$ and $B = 0$. This makes sense, because the individual fields of the two wires are everywhere equal in magnitude and opposite in direction, so by the principle of superposition the net field is zero.

Example 25.2. Consider an infinitely long cylindrical wire with some nonzero radius R. The wire carries a uniform current I_0 that is spread uniformly over the cross section of the wire; in other words, the current density is constant throughout the wire. Use Ampère's law to find the magnetic field at all locations inside and outside the wire. Express your answer as a function of r, the perpendicular distance from the center of the wire.

Solution. Based on the cylindrical symmetry of the wire, the magnitude of the field does not vary as we move around a circle of radius r. Further, based on our experience with the infinitely long thin wire, the magnetic field lines form concentric circles around the center of the wire. This fact can be verified by appealing to the Biot-Savart law and applying the right-hand rule to determine the direction of the magnetic field caused by each small element of the wire.

Starting from a point outside the wire [Figure 25.6(a)], let's integrate over a closed circular path in the direction of the magnetic field. Thus,

$$\vec{B} \cdot d\vec{s} = (B)(ds)\cos(0) = B\, ds,$$

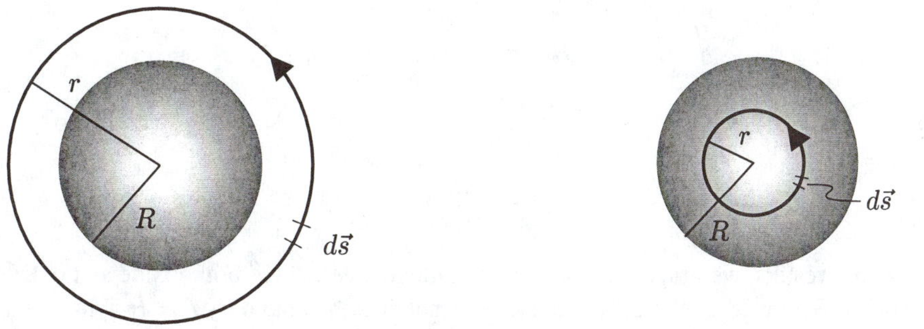

(a) Finding the magnetic field at a point outside the wire

(b) Finding the magnetic field at a point inside the wire

Figure 25.6. Schematic for Example 25.2.

and the line integral in Ampère's law becomes

$$\oint_C \vec{B} \cdot d\vec{s} = \oint_C B\, ds = B \oint_C ds.$$

Because B does not vary as we move around a circle of radius r, we are able to factor B out of the integral. The remaining integral is again just the circumference of the circle, or $2\pi r$. On the right side of Ampère's law, note that the entire current I_0 passes through this path, and so $I_C = I_0$. Combining these results,

$$\oint_C \vec{B} \cdot d\vec{s} = B(2\pi r) = \mu_0 I_0.$$

Solving for B,

$$B = \frac{\mu_0 I_0}{2\pi r}.$$

This result is essentially the same as if the wire were very thin; in fact, the thickness of the wire does not matter as long as $R < r$. This makes sense, because according to Ampère's law it is the total current passing through the surface outlined by the curve C that determines the magnetic field.

Similarly, for a point inside the wire, we integrate around a circular path as shown in Figure 25.6(b). The integration still follows the direction of the magnetic field line, so the left side of Ampère's law remains

$$\oint_C \vec{B} \cdot d\vec{s} = B(2\pi r).$$

On the right side, however, the current passing through C is not the entire I_0. Rather, we can use the fact that the current density is constant to find I_C:

$$\text{current density} = \frac{I_C}{\pi r^2} = \frac{I_0}{\pi R^2}.$$

Thus,

$$I_C = I_0 \frac{r^2}{R^2}.$$

Using this result in Ampère's law,

$$\oint_C \vec{B} \cdot d\vec{s} = B(2\pi r) = \mu_0 I_C = \mu_0 I_0 \frac{r^2}{R^2}.$$

Solving for B,

$$B = \frac{\mu_0 I_0 r}{2\pi R^2}.$$

To summarize our results, we graph the magnitude of the magnetic field both inside and outside the wire (as a function of r) in Figure 25.7. Notice that the results match at the boundary $r = R$, which is a good sign that we have solved the problem correctly. You should see another striking similarity between electricity and magnetism here. The graph in Figure 25.7 bears a striking resemblance to the graph of the electric field as a function of the radius of a uniformly charged solid (nonconducting) sphere, which we found using Gauss's law in Chapter 23. △

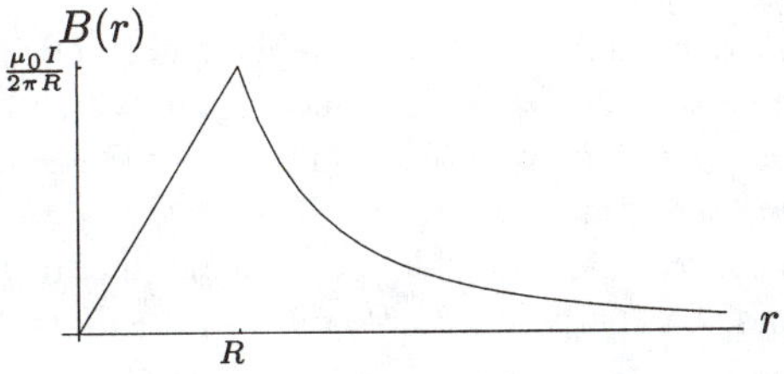

Figure 25.7. Magnitude of the magnetic field vs. radius for a current-carrying wire of radius R.

Consider again the single circular loop of current, shown in Figure 25.4. We used the Biot-Savart law to find the magnetic field at the center of the loop. Because of the curved magnetic field lines (Figure 25.5) caused by this distribution of current, it is impractical to use Ampere's law to find the magnetic field at an arbitrary point in space. Notice, however, that in the flat disk (with the loop as its boundary) the field lines are all perpendicular to that disk. This suggests that we might make a more uniform field by stacking identical current loops atop one another [Figure 25.8(a)]. In practice, this situation is well approximated by winding a single wire into a coil that forms a helix of constant radius, with the wires packed together as closely as possible. Such a device, shown in Figure 25.8(b), is called a **solenoid**.

(a) Circular loops of wire in a stack

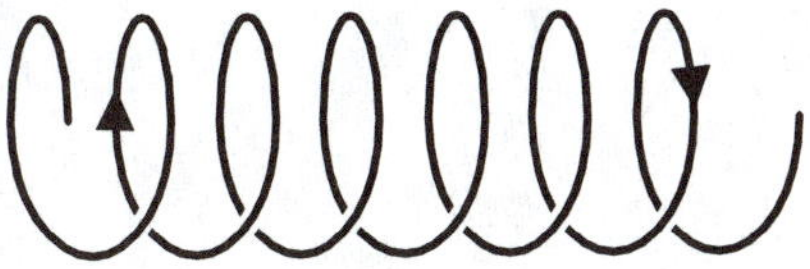

(b) A coil of wire wound into a helix forms a solenoid

Figure 25.8. Developing the concept of a solenoid.

We can use Ampère's law to find an approximate value for the magnetic field inside a solenoid. We restrict ourselves to points inside the solenoid that are not too close to the ends, where we see the field lines bowing outward, as in the single loop field in Figure 25.5. Near the geometric center of the solenoid, the field lines tend to be fairly straight and parallel. We will take this as an experimental fact, although you can convince yourself why this should be so if you (1) use the principle of superposition, adding the fields of each single circular loop, or (2) use the Biot-Savart law, picking representative bits of current distributed symmetrically around the center of the solenoid.

A schematic of the magnetic field of the solenoid is shown in Figure 25.9(a). We apply Ampère's law by integrating over a *rectangular* curve C shown in Figure 25.9(b).

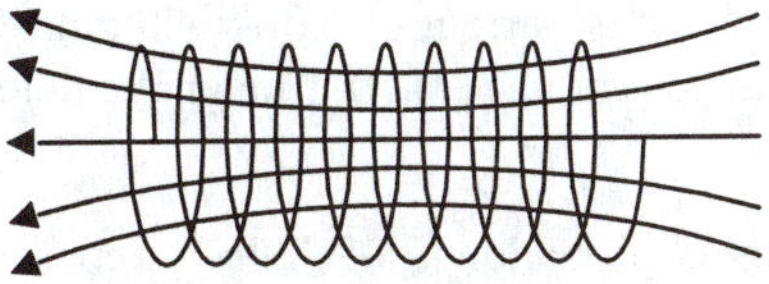

(a) Magnentic field lines

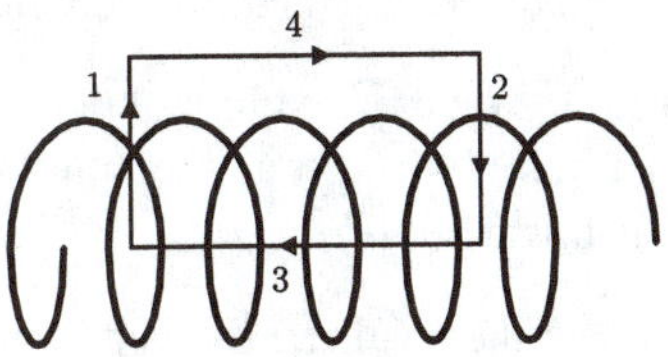

(b) Applying Ampère's law to a rectangular loop

Figure 25.9. Analyzing the magnetic field of a solenoid.

One side of the curve is just outside the solenoid, where the magnetic field is approximately zero as long as we are not too close to the ends of the solenoid. Once again, we will take this as an experimental fact, but you can use symmetry arguments and/or the Biot-Savart law to convince yourself it is true. Because the magnetic field in that region is approximately zero, the line integral $\int \vec{B} \cdot d\vec{s} = 0$ along that side [labeled 1 in Figure 25.9(b)]. Along the part of side 2 outside the solenoid, the magnetic field is also zero. Along the part of side 2 inside the solenoid, $\vec{B}$ is perpendicular to $d\vec{s}$, so $\int \vec{B} \cdot d\vec{s} = 0$ along side 2. The same argument can be made for side 4. This leaves side 3, where $\vec{B}$ and $d\vec{s}$ are in the same direction. With the magnetic field constant over that side (which we arbitrarily assign a length L), we have

$$\int_{\text{side 3}} \vec{B} \cdot d\vec{s} = \int_{\text{side 3}} B\, ds = B \int_{\text{side 3}} ds = BL.$$

Recalling that the line integral over each of the other three sides is zero, we can apply Ampère's law. But what is I_C, the current enclosed by our rectangular path C? If N loops of wire pass through the rectangular region outlined by C, then $I_C = NI$, where I is the current flowing through the solenoid. (The geometry of the solenoid guarantees that the current in each loop passes through the rectangle going in the same direction.) Therefore, Ampère's law says

$$\oint_C \vec{B} \cdot d\vec{s} = \int_{\text{side 3}} \vec{B} \cdot d\vec{s} = BL = \mu_0 I_C = \mu_0 NI$$

or

$$B = \frac{\mu_0 NI}{L}.$$

In this result, N and L are adjustable parameters, because they are based on the size of the curve C over which the integration takes place. However, for a uniformly wound solenoid, the ration N/L is constant. Let's call this constant n; it is simply the number of turns of wire per unit length. The magnitude of the magnetic field near the center of the solenoid is thus expressed more succinctly as

$$B = \mu_0 nI. \tag{25.8}$$

You should verify that Equation (25.8) is dimensionally correct. It makes physical sense that B increases when n or I increases. Our analysis using Ampère's law has shown that B is directly proportional to both of those parameters.

Solenoids are useful for providing a fairly uniform magnetic field. Constructing a uniform magnetic field is not a trivial matter; it can't be done the way we made uniform static electric fields, because there are no magnetic monopoles. We assumed throughout Chapter 24 that it is possible to make a nearly uniform magnetic field, so it's good to know now how it's done! The magnetic field outside a solenoid is a fair approximation of a dipole field, resembling the field of a bar magnet of similar dimensions.

Example 25.3. A wire used in making a solenoid can carry a steady current of 6.0 A without overheating. How many turns of wire per unit length should there be in order to produce a 0.10-T magnetic field inside the solenoid? Is this feasible?

Solution. From the result we derived in our study of the solenoid [Equation (25.8)], the number of turns per unit length is

$$n = \frac{B}{\mu_0 I}.$$

Inserting numerical values,

$$n = \frac{0.10 \text{ T}}{(4\pi \times 10^{-7} \text{ T·m/A})(6.0 \text{ A})} = 1.3 \times 10^4 \text{ m}^{-1}$$

or 1.3×10^4 turns per meter.

To analyze this result, notice that it corresponds to 1.3×10^1 turns per mm, or 13 turns per mm. For a singly wrapped coil, this means that the diameter of the wire could be no more than

$$\frac{1}{13 \text{ mm}^{-1}} = 0.077 \text{ mm}.$$

This is rather thin for wire that is expected to carry 6.0 A. One way around this is to make the solenoid many layers of wire thick. This will work, because as you can see from Equation (25.8), the magnetic field strength inside the solenoid does not depend on the diameter of each coil. Thus, the solenoid can be many layers thick. If it is (say) 10 layers thick, this effectively increases n by a factor of 10, and so the wire could be 0.77 mm in diameter—a much more reasonable value. △

25.3 Parallel Current-Carrying Wires

In this section, we apply our understanding of magnetic fields to the problem of how two parallel current-carrying wires affect one another. Let's assume for now that the wires are long enough that we can consider them infinite. That is, we will ignore the end effects and concentrate of the forces near the middle of the wires.

In Figure 25.10, we show two long parallel wires, carrying currents I_1 and I_2 (not necessarily the same) in the same direction, with a distance a between the wires. We know that the magnetic field generated by a long straight wire is given by Equation (25.3) and pictured in Figures 25.2(a) and 25.2(b). (Remember also that the field lines are concentric circles, curling around in the direction the fingers of your right hand curl when your right thumb points in the direction of the current flow.)

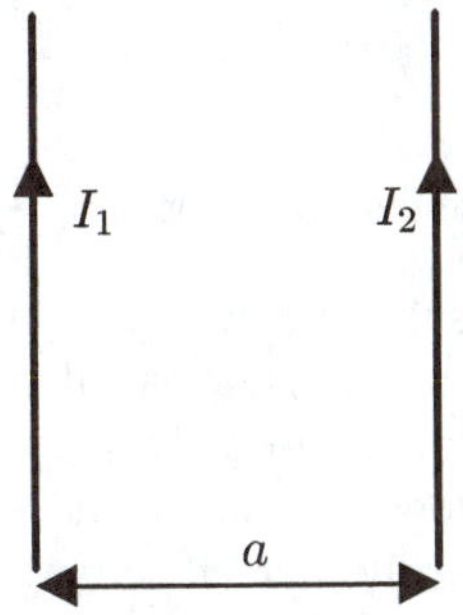

Figure 25.10. Two long parallel wires, carrying currents I_1 and I_2.

Subject to the approximations we have stated, the magnetic field anywhere on the wire on the right in Figure 25.10 (the one carrying I_2) due to the other wire (carrying I_1) has a magnitude

$$B_1 = \frac{\mu_0 I_1}{2\pi a}$$

and a direction perpendicular to the wire and going straight into the page. Then by Equation (24.21), the force on a section of length L of the wire on the right is

$$\vec{F}_2 = I_2\vec{L} \times \vec{B}_1,$$

where as usual the direction of $\vec{L}$ is in the direction of positive current flow. In the diagram, $\vec{L}$ is straight up and $\vec{B}_1$ is straight into the page, so the cross product $\vec{L} \times \vec{B}_1$ is perpendicular to both and to the *left*. This tells us that the wire carrying I_2 is attracted toward the other wire. Because our choice of L was arbitrary, it is more meaningful to compute the force per unit length on the wire on the right. It is

$$\text{force per unit length} = \frac{F_2}{L}.$$

Because $\vec{L}$ is perpendicular to $\vec{B}$, we have $F_2 = \|\vec{F}_2\| = I_2 L B_1 \sin(\pi/2) = I_2 L B_1$. Therefore,

$$\text{force per unit length} = \frac{I_2 L B_1}{L} = I_2 B_1 = I_2 \left(\frac{\mu_0 I_1}{2\pi a} \right)$$

or

$$\text{force per unit length} = \frac{\mu_0 I_1 I_2}{2\pi a}. \tag{25.9}$$

Now let's ask the question: What is the force per unit length on the other wire (the one on the left, carrying I_1)? Using the same method as before, the magnetic field due to the other wire has a magnitude

$$B_2 = \frac{\mu_0 I_2}{2\pi a}$$

and a direction perpendicular to the wire and going straight out of the page. The force on a section of length L of the wire on the left is

$$\vec{F}_1 = I_1 \vec{L} \times \vec{B}_2$$

where the direction of $\vec{L}$ is in the direction of positive current flow. Now the direction of $\vec{L} \times \vec{B}_2$ is to the *right*, so the wire on the left is also attracted toward the other wire. As before, we can find the force per unit length on the wire carrying I_1:

$$\text{force per unit length} = \frac{I_1 L B_2}{L} = I_1 B_2 = I_1 \left(\frac{\mu_0 I_2}{2\pi a} \right)$$

or

$$\text{force per unit length} = \frac{\mu_0 I_1 I_2}{2\pi a}. \tag{25.10}$$

We should not be surprised by this result. The two wires interact with each other through a distance via a magnetic field, and no other forces are present. Newton's third law tells us that the forces on the two wires should have equal magnitudes and opposite directions. This is just what we have found. The mutual attraction of these two current-carrying wires is analogous to the mutual gravitational attraction of two isolated masses, or the electrostatic attraction of two oppositely charged bodies.

To carry this last analogy further, consider what happens if the direction of one of the currents (say I_2) is reversed, but the geometry is left the same. The magnetic field $\vec{B}_1$ due to the wire I_1 is the same as before. But now the direction of the vector $\vec{L}$ in the wire carrying I_2 is reversed. As a result, the force on that wire is to the *right*, with the same magnitude per unit length as we found before. A similar analysis for the other wire shows that the force on it is to the *left*, with the same magnitude per unit length as before. Thus, the two wires repel each other in this configuration; the forces on them have equal magnitudes and opposite directions, as required by Newton's third law. You can consider this case to be a magnetostatic analog to the electrostatic repulsion of two like charges.

This analysis serves as the basis for the definition of the SI unit ampere (A). If two long parallel wires carrying equal currents experience a force per unit length of exactly 2×10^{-7} N, then the current is 1 ampere. This definition of the ampere, along with the second (s), provides the definition of the coulomb as 1 A·s.

25.4 Maxwell's Displacement Current

Throughout our development of Ampère's law in Section 25.2, we assumed a *constant* current I_C passing through a surface with a closed boundary C. Thus, the form of Ampère's law given in Equation (25.7)

$$\oint_C \vec{B} \cdot d\vec{s} = \mu_0 I_C \tag{25.7}$$

is valid only in magnetostatics. A constant current I_C gives rise to a constant magnetic field $\vec{B}$.

It was the Scottish physicist James Clerk Maxwell (1831–1879) who not only realized this drawback to the form of Ampère's law presented in Equation (25.7), but also found how to correct it. The problem with Ampère's law is often demonstrated using a simple parallel-plate capacitor, as shown in Figure 25.11(a). A wire is attached to each capacitor plate, which we can imagine connecting in a closed circuit to a battery and perhaps a resistor, so that a current I will flow to charge the capacitor. We have drawn in a curve C for the purpose of applying Ampère's law. Notice that C encloses the wire. Thus, we should expect the magnetic field along C to be governed by Equation (25.7) with $I_C = I$ (the current in the wire). This seems clear enough if we think of the current I passing through the rectangular region with boundary C, as in Figure 25.11(b). But here is the problem: We could just as well pick another region [Figure 25.11(c)] bounded by C, which passes through the capacitor plates. You know that no current flows between the plates, so this would tell us $I_C = 0$, in contradiction to our finding using the surface in Figure 25.11(b). Ampère's law is supposed to be perfectly general and should not allow such a contradiction.

Maxwell found a way around this dilemma. Surely to be consistent with our previous form of Ampère's law, the analysis with the current $I_C = I$ passing through the rectangular region in Figure 25.11(b) must be correct. To produce the same effect (i.e., the same magnetic field along C) when the surface shown in Figure 25.11(c) is used, Maxwell proposed that another term,

$$\mu_0 \epsilon_0 \frac{d\Phi_E}{dt},$$

be added to the right side of Ampère's law, where Φ_E is the flux of the electric field passing through the surface with boundary C. An electric field flux Φ_E passes through the surface in Figure 25.11(c), because

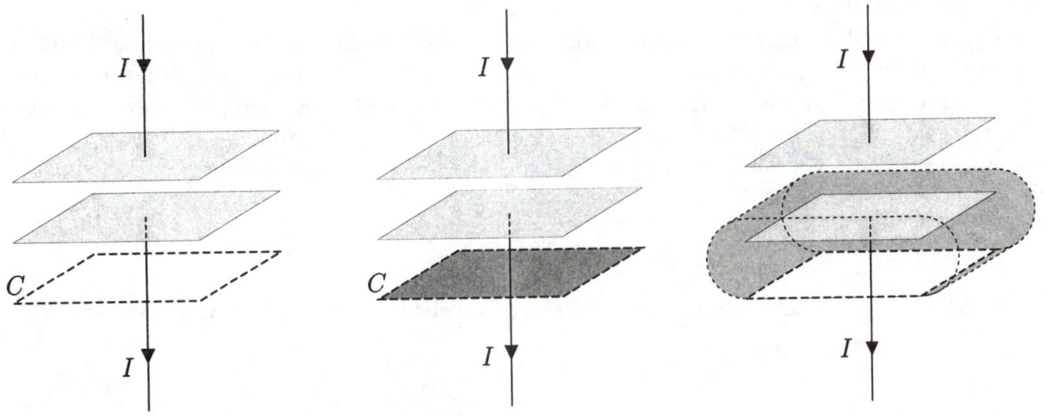

(a) Curve C near parallel-plate capacitor (b) One surface with C as boundary (c) A second surface with C as boundary

Figure 25.11. The argument for Maxwell's displacement current.

there is an electric field between the capacitor plates. Therefore, it is plausible that this added term can remedy the problem. We will have to look at the situation more carefully to see how it works. First, however, notice that Maxwell's hypothesis turns Ampère's law into

$$\oint_C \vec{B} \cdot d\vec{s} = \mu_0 I_C + \mu_0 \epsilon_0 \frac{d\Phi_E}{dt}. \tag{25.11}$$

If we let the portion of the surface between the capacitor plates in Figure 25.11(c) be perpendicular to the electric field, then by the definition of flux

$$\Phi_E = EA$$

where E is the magnitude of the electric field between the plates (we can use the infinite-plate approximation and assume E is constant) and A is the surface area of the plates. From our study of capacitors [Equation (19.3)], we know that $EA = Q/\epsilon_0$, where Q is the magnitude of the charge on each plate. The reformed version of Ampère's law tells us to take the time derivative of the electric flux, which is

$$\frac{d\Phi_E}{dt} = \frac{d}{dt}[EA] = \frac{d}{dt}\left[\frac{Q}{\epsilon_o}\right] = \frac{1}{\epsilon_0}\frac{dQ}{dt}.$$

Note that dQ/dt is the rate at which the capacitor is being charged; it is also equal to the current in the wire, because that current supplies the charge on the capacitor. Therefore,

$$\frac{d\Phi_E}{dt} = \frac{1}{\epsilon_0} I.$$

We now insert this result into our new version of Ampère's law. Recall that along the surface defined in Figure 25.11(c), the current $I_C = 0$. Therefore,

$$\oint_C \vec{B} \cdot d\vec{s} = \mu_0(0) + \mu_0\epsilon_0 \frac{d\Phi_E}{dt} = \mu_0\epsilon_0\left(\frac{1}{\epsilon_0} I\right) = \mu_0 I.$$

This is in perfect agreement with the result we obtain using the rectangular surface in Figure 25.11(b), where $I = I_C$ and $d\Phi_E/dt = 0$.

We have shown that Maxwell's improved form of Ampère's law [Equation (25.11)] works for the specific geometry of the parallel-plate capacitor. It turns out to be true in general, regardless of the geometry of the current-carrying wires and electric fields involved (we will not prove this fact here). Notice that the two terms on the right side of Equation (25.11) each contain the factor μ_0. By comparing the two terms, it is apparent (you should verify this yourself—see the problems) that the factor

$$\epsilon_0 \frac{d\Phi_E}{dt}$$

has the dimensions of electric current. For that reason, it is called **Maxwell's displacement current** I_d. That is,

$$I_d = \epsilon_0 \frac{d\Phi_E}{dt}. \tag{25.12}$$

In our example with the parallel-plate capacitor, the displacement current plays the role of the conduction current I in that region of space where the conducting wire is absent (inside the capacitor). In fact, we showed above that $I_d = I$. We can write Ampère's law (the corrected version written by Maxwell) in terms of the displacement current as

$$\oint_C \vec{B} \cdot d\vec{s} = \mu_0 I_C + \mu_0 I_d$$

or

$$\oint_C \vec{B} \cdot d\vec{s} = \mu_0 (I_C + I_d). \tag{25.13}$$

From this analysis, we must amend our earlier statement that magnetic fields are caused by currents. Rather, magnetic field are caused by currents *and* by changing electric fields.

Example 25.4. In the circuit shown in Figure 25.12, $\mathcal{E} = 9.0$ V, $R = 10$ kΩ, $C = 25$ μF, and the capacitor is initially uncharged. At 0.50 s after the switch S is closed, find the current that flows in the circuit. Then find $d\Phi_E/dt$ at that time.

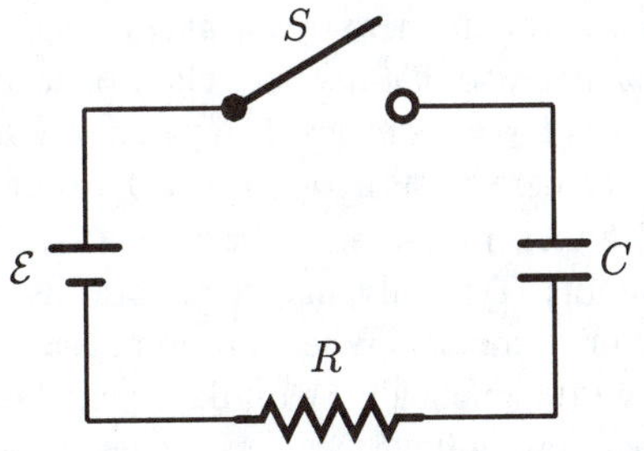

Figure 25.12. The circuit for Example 25.4.

Solution. The current in the circuit is given by Equation (21.35):

$$I = \frac{\mathcal{E}}{R} \exp\left(-\frac{t}{RC}\right).$$

Inserting the given numerical values, we find

$$I = \frac{9.0 \text{ V}}{10^4 \, \Omega} \exp\left[-\frac{0.50 \text{ s}}{(10^4 \, \Omega)(25 \times 10^{-6} \text{ F})}\right] = 1.22 \times 10^{-4} \text{ A}$$

or 0.122 mA. We know from the analysis in the text that this current equals the displacement current. Thus,

$$I = I_d = \epsilon_0 \frac{d\Phi_E}{dt}.$$

The rate of change of electric flux is

$$\frac{d\Phi_E}{dt} = \frac{I}{\epsilon_0}.$$

Inserting numerical values (with $I = 0.122$ mA at $t = 0.50$ s),

$$\frac{d\Phi_E}{dt} = \frac{1.22 \times 10^{-4} \text{ A}}{8.85 \times 10^{-12} \text{ C}^2/(\text{N}\cdot\text{m}^2)} = 1.38 \times 10^7 \text{ A}\cdot\text{N}\cdot\text{m}^2/\text{C}^2 = 1.38 \times 10^7 \text{ (N/C)}\cdot\text{m}^2/\text{s}.$$

By writing the units as (N/C)$\cdot$m^2/s, it can be seen that we have an electric flux (field times area) per unit time. △

25.5 Magnetism in Matter

Earlier, we saw how Ampère gave a successful qualitative explanation for magnetism in matter: There are small currents within magnetic materials that, when combined, produce the effects we see macroscopically. We now know that it is the electrons in atoms that are primarily responsible for the magnetic effects we see, because their orbital and spin angular momenta give individual atoms magnetic dipole moments. In this section, we describe the three classes of magnetic behavior into which almost all materials can be placed, and we will give a (mostly qualitative) explanation of how these behaviors follow from the various ways of combining atomic magnetic moments.

In **ferromagnetic** materials (the root *ferro-* referring to iron) a bulk sample of the material maintains a magnetic dipole moment that is nearly constant in time. An iron magnet is an example of a ferromagnetic material. In addition to iron, only four other elements (the most common ones are nickel and cobalt) are ferromagnetic, although many other metal alloys are ferromagnetic. We have already seen how a single current loop (Figure 25.5) generates a magnetic field that resembles that of a bar magnet, and how a number of parallel loops effectively stacked together in a solenoid generate a magnetic field that looks even more like a bar magnet's. In a bar magnet made of iron (or another ferromagnetic material), the individual atomic magnetic moments tend to line up with one another, so that the net field produced resembles that of a solenoid. The lining up of the magnetic moments over large distances (compared with the size of an atom) is characteristic of ferromagnetic material. We say that there exists a *long-range order* of the magnetic moments.

Ferromagnetic materials establish their long-range order because of the combined action of smaller regions of magnetic order, called **domains**. Typically, magnetic domains have linear dimensions on the order of micrometers (10^{-6} m) to tens of micrometers. When a ferromagnetic material is subjected to an applied magnetic field, domains tend to align (magnetically) with their neighboring domains (Figure 25.13). That tendency makes the alignment spread throughout the piece of ferromagnetic material and helps keep the alignment stable over time.

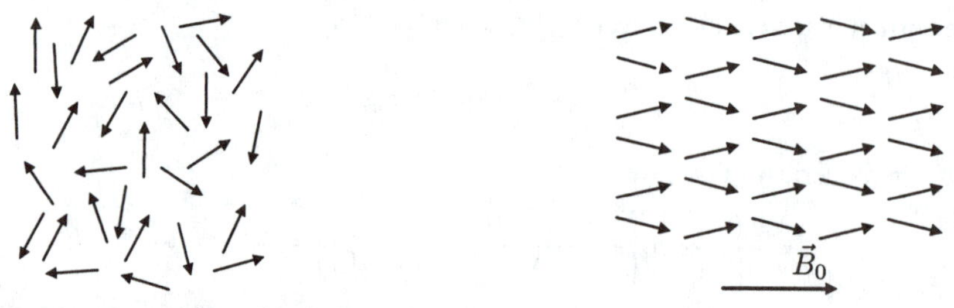

(a) Initially random orientation of domains in zero applied field

(b) Alignment occurs when a magnetic field is applied

Figure 25.13. Alignment of magnetic domains with an applied magnetic field.

Competing with the tendency to align is the random thermal motion of atoms or molecules in the solid. In ferromagnetic materials, the alignment can last many years, but eventually the orientation of the domains becomes more random. How long it takes for this to occur varies, depending on the type of material, and the size and history of the particular sample. For example, if a bar magnet is frequently used in the presence of other magnetic fields, it can lose its alignment rapidly, unless the applied fields happen to line up with the desired alignment. Large magnets stored outside other magnetic fields can retain a high degree of alignment for many years, but not forever. The common reference to ferromagnetic materials as "permanent magnets" is an exaggeration.

Because thermal motion eventually breaks down ferromagnetic alignment, the breakdown process can be

Table 25.1. Magnetic susceptibility of selected paramagnetic materials ($\chi > 0$) and diamagnetic materials ($\chi < 0$) at $T = 293$ K

Material	$\chi(\times 10^{-6})$	Material	$\chi(\times 10^{-6})$
Aluminum	$+16.5$	Lead	-23.0
Argon	-5.5	Manganese	$+529$
Calcium	$+40$	Mercury	-24.1
Cobalt Chloride	$+12,660$	Molybdenum	$+89.0$
Copper	-5.46	Oxygen	$+3449$
Gold	-28.0	Vanadium	$+255$

Source: CRC Handbook of Chemistry and Physics

enhanced by raising the temperature of the sample. In fact, each kind of ferromagnetic material has a specific temperature called the **Curie temperature**, above which it the material is not ferromagnetic (and hence there is no alignment of magnetic dipoles in the material, unless an external magnetic field is applied). The Curie temperature for iron is about 1040 K; for most ferromagnetic materials it is lower, and for some it is even below room temperature.

Most materials fall into one of the other two major categories: either **paramagnetic** or **diamagnetic**. In paramagnetic materials, there is no bulk magnetic moment, but a weak magnetic moment is obtained when the material is placed in a magnetic field, with the alignment in the direction of the applied field. In diamagnetic materials, no bulk magnetic moment exists either, but in the presence of an applied magnetic field, a weak magnetic moment is established in a direction opposite to the applied field. For studying paramagnetic and diamagnetic materials, it is useful to introduce the **magnetization** M, defined as the net magnetic moment per unit volume. Then the **magnetic susceptibility** χ of a material is defined by

$$\chi = \frac{\mu_0 M}{B_0} \tag{25.14}$$

where M is the magnetization induced by an applied field of magnitude B_0. Magnetic susceptibility is a scalar quantity. It is convenient to have $\chi > 0$ for paramagnetic materials and $\chi < 0$ for diamagnetic materials. This way, if the magnetization and applied magnetic fields are expressed as vectors (we choose not to do so here), the sign on the susceptibility makes the vectors parallel or antiparallel, as appropriate.

Before discussing diamagnetic and paramagnetic materials individually, let's examine a classical model of the magnetic moment generated by an orbiting electron (Figure 25.14). Assuming an electron in uniform circular motion, we apply the rules of angular momentum (Chapter 10) to find that the angular momentum $\vec{L} = \vec{r} \times \vec{p}$ of the electron is in the direction shown in the diagram. The current loop (the electron in orbit)

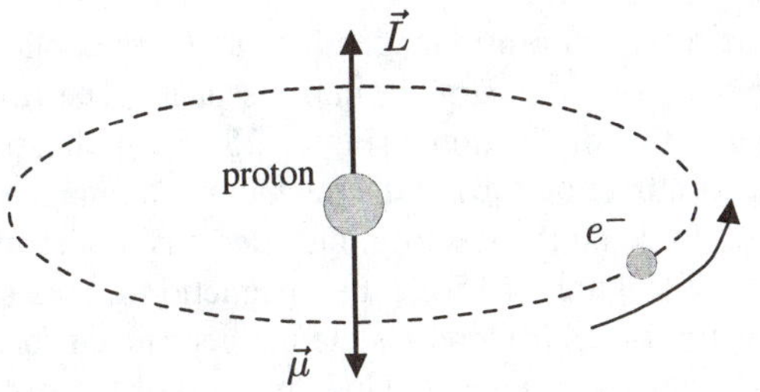

Figure 25.14. A model of an atom used to understand magnetic moments.

has a magnetic moment of magnitude $\mu = IA$, where $A = \pi r^2$ is the area of the loop. The magnitude of the current I is the absolute value (e) of the electron's charge divided by the period T for one orbit. Using the fact that the electron's speed is $v = 2\pi r/T$, we find

$$\mu = IA = \frac{e}{T}A = \frac{e}{2\pi r/v}(\pi r^2) = \frac{erv}{2}.$$

The magnitude of the angular momentum of a particle of mass m (in this case, m is the mass of the electron) in uniform circular motion is just $L = mvr$, so we can write

$$\mu = \frac{e}{2m}L.$$

Both the angular momentum and magnetic moment are vector quantities, and in terms of these vectors,

$$\vec{\mu} = \frac{e}{2m}\vec{L}. \qquad (25.15)$$

As shown in the diagram, $\vec{\mu}$ points in the opposite direction of $\vec{L}$.

In quantum mechanics, the magnitude of electron's angular momentum is *quantized* in units of $\hbar$ (defined as Planck's constant h divided by 2π). Therefore, the base unit of magnetic moment in quantum mechanics is

$$\mu_B = \frac{e\hbar}{2m} \qquad (25.16)$$

where as before m is the mass of the electron. Because all the factors on the right side of Equation (25.16) are constants, the resultant μ_B is also a constant. It is called the **Bohr magneton** and has a numerical value

$$\mu_B = \frac{(1.602 \times 10^{-19}\ \text{C})(1.055 \times 10^{-34}\ \text{J·s})}{2(9.109 \times 10^{-31}\ \text{kg})} = 9.27 \times 10^{-24}\ \text{J/T}.$$

The alignment of the atomic dipoles of a paramagnet in an applied field seems natural enough, like a compass needle swinging toward the north in Earth's magnetic field. How is it then that in a diamagnet the alignment is opposite to the applied field? Diamagnetic materials normally have even numbers of electrons. By the rules of quantum mechanics, the electronic spins and orbital angular momenta within an atom then combine roughly in pairs in such a way as to cancel, leaving no net magnetic moment. We show a classical approximation of this process in Figure 25.15, with two electrons in the same atom orbiting the atomic nucleus in circles of equal size. The orbits are in opposite directions, so that the electrons have angular momenta of equal magnitude and opposite directions. Hence, the electrons' orbital motions produce magnetic moments of equal magnitude and opposite directions, which add to zero. Note that the electrostatic force attracting the electron to the (positive) nucleus is the same for each electron. Also note the directions of the two individual magnetic moments, as stated in the diagram.

Now consider what happens when a uniform magnetic field $\vec{B}_0$ is applied to our model atom. For convenience, let's assume the direction of $\vec{B}_0$ is straight into the page. The resulting magnetic force on each electron is $\vec{F} = q\vec{v} \times \vec{B} = -e\vec{v} \times \vec{B}$. For electron 1 [Figure 25.15(a)], that force is directed straight up (opposite to the electrostatic force of attraction toward the nucleus). The magnetic force is small, but its effect is to decrease the net centripetal force on the electron, thus decreasing the magnitude of the corresponding magnetic moment. For electron 2 [Figure 25.15(b)], the magnetic force is straight down, and thus it adds a small amount to the electrostatic force. It increases the net centripetal force, which in turn increases the magnitude of the corresponding magnetic moment. Now the sum of the two electrons' magnetic moments is no longer zero; the magnetic moment of electron 2 wins out, just barely. The observed effect is a weak induced magnetic moment pointing out of the page, opposite to the direction of our applied magnetic field.

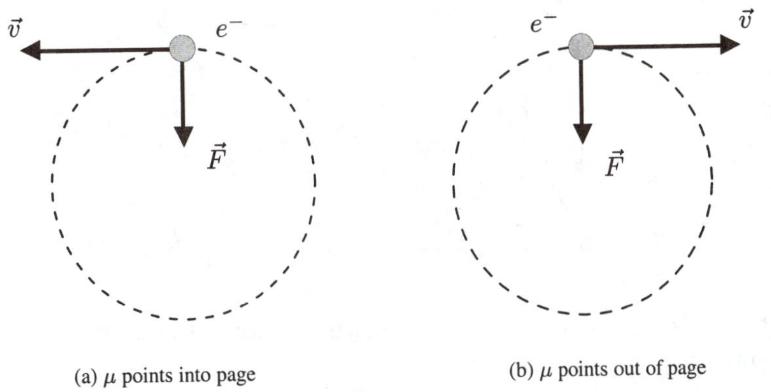

Figure 25.15. Crude model of an atom, in which two electrons in the same atom orbiting the atomic nucleus in circles of equal size.

In paramagnetic materials, Equation (25.14) predicts a magnetization

$$M = \frac{\chi}{\mu_0} B_0. \tag{25.17}$$

Typical values of the magnetic susceptibility χ (a dimensionless number) range from 10^{-6} to 10^{-2}. Remember that $\chi > 0$ for paramagnetic materials. By comparison, the range of χ for diamagnetic materials (where $\chi < 0$) is -10^{-6} to -10^{-4} (see Table 25.1). In paramagnetic materials, the tendency for the magnetic dipoles to align in the magnetic field is always in competition with the dipoles' random thermal motion, which tends to reduce the alignment and hence limit the magnetization of a sample. In 1895, the French physicist Pierre Curie (1859–1906) discovered an inverse relationship between magnetic susceptibility and temperature—that is, χ is proportional to $1/T$. With a proportionality constant C (called the **Curie constant**, which is really a parameter with a different numerical value for each different paramagnetic material), we can write

$$\chi = \frac{C}{T}. \tag{25.18}$$

Thus, Equation (25.17) can be rewritten as

$$M = \frac{C}{\mu_0} \frac{B_0}{T}. \tag{25.19}$$

Equation (25.19) is commonly called the **Curie law** for paramagnetism.

The Curie law predicts that the magnetization of a paramagnetic sample is proportional to the fraction B_0/T. A typical graph of experimental data is shown in Figure 25.16. Notice that the Curie law is consistent with the data only for lower values of B_0/T. At higher values of B_0/T, the magnetization levels off, indicating that the magnetic dipoles have become as well aligned as possible. It should make sense that this occurs when the magnetic field magnitude B_0 is high. It also makes sense that it occurs when T is relatively *low*, because then the thermal motion is minimized, and it is easier for even a modest applied field to align the magnetic dipoles.

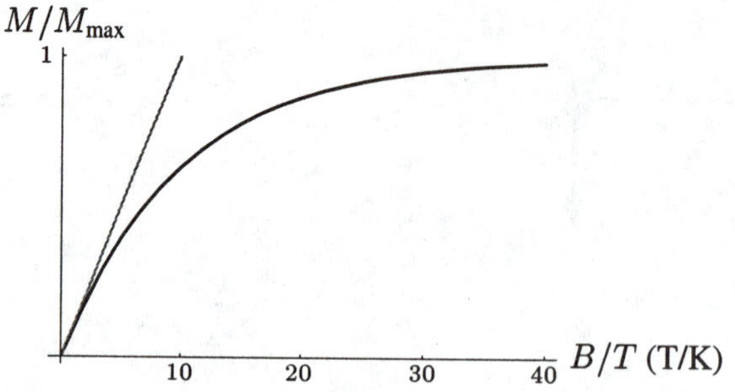

Figure 25.16. Magnetization vs. B/T in a paramagnetic material. The Curie law prediction is the gray line.

25.6 Divergence

25.6.1 Flux Density

In Chapter 23, we defined the flux of a vector field $\vec{F} : \mathbb{R}^3 \to \mathbb{R}^3$ for a surface S as

$$\Phi_F = \iint_S \vec{F} \cdot \hat{n}\, dA.$$

This scalar quantity depends on both the field $\vec{F}$ and the chosen surface. In many situations, it is convenient to define and understand a related quantity that is independent of the choice of surface. We define the **flux density** as follows. At a point (x, y, z), consider a solid region of volume V with a surface S as its boundary. The flux density is defined as

$$\lim_{V\to 0} \frac{\Phi_F}{V} = \lim_{V\to 0} \frac{1}{V} \iint_S \vec{F} \cdot \hat{n}\, dA$$

with the convention that the surface normals point outward. From this definition, it is not clear whether the flux density might depend on the shape of the solid region used in the limit. We do not discuss this point here. Instead, we pick a specific geometry that allows us to find a convenient relation between the flux density and the vector field $\vec{F}$.

For the point (x, y, z), take the solid region to be a rectangular box with its faces parallel to the coordinate planes, as shown in Figure 25.17. Let one corner lie at (x, y, z), and the diagonal corner at $(x + \Delta x, y + \Delta y, z + \Delta z)$. The volume of the box is $\Delta x \Delta y \Delta z$. For this geometry, the surface integral over the boundary of the box is approximated by a sum of six terms:

$$\iint_S \vec{F} \cdot \hat{n}\, d\vec{A} \approx \sum_{i=1}^{6} \vec{F_i} \cdot \vec{A_i}$$

with one term for each face of the box. Let $i = 1$ correspond to the face parallel to the yz-plane containing the point (x, y, z). For this face, $\vec{A_1} = -\Delta y \Delta z\, \hat{\imath}$, so

$$\vec{F_1} \cdot \vec{A_1} = -\vec{F}(x, y, z) \cdot \Delta y \Delta z\, \hat{\imath} = -F_x(x, y, z)\, \Delta y \Delta z.$$

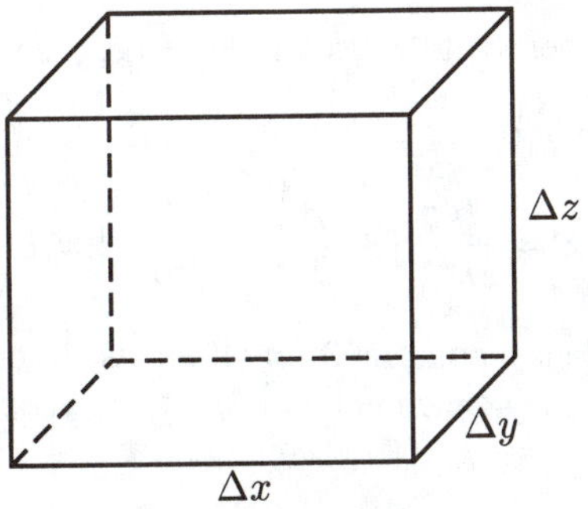

Figure 25.17. Rectangular box used to illustrate flux density.

If we let $i = 2$ correspond to the opposite face, we have $\vec{A_2} = +\Delta y \Delta z\, \hat{\imath}$, and

$$\vec{F_2} \cdot \vec{A_2} = \vec{F}(x + \Delta x, y, z) \cdot \Delta y \Delta z\, \hat{\imath} = F_x(x + \Delta x, y, z)\, \Delta y \Delta z\,.$$

Pairing the opposite sides parallel to the xz-plane and xy-plane in a similar fashion, we have

$$\iint_S \vec{F} \cdot \hat{n}\, dA \approx [F_x(x + \Delta x, y, z) - F_x(x, y, z)]\, \Delta y \Delta z + [F_y(x, y + \Delta y, z) - F_y(x, y, z)]\, \Delta x \Delta z$$
$$+ [F_z(x, y, z + \Delta z) - F_z(x, y, z)]\, \Delta x \Delta y.$$

To get the flux density at (x, y, z), we divide by the volume $V = \Delta x \Delta y \Delta z$ and take a limit as $V \to 0$. This is equivalent to $\Delta x, \Delta y, \Delta z \to 0$. The flux density is thus

$$\begin{aligned}
&\lim_{V \to 0} \frac{1}{V} \iint_S \vec{F} \cdot \hat{n}\, dA \\
&= \lim_{\Delta x, \Delta y, \Delta z \to 0} \frac{1}{\Delta x \Delta y \Delta z} \{[F_x(x + \Delta x, y, z) - F_x(x, y, z)]\, \Delta y \Delta z \\
&\qquad + [F_y(x, y + \Delta y, z) - F_y(x, y, z)]\, \Delta x \Delta z + [F_z(x, y, z + \Delta z) - F_z(x, y, z)]\, \Delta x \Delta y\} \\
&= \lim_{\Delta x, \Delta y, \Delta z \to 0} \left[\frac{F_x(x + \Delta x, y, z) - F_x(x, y, z)}{\Delta x} + \frac{F_y(x, y + \Delta y, z) - F_y(x, y, z)}{\Delta y} \right. \\
&\qquad \left. + \frac{F_z(x, y, z + \Delta z) - F_z(x, y, z)}{\Delta z} \right] \\
&= \frac{\partial F_x}{\partial x} + \frac{\partial F_y}{\partial y} + \frac{\partial F_z}{\partial z}.
\end{aligned}$$

Note that the final result is fairly simple to write and depends *only* on the field $\vec{F}$, and not on the choice of surface. The expression we just derived is important enough to deserve its own name, which is given in the following definition.

Definition 25.1. Let $\vec{F} : \mathbb{R}^3 \to \mathbb{R}^3$ be a vector field with $\vec{F}(x, y, z) = \langle F_x(x, y, z), F_y(x, y, z), F_z(x, y, z) \rangle$. The **divergence of $\vec{F}$** is denoted div $\vec{F}$ and is defined as

$$\operatorname{div} \vec{F}(x, y, z) = \frac{\partial F_x}{\partial x} + \frac{\partial F_y}{\partial y} + \frac{\partial F_z}{\partial z}.$$

Example 25.5. Compute the divergence of the vector field $\vec{F}(x, y, z) = \langle x, y, z \rangle$.

Solution. The computation is straightforward:

$$\text{div}\langle x, y, z \rangle = \frac{\partial}{\partial x}[x] + \frac{\partial}{\partial y}[y] + \frac{\partial}{\partial z}[z] = 1 + 1 + 1 = 3.$$

For this vector field, the divergence is a constant 3 for all inputs. In other terms, the flux density is 3 for all input points. Roughly, this means that at each point (x, y, z), the vector field outputs point more away from the point than in toward the point. △

Example 25.6. Compute the divergence of the vector field $\vec{F}(x, y, z) = \langle x^2y, y + z, yz^2 \rangle$.

Solution. The computation is again straightforward:

$$\text{div}\langle x^2y, y + z, yz^2 \rangle = \frac{\partial}{\partial x}[x^2y] + \frac{\partial}{\partial y}[y + z] + \frac{\partial}{\partial z}[yz^2] = 2xy + 1 + 2yz.$$

For this vector field, the divergence is not constant. △

25.6.2 The Operator View of Divergence

In Section 18.3, we studied the gradient for a function $f : \mathbb{R}^n \to \mathbb{R}$. The components of a gradient vector are the n partial derivatives of f. We saw in Chapter 18 that the gradient has some geometric and physical applications, including measuring the direction and magnitude of steepest ascent along a hill and relating the electric field to the electric potential. You may wish to review Chapter 18 for other facts about the gradient and other applications.

In our study of the gradient, the symbol $\vec{\nabla}$ told you to construct the vector described above. For example, in three dimensions

$$\vec{\nabla} f = \frac{\partial f}{\partial x}\hat{\imath} + \frac{\partial f}{\partial y}\hat{\jmath} + \frac{\partial f}{\partial z}\hat{k} = \left\langle \frac{\partial f}{\partial x}, \frac{\partial f}{\partial y}, \frac{\partial f}{\partial z} \right\rangle. \tag{25.20}$$

It is often useful to think of the vector **operator** $\vec{\nabla}$ as a separate entity, which can be written

$$\vec{\nabla} = \hat{\imath}\frac{\partial}{\partial x} + \hat{\jmath}\frac{\partial}{\partial y} + \hat{k}\frac{\partial}{\partial z} = \left\langle \frac{\partial}{\partial x}, \frac{\partial}{\partial y}, \frac{\partial}{\partial z} \right\rangle \tag{25.21}$$

in three dimensions. In general, the operator $\vec{\nabla}$ contains the n partial derivatives of the n-dimensional space. The operator by itself is not a function, but it produces a function when it operates on a function f, as in Equation (25.20). This is just like an ordinary derivative operator d/dx, which produces a function df/dx when it operates on a function f.

In this section, we examine another way in which the operator $\vec{\nabla}$ can be used. Consider a vector field $\vec{F} : \mathbb{R}^3 \to \mathbb{R}^3$ with

$$\vec{F} = \langle F_x, F_y, F_z \rangle .$$

Think now of taking a dot product of the vector operator $\vec{\nabla}$ as written in Equation (25.21) and the vector $\vec{F}$:

$$\vec{\nabla} \cdot \vec{F} = \left\langle \frac{\partial}{\partial x}, \frac{\partial}{\partial y}, \frac{\partial}{\partial z} \right\rangle \cdot \langle F_x, F_y, F_z \rangle .$$

If we simply follow the usual definition of dot product, the result is

$$\vec{\nabla} \cdot \vec{F} = \frac{\partial F_x}{\partial x} + \frac{\partial F_y}{\partial y} + \frac{\partial F_z}{\partial z}. \tag{25.22}$$

The result is a *scalar function* (naturally, because we took a dot product), consisting of the sum of the partial derivatives of each component of $\vec{F}$ with respect to the corresponding input variable $(x, y,$ and $z)$. We recognize $\vec{\nabla} \cdot \vec{F}$ as the divergence of $\vec{F}$, which we defined previously using the concept of flux density.

There is nothing to restrict the definition of divergence to three dimensions. We will now generalize the procedure with a definition.

Definition 25.2. Let $\vec{F} : \mathbb{R}^n \to \mathbb{R}^n$ be a vector field with components $\vec{F}(\vec{r}) = \langle F_1(\vec{r}), F_2(\vec{r}), ..., F_n(\vec{r})\rangle$ for $\vec{r} = \langle x_1, x_2, \cdots, x_n\rangle$. The **divergence of $\vec{F}$** is denoted div $\vec{F}$ and defined as

$$\text{div}\,\vec{F}(\vec{r}) = \vec{\nabla} \cdot \vec{F}(\vec{r}) = \frac{\partial F_1}{\partial x_1} + \frac{\partial F_2}{\partial x_2} + \cdots + \frac{\partial F_n}{\partial x_n}.$$

For the divergence to be defined, each of the component functions F_i must be have a first partial derivative with respect to the corresponding variable x_i.

Let's make some observations about the divergence before going on to see how it is used. First, we emphasize that the divergence of a vector is a scalar function of all the input components x_i. That is, $\vec{\nabla} \cdot \vec{F} : \mathbb{R}^n \to \mathbb{R}$. Second, it should be clear from the definition that the divergence is only defined for functions of the type $\mathbb{R}^n \to \mathbb{R}^n$, because each output component F_i must have a corresponding input component x_i.

We will use the two notations div $\vec{F}$ and $\vec{\nabla} \cdot \vec{F}$ interchangeably. Physicists and engineers often refer to the vector operator $\vec{\nabla}$ as "del" and say "del dot F" to refer to the divergence of $\vec{F}$ (because of the way $\vec{\nabla} \cdot \vec{F}$ appears on the page). That is, they use "del dot F" instead of, or interchangeably with, "divergence of F." Mathematicians are more consistent about using the word "divergence."

Example 25.7. Find the divergence of the vector fields

$$\vec{A}(x, y, z) = \left(2xy^2 + y\right)\hat{\imath} - \left(2x^3y - 4y^2 - 7\right)\hat{\jmath}$$

and

$$\vec{B}(x, y, z) = \left\langle 3xz, 4x^3 - x, x^2 + y^2 + z^2\right\rangle.$$

Solution. Notice that $\vec{A}$ is a vector field of the type $\vec{A} : \mathbb{R}^2 \to \mathbb{R}^2$. Therefore, from the definition of divergence,

$$\vec{\nabla} \cdot \vec{A} = \frac{\partial A_x}{\partial x} + \frac{\partial A_y}{\partial y}.$$

Computing the necessary partial derivatives,

$$\vec{\nabla} \cdot \vec{A}(x, y, z) = 2y^2 - \left(2x^3 - 8y\right) = 2y^2 - 2x^2 + 8y.$$

The vector field $\vec{B}$ is of the type $\vec{B} : \mathbb{R}^3 \to \mathbb{R}^3$. Therefore, from the definition of divergence,

$$\text{div}\,\vec{B} = \frac{\partial B_x}{\partial x} + \frac{\partial B_y}{\partial y} + \frac{\partial B_z}{\partial z}.$$

Computing the necessary partial derivatives,

$$\text{div}\,\vec{B}(x, y, z) = 3z + 0 + 2z = 5z.$$

In each case, the resulting divergence is a scalar function of the input variables. △

25.6.3 Application to Fluid Flow

We now use a familiar physics example (fluid flow) to illustrate a physical interpretation of the divergence. The velocity of a flowing fluid can described by a three-dimensional vector field (shown schematically in Figure 25.18). Like any vector field, this means that every point in a well-defined region of space (in this case, inside the fluid) has a velocity vector describing the velocity of the fluid at that point. Thus, the velocity field $\vec{v}(x, y, z)$ is a function of position. It may also vary as a function of time, if the velocity field is changing, but for the moment we will ignore that consideration to concentrate on the dependence of $\vec{v}$ on x, y, and z.

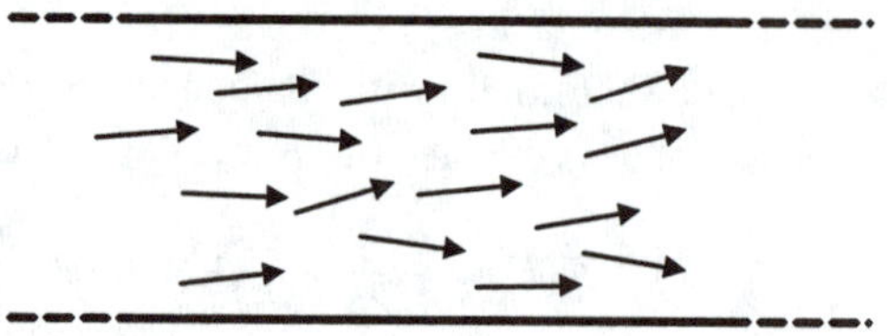

Figure 25.18. Velocity vector field for fluid flow.

It is convenient to define another vector $\vec{R} = \rho\vec{v}$, where ρ is the density (mass per unit volume) of the fluid. Note that in a fluid ρ may also vary with position and time. Using the dimensions of density and velocity, we can see the dimensions of $\vec{R}$ are

$$\text{dimensions of } \vec{R} = \left(\frac{\text{mass}}{\text{volume}}\right)\left(\frac{\text{distance}}{\text{time}}\right) = \frac{\text{mass}}{(\text{area})(\text{time})}$$

or mass per unit area per unit time. Remember that $\vec{R}$ is a vector, with components R_x, R_y, and R_z. Now consider a very small rectangular box, with dimensions Δx by Δy by Δz, somewhere in the fluid (shown schematically in Figure 25.17). By the linear approximation, the difference between the two R_x outputs on the right and left sides of the box is

$$\Delta R_x \approx \frac{\partial R_x}{\partial x}\Delta x.$$

Given the dimensions of $\vec{R}$ we found above, the net mass per unit time flowing through the left and right sides is (ΔR_x)(area of left and right sides) $= (\Delta R_x)(\Delta y\, \Delta z)$. Therefore, the net mass per unit time flowing through the left and right sides of the box is

$$\frac{\partial R_x}{\partial x}\Delta x\Delta y\Delta z.$$

Following the same procedure for the top/bottom and front/back of the box, the net change of mass per unit time for the whole box is

$$\begin{aligned}\frac{\text{change in mass}}{\text{unit time}} &\approx \frac{\partial R_x}{\partial x}\Delta x\,\Delta y\,\Delta z + \frac{\partial R_y}{\partial y}\Delta x\,\Delta y\,\Delta z + \frac{\partial R_z}{\partial z}\Delta x\,\Delta y\,\Delta z \\ &= \left(\frac{\partial R_x}{\partial x} + \frac{\partial R_y}{\partial y} + \frac{\partial R_z}{\partial z}\right)\Delta x\,\Delta y\,\Delta z.\end{aligned}$$

You should recognize the divergence in the above expression, so we can write

$$\frac{\text{change in mass}}{\text{unit time}} \approx (\vec{\nabla}\cdot\vec{R})\Delta x\Delta y\Delta z. \qquad (25.23)$$

In addition to the outward flow calculated in Equation (25.23), we need to consider the possibility that mass is being added to or taken out of the fluid. When mass is added, we say there is a *source*, and when mass is removed, we say there is a *sink* (think of these as a faucet and a drain, respectively). Let k be the rate (per unit time) of new mass per unit volume in the fluid, so that for a source $k > 0$, and for a sink $k < 0$. Then for our small box, we can say

$$\frac{\text{net change in mass}}{\text{unit time}} = \text{change due to sources and sinks} - \text{outflow in Equation (25.23).} \tag{25.24}$$

From dimensional analysis, we can see that the left side of Equation (25.24) can be written

$$\frac{\text{net change in mass}}{\text{unit time}} = \frac{\partial \rho}{\partial t}\Delta x \Delta y \Delta z,$$

where we have used the fact that density is mass/volume. Notice that we have used the *partial derivative* of the density with respect to time, because the density can also depend on position. The change in mass per unit time due to sources and sinks is just k times the volume. Putting these results and the result of Equation (25.23) together in Equation (25.24), we have for the box

$$\frac{\partial \rho}{\partial t}\Delta x \Delta y \Delta z \approx (k)\Delta x \Delta y \Delta z - (\vec{\nabla} \cdot \vec{R})\Delta x \Delta y \Delta z.$$

Dividing by the common factor $\Delta x \Delta y \Delta z$ and rearranging, we have

$$\vec{\nabla} \cdot \vec{R} = k - \frac{\partial \rho}{\partial t}. \tag{25.25}$$

Now let's interpret the result expressed in Equation (25.25), and thereby give another physical interpretation to the divergence. To simplify things, suppose there are no sources or sinks, so we can take $k = 0$. Then for this case,

$$\vec{\nabla} \cdot \vec{R} = -\frac{\partial \rho}{\partial t} \qquad \text{(no sources or sinks).} \tag{25.26}$$

Stated in words, the divergence of $\vec{R}$ is the negative of the rate of change of density of the fluid. We have taken the mass of the fluid to be constant. Therefore, a positive divergence of $\vec{R}$ corresponds to a decreasing density, which implies an *expansion of the fluid*. Conversely, a negative divergence of $\vec{R}$ implies an increase in density or a *compression of the fluid*. Remember that $\vec{R}$ is proportional to $\vec{v}$, and in fact the vectors $\vec{R}$ and $\vec{v}$ point in the same direction at every point in space. Thus, we can claim that a positive divergence of the velocity field ($\vec{\nabla} \cdot \vec{v}$) also implies an expansion of the fluid, and a negative divergence of the velocity field implies a compression of the fluid.

Based on this analysis, we say that a fluid is **incompressible** if $\partial \rho / \partial t = 0$. This makes physical sense, because if a fluid undergoes neither compression nor expansion its density is constant.

Equation (25.26), valid when there are no sources or sinks, is often referred to as the **equation of continuity**. A similar equation comes up quite frequently in dealing with applications of vector fields in physics and engineering, including those applications in which the vector field being studied is not the velocity field of a fluid. In Chapter 27, we examine the physical significance of the divergence of electric and magnetic fields.

It is often useful to make a graphical interpretation of the divergence of a vector field. Consider the two-dimensional vector field shown in Figure 25.19. Think of computing the divergence at a point as we did for the velocity field of the fluid: Draw a small rectangle, and find the net difference between the incoming and outgoing vectors along the sides of the rectangle. Then taking a limit that shrinks the rectangle's size to zero gives the divergence of the vector field at a single point. If you try this for various points in the vector fields shown in Figure 25.19, you should be able to convince yourself that a nonzero divergence exists everywhere in the picture, because any small rectangle you draw has a greater flux of the vector field going out than going in. In fact, it is the net *flux* of the vector field that determines the divergence; we will develop this point more carefully in Chapter 27, and thereby make the connection to electric and magnetic fields (you have already seen the flux of the electric field in Gauss's law).

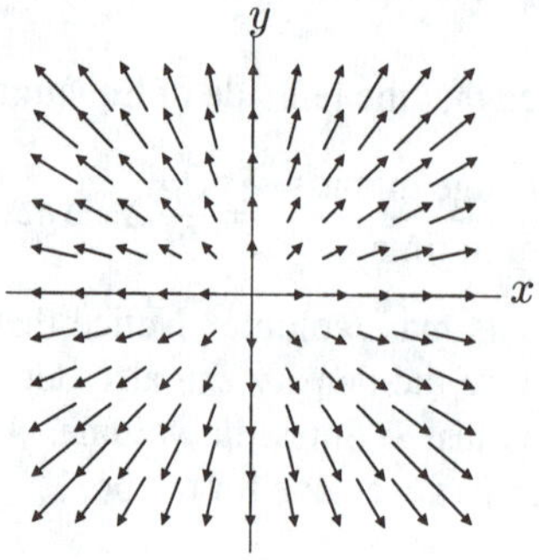

Figure 25.19. A vector field.

Example 25.8. Consider a vector field $\vec{R} = \rho\vec{v}$ that is described by the function

$$\vec{R} = \frac{1}{x+2}\,\hat{\imath}$$

for all points in space with $x > 0$. The function outputs are in units kg/(m^2·s) for inputs in m. (a) Graph the vector field. (b) Compute the divergence of the vector field, and interpret the result, assuming $k = 0$. (c) Reconcile your answer in (b) with an interpretation of the graph in (a).

Solution. (a) A graph is shown in Figure 25.20, with the view in the xy-plane. (The field looks the same for any plane parallel to the xy-plane.)

(b) From the definition of divergence,

$$\vec{\nabla}\cdot\vec{R} = \frac{\partial R_x}{\partial x} + \frac{\partial R_y}{\partial y} + \frac{\partial R_z}{\partial z} = -\frac{1}{(x+2)^2} + 0 + 0 = -\frac{1}{(x+2)^2}.$$

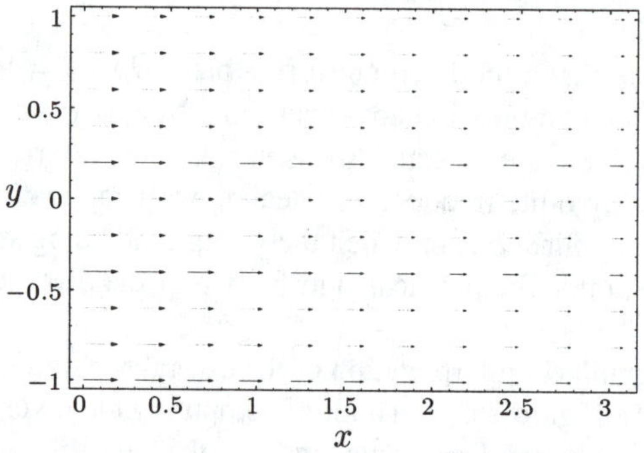

Figure 25.20. Vector field of Example 25.8.

The divergence of $\vec{R}$ is negative for all x, and therefore we conclude that the fluid described by this field is being compressed.

(c) If we imagine a rectangle (with its sides parallel to the x- and y-axes) anywhere in the field in Figure 25.20, clearly a greater flux goes into the rectangle than emerges from it. This is consistent with a negative divergence. If you think of flowing along with the fluid (to the right in the diagram), it must be compressed for the magnitude of the velocity to decrease as we move to the right. Again, the rectangle can help you to visualize this. With more fluid entering from the left (per unit time) than leaving from the right of the rectangle, there has to be a compression of the fluid within the rectangle. △

25.7 Curl

Think again of the vector operator $\vec{\nabla}$ as a separate entity, as we wrote it in Equation (25.21):

$$\vec{\nabla} = \left\langle \frac{\partial}{\partial x}, \frac{\partial}{\partial y}, \frac{\partial}{\partial z} \right\rangle$$

in three dimensions. In the preceding section, we developed the divergence by taking the dot product of the vector operator $\vec{\nabla}$ with a vector field. Now let's try taking the *cross product* of the operator $\vec{\nabla}$ with a vector field (remember that the cross product is only defined for three dimensions). Doing so leads to the following definition.

Definition 25.3. Let $\vec{F} : \mathbb{R}^3 \to \mathbb{R}^3$ be a vector field with $\vec{F}(x,y,z) = \langle F_x(x,y,z), F_y(x,y,z), F_z(x,y,z)\rangle$. The **curl of $\vec{F}$** is denoted curl $\vec{F}$ and defined as the vector field

$$\operatorname{curl} \vec{F}(x,y,z) = \vec{\nabla} \times \vec{F} = \left\langle \frac{\partial F_z}{\partial y} - \frac{\partial F_y}{\partial z}, \frac{\partial F_x}{\partial z} - \frac{\partial F_z}{\partial x}, \frac{\partial F_y}{\partial x} - \frac{\partial F_x}{\partial y} \right\rangle.$$

For the curl to be defined, each of the component functions F_i must have first partial derivatives with respect to x,y, and z.

We emphasize that the curl of $\vec{F}$ is a three-dimensional vector field, with curl $\vec{F} : \mathbb{R}^3 \to \mathbb{R}^3$. Like Equation (10.21) that defines the cross product, the formula for the curl of $\vec{F}$ in Definition 25.3 is long, but the fact that all the variables recur in cyclic permutations makes it simpler to remember.

Just as with divergence, different words and notations are used for the curl. Physicists and engineers often say "del cross F" to refer to the curl of $\vec{F}$ (because of the way $\vec{\nabla} \times \vec{F}$ appears on the page). That is, they use "del cross F" instead of or interchangeably with "curl of F."

Example 25.9. Find the curl of the vector fields

$$\vec{A}(x,y,z) = \left(2xy^2 + y\right)\hat{\imath} - \left(2x^3y - 4y^2 - 7\right)\hat{\jmath}$$

and

$$\vec{B}(x,y,z) = \left\langle 3xz, 4x^3 - x, x^2 + y^2 + z^2 \right\rangle.$$

Solution. We use Definition 25.3 to evaluate the curl of $\vec{A}$. Notice that $F_z = 0$.

$$\begin{aligned}
\vec{\nabla} \times \vec{A} &= \left(\frac{\partial A_z}{\partial y} - \frac{\partial A_y}{\partial z}\right)\hat{\imath} + \left(\frac{\partial A_x}{\partial z} - \frac{\partial A_z}{\partial x}\right)\hat{\jmath} + \left(\frac{\partial A_y}{\partial x} - \frac{\partial A_x}{\partial y}\right)\hat{k} \\
&= (0-0)\,\hat{\imath} + (0-0)\,\hat{\jmath} + \left(-6x^2y - (4xy+1)\right)\hat{k} \\
&= \left(-6x^2y - 4xy - 1\right)\hat{k}.
\end{aligned}$$

Similarly,

$$\begin{aligned}\vec{\nabla}\times\vec{B} &= \left\langle \frac{\partial}{\partial y}[x^2+y^2+z^2]-\frac{\partial}{\partial z}[4x^3-x], \frac{\partial}{\partial z}[3xz]-\frac{\partial}{\partial x}[x^2+y^2+z^2], \frac{\partial}{\partial x}[4x^3-x]-\frac{\partial}{\partial y}[3xz]\right\rangle \\ &= \langle 2y-0, 3x-2x, 12x-1\rangle = \langle 2y, x, 12x-1\rangle.\end{aligned}$$

△

Because the cross product is defined only for vectors in $\mathbb{R}^3$, this definition of curl is sensible only for vector fields $\vec{F}:\mathbb{R}^3\to\mathbb{R}^3$. However, there is a natural way to extend the idea of curl to a vector field $\vec{F}:\mathbb{R}^2\to\mathbb{R}^2$. Let $\vec{F}(x,y)=\langle F_x(x,y),F_y(x,y)\rangle$. Now build a related vector field $\vec{G}:\mathbb{R}^3\to\mathbb{R}^3$ by tacking on a third component with value 0 for every input. That is, let $\vec{G}(x,y,z)=\langle F_x(x,y),F_y(x,y),0\rangle$. Compute the curl of $\vec{G}$ to get

$$\begin{aligned}\operatorname{curl}\vec{G}(x,y,z) &= \left\langle \frac{\partial}{\partial y}[0]-\frac{\partial}{\partial z}[F_y(x,y)], \frac{\partial}{\partial z}[F_x(x,y)]-\frac{\partial}{\partial x}[0], \frac{\partial}{\partial x}[F_y(x,y)]-\frac{\partial}{\partial y}[F_x(x,y)]\right\rangle \\ &= \left\langle 0,0,\frac{\partial F_y}{\partial x}-\frac{\partial F_x}{\partial y}\right\rangle.\end{aligned}$$

Note that the partial derivatives with respect to z are zero because F_x and F_y depend only on x and y. The final result has only one nonzero component. With this as motivation, we have the following definition.

Definition 25.4. Let $\vec{F}:\mathbb{R}^2\to\mathbb{R}^2$ be a vector field with $\vec{F}(x,y)=\langle F_x(x,y),F_y(x,y)\rangle$. The **scalar curl of $\vec{F}$** is denoted $\operatorname{curl}\vec{F}$ and defined as

$$\operatorname{curl}\vec{F}(x,y)=\frac{\partial F_y}{\partial x}-\frac{\partial F_x}{\partial y}.$$

In practice, we often omit the adjective "scalar" and simply refer to the curl of $\vec{F}$. Context will always tells us whether we are using the vector curl or the scalar curl.

Example 25.10. Compute the curl of the $\vec{E}(x,y)=\langle xy^2,x+y\rangle$.

Solution. Since the vector field has inputs and outputs in $\mathbb{R}^2$, we compute the scalar curl to get

$$\operatorname{curl}\vec{E}(x,y)=\frac{\partial}{\partial x}[x+y]-\frac{\partial}{\partial y}[xy^2]=1+2xy.$$

△

In Chapter 22, we proved Theorem 22.3: If $\vec{F}(x,y)=\langle P(x,y),Q(x,y)\rangle$ has components with continuous partial derivatives and there exists a potential function for $\vec{F}$, then

$$\frac{\partial P}{\partial y}=\frac{\partial Q}{\partial x}.$$

We can write the conclusion as

$$\frac{\partial Q}{\partial x}-\frac{\partial P}{\partial y}=0.$$

That is, the conclusion is equivalent to saying that the scalar curl of the vector field $\vec{F}$ is zero if there is a potential function for $\vec{F}$. Recall that $V:\mathbb{R}^2\to\mathbb{R}$ is a potential function for $\vec{F}$ if $\vec{\nabla}V(x,y)=\vec{F}(x,y)$. This result can be generalized to vector fields with inputs and outputs in $\mathbb{R}^3$. The key is the identity given in the following theorem.

Theorem 25.1. *If $V : \mathbb{R}^3 \to \mathbb{R}$ has continuous partial derivatives, then the curl of the gradient of V is zero. That is,*

$$\vec{\nabla} \times (\vec{\nabla} V) = \vec{0}.$$

Proof. First, we have from the definition of the gradient

$$\vec{\nabla} V = \left\langle \frac{\partial V}{\partial x}, \frac{\partial V}{\partial y}, \frac{\partial V}{\partial z} \right\rangle.$$

Using the definition of curl,

$$\vec{\nabla} \times (\vec{\nabla} V) = \left\langle \frac{\partial^2 V}{\partial y \partial z} - \frac{\partial^2 V}{\partial z \partial y}, \frac{\partial^2 V}{\partial x \partial z} - \frac{\partial^2 V}{\partial z \partial x}, \frac{\partial^2 V}{\partial x \partial y} - \frac{\partial^2 V}{\partial y \partial x} \right\rangle.$$

By the equality of mixed partial derivatives, each of the expressions in square brackets is zero, and

$$\vec{\nabla} \times (\vec{\nabla} V) = \langle 0, 0, 0 \rangle = \vec{0}.$$

This is the conclusion of the theorem. □

The generalization of Theorem 25.2 now follows.

Theorem 25.2. *If $\vec{F} : \mathbb{R}^3 \to \mathbb{R}^3$ has components with continuous partial derivatives and there exists a potential function for $\vec{F}$, then*

$$\operatorname{curl} \vec{F}(x, y, z) = \vec{\nabla} \times \vec{F} = \vec{0}.$$

Proof. By hypothesis, there is a potential function for $\vec{F}$ so $\vec{F}(x, y, z) = \vec{\nabla} V(x, y, z)$ for some function $V : \mathbb{R}^3 \to \mathbb{R}$. Using this fact together with Theorem 25.1, we have

$$\operatorname{curl} \vec{F}(x, y, z) = \vec{\nabla} \times \vec{F} = \vec{\nabla} \times (\vec{\nabla} V) = \vec{0}.$$

This is the conclusion we seek. □

We have seen in previous chapters that it is possible to express a *conservative* force field as the gradient of a scalar function. For example, from Chapter 18, a static electric field is $\vec{E} = -\vec{\nabla} V$, where V is the electric potential. From the result of Theorem 25.2, we can infer that the curl of a static electric field $\vec{E}$ is zero, or stated symbolically

$$\vec{\nabla} \times \vec{E} = \vec{0}. \tag{25.27}$$

Also, in Chapter 27 we will see that the converse of Theorem 25.2 is true. That is, if $\operatorname{curl} \vec{F} = \vec{0}$ on a reasonable domain, then a potential function exists for $\vec{F}$.

As the name *curl* implies, the curl has an interpretation in terms of rotational motion. The following example illustrates this.

Example 25.11. A solid uniform circular disk of radius R lies in the xy-plane with its center at the origin. The disk rotates counterclockwise (viewed from above) with a constant angular speed ω. Show that the curl of the velocity field for points in the disk is $2\vec{\omega}$, where $\vec{\omega}$ is the angular velocity vector.

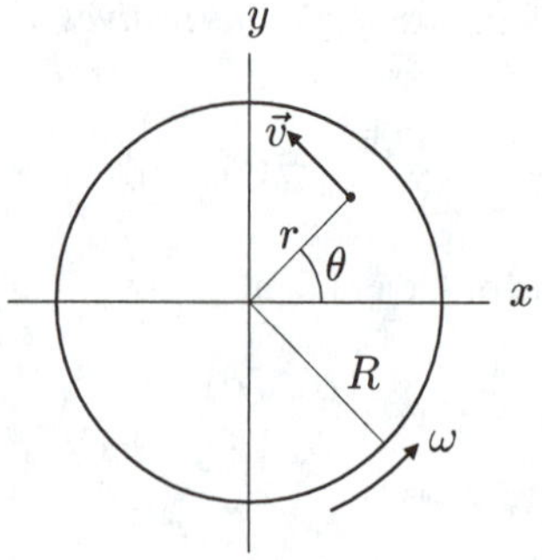

Figure 25.21. Rotating disk.

Solution. By the convention developed in Section 10.1, the direction of the angular velocity vector is in the $+z$-direction (perpendicular to the disk and given by the right-hand rule). That is, $\vec{\omega} = \omega\,\hat{k}$. For an arbitrary point P in the disk a distance r from the center of rotation (Figure 25.21), the speed of that point is $v = \omega r$. Thus, the velocity components are

$$v_x = -v\,\sin\theta = -\omega r\,\sin\theta \qquad \text{and} \qquad v_y v\cos\theta = \omega r\,\cos\theta.$$

From trigonometry, $\sin\theta = y/r$ and $\cos\theta = x/r$, so

$$v_x = -\omega y \qquad \text{and} \qquad v_y = \omega x.$$

With these velocity components, we can compute the curl of $\vec{v}$:

$$\vec{\nabla} \times \vec{v} = \hat{\imath}\,(0-0) + \hat{\jmath}\,(0-0) + \hat{k}(\omega - (-\omega)) = 2\omega\,\hat{k} = 2\vec{\omega}.$$

The curl of the velocity field is $2\vec{\omega}$. △

It is true that in general, a velocity field $\vec{v}$ for rigid rotation about an axis has the property that

$$\vec{\nabla} \times \vec{v} = 2\vec{\omega} \tag{25.28}$$

where $\vec{\omega}$ is the angular velocity vector, as we defined it in Chapter 10 (Section 10.1). A nonzero curl implies that some rotation of the physical entity is being described by the velocity field, while if the curl is zero, it implies no rotation. When the curl of the field is zero, we call the field **irrotational**, and when the curl is nonzero, we call the field **rotational**. In the preceding example, we considered the velocity field of a rigid rotator, in which the angular velocity is the same for every point. In a fluid, however, different points in the fluid may have different angular velocities (and hence different values of $\vec{\nabla} \times \vec{v}$). In this case, you should consider the output of $\vec{\nabla} \times \vec{v}$ at each point in the field to represent the local value of $2\vec{\omega}$.

Example 25.12. Consider a fluid with the two-dimensional velocity field

$$\vec{v} = 3y\,\hat{\imath}$$

with output in m/s for input in m. Find the curl of the velocity, and thereby find the angular velocity in the plane. Graph the velocity field, and reconcile the graph with your computed value of angular velocity.

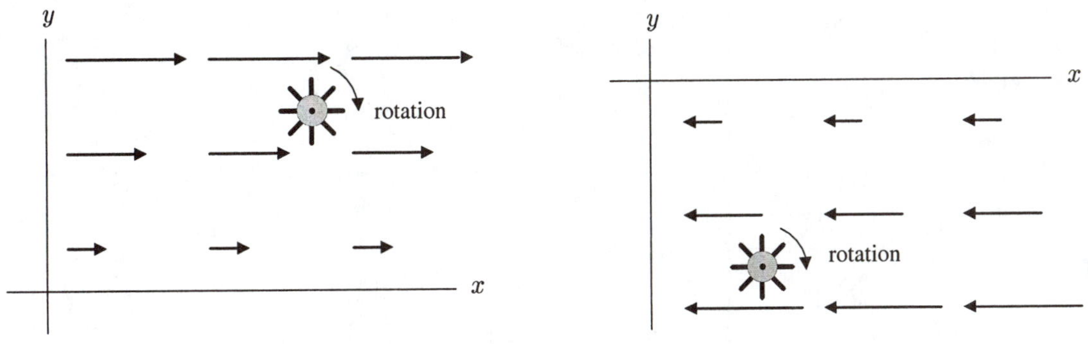

(a) Velocity field above the x-axis

(b) Velocity field below the x-axis

Figure 25.22. Paddle wheels to visualize curl in Example 25.12.

Solution. The curl of the velocity is

$$\vec{\nabla} \times \vec{v} = \hat{\imath}\,(0-0) - \hat{\jmath}\,(0-0) + \hat{k}\,(0-3) = -3\,\hat{k}.$$

The curl of $\vec{v}$ is a constant $-3\,\hat{k}$ everywhere in the plane. We used SI units for velocity, and the units for angular velocity will be rad/s, with

$$\vec{\omega} = \frac{1}{2}(\vec{\nabla} \times \vec{v}) = -\frac{3}{2}\,\hat{k} \text{ rad/s}.$$

The graph of the velocity field is shown in Figures 25.22(a) and 25.22(b). At first glance, it may appear strange that there is a nonzero *curl* and that the fluid has *angular velocity*, even though all the velocity field lines are straight. A common test we use is to imagine placing at various locations in the plane a paddle wheel, which is free to turn under the influence of the fluid. In Figure 25.22(a), we show a paddle wheel placed in the first quadrant of the field $\vec{v} = 3y\,\hat{\imath}$. The fluid velocity is higher on the top of the wheel than on the bottom, causing it to turn clockwise. By the right-hand rule, a clockwise rotation in the xy-plane corresponds to an angular velocity in the $-z$-direction, in agreement with our computation using the curl. Similarly, a paddle wheel placed in the fourth quadrant [Figure 25.22(b)] experiences a greater fluid velocity on the bottom of the wheel, again resulting in a clockwise rotation. △

Example 25.13. Repeat the analysis of the previous example for a two-dimensional velocity field

$$\vec{v} = y^2\,\hat{\imath}.$$

Solution. The curl of the velocity is

$$\vec{\nabla} \times \vec{v} = \hat{\imath}\,(0-0) - \hat{\jmath}\,(0-0) + \hat{k}\,(0-2y) = -2y\,\hat{k}.$$

The curl is not constant throughout the plane, so the angular velocity varies with position. The angular velocity is

$$\vec{\omega} = \frac{1}{2}(\vec{\nabla} \times \vec{v}) = -y\,\hat{k} \text{ rad/s}.$$

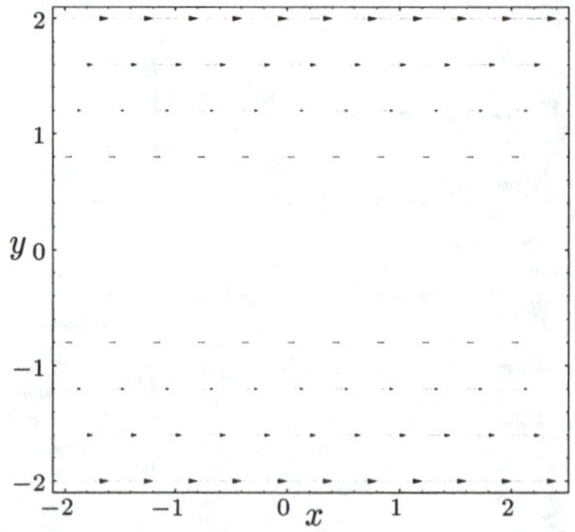

Figure 25.23. The vector field of Example 25.13.

A graph of this velocity field is shown in Figure 25.23. According to our calculation, the angular velocity along the x-axis ($y = 0$) is zero. This makes sense, because a paddle wheel placed on the x-axis has fluid with equal angular velocity above and below. A paddle wheel placed at some point $y > 0$ has a greater fluid velocity above than below, so it turns clockwise, and $\vec{\omega}$ is in the $-z$-direction, in agreement with the formula above. Similarly, a paddle wheel placed at $y < 0$ turns counterclockwise, and $\vec{\omega}$ is in the $+z$-direction, as it should be. △

25.8 Problems

25.1 The Biot-Savart Law

1. Show that if SI units are used for all the quantities on the right side of Equation (25.4) (the Biot-Savart law), the correct unit of magnetic field (T) is generated.

Problems 2–7 refer to the circular conducting wire carrying a constant current I, as described in Example 25.1. Assume that the circle is in the xy-plane with its center at the origin.

2. Use symmetry arguments to find the direction of the magnetic field at a point on the z-axis.

3. Using the Biot-Savart law, write an expression containing a definite integral that gives the magnitude of the magnetic field at a point on the z-axis.

4. Evaluate the definite integral you obtained in the preceding problem and thus determine the magnetic field on the z-axis.

5. Show that your answer in the preceding problem agrees with the result of Example 25.1 and that it approaches zero in the limit as the distance from the circle becomes infinite. What is the approximate functional dependence of the magnetic field on distance from the circle as the distance becomes much larger than the radius of the circle?

6. Using a computer program, obtain a numerical estimate of the magnetic field at the following places and compare with a numerical value obtained using the exact result in Problem 4: (a) the origin; (b) $(0, 0, R)$; and (c) $(0, 0, -10R)$. Assume reasonable numerical values, such as $I = 0.1$ A and $R = 25$ cm.

7. Using a computer program, obtain a numerical estimate of the magnetic field at $(R, 0, R)$. Assume reasonable numerical values as in the preceding problem. Does the *direction* of the magnetic field you computed make sense, given what you know of the magnetic field of a circular loop?

In Problems 8–11, consider a *finite* straight wire running along the x-axis from $x = -L$ to $x = +L$. The wire carries a steady current I.

8. Use symmetry arguments to find the direction of the magnetic field at points (not at the origin) on the y-axis and z-axis.

9. Use the Biot-Savart law to find the magnetic field at points on the y-axis.

10. Using a computer program, obtain a numerical estimate of the magnetic field at the point $(0, L, 0)$. Assume reasonable numerical values, for example, $I = 0.1$ A and $L = 25$ cm.

11. Using the same numerical values as in the previous problem, obtain a numerical estimate of the magnetic field at the point $(L, L, 0)$.

12. A *semi-infinite* wire carries a steady current I starting at the origin and traveling along the $+y$-axis in the $+y$-direction. Find the magnetic field at a point (x_0, y_0) in the first quadrant of the xy-plane.

13. A current I travels along the $+x$-axis from $x \gg 0$, traveling in the $-x$-direction and terminating at the origin. Find the magnetic field at a point (x_0, y_0) in the first quadrant of the xy-plane.

14. Using the results of Problems 12 and 13, find the magnetic field at a point (x_0, y_0) in the first quadrant of the xy-plane for the infinite wire shown below.

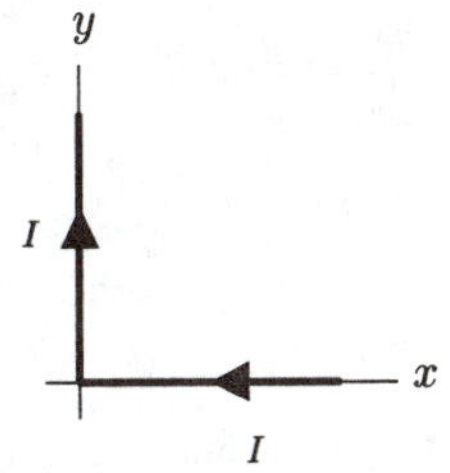

15. A wire in the shape of a semicircle of radius R lies in the upper half-plane $(y > 0)$ of the xy-plane. It carries a constant current I counterclockwise as viewed from above. Find the magnetic field at the origin.

16. Use the result of the preceding problem to find the magnetic field at the origin for the current shown in the diagram below. The curved portion of the wire is a semicircle of radius R.

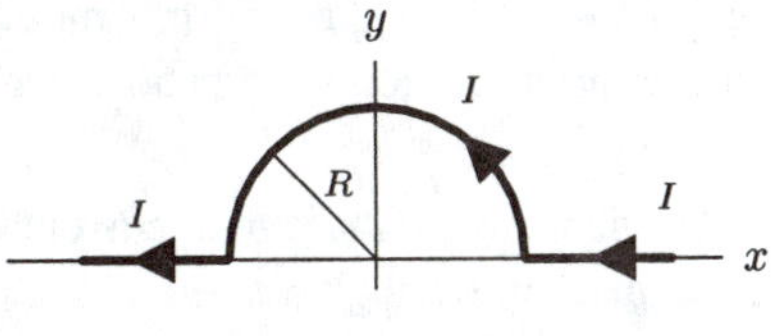

17. An infinite wire carrying a constant current I is bent into a "hairpin" shape as shown in the diagram below. The curved portion of the wire is a semicircle of radius R. Find the magnetic field at the origin.

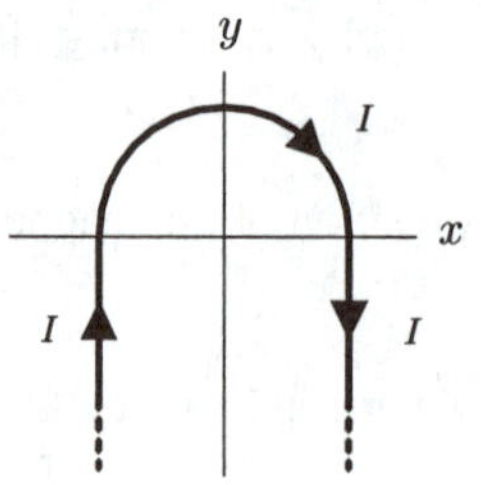

18. Two very long straight wires each carry the same current I_0 as shown below. What is the magnetic field at the point P?

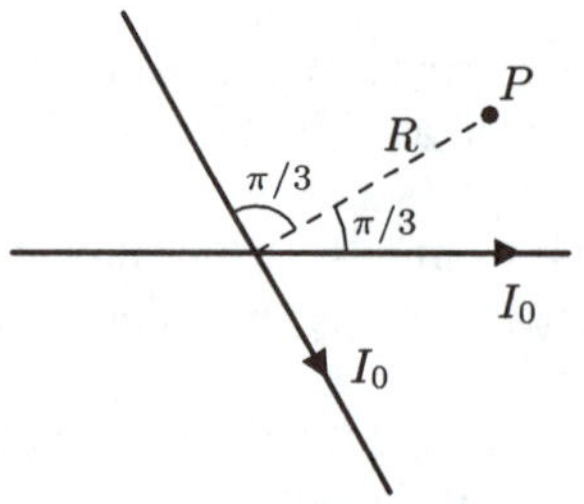

19. A conducting wire is bent into a square with side L. If the wire carries a steady current I, what is the magnetic field at the center of the square?

25.2 Ampère's Law

1. Two very long parallel wires are very close together and carry currents of 2.95 A and 1.34 A in opposite directions. Find the magnetic field a perpendicular distance of 1.00 cm from the wires.

2. A long cylindrical wire 1.25 mm in radius has a uniform current density. At what point outside the wire is the magnetic field the same as at a point inside the wire a perpendicular distance 1.00 mm from the center of the wire?

3. A long solenoid is wound with 25 turns of wire per centimeter of length. The wire carries a current of 3.75 A. (a) What is the magnetic field near the center of the solenoid? (b) What is the trajectory of a proton inside the solenoid that begins with a speed of 10^4 m/s moving parallel to the solenoid's axis of symmetry? (c) Repeat part (b) if the initial motion is perpendicular to the symmetry axis.

4. You have current-carrying wire that can carry a maximum current of 5.00 A. Design a solenoid using this wire that has a magnetic flux (passing through a surface perpendicular to the solenoid's axis) of 1.50×10^{-4} T·m^2.

Problems 5–7 refer to a **toroid** (shown in the figure), consisting of a number of coils of uniform size stacked together and then bent into the shape of a torus (a "bagel" or "donut" shape). This means the coil density is actually a little higher near the center of the torus and lower near the outside.

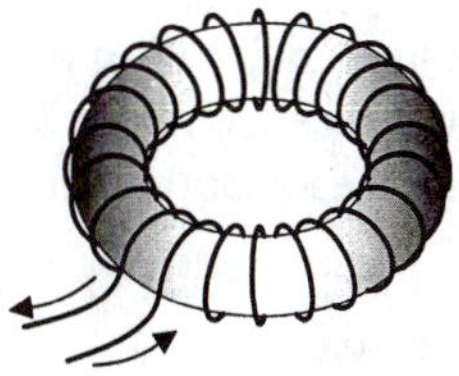

5. Using symmetry arguments, describe the shape of the magnetic field lines inside the toroid (i.e., in the interior of the coils, not the geometric center of the torus).

6. Use Ampere's law to show that if the toroid carries a constant current I and has a total of N coils, then the magnitude of the magnetic field inside the toroid a distance r from the geometric center of the torus is approximately

$$B = \frac{\mu_0 I N}{2\pi r}.$$

7. Describe a possible trajectory of a charged particle moving inside the toroid. How could a toroid be used in "magnetic confinement" of charged particles?

8. A long cylindrical wire of radius R carries a current density that is not uniform but varies according to the function

$$J = I_0 \frac{r^2}{R^4}$$

where I_0 is a constant and r is the distance from the center of the wire. Find the magnetic field everywhere, inside and outside the wire.

9. A long cylindrical wire of radius R carries a current I_0 spread uniformly over its outer surface. Find the magnetic field everywhere, inside and outside the wire.

10. The diagram below shows a cross section of *coaxial wire*, so-called because the inner conducting wire and outer conducting shell have a common axis along the center of the wire. Suppose the inner conducting wire and outer conducting shell each carry a uniform current density and the same net current I_0, but those currents travel in opposite directions. Find the magnetic field a distance r from the center of the wire. You will need to consider four specific cases: $r < a$, $a < r < b$, $b < r < c$, and $r > c$. Explain why coaxial wire is used in applications in which it is important that there be minimal magnetic fields outside the wire.

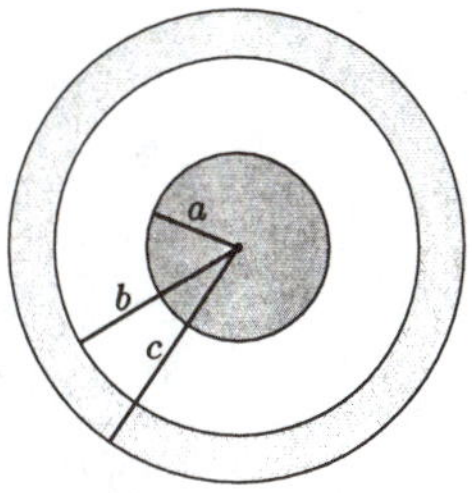

25.3 Parallel Current-Carrying Wires

1. Two long parallel conductors each carry the same current $I = 1.45$ A in the $+z$-direction. One conductor lies along the z-axis and the other is in the xz-plane at $x = 20$ cm. Find the magnetic field along each of these lines in the xz-plane: (a) $x = -10$ cm; (b) $x = 10$ cm; and (c) $x = 30$ cm.

2. Repeat the preceding problem if the conductor along the line $x = 20$ cm carries a current I in the $-z$-direction.

3. For each of Problems 1 and 2, what is the force per unit length on each wire?

Problems 4–7 refer to the diagram below, with four long straight wires are parallel to the z-axis as shown. Each wire carries a current in the $+z$- or $-z$-direction as indicated, with each current equal to 500 mA.

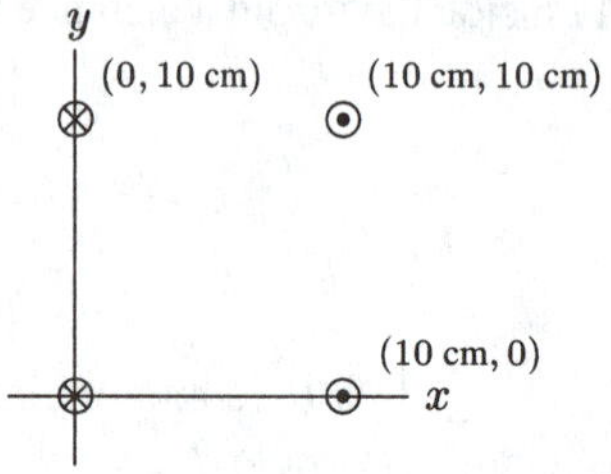

4. Find the magnetic field at the point (5 cm, 5 cm, 0).

5. Find the magnetic field at the point (0, 5 cm, 0).

6. Find the force per unit length on the wire that lies on the z-axis.

7. Use symmetry arguments with your answer to the preceding problem to find the force per unit length on each of the other three wires.

For Problems 8–11 refer to the previous diagram, but assume now that all the currents travel in the $+z$-direction.

8. Find the magnetic field at the point (5 cm, 5 cm, 0).

9. Find the magnetic field at the point (0, 5 cm, 0).

10. Find the force per unit length on the wire that lies on the z-axis.

11. Use symmetry arguments with your answer to the preceding problem to find the force per unit length on each of the other three wires.

12. A long straight wire carries a 9.50-A current along the y-axis in the $+y$-direction. Find the net force on the rectangular loop in the xy-plane as shown below, if the loop carries a constant current $I = 2.75$ A.

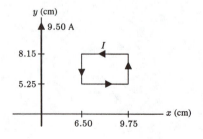

25.4 Maxwell's Displacement Current

1. Show that if SI units are used on the right side of Equation (25.12), the correct SI units (A) are obtained for the displacement current.

2. In the RC circuit shown in Figure 25.12, the electric flux through the capacitor is changing at a rate

$$\frac{d\Phi_E}{dt} = 5.93 \times 10^8 \text{ (N/C)}\cdot\text{m}^2/\text{s}.$$

(a) What is the current in the circuit when the flux is changing at this rate? (b) Assume the capacitor was initially uncharged. If $\mathcal{E} = 36.0$ V, $R = 150\,\Omega$, and $C = 1500\,\mu$F, for how much time has the circuit been charging when the electric flux is changing at that rate?

3. Show that the displacement current across a parallel-plate capacitor with capacitance C is

$$I_d = C\frac{dV}{dt}.$$

4. A 0.25-μF capacitor has an applied potential difference across its parallel plates that varies sinusoidally with time t as

$$V = (16.0 \text{ V}) \cos(\omega t)$$

where $\omega = 60 \text{ s}^{-1}$. (a) Find the displacement current as a function of time. (b) At what times is the displacement current equal to zero?

25.5 Magnetism in Matter

1. Show that if SI units are used on the right side of Equation (25.16), the correct units (J/T) are obtained for the Bohr magneton.

2. Estimate the magnetization of iron, assuming an average magnetic moment of one Bohr magneton per atom. The density of iron is 7870 kg/m^3 and its average atomic mass is 55.85 g/mol.

3. (a) What is the magnetization of copper (at room temperature) when a 150-mT magnetic field is applied? (b) What is the net magnetic moment of a 150-g sample of copper under those conditions?

4. Show that magnetic susceptibility is a dimensionless quantity.

5. Consider again the model for diamagnetism. Using the hydrogen atom as a model (in which the electron orbits the proton at a radius of 5.29×10^{-11} m), estimate numerically the fractional change in the magnetic moment (as a fraction of the Bohr magneton) when a 1.0-T magnetic field is applied to a hydrogen atom. Does your answer seem realistic, based on what you know of diamagnetism?

6. A 1.50-kg cube of aluminum is immersed in a uniform 0.430-T magnetic field with one face perpendicular to the field. What is the net magnetic moment?

7. Estimate the magnetic susceptibility of manganese immersed in liquid nitrogen at a temperature of 77 K.

25.6 Divergence

In Problems 1–6, compute the divergence of the given vector field.

1. $\vec{g}(x,y) = (x^2+y^2)\,\hat{\imath} - \left(\dfrac{1}{x^2+y^2}\right)\hat{\jmath}$

2. $\vec{A}(x,y) = -2x\,\hat{\imath} + \left(4x^2y^2 - e^y\right)\hat{\jmath}$

3. $\vec{B}(x,y,z) = \left\langle 3xy, -2x^2z^2+4yz, 6xz^4 \right\rangle$

4. $\vec{F}(x,y,z) = \left\langle \sin^2 x + \sin^2 y, \cos^2 x + \cos^2 y, 3z \right\rangle$

5. $\vec{A}(x,y,z) = \left\langle e^{x^2}, \ln(4xy), \sin^{-1}(xyz) \right\rangle$

6. $\vec{G}(x_1,x_2,x_3,x_4) = \left\langle 3x_1x_2, -4x_1^2 + 5x_2x_3x_4, 9x_2x_4, -x_3x_4^4 \right\rangle$

7. Consider a function $f : \mathbb{R}^3 \to \mathbb{R}$ and a vector field $\vec{F} : \mathbb{R}^3 \to \mathbb{R}^3$. Let $\vec{r} = \langle x,y,z \rangle$. Prove the following product rule:

$$\vec{\nabla} \cdot \left(f(\vec{r})\vec{F}(\vec{r})\right) = f(\vec{r})\vec{\nabla} \cdot \vec{F}(\vec{r}) + \left(\vec{\nabla} f(\vec{r})\right) \cdot \vec{F}(\vec{r}).$$

[*Hint:* Write out components for each side of the identity.]

In Problems 8–11, compute the divergence of the given vector field where $\vec{r} = \langle x,y,z \rangle$ and $r = \|\vec{r}\|$.

8. $\vec{F}(\vec{r}) = \vec{r}$

9. $\vec{F}(\vec{r}) = \hat{r}$

10. $\vec{F}(\vec{r}) = \dfrac{\vec{r}}{r^2}$

11. $\vec{F}(\vec{r}) = \dfrac{\vec{r}}{r^3}$

12. The divergence of the gradient of a function $f : \mathbb{R}^n \to \mathbb{R}$ is called the **Laplacian** of f and is written $\nabla^2 f = \vec{\nabla} \cdot (\vec{\nabla} f)$. Show that in Cartesian coordinates

$$\nabla^2 f = \frac{\partial^2 f}{\partial x^2} + \frac{\partial^2 f}{\partial y^2} + \frac{\partial^2 f}{\partial z^2}.$$

In Problems 13–18, compute the Laplacian of the given function. The Laplacian is defined in Problem 7.

13. $f(x,y) = 3x^2y^2 - \sin^2 x$

14. $g(x,y,z) = -5\dfrac{x^2}{z^2} + 6x^3$

15. $V(x,y,z) = x^2 + y^2 + z^2$

16. $f(x,y,z) = \dfrac{1}{x} + 4y^{10}z$

17. $f(x,y) = \sin x \cos y$

18. $g(y,z) = -9y^3z^4$

In Problems 19–22, the vector field $\vec{R} = \rho\vec{v}$ is defined for all points in space unless specified otherwise. Graph the vector field $\vec{R}$, find the divergence of $\vec{R}$, and give a physical interpretation of the divergence based on the graph.

19. $\vec{R} = \dfrac{1}{x^2+y^2}\,\hat{\jmath} \qquad x^2+y^2 \neq 0$

20. $\vec{R} = y\,\hat{\imath} + x\,\hat{\jmath}$

21. $\vec{R} = 3x\;(\hat{\imath} + \hat{\jmath})$

22. $\vec{R} = -2x\,\hat{\imath} + 3xy\,\hat{\jmath}$

25.7 Curl

In Problems 1–10, find the curl of the given vector field.

1. $\vec{A}(x,y,z) = \langle xy, 2y^2z^2, 4e^x \rangle$

2. $\vec{B}(x,y,z) = (\sin 4z)\,\hat{\imath} + xe^z\,\hat{\jmath}$

3. $\vec{A}(x,y,z) = \left\langle y, -x, \dfrac{3}{x^2+y^2+z^2} \right\rangle$

4. $\vec{R}(x,y,z) = (-3\ln y)\,\hat{\imath} + 5xyz\,\hat{k}$

5. $\vec{E}(x,y,z) = \langle x^2+y^2+z^2, 4e^{-x^2}, \cos^{-1}(x) \rangle$

6. $\vec{F}(x,y) = \langle x, y \rangle$

7. $\vec{F}(x,y) = \langle y, x \rangle$

8. $\vec{F}(x,y) = \langle y, -x \rangle$

9. $\vec{F}(x,y) = \langle 3xy^2, y^2 - x^2 \rangle$

10. $\vec{F}(x,y) = \langle \sin(xy), \cos(xy) \rangle$

11. Consider a function $f : \mathbb{R}^3 \to \mathbb{R}$ and a vector field $\vec{F} : \mathbb{R}^3 \to \mathbb{R}^3$. Let $\vec{r} = \langle x, y, z \rangle$. Prove the following product rule:

$$\vec{\nabla} \times \left(f(\vec{r})\vec{F}(\vec{r})\right) = f(\vec{r})\vec{\nabla} \times \vec{F}(\vec{r}) + \left(\vec{\nabla} f(\vec{r})\right) \times \vec{F}(\vec{r}).$$

[*Hint:* Write out components for each side of the identity.]

In Problems 12–15, compute the divergence of the given vector field where $\vec{r} = \langle x, y, z \rangle$ and $r = \|\vec{r}\|$.

12. $\vec{F}(\vec{r}) = \vec{r}$

13. $\vec{F}(\vec{r}) = \hat{r}$

14. $\vec{F}(\vec{r}) = \dfrac{\vec{r}}{r^2}$

15. $\vec{F}(\vec{r}) = \dfrac{\vec{r}}{r^3}$

16. Consider vector fields $\vec{A} : \mathbb{R}^3 \to \mathbb{R}^3$ and $\vec{B} : \mathbb{R}^3 \to \mathbb{R}^3$. Prove the following identity :

$$\vec{\nabla} \times (\vec{A} \times \vec{B}) = (\vec{\nabla} \cdot \vec{B})\vec{A} - (\vec{\nabla} \cdot \vec{A})\vec{B} + (\vec{B} \cdot \vec{\nabla})\vec{A} - (\vec{A} \cdot \vec{\nabla})\vec{B}$$

where $\vec{F} \cdot \vec{\nabla}$ is an operator defined as

$$(\vec{F} \cdot \vec{\nabla})\vec{G} = \langle \vec{F} \cdot (\vec{\nabla} G_x), \vec{F} \cdot (\vec{\nabla} G_y), \vec{F} \cdot (\vec{\nabla} G_z) \rangle.$$

17. For a rigid rotator with constant angular velocity $\vec{\omega}$, the velocity of a point P in the rigid rotator is given by

$$\vec{v} = \vec{\omega} \times \vec{r}$$

where $\vec{r}$ is a displacement vector from a point on the rotation axis to P. Use this fact along with the result of Problem 16 to evaluate

$$\vec{\nabla} \times \vec{v} = \vec{\nabla} \times (\vec{\omega} \times \vec{r})$$

for a rigid rotator. Verify that

$$\vec{\nabla} \times \vec{v} = 2\vec{\omega}$$

18. At the beginning this chapter, we mentioned that a static magnetic field can be expressed as the curl of a *vector potential* $\vec{A}$, that is,

$$\vec{B} = \vec{\nabla} \times \vec{A}.$$

What is the vector potential for a uniform magnetic field $\vec{B} = B_0\,\hat{k}$?

Problems 19–21 refer to the force field in the xy-plane described by $\vec{F} = y\,\hat{\imath} + x\,\hat{\jmath}$.

19. Graph the force field.

20. Compute the curl of the force. Is the force conservative?

21. Check to see if the force is conservative by computing the work done by the force in moving a particle from one point to another along two different paths. Let path 1 be a straight line from the origin to $(1, 0)$. Let path 2 be made up of three straight line segments: from the origin to $(0, 1)$; from $(0, 1)$ to $(1, 1)$; from $(1, 1)$ to $(1, 0)$. Is the work the same along the two paths? Reconcile your answer with whether the field is conservative.

Problems 22–24 refer to the force field in the xy-plane described by $\vec{F} = -y\,\hat{\imath} + x\,\hat{\jmath}$.

22. Graph the force field.

23. Compute the curl of the force. Is the force conservative?

24. Check to see if the force is conservative by computing the work done by the force in moving a particle from one point to another along two different paths. Let path 1 be a straight line from the origin to $(1, 0)$. Let path 2 be made up of three straight line segments: from the origin to $(0, 1)$; from $(0, 1)$ to $(1, 1)$; from $(1, 1)$ to $(1, 0)$. Is the work the same along the two paths? Reconcile your answer with whether the field is conservative.

Chapter 26

Faraday's Law and Inductance

We now turn our attention to another aspect of electromagnetism, which completes the laws of electromagnetism known as Maxwell's equations. In studying Maxwell's displacement current in Chapter 25, you saw how a changing electric flux induces a magnetic field. Given the similarities we have seen between electricity and magnetism, it should come as no surprise that a changing magnetic flux induces an electric field. *Faraday's law* describes this process mathematically. Our study of magnetic flux also helps us understand an important element in electric circuits, the *inductor*, and the general property of inductance. We consider some examples of how inductors are used in combination with resistors and capacitors in electric circuits.

26.1 Faraday's Law

26.1.1 Magnetic Flux and Gauss's Law for Magnetism

Before beginning this section, you may want to review briefly the earlier sections on electric flux (specifically Sections 23.1 and Section 25.4). Magnetic flux through a surface (not necessarily a closed surface) S is defined like the flux of any other vector. We use Φ_B to denote the flux of a magnetic field $\vec{B}$, and in keeping with our previous definition of flux

$$\Phi_B = \iint_S \vec{B} \cdot \hat{n}\, dA \tag{26.1}$$

where dA is an area element on the surface and $\hat{n}$ is a unit vector normal to the surface at that point. As illustrated in Figure 26.1, magnetic flux is due to a magnetic field passing through the surface. Magnetic flux has dimensions of magnetic field times area, and therefore its SI units are T·m^2.

Example 26.1. A solenoid has 1500 circular turns of wire per meter. The diameter of the circular windings is 2.50 cm, and the solenoid carries a steady current of 9.45 A. What is the magnetic flux passing through a planar surface area that has as its boundary one of the circular windings near the center of the solenoid?

Solution. The relation of the surface to the magnetic field is depicted in Figure 26.2. We know that near the center of a solenoid, the magnetic field is nearly uniform, in the direction of the solenoid's symmetry axis, and has a magnitude

$$B = \mu_0 n I \tag{25.8}$$

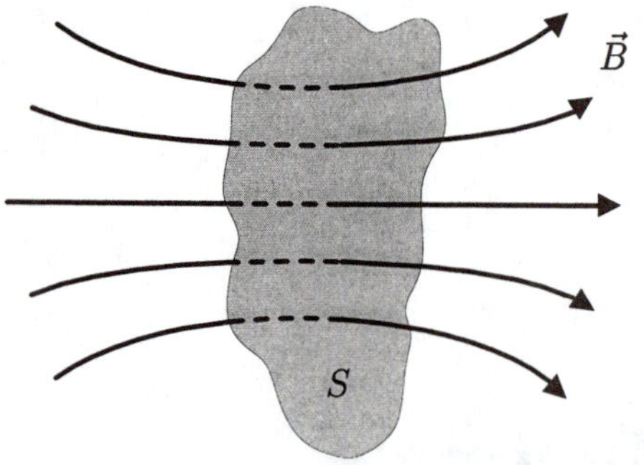

Figure 26.1. Magnetic flux is due to a magnetic field passing through the surface.

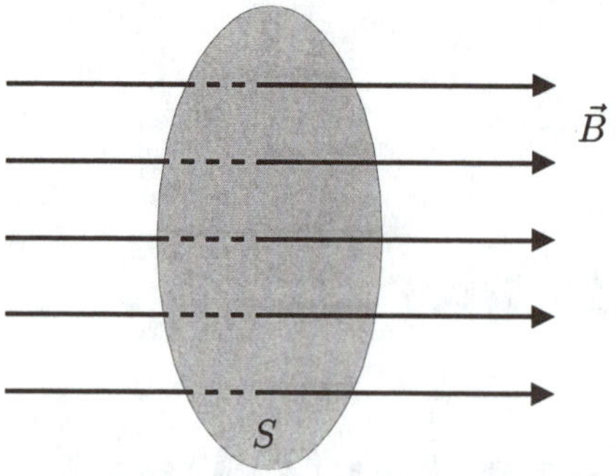

Figure 26.2. Magnetic field and surface used in for Example 26.1.

where n is the number of turns of coil per unit length, and I is the current in the wire. The direction of the magnetic field is perpendicular to the surface over which we integrate. Thus,

$$\vec{B} \cdot \hat{n} = (B)(1)\cos(0) = B.$$

With the magnitude of the field B constant over the surface S, we can factor B out of the flux integral:

$$\Phi_B = \iint_S \vec{B} \cdot \hat{n}\, dA = \iint_S B\, dA = B \iint_S dA.$$

The remaining surface integral is simply the surface enclosed by the circle, with area πr^2. Then

$$\Phi_B = BA = (\mu_0 n I)(\pi r^2) = \pi \mu_0 n I r^2.$$

Inserting numerical values,

$$\Phi_B = \pi (4\pi \cdot 10^{-7}\ \text{T·m/A})(1500\ \text{m}^{-1})(9.50\ \text{A})(0.0250\ \text{m})^2 = 3.5 \times 10^{-5}\ \text{T·m}^2.$$

Note that magnetic flux is a scalar quantity, even though a direction is associated with the magnetic field. △

Gauss's law for magnetism. An important result arises if we compute the magnetic flux through a *closed* surface S. Remember that the *electric* flux through a closed surface is an essential component of Gauss's law:

$$\oiint \vec{E} \cdot \hat{n}\, dA = \frac{q_{\text{enc}}}{\epsilon_0}.$$

In words, the electric flux through a closed surface equals the net charge enclosed by the surface divided by ϵ_0. What is the equivalent of Gauss's law in magnetism? Because there are no magnetic monopoles (the magnetic analog of the charge q_{enc} in Gauss's law), a closed surface can enclose no net magnetic "charge." Therefore, the magnetic flux through a closed surface is always zero. Symbolically,

$$\oiint \vec{B} \cdot \hat{n}\, dA = 0 \tag{26.2}$$

which is known as **Gauss's law for magnetism**. Equation (26.2) is actually a succinct mathematical statement of the nonexistence of magnetic monopoles.

Let's illustrate Gauss's law for magnetism by thinking about the magnetic flux through several closed surfaces in the vicinity of a bar magnet (Figure 26.3). Surface S_1 encloses the entire magnet. Field lines emerging from the north pole of the magnet terminate at the south pole. Those lines that emerge from S_1 are associated with some positive flux through the surface, because by convention the unit vector $\hat{n}$ points outward from a closed surface. However, all those lines eventually reenter S_1 on the way back into the south pole, making some negative flux. Therefore, it is plausible that the net flux through S_1 is zero, as stated in Equation (26.2). The net magnetic "charge" enclosed is also zero, because S_1 encloses both north and south poles. Similarly, for a closed surface S_2 that does not enclose any of the bar magnet, any field line that enters the surface also leaves it, and a similar argument can be made.

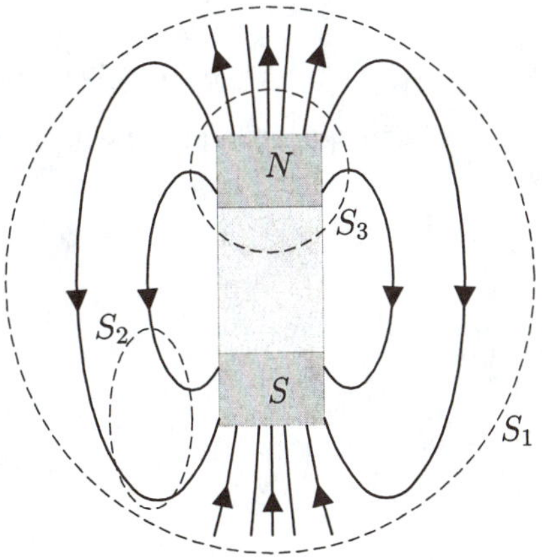

Figure 26.3. Magnetic field lines and several closed surfaces (in cross section) near a bar magnet.

The surface S_3 is a clever attempt to enclose only one of the poles with a surface. Once again, all the field lines that emerge from the surface eventually reenter it (some of these may be in the metal of the magnet itself). And the net magnetic "charge" is still zero—remember that a smaller piece of a magnet still contains both poles. Gauss's law for magnetism is still valid in this case.

26.1.2 Introduction to Faraday's Law

As another example of magnetic flux, consider the bar magnet and solenoid in Figure 26.4(a). The north pole of the bar magnet is just inside the solenoid. Given the shape of the magnetic field generated by the bar magnet (see, e.g., Figure 26.3), with field lines emerging from the north pole, a net nonzero magnetic flux passes through the solenoid coils. Notice that the solenoid is not connected to any source of emf, and no current flows though it. In Figure 26.4(b), we show the solenoid connected to an ammeter, which reads zero, to illustrate that the existence of a constant magnetic flux does not give rise to any current in the solenoid.

In about 1830, the English physicist Michael Faraday (1791–1867) noticed that a *moving* magnet induces an electric current in a coil. We can illustrate this schematically with our bar magnet and solenoid. In Figure 26.5(a), we show the north pole of the bar magnet being pushed into the solenoid, resulting in a flow of current in the solenoid. We will discuss the direction of the induced current later, but for the moment note that when the magnet is moved in the opposite direction—that is, when the north pole is withdrawn from the solenoid as in Figure 26.5(b)—the induced current is in the opposite direction.

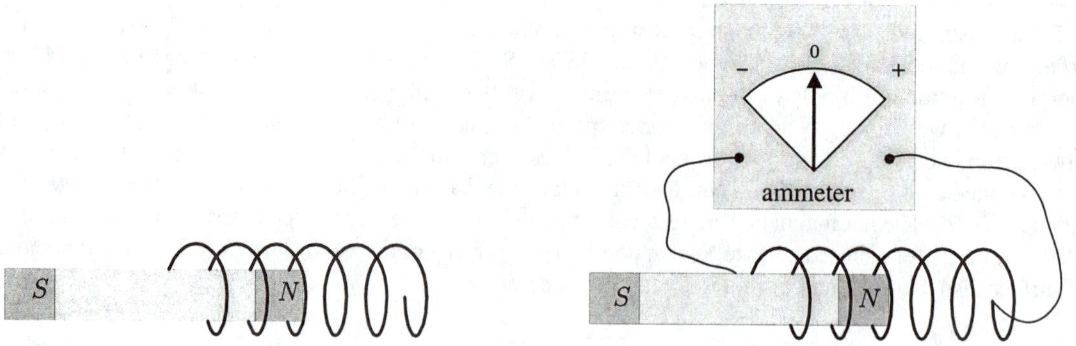

(a) North pole at rest in the solenoid

(b) Ammeter connected to the solenoid records zero current

Figure 26.4. A bar magnet interacting with a solenoid.

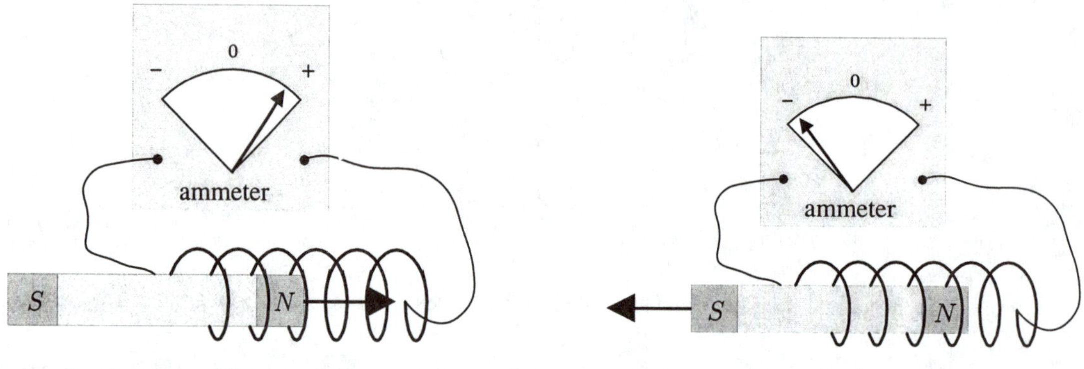

(a) North pole moving into the solenoid

(b) North pole moving out of the solenoid

Figure 26.5. Current induced by a changing magnetic flux. The direction of the current depends on the direction in which the magnet is moving.

That there is induced current and that it changes direction when the magnet's motion is reversed is analogous to Maxwell's displacement current. Maxwell's displacement current can be explained by thinking of a magnetic field arising from a change in electric flux in a region. Similarly, a changing magnetic flux gives rise to an electric field, which causes electric current to flow in the coil. (Remember that electric current results from the presence of an electric field in the conducting wire.) This is another example of the remarkable symmetry between electric and magnetic effects.

The similarity is even more striking when we write the mathematical relationship between the changing magnetic flux Φ_B and the induced electric field $\vec{E}$:

$$\oint_C \vec{E} \cdot d\vec{s} = -\frac{d\Phi_B}{dt}. \tag{26.3}$$

In Equation (26.3), known appropriately as **Faraday's law**, the line integral is carried out over a closed curve (hence the symbol $\oint$), and the magnetic flux is the net flux over a surface having the integration path as its boundary, just as with Maxwell's displacement current. Notice that Faraday's law and that part of

Equation (25.11) for Maxwell's displacement current are almost the same, with the electric and magnetic fields interchanged [there is an extra dimensional factor $\mu_0\epsilon_0$ in Equation (25.11)]. The extra minus sign on the right side of Faraday's law is important and will be explained in the next section.

First, however, we give an alternative form of Faraday's law appropriate for the magnet and solenoid we have been considering. The emf $\mathcal{E}$ around one loop of wire in the solenoid can be related to the electric field in the wire. Using the relationship between potential difference and electric field from Chapter 18,

$$\mathcal{E} = \oint_C \vec{E} \cdot d\vec{s}. \tag{26.4}$$

Therefore, the emf in the loop is given by Faraday's law as

$$\mathcal{E} = -\frac{d\Phi_B}{dt}. \tag{26.5}$$

26.1.3 Lenz's Law and Motional EMF

Lenz's law is not a separate rule, but part of Faraday's law; symbolically, it is the minus sign on the right side of Equation (26.3). In words, Lenz's law says that the current produced in the solenoid (in our example in the previous section) generates a magnetic flux that *opposes* the change in flux that gives rise to that current. We illustrate this principle with an example.

For simplicity, let's consider just a single closed circular loop of conducting wire, with a bar magnet moving toward it [Figure 26.6(a)]. With the north pole approaching the loop, the number of field lines going through the loop increases, and hence the magnetic flux through the loop is increasing with time. Then the induced current is in the direction shown, because Lenz's law tells us that the magnetic flux in the loop generated by the induced current opposes the increasing flux due to the moving magnet.

Now consider what happens if the same magnet is moved away from the loop [Figure 26.6(b)]. In this case, the magnetic flux through the loop is decreasing, so Lenz's law dictates that in order to oppose this change, the induced current tends to restore the decreasing magnetic flux. Thus, the induced current in the loop is in the opposite direction as in Figure 26.6(a), where the magnet moves toward the loop. The important thing to remember is that Lenz's law says the induced current generates a magnetic field that opposes not the magnetic flux, but the *change* in the magnetic flux. We illustrate this point again with another example.

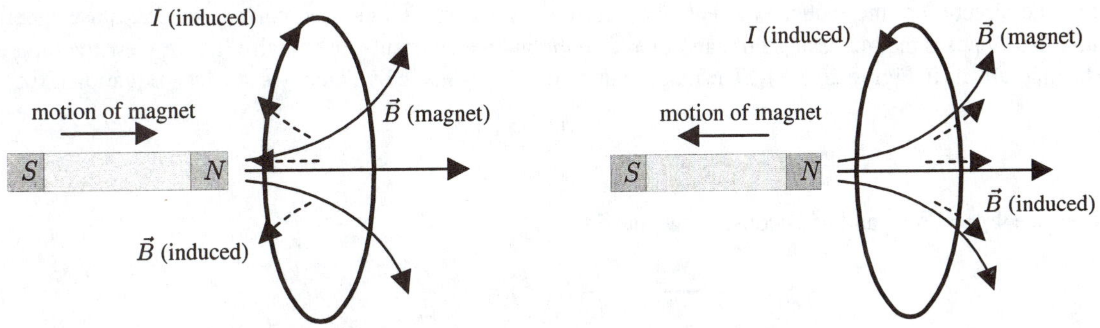

(a) North pole moves toward the loop

(b) North pole moves away from the loop

Figure 26.6. Direction of current induced in a conducting loop of wire is determined by Lenz's law.

Example 26.2. A square conducting loop with each side $a = 15.0$ cm lies in the xy-plane, as shown in Figure 26.7(a). A uniform magnetic field $\vec{B} = B_0\,\hat{k}$ exists throughout the region of the loop, with $B_0 = 350$ mT. (a) What is the magnetic flux through the loop? (b) Find the direction of the current induced in the loop and the emf induced in the loop if B_0 is increasing at a rate of 25 mT/s. (c) Repeat if B_0 is *decreasing* at a rate of 25 mT/s.

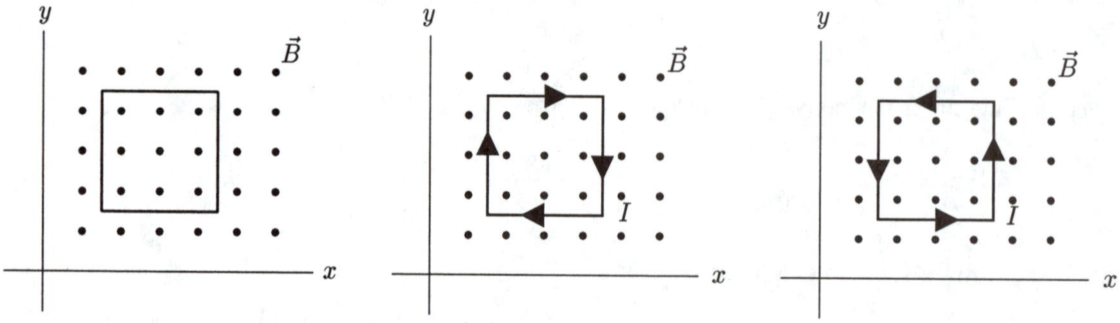

(a) Square loop with magnetic flux through it

(b) Induced current with B increasing in time

(c) Induced current with B decreasing in time

Figure 26.7. Analyzing the induced current in a square conducting loop in a uniform magnetic field that varies in time.

Solution. (a) Because the magnetic field is uniform and normal to the plane of the loop, $\vec{B}$ and $\hat{n}$ are in the same direction, and the magnetic flux is

$$\Phi_B = \iint_S \vec{B}\cdot\hat{n}\,dA = \iint_S B_0\,dA = B_0 \iint_S dA.$$

The remaining double integral is just the area of the square, or a^2. Thus, $\Phi_B = B_0 a^2$. Inserting numerical values,

$$\Phi_B = (350 \times 10^{-3}\text{ T})(0.150\text{ m})^2 = 7.88 \times 10^{-3}\text{ T·m}^2.$$

(b) The direction of the induced current is given by Lenz's law. The induced current generates a magnetic flux that opposes the increasing flux, and therefore the induced current is clockwise [as we view the loop in the diagram, as in Figure 26.7(b)]. The magnitude of the emf is given by Faraday's law [see Equation (26.5)]:

$$|\mathcal{E}| = \left| -\frac{d\Phi_B}{dt} \right|.$$

Because $\Phi_B = B_0 a^2$ and a^2 is constant, we have

$$\frac{d\Phi_B}{dt} = \frac{d}{dt}(B_0 a^2) = a^2 \frac{dB_0}{dt}.$$

Thus,

$$|\mathcal{E}| = a^2 \frac{dB_0}{dt} = (0.150\text{ m})^2 (25 \times 10^{-3}\text{ T/s}) = 5.63 \times 10^{-4}\text{ m}^2\text{·T/s} = 5.63 \times 10^{-4}\text{ V}.$$

(c) Because the absolute value of the rate of change of B_0 is the same as in (b), the emf is the same as before, or 5.63×10^{-4} V. But now Lenz's law tells us that the direction of the induced current is counterclockwise [Figure 26.7(c)]. Because the magnetic flux is decreasing, the induced current is attempting to restore the lost flux, which is just what a counterclockwise current will do. △

In the preceding example, we found the *magnitude* of the emf $\mathcal{E}$, essentially not using the minus sign in Faraday's law [Equation (26.5)]. In using Equation (26.5), it will usually work this way. You can think of the minus sign as a reminder to apply Lenz's law to find the direction of the induced current. In the more general form of Faraday's law [Equation (26.3)], the minus sign has a more specific meaning, which we now explain with an example.

Suppose a uniform magnetic field exists in the $+z$-direction, as shown in Figure 26.8. The magnetic field passes through the xy-plane in a circular region of radius R, as shown. What is the electric field along the circle that circumscribes the magnetic field, if the magnitude of the magnetic field is increasing at a rate dB/dt? From our previous examples, we expect the direction of the electric field to be in the xy-plane and tangent to the circle, as shown in the diagram. We know this, because if there were a conducting wire along the circle, the induced current would flow in this direction, according to Lenz's law, in order to oppose the increasing magnetic flux. In the absence of the wire, an electric field still exists in the xy-plane, due to the changing magnetic flux through the plane. Faraday's law says that the line integral of that electric field $\vec{E}$ around a closed path equals the negative of the rate of change of magnetic flux through that path:

$$\oint_C \vec{E} \cdot d\vec{s} = -\frac{d\Phi_B}{dt}. \tag{26.3}$$

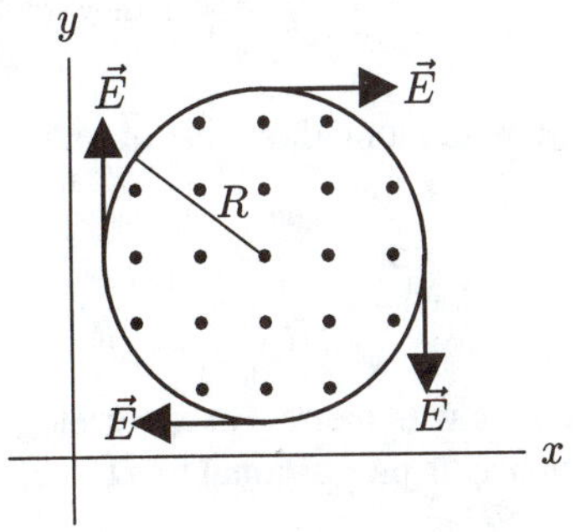

Figure 26.8. Electric field along a circle circumscribing a magnetic field, with magnitude of the magnetic field increasing in time.

We choose the circle that circumscribes the magnetic field as our path. Then because $\vec{E}$ is tangent to the circle, $\vec{E}$ and $d\vec{s}$ are in the same direction at every point on the circle (we integrate clockwise around the circle). Thus, $\vec{E} \cdot d\vec{s} = (E)(ds)(\cos 0) = E\, ds$, and we can easily evaluate the line integral:

$$\oint_C \vec{E} \cdot d\vec{s} = \oint_C E\, ds = E \oint_C ds$$

where we have factored E out of the integral, because by symmetry we know it is constant along the circle. The remaining line integral $\oint_C ds$ is simply the circumference of the circle, or $2\pi R$. Therefore, in this example,

$$\oint_C \vec{E} \cdot d\vec{s} = 2\pi R E. \tag{26.6}$$

Turning our attention to the other side of Faraday's law, we need to know the magnetic flux:

$$\Phi_B = \iint_S \vec{B} \cdot \hat{n}\, dA.$$

It is clear that we want to integrate over the circular area containing the magnetic field. But should we take $\hat{n}$ to be in the $+z$- or $-z$-direction? This is where the sign convention comes into play. By convention, we use another right-hand rule: Curl the fingers of your right hand in the direction of the path followed in computing the line integral (as we did above); then your right thumb points in the direction of $\hat{n}$. In this case, $\hat{n}$ points in the $-z$-direction, and so $\vec{B} \cdot \hat{n} = (B)(1)(\cos\pi) = -B$. Thus, we see that at any instant

$$\Phi_B = \iint_S \vec{B} \cdot \hat{n}\, dA = \iint_S -B\, dA = -B \iint_S dA$$

because the magnitude B is the same at all points in the circular region. The remaining surface integral is just the area of the circle, πR^2, and $\Phi_B = -\pi R^2 B$. Because R is constant,

$$\frac{d\Phi_B}{dt} = \frac{d}{dt}(-\pi R^2 B) = -\pi R^2 \frac{dB}{dt}. \qquad (26.7)$$

Combining Equations (26.6) and (26.7) in Faraday's law [Equation (26.3)], we have

$$2\pi R E = -\left(-\pi R^2 \frac{dB}{dt}\right) = \pi R^2 \frac{dB}{dt}.$$

Therefore, the magnitude of the electric field along the circle of radius R is

$$E = \frac{\pi R^2 \dfrac{dB}{dt}}{2\pi R} = \frac{R}{2}\frac{dB}{dt}. \qquad (26.8)$$

It should not be surprising that E increases with a circle of increasing radius, because more flux is contained in a larger circle. It also makes sense that E is proportional to dB/dt, because a more rapidly increasing flux should generate a larger electric field.

Example 26.3. In the example just considered (Figure 26.8), explain why the same result [Equation (26.8)] is obtained if we do the line integral around the circle in the *counterclockwise* direction, given the convention established for choosing $\hat{n}$.

Solution. With the path reversed for the line integral, the vector $\vec{E}$ now points in the opposite direction of $d\vec{s}$ everywhere on the circle. Then $\vec{E} \cdot d\vec{s} = (E)(ds)(\cos\pi) = -E\, ds$. Carrying out the line integral as in Equation (26.6) leads to

$$\oint_C \vec{E} \cdot d\vec{s} = -2\pi R E.$$

The convention established for choosing $\hat{n}$ tells us that $\hat{n}$ is now in the $+z$-direction. This is the same direction as $\vec{B}$, so now $\vec{B} \cdot \hat{n} = (B)(1)(\cos 0) = B$. The rate of change of magnetic flux becomes

$$\frac{d\Phi_B}{dt} = \frac{d}{dt}(\pi R^2 B) = \pi R^2 \frac{dB}{dt}.$$

By Faraday's law,

$$\oint_C \vec{E} \cdot d\vec{s} = -\frac{d\Phi_B}{dt},$$

so

$$-2\pi RE = -\left(\pi R^2 \frac{dB}{dt}\right),$$

and the same result

$$E = \frac{-\pi R^2 \frac{dB}{dt}}{-2\pi R} = \frac{R}{2}\frac{dB}{dt}$$

is obtained. △

Example 26.4. In the preceding example (Figure 26.8), find the magnitude of the electric field using numerical values $R = 18.50$ cm and $dB/dt = 4.50$ T/s. Show carefully that the correct SI units of electric field are obtained.

Solution. Inserting the numerical values in Equation (26.8) for the electric field magnitude,

$$E = \frac{0.1850 \text{ m}}{2}(4.50 \text{ T/s}) = 0.416 \text{ m·T/s}.$$

To put the units into a more recognizable form for the electric field, recall that

$$1 \text{ T} = 1 \frac{\text{kg}}{\text{C·s}}.$$

Therefore,

$$1 \text{ m·T/s} = 1 \frac{\text{m}}{\text{s}}\frac{\text{kg}}{\text{C·s}} = 1 \frac{\text{kg} \cdot \text{m}}{\text{s}^2}\frac{1}{\text{C}}.$$

However, from mechanics we know 1 kg·m/s^2 = 1 N. Thus,

$$1 \text{ m·T/s} = 1 \text{ N/C},$$

and in this example

$$E = 0.416 \text{ N/C},$$

as expected for an electric field. △

Motional emf Motional emf is the emf induced in a conductor when there is relative motion between the conductor and a magnetic field. You have already seen examples of motional emf, in the situations depicted in Figures 26.5, 26.6(a), and 26.6(b). In the examples that follow, we look at the motional emf that arises when a conductor moves through a fixed magnetic field.

In the situation shown in Figure 26.9(a), a square loop of conducting wire (with side a) moves through the xy-plane in the $+x$-direction with a constant speed v. As shown, a uniform magnetic field $\vec{B} = -B_0\,\hat{k}$ exists in the region of space where $x > 0$. When the entire loop is in the region $x < 0$, there is a constant magnetic flux (zero) through it, and therefore no induced current. But after the leading edge of the loop crosses the

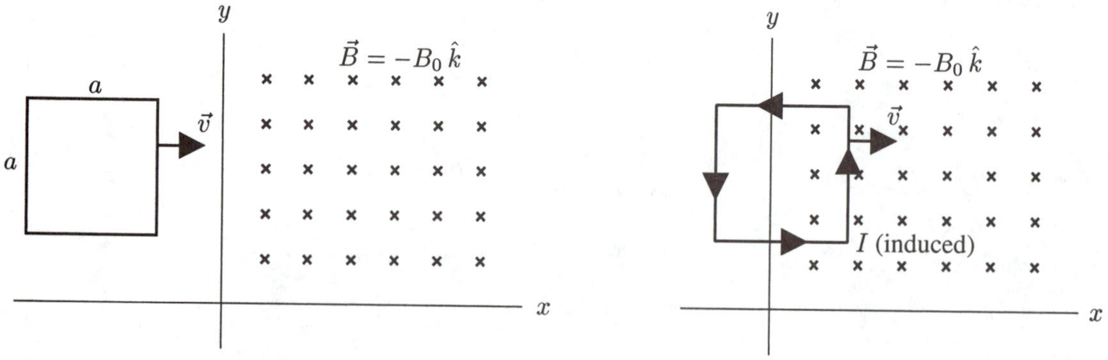

(a) Entire loop is in the region $x < 0$

(b) Leading edge of the loop has crossed the line $x = 0$

Figure 26.9. A square loop of conducting wire moves through the xy-plane in the $+x$-direction with a constant speed v. A uniform magnetic field $\vec{B} = -B_0\,\hat{k}$ exists in the region of space where $x > 0$.

line $x = 0$, the flux through the loop increases as it moves to the right. By Faraday's law, we know that a changing magnetic flux induces a current in the wire. Because the flux is *increasing*, Lenz's law says that the induced current opposes this change. Therefore, the induced current in the loop is counterclockwise, as shown in Figure 26.9(b).

You should expect that the amount of motional emf induced depends on the size of the loop and how fast it is moving. We now find the emf using Faraday's law. The amount of flux through the loop depends on the loop's position as a function of time; we define $t = 0$ as the time when the leading (right) edge of the loop reaches $x = 0$. Then the position of that leading edge is $x = vt$. The amount of magnetic flux through the loop when it is in the position shown in Figure 26.9(b) is therefore (B_0)(area of loop in magnetic field) $=$ $(B_0)[(a)(vt)] = B_0avt$. This is really the absolute value of the flux—as before, we won't worry about the sign, because we have already figured out the direction of the current in the loop and only need to know the magnitude of the emf.

Applying Faraday's law,

$$|\mathcal{E}| = \left|-\frac{d\Phi_B}{dt}\right| = \left|-\frac{d}{dt}(B_0avt)\right|.$$

With B_0, a, and v all constant, the derivative is straightforward:

$$|\mathcal{E}| = |-B_0av| = B_0av. \tag{26.9}$$

As expected, the magnitude of the induced emf depends on the magnitude of the magnetic field, the size of the loop, and the speed with which the loop is moving through the field. The following numerical example will give you an idea of how much emf and current can be generated this way with ordinary values of the various parameters.

Example 26.5. For the square conducting loop moving through the field in Figure 26.9(b), find the magnitude of the induced emf and the current with the following parameters: $a = 5.0$ cm, $B_0 = 2.50$ T, $v = 60$ cm/s, and the resistance of the wire (measured the entire way around) is 0.65 Ω.

Solution. From Equation (26.9), the magnitude of the emf is

$$|\mathcal{E}| = B_0 a v = (2.50\ \text{T})(0.05\ \text{m})(0.60\ \text{m/s}) = 0.075\ \text{T}\cdot\text{m}^2/\text{s}\ = 0.075\ \text{V}.$$

The induced current is

$$I = \frac{|\mathcal{E}|}{R} = \frac{0.075\ \text{V}}{0.65\ \Omega} = 0.12\ \text{A}.$$

△

What happens if the loop shown in Figure 26.9(b) continues to move to the right, until it is entirely in the region $x > 0$? Once it is entirely in the region $x > 0$, the same amount of the field is always passing through the loop, so the magnetic flux through the loop is constant, even though it continues to move. When this situation is reached, $d\Phi_B/dt = 0$, and Faraday's law dictates that there is no induced current in the wire.

Let's see how our previous study of the force on a current-carrying conductor in a magnetic field (Chapter 24) applies to the problem we have been considering. Returning again to Figure 26.9(b), you can see that parts of the loop carry current through a magnetic field. Remember that the force on a straight segment of wire carrying a current I through a magnetic field $\vec{B}$ is

$$\vec{F} = I\vec{L} \times \vec{B} \tag{24.21}$$

where $\vec{L}$ is a displacement vector with magnitude equal to the length of the wire segment and direction in the direction of current flow. Let's use Equation (24.21) to analyze the forces on the different parts of the square loop in Figure 26.9(b). On that portion of the top of the loop that is in the field, applying the right-hand rule gives a force in the $-y$-direction. Doing the same to that portion of the bottom of the loop in the field gives a force in the $+y$-direction. By symmetry, the magnitudes of these forces are equal, and therefore their (vector) sum is zero.

This leaves the right edge of the square for us to consider. By the right-hand rule, the force in Equation (24.21) is in the $-x$-direction, or to the left in the diagram. This means that the loop does not continue to the right with a constant speed. The magnetic field acts as a "brake" that will slow the loop as it enters the magnetic field. If we wish to keep the loop moving to the right with a constant speed, a balancing force to the right (the $+x$-direction) is required from an external agent in order to make the net force on the loop zero. The magnitude of the external force equals the magnitude of the magnetic force on that segment of the wire, or

$$F = \|\vec{F}\| = ILB_0 \sin\frac{\pi}{2} = IaB_0 \tag{26.10}$$

where we have used the fact that $L = a$. If the loop has resistance R, the current is

$$I = \frac{|\mathcal{E}|}{R} = \frac{B_0 a v}{R},$$

so the magnitude of the force becomes

$$F = \left(\frac{B_0 a v}{R}\right) a B_0 = \frac{a^2 v B_0^2}{R}. \tag{26.11}$$

Continuing this analysis, let's consider the power expended by the external force in moving the loop with a constant speed v. In general, the instantaneous power is $P = \vec{F} \cdot \vec{v}$ [Equation (7.24)], and with $\vec{F}$ and $\vec{v}$ both in the $+x$-direction, we have $P = Fv$. Thus,

$$P = Fv = \frac{a^2 v^2 B_0^2}{R}. \tag{26.12}$$

The work done by the external force does not show up as kinetic energy, because the loop moves with constant speed. What happens to it then? You know that an electrical circuit with current I and resistance R uses energy at a rate $P = I^2R$ [Equation (19.33)]. In our square loop,

$$P = I^2R = \left(\frac{B_0 av}{R}\right)^2 (R) = \frac{a^2v^2B_0^2}{R}. \qquad (26.13)$$

Comparing Equations (26.12) and (26.13), we see that the power supplied by the external force eventually shows up in the heating of the resistive wire. Or put another way, the mechanical energy supplied by the external force is converted to electrical energy, which is in turn converted to heat energy. Equations (26.12) and (26.13) show that we can account exactly for all the energy in this way. This example has proved to be a useful way for us to connect a number of the topics we have studied throughout this book!

26.1.4 Applications

We just saw how a conducting loop moving in a plane perpendicular to a magnetic field is slowed because of the action of the magnetic field on the induced current. This is one type of magnetic brake. Another, similar type is shown in Figure 26.10(a). A thin solid conducting plate lies parallel to the xy-plane and moves in the $+x$-direction. A uniform magnetic field $\vec{B} = B_0\,\hat{k}$ exists in the region $x < 0$. Let's assume as before that the conducting plate started entirely in the region $x < 0$ and then entered the magnetic field, to reach the position shown in the diagram.

As the plate moves to the right, moving farther into the magnetic field, the magnetic flux through it increases. In a conductor, charges are free to flow, and Faraday's law (with Lenz's law) tells us that currents will be induced in the conductor in such a way as to oppose the flux increase. To oppose the increasing flux, the induced current must generate a magnetic field in the $-z$-direction, and it does so with the circulating induced currents shown schematically in Figure 26.10(a). These circulating currents are called **eddy currents**.

Just like the induced current in the single loop in Figure 26.9(b), the effect of the eddy currents is to retard the passage of the conductor as it enters the magnetic field. This is because the moving charges that make

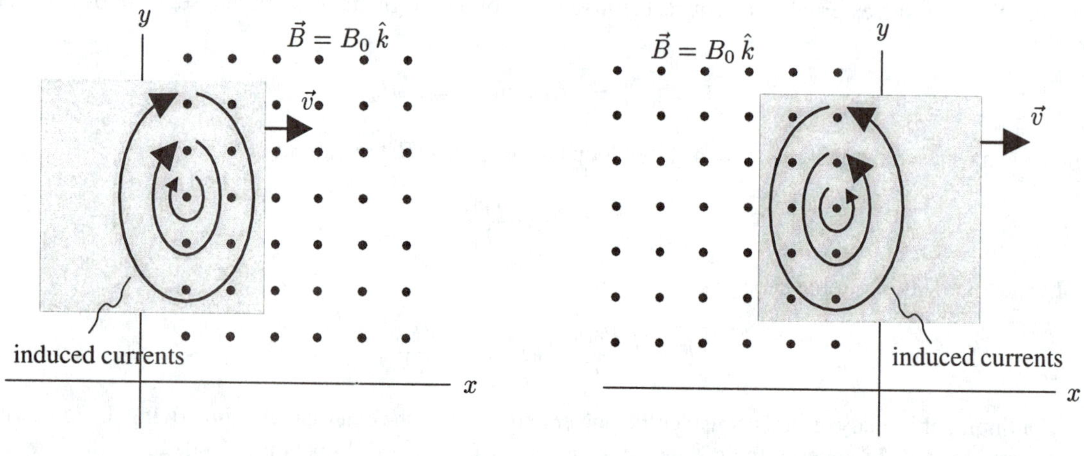

(a) Conducting plate moving into a uniform magnetic field $\vec{B} = B_0\,\hat{k}$ in the region $x < 0$

(b) Conducting plate moving out of a uniform magnetic field $\vec{B} = B_0\,\hat{k}$ in the region $x > 0$

Figure 26.10. A thin solid conducting plate lies parallel to the xy-plane and moves in the $+x$-direction.

up the eddy currents experience a force $\vec{F} = q\vec{v} \times \vec{B}$ in the magnetic field, and the same reasoning as we applied with the single conducting loop can be used to show that the net force on the conducting plate in Figure 26.10(a) is in the $-x$-direction, that is, opposing the motion.

A similar thing happens if the conducting plate is *leaving* the magnetic field, as in Figure 26.10(b). Now the plate, coming from the region $x < 0$ that contains a magnetic field $\vec{B} = B_0\,\hat{k}$, is moving into the field-free region $x > 0$. With a decreasing magnetic flux through the plate, Lenz's law says that the induced current will tend to restore the lost flux, so now the eddy currents are counterclockwise, as shown in Figure 26.10(b). Once again, it is possible (see the problems) to use the force law $\vec{F} = q\vec{v} \times \vec{B}$ to show that a net force acts on the plate in the $-x$-direction. That is, its motion is decelerated both when it enters and leaves the magnetic field. It's possible to use this principle to design a magnetic brake, say by having a metal plate fixed to a rolling wheel pass in and out of a magnetic field as it rotates. Magnetic brakes are used on train and truck wheels, as well as in numerous machines.

In Figure 26.11(a), we show a rectangular coil of wire, which rotates about a fixed axis (normally its symmetry axis) in a magnetic field. We will show that such a device can be used to generate electrical current, and therefore it is called a **generator**. In practical applications, some external agent is used to turn the coil. For example, in a hydroelectric power plant, falling water turns the wheels of a turbine, which can have such a coil attached to it. A generator is therefore useful in converting mechanical energy to electrical energy.

The basic principle underlying the operation of the generator is Faraday's law. As the coil rotates about the y-axis [Figure 26.11(a)], the amount of magnetic flux through it changes, and the changing magnetic flux results in an emf in the coil. Let's calculate the flux and use Faraday's law to find the emf. As usual, we define a unit vector $\hat{n}$ to be perpendicular to the plane of the coil. With the magnitude of the field and the area A of the coil constant, we can think of the variation in the flux as due entirely to the changing angle. From the definition of magnetic flux,

$$\Phi_B = \iint_S \vec{B} \cdot \hat{n}\, dA$$

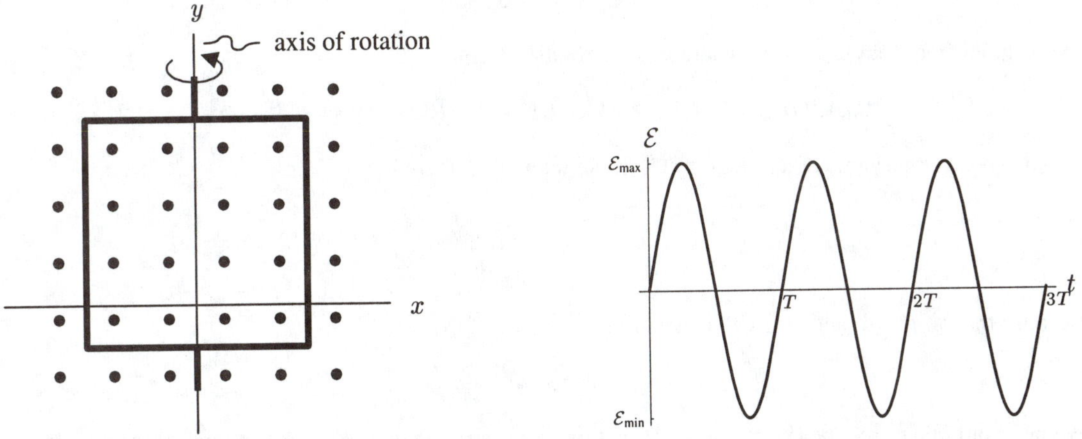

(a) Coil rotating around the y-axis in a uniform magnetic field

(b) Induced emf as a function of time

Figure 26.11. Induced emf in a coil rotating in a magnetic field.

where in this case the integration is to be carried out over the rectangular surface circumscribed by the coil. The scalar product in the surface integral is

$$\vec{B} \cdot \hat{n} = (B_0)(1)\cos\theta = B_0 \cos\theta$$

where θ is the angle between the normal vector $\hat{n}$ and the magnetic field. From rotational kinematics, we can write $\theta = \omega t$, where we will assume that the angular speed of the coil is a constant ω. Combining these results, we write the magnetic flux through the coil as a function of time:

$$\Phi_B = B_0 \cos(\omega t) \iint_S dA = B_0 A \cos(\omega t).$$

Now applying Faraday's law [Equation (26.5)], we have for a coil containing N identical loops

$$\mathcal{E} = -N \frac{d\Phi_B}{dt} = -N B_0 A[-\omega \sin(\omega t)]$$

or

$$\mathcal{E} = N B_0 A \omega \sin(\omega t). \tag{26.14}$$

The emf varies sinusoidally, as shown in Figure 26.11(b). The result is alternating current, which varies sinusoidally, and thus reverses its direction each half-cycle.

Example 26.6. For the generating coil shown in Figure 26.11(a), assume the following parameters: The coil has 100 turns, $B_0 = 3.45$ T, the coil is a square 10 cm on a side, and the coil rotates 60 times per second (this is the frequency of the alternating current produced by U.S. utility companies). (a) What is the maximum emf (magnitude) produced? (b) What is the average emf (magnitude) produced? (c) What is the average current if the coil's total resistance is 5.2 Ω?

Solution. Because the sine function varies between -1 and $+1$, the maximum magnitude of emf is given by Equation (26.14) as

$$\mathcal{E}_{\text{max}} = N B_0 A \omega.$$

The angular frequency $\omega = 2\pi\nu$, where ν is the rotation frequency. Thus,

$$\mathcal{E}_{\text{max}} = N B_0 A(2\pi\nu) = (100)(3.45\text{ T})(0.01\text{ m})^2[2\pi(60\text{ s}^{-1})] = 13.0\text{ T·m}^2\text{/s} = 13.0\text{ V}.$$

(b) The average value of a function $f(t)$ on an interval $t : [a, b]$ is

$$\overline{f(t)} = \frac{\int_a^b f(t)\,dt}{\int_a^b dt}.$$

Because the N, B_0, A, and ω are all constant, the average of the emf is

$$\overline{\mathcal{E}} = N B_0 A \omega \overline{\sin(\omega t)}.$$

We need only take the average of the sine function from 0 to π, because the magnitude of the sine function repeats itself with that period. Thus,

$$\overline{\sin t} = \frac{\int_0^\pi \sin t\,dt}{\int_0^\pi dt} = \frac{2}{\pi}$$

and the average magnitude of the emf is

$$\overline{\mathcal{E}} = \frac{2}{\pi} N B_0 A \omega = \frac{2}{\pi} \mathcal{E}_{\max} .$$

Inserting numerical values,

$$\overline{\mathcal{E}} = \frac{2}{\pi}(13.0\ \text{V}) = 8.3\ \text{V}.$$

(c) The average current follows from the average emf:

$$\overline{I} = \frac{\overline{\mathcal{E}}}{R} = \frac{8.3\ \text{V}}{5.2\,\Omega} = 1.6\ \text{A}.$$

△

An **electric motor** is essentially a generator in reverse. Electric current is now the input, rather than end product. Current flows through a coil [such as the one we described for the generator, in Figure 26.11(a)], with the coil immersed in a magnetic field. As you saw in Section 24.4, a current-carrying loop in a magnetic field experiences a torque. (Note that a device called a *commutator* must be used to switch the direction of the current twice per rotation, in order to keep the torque in the same direction throughout the coil's rotation.) The turning loop can then be used to do mechanical work, which is the basic principle in an electric motor.

26.2 Inductance

26.2.1 Inductors and Self-Inductance

Faraday's law has important implications for a loop or coil of wire (single or multiple) that is a part of an electric circuit. To understand how the presence of a coil of wire changes the behavior of a simple circuit, consider the circuit shown in Figure 26.12, in which a complete circuit contains just a battery and a variable resistor. If you change the resistance continuously, the current in the circuit changes too, because the current at any time is given by $I = \mathcal{E}/R$. A sudden change in the resistance causes just as sudden a change in the current. We can say that the response of the current to changes in resistance is virtually instantaneous.

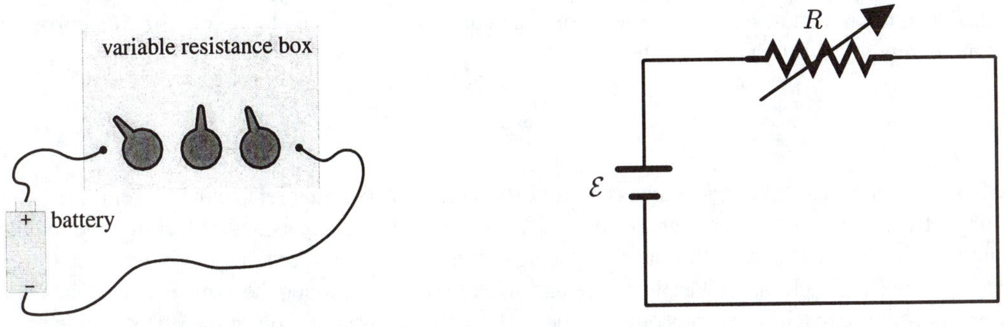

(a) Schematic of a battery and resistance box

(b) The equivalent circuit diagram

Figure 26.12. Circuit containing a battery and a variable resistor.

Now suppose we add a coil consisting of a single loop of wire in series with the resistor [Figure 26.13(a)]. A steady current gives rise to a constant magnetic flux through the loop. When we change the current going through the coil, this changes the magnetic flux through the coil. Faraday's law (with Lenz's law) dictates that an added current is induced in the coil, in such a way that it opposes the change in flux. Let's consider what effect this has qualitatively, and later we'll make it more quantitative. Suppose we start with a steady current, and then increase the current (by lowering the resistance, say from R_0 to R_1). The (formerly steady) magnetic flux through the coil increases, but an induced current in the opposite direction arises to oppose the flux increase. The observed effect of this is to retard the current increase: The net current in the circuit increases, but at a slower rate than if the coil were not present. We will see later that the current *eventually* rises from $\mathcal{E}/R_0$ to $\mathcal{E}/R_1$, but that this change does not happen instantaneously. Similarly, what if the resistance is *increased* from R_0 to R_2? Now the current in the circuit drops, but an induced current in the coil tends to oppose this change by restoring some of the lost current. Again, the effect is that the current change is retarded, so that the drop in current from $\mathcal{E}/R_0$ to $\mathcal{E}/R_2$ does not happen instantaneously.

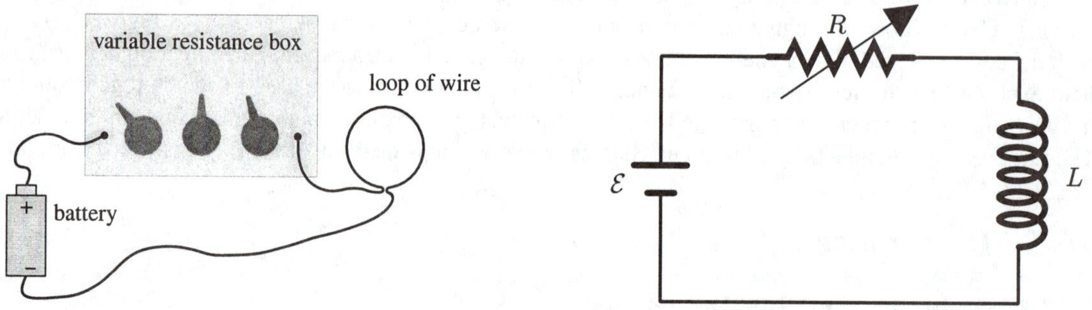

(a) Schematic of a battery, resistance box, and loop of wire

(b) The equivalent circuit diagram

Figure 26.13. Circuit containing a battery, a variable resistor, and a loop of wire, used to illustrate inductance.

We say that the circuit (in particular the coil) just described has **self-inductance**, or simply **inductance**. A device specifically designed to have inductance is called an **inductor**. Notice that the symbol L is used for inductance, and we introduce the schematic symbol for an inductor in Figure 26.13(b). The definition of the physical quantity inductance comes in the form of an equation. For a single loop of wire, if a current I causes a magnetic flux Φ_B through the coil, then

$$\Phi_B = LI \tag{26.15}$$

and L is the inductance. The inductance is a proportionality constant that relates the current through a coil to the magnetic flux through the coil, and Equation (26.15) reflects the experimental fact that the magnetic flux through a coil is proportional to the current flowing in the wire. We will see that the inductance thus depends only on geometrical factors, such as the shape and size of the coil, in much the same way that the capacitance of a parallel-plate capacitor depends only on the surface area and separation of the plates and is constant for a capacitor of fixed dimensions.

Let's find the inductance of inductors with some specific geometric shapes in worked examples. First, however, let's use Faraday's law to find the emf in an inductor, as a function of how rapidly the current in the inductor is changing. It is important to know the emf in the inductor, so that we can then use Kirchoff's rules to analyze electric circuits containing inductors. From Faraday's law $\mathcal{E} = -d\Phi_B/dt$, so using Equation (26.15)

$$\mathcal{E} = -\frac{d}{dt}(LI).$$

For a particular inductor, the inductance is fixed, so L is treated as a constant in taking the derivative. Thus,

$$\mathcal{E} = -L\frac{dI}{dt}. \tag{26.16}$$

Once again, the minus sign in Equation (26.16) represents Lenz's law; it tells us that the effect of the emf in the inductor is to oppose the change in the current passing through it. We will see how this works when we analyze electric circuits in the following sections.

Dimensions and units for inductance Going back to the definition of inductance in Equation (26.15), we see that it has dimensions of magnetic flux divided by current. Then the correct SI units are

$$\frac{\mathrm{T{\cdot}m^2}}{\mathrm{A}}.$$

For convenience, the SI unit **henry** (H) is defined

$$1\,\mathrm{H} = = 1\,\frac{\mathrm{T{\cdot}m^2}}{\mathrm{A}}$$

and is used as the unit for inductance. The henry is named for the American physicist Joseph Henry (1797–1878), who made outstanding contributions toward the understanding of electromagnetism. From Equation (26.16), it should also be evident that 1 H is equivalent to 1 V·s/A.

Example 26.7. Use the definition of inductance to find the inductance of a solenoid of length a and cross-sectional area A having a uniform n turns of wire per unit length.

Solution. We use the approximation that the magnetic field inside a solenoid is parallel to the symmetry axis and has a magnitude $B = \mu_0 nI$. Some error results in using this approximation throughout the solenoid, because it ignores end effects, but it is the best we can do symbolically. The total flux through the solenoid is the product of the number of loops (number of loops per unit length times the length), the magnitude of the field B, and the cross-sectional area A. That is,

$$\Phi_B = NBA = (na)(\mu_0 nI)(A).$$

From the definition of inductance, the inductance L is simply this flux divided by the current, or

$$L = \frac{\Phi_B}{I} = \mu_0 n^2 aA. \tag{26.17}$$

As mentioned in the text, the inductance of a particular type of inductor (in this case, a solenoid) depends only on the geometrical factors n, a, and A. It doesn't depend on the current or the inductor's placement in a circuit, but rather is constant for a solenoid of fixed dimensions. It should make physical sense that the inductance (a measure of the inductor's response to a changing current) increases whenever n, a, or A increases. △

Example 26.8. Find the numerical value of the inductance of a solenoid having 1800 turns of wire per meter, a circular cross section of radius 1.50 cm, and a length of 12.5 cm.

Solution. From the result of the previous example, the numerical value is

$$L = (4\pi \times 10^{-7}\ \mathrm{T \cdot m/A})(1500\ \mathrm{m}^{-1})^2(0.125\ \mathrm{m})[\pi(0.015\ \mathrm{m})^2]$$
$$= 2.50 \times 10^{-4}\ \mathrm{T \cdot m^2/A} = 2.50 \times 10^{-4}\ \mathrm{H}$$

or 0.250 mH. Inductances on the order of mH are fairly typical in electric circuits. △

Example 26.9. Find the inductance of a toroid with a total of N turns of wire and the following dimensions: Its cross-section is rectangular, with inner radius a, outer radius b, and constant height c. Evaluate numerically with the values $N = 2000$, $a = c = 10$ cm, $b = 20$ cm.

Solution. From the problems for Section 25.2, we know that the magnetic field inside a toroid is not uniform, but varies with the perpendicular distance r from the symmetry axis as

$$B = \frac{\mu_0 I N}{2\pi r}$$

with the direction of the magnetic field tangent to a circle of radius r having its center on the symmetry axis. Because the magnetic field varies as a function of r, we must construct a definite integral to find the total magnetic flux through a cross section of the toroid, as shown in Figure 26.14. The magnetic flux through a thin strip of width Δr as shown at a distance r from the central axis is

$$\text{magnetic flux} = (B)(\text{area of strip}) = B(c\,\Delta r)$$

where we have used the fact that the magnetic field is in the direction of the normal for this surface to evaluate the dot product $\vec{B} \cdot \hat{n}$ used to compute magnetic flux. Then the net flux for the entire cross section is found by constructing a definite integral over all such strips from $r = a$ to $r = b$. This is accompished by taking the limit as Δr approaches zero to get

$$\Phi_B = \int_a^b B c\, dr.$$

Substituting for B and factoring out constants,

$$\Phi_B = \int_a^b \left(\frac{\mu_0 I N}{2\pi r}\right) c\, dr = \frac{\mu_0 I N c}{2\pi} \int_a^b \frac{1}{r}\, dr.$$

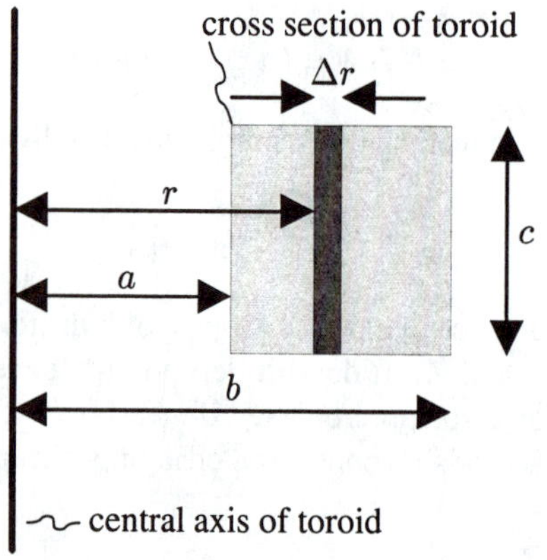

Figure 26.14. Cross section of the toroid, showing its physical dimensions.

This is the flux through *one* loop, so for all N loops in the toroid we multiply by another factor of N to find a total flux of

$$\Phi_B = \frac{\mu_0 I N^2 c}{2\pi} \int_a^b \frac{1}{r}\, dr.$$

The evaluation of the definite integral is straightforward, with

$$\int_a^b \frac{1}{r}\, dr = (\ln r)\Big|_a^b = \ln b = \ln a = \ln\left(\frac{b}{a}\right)$$

giving

$$\Phi_B = \frac{\mu_0 I N^2 c}{2\pi} \ln\left(\frac{b}{a}\right).$$

Finally, the definition of inductance tells us that

$$L = \frac{\Phi_B}{I} = \frac{\mu_0 N^2 c}{2\pi} \ln\left(\frac{b}{a}\right).$$

Once again, you should be able to justify why the inductance increases whenever N, c, or b increases or when a decreases.

Evaluating numerically,

$$L = \frac{(4\pi \times 10^{-7}\ \mathrm{T{\cdot}m/A})(2000)^2(0.10\ \mathrm{m})}{2\pi} \ln\left(\frac{0.20\ \mathrm{m}}{0.10\ \mathrm{m}}\right) = 5.5 \times 10^{-2}\ \mathrm{T{\cdot}m^2/A} = 0.055\ \mathrm{H} = 55\ \mathrm{mH}.$$

This is also a typical inductance value. △

26.2.2 *RL* Circuits

In this section, we use our knowledge calculus to understand quantitatively what happens when an inductor and resistor are put together in a series circuit, and we relate the results to our previous qualitative understanding of what an inductor does according to Faraday's law. We start by analyzing the circuit shown in Figure 26.15. For obvious reasons, this is known as a ***RL*-series circuit**. Initially, the switch S is open as shown, but at time $t = 0$ the switch will be closed. As in other simple series circuits, we wish to know the current as a function of time. Because this is a series circuit, you know that the current is the same everywhere in the circuit at any time. But how does it vary as a function of time?

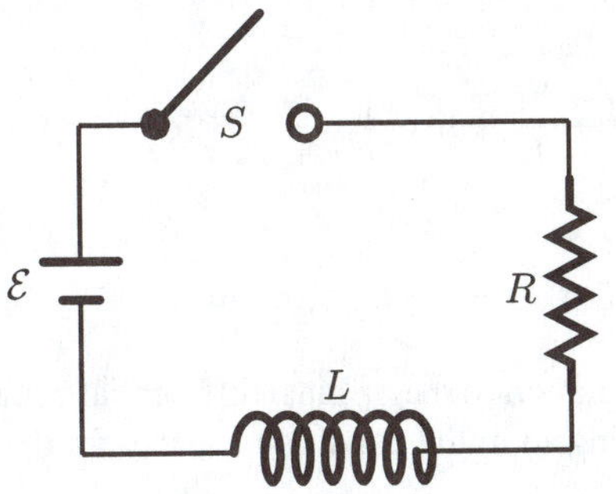

Figure 26.15. *RL*-series circuit with an open switch.

To answer that question, we apply Kirchoff's loop rule. When current flows, it flows clockwise around the loop, from the positive terminal to the negative terminal of the battery. Starting in the lower-left corner of the diagram and proceeding around the loop in a clockwise direction, the sum of the potential differences across the circuit elements is

$$\mathcal{E} - IR - L\frac{dI}{dt} = 0. \tag{26.18}$$

In Equation (26.18), the sign on the term IR is negative, because we are going around the loop with the current. The term $-L\,dI/dt$ is the potential difference across the inductor, as dictated by Equation (26.16).

With $\mathcal{E}$, R, and L constants, you should recognize that Equation (26.18) is a linear, first-order differential equation for $I(t)$, which can be solved using the technique of separation of variables (Section 6.2). First, we rearrange to separate the factors and terms containing I and t:

$$L\frac{dI}{dt} = \mathcal{E} - IR$$

or

$$\left(\frac{L}{\mathcal{E} - IR}\right) dI = dt.$$

A substitution $u = \mathcal{E} - IR$ leads to $du/dI = -R$, so

$$\left(\frac{L}{u}\right)\left(-\frac{du}{R}\right) = dt$$

or, rearranging,

$$\frac{1}{u}\,du = -\frac{R}{L}\,dt.$$

Because R and L are constants, the separation of variables has been accomplished. Now we take the indefinite integral of both sides, collecting the constant of integration on the right:

$$\int \frac{1}{u}\,du = -\frac{R}{L}\int dt + \text{ constant.}$$

Thus,

$$\ln u = -\frac{Rt}{L} + \ln C.$$

For convenience, we have called the constant $\ln C$, anticipating that the next step is to exponentiate both sides to recover u, and then I:

$$u = \exp\left(-\frac{Rt}{L} + \ln C\right) = e^{-Rt/L}e^{\ln C} = Ce^{-Rt/L}.$$

Recalling that $u = \mathcal{E} - IR$, we have

$$\mathcal{E} - IR = Ce^{-Rt/L}. \tag{26.19}$$

Equation (26.19) is the general solution to the original differential equation [Equation (26.18)]. As usual, we can find a specific solution by using an initial condition to solve for the constant C. In this case, we know the current $I = 0$ at $t = 0$. Thus,

$$\mathcal{E} - 0 = Ce^0 = C,$$

so

$$C = \mathcal{E}.$$

The specific solution is

$$\mathcal{E} - IR = \mathcal{E}e^{-Rt/L}$$

or, rearranging to solve for I,

$$I = \frac{\mathcal{E}}{R}\left(1 - e^{-Rt/L}\right). \tag{26.20}$$

Let's analyze the specific solution [Equation (26.20)] to see if it makes sense physically. The graph of the current as a function of time (Figure 26.16) aids our analysis. Just after the switch is closed, the current does not rise instantaneously to a value of $\mathcal{E}/R$, as it would in a circuit without an inductor. Rather, it increases gradually. This makes sense, because the current induced in the inductor fights against the current increase the battery is trying to impose on the circuit. As the current increases, the slope of the graph dI/dt becomes smaller, and the emf induced in the inductor becomes proportionally less. Eventually (after a long time) the graph flattens out, and the slope dI/dt asymptotically approaches zero. In this limit, a steady-state situation is approached. When the rate of change of current becomes negligible, the induced emf in the inductor is also negligible. At this point, the inductor acts just like a straight conducting wire, because if there is no change in the current, there is no change in the magnetic flux through the inductor, and hence no emf induced in the inductor. The asymptotic value of current is $I = \mathcal{E}/R$, just as it would be with no inductor.

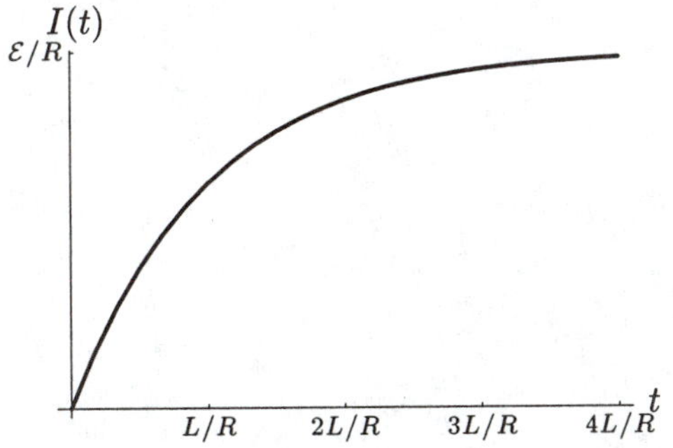

Figure 26.16. Current as a function of time in the RL-series circuit.

It is useful to compare the behavior of this circuit with that of the charging RC circuit we studied in Section 21.5. We found that an initially uncharged capacitor connected in series with a resistor and battery has a charge that varies as a function of time according to the function

$$Q = \mathcal{E}C\left(1 - e^{-t/RC}\right). \tag{21.32}$$

The similarity of Equations (21.32) and (26.20) is striking. In studying the RC circuit, we found it convenient to define a time constant $\tau = RC$ and measure time in those units. In our RL series circuit, it is apparent [by looking at the exponential term in Equation (26.20)] that an ***RL* time constant**

$$\tau = \frac{L}{R} \tag{26.21}$$

will be useful in this context. For example, after one time constant has elapsed, $t = \tau = L/R$ and the current is

$$I(\tau) = \frac{\mathcal{E}}{R}\left(1 - e^{-R\tau/L}\right) = \frac{\mathcal{E}}{R}\left(1 - e^{-1}\right).$$

The factor $\left(1 - e^{-1}\right)$ has a decimal approximation 0.632. This means that after one time constant has elapsed, the current has reached about 63.2% of its final value. This is consistent with what we see on the graph in Figure 26.16.

Example 26.10. In the circuit shown in Figure 26.15, assume the following numerical values: $\mathcal{E} = 12.0$ V, $R = 250\,\Omega$, and $L = 4.50$ mH. (a) Find the RL time constant. (b) What is the current in the circuit after two time constants have elapsed? (c) How much time does it take for the current to reach exactly one-half of its final (steady-state) value?

Solution. (a) From the numerical values given,

$$\tau = \frac{L}{R} = \frac{4.50 \times 10^{-3}\ \mathrm{H}}{250\,\Omega} = 1.80 \times 10^{-5}\ \mathrm{H}/\Omega = 1.80 \times 10^{-5}\ \mathrm{s}.$$

(b) At $t = 2\tau$, the current is

$$I = \frac{\mathcal{E}}{R}\left(1 - e^{-Rt/L}\right) = \frac{\mathcal{E}}{R}\left(1 - e^{-(2\tau/\tau)}\right) = \frac{12.0\ \mathrm{V}}{250\,\Omega}\left(1 - e^{-2}\right) = 4.15 \times 10^{-2}\ \mathrm{A} = 41.5\ \mathrm{mA}.$$

(c) From the graph of current vs. time in the text (Figure 26.16), we should expect that this value will be reached shortly before one time constant. Because the steady-state value of the current is $\mathcal{E}/R$, we desire a value $I = \mathcal{E}/2R$. Thus,

$$I = \frac{\mathcal{E}}{2R} = \frac{\mathcal{E}}{R}\left(1 - e^{-Rt/L}\right).$$

Solving for t,

$$\frac{1}{2} = 1 - e^{-Rt/L}$$

or

$$e^{-Rt/L} = \frac{1}{2}.$$

Taking the logarithm of both sides:

$$-\frac{Rt}{L} = \ln\left(\frac{1}{2}\right).$$

Solving for t,

$$t = -\frac{L}{R}\ln\left(\frac{1}{2}\right).$$

Inserting numerical values,

$$t = -\frac{4.50 \times 10^{-3}\ \mathrm{H}}{250\,\Omega}\ln\left(\frac{1}{2}\right) = 1.25 \times 10^{-5}\ \mathrm{s}.$$

This is certainly in the range we expected. △

26.2.3 LC Circuits

You have just seen how the time-varying current in an RL circuit is similar to the variation in the charge on a capacitor with time in an RC circuit. In this section, we'll explore what happens when we put an inductor in series with a capacitor in an **LC circuit**. The simplest possible combination of these two circuit elements is shown in Figure 26.17(a); the inductor and capacitor are connected in series with no battery or resistor. For current to flow in this circuit when we close the switch, we suppose that some initial charge Q_0 is stored on the capacitor.

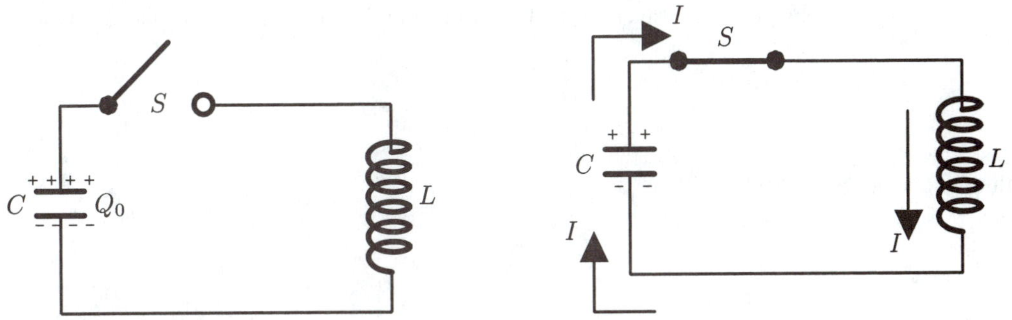

(a) Initially the capacitor is charged, and the switch is open

(b) The switch is closed, and current flows

Figure 26.17. An LC circuit.

Let's first consider what happens qualitatively, based on our knowledge of how capacitors and inductors work. Remember that the charged capacitor carries charges of equal magnitude and opposite sign on its two plates. This means that the positively charged plate is at a higher electric potential than the negatively charged plate. When the switch is closed, current flows from a higher to lower potential, in the direction shown in Figure 26.17(b). The charge on the capacitor is the source of the current, so as current begins to flow, the charge on the capacitor decreases. (Because we like to think in terms of positive current, think of the positive charge leaving the top plate of the capacitor and heading through the circuit to the lower plate.) Based on our previous study of inductors, we expect the current to rise gradually from zero, as the inductor opposes the increasing current from the capacitor. At some point, all the charge will have drained off the capacitor, but there will still be current through the inductor at that time, because the inductor opposes any sudden drop in the current. This means that positive charges are still traveling toward the lower plate of the capacitor. When they arrive there, the capacitor will again be charged, this time with positive charges on the bottom plate and negative charges on the top.

The capacitor and inductor are completely symmetric, so the process just described will repeat itself, with the positive charge returning from the bottom plate through the inductor to the top plate. The result is an *oscillation* of the charge from one capacitor plate to the other and back again. In a perfectly resistance-free circuit, no energy is lost, so the energy initially stored in the electric field in the capacitor (see Section 19.1) remains in the circuit, and the oscillation of charge continues indefinitely. In Section 26.2.4, we discuss the energy stored in a magnetic field, so that we can understand how the energy is stored when the capacitor loses its charge and current is flowing through the inductor. But first, we establish the quantitative basis for the oscillations just described.

Let's just apply Kirchoff's loop rule to the circuit in Figure 26.17(b). Starting from the lower-left corner and proceeding around the loop clockwise, we find

$$\frac{Q}{C} - L\frac{dI}{dt} = 0. \tag{26.22}$$

In this analysis, we have used the fact (see Chapter 19) that the potential difference across the capacitor plates is Q/C, where Q is the magnitude of the charge on each plate. Based on our qualitative analysis above, we expect that Q will vary as a function of time. As usual, the potential difference across the inductor is $-L\,dI/dt$, as given by Faraday's law. A connection exists between the current I through the inductor and charge Q on the capacitor. The charge on the capacitor is the source of the current, and therefore by conservation of charge, $I = -dQ/dt$. The minus sign here reflects the fact that we have defined the current to be positive when the charge on the capacitor is decreasing. Taking a derivative of this expression for I gives

$$\frac{dI}{dt} = -\frac{d^2Q}{dt^2}. \tag{26.23}$$

Substituting this into Equation (26.22) gives

$$\frac{Q}{C} + L\frac{d^2Q}{dt^2} = 0$$

or, rearranging,

$$L\frac{d^2Q}{dt^2} = -\frac{1}{C}Q. \tag{26.24}$$

Look closely at Equation (26.24). It is a second-order, linear differential equation for the function $Q(t)$ (with L and C constant). In fact, it closely resembles a second-order differential equation we studied extensively in Chapter 12, the equation of motion for the simple harmonic oscillator:

$$m\frac{d^2x}{dt^2} = -kx. \tag{12.4}$$

Recall that a solution to this equation of motion is $x = A\cos(\omega t + \delta)$ where A and δ are arbitrary constants in the general solution and can be determined if initial conditions are known. The constant ω depends on the parameters m and k:

$$\omega = \sqrt{\frac{k}{m}}. \tag{12.7}$$

The constant ω is also the angular frequency of simple harmonic motion, and is related to the frequency ν and period T of the oscillation:

$$\omega = 2\pi\nu = \frac{2\pi}{T}.$$

We use our experience with the simple harmonic oscillator to solve the differential equation [Equation (26.24)] for the charge on the capacitor as a function of time. Because of the similar form of the two differential equations, we need only replace the constants m and k with L and $1/C$, respectively. Then the charge on the capacitor as a function of time is

$$Q(t) = Q_0\cos(\omega t + \delta) \tag{26.25}$$

where now the angular frequency of oscillation is

$$\omega = \sqrt{\frac{1}{LC}}. \tag{26.26}$$

Equation (26.25) tells us that the charge on the capacitor does indeed oscillate from one plate to the other and that it does so sinusoidally. Notice that because the output of the cosine function varies from -1 to $+1$, the output of $Q(t)$ is negative some of the time. This goes against our previous convention of letting Q represent just the magnitude of the charge on a capacitor. Now $Q(t)$ carries the added information of the sign of the charge on each plate. Referring to Figure 26.17(b), we'll say that $Q > 0$ when the positive charge is on the top plate, and $Q < 0$ when negative charge is on the top plate.

Once again, Q_0 and δ are constants that can be determined if initial conditions are known. The constant Q_0 represents the maximum charge on the capacitor, and δ is a phase angle that depends on the initial charge and current. With the initial conditions shown in Figure 26.17(a), the initial charge is the maximum charge on the capacitor, so in this case

$$Q(t) = Q_0 \cos(\omega t) \tag{26.27}$$

because the maximum value of the cosine function is $\cos(0) = 1$.

Assuming the initial conditions as we have described them previously [Figure 26.17(a)], so that $Q(t)$ is given by Equation (26.27), we can easily find the current in the inductor as a function of time. Recalling that $I = -dQ/dt$,

$$I(t) = -\frac{d}{dt}[Q_0 \cos(\omega t)] = -Q_0 \frac{d}{dt}[\cos(\omega t)] = Q_0\omega \sin(\omega t).$$

Defining $I_{\max} = Q_0\omega$ [which is reasonable, because this is the maximum output of the function $I(t)$], we have

$$I(t) = I_{\max} \sin(\omega t). \tag{26.28}$$

The output of this function together with the output of $Q(t)$ is shown in Figure 26.18. The function $I(t)$ in Equation (26.28) is in good agreement with our previous qualitative analysis. The action of the inductor causes the current to rise gradually from $I = 0$ at time $t = 0$. The current in the inductor reaches a maximum when the charge on the capacitor is zero. When the positive charge is all on the bottom plate of the capacitor (corresponding to $Q = -Q_0$), the current in the inductor is again zero—equivalent to the initial situation with the charges reversed. Then the charge flow reverses, so the current becomes negative as the cycle repeats, with the positive charge flowing in the other direction until the initial situation is repeated.

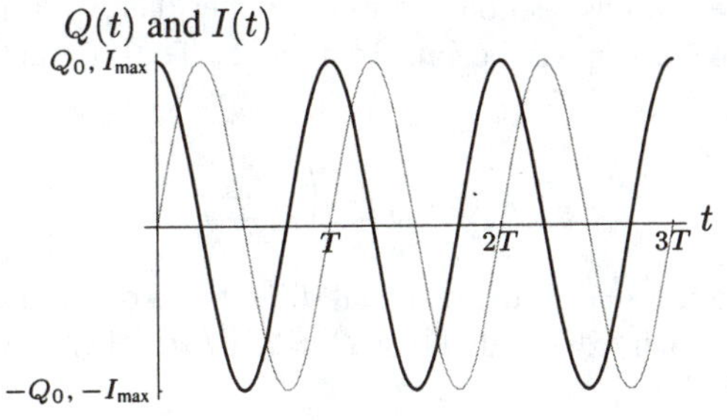

Figure 26.18. Current (light curve) and charge (dark curve) as functions of time.

Example 26.11. You are given a 15.0-mH inductor. (a) With what value capacitor should the inductor be combined (Figure 26.17(b)) in order to make the oscillation frequency exactly 1 MHz (i.e., 10^6 oscillations per second)? (b) Assuming the capacitance you found in (a), what is the maximum current in the inductor if the maximum charge on the capacitor is 25 nC?

Solution. (a) We can use the relationship between frequency and angular frequency, along with the relationship between angular frequency, capacitance, and inductance [Equation (26.26)]:

$$2\pi\nu = \omega = \sqrt{\frac{1}{LC}}.$$

Thus,

$$(2\pi\nu)^2 = 4\pi^2\nu^2 = \frac{1}{LC}$$

so

$$C = \frac{1}{4\pi^2\nu^2 L}.$$

Inserting numerical values,

$$C = \frac{1}{4\pi^2(10^6\ \mathrm{s}^{-1})^2(1.50 \times 10^{-2}\ \mathrm{H})} = 1.7 \times 10^{-12}\ \mathrm{F} = 1.7\ \mathrm{pF}.$$

(b) From the text $I_{\max} = Q_0\omega$, so

$$I_{\max} = Q_0(2\pi\nu) = 2\pi Q_0\nu.$$

Inserting numerical values,

$$I_{\max} = 2\pi(25 \times 10^{-9}\ \mathrm{C})(10^6\ \mathrm{s}^{-1}) = 0.16\ \mathrm{C/s} = 0.16\ \mathrm{A}.$$

△

26.2.4 Energy in a Magnetic Field

In this section, we use the inductor as a model to discuss how energy is stored in a magnetic field. This discussion is analogous to how we used a capacitor as a model to discuss how energy is stored in an electric field, in Section 19.1, and you should pay attention to the similarities.

Recall from our study of dc circuits that the power delivered to a circuit element is the product of the current and potential difference. In this section, we use i (rather than I) to represent the variable current passing through the inductor (the reason will become clear later). Thus, for an inductor, the power delivered is

$$P = i|\Delta V| = i\left(L\frac{di}{dt}\right)$$

where we have taken the absolute value of the potential difference so that the power will be positive. By definition, power is the rate at which work is done, or $P = dW/dt$. Then the work done in a time interval $t : [0, t_1]$ is the definite integral

$$W = \int_0^{t_1} P\,dt$$

or in this case

$$W = \int_0^{t_1} iL\,\frac{di}{dt}\,dt.$$

Simplifying,

$$W = L\int_0^I i\,di$$

where we have assumed $i = 0$ at $t = 0$ and $i = I$ at $t = t_1$ (the end of the interval). The work done is

$$W = L\left(\frac{1}{2}i^2\right)\Big|_0^I = \frac{1}{2}LI^2.$$

If there are no energy losses, the inductor is a conservative system, so the change in potential energy associated with the inductor's magnetic field in this process is

$$\Delta U_B = W = \frac{1}{2}LI^2.$$

It is convenient to define the potential energy to be zero when the magnetic field is zero (and hence when $I = 0$), so we can write a potential energy function as

$$U_B = \frac{1}{2}LI^2. \qquad (26.29)$$

Before going on, it is useful to compare this with the potential energy we associated with the electric field in a capacitor:

$$U = \frac{Q^2}{2C}. \qquad (19.15)$$

The similarity is striking, with the square of the current (inductor) replacing the square of the charge (capacitor), and L replacing $1/C$. The similarities continue to appear if we consider the energy density (the energy per unit volume) u_B associated with the magnetic field. Recall that the energy density in the electric field E is

$$u = \frac{1}{2}\epsilon_0 E^2. \qquad (19.17)$$

We use a solenoid as a model inductor to find energy density. Ignoring end effects, the magnetic field inside a solenoid is uniform. The volume of the interior of the solenoid is the product of its length a and cross-sectional area A, so if the energy stored in the magnetic field is U_B, the energy density is

$$u_B = \frac{U_B}{aA}.$$

Using the energy U_B of the inductor from Equation (26.29),

$$u_B = \frac{\frac{1}{2}LI^2}{aA}.$$

Earlier we found that the inductance of an ideal solenoid is

$$L = \mu_0 n^2 aA \qquad (26.17)$$

where the solenoid has n turns of wire per unit length. With this, the energy density in the solenoid becomes

$$u_B = \frac{\frac{1}{2}(\mu_0 n^2 a A) I^2}{aA} = \frac{1}{2}\mu_0 n^2 I^2.$$

Because the magnitude of the (assumed uniform) magnetic field inside the solenoid is $B = \mu_0 n I$, we can rewrite the energy density in terms of B as

$$u_B = \frac{B^2}{2\mu_0}. \tag{26.30}$$

We have derived Equation (26.30) for the specific geometry of the solenoid, but it turns out that it is perfectly general. The energy density of any magnetic field with magnitude B is given by that equation. Notice how similar it is to Equation (19.17), the energy density of an electric field. In each case, the energy density is proportional to the square of the magnitude of the field.

26.2.5 *RLC* Circuits and Harmonic Motion

In Section 26.2.3, we saw how the oscillation of charge through an LC circuit resembled the oscillation of a mass on a spring from Chapter 12. In this section, we explore more of the connections and analogies between linear oscillatory motion and the behavior of electric circuits containing inductors, capacitors, and resistors.

To begin, let's look more closely at the example from Section 26.2.3 [the LC circuit in Figure 26.17(b)], thinking of the linear harmonic oscillator as the mechanical analog. The analogy is most successful if we associate the electric energy in the capacitor with energy stored in the spring in the mechanical oscillator, and associate the magnetic energy in the inductor with the kinetic energy of the mass attached to the spring. At the start of a cycle, all the energy of the LC circuit is stored in the capacitor (with no magnetic flux in the inductor). The mechanical analog of this is an initial situation with the spring (with force constant k) compressed and the mass m at rest, so that all the energy of the system is stored in the spring. Then current begins to flow in the LC circuit, and some of the electrical energy in the capacitor is converted into magnetic energy in the inductor. Similarly, as the mass begins to move, some of the spring's potential energy is converted into kinetic energy. After one-fourth of a period in the LC circuit, all the energy is magnetic, because no charge is left on the capacitor. After one-fourth cycle of the mechanical oscillator, the spring is in its equilibrium position, and all the system's energy is in the kinetic energy of the mass. After one-half of a period in the LC circuit, the charge is once again all on the capacitor (although the signs are reversed). After one-half period in the mechanical oscillator, the spring is extended to its maximum displacement and the mass is momentarily at rest.

In both the LC circuit (without resistance) and the mechanical oscillator (without friction), the total energy of the system is conserved. The total energy of the mechanical oscillator is the total mechanical energy E, given by $E = K + U$, or in this case

$$E = K + U = \frac{1}{2}mv^2 + \frac{1}{2}kx^2.$$

In the LC circuit, the total energy is the sum of the electric and magnetic energy:

$$\text{total energy} = U + U_B = \frac{Q^2}{2C} + \frac{1}{2}LI^2. \tag{26.31}$$

To summarize, we show the analogous energy relationships in Table 26.1.

There is a clear correspondence between the mechanical displacement x and capacitor charge Q, and between the mechanical velocity v (the derivative of the displacement) and the electric current I (the derivative of the

Table 26.1. Analogous energy relationships

Mechanical oscillator	LC circuit
$U = \frac{1}{2}kx^2$	$U = \frac{Q^2}{2C}$
$K = \frac{1}{2}mv^2$	$U_B = \frac{1}{2}LI^2$

charge). Judging from the corresponding entries in the table, it appears that the role of the spring constant k is taken in the LC circuit by the factor $1/C$. This is surely in keeping with the roles these constants play in the second-order differential equations governing the respective situations (see Section 26.2.3). It even makes sense that the inverse of capacitance is the quantity analogous to the spring constant. Because capacitance is charge stored per unit of potential difference, a lower-valued capacitor holds less charge per unit potential energy than a higher-valued capacitor. This is like having a stiffer spring, which has less displacement for a given amount of potential energy than a weaker spring. Thus, a *lower* capacitance corresponds to a *higher* spring constant. The inductance L plays the same role as mass m. Inductance is the tendency to keep the current where it is, fighting off any changes. It's a sort of electrical inertia, if you will, just as mass is a measure of inertia in the mechanical system.

Now let's add a resistor to the circuit, as shown in Figure 26.19 (for obvious reasons this is now called an ***RLC* series circuit**). As before, we assume that the capacitor has an initial charge Q_0 and the switch is open, so that no current flows. When the switch, is closed current begins to flow in the clockwise direction around the loop. We apply Kirchoff's loop rule as we did in Equation (26.22). Now an additional potential difference of $-IR$ exists across the resistor, so the loop rule gives

$$\frac{Q}{C} - L\frac{dI}{dt} - IR = 0.$$

Recalling that $I = -dQ/dt$, this equation can be rewritten

$$L\frac{d^2Q}{dt^2} = -\frac{Q}{C} - R\frac{dQ}{dt}. \tag{26.32}$$

The extra term $R\,dQ/dt$ in this equation makes its solution more difficult. Fortunately, we have prior experience with a similar equation in mechanics. It is essentially the same as Equation (12.36), which describes

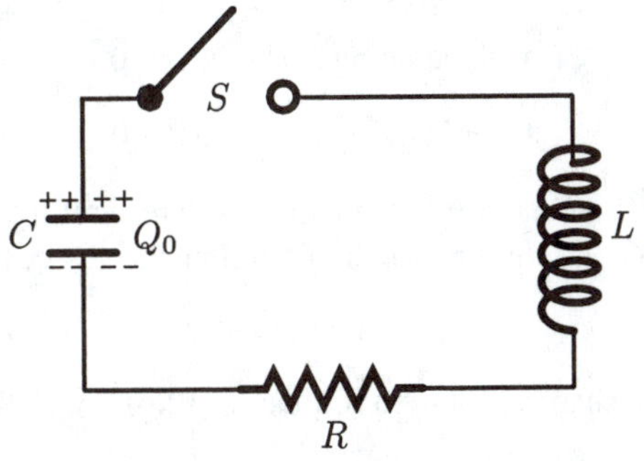

Figure 26.19. An RLC circuit with an open switch.

a simple harmonic oscillator with a frictional force proportional to the speed of the oscillating particle:

$$m\,\frac{d^2x}{dt^2} = -kx - b\,\frac{dx}{dt}. \tag{12.36}$$

The constants in the two equations (mechanical and electrical) bear a familiar correspondence: L corresponds to m, $1/C$ to k, Q to x, and now the resistance R corresponds to the damping constant b. There is also a physical correspondence between R and b, because we know the effect of mechanical damping and electrical resistance is to dissipate energy from the system in the form of heat.

Before looking at solutions of Equation (26.32), let's explore the fact that energy is not conserved in the RLC circuit. We know the total electric energy in the electric and magnetic fields [Equation (26.31)]. The rate of change in the total energy is the derivative of this quantity with respect to time:

$$\text{rate of change of total energy} = \frac{d}{dt}\left[\frac{Q^2}{2C} + \frac{1}{2}LI^2\right].$$

The derivative of each term requires the chain rule:

$$\text{rate of change of total energy} = \left(\frac{1}{2C}\right)(2Q)\left(\frac{dQ}{dt}\right) + \frac{1}{2}L(2I)\left(\frac{dI}{dt}\right) = \frac{Q}{C}\frac{dQ}{dt} + LI\,\frac{dI}{dt}.$$

Recalling that $I = -dQ/dt$, we have

$$\text{rate of change of total energy} = -I\frac{Q}{C} - IL\,\frac{d^2Q}{dt^2} = -I\left(\frac{Q}{C} + L\,\frac{d^2Q}{dt^2}\right).$$

The quantity in brackets is given by Equation (26.32) for the RLC circuit as $-R\,dQ/dt$, so

$$\text{rate of change of total energy} = -I\left(-R\,\frac{dQ}{dt}\right) = -I^2R.$$

We conclude that energy is being *lost* at a rate of I^2R. This makes perfect sense, because you know from our study of dc circuits [Equation (19.34)] that the resistor in the circuit dissipates energy at a rate I^2R. Our energy bookkeeping is therefore consistent with what we know about resistors.

Returning to the differential equation [Equation (26.32)], we can use our study of damped oscillations in Section 12.4 to understand how the circuit behaves. To summarize briefly the results of that section, we first defined the parameters $\beta = b/2m$ and $\omega_0^2 = k/m$. Then we characterized the oscillator's behavior three ways:

$$\text{Underdamping}: \beta^2 - \omega_0^2 < 0$$

$$\text{Critical Damping}: \beta^2 - \omega_0^2 = 0$$

$$\text{Overdamping}: \beta^2 - \omega_0^2 > 0$$

We can make the same characterization in the RLC circuit, if we are careful to define the parameters accordingly: $\beta = R/2L$ and $\omega_0^2 = 1/LC$. Using the results of Section 12.4, we can then write analytic solutions appropriate for each case.

Underdamping Rewriting the mechanical solution [Equation (12.41)] with the appropriate electromagnetic quantities, we have

$$Q(t) = Q_0 e^{-\beta t}\cos(\omega_u t) \tag{26.33}$$

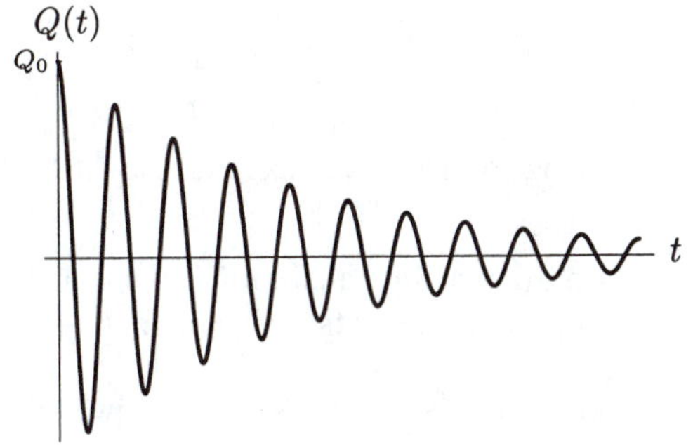

(a) Underdamping

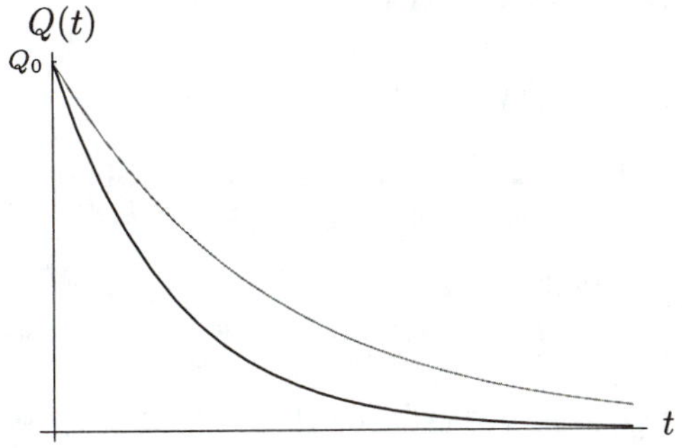

(b) Critical damping (dark curve) and overdamping (light curve)

Figure 26.20. Charge on the capacitor as a function of time in the RLC circuit.

where the underdamped angular frequency is $\omega_u = \sqrt{\omega_0^2 - \beta^2}$. A graph of the resulting charge oscillation is shown in Figure 26.20(a). [Remember that $Q(t)$ represents the charge on the capacitor as a function of time.] The factor $e^{-\beta t}$ causes an exponential decay of the oscillation "amplitude" (in this case, the maximum charge on the capacitor in each cycle). After a long time, the maximum charge in a cycle approaches zero, because the resistor has dissipated all the original energy.

Critical damping With $\beta^2 = \omega_0^2$, the solution resembles that of the critically damped mechanical oscillator (Section 12.4):

$$Q(t) = (Q_0 + At)e^{-\beta t} \tag{26.34}$$

where Q_0 is the initial charge and A is another constant that depends on initial conditions. The charge on the capacitor simply decays exponentially without oscillating at all. the charge on the capacitor approaches zero asymptotically after a long time, as shown in Figure 26.20(b).

Overdamping Now the solution follows by putting the analogous electromagnetic quantities into Equation (12.42):

$$Q(t) = Q_1 \exp\left[\left(\sqrt{\beta^2 - \omega_0^2} - \beta\right)t\right] + Q_2 \exp\left[\left(-\sqrt{\beta^2 - \omega_0^2} - \beta\right)t\right] \tag{26.35}$$

where Q_1 and Q_2 are constants, with the restriction $Q_1 + Q_2 = Q_0$. This solution is essentially the sum of two decaying exponentials, and again no oscillations result, as seen in Figure 26.20(b).

26.3 Problems

26.1 Faraday's Law

1. Show that (as in Example 26.2) the SI units $m^2 \cdot T/s$ for rate of change of magnetic flux reduce to the SI unit for emf (volts).

   In Problems 2–4, consider a straight conducting wire of length a aligned parallel to the y-axis and moving in the $+x$-direction with constant speed v. There is a uniform magnetic field $\vec{B} = -B_0\,\hat{k}$.

2. By considering the flux through the area swept out by the moving wire, use Faraday's law to show that a motional emf of magnitude $B_0 av$ exists between the two ends of the wire.

3. Explain why current flows in the wire for only a short time, with the steady-state situation being charges (of opposite sign) built up on the two ends of the wire. On which end of the wire are the positive charges, and on which end are the negative charges?

4. Use the fact that moving charges experience a force $\vec{F} = q\vec{v} \times \vec{B}$ in a magnetic field to obtain the same results as in the previous two problems. (This shows that in this case the laws of electromagnetism allow two different but equivalent approaches to the problem.)

5. Show why a conducting plate moving transverse to a uniform magnetic field is decelerated both when it enters the field [Figure 26.10(a)] and leaves the field [Figure 26.10(b)].

6. A bar magnet is moved along the symmetry axis of a solenoid. The wires from the two ends of the solenoid are connected, forming a closed circuit. Initially, the magnet is outside the solenoid. It is then pulled straight through the solenoid (with the north pole entering first) at constant speed, emerging from the other side. Describe qualitatively what happens to the current in the solenoid as a function of time. Assume that the solenoid is longer than the magnet.

Problems 7–9 refer to the following situation. A square conducting loop with each side $a = 8.50$ cm lies in the yz-plane. Throughout the yz-plane, a time-varying magnetic field $\vec{B} = B_0 t\,\hat{\imath}$ exists, with $B_0 = 0.150$ mT/s.

7. What is the magnetic flux through the loop as a function of time? Evaluate numerically the magnetic flux at $t = 1.0$ s and $t = 10.0$ s.

8. What is the direction of the current induced in the loop?

9. The conducting wire has a resistance per unit length of 1.25 Ω/m. Find the electric current in the loop at $t = 5.0$ s.

10. A uniform magnetic field $\vec{B} = B_0\,\hat{k}$ (with $B_0 = 90$ mT) exists in a circular region of the xy-plane, centered at the origin, with a radius 25 cm. If the parameter B_0 is changing at a rate +25 mT/s, find the induced electric field at the following points in the xy-plane: (a) the origin; (b) (10 cm, 0); (c) (15 cm, 15 cm); and (d) (−15 cm, 20 cm).

11. Consider again the square conducting loop in Figure 26.9(b). Assume that its front edge reaches the y-axis at time $t = 0$ with a speed v_0 and that it is accelerating at a constant rate a_0. Find the emf induced in the loop as a function of time.

Problems 12–14 refer to the diagram below, in which a straight conducting rail slides to the right with a constant speed v over the conducting rails as shown. Assume the magnetic field is uniform and has a magnitude B, and a fixed resistor R is situated on the left end of the rail section. Except for the resistor, assume negligible resistance in all conductors.

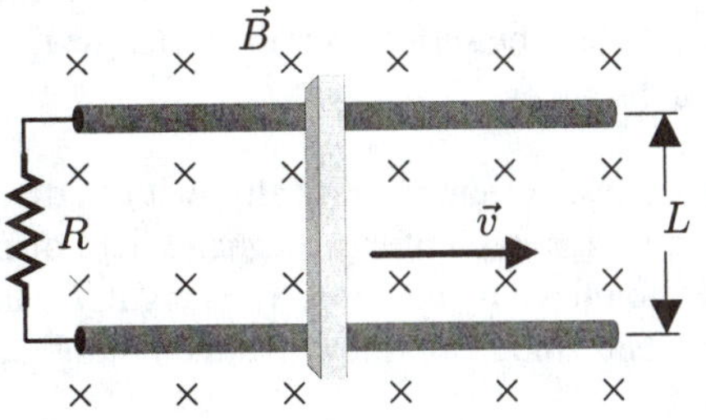

12. Find the current induced in the closed loop (formed by the fixed rails and moving rail). What is the direction of that current?

13. How much force must an external agent supply to keep the sliding rail moving with a constant speed?

14. Compare the rate at which work is done by the external agent (described in the previous problem) with the rate at which energy is dissipated in the resistor.

In Problems 15–16, consider the following situation. A long straight wire lies along the y-axis, with current flowing in the $+y$-direction. At a certain time, the current has a magnitude 4.50 A and is decreasing at a rate of 1.00 A/s. A square loop of wire 5.00 cm on a side lies in the xy-plane. The edges of the square are parallel to the x- and y-axes, with the two vertical edges along the lines $x = 2.50$ cm and $x = 7.50$ cm.

15. Find the magnetic flux through the square when the current in the straight wire is 4.50 A.

16. At what rate is the flux changing? Use this result to find the emf induced in the square and the direction of the induced current.

17. A long straight wire lies along the y-axis, with constant current I_0 flowing in the $+y$-direction. A conducting rod parallel to the x-axis has its two ends at $x = x_0$ and $x = x_0 + a$. It moves in the $+y$-direction with a constant speed v. What is the emf induced between the two ends of the rod?

18. An airplane with a wingspan of 35 m flies horizontally at a speed of 800 km/h in a region where Earth's magnetic field is vertical and has a magnitude 5.0×10^{-5} T. What is the emf induced between the wing tips? Is this something that should concern the pilot? Explain.

In Problems 19–20, consider a single circular loop of conducting wire, with area A, completely inside a solenoid, with the loop perpendicular to the solenoid's axis. The solenoid has n turns of wire per unit length.

19. If the current in the solenoid follows the relation $I = I_0 \cos(\omega t)$ (where I_0 and ω are constants), what is the induced emf in the loop as a function of time?

20. Now assume a steady current I_0 in the solenoid. Find the induced emf in the loop as a function of time if it is rotated about its diameter with angular velocity ω.

21. Consider the magnetic field given by

$$\vec{B}(x,y,z) = \left\langle z^3 e^{-(x^2+y^2)}, x+y+z, xyz \right\rangle .$$

Compute the magnetic flux for this field for each of the following surfaces: (a) The surface S is the triangle with vertices $(1,0,2)$, $(1,1,2)$, and $(0,1,2)$. (b) The surface A is the circle in the xz-plane of radius 2 centered at the origin. (c) The surface S is the rectangle in the yz-plane given by $[1,3]\times[-2,0]$. Assume SI units for $\vec{B}$ and distances.

22. A rectangular conducting loop is oriented vertically, with vertical height h and width a. The loop is dropped from rest into a horizontal uniform magnetic field of magnitude B, with the field aligned perpendicular to the surface outlined by the loop. The height h is large enough so that the trailing (top) edge of the loop remains outside the field. Show that the falling loop reaches a terminal speed

$$v_t = \frac{mgR}{B^2 a^2}$$

where R is the resistance of the loop and m is its mass.

23. Assume the generating coil shown in Figure 26.11(a) has 150 turns of wire and is turned at 100 Hz. The coil is a square 25 cm on a side. Find the maximum emf that can be induced in this coil in (a) Earth's magnetic field, with a magnitude 5.0×10^{-5} T, and (b) the field of a strong laboratory electromagnet, with a field magnitude 7.5 T.

24. Assuming a uniform magnetic field $\vec{B} = B_0\,\hat{\jmath}$, show that Gauss's law for magnetism holds for the following surfaces: (a) a sphere immersed in the field, (b) a sphere half in the field and half out of the field.

25. Assuming a uniform magnetic field $\vec{B} = B_0\,\hat{\imath}$, show that Gauss's law for magnetism holds for the following surfaces: (a) a cylinder with its axis parallel to the x-axis, (b) a cylinder with its axis parallel to the y-axis.

26. The magnetic field of a real object (say a bar magnet) is difficult to define symbolically. Consider instead a rough equivalent in electrostatics: a point charge $+Q$ at $(0,a,0)$ and a point charge $-Q$ at $(0,-a,0)$. Using numerical methods, integrate the electric flux over a spherical surface, of radius $2a$ centered at the origin, and show that Gauss's law holds.

27. A circular coil of area 0.05 m^2 and 50 turns of wire is rotated 20 times per second about a diameter parallel to the x-axis in a uniform magnetic field

$$\vec{B} = B_0(\hat{\imath} + \hat{\jmath})$$

where $B_0 = 850$ mT. What is the maximum emf induced in the coil?

28. The coil of an electric motor has 120 turns of wire and a cross-sectional area 0.095 m^2. The coil carries a current of 12.0 A and lies with its axis parallel to a uniform magnetic field of 600 mT. What is the maximum torque produced? What is the average torque produced?

26.2 Inductance

1. Show from the basic relation $\Phi_B = LI$ [not from Equation (26.16)] that the SI units of inductance are equivalent to

$$1\,\text{H} = 1\,\text{V·s/A} = 1\,\text{V·s}^2/\text{C}.$$

2. Show that the SI units for the RL time constant $\tau = L/R$ reduce to seconds.

3. For the RL circuit shown in Figure 26.15, show that the time required for the current to reach a fraction f of its final (steady-state) value is $t = -\tau \ln(1 - f)$.

4. You require a 1.0-mH inductor in the form of a solenoid of length 10 cm. Specify the other parameters of the solenoid. [*Note*: more than one answer is possible.]

5. An electrical circuit must form a closed loop, and therefore there is some inductance in the circuit, intended or not. Consider a simple circuit with just a 18.0-V battery and 100-Ω resistor. If the circuit lies flat on the table in a circular loop of radius 20 cm, estimate the inductance of the loop. How does your estimate compare with a 1-mH inductor?

6. You want to make a 0.50-mH inductor using a solenoid with 900 turns of wire per meter. (a) What should the length of the solenoid be if its radius is 3.50 cm? (b) What should the radius of the solenoid be if its length is 25 cm?

7. You wish to make a toroid with an inductance of 40 mH. The toroid's cross section is square, 10 cm by 10 cm, and it is to have 1600 turns. What is the toroid's inner radius?

Problems 8–10 refer to the RL circuit shown in Figure 26.15. Assume $\mathcal{E} = 25.0$ V, $R = 150\,\Omega$, and $L = 25.0$ mH.

8. What is the maximum current in the circuit?

9. At what time (after the switch is closed) is the current in the circuit (a) 10% (b) 50% (c) 90 % (d) 99% of its maximum value?

10. What is the current in the circuit 10 μs after the switch is closed?

11. In the diagram below, the switch is initially in the position shown and then thrown to position 1. With $\mathcal{E} = 12.0$ V, $R = 105\,\Omega$ and $L = 145$ mH, what is the current after (a) 1 ms? (b) 10 s? (c) After 10 s has elapsed, the switch is thrown to position 2. The switch used is a "make before break" switch, which completes the circuit in position 2 before disconnecting the circuit in position 1. What is the current in the inductor as a function of time after the switch is thrown to position 2?

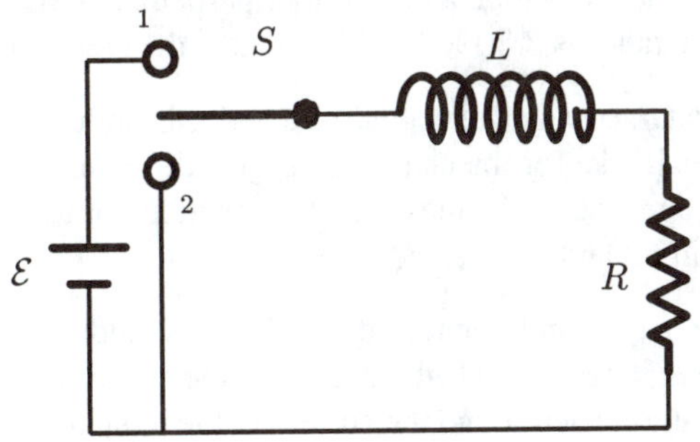

Use the figure below for Problems 12–13. Assume $\mathcal{E} = 36.0$ V, $R_1 = R_2 = 100\,\Omega$, $R_3 = 250\,\Omega$, and $L = 250$ mH.

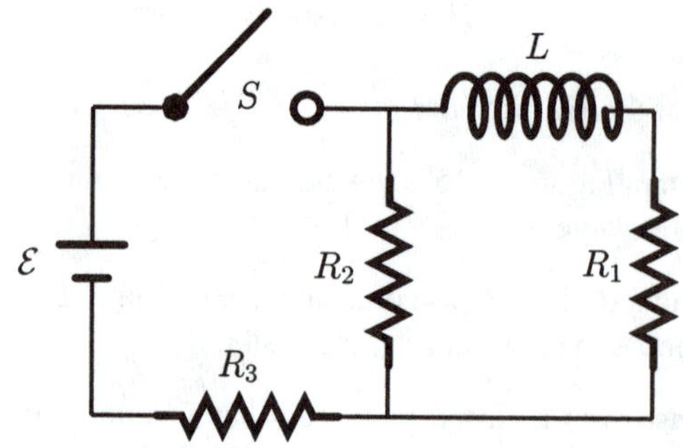

12. Initially there is no current in the circuit. Find the current in each resistor immediately after the switch S is closed and a long time later.

13. After the switch has been closed for a long time, it is opened. Find the current in each resistor immediately after the switch is opened.

14. A 45-μF capacitor carries a 240-μC charge. If this capacitor is discharged through a 1.2-H inductor, (a) how much time after the connection is made is the maximum current reached in the inductor, and (b) what is the value of the maximum current?

15. A 250-nF capacitor is discharged across a 1.28-mH inductor. What is the period of the charge oscillations?

16. A 0.500-μF capacitor is charged and connected to an inductor. The resulting current in the inductor is given by the function

$$I = I_0 \sin(\omega t)$$

with $I_0 = 5.90$ mA and $\omega = 2.54 \times 10^4\ \text{s}^{-1}$. Find (a) the initial potential difference and charge on the capacitor, and (b) the inductance of the inductor.

17. In an LC circuit, the values of the capacitance and inductance are 2.50 μF and 1.04 H, respectively. At a certain time, the charge on the capacitor carries its maximum value of 20.0 μC. What is the charge on the capacitor exactly 1.00 ms later?

18. LC circuits are used in radios to tune a circuit to a particular frequency. If the desired frequency is 107.7 MHz and the inductance is 3.50×10^{-6} H, what is the capacitance?

19. A 40-nC capacitor is charged and connected to a 3.5 mH inductor. (a) How much time after the connection is made does it take for the charge to reach half its initial value? (b) How much time after the connection is made does it take for the current in the inductor to reach half its maximum value? (c) Explain why the times in (a) and (b) are not the same.

20. A 240-μF capacitor is charged and connected to a 2.56-mH inductor. (a) When is the first time after the connection is made that exactly half the energy in the circuit is magnetic and half is electric? (b) What are the charge on the capacitor and the current in the inductor at that time?

21. (a) What are the SI units for energy density? (b) Show that the SI units on the right side of the equation

$$u_B = \frac{B^2}{2\mu_0}$$

reduce to the proper SI units for energy density.

22. What is the energy density in a uniform magnetic field of 10 T? If a cubical room 5.0 m on a side were filled with this magnetic field, what is the total energy in the field? (Note that this is an extreme and not very realistic situation.)

23. A solenoid has 2400 turns per meter, a length of 25.0 cm, and a diameter of 3.45 cm. the solenoid carries a current of 7.65 A. (a) Find the magnetic field inside the solenoid. (b) Use your answer to (a) to find the energy density in the solenoid, and use this result to find the total energy stored in the magnetic field. (c) Compute the energy in the solenoid using $U_B = \frac{1}{2}LI^2$, and show that your answer agrees with the answer in (b).

24. In an LC circuit, a 500-μF capacitor is connected to a 1.25-H inductor. At a certain time, the charge on the capacitor is zero and the current in the inductor is 0.342 A. (a) How much time after this will the charge on the capacitor reach its maximum value? (b) What is the total energy in the circuit? (c) What is the maximum charge on the capacitor?

25. A solenoid with 1200 turns of wire per meter has wire that can handle a maximum current of 3.50 A. The solenoid's length is 12.0 cm and its cross-sectional area is 0.015 m^2. (a) What is the solenoid's inductance? (b) What is the maximum amount of magnetic energy that can be stored in the solenoid? (c) What is its energy density when the maximum energy is stored?

26. In an RLC circuit, the 10.0 μF capacitor initially carries a charge of 250 μC, and there is no current in the circuit. The other circuit elements carry numerical values $L = 0.150$ H and $R = 210\,\Omega$. (a) Is the resulting charge flow underdamped, overdamped, or critically damped? (b) What is the charge on the capacitor 1.00 ms and 100 ms after initial configuration described above?

27. Repeat part (a) of the preceding problem if R is changed to 350 Ω.

28. Find an expression for the current as a function of time in an underdamped RLC circuit.

29. Find an expression for the current as a function of time in a critically damped RLC circuit.

30. In a real LC circuit, some small but unavoidable resistance exists. Suppose we start an LC oscillation with a charge of 1.35 mC on a 400-μF capacitor with a 0.150-H inductor, and the total resistance in the circuit is 0.950 Ω. What fraction of the energy in the circuit is last after (a) 100 oscillations? (b) 10 seconds?

31. Show that the ratio of the angular frequency ω_u in an underdamped RLC circuit to the angular frequency ω_0 of the same circuit without resistance is

$$\frac{\omega_u}{\omega_0} = \sqrt{1 - \frac{CR^2}{4L}}.$$

Explain what happens to this ratio (and what happens physically in the circuit) when

$$\text{(a)} \quad \frac{CR^2}{4L} = 1 \quad \text{and} \quad \text{(b)} \quad \frac{CR^2}{4L} > 1.$$

Chapter 27

Vector Calculus and Maxwell's Equations

In the first half of this chapter, we pull together many of the mathematical concepts from previous chapters into generalizations of the fundamental theorem of calculus. These generalizations involve line integrals, normal line integrals, double integrals, and surface integrals along with gradient, divergence, and curl. In the first section, we deal with results relevant to working in the plane, and in the second section, we look at results relevant to working in space.

In the second half of the chapter, we pull together many of the physical concepts concerning electricity and magnetism into *Maxwell's equations*. In Section 27.3, we use the results of Sections 27.1 and 27.2 to recast the integral statements of Gauss's law, Ampère's law, and Faraday's law as differential equations. These forms of the fundamental relations about electricity and magnetism allow us to explore *electromagnetic waves*. In the final section, we briefly survey two major areas of modern physics, specifically *relativity* and *quantum physics*. Each of these has roots in electromagnetism.

27.1 Vector Calculus in the Plane

27.1.1 Overview

The second fundamental theorem for definite integrals relates a definite integral of a continuous function $g : [a, b] \to \mathbb{R}$ to the outputs of an antiderivative G at the endpoints a and b. Specifically, if $G'(x) = g(x)$ for all x in $[a, b]$, then

$$\int_a^b g(x)\,dx = G(b) - G(a).$$

We can write this without reference to g as

$$\int_a^b G'(x)\,dx = G(b) - G(a).$$

The fundamental theorem for line integrals has an analogous form: If $\vec{\nabla} V(\vec{r}) = \vec{F}(\vec{r})$ for all $\vec{r}$ in a region containing the curve C, then

$$\int_C \vec{F} \cdot d\vec{s} = V(\vec{r}_1) - V(\vec{r}_2)$$

where $\vec{r}_1$ and $\vec{r}_2$ are the endpoints of the curve. Stated without reference to $\vec{F}$, this is

$$\int_C \vec{\nabla} V \cdot d\vec{s} = V(\vec{r}_1) - V(\vec{r}_2).$$

There are analogs of the fundamental theorem for double integrals, triple integrals, and surface integrals. In finding the analogs of these theorems for other types of integrals, we need to consider five things:

- the set of inputs over which we are integrating;
- the boundary of the set of inputs;
- the type of function in the integrand;
- the relevant notion of derivative; and
- the corresponding type of function for an "antiderivative ."

For example, a double integral involves integrating over a region in the plane. For the "nice" region shown in Figure 27.1, the boundary is closed curve. The integrand is a function of two variables $f : \mathbb{R}^2 \to \mathbb{R}$. Two options exist for the right notion of derivative, namely the scalar curl and the divergence. Each of these involves a vector field as the relevant notion of antiderivative. The biggest question we face is how to deal with the fact that the boundary of the integration region is a curve, not just two endpoints. The answer is that the correct fundamental theorems involve integrals on both sides of the equality. For a double integral there are two fundamental theorems. The first involves the line integral of a scalar curl over the boundary curve; this result is known as *Green's theorem.* The second involves the normal line integral of a divergence over the boundary curve; this result is known as *the planar divergence theorem.*

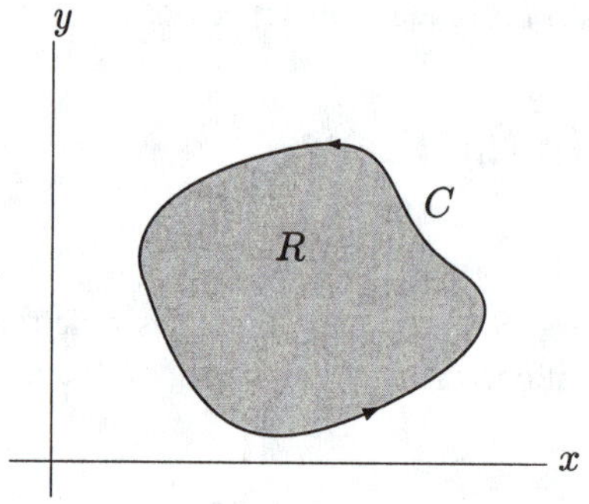

Figure 27.1. Plane region as the region of integration for a double integral.

27.1.2 Green's Theorem

Green's theorem relates the double integral of a function $f : \mathbb{R}^2 \to \mathbb{R}$ over a planar region to the line integral of an "antiderivative" over the closed curve that is the boundary of the region. Let's start with a version of the theorem relevant to a "nice" region R, such as the one shown in Figure 27.1. By "nice," we mean that the boundary is a simple closed curve C. We orient C in the counterclockwise direction as indicated in the figure.

Theorem 27.1 (Green's theorem). *Let R be a planar region bounded by a simple closed curve C oriented counterclockwise. Let $f : \mathbb{R}^2 \to \mathbb{R}$ be continuous. If $\vec{F} : \mathbb{R}^2 \to \mathbb{R}^2$ is a vector field such that curl $\vec{F}(x,y) = f(x,y)$ for all (x,y) in R, then*

$$\iint_R f(x,y)\, dA = \oint_C \vec{F} \cdot d\vec{s}.$$

In terms of the components $\vec{F}(x,y) = \langle P(x,y), Q(x,y)\rangle$, this is equivalent to

$$\iint_R \left(\frac{\partial Q}{\partial x} - \frac{\partial P}{\partial y}\right) dA = \oint_C \langle P, Q\rangle \cdot d\vec{s}.$$

Proof. We give a proof for a special case in order to simplify some of the details. Consider a plane region R, for which the boundary curve C can be split up in two different ways as indicated in Figure 27.2. In one view, the curve C comprises the graphs of two functions $y = \alpha(x)$ and $y = \beta(x)$ for x in the interval $[a, b]$. The region R can thus be described by the bounds

$$a \le x \le b \qquad \text{and} \qquad \alpha(x) \le y \le \beta(x). \tag{27.1}$$

In the other view, the curve C comprises the graphs of the two functions $x = \gamma(y)$ and $x = \delta(y)$ for y in the interval $[c, d]$. This gives a description of the region R by the bounds

$$\gamma(y) \le x \le \delta(y) \qquad \text{and} \qquad c \le y \le d. \tag{27.2}$$

To make use of the two decriptions of R, we invoke linearity of double integrals to write

$$\iint_R \left(\frac{\partial Q}{\partial x} - \frac{\partial P}{\partial y}\right) dA = \iint_R \frac{\partial Q}{\partial x}\, dA - \iint_R \frac{\partial P}{\partial y}\, dA. \tag{27.3}$$

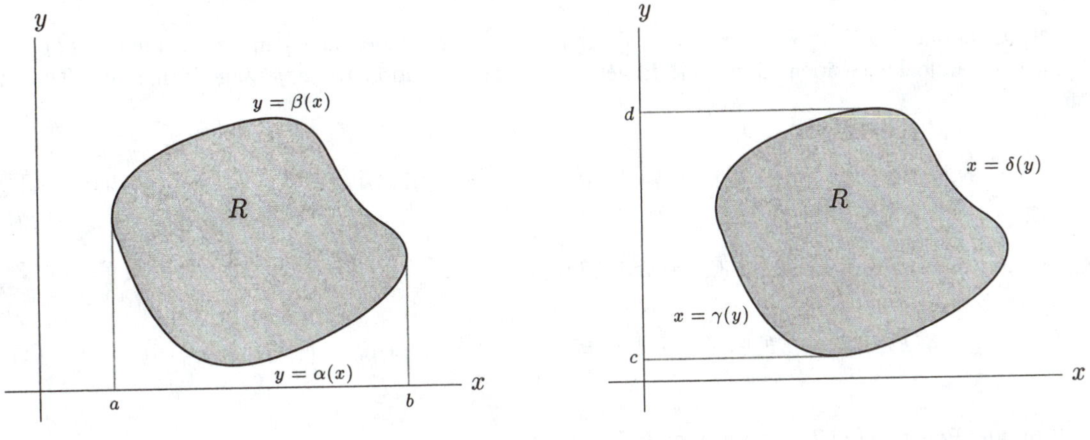

(a) Boundary curve C viewed as the graphs of $\alpha(x)$ and $\beta(x)$

(b) Boundary curve C viewed as the graphs of $\gamma(y)$ and $\delta(y)$

Figure 27.2. Two views of the boundary of R.

We analyze each of the integrals on the right side of Equation (27.3) in turn. Our strategy is to show that

$$\iint_R \frac{\partial Q}{\partial x}\, dA = \oint_C \vec{F_1} \cdot d\vec{s}$$

where $\vec{F_1}(x, y) = \langle 0, Q(x, y)\rangle$ and

$$\iint_R \frac{\partial P}{\partial y}\, dA = -\oint_C \vec{F_2} \cdot d\vec{s}$$

where $\vec{F_2}(x, y) = \langle P(x, y), 0\rangle$.

Using the bounds for R given in Equation (27.2) in an iterated integral,

$$\iint_R \frac{\partial Q}{\partial x}\, dA = \int_c^d \int_{\gamma(y)}^{\delta(y)} \frac{\partial Q}{\partial x}\, dx\, dy.$$

Using the second fundamental theorem for definite integrals on the x-integration, we have

$$\iint_R \frac{\partial Q}{\partial x}\, dA = \int_c^d (Q(\delta(y), y) - Q(\gamma(y), y))\, dy = \int_c^d Q(\delta(y), y)\, dy - \int_c^d Q(\gamma(y), y)\, dy. \tag{27.4}$$

The two definite integrals here are equivalent to the line integral of the vector field $\vec{F_1}(x, y) = \langle 0, Q(x, y)\rangle$ over the curve C. To see this, think of splitting C into two pieces, C_1 as the graph of $\delta(y)$, and C_2 as the graph of $\gamma(y)$. Parametrize C_1 by $\vec{r_1}(y) = \langle \delta(y), y\rangle$ for y in $[c, d]$. Note that this traces out C_1 in an orientation consistent with the counterclockwise orientation of C. By defintion, the line integral of $\vec{F_1}$ over C_1 is

$$\int_{C_1} \vec{F_1} \cdot d\vec{s} = \int_c^d \langle 0, Q(\delta(y), y)\rangle \cdot \langle \delta'(y), 1\rangle\, dy = \int_c^d Q(\delta(y), y)\, dy. \tag{27.5}$$

Next, parametrize C_2 by $\vec{r_2}(y) = \langle \gamma(y), y\rangle$ for y in $[c, d]$. This traces out C_2 in a direction that is opposite to the counterclockwise orientation of C. To get the correct orientation, we let y range from d to c. Thus, we have

$$\int_{C_2} \vec{F_1} \cdot d\vec{s} = \int_d^c \langle 0, Q(\gamma(y), y)\rangle \cdot \langle \gamma'(y), 1\rangle\, dy = \int_d^c Q(\gamma(y), y)\, dy = -\int_c^d Q(\gamma(y), y)\, dy. \tag{27.6}$$

Combining the results of Equations (27.5) and (27.6), we have

$$\oint_C \vec{F_1} \cdot d\vec{s} = \int_{C_1} \vec{F_1} \cdot d\vec{s} + \int_{C_2} \vec{F_1} \cdot d\vec{s} = \int_c^d Q(\delta(y), y)\, dy - \int_c^d Q(\gamma(y), y)\, dy. \tag{27.7}$$

Substituting Equation (27.7) into Equation (27.4) gives

$$\iint_R \frac{\partial Q}{\partial x}\, dA = \int_C \vec{F_1} \cdot d\vec{s}. \tag{27.8}$$

Now return to Equation (27.3), and consider the second double integral on the right side. With the description of R given by the bounds in Equation (27.1), the double integral can be expressed as

$$\iint_R \frac{\partial P}{\partial y}\, dA = \int_a^b \int_{\alpha(x)}^{\beta(x)} \frac{\partial P}{\partial y}\, dy\, dx. \tag{27.9}$$

Using an argument similar to that above, we can show

$$\iint_R \frac{\partial P}{\partial y}\, dA = -\oint_C \vec{F}_2 \cdot d\vec{s} \tag{27.10}$$

where $\vec{F}_2(x, y) = \langle P(x, y), 0\rangle$.

Substituting from Equations (27.8) and (27.8) into Equation (27.3) gives

$$\iint_R \left(\frac{\partial Q}{\partial x} - \frac{\partial P}{\partial y}\right) dA = \oint_C \vec{F}_1 \cdot d\vec{s} + \oint_C \vec{F}_2 \cdot d\vec{s} = \oint_C \left(\vec{F}_1 + \vec{F}_2\right) \cdot d\vec{s} = \oint_C \langle P(x,y), Q(x,y)\rangle \cdot d\vec{s}.$$

This completes the proof for the relatively simple type of region we have assumed. □

Example 27.1. Verify Green's theorem for $\vec{F}(x,y) = \langle P(x,y), Q(x,y)\rangle = \langle xy, y^2\rangle$ and the rectangle $R = [0,2] \times [0,2]$ by directly computing

$$\iint_R \left(\frac{\partial Q}{\partial x} - \frac{\partial P}{\partial y}\right) dA \qquad \text{and} \qquad \oint_C \langle P(x,y), Q(x,y)\rangle \cdot d\vec{s}.$$

Solution. For the double integral, first compute

$$\frac{\partial Q}{\partial x} - \frac{\partial P}{\partial y} = 0 - x = -x.$$

Thus,

$$\iint_R \left(\frac{\partial Q}{\partial x} - \frac{\partial P}{\partial y}\right) dA = \int_0^2 \int_0^2 (-x)\, dx\, dy = -\int_0^2 dy \int_0^2 x\, dx = -4.$$

For the line integral, the boundary consists of four line segments as shown in Figure 27.3.

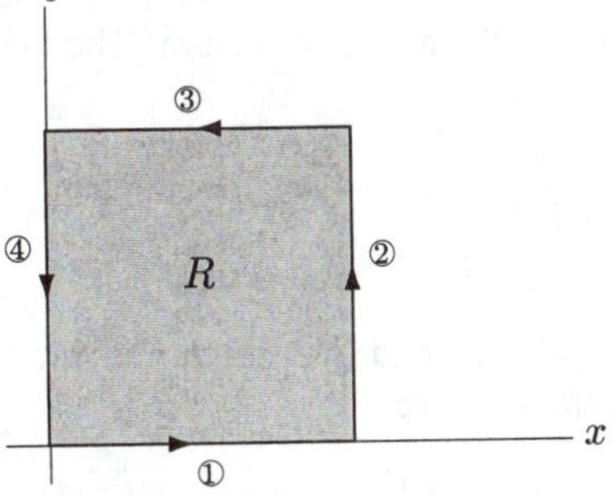

Figure 27.3. Region and boundary curve for Example 27.1.

Let's evaluate a line integral for each segment and then sum the four results for the total:

$$\oint_C \vec{F} \cdot d\vec{s} = \int_{C_1} \vec{F} \cdot d\vec{s} + \int_{C_2} \vec{F} \cdot d\vec{s} + \int_{C_3} \vec{F} \cdot d\vec{s} + \int_{C_4} \vec{F} \cdot d\vec{s}.$$

We do one of these explicitly, and quote results for the other three. Consider the line segment labeled 2 from $(2, 0)$ to $(2, 2)$. This is parametrized by $\vec{r}_2(y) = \langle 2, y \rangle$ for y in $[0, 2]$. The line segment is traced out in the direction consistent with a counterclockwise orientation for the boundary. We compute $\vec{r}_2'(y) = \langle 0, 1 \rangle$ and $\vec{F}(2, y) = \langle 2y, y^2 \rangle$. Thus,

$$\int_{C_2} \vec{F} \cdot d\vec{s} = \int_0^2 \langle 2y, y^2 \rangle \cdot \langle 0, 1 \rangle \, dy = \int_0^2 y^2 \, dy = \frac{8}{3}.$$

Similar computations give us

$$\int_{C_1} \vec{F} \cdot d\vec{s} = 0, \qquad \int_{C_3} \vec{F} \cdot d\vec{s} = -4, \qquad \text{and} \qquad \int_{C_4} \vec{F} \cdot d\vec{s} = -\frac{8}{3}.$$

With these results, we have

$$\oint_C \vec{F} \cdot d\vec{s} = 0 + \frac{8}{3} - 4 - \frac{8}{3} = -4.$$

The double integral and the line integral both have value -4, as Green's theorem tells us they must. △

In Example 27.1, the line integral is more difficult to compute than the double integral. This is often the case, so Green's theorem has practical value as a tool for evaluating line integrals along closed curves.

Green's thereom is also one key to our argument for the converse of Theorem 22.3. This converse is given in the following theorem.

Theorem 27.2. *Let $\vec{F} : \mathbb{R}^2 \to \mathbb{R}^2$ have continuous partial derivatives for all (x, y) in $\mathbb{R}^2$. If $\operatorname{curl} \vec{F}(x, y) = 0$ for all (x, y) in $\mathbb{R}^2$, then there exists a potential function for $\vec{F}$. That is, there is a function $V : \mathbb{R}^2 \to \mathbb{R}$ such that $\vec{\nabla} V(x, y) = \vec{F}(x, y)$ for all (x, y) in $\mathbb{R}$.*

Proof. This proof has two pieces. In the first piece, we establish that $\vec{F}$ is *conservative* in the sense defined in Chapter 7. This means that line integrals of $\vec{F}$ are *path-independent*. In the second piece, we take advantage of path-independence to construct a potential function V in terms of a line integral.

Let (x_1, y_1) and (x_2, y_2) be any two points in $\mathbb{R}^2$. Let C_1 and C_2 be two curves, each starting at (x_1, y_1) and ending at (x_2, y_2). Define the closed curve $\tilde{C}$ as starting at (x_1, y_1), following C_1 to (x_2, y_2), and then returning to (x_1, y_2) by following C_2 in the reverse direction. The line integral of $\vec{F}$ on $\tilde{C}$ can be split into two line integrals as

$$\oint_{\tilde{C}} \vec{F} \cdot d\vec{s} = \int_{C_1} \vec{F} \cdot d\vec{s} - \int_{C_2} \vec{F} \cdot d\vec{s}. \tag{27.11}$$

On the other hand, we can use Green's theorem to evaluate the line integral over $\tilde{C}$. Let R be the region bounded by $\tilde{C}$. From Green's theorem, we conclude

$$\oint_{\tilde{C}} \vec{F} \cdot d\vec{s} = \iint_R \operatorname{curl} \vec{F} \, dA = \iint_R 0 \, dA = 0. \tag{27.12}$$

Using Equation (27.12) in Equation (27.11), we conclude

$$\int_{C_1} \vec{F} \cdot d\vec{s} = \int_{C_2} \vec{F} \cdot d\vec{s}.$$

This establishes the path-independence of line integrals for $\vec{F}$, so $\vec{F}$ is conservative.

To define a function $V : \mathbb{R}^2 \to \mathbb{R}$, we first declare $V(0,0) = 0$. (We are essentially constructing an antiderivative here. The choice $V(0,0) = 0$ is akin to choosing a particular value for the choice of constant we have in any antiderivative problem.) For a generic input (x, y), we let C be a path from $(0,0)$ to (x, y), and then define

$$V(x,y) = \int_{C} \vec{F} \cdot d\vec{s}.$$

An important point here is that the output $V(x, y)$ depends only on the input (x, y) and *not* on the choice of the curve C, because of the path-independence we established in the previous paragraph.

It remains to show that the gradient of V is equal to $\vec{F}$. Let $\vec{F}(x,y) = \langle P(x,y), Q(x,y)\rangle$. We must show that

$$\vec{\nabla} V(x,y) = \left\langle \frac{\partial V}{\partial x}, \frac{\partial V}{\partial y} \right\rangle = \langle P(x,y), Q(x,y)\rangle = \vec{F}(x,y)$$

or

$$\frac{\partial V}{\partial x} = P(x,y) \qquad \text{and} \qquad \frac{\partial V}{\partial y} = Q(x,y).$$

We verify the first equation holds, and leave the second as an exercise.

Consider a specific choice for a curve C from $(0,0)$ to (x, y). First, go from $(0,0)$ to $(0, y)$ along the y-axis, and then go from $(0, y)$ to (x, y) along a line segment parallel to the x-axis. The first line segment can be parametrized by $\langle 0, t\rangle$ for t in $[0, y]$, and the second line segment can be parametrized by $\langle t, y\rangle$ for t in $[0, x]$. With these parametrizations, we have

$$\begin{aligned} V(x,y) = \int_{C} \vec{F} \cdot d\vec{s} &= \int_0^y \langle P(0,t), Q(0,t)\rangle \cdot \langle 0,1\rangle\, dt + \int_0^x \langle P(t,y), Q(t,y)\rangle \cdot \langle 1,0\rangle\, dt \\ &= \int_0^y Q(0,t)\, dt + \int_0^x P(t,y)\, dt. \end{aligned}$$

Now consider the partial derivative with respect to x of this last expression. The derivative of the first term is zero because that term is constant in x. For the second term, apply the first fundamental theorem of calculus to conclude that the derivative is $P(x, y)$. We thus have

$$\frac{\partial V}{\partial x} = P(x,y).$$

A similar argument gives

$$\frac{\partial V}{\partial y} = Q(x,y).$$

Together, these give us $\vec{\nabla} V(x,y) = \vec{F}(x,y)$. Thus, V is a potential function for $\vec{F}$.

□

27.1.3 The Planar Divergence Theorem

The planar divergence theorem relates the double integral of a function $f : \mathbb{R}^2 \to \mathbb{R}$ over a planar region to the normal line integral of an "antiderivative" over the closed curve that is the boundary of the region. This is quite similar to Green's theorem, but differs in that a *normal* line integral is involved and, as we shall see, the relevant notion of derivative is the divergence rather that the scalar curl.

Theorem 27.3 (Planar divergence theorem). *Let R be a planar region bounded by a simple closed curve C. Let $f : \mathbb{R}^2 \to \mathbb{R}$ be continuous. If $\vec{F} : \mathbb{R}^2 \to \mathbb{R}^2$ is a vector field such that div $\vec{F}(x,y) = f(x,y)$ for all (x,y) in R, then*

$$\iint\limits_R f(x,y)\,dA = \iint\limits_R \vec{\nabla}\cdot\vec{F}\,dA = \oint\limits_C \vec{F}\cdot\hat{n}\,ds.$$

In terms of the components $\vec{F}(x,y) = \langle P(x,y), Q(x,y)\rangle$, this is equivalent to

$$\iint\limits_R \left(\frac{\partial P}{\partial x} + \frac{\partial Q}{\partial y}\right) dA = \oint\limits_C \langle P, Q\rangle\cdot\hat{n}\,ds.$$

Proof. Our proof uses Green's theorem applied to the new vector field $\vec{G}(x,y) = \langle -Q(x,y), P(x,y)\rangle$. Note that at each input (x,y), the output $\vec{F}(x,y)$ is perpendicular to the output $\vec{G}(x,y)$ because $\vec{F}(x,y)\cdot\vec{G}(x,y) = 0$. Thus, for each input on a curve, the normal component of $\vec{F}(x,y)$ is equal to the tangential component of $\vec{G}(x,y)$, as illustrated in Figure 27.4. Hence, the normal line integral of $\vec{F}$ on a curve is equal to the line integral of $\vec{G}$ for the same curve. That is,

$$\int\limits_C \vec{F}\cdot\hat{n}\,ds = \int\limits_C \vec{G}\cdot d\vec{s}. \tag{27.13}$$

Also note that that

$$\operatorname{curl}\vec{G}(x,y) = \frac{\partial}{\partial x}(P) - \frac{\partial}{\partial y}(-Q) = \frac{\partial P}{\partial x} + \frac{\partial Q}{\partial y} = \vec{\nabla}\cdot\vec{F}(x,y). \tag{27.14}$$

Applying Green's theorem to $\vec{G}(x,y)$ gives

$$\iint\limits_R \operatorname{curl}\vec{G}\,dA = \oint\limits_C \vec{G}\cdot d\vec{s}.$$

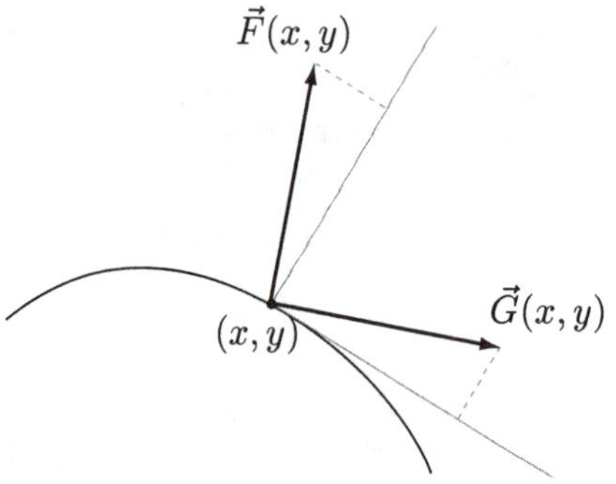

Figure 27.4. At an input on a curve, the normal component of $\vec{F}$ is equal to the tangential component of $\vec{G}$.

Using the result in Equation (27.13) on the right side and the result in Equation (27.14) to substitute for the left side, we can rewrite this as

$$\iint_R \left(\frac{\partial P}{\partial x} + \frac{\partial Q}{\partial y} \right) dA = \oint_C \vec{F} \cdot \hat{n}\, ds.$$

This gives us the result we sought to prove. □

Recall that the normal line integral of a vector field $\vec{F}$ in the plane has a physics interpretation as the flux of $\vec{F}$ throught the relevant curve. The planar divergence theorem can thus be interpreted as saying that the flux of a vector field $\vec{F}$ through a closed curve is equal to the double integral of the divergence of $\vec{F}$ over the region enclosed by the curve. If the divergence is positive at all points in the region, then the double integral has a positive value, and hence the flux must be positive. If the divergence is negative at all points in the region, then the flux must be negative. Finally, if the divergence is zero at all points in the region, then the flux is zero.

27.2 Vector Calculus in Space

Both Green's theorem and the planar divergence theorem can be generalized to vector fields $\vec{F} : \mathbb{R}^3 \to \mathbb{R}^3$. The generalization of Green's theorem is known as *Stokes's theorem.* The generalization of the planar divergence theorem is known as the (usual) *divergence theorem.*

27.2.1 Stokes's Theorem

The setup for Stokes's theorem can be viewed as a "lifting out of the plane" of the various elements in Green's theorem. The planar region R is lifted into a surface S, with the double integral over R translating into a surface integral over S. The boundary of the planar region R is lifted into a curve C in space that forms the edge of the surface S. The relevant type of vector field $\vec{F}$ has inputs in $\mathbb{R}^3$ and outputs in $\mathbb{R}^3$. The scalar curl translates into the usual curl, $\vec{\nabla} \times \vec{F}$.

To get the correct signs in Stokes's theorem, we need to specify an orientation for the curve C that forms the edge of the surface S. We say that C has **positive orientation** if the thumb of a right hand points to the same side of the surface as the surface normals when the fingers wrap around the curve in the orientation of the curve. Figure 27.5 shows an example of a curve that has positive orientation.

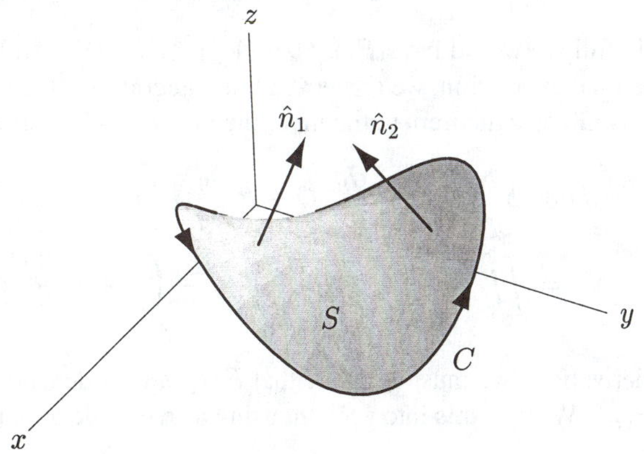

Figure 27.5. The curve C forming the edge of the surface S is shown with a positive orientation.

Theorem 27.4 (Stokes's theorem). *Let S be a surface in $\mathbb{R}^3$, with an edge given by a closed curve C with positive orientation. If the vector field $\vec{F} : \mathbb{R}^3 \to \mathbb{R}^3$ has components with continuous first partial derivatives, then*

$$\iint_S (\vec{\nabla} \times \vec{F}) \cdot \hat{n}\, dA = \oint_C \vec{F} \cdot d\vec{s}.$$

Proof. We give a proof for the special case in which the surface is the graph of a function $g : \mathbb{R}^2 \to \mathbb{R}$ for inputs in some region R of the plane. The graph of g can be parametrized as $\vec{r}(x, y) = \langle x, y, g(x, y)\rangle$ for (x, y) in the region R. It is straightforward to compute the normal vectors as

$$\vec{n}(x, y) = \vec{T}_x(x, y) \times \vec{T}_y(x, y) = \left\langle -\frac{\partial g}{\partial x}, -\frac{\partial g}{\partial y}, 1 \right\rangle.$$

The surface integral of $\vec{\nabla} \times \vec{F} = \langle P, Q, R\rangle$ over the surface thus becomes

$$\iint_S (\vec{\nabla} \times \vec{F}) \cdot \hat{n}\, dA = \iint_R \left[-\left(\frac{\partial R}{\partial y} - \frac{\partial Q}{\partial z}\right)\frac{\partial g}{\partial x} - \left(\frac{\partial P}{\partial z} - \frac{\partial R}{\partial x}\right)\frac{\partial g}{\partial y} + \left(\frac{\partial Q}{\partial x} - \frac{\partial P}{\partial y}\right)\right] dA. \quad (27.15)$$

Now turn to the line integral over the edge of the graph as given by the curve C. Let C_1 denote the projection of C into the xy-plane. The curve C_1 will bound the region R. If $\vec{r}_1(t) = \langle x(t), y(t)\rangle$ parametrizes C_1 for t in $[a, b]$, then $\vec{r}(t) = \langle x(t), y(t), g(x(t), y(t))\rangle$ will parametrize C, also for t in $[a, b]$. Using a chain rule, we compute

$$\vec{r}'(t) = \left\langle x'(t), y'(t), \frac{\partial g}{\partial x}x'(t) + \frac{\partial g}{\partial y}y'(t) \right\rangle.$$

Thus, the line integral of $\vec{F}$ over C is

$$\begin{aligned}
\oint_C \vec{F} \cdot d\vec{s} &= \int_a^b \left[Px'(t) + Qy'(t) + R\left(\frac{\partial g}{\partial x}x'(t) + \frac{\partial g}{\partial y}y'(t)\right)\right] dt \\
&= \int_a^b \left[\left(P + R\frac{\partial g}{\partial x}\right)x'(t) + \left(Q + R\frac{\partial g}{\partial y}\right)y'(t)\right] dt \\
&= \oint_{C_1} \langle P + R\frac{\partial g}{\partial x}, Q + R\frac{\partial g}{\partial y}\rangle \cdot d\vec{s}. \qquad (27.16)
\end{aligned}$$

[Note that if written out in full we would have $P(x(t), y(t), g(x(t), y(t)))$, with similar expressions for the inputs of Q and R.] In the final expression, we arrive at a line integral over the closed curve C_1 in the plane.

The next step is to apply Green's theorem to the final line integral in Equation (27.16). This gives

$$\begin{aligned}
\oint_C \vec{F} \cdot d\vec{s} &= \oint_{C_1} \left\langle P + R\frac{\partial g}{\partial x}, Q + R\frac{\partial g}{\partial y} \right\rangle \cdot d\vec{s} \\
&= \iint_R \left[\frac{\partial}{\partial x}\left(Q + R\frac{\partial g}{\partial y}\right) - \frac{\partial}{\partial y}\left(P + R\frac{\partial g}{\partial x}\right)\right] dA. \qquad (27.17)
\end{aligned}$$

In computing the partial derivatives, we must be aware that P, Q, and R depend on x and y both directly and indirectly through the $g(x, y)$. We take this into account using a chain rule to get

$$\frac{\partial}{\partial x}\left(Q + R\frac{\partial g}{\partial y}\right) = \frac{\partial Q}{\partial x} + \frac{\partial Q}{\partial z}\frac{\partial g}{\partial x} + R\frac{\partial^2 g}{\partial x \partial y} + \frac{\partial R}{\partial x}\frac{\partial g}{\partial y} + \frac{\partial R}{\partial z}\frac{\partial g}{\partial x}\frac{\partial g}{\partial y} \qquad (27.18)$$

and

$$\frac{\partial}{\partial y}\left(P + R\frac{\partial g}{\partial x}\right) = \frac{\partial P}{\partial y} + \frac{\partial P}{\partial z}\frac{\partial g}{\partial y} + R\frac{\partial^2 g}{\partial y \partial x} + \frac{\partial R}{\partial y}\frac{\partial g}{\partial x} + \frac{\partial R}{\partial z}\frac{\partial g}{\partial y}\frac{\partial g}{\partial x}. \tag{27.19}$$

For the integrand of the double integral in Equation (27.17), we need the difference between the right sides of Equations (27.18) and (27.19). Note that four terms cancel, leaving an integrand that consists of six terms. Furthermore, these six terms are precisely those in the double integral on the right side of Equation (27.15). Thus, we have shown that the surface integral is equal to the line integral. □

Example 27.2. Verify Stokes's theorem for $\vec{F}(x, y, z) = \langle z, x, y\rangle$ and the surface S as the upper hemisphere of radius 1 centered at the origin.

Solution. To verify Stokes's theorem, we compute both the surface integral and the line integral.

To compute the surface integral, we use the usual parametrization in terms of spherical coordinates:

$$\vec{r}(\phi, \theta) = \langle \sin\phi\cos\theta, \sin\phi\sin\theta, \cos\phi\rangle$$

with ϕ in the interval $[0, \pi/2]$ and θ in the interval $[0, 2\pi]$. From Example 23.13, we know the normal vectors for this parametrization are given by

$$\vec{n}(\phi, \theta) = \langle \sin^2\phi\,\cos\theta, \sin^2\phi\,\sin\theta, \cos\phi\,\sin\phi\rangle.$$

The surface integral involves the curl of $\vec{F}$, which we compute to be

$$\vec{\nabla} \times \vec{F}(x, y, z) = \langle 1, 1, 1\rangle.$$

The surface integral is thus

$$\iint_S (\vec{\nabla} \times \vec{F}) \cdot \hat{n}\, dA = \int_0^{2\pi}\int_0^{\pi/2} \langle 1, 1, 1\rangle \cdot \langle \sin^2\phi\,\cos\theta, \sin^2\phi\,\sin\theta, \cos\phi\,\sin\phi\rangle\, d\phi\, d\theta$$

$$= \int_0^{2\pi}\int_0^{\pi/2} \left(\sin^2\phi\,\cos\theta + \sin^2\phi\,\sin\theta + \cos\phi\,\sin\phi\right)\, d\phi\, d\theta.$$

It is a straightforward exercise to compute this double integral. The result is

$$\iint_S (\vec{\nabla} \times \vec{F}) \cdot \hat{n}\, dA = \pi.$$

To compute the line integral, we first note that the edge of the surface is the unit circle in the xy-plane. This is easily parametrized as $\vec{r}(\theta) = \langle \cos\theta, \sin\theta, 0\rangle$ for θ in $[0, 2\pi]$. For the line integral, we need the derivative $\vec{r}\,'(\theta) = \langle -\sin\theta, \cos\theta, 0\rangle$. We also need the outputs $\vec{F}(\vec{r}(\theta))$ along the curve:

$$\vec{F}(\vec{r}(\theta)) = \vec{F}(\cos\theta, \sin\theta, 0) = \langle 0, \cos\theta, \sin\theta\rangle.$$

The line integral is thus

$$\oint_C \vec{F} \cdot d\vec{s} = \int_0^{2\pi} \langle 0, \cos\theta, \sin\theta\rangle \cdot \langle -\sin\theta, \cos\theta, 0\rangle\, d\theta = \int_0^{2\pi} \cos^2\theta\, d\theta = \pi.$$

The surface integral and the line integral are equal which provides some verification of Stokes's theorem. △

We can make use of Stokes's theorem to elaborate on our interpretation of the curl in terms of rotation. Consider a velocity vector field $\vec{v} : \mathbb{R}^3 \to \mathbb{R}^3$ and the curl $\vec{\nabla} \times \vec{v}$ at a particular input (x_0, y_0, z_0). Let S be a small disk centered at the point (x_0, y_0, z_0) and oriented with normal vector $\hat{n}$ aligned to the curl $\vec{\nabla} \times \vec{v}(x_0, y_0, z_0)$, as shown in Figure 27.6. Take A to be the area of the disk. If the area is small, then the surface integral of $\vec{\nabla} \times \vec{v}$ over S can be approximated with one term:

$$\iint_S (\vec{\nabla} \times \vec{v}) \cdot \hat{n}\, dA \approx (\vec{\nabla} \times \vec{v}) \cdot \hat{n} A.$$

Because $\hat{n}$ is aligned with $\vec{\nabla} \times \vec{v}$, the dot product results in just $\|\vec{\nabla} \times \vec{v}\|$. Thus, we have

$$\iint_S (\vec{\nabla} \times \vec{v}) \cdot \hat{n}\, dA \approx \|\vec{\nabla} \times \vec{v}\| A$$

for a small disk of area A oriented with a normal aligned in the direction of the curl field. On the other hand, we can apply Stokes's theorem to get

$$\iint_S (\vec{\nabla} \times \vec{v}) \cdot \hat{n}\, dA = \oint_C \vec{v} \cdot d\vec{s}$$

where C is the circle that forms the edge of the disk. Equating (at least approximately) the right sides of the previous two equations gives

$$\|\vec{\nabla} \times \vec{v}\| A \approx \oint_C \vec{v} \cdot d\vec{s}.$$

Dividing by A, we have

$$\|\vec{\nabla} \times \vec{v}\| \approx \frac{\oint_C \vec{v} \cdot d\vec{s}}{A}.$$

The line integral on the right side sums the velocity vector field around a small circle, and can thus be interpreted as a **circulation**. With this, we interpret the magnitude of the curl as a **circulation density**.

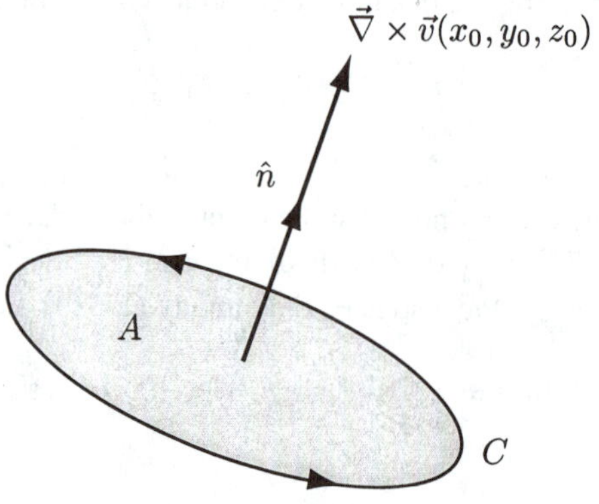

Figure 27.6. A small disk aligned with normal $\hat{n}$ in the direction of the curl $\vec{\nabla} \times \vec{v}$.

Note that we have arrived at a circulation density for a disk aligned with the direction of the curl at the chosen point (x_0, y_0, z_0). If we consider a disk with any other orientation, the dot product contributes a factor of $\cos\theta < 1$. This implies that the *direction* of the curl gives the axis direction around which the circulation density is maximum. The magnitude of the curl gives that maximum circulation density. There is an analogy here to the gradient vector field $\vec{\nabla} f$ for a function $f : \mathbb{R}^3 \to \mathbb{R}$. At each point in the domain of f, the direction of the gradient gives the direction of maximum rate of change in f. The magnitude of the gradient gives that maximum rate of change.

Another use of Stokes's theorem comes in proving the converse of Theorem 25.2. This converse is given in the following theorem (which is a direct generalization of Theorem 27.2).

Theorem 27.5. *Let $\vec{F} : \mathbb{R}^3 \to \mathbb{R}^3$ have continuous partial derivatives for all inputs (x, y, z) in $\mathbb{R}^3$. If* $\operatorname{curl}\vec{F}(x, y, z) = \vec{0}$ *for all (x, y, z) in $\mathbb{R}^3$, then there exists a potential function for $\vec{F}$. That is, there is a function $V : \mathbb{R}^3 \to \mathbb{R}$ such that $\vec{\nabla} V(x, y, z) = \vec{F}(x, y, z)$ for all (x, y, z) in $\mathbb{R}^3$.*

Proof. The proof here is very similar to the proof of Theorem 27.2, so we only outline the major steps. First, establish that $\vec{F}$ is conservative with the key idea that for any closed curve $\tilde{C}$ in space, Stokes's theorem gives

$$\int_{\tilde{C}} \vec{F} \cdot d\vec{s} = \iint_S (\vec{\nabla} \times \vec{F}) \cdot \hat{n}\, dA = \iint_S 0\, dA = 0$$

where S is any surface that has $\tilde{C}$ as an edge. Next, define $V : \mathbb{R}^3 \to \mathbb{R}$ by the conditions $V(0, 0, 0) = 0$ and

$$V(x, y, z) = \int_C \vec{F} \cdot d\vec{s}$$

for any curve C from $(0, 0, 0)$ to (x, y, z). Finally, show that $\vec{\nabla} V(x, y, z) = \vec{F}(x, y, z)$ one component at a time. To accomplish this for each component, choose the correct path consisting of line segments parallel to the coordinate axes. For example, to show that $\partial V/\partial x = P(x, y, z)$, use the line segment that goes from $(0, 0, 0)$ to $(0, y, 0)$ along the y-axis, followed by the line segment parallel to the z-axis that goes from $(0, y, 0)$ to $(0, y, z)$, and then the line segment parallel to the x-axis that goes from $(0, y, z)$ to (x, y, z). □

The hypotheses of Theorem 27.5 can be weakened considerably. Suppose the vector field $\vec{F}$ is not defined for one specific input, say $(0, 0, 0)$. This is the case for many physically interesting vector fields, such as $\vec{F}(\vec{r}) = \hat{r}/r^2$. The proof can be modified to avoid this "bad point." First, we can always choose the surface S used in the application of Stokes's theorem to avoid the point $(0, 0, 0)$. Second, we can choose some other point (x_0, y_0, z_0) to declare $V(x_0, y_0, z_0) = 0$. Third, we define $V(x, y, z)$ in terms of a line integral from (x_0, y_0, z_0) to (x, y, z) along any curve that avoids $(0, 0, 0)$. Finally, argue that $\vec{\nabla} V(x, y, z) = \vec{F}(x, y, z)$ by using curves consisting of line segments parallel to the coordinate axes connecting (x_0, y_0, z_0) to (x, y, z).

The proof can be suitably modified if $\vec{F}$ is not defined for even larger pieces of $\mathbb{R}^3$. However, there are some limitations. For example, the proof cannot be modified if $\vec{F}$ is not defined for all points on the z-axis. The obstacle comes in considering a closed curve that wraps around the z-axis such as the unit circle centered at the origin in the xy-plane. Every surface for which this curve is the edge must contain at least one point on the z-axis. Stokes's theorem cannot be applied because $\vec{F}$ is not defined for that one point.

27.2.2 The Divergence Theorem

To generalize the planar divergence theorem to $\mathbb{R}^3$, we start by noting that the analog of the normal line integral for a vector field in the plane is the surface integral for a vector field in space. Thus, we consider a solid region D that has a closed surface S as its boundary. A generic example is shown in Figure 27.7. As usual, we use an outward pointing normal for the closed surface. The divergence of a vector field in the plane naturally generalizes to the divergence of a vector field in space.

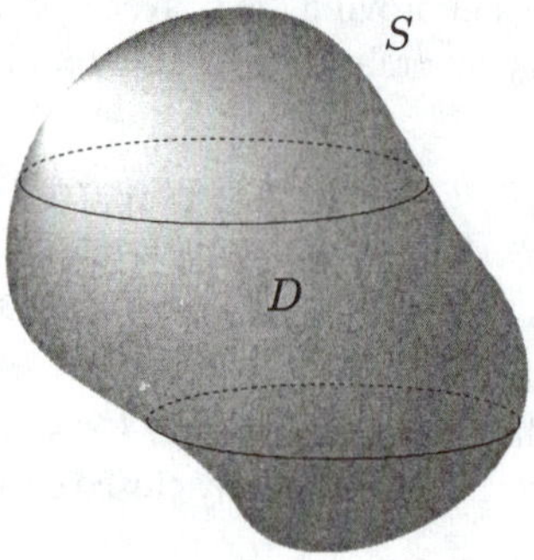

Figure 27.7. Solid region D bounded by a closed surface S.

Theorem 27.6 (Divergence theorem). *Let D be a solid region in* $\mathbb{R}^3$ *with a closed surface S as its boundary. If the vector field* $\vec{F} : \mathbb{R}^3 \to \mathbb{R}^3$ *has components with continuous first partial derivatives, then*

$$\iiint_D (\vec{\nabla} \cdot \vec{F})\, dV = \oiint_S \vec{F} \cdot \hat{n}\, dA.$$

A proof of the divergence theorem is difficult, and we do not include one here. As usual, we verify the theorem for a specific example.

Example 27.3. Verify the divergence theorem for $\vec{F}(x, y, z) = \langle x, y, z \rangle$ and the ball of radius R centered at the origin.

Solution. We will determine the values of the triple integral and the surface integral by geometric arguments.

The divergence of the given vector field is straightforward to compute as

$$\vec{\nabla} \cdot \vec{F}(x, y, z) = 3.$$

Thus, the triple integral is simply

$$\iiint_D (\vec{\nabla} \cdot \vec{F})\, dV = 3 \iiint_D dV = 3\left(\frac{4}{3}\pi R^3\right) = 4\pi R^3.$$

For the surface integral, we note that the vector field points radially out from the origin at all points in $\mathbb{R}^3$. Thus, at each point on the sphere that is the boundary of the ball, the vector field output $\vec{F}(x, y, z)$ is aligned with the unit normal vector $\hat{n}(x, y, z)$. The dot product in the surface integral is thus

$$\vec{F} \cdot \hat{n} = \|F\|(1)(1) = \|F\|.$$

For inputs on the sphere, the vector field outputs have magnitude equal to R. The surface integral then reduces to

$$\oiint_S \vec{F} \cdot \hat{n}\, dA = R \oiint_S dA = R(4\pi R^2) = 4\pi R^3.$$

The triple integral and the surface integral are equal, verifying the divergence theorem for this case. △

In some cases, we can use the divergence theorem to evaluate a surface integral by evaluating the related triple integral explicitly. This can be easier than directly evaluating the surface integral through the definition because we need not parametrize the surface explicitly. We illustrate in the following example.

Example 27.4. Compute the surface integral of $\vec{F} = \langle xy, yz, xz \rangle$ for the surface of the rectangular solid $[1, 2] \times [0, 4] \times [-1, 1]$.

Solution. By the divergence theorem, we have

$$\oiint_S \langle xy, yz, xz \rangle \cdot \hat{n}\, dA = \iiint_D (\vec{\nabla} \cdot \langle xy, yz, xz \rangle)\, dV$$

where D is the rectangular solid and S is the surface of the solid. We compute the divergence

$$\vec{\nabla} \cdot \langle xy, yz, xz \rangle = y + z + x$$

and thus have

$$\oiint_S \langle xy, yz, xz \rangle \cdot \hat{n}\, dA = \iiint_D (x + y + z)\, dV.$$

As usual, we evaluate the triple integral through an iterated integral, giving

$$\begin{aligned}\oiint_S \langle xy, yz, xz \rangle \cdot \hat{n}\, dA &= \int_{-1}^{1} \int_{0}^{4} \int_{1}^{2} (x + y + z)\, dx\, dy\, dz \\ &= \int_{-1}^{1} \int_{0}^{4} \int_{1}^{2} x\, dx\, dy\, dz + \int_{-1}^{1} \int_{0}^{4} \int_{1}^{2} y\, dx\, dy\, dz + \int_{-1}^{1} \int_{0}^{4} \int_{1}^{2} z\, dx\, dy\, dz \\ &= 12 + 16 + 0 = 28.\end{aligned}$$

The value of the surface integral is thus 28. △

In Table 27.1 (on the following page), we summarize the fundamental theorems of calculus developed throughout the text. The conclusion of each theorem relates an integral of some type of derivative over some region of integration to a sum or integral of the function being differentiated over the "edge" of the integration region. In some situations, the conclusions give us an efficient way to evaluate one integral by "trading in" for a simpler sum or integral. In the next section, we will make use of the theorems in the context of physics. Specifically, we will use several of the theorems to pass from the integral forms of the fundamental laws describing electromagnetism to differential equation forms. On a very different level, the theorems provide a unified view of the mathematics we have covered in this text.

Table 27.1. Fundamental theoreoms of calculus

	Type of function	Integration region	Edge of region	Derivative	Conclusion
Second FTC	$F:\mathbb{R}\to\mathbb{R}$	interval $[a,b]$	endpoints a and b	ordinary	$\int_a^b F'(x)\,dx = F(b)-F(a)$
FTC for Line Integrals	$f:\mathbb{R}^n\to\mathbb{R}$	curve in $\mathbb{R}^n$	endpoints $\vec{r}_a$ and $\vec{r}_b$	gradient	$\int_{\vec{r}_a}^{\vec{r}_b} \vec{\nabla} f\cdot d\vec{s} = f(\vec{r}_b)-f(\vec{r}_a)$
Green's Theorem	$\vec{F}:\mathbb{R}^2\to\mathbb{R}^2$	region R in plane	simple closed curve C in plane	scalar curl	$\iint_R \left(\frac{\partial Q}{\partial x}-\frac{\partial P}{\partial y}\right) dA = \oint_C \langle P,Q\rangle\cdot d\vec{s}$
Planar Divergence Theorem	$\vec{F}:\mathbb{R}^2\to\mathbb{R}^2$	region R in plane	simple closed curve C in plane	divergence	$\iint_R \left(\frac{\partial P}{\partial x}+\frac{\partial Q}{\partial y}\right) dA = \oint_C \langle P,Q\rangle\cdot \hat{n}\,ds$
Stokes's Theorem	$\vec{F}:\mathbb{R}^3\to\mathbb{R}^3$	surface S in space	simple closed curve C in space	curl	$\iint_S (\vec{\nabla}\times\vec{F})\cdot\hat{n}dA = \oint \vec{F}\cdot d\vec{s}$
Divergence Theorem	$\vec{F}:\mathbb{R}^3\to\mathbb{R}^3$	solid region D in space	closed surface S in space	divergence	$\iiint_D (\vec{\nabla}\cdot\vec{F})\,dV = \oiint_S \vec{F}\cdot\hat{n}\,dA$

27.3 Maxwell's Equations

We saw in Section 25.4 how Maxwell generalized Ampère's law to include situations in which there is not only a constant electric current, but also a changing electric flux. This adds a term to Ampère's law and makes it more symmetric with Faraday's law, which relates a changing magnetic flux to an induced electric field. In Section 27.4, we'll see how Maxwell used the generalized form of Ampère's law together with Faraday's law to predict the existence of electromagnetic waves. This was one of the major contributions of nineteenth-century physics, and with its experimental confirmation have come numerous applications in the twentieth century.

Because of Maxwell's role as just described above, the four fundamental equations of electromagnetism (Gauss's law, Gauss's law for magnetism, Ampère's law, and Faraday's law) are called **Maxwell's equations**. In this section, we examine Maxwell's equations as a group and use Stokes's theorem and the divergence theorem to change them from integral form to the (often more useful) differential form.

27.3.1 Maxwell's Equations in Integral Form

To begin, let's collect the four Maxwell's equations as named above. We present them not in the order we have encountered them in this book, but rather group them in pairs by type. First, we have the integrals of the electric and magnetic fields over a closed surface:

$$\oiint \vec{E}\cdot\hat{n}\,dA = \frac{q_{\text{enc}}}{\epsilon_0} \qquad \text{Gauss's law} \qquad (23.7)$$

and

$$\oiint \vec{B} \cdot \hat{n}\, dA = 0 \qquad \text{Gauss's law for magnetism.} \tag{26.2}$$

The second pair of equations are the integrals of the electric and magnetic fields over a closed path:

$$\oint_C \vec{B} \cdot d\vec{s} = \mu_0 I_C + \mu_0 \epsilon_0 \frac{d\Phi_E}{dt} \qquad \text{Ampère's law} \tag{25.11}$$

and

$$\oint_C \vec{E} \cdot d\vec{s} = -\frac{d\Phi_B}{dt} \qquad \text{Faraday's law.} \tag{26.3}$$

These four equations are Maxwell's equations in integral form. The pairs of equations each possess a high degree of symmetry, though this symmetry is not complete (an issue we will address in Section 27.3.3).

In the form in which they are written above, Maxwell's equations are valid only for a vacuum. As you saw when we studied capacitors with dielectric materials (Section 19.2), it is easy to "fix" the electrostatic equations valid in a vacuum by replacing the permittivity of free space ϵ_0 with a modified permittivity $\epsilon = \kappa\epsilon_0$, where κ is a dimensionless parameter with a numerical value greater than one. A similar strategy can be applied to Maxwell's equations, although we will not pursue it, because we will not study electric fields in materials in any more detail in this course. It would also be necessary to modify the permeability of free space μ_0 to obtain the correct magnetic field in materials.

Maxwell's equations provide a compete description of the electric and magnetic fields in free space. To form a complete theory of electromagnetism in a vacuum, we need only add the Lorentz force law

$$\vec{F} = q(\vec{E} + \vec{v} \times \vec{B}) \tag{24.2}$$

so that we can tell what happens to a charged particle that exists in an electric and/or magnetic field. Virtually all the electromagnetism we have studied in this book is based on one or more of Maxwell's equations and the Lorentz force law, or could be re-derived from this basis. For example, Coulomb's law follows easily from Gauss's law.

27.3.2 Maxwell's Equations in Differential Form

The vector calculus theorems we studied earlier in this chapter can be applied directly to Maxwell's equations to change them from integral equations to differential equations. First, let's consider Gauss's law:

$$\oiint \vec{E} \cdot \hat{n}\, dA = \frac{q_{\text{enc}}}{\epsilon_0}.$$

Remember that the charge q_{enc} on the right side of the equation is the *net charge* enclosed in the volume bounded by the surface, over which the surface integral on the left side of the equation is carried out. The charge within that volume can be distributed according to some charge density function ρ. The function ρ is a function of the coordinates being used. For example, in Cartesian coordinates $\rho = \rho(x, y, z)$. We'll leave off the functional dependence in order to express the result in the most general from possible. From the definition of charge density (charge per unit volume), the net charge q_{enc} enclosed in a volume is the triple integral of the charge density over the entire volume. That is,

$$q_{\text{enc}} = \iiint_D \rho\, dV. \tag{27.20}$$

From the divergence theorem, we know that the surface integral in Gauss's law can be expressed as a triple integral (over the enclosed volume) of the divergence of the electric field $\vec{E}$. Symbolically, the divergence theorem says

$$\oiint \vec{E} \cdot \hat{n}\, dA = \iiint_D (\vec{\nabla} \cdot \vec{E})\, dV.$$

But from Gauss's law, the surface integral is also $q_{\text{enc}}/\epsilon_0$, so

$$\frac{q_{\text{enc}}}{\epsilon_0} = \iiint_D (\vec{\nabla} \cdot \vec{E})\, dV$$

or, upon multiplying through by the constant ϵ_0,

$$q_{\text{enc}} = \iiint_D \epsilon_0 (\vec{\nabla} \cdot \vec{E})\, dV. \tag{27.21}$$

The charge q_{enc} is the same in Equations (27.20) and (27.21), and the integrals are taken over the same volume. Therefore, the integrands must be equal, so

$$\epsilon_0 (\vec{\nabla} \cdot \vec{E}) = \rho.$$

Or, as it is more commonly written,

$$\vec{\nabla} \cdot \vec{E} = \frac{\rho}{\epsilon_0}. \tag{27.22}$$

This is Gauss's law in differential form.

It is straightforward to derive Gauss's law for magnetism in differential form by following the same steps. Alternately, we may simply observe that there is no magnetic "charge" (there are no magnetic monopoles), and hence the density of magnetic "charge" is always zero. From this it follows that Gauss's law for magnetism reads

$$\vec{\nabla} \cdot \vec{B} = 0. \tag{27.23}$$

Next, we turn to Ampère's law

$$\oint_C \vec{B} \cdot d\vec{s} = \mu_0 I_C + \mu_0 \epsilon_0 \frac{d\Phi_E}{dt}.$$

Both terms on the right side of Ampère's law involve surface integrals over a surface. The current I_C is the net current passing through any surface bounded by the closed curve C in the line integral on the left side of the equation. When studying current flow in cylindrical wires in Section 19.3, we assumed a uniform flow of current throughout the wire's cross section. But in general, the flow of current can vary in both magnitude and direction over an arbitrary surface. To take this into account, we redefine the **current density** as a vector field $\vec{J}$, with a direction corresponding to the direction of current flow at each point and magnitude equal to the local current per unit area at each point on the surface. Then the net current I_C is simply the flux of the current density through the surface. Symbolically,

$$I_C = \iint \vec{J} \cdot \hat{n}\, dA \tag{27.24}$$

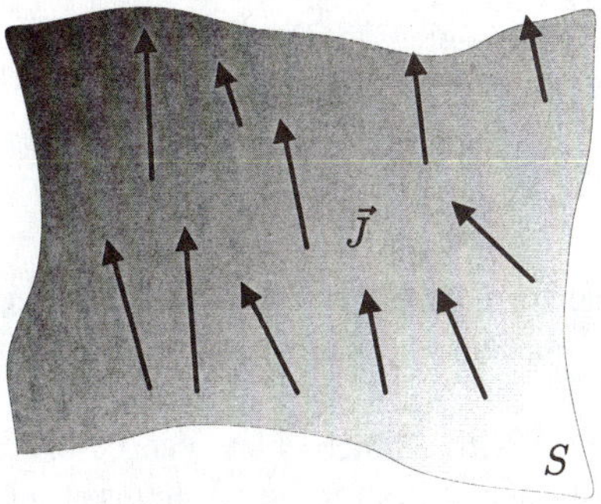

Figure 27.8. Generic vector field $\vec{J}$ along a surface S.

where the unit vector $\hat{n}$ is normal to the surface at each point, as usual. In Figure 27.8, we show a generic vector field $\vec{J}$ along a surface S.

The second term of the right side of Ampère's law also contains a surface integral in the electric flux. Using the definition of electric flux, that term can be rewritten

$$\mu_0 \epsilon_0 \frac{d\Phi_E}{dt} = \mu_0 \epsilon_0 \frac{d}{dt} \left(\iint \vec{E} \cdot \hat{n}\, dA \right)$$

where again the surface of integration is any one bounded by the closed curve C described on the left side of Ampère's law. Now we can interchange the order of integration and differentiation, and since the unit vector $\hat{n}$ is constant, we have

$$\mu_0 \epsilon_0 \frac{d\Phi_E}{dt} = \mu_0 \epsilon_0 \iint \frac{d\vec{E}}{dt} \cdot \hat{n}\, dA. \tag{27.25}$$

Collecting the results of Equations (27.24) and (27.25), Ampère's law becomes

$$\oint_C \vec{B} \cdot d\vec{s} = \mu_0 \iint \vec{J} \cdot \hat{n}\, dA + \mu_0 \epsilon_0 \iint \frac{d\vec{E}}{dt} \cdot \hat{n}\, dA.$$

The two surface integrals are over the same surface, so we can combine those integrals to give

$$\oint_C \vec{B} \cdot d\vec{s} = \iint \left(\mu_0 \vec{J} \cdot \hat{n} + \mu_0 \epsilon_0 \frac{d\vec{E}}{dt} \cdot \hat{n} \right) dA.$$

We can now factor out the unit vector $\hat{n}$:

$$\oint_C \vec{B} \cdot d\vec{s} = \iint \left(\mu_0 \vec{J} + \mu_0 \epsilon_0 \frac{d\vec{E}}{dt} \right) \cdot \hat{n}\, dA. \tag{27.26}$$

It should be apparent that Stokes's theorem can be applied to Equation (27.26). Symbolically we can write Stokes's theorem with the magnetic field vector $\vec{B}$ as

$$\oint_C \vec{B} \cdot d\vec{s} = \iint (\vec{\nabla} \times \vec{B}) \cdot \hat{n}\, dA. \tag{27.27}$$

If the surface integrals in Equations (27.26) and (27.27) are taken over the same surface area, the integrands must be the same, in order to give the same result on the left side of the equations. Therefore,

$$\vec{\nabla} \times \vec{B} = \mu_0 \vec{J} + \mu_0 \epsilon_0 \frac{\partial \vec{E}}{\partial t} \tag{27.28}$$

which is Ampère's law in differential form. Notice that the time derivative of the electric field has been changed to a partial derivative. This is because the derivative is now being carried out over a general region of space, rather than a fixed surface.

The approach we have just followed with Ampère's law can be followed to change Faraday's law into differential form. We will leave the full derivation for the problems at the end of the chapter. Faraday's law in differential form is

$$\vec{\nabla} \times \vec{E} = -\frac{\partial \vec{B}}{\partial t}. \tag{27.29}$$

As a group, Equations (27.22), (27.23), (27.28), and (27.29) constitute Maxwell's equations in differential form.

You may well wonder why physicists find it preferable to express Maxwell's equations in differential form rather than integral form. It's impossible to give a complete answer to this question here, because these are *partial differential equations*. Because we have not given a full treatment of partial differential equations in this course, we cannot get involved in significant applications. But from your experience with ordinary differential equations throughout the course, you should have developed some sense that it is generally easier to deal with derivatives than with integrals. Further, your experience with Maxwell's equations in integral form (particularly Gauss's law and Ampère's law) has been that these equations are only useful when a high degree of symmetry is present. All four of Maxwell's equations in integral form require a surface integral, and two require a line integral as well. By converting to differential form, we essentially trade these for the divergence and curl of the electric and magnetic fields, which are generally easier to compute.

27.3.3 Symmetry and Lack of Symmetry in Maxwell's Equations

Let's return to Maxwell's equations in their more familiar integral form, as presented at the beginning of Section 27.3.1. Thinking of the equations in pairs (Gauss's law and Gauss's law for magnetism forming one pair, and Ampère's law and Faraday's law forming the other), rather clear symmetries show that electric and magnetic fields are both similar and connected to one another.

Gauss's law relates the integral of the electric field integrated over a closed surface to the charge enclosed by that surface. Gauss's law for magnetism relates the magnetic field integrated over a closed surface to the "magnetic charge" enclosed by that surface. (Because there are no magnetic monopoles, the magnetic charge is always zero.)

Ampère's law relates the magnetic field along a closed path to the electric current and changing electric flux through a surface enclosed by that path. Faraday's law relates the electric field along a closed path to the changing magnetic flux through a surface enclosed by that path. The apparent lack of symmetry due to the presence of the constant factor $\mu_0\epsilon_0$ with the changing electric flux in Ampère's law is an illusion. The fact that these constants appear here is due to our choice of the SI system of units. The fact that a minus sign appears with the changing magnetic flux in Faraday's law (but not the changing electric flux in Ampère's law) is also not a lack of symmetry, but simply a statement of how changing electric and magnetic fields create one another. Indeed, the minus sign is *necessary* in order for the changing fields to create one another in a self-consistent way.

The most glaring lack of symmetry in Maxwell's equations is in the fact that Ampère's law contains the term $\mu_0 I_C$, for which Faraday's law has no counterpart. Let's examine this extra term arises. The quantity

I_C is the rate of flow of electric charge through the closed curve C. The symmetry of Maxwell's equations (with electric and magnetic quantities reversing roles in each pair of equations) tells us that the corresponding quantity in Faraday's law would be the rate of flow of *magnetic charges* through the closed curve. Since magnetic charges don't exist, this term is absent in Faraday's law, just as there is no magnetic charge in Gauss's law for magnetism.

For years, theoretical physicists have speculated on whether magnetic monopoles exist, even though we don't see them in normal matter. Several reasons lead us to believe that magnetic monopoles exist but are extremely rare in the universe and hard to detect. For *some* physicists, one reason is that the existence of magnetic monopoles would make Maxwell's equations more symmetric! There would be a nonzero magnetic monopole charge density on the right side of Gauss's law for magnetism, and an added term on the right side of Faraday's law, containing the magnetic monopole *current*. Indeed, this would make Maxwell's equations as a whole appear more symmetric. However, as of this writing, experimental evidence does not yet indicate the existence of magnetic monopoles.

27.4 Electromagnetic Waves

We have seen on numerous occasions how electric and magnetic phenomena are similar and related to each other. In this section, we explore the phenomenon of electromagnetic waves, which is perhaps the most intimate and significant connection between electricity and magnetism. First, we develop a qualitative understanding (based on Maxwell's equations) of how it is possible for electromagnetic waves to occur. Then we look more closely at the details of how the mathematical description of electromagnetic waves follows directly from Maxwell's equations. Finally, we consider some of the experimental evidence and examine the different types of electromagnetic waves on the electromagnetic spectrum.

27.4.1 Introduction and Qualitative Description

Consider again two of Maxwell's equations: Ampère's law and Faraday's law. Ampère's law says that a changing electric flux gives rise to a magnetic field. In most of the examples we have considered, the electric flux changes at a constant rate. But what if the rate of change of electric flux varies? Then the induced magnetic field also varies in time. From Faraday's law, we know that the changing magnetic field in turn induces another electric field. This process can be continued indefinitely, as long as the subsequent derivatives of the induced electric or magnetic fields are not constant. As a result, changing electric and magnetic fields coexist in space, each one being responsible for the other. An electromagnetic wave is one possible outcome of the process just described qualitatively.

An electromagnetic wave is shown in Figure 27.9. One important feature of electromagnetic waves is that they are **transverse waves**, meaning that the disturbance (in this case the electric and magnetic fields) is

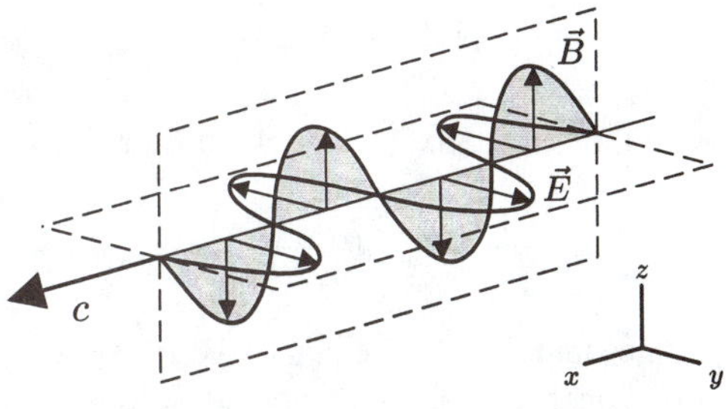

Figure 27.9. An electromagnetic wave travels in the x-direction. The electric field is in the $\pm y$-direction, and the magnetic field is in the $\pm z$-direction.

perpendicular to the direction in which the wave travels. In an electromagnetic wave, the electric and magnetic fields are also perpendicular to each other; therefore, the electric field, magnetic field, and direction of travel are all perpendicular to one another. This feature is shown in the diagram, in which the electromagnetic wave travels in the x-direction, the electric field is in the $\pm y$-direction, and the magnetic field is in the $\pm z$-direction.

Another important property of electromagnetic waves is that they all travel with the same speed

$$c = \frac{1}{\sqrt{\mu_0 \epsilon_0}} = 299{,}792{,}458 \text{ m/s} \tag{27.30}$$

in a vacuum, independent of their wavelength. Notice how the speed of light depends on the permitivity constant from electricity and the permeability constant from magnetism. In any medium other than a vacuum, electromagnetic waves travel with a speed

$$v = \frac{c}{n}$$

where n is a dimensionless number ($n > 1$) known as the **index of refraction** for that medium. The index of refraction is easily measured for different media by measuring the angle at which light refracts (bends) when entering a particular medium (visible light is one kind of electromagnetic wave). For water, the index of refraction is about 1.3, and for many types of glass, it is about 1.5. The index of refraction is normally wavelength-dependent, so different colors of light are bent by different amounts when entering a particular medium. This is why the colors of light are separated by a glass prism.

27.4.2 Electromagnetic Waves and Maxwell's Equations

We now obtain a mathematical description of an electromagnetic wave by applying the differential forms of Faraday's law and Ampère's law to the wave shown in Figure 27.9. This picture is based on certain assumptions that simplify the situation. The electric field is assumed to have a nonzero y-component that depends on x and t. We assume the other two components are identically zero. With this, we can write

$$\vec{E}(\vec{r}, t) = \langle 0, E_y(x, t), 0 \rangle. \tag{27.31}$$

In a similar fashion, the magnetic field is assumed to have a nonzero z-component that depends on x and t, so we can write

$$\vec{B}(\vec{r}, t) = \langle 0, 0, B_z(x, t) \rangle. \tag{27.32}$$

A more general analysis that removes these assumptions goes beyond the scope of this text.

We now write the differential form of Faraday's law for an electric field and a magnetic field in the forms given by Equations (27.31) and (27.32). Using the definition of curl for the field given in Equation (27.31), we get

$$\vec{\nabla} \times \vec{E} = \left\langle 0, 0, \frac{\partial E_y}{\partial x} \right\rangle.$$

The partial derivative with respect to time of the magnetic field in Equation (27.32) is

$$\frac{\partial \vec{B}}{\partial t} = \left\langle 0, 0, \frac{\partial B_z}{\partial t} \right\rangle.$$

Faraday's law [Equation (27.29)] states that $\vec{\nabla} \times \vec{E}$ is equal to the negative of $\partial \vec{B} / \partial t$. With the assumptions we have made, the only nonzero components are the z-components, so we get the scalar equation

$$\frac{\partial E_y}{\partial x} = -\frac{\partial B_z}{\partial t}. \tag{27.33}$$

A similar series of steps using the differential form of Ampère's law [Equation (27.28)] leads to a second relation between E_x and B_z. First, note that the conduction current I_C is zero, and thus the current density $\vec{J}$ is zero. For the field in Equation (27.32), the curl is given by

$$\vec{\nabla} \times \vec{B} = \left\langle 0, -\frac{\partial B_z}{\partial x}, 0 \right\rangle.$$

The partial derivative with respect to time of the electric field in Equation (27.31) is

$$\frac{\partial \vec{E}}{\partial t} = \left\langle 0, \frac{\partial E_y}{\partial t}, 0 \right\rangle.$$

With these equations, the only nonzero component is the y-component, so Ampère's law reduces to the single scalar equation

$$\frac{\partial B_z}{\partial x} = -\mu_0 \epsilon_0 \frac{\partial E_y}{\partial t}. \tag{27.34}$$

Taking the partial derivative with respect to t on both sides of Equation (27.34) gives

$$\frac{\partial^2 B_z}{\partial t\, \partial x} = -\mu_0 \epsilon_0 \frac{\partial^2 E_y}{\partial t^2}. \tag{27.35}$$

Taking the partial derivative with respect to x on both sides of Equation (27.33) gives

$$\frac{\partial^2 E_y}{\partial x^2} = -\frac{\partial^2 B_z}{\partial x\, \partial t}. \tag{27.36}$$

Because the order of mixed second partial derivatives does not matter, the expressions in Equations (27.35) and (27.36) are equivalent, so

$$\frac{\partial^2 E_y}{\partial x^2} = \mu_0 \epsilon_0 \frac{\partial^2 E_y}{\partial t^2}. \tag{27.37}$$

Equation (27.37) is precisely the classical wave equation for a transverse wave $E_y(x,t)$ traveling with a speed $c = 1/\sqrt{\mu_0 \epsilon_0}$. We have therefore shown how the electric part of the electromagnetic wave follows directly from Maxwell's equations. By a similar procedure [taking the partial derivative of Equation (27.33) with respect to time and the partial derivative of Equation (27.34) with respect to x], we get a wave equation for $B_z(x,t)$:

$$\frac{\partial^2 B_z}{\partial x^2} = \mu_0 \epsilon_0 \frac{\partial^2 B_z}{\partial t^2}. \tag{27.38}$$

This shows that the magnetic part of the electromagnetic wave also travels with a speed c. Solutions to the wave equations [Equations (27.37) and (27.38)] are of the form

$$E_y(x,t) = E_0 \cos(kx - \omega t + \phi) \tag{27.39}$$

and

$$B_z(x,t) = B_0 \cos(kx - \omega t + \phi). \tag{27.40}$$

Before going on, let's briefly examine the parameters in these solutions. The constant ϕ is arbitrary (one of the arbitrary constant in the solution of a second-order differential equation; k is the other arbitrary

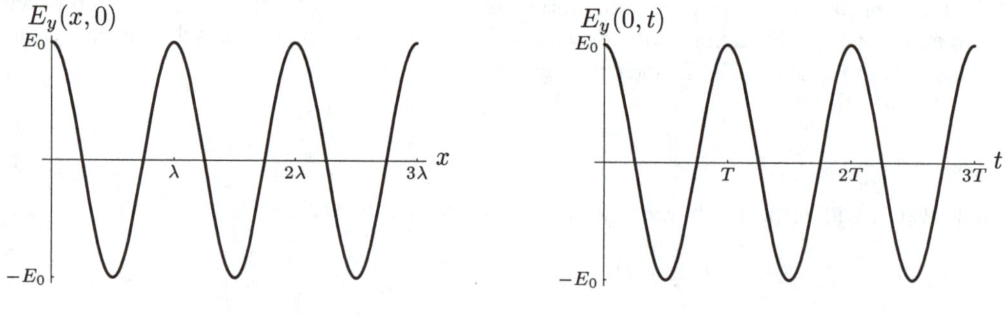

(a) The spatial dependence with $t = 0$

(b) The time dependence with $x = 0$

Figure 27.10. Constant t and constant x views of the solution in Equation (27.39).

constant), so for the time being let $\phi = 0$. Now consider a "snapshot" of the wave taken at $t = 0$. With both $t = 0$ and $\phi = 0$, the solution for the electric part of the wave (we could just as well use the magnetic part) reduces to

$$E_y(x,0) = E_0 \cos(kx)$$

as shown in Figure 27.10(a). Notice that one complete cycle of the cosine function (the argument of the cosine ranging from 0 to 2π) corresponds to one wavelength λ of the wave, that is, the distance from wave crest to wave crest. Therefore,

$$k\lambda = 2\pi$$

or

$$k = \frac{2\pi}{\lambda}.$$

Since k is an arbitrary constant in the solution of the differential equation, this means that there is no physical restriction on the wavelength λ of an electromagnetic wave.

Similarly, we can look at how the wave varies in time for a fixed value of x, for example, $x = 0$. Keeping $\phi = 0$, the solution to the wave equation now reduces to

$$E_y(0,t) = E_0 \cos(-\omega t)$$

as shown in Figure 27.10(b). Now a complete cycle from 0 to 2π in the cosine function corresponds to one complete period T for the wave, so $\omega T = 2\pi$ or

$$\omega = \frac{2\pi}{T}.$$

If we think of k as arbitrary, then ω is not arbitrary, because ω and k are tied to each other by the speed of the wave. We can show this by considering the ratio ω/k:

$$\frac{\omega}{k} = \frac{2\pi/T}{2\pi/\lambda} = \frac{\lambda}{T}.$$

But for a traveling wave, λ/T is the speed of the wave, which in this case we are calling c:

$$c = \frac{\lambda}{T} = \frac{\omega}{k}. \tag{27.41}$$

It remains to be shown that the speed of the electromagnetic wave is in fact the speed of light. We can do this while verifying that Equation (27.39) is a solution to the electric wave equation. (Again a similar verification can be done for the magnetic wave.) We need the second partial derivatives of the electric wave [Equation (27.39)] with respect to time and x. These are

$$\frac{\partial^2 E_y}{\partial x^2} = -k^2 E_0 \cos(kx - \omega t + \phi)$$

and

$$\frac{\partial^2 E_y}{\partial t^2} = -\omega^2 E_0 \cos(kx - \omega t + \phi).$$

Substituting these second derivatives into the original wave equation [Equation (27.37)]:

$$-k^2 E_0 \cos(kx - \omega t + \phi) = \mu_0 \epsilon_0 \big[-\omega^2 E_0 \cos(kx - \omega t + \phi)\big]$$

which can be simplified to

$$k^2 = \mu_0 \epsilon_0 \omega^2.$$

Rearranging gives

$$\frac{\omega}{k} = \sqrt{\frac{1}{\mu_0 \epsilon_0}}.$$

Thus [comparing with Equation (27.41)], we see that the speed of the wave is in fact

$$c = \frac{1}{\sqrt{\mu_0 \epsilon_0}} = 299,792,458 \text{ m/s}. \tag{27.30}$$

Following the same procedure for the magnetic wave, you can see that it has the same speed (c), so the electric and magnetic parts of the wave travel together with the same speed. We can also show that they travel with the same phase ϕ (as shown in Figure 27.9). Taking the derivatives indicated in Equation (27.33), we have

$$\frac{\partial E_y}{\partial x} = -kE_0 \sin(kx - \omega t + \phi) = (-\omega)\, B_0 \sin(kx - \omega t + \phi) = -\frac{\partial B_z}{\partial t}.$$

For these expressions to be equal for all time t and position x, the phase ϕ must be equal for the electric and magnetic parts of the wave. Solving the remaining expression gives $kE_0 = \omega B_0$ or

$$\frac{\omega}{k} = \frac{E_0}{B_0}.$$

But we already know that $\omega/k = c$, the speed of the electromagnetic wave, so

$$c = \frac{E_0}{B_0}. \tag{27.42}$$

Equation (27.42) expresses another important feature of electromagnetic waves: The speed (equal to the speed of light) equals the ratio of the amplitudes of the electric and magnetic parts of the wave.

Example 27.5. Using SI units, show that the ratio of the electric field to the magnetic field [Equation (27.42)] has units of meters per second.

Solution. The units of electric field are N/C, and the units of magnetic field are T. Recalling (Chapter 24) that 1 T = 1 N/(A·m), this gives

$$\text{units for } \left[\frac{E_0}{B_0}\right] = \frac{\text{N/C}}{\text{T}} = \frac{\text{N/C}}{\text{N/(A·m)}} = \frac{\text{A·m}}{\text{C}}.$$

But since 1 A = 1 C/s,

$$\text{units for } \left[\frac{E_0}{B_0}\right] = \frac{\text{(C/s)m}}{\text{s}} = \frac{\text{m}}{\text{s}}$$

as required. △

Example 27.6. Consider the electromagnetic wave for a red helium-neon laser light, with wavelength $\lambda =$ 632.8 nm and the amplitude of the magnetic wave $B_0 = 0.350\ \mu$T. (a) Find the angular frequency ω and the frequency of the light wave. (b) Find the amplitude of the electric part of the wave.

Solution. (a) From the analysis of the electromagnetic wave in the text, we know that $\omega/k = c$, where the speed c of the wave is constant. Because $k = 2\pi/\lambda$, we have

$$\omega = ck = \frac{2\pi c}{\lambda}.$$

Inserting numerical values,

$$\omega = \frac{2\pi(2.998 \times 10^8 \text{ m/s})}{632.8 \times 10^{-9} \text{ m}} = 2.977 \times 10^{15} \text{ s}^{-1}.$$

The frequency of a wave is $\nu = 1/T = \omega/2\pi$, or

$$\nu = \frac{ck}{2\pi} = \frac{c}{\lambda}.$$

Inserting numerical values,

$$\nu = \frac{2.998 \times 10^8 \text{ m/s}}{632.8 \times 10^{-9} \text{ m}} = 4.738 \times 10^{14} \text{ s}^{-1}.$$

(b) From Equation (27.42), we know the relation between the amplitudes of the electric and magnetic waves:

$$E_0 = cB_0 = (2.998 \times 10^8 \text{ m/s})(0.350 \times 10^{-6} \text{ T}) = 105 \text{ m·T/s} = 105 \text{ N/C}.$$ △

The relationship $\nu = c/\lambda$ (derived in the last example) between the frequency and wavelength is a fundamental one for all electromagnetic waves in a vacuum. It is important to know that an inverse relationship exists between the frequency and wavelength: Longer wavelengths correspond to lower frequencies, and shorter wavelengths correspond to higher frequencies.

27.4.3 Experimental Evidence and the Electromagnetic Spectrum

When Maxwell developed his electromagnetic theory in the 1860s, the speed of visible light was known to be about 3×10^8 m/s, with an uncertainty of less than 10%. Therefore, Maxwell's prediction of $c = \sqrt{1/\mu_0\epsilon_0}$ for the speed of electromagnetic waves was consistent with the known speed of light. In 1879, the American physicist Albert A. Michelson (1852–1931) made an extremely precise measurement of the speed of light, good to a fraction of 1%. Michelson's results were again consistent with Maxwell's theory.

What remained to be found was experimental confirmation of electromagnetic waves outside the visible spectrum. (Recall that Maxwell's theory places no restriction on the wavelength of electromagnetic waves.) This confirmation was provided in 1887 by the German physicist Heinrich Hertz (1857–1894), in whose honor the SI unit of frequency is named. Hertz produced electromagnetic waves using time-varying electrical discharges. The resulting waves induced time-varying currents to flow in a nearby coil of wire.

In Hertz's experiments, the oscillation frequency was extremely high—on the order of 10^8 Hz. Since the speed of electromagnetic waves is the speed of light (3×10^8 m/s, which Hertz confirmed by experiment), we can use our wavelength versus frequency relationship to find the wavelength:

$$\lambda = \frac{c}{\nu} \cong \frac{3 \times 10^8 \text{ m/s}}{10^8 \text{ s}^{-1}} = 3 \text{ m}.$$

We now use the term **radio waves** to refer to electromagnetic waves with wavelengths on this order of magnitude. In fact, electromagnetic waves with wavelengths greater than about 1 mm are generically called radio waves, even though standard AM, FM, television, and short wave radio signals each use only small ranges of these wavelengths.

Hertz subsequently did more experiments with his radio waves, showing that the speed was constant (independent of wavelength) and that radio waves exhibit the same wave properties as visible light: interference, diffraction, and polarization. Thus, he provided convincing evidence that Maxwell's predicted electromagnetic waves exist for wavelengths radically different than the wavelength of visible light. Of course, the discovery of radio waves had tremendous practical use too. By 1900, Gugliemo Marconi had used radio waves to transmit Morse code messages over vast distances. In 1920, commercial radio broadcasts began in the United States, and in the 1930s television broadcasts began (also using radio frequency waves to transmit signals). Today, many types of telecommunications use radio waves, including cell phones and satellite transmission of data and voice information.

Electromagnetic radiation has now been detected throughout a huge range of wavelengths, from radio waves many meters long to **gamma rays** (produced in nuclear transitions) shorter than 10^{-15} m! Between these extremes lie, in order of increasing wavelength, **x rays** (high-energy waves produced in atomic transitions or by the slowing of high-energy electrons), **ultraviolet waves** (normally produced by the same mechanisms as visible light), visible light, and **infrared waves**, which can be given off by objects radiating heat energy). In Figure 27.11, we show the **electromagnetic spectrum**, on which each type of electromagnetic wave is placed in its own range of wavelengths. You should not think of the borders between the various types of electromagnetic waves as sharp boundaries. The electromagnetic spectrum is continuous, with one type of wave gradually changing into another as wavelength changes. The boundaries are not even agreed upon by all physicists; our choice is a fairly conventional one. Notice that visible light takes up the narrowest band on the logarithmically plotted spectrum, ranging in wavelength from just 400 nm to 700 nm. Even these numbers are chosen by convention, since the ability to perceive light of different wavelengths differs from person to person.

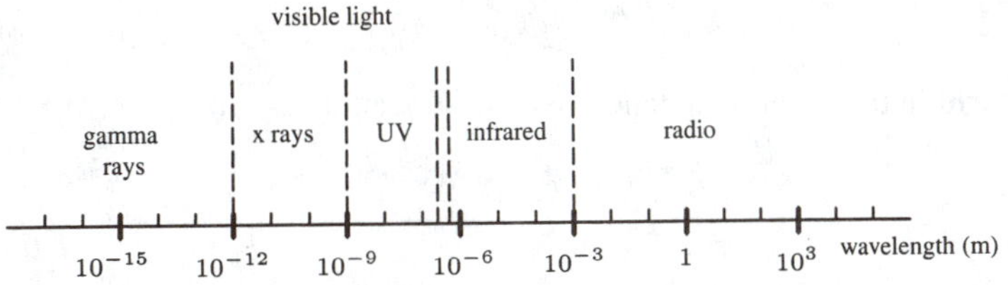

Figure 27.11. The electromagnetic spectrum.

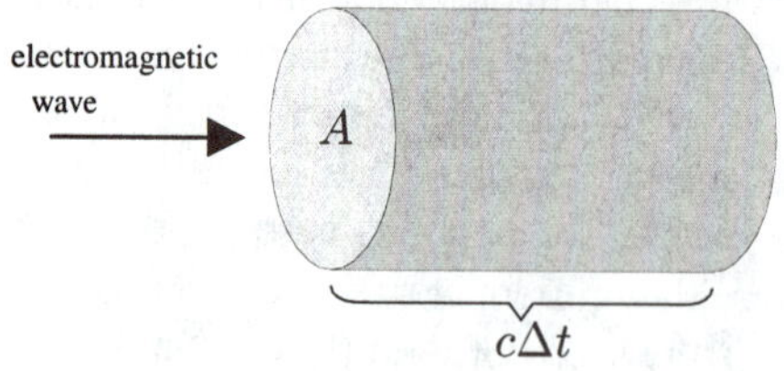

Figure 27.12. Continuous electromagnetic wave passing through a flat circular surface area A.

27.4.4 Energy and Momentum of Electromagnetic Waves

You know from experience that electromagnetic waves (such as light from the Sun) carry energy. In this section, we develop a quantitative model for energy transport in electromagnetic waves. Figure 27.12 shows a continuous electromagnetic wave passing through a flat circular surface area A. The wave's direction of travel is perpendicular to that surface. The energy ΔU passing through the surface in a time Δt can be found by dimensional analysis:

$$\frac{\Delta U}{\Delta t} = \frac{\text{energy}}{\text{time}} = \frac{\text{energy}}{\text{volume}}\,\frac{\text{volume}}{\text{time}}.$$

The energy per unit volume is the net electric and magnetic energy density, which for now we will simply call u. The volume swept out by the wave in time Δt is the volume of the cylinder as shown, with cross-sectional area A and length $c\,\Delta t$, since the electromagnetic wave's speed is c. Combining these results,

$$\frac{\Delta U}{\Delta t} = u\frac{A\,(c\,\Delta t)}{\Delta t} = uAc.$$

Because the area A is arbitrary, physicists prefer to think not of this quantity but of a quantity S, defined as the energy per unit area per unit time, which is simply $\Delta U/\Delta t$ divided by A. Thus,

$$S = \frac{\Delta U}{A\,\Delta t}$$

or

$$S = uc.$$

Some further insight is gained if we relate the quantity S to the electric and magnetic fields within the electromagnetic wave. The total electromagnetic energy density is the sum of the electric energy density and magnetic energy density:

$$u = \frac{1}{2}\epsilon_0 E^2 + \frac{B^2}{2\mu_0}. \tag{27.43}$$

We also know that within the electromagnetic wave [see Equation (27.42)]

$$c = \frac{E_0}{B_0} = \frac{E_0\cos(kx - \omega t + \phi)}{B_0\cos(kx - \omega t + \phi)}$$

or

$$c = \frac{E}{B}. \tag{27.44}$$

Using this fact, we can examine the two terms on the right side of Equation (27.43) separately. The term (containing the electric field) is

$$\frac{1}{2}\epsilon_0 E^2 = \frac{1}{2}\epsilon_0 (cB)^2 = \frac{1}{2}\epsilon_0 c^2 B^2.$$

But $c^2 = 1/\mu_0\epsilon_0$, so

$$\frac{1}{2}\epsilon_0 E^2 = \frac{1}{2}\epsilon_0 \left(\frac{1}{\mu_0\epsilon_0}\right) B^2 = \frac{B^2}{2\mu_0}.$$

Therefore, the two terms $\frac{1}{2}\epsilon_0 E^2$ and $B^2/2\mu_0$ are equal, and the electric and magnetic fields make equal contributions to the energy density u of the electromagnetic wave. So, in terms of both E and B, we can write

$$u = \frac{1}{2}\epsilon_0 E^2 + \frac{1}{2}\epsilon_0 E^2 = \epsilon_0 E^2 \tag{27.45}$$

or

$$u = \frac{B^2}{2\mu_0} + \frac{B^2}{2\mu_0} = \frac{B^2}{\mu_0}.$$

Now returning to the quantity S defined above as the energy per unit area per unit time transported by the electromagnetic wave, we can use Equation (27.45) to write

$$S = uc = \epsilon_0 E^2 c.$$

From Equation (27.44), $E = cB$, so

$$S = \epsilon_0 E\,(cB)\,c = \epsilon_0 c^2 EB.$$

But $\epsilon_0 c^2 = 1/\mu_0$, so

$$S = \frac{1}{\mu_0} EB. \tag{27.46}$$

From this result, we define the **Poynting vector** $\vec{S}$ [named for the English physicist John H. Poynting (1852–1914)] to have a magnitude given by Equation (27.46) and a direction in the electromagnetic wave's direction of propagation. Based on the mutual orthogonality of the electric field, magnetic field, and direction of propagation of the electromagnetic wave (Figure 27.9), we express the Poynting vector in its most compact form as

$$\vec{S} = \frac{1}{\mu_0}\vec{E} \times \vec{B}.$$

The **intensity** I of an electromagnetic wave is the time-averaged value of the magnitude S of the Poynting vector. We find an expression for the intensity by computing the time average of the electric field (or the magnetic field). The electric field in the electromagnetic wave has a magnitude

$$E = E_0 \cos(kx - \omega t + \phi).$$

The average value of the cosine-squared function over a whole period is $\frac{1}{2}$, so the average value of E^2 is $\frac{1}{2}E_0^2$. Using the expression $S = \epsilon_0 E^2 c$ from above, we see that

$$I = S_{\text{avg}} = \epsilon_0 c \left(\frac{1}{2}E_0^2\right) = \frac{1}{2}\epsilon_0 E_0^2 c.$$

It may not seem so obvious that electromagnetic waves carry momentum as well as energy. But this fact also follows from the laws of electromagnetism and has been confirmed experimentally. Consider again an electromagnetic wave as pictured in Figure 27.9. With the electric and magnetic fields restricted to the y-direction and z-direction (respectively), we can write

$$\vec{E} = E_y\,\hat{\jmath} \qquad \text{and} \qquad \vec{B} = B_z\,\hat{k}.$$

If the electromagnetic wave encounters a charge q, then the charge experiences a Lorentz force. The x-component of that force is

$$F_x = qE_x + q(\vec{v} \times \vec{B})_x.$$

We know that $E_x = 0$, and with the given magnetic field $\vec{B} = B_z\,\hat{k}$, the x-component of the cross product is

$$(\vec{v} \times \vec{B})_x = v_y B_z - v_z B_y = v_y B_z.$$

Therefore,

$$F_x = q v_y B_z.$$

From mechanics, $F_x = dp_x/dt$ and from our study of electromagnetic waves $B_z = E_y/c$, so

$$\frac{dp_x}{dt} = q v_y \frac{E_y}{c}. \tag{27.47}$$

Now let's consider the *power* generated by the wave striking the charged particle. Power is the rate at which energy is transferred (dU/dt), and in mechanics we found that the instantaneous power is also given by the dot product $\vec{F} \cdot \vec{v}$. In this case, the force is the Lorentz force on the charged particle q. Thus,

$$\text{Power } = \frac{dU}{dt} = \vec{F} \cdot \vec{v} = [q\vec{E} + q(\vec{v} \times \vec{B})] \cdot \vec{v} = q\vec{E} \cdot \vec{v} + q(\vec{v} \times \vec{B}) \cdot \vec{v}.$$

But since $\vec{v} \times \vec{B}$ is perpendicular to $\vec{v}$, we know $(\vec{v} \times \vec{B}) \cdot \vec{v} = 0$. Therefore,

$$\frac{dU}{dt} = q\vec{E} \cdot \vec{v}$$

or, with $\vec{E} = E_y\,\hat{\jmath}$,

$$\frac{dU}{dt} = qE_y v_y. \tag{27.48}$$

Combining the results of Equations (27.47) and (27.48), we have

$$\frac{dp_x}{dU} = \frac{1}{c}. \tag{27.49}$$

Equation (27.49) expresses a striking and important result: *In an electromagnetic wave, the ratio of the momentum to the energy is* $1/c$. Thus, for an electromagnetic wave traveling in the x-direction with some total energy U and momentum (magnitude) p_x, the momentum is

$$p_x = \frac{U}{c}. \tag{27.50}$$

[Note: More formally, we could take the indefinite integral of Equation (27.49) to find Equation (27.50).] This result was first confirmed experimentally by E. Nichols and G. F. Hull in 1901. An electromagnetic wave does transfer momentum to an object it strikes. If all the electromagnetic energy is absorbed, the amount of momentum transferred is given by Equation (27.50). But if the object is perfectly reflecting (a perfect mirror), conservation of momentum dictates that *twice* as much momentum is transferred to the mirror—that is, $\Delta p_x(\text{mirror}) = 2U/c$, where U is the energy of the electromagnetic wave. Physicists often refer to the ability of light (or other electromagnetic radiation) to transfer momentum as "radiation pressure," because with a transfer of momentum there is associated a force, and pressure is force per unit area.

The momentum transfer is generally small and difficult to detect. The factor $1/c$ between the momentum and energy makes the momentum transfer small numerically compared with the energy of the electromagnetic wave. We illustrate this relationship with an example.

Example 27.7. Find the maximum possible (i.e., ignoring frictional effects) momentum transfer of a 5.0-mW laser beam striking a perpendicular, perfectly reflecting surface for exactly one hour.

Solution. The energy of the electromagnetic wave is the power multiplied by the time, or $U = P\,\Delta t$. For a perfectly reflecting surface, the maximum momentum transfer is

$$p_x = \frac{2U}{c} = \frac{2P\,\Delta t}{c}.$$

Inserting numerical values (with $\Delta t = 1\text{ h} = 3600\text{ s}$),

$$p_x = \frac{2(5.0 \times 10^{-3}\text{ J/s})(3600\text{ s})}{3.00 \times 10^{8}\text{ m/s}} 1.2 \times 10^{-7}\text{ J/m} = 1.2 \times 10^{-7}\text{ kg·m/s}$$

where we have converted into the more familiar SI unit for momentum. This is an extremely small amount of momentum and would be difficult to detect! △

Example 27.8. The laser beam of the preceding example is confined to a circular cross section of diameter 0.080 mm. Find the electromagnetic intensity I and the maximum amplitudes of the electric and magnetic fields in the wave.

Solution. The intensity is the average magnitude of the Poynting vector, which is the energy per unit area per unit time. Thus,

$$I = S_{\text{avg}} = \frac{\Delta U/\Delta t}{A}.$$

Inserting the numerical values for this problem, with $\Delta U/\Delta t = 5.0\text{ mW}$,

$$I = \frac{5.0 \times 10^{-3}\text{ W}}{\pi\,(4.0 \times 10^{-5}\text{ m})^2} = 9.95 \times 10^{5}\text{ W/m}^2.$$

This high intensity is not unusual for a laser beam, in which a modest amount of power is concentrated in a very small area. △

27.5 Electromagnetism, Relativity, and Quantum Physics

In 1905, Albert Einstein published his theory of special relativity, governing the measurements made in different reference frames that are in constant motion relative to one another. In special relativity, the laws of

Newtonian mechanics are no longer valid. Although Newton's laws still serve us well at speeds much less than the speed of light, they are increasingly inaccurate as speeds approach the speed of light.

However, the laws of electromagnetism (specifically Maxwell's equations and the Lorentz force law) are still valid in special relativity. In fact, it was Einstein's belief in the invariance of these laws that served as the theoretical foundation for special relativity. The title of Einstein's 1905 paper "The Electrodynamics of Moving Bodies" shows this was very much on his mind.

In special relativity, different observers naturally make different measurements of electric and magnetic fields. For example, consider an observer A, who measures a charge q to be at rest. This observer sees a static electric field of that single charge and no magnetic field. How does the situation appear to a second observer B, who is moving with a constant velocity $\vec{v}$ relative to A? In B's frame of reference, the charge is moving with a constant velocity $-\vec{v}$. Therefore, B sees a steady current, and hence a magnetic field, as dictated by Ampère's law. One of Einstein's achievements was to show how to transform Maxwell's equations from one reference frame to another. The transformations he derived are consistent with the example we just gave, both qualitatively and quantitatively. Maxwell's equations are therefore consistent from one reference frame to another, even though different observers measure different values for their electric and magnetic fields. In this way, Einstein helped physicists better understand the connections between electricity and magnetism.

The relativistic transformation of Maxwell's equations is beyond the scope of this book. But we can show another important result of Einstein's theory here, based on our understanding of the momentum and energy of electromagnetic waves from Section 26.5. We'll make use of a heuristic device known as "Einstein's box" pictured in Figure 27.13. The box is initially at rest. Then a short flash of light is sent from the left side of the box toward the right side as shown. We know from Section 27.4 that the light carries momentum $p_x = E/c$, where E is the energy of the light flash. (We use E for energy in this section, since we are no longer concerned with the electric field.) Because linear momentum is conserved, the box must recoil with a

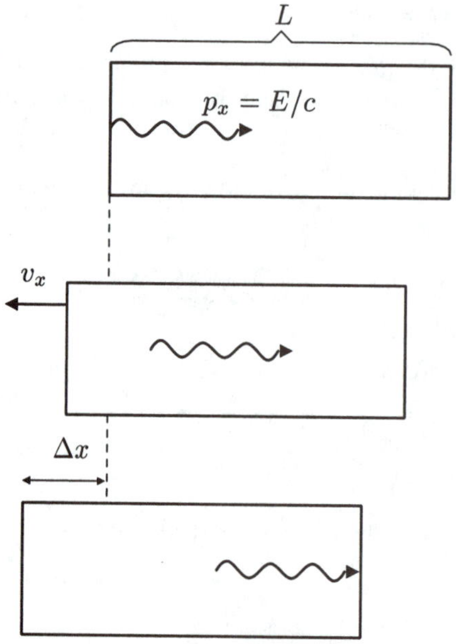

Figure 27.13. Einstein's box at three instances of time. The top frame shows the light as it is emitted at one end. The middle frame shows the light in transit. The bottom frame shows the light as it is absorbed at the other end. In the process, the box recoils a distance Δx.

linear velocity component v_x as shown. If the mass of the box is M, then conservation of momentum gives

$$\frac{E}{c} + Mv_x = 0$$

because before the light emission the momentum of the system was zero. Solving for v_x,

$$v_x = -\frac{E}{Mc}.$$

A short time Δt later, the light is absorbed on the right side of the box, and the system is again at rest. During that time, the box has moved a distance $\Delta x = v_x\,\Delta t$. Since we know the momentum of the light is small, we will assume $\Delta x << L$ (the length of the box), so to an excellent approximation $\Delta t = L/c$. Thus,

$$\Delta x = v_x\,\Delta t = \left(-\frac{E}{Mc}\right)\left(\frac{L}{c}\right) = -\frac{EL}{Mc^2}. \qquad (27.51)$$

(*Note*: the derivation still works without the approximation $\Delta t = L/c$, as you are asked to show in the problems at the end of the chapter.)

After the light is absorbed on the right side of the box, the entire box has moved by Δx. But in an isolated system, the center of mass cannot move. Therefore, the light should have carried a small amount of mass m with it. This is consistent with the light carrying momentum. The equation of the center of mass (assuming the center of the box started at $x = 0$) is

$$mL + M\Delta x = 0. \qquad (27.52)$$

(Again, this is an approximation, based on the fact that $L >> \Delta x$.) Substituting the result of Equation (27.51) for Δx into Equation (27.52), we find

$$mL + M\left(-\frac{EL}{Mc^2}\right)$$

which after some algebra reduces to

$$E = mc^2. \qquad (27.53)$$

Equation (27.53) is probably the most widely recognized equation in all of physics! It is a statement of the equivalence of mass and energy. In the context of our example, we may interpret it as follows: The light sent across the box carried an equivalent mass $m = E/c^2$ with it. This amount of mass was converted into energy $E = mc^2$ at the left side of the box and then converted back into mass when it was absorbed on the right side. It is striking to see how this important result follows from our study of electromagnetism.

Example 27.9. Assume a box as described above with $M = 1.00$ kg and $L = 10.0$ cm. If the energy content of the light burst is 1000 J, what is the equivalent mass m associated with the light, and approximately how far does the box move while the light is in transit?

Solution. Using the result $E = mc^2$, the equivalent mass is

$$m = \frac{E}{c^2} = \frac{1000\ \text{J}}{(3.00 \times 10^8\ \text{m/s})^2} = 1.11 \times 10^{-14}\ \text{J·s}^2/\text{kg}^2 = 1.11 \times 10^{-14}\ \text{kg}.$$

The distance traveled by the box is approximately

$$\Delta x = -\frac{EL}{Mc^2} = -\frac{(1000\ \text{J})(0.10\ \text{m})}{(1.00\ \text{kg})(3.00 \times 10^8\ \text{m/s})^2} = -1.11 \times 10^{-15}\ \text{m}.$$

△

The numerical results of this example show that the approximations we made in the text are justified. On the other hand, they discourage us from believing that this experiment could really be carried out! Einstein was famous for developing such *gedankenexperiments* (thought experiments) that would illustrate a theoretical point, even if the experiment could not be done in practice. Fortunately, since 1905, the equivalence of mass and energy has been examined and tested so many times (particularly in experiments in atomic and nuclear physics) that it is now on a solid experimental basis.

Another experimental fact about light is that it is not a continuous wave phenomenon, but rather consists of extremely small, discrete particlelike bundles of energy called **photons**. The existence of photons was first postulated by the German physicist Max Planck (1858–1947) in 1900 to explain the distribution of energy radiated by a blackbody. In 1905, Einstein (in a separate work from his theory of relativity) used Planck's hypothesis to explain the emission of electrons from metals bombarded with visible and ultraviolet light (a phenomenon known as the **photoelectric effect**).

We say that electromagnetic energy is **quantized** in photons. In spite of their particle-like nature, photons retain wave properties, because we still observe the wave phenomena of interference, diffraction, and polarization. How photons can possess wave and particle properties simultaneously was a puzzle to early twentieth-century physicists; it took many years for them to become accustomed to this duality. The quantum of energy of a photon is based on Planck's constant h, with a numerical value of approximately

$$h = 6.626 \times 10^{-34}\ \text{J·s}.$$

The energy of photon in terms of its wave frequency ν is

$$E = h\nu$$

or, since $\nu = c/\lambda$ for electromagnetic waves,

$$E = \frac{hc}{\lambda}. \tag{27.54}$$

Example 27.10. The red light from a helium-neon laser has a wavelength $\lambda = 632.8$ nm. For a 2.0-mW laser find (a) the energy of each photon, (b) the momentum of each photon, and (c) the number of photons emitted per second.

Solution. (a) From Equation (27.54), the energy of each photon is

$$E = \frac{hc}{\lambda} = \frac{(6.626 \times 10^{-34}\ \text{J·s})(3.00 \times 10^{8}\ \text{m/s})}{632.8 \times 10^{-9}\ \text{m}} = 3.14 \times 10^{-19}\ \text{J}.$$

(b) For light, the relationship between momentum and energy is $p = E/c$, so

$$p = \frac{E}{c} = \frac{3.14 \times 10^{-19}\ \text{J}}{3.00 \times 10^{8}\ \text{m/s}} = 1.05 \times 10^{-27}\ \text{J·s/m} = 1.05 \times 10^{-27}\ \text{kg·m/s}.$$

(c) Because 2.0 mW = 2.0×10^{-3} J/s, the number of photons emitted in one second is

$$\frac{\text{number}}{\text{s}} = \frac{\text{energy}}{\text{s}} \frac{1\ \text{photon}}{\text{energy}}.$$

The energy of one photon was found in part (a), so we can insert numerical values:

$$\frac{\text{number}}{\text{s}} = 2.0 \times 10^{-3}\ \text{J/s}\ \frac{1}{3.14 \times 10^{-19}\ \text{J}} = 6.37 \times 10^{15}\ \text{s}^{-1}.$$

The large number of photons shows why we don't notice the particle nature of light in our everyday experience. △

In this section, we have only been able to provide a small taste of relativity and quantum physics, two areas of study that have dominated physics in the twentieth century. However, even this brief introduction will allow you appreciate that electromagnetic theory is closely tied with both relativity and quantum theory. In Maxwell's theory of the 1860s, there were already to be found the seeds that grew into modern physics. As you go on to study modern physics, you will see electromagnetism as part and parcel of the unified whole that is modern physics.

27.6 Problems

27.1 Vector Calculus in the Plane

1. Verify Green's theorem for $\vec{F}(x, y) = \langle y^2, x\rangle$ and the triangular region with vertices at $(0,0)$, $(4,0)$, and $(3,2)$ by computing both the relevant double integral and the relevant line integral explicitly.

2. Use Green's theorem to compute the line integral of $\vec{F}(x,y) = \langle x + y, x - y\rangle$ for the triangle with vertices at $(0,0)$, $(4,0)$ and $(3,2)$ oriented counterclockwise.

3. Use Green's theorem to show that the line integral $\frac{1}{2}\oint_C \langle -y, x\rangle \cdot d\vec{s}$ gives the area of the region bounded by the closed curve C.

4. Use the result of Problem 3 to compute the area of a circle of radius r.

5. Write the details of the argument that goes from Equation (27.9) to Equation (27.10).

6. Verify the planar divergence theorem for $\vec{F}(x,y) = \langle y^2, x\rangle$ and the triangular region with vertices at $(0,0)$, $(4,0)$, and $(3,2)$ by computing both the relevant double integral and the relevant normal line integral explicitly.

7. Use the planar divergence theorem to compute the normal line integral of $\vec{F}(x,y) = \langle x+y, x-y\rangle$ for the triangle with vertices at $(0,0)$, $(4,0)$, and $(3,2)$.

8. Complete the proof of Theorem 27.2 by showing that $\partial V/\partial y = Q$.

27.2 Vector Calculus in Space

1. Verify Stokes's theorem for the vector field $\vec{F}(x,y,z) = \langle xy, yz, xz\rangle$ with the upper hemisphere of radius one centered at the origin as the surface S.

2. Use Stokes's theorem to compute the line integral $\oint \vec{F}\cdot d\vec{s}$ for the vector field $\vec{F}(x,y,z) = \langle yz, xz, zy\rangle$ over the closed curve consisting of line segments joining the points $(1,0,0)$, $(0,2,0)$, and $(0,0,3)$.

3. Prove that if f and g have continuous partial derivatives, then

$$\iint_S (\vec{\nabla} f \times \vec{\nabla} g)\cdot \hat{n}\, dA = \oint_C f\vec{\nabla} g \cdot d\vec{s}$$

where C is a closed curve forming the edge of the surface S.

4. Use the divergence theorem to compute the surface integral $\oiint_S \vec{F} \cdot \hat{n}\, dA$ for $\vec{F}(x, y, z) = \langle 2x, y^2, z^2 \rangle$ over the unit sphere centered at the origin.

5. Use the divergence theorem to compute the flux of the vector field $\vec{F}(x, y, z) = \langle 3xy, y^2, -x^2y^4 \rangle$ for the surface of the cube bounding the solid region $[0, 1] \times [0, 2] \times [0, 3]$.

6. Use the divergence theorem to compute the flux of the vector field $\vec{F}(x, y, z) = \langle 3xy, y^2, -x^2y^4 \rangle$ for the surface of the tetrahedron with vertices at $(0, 0, 0)$, $(1, 0, 0)$, $(0, 1, 0)$, and $(0, 0, 1)$.

7. Prove that

$$\oiint_S (\vec{\nabla} \times \vec{G}) \cdot \hat{n}\, dA = 0$$

for any vector field $\vec{G}$ with continuous first partial derivatives and any closed surface S.

8. In Section 25.6, we introduced the idea of *flux density* to motivate the definition of *divergence*. We made an argument relating the flux density to the divergence by considering a rectangular box. Use the divergence theorem to make a more general argument that does not use any particular shape for the surface over which the flux is computed.

27.3 Maxwell's Equations

1. Beginning with Gauss's law for magnetism in integral form, apply the divergence theorem to show that in differential form

$$\vec{\nabla} \cdot \vec{B} = 0.$$

2. Beginning with Faraday's law in integral form, apply Stokes's theorem to show that in differential form

$$\vec{\nabla} \times \vec{E} = -\frac{\partial \vec{B}}{\partial t}.$$

3. In this problem, assume that magnetic monopoles exist, so that Gauss's law for magnetism can be rewritten

$$\oiint \vec{B} \cdot \hat{n}\, dA = \mu_0 Q_{\mathrm{m}}$$

where Q_{m} is the net "magnetic charge" enclosed in the surface integral. (a) What are the SI units for magnetic charge? (b) We define the magnetic monopole current I_{m} as the flow rate of magnetic charge—that is, $I_{\mathrm{m}} = dQ_{\mathrm{m}}/dt$. Show that if Faraday's law is modified to include this monopole current in the form

$$\oint_C \vec{E} \cdot d\vec{s} = -\mu_0 I_{\mathrm{m}} - \frac{d\Phi_B}{dt}$$

then this expression is dimensionally correct.

4. Assuming magnetic monopoles exist as described in the preceding problem, rewrite Gauss's law for magnetism and Faraday's law in differential form.

5. Consider a rectangular solid with sides Δx, Δy, and Δz. In the linear approximation, the difference between the x-component of the magnetic field B_x at the two faces parallel to the x-axis is

$$B_x(x+\Delta x, y, z) - B_x(x, y, z) = \frac{\partial B_x}{\partial x}\Delta x.$$

Similar relationships exist for the faces separated by Δy and Δz. Show that in the limit as Δx, Δy, and Δz approach zero Gauss's law for magnetism (in integral form) leads to the result

$$\frac{\partial B_x}{\partial x} + \frac{\partial B_y}{\partial y} + \frac{\partial B_z}{\partial z} = \vec{\nabla} \cdot \vec{B} = 0.$$

Note: This is an alternative method of deriving Gauss's law in differential form.

27.4 Electromagnetic Waves

1. Show that the pressure (force per unit area) exerted on a flat, perfectly absorbing surface by an electromagnetic wave perpendicular to the surface is S/c where S is the magnitude of the Poynting vector.

2. Use Faraday's law to show that for an electromagnetic wave propogating in the x-direction

$$\frac{\partial E}{\partial x} = -\frac{\partial B}{\partial t}.$$

You will need to compute the line integral of $\vec{E} \cdot d\vec{s}$ around a closed rectangle in the xy-plane (assuming the magnetic field is in the z-direction) and use the linear approximation to estimate the magnetic flux Φ_B through the rectangle.

3. Use Ampère's law (with conduction current $I_C = 0$) to show that for an electromagnetic wave propagating in the x-direction

$$\frac{\partial B}{\partial x} = -\mu_0 \epsilon_0 \frac{\partial E}{\partial t}$$

You will need to compute the line integral of $\vec{B} \cdot d\vec{s}$ around a closed rectangle in the xz-plane (assuming the magnetic field is in the z-direction) and use the linear approximation to estimate the electric flux Φ_E through the rectangle.

4. Show carefully that the SI units in the expression

$$\frac{1}{\sqrt{\mu_0 \epsilon_0}}$$

reduce to meters per second.

5. The amplitude of the electric field in an electromagnetic wave is 55.0 N/C. (a) Find the amplitude of the magnetic field. (b) Find the intensity of the electromagnetic wave. (c) Find the energy density in the electromagnetic wave.

6. At a given point in an electromagnetic wave, the electric field points in the direction of a unit vector

$$\hat{u}_E = \frac{1}{\sqrt{2}}\langle -1, 1, 0\rangle$$

and the magnetic field points in the direction of a unit vector

$$\hat{u}_B = \frac{1}{\sqrt{2}} \langle 1, 1, 0 \rangle$$

In what direction is the wave traveling? *Hint*: consider the Poynting vector.

7. A cylindrical wire has a cross-sectional area A. The wire carries a current I and has resistance R. Show that the Poynting vector at the surface of the wire has a magnitude

$$S = \frac{I^2 R}{A} .$$

What is the direction of the Poynting vector?

8. The magnetic field of an electromagnetic wave is given by

$$\vec{B} = \langle B_0 \sin(kz - \omega t), 0, 0 \rangle .$$

(a) In which direction is the wave traveling? (b) What is the corresponding electric field?

9. An electromagnetic wave has electric and magnetic amplitudes of E_0 and B_0, respectively. Write the equations for $\vec{E}(x, y, z, t)$ and $\vec{B}(x, y, z, t)$ if the wave is traveling in the $-y$-direction and $\vec{E}$ is in the $+x$-direction.

In Problems 10–15, use the fact that the intensity of solar radiation reaching Earth is 1400 W/m^2.

10. What is the total power (in watts) of the sunlight incident on Earth?

11. How much energy does Earth receive from the Sun in one day?

12. What is the power output of the Sun? [*Hint:* Consider what fraction of the Sun's radiation strikes Earth.]

13. The United States uses about 1×10^{19} J of electrical energy per year. If a solar collector could work with an average efficiency of 5% (including cloudy days and night time!), how large would the side of a square collector have to be to supply the United States with all its electricity?

14. What is the force of the Sun's radiation pressure on Earth? Compare your result with the gravitational force between the Sun and Earth.

15. It has been proposed to use a perfectly reflecting sheet of metal as a "solar sail" to send spacecraft from Earth to the outer parts of the solar system. How thick should a sheet of aluminum (aluminum's density = 2700 kg/m^3) be so that the radiation pressure from the Sun just balances the Sun's gravitational attraction? Does the aluminum sheet's surface area matter? Does its distance from the Sun matter? Explain.

27.5 Electromagnetism, Relativity, and Quantum Physics

1. How many photons does it take to make 1 J of energy for the following kinds of electromagnetic radiation: (a) radio waves, $\lambda = 3.5$ m; (b) visible light, $\lambda = 540$ nm; and (c) gamma rays, $\lambda = 5.2 \times 10^{-14}$ m?

2. How much energy is released in an atomic bomb if 1.0 g of matter is converted into energy?

3. Consider again the "Einstein's box" gedankenexperiment from the text. This time take into account that the light does not travel the full distance L, but due to the box's recoil Δx the light travels only $L - \Delta x$. Also, note that if the box begins with mass M the recoiling box has mass $M - m$. With these changes, show that the same result ($E = mc^2$) is found.

4. Find the range of (a) energies and (b) frequencies corresponding to the visible range of wavelength, 400–700 nm.

5. Using the data from the problems in Section 27.5 and assuming a uniform wavelength of 540 nm, estimate the number of photons per second released by the Sun.

Chapter 28

The Bohr Atom

In this concluding chapter, we use many of the physics concepts we have studied throughout the book, including the dynamics of circular motion, electrostatic forces, potential energy and conservation of energy, and electromagnetic radiation. However, classical physics alone is unable to explain the structure of the atom. We show how Bohr was able to overcome these difficulties and develop a successful model of the hydrogen atom, by combining classical physics with some radical new assumptions. Finally, we examine the subsequent confirmation of Bohr's theory with more experimental evidence and consider the limitations of the theory that eventually led to a more complete quantum mechanics.

28.1 Background: Early Theories and Experiments

The idea of an atom—an elementary, indivisible building clock of matter—was conceived by the Greek philosophers Democritus and Leucippus in the fifth century B.C.E. By the early fourth century B.C.E., the Greeks had identified earth, water, air, and fire as the four atomic "elements," and the philosopher Plato (428–384 B.C.E.) associated each of the elements with regular geometric solids: The tetrahedron (pyramid), the cube, the dodecahedron, and the icosahedron. We can regard this as one of the earliest efforts to integrate physics and mathematics!

Because atoms are so small, ancient philosophers could only speculate about the properties of atoms. Relatively little progress had been made by the late eighteenth century, except that by then the four Greek elements had been discarded in favor of the many elements we recognize today (hydrogen, oxygen, iron, etc.). Around 1800, chemists (including Dalton and Avogadro) began to do experiments that showed how different atoms combine with each other. By 1869, enough experimental evidence had been accumulated for the Russian chemist Dmitri Mendeleev (1834–1907) to develop the periodic table of the elements. Mendeleev's table is a good approximation of the one we still use, although many elements were not yet discovered in 1869 (in particular, the ideal gases and many rare earth metals). The elements in a column of the periodic table exhibit similar chemical properties. But as of the end of the nineteenth century, what made different elements exhibit similar properties was unknown, because the structure of atoms was not yet understood. Any successful atomic theory would have to explain the regular behavior of atoms in the periodic table.

In the middle of the nineteenth century, the German chemists Gustav Kirchoff (1824–1887) and Robert Bunsen (1811–1899) developed the field of **spectroscopy**, identifying the emission spectra corresponding to different atoms. Kirchoff and Bunsen found that each atom has a **characteristic spectrum**—that is, emission of a set of wavelengths unique to that kind of atom. For example, we show the characteristic spectrum of hydrogen in Figure 28.1. Today we can easily obtain such a spectrum by taking a glass tube filled with

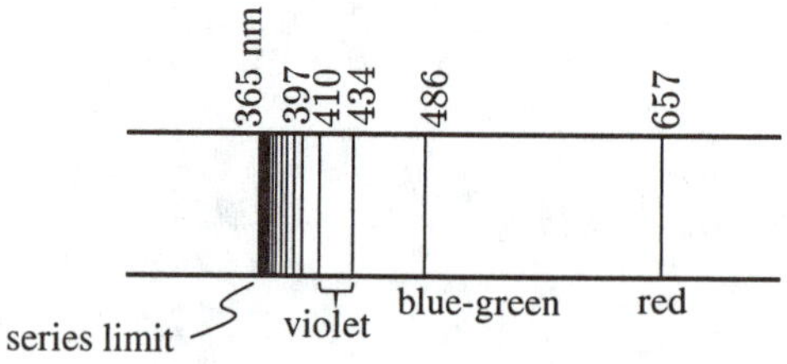

Figure 28.1. Characteristic spectrum of hydrogen.

hydrogen gas and putting a potential difference across the two ends of the tube. The electrical energy put into the tube forces some of the hydrogen atoms into higher energy states. When the atoms naturally fall back to lower energy states, the excess energy is emitted in the form of light. The emitted light is a mixture of the colors of the spectrum in Figure 28.1; we can then separate the light into the distinct colors shown with a prism or diffraction grating.

Any successful atomic theory would have to explain the characteristic emission spectra of atoms. One clue that helped advance the theory was discovered by the Swiss school teacher Johann Balmer in 1885. He noticed that the spectral lines of hydrogen (Figure 28.1) follow a regular pattern that can be expressed in a fairly simple mathematical formula. In terms of SI units, Balmer's formula for the wavelength λ of each hydrogen spectrum line is

$$\lambda = \left(\frac{s^2}{s^2 - 4}\right) 364.56 \text{ nm}, \tag{28.1}$$

with s equal to any integer greater than 2. You should check for yourself that the red line in hydrogen ($\lambda = 656.5$ nm) corresponds to $s = 3$, the blue-green line ($\lambda = 486.3$ nm) corresponds to $s = 4$, and so on. In the limit as s approaches infinity, the spectral lines get very close together, approaching a **series limit** at $\lambda = 364.54$ nm.

The mathematical regularity of Balmer's formula is a sign that the hydrogen atom has a regular physical structure. But in the late nineteenth century, physicists and chemists could only guess at that structure. Another crucial part of the puzzle was solved with J. J. Thomson's measurement of the charge to mass ratio for the electron in 1897 (see Section 24.2). Because the charge to mass ratio of the electron is nearly 2000 times greater than that of the ionized hydrogen atom (a proton), this implies that the electron is nearly 2000 times lighter than the proton (assuming they have equal charge). The hydrogen atom is composite, with the proton and electron joined together.

But what is the structure of the hydrogen atom—that is, what are the relative positions of the proton and electron? And what happens to that structure when the atom absorbs energy, and subsequently gives off energy by emitting light? It seems plausible enough the positive proton and negative electron remain bound to each other through electrostatic attraction, such as the Sun and Earth are bound together by their gravitational attraction. For a time, Thomson believed that the hydrogen atom was like a "plum pudding," with most of the hydrogen atom consisting of a lump (the pudding) of positive charge, with the smaller electron (a plum) floating around inside. Larger atoms could then be modeled as larger lumps of pudding with more electrons (Figure 28.2).

Unfortunately, Thomson and other physicists were unable to show how the dynamics of the plum pudding model would work to form stable atoms, let alone understand the transitions and emission spectra. Thomson's model, along with the "solar system" model in which electrons orbit the atom's positive charge as the planets orbit the Sun, suffered from a significant flaw. In Maxwell's electromagnetic theory, accelerated charges radiate energy in the form of electromagnetic radiation. Any stable configuration would require the electrons

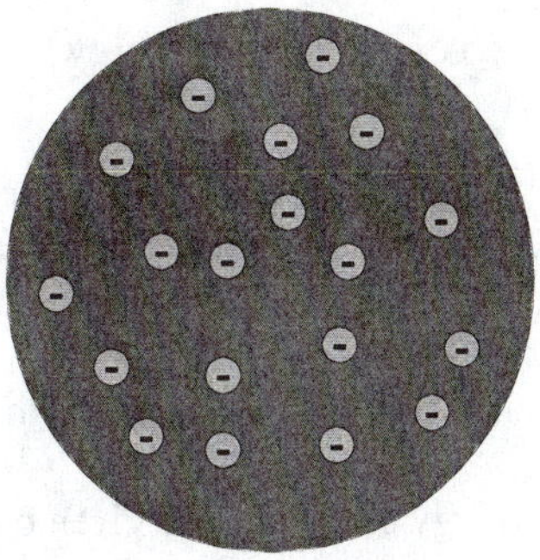

Figure 28.2. Thomson's "plum pudding" model of the atom.

to be in motion. But this in turn requires that the electrons accelerate (e.g., toward the center of a circular orbit) in order to remain in the atom. In doing so, the electrons would radiate energy, and thus stable atoms should not be possible! On the other hand, we have the evidence of our own eyes that atoms are stable for many years. In the first decade of the twentieth century, this contradiction greatly puzzled physicists.

In 1909, another crucial experiment was proposed by the great New Zealand–born experimental physicist Ernest Rutherford (1871–1937), who was working in England. The experiment consisted of firing a beam of alpha particles (helium atoms with the two electrons stripped away) toward a thin metal foil. The positively charged alpha particles are scattered by their interactions with the positive and negative charges in the atoms in the foil. Rutherford reasoned that the pattern of scattered alpha particles could reveal something about the internal structure of the atoms. Rutherford's colleagues Hans Geiger and Ernest Marsden carried out a series of experiments, and the results were reported in 1911. They were surprised to find that a small but significant number of alpha particles scattered at large angles, even directly backwards. The only way Rutherford could explain these results was to assume that the positive charge of the atoms is not spread out in the form of Thomson's pudding, but rather is concentrated in a small **nucleus**, with dimensions of the order of 10^{-15} m to 10^{-14} m. This is *much* smaller than the atom as a whole, which was known to be on the order of 10^{-10} m across.

28.2 The Bohr Model

The results of the alpha particle scattering experiments (now appropriately called **Rutherford scattering**) seem to lead back toward a "solar system" model of the atom, with the relatively heavy nucleus acting like the Sun for the lighter orbiting "planetary" electrons. But how could such an atom be stable, when the orbiting electrons have a centripetal acceleration, and thus according to Maxwell's electromagnetic theory should lose energy continuously? This was the situation faced by the young Danish theoretical physicist Niels Bohr (1885–1962) in 1913. Bohr used the fact that stable atoms *do* exist, combined with the quantum hypothesis used previously by Planck and Einstein (Section 27.5), to form the first successful quantum theory of the hydrogen atom.

Specifically, Bohr made four "general assumptions" upon which he based his model of the hydrogen atom. These assumptions are as follows:

1. There exist "stationary states," in which the atom is stable and does not radiate energy continuously as predicted by classical electromagnetic theory. These states do not occur randomly, but rather correspond to specific energy levels.

2. When a transition occurs between energy states E_ℓ and E_u (ℓ for lower and u for upper), the transition

is accompanied by the emission or absorption of a photon of energy

$$\Delta E = E_u - E_\ell = h\nu, \tag{28.2}$$

where h is Planck's constant (see Section 27.5) and $E_u > E_\ell$. That is, an atom with energy E_ℓ can absorb a photon of energy $h\nu$ in order to reach an energy state E_u, or an atom with energy E_u can spontaneously emit a photon of energy $h\nu$, leaving the atom with energy E_ℓ.

3. The dynamics of the stationary states can be understood with classical mechanics, but transitions between states cannot.

4. The mean value of the total kinetic energy of the atom is given by

$$K = \frac{1}{2} n h \nu_{\text{orb}}$$

where n is an integer and ν_{orb} is the orbital frequency of the electron's revolution around the proton (the nucleus). This assumption is equivalent to assuming that the electron's orbital angular momentum is

$$L = n\frac{h}{2\pi}. \tag{28.3}$$

Note: Because the combination of constants $h/2\pi$ occurs so often in quantum mechanics, it is customary to define a new constant "h-bar" as

$$\hbar = \frac{h}{2\pi}.$$

With this definition, Equation (28.3) can be rewritten

$$L = n\hbar$$

where n is an integer.

Let's begin looking at the dynamics of the electron's orbit around the proton in hydrogen, which according to Bohr's third assumption can be done with classical mechanics. We assume a circular orbit (of radius r) for the electron, and since the proton is nearly 2000 times heavier than the electron, we can make the approximation that the proton is fixed, while the electron orbits around it (Figure 28.3). From electrostatics (Coulomb's law), we know that the force on the electron is directly toward the proton and has a magnitude

$$F = k\frac{(e)(e)}{r^2} = \frac{ke^2}{r^2}$$

because the charges on the proton and electron are $+e$ and $-e$, respectively. Because this is the *only* force on the electron, it is also equal to the centripetal force that keeps the electron in its circular path. In general, we know from mechanics that the magnitude of the centripetal force is mv^2/r. Equating this with the Coulomb force, we have

$$\frac{mv^2}{r} = \frac{ke^2}{r^2} \tag{28.4}$$

where m is the electron's mass.

We can use Equation (28.4) to say something about the energy of the hydrogen atom. The electron has kinetic energy $K = \frac{1}{2}mv^2$, and the system has potential energy

$$U = k\frac{q_1 q_2}{r}.$$

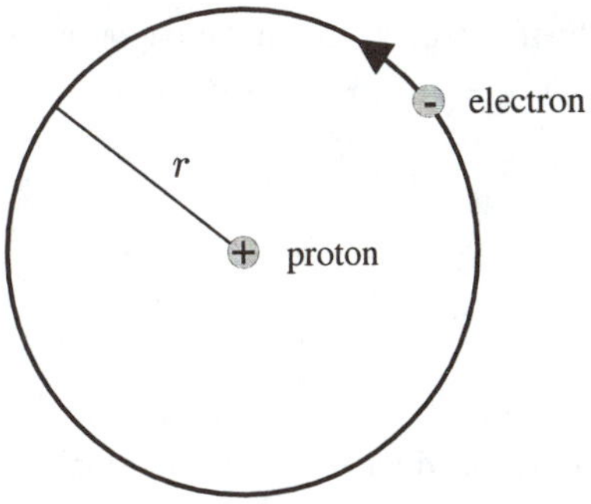

Figure 28.3. Electron in orbit about a fixed proton.

With $q_1 = e$ and $q_2 = -e$,

$$U = -\frac{ke^2}{r}.$$

The total mechanical energy of the atom is

$$E = K + U = \frac{1}{2}mv^2 - \frac{ke^2}{r}.$$

But from Equation (28.4), the kinetic energy can be expressed

$$K = \frac{1}{2}mv^2 = \frac{ke^2}{2r}.$$

Therefore, the total energy of the atom is

$$E = \frac{ke^2}{2r} - \frac{ke^2}{r} = -\frac{ke^2}{2r}. \tag{28.5}$$

Before going on, we should comment briefly on the fact that the total energy is negative. This situation should remind you of the total energy of a satellite of mass m in orbit around a fixed mass M in a circular orbit of radius r:

$$E = -\frac{GMm}{2r}. \tag{15.32}$$

The similarity between Equations (28.5) and (15.32) is striking, but it should not be surprising, given the similarities between Newton's law of gravitation and Coulomb's law and the fact that we have assumed a planetary model of the atom. The fact that the total energy is negative results from our (arbitrary) choice of defining the potential energy U to be zero when r approaches infinity. Given that choice the negative total energy indicates a bound system—that is, one in which the electron is bound to the proton to form a stable atom.

Using Bohr's other assumptions, we can learn more about the hydrogen atom. The magnitude of the angular momentum of a particle of mass m in uniform circular motion with speed v and radius r is

$$L = \|\vec{r} \times \vec{p}\| = rp = rmv$$

where we have used the fact that in uniform circular motion the angle between the vectors $\vec{r}$ and $\vec{p}$ is always $\pi/2$. Bohr's fourth assumption tells us that $L = n\hbar$ (where n is an integer), so

$$rmv = n\hbar$$

which can be solved for the speed:

$$v = \frac{n\hbar}{rm}. \tag{28.6}$$

Equations (28.4) and (28.6) are two independent equations relating v and r. Thus, we can eliminate v from this pair of equations and solve for r, thereby finding the radius of the electron's orbit in the hydrogen atom. Substituting v from Equation (28.6) into Equation (28.4) and rearranging, we find

$$\frac{m}{r}\left(\frac{n\hbar}{rm}\right)^2 = \frac{ke^2}{r^2}$$

so

$$\frac{n^2\hbar^2}{mr^3} = \frac{ke^2}{r^2}.$$

Solving for r gives

$$r = \frac{n^2\hbar^2}{kme^2}.$$

Each different integer value of n corresponds to a different allowed value of r for the hydrogen atom. The quantized radii (and as we will see, the quantized energy values) depend on n, so n is called a **quantum number**, and we call the allowed radii r_n to show this dependence. Thus,

$$r_n = \frac{n^2\hbar^2}{kme^2}. \tag{28.7}$$

The smallest radius corresponds to a quantum number $n = 1$ and is called the **Bohr radius** a_0. Notice, that apart from n, the right side of Equation (28.7) is made up entirely of physical constants, so we can easily find a numerical value of the Bohr radius:

$$a_0 = \frac{\hbar^2}{kme^2} = \frac{(1.055 \times 10^{-34}\ \text{J})^2}{(8.988 \times 10^9\ \text{N}\cdot\text{m}^2/\text{C}^2)(9.109 \times 10^{-31}\text{kg})(1.602 \times 10^{-19}\ \text{C})^2} = 5.29 \times 10^{-11}\ \text{m}.$$

Therefore, the diameter of the electron's orbit is just a little over 1×10^{-10} m, in agreement with our previous statements about the size of the atom. The radii of the other Bohr orbits increase as the square of the quantum number n. Rewriting Equation (28.7) in terms of the Bohr radius,

$$r_n = n^2 a_0. \tag{28.8}$$

The radii of the allowed electron orbits are a_0, $4a_0$, $9a_0$, and so on.

Now that we know the radius of all the allowed orbits, we can find the corresponding energies. Substituting the radius [Equation (28.8)] into our expression for the total energy [Equation (28.5)], we find for the total energy E_n of the nth quantum level

$$E_n = -\frac{ke^2}{2r_n} = -\frac{ke^2}{2\left(n^2 a_0\right)}$$

or

$$E_n = -\frac{ke^2}{2a_0 n^2}. \tag{28.9}$$

It is convenient to collect the constants in Equation (28.9) and define another constant E_0 (with dimensions of energy) as

$$E_0 = \frac{ke^2}{2a_0}. \tag{28.10}$$

Inserting the physical constants yields a numerical value

$$E_0 = 2.18 \times 10^{-18}\text{ J} = 13.6\text{ eV}.$$

Physicists often prefer to use electron-volts as units of energy when dealing with atoms, and we will follow that practice here. From Equations (28.9) and (28.10), you can see that in terms of the constant E_0, the energy of the hydrogen atom is

$$E_n = -\frac{E_0}{n^2}. \tag{28.11}$$

The lowest possible energy (called the **ground-state energy**) corresponds to a quantum number $n = 1$. We say that the hydrogen atom is in the **ground state** when $n = 1$ and in an **excited state** when $n > 1$. The ground state-energy is clearly

$$E_1 = -\frac{E_0}{1^2} = -E_0 = -13.6\text{ eV}.$$

The next lowest energy corresponds to the $n = 2$ quantum state (the first excited state), so

$$E_2 = -\frac{E_0}{2^2} = -\frac{E_0}{4} = -3.40\text{ eV}.$$

In theory, nothing prevents a hydrogen atom from having any quantum number n, and so there are an infinite number of energy levels E_n. We show the relationship between the energy levels with an **energy-level diagram** in Figure 28.4. How do we know we have the right energies? This is where Bohr's second assumption comes in. Atoms in excited states tend to fall spontaneously into lower energy states, and eventually into the ground state (although the Bohr theory does not predict why or how this happens). Bohr's second assumption [specifically Equation (28.2)] then allows us to find the frequency or wavelength of a photon emitted when such a transition occurs.

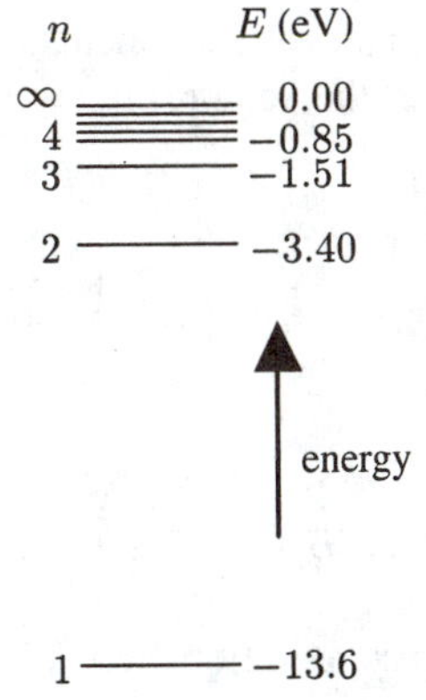

Figure 28.4. Energy-level diagram for the hydrogen atom.

For example, consider a transition from the $n = 3$ state to the $n = 2$ state. We can use Equation (28.2), together with the fact that for an electromagnetic wave the frequency is $\nu = c/\lambda$ (where c is the speed of light), to find the wavelength of the emitted photon. We have

$$E_u - E_\ell = h\nu = \frac{hc}{\lambda}.$$

Thus

$$\frac{hc}{\lambda} = -\frac{E_0}{n_u^2} - \left(-\frac{E_0}{n_\ell^2}\right) = E_0\left(\frac{1}{n_\ell^2} - \frac{1}{n_u^2}\right). \tag{28.12}$$

In this example, $n_\ell = 2$ and $n_u = 3$. Inserting these and the other numerical values, we find

$$\lambda = \frac{hc}{E_0\left(\frac{1}{n_\ell^2} - \frac{1}{n_u^2}\right)} = \frac{(4.136 \times 10^{-15}\ \text{eV}\cdot\text{s})(2.998 \times 10^8\ \text{m/s})}{(13.6\ \text{eV})\left(\frac{1}{2^2} - \frac{1}{3^2}\right)} = 6.56 \times 10^{-7}\ \text{m} = 656\ \text{nm}.$$

This wavelength corresponds to the red spectral line observed in a hydrogen discharge.

We can generalize this procedure and thereby show that all the observed hydrogen spectrum is predicted by the Bohr theory. From Equation (28.12), we see that whenever a hydrogen atom makes a transition from a higher energy state with quantum number n_u to a lower energy state with a quantum number n_ℓ, the wavelength λ of the photon emitted is given by

$$\frac{1}{\lambda} = \frac{E_0}{hc}\left(\frac{1}{n_\ell^2} - \frac{1}{n_u^2}\right).$$

For convenience, the constants h, c, and E_0 are collected together in the **Rydberg constant** R defined as

$$R = \frac{E_0}{hc} = 1.097 \times 10^7\ \text{m}^{-1}. \tag{28.13}$$

In terms of the Rydberg constant,

$$\frac{1}{\lambda} = R\left(\frac{1}{n_\ell^2} - \frac{1}{n_u^2}\right). \tag{28.14}$$

Let's examine a special case of this formula for the wavelength of a photon emitted by hydrogen: all the wavelengths that can be produced with $n_\ell = 2$ (this requires $n_u > 2$). Inserting $n_\ell = 2$ into Equation (28.14) and solving for λ, we have

$$\frac{1}{\lambda} = R\left(\frac{1}{2^2} - \frac{1}{n_u^2}\right) = R\left(\frac{n_u^2 - 4}{4n_u^2}\right)$$

so

$$\lambda = \frac{4}{R}\left(\frac{n_u^2}{n_u^2 - 4}\right).$$

Using the numerical value of the Rydberg constant in Equation (28.13), this reduces to

$$\lambda = \frac{4}{1.097 \times 10^7\ \text{m}^{-1}}\left(\frac{n_u^2}{n_u^2 - 4}\right) = 3.646 \times 10^{-7}\ \text{m}\left(\frac{n_u^2}{n_u^2 - 4}\right) = 364.6\ \text{nm}\left(\frac{n_u^2}{n_u^2 - 4}\right).$$

This matches Balmer's formula [Equation (28.1)] for the optical spectrum of hydrogen (with $n_u = s$). This correspondence between theory and experiment is one of the major successes of Bohr's atomic theory.

What about all the other electron transitions in hydrogen, with $n_\ell \neq 2$? It turns out that these can also be observed experimentally, although they are not as obvious, because the photons emitted are not in the visible part of the spectrum. For example, consider the transition from $n_u = 2$ to $n_\ell = 1$. The wavelength of the photon emitted when this transition takes place can be computed using Equation (28.14):

$$\frac{1}{\lambda} = R\left(\frac{1}{n_\ell^2} - \frac{1}{n_u^2}\right) = \left(1.097 \times 10^7 \text{ m}^{-1}\right)\left(\frac{1}{1^2} - \frac{1}{2^2}\right) = 8.228 \times 10^6 \text{ m}^{-1}.$$

Taking the reciprocal,

$$\lambda = 1.22 \times 10^{-7} \text{ m } = 122 \text{ nm}.$$

This is an ultraviolet photon. In fact, the entire series of photons (called the **Lyman series**) emitted with $n_\ell = 1$ lies in the ultraviolet region. In Figure 28.5, we show some transitions in the Lyman series, along with some for the **Balmer series** ($n_\ell = 2$) and the **Paschen series** ($n_\ell = 3$). In every series, there is a series limit—that is, the shortest possible wavelength, corresponding to the largest possible transition, a transition from $n_u = \infty$. For example, the series limit for the Paschen series means a transition from $n_u = \infty$ to $n_\ell = 3$. The wavelength of the photon emitted in this transition is given by

$$\frac{1}{\lambda} = R\left(\frac{1}{n_\ell^2} - \frac{1}{n_u^2}\right) = \left(1.097 \times 10^7 \text{ m}^{-1}\right)\left(\frac{1}{3^2} - 0\right)$$

where we have taken the limit of $1/n_u^2$ as n_u approaches infinity. Then

$$\frac{1}{\lambda} = \left(1.097 \times 10^7 \text{ m}^{-1}\right)\left(\frac{1}{9}\right) = 1.219 \times 10^6 \text{ m}^{-1}$$

or

$$\lambda = 8.20 \times 10^{-7} \text{ m } = 820 \text{ nm},$$

which is an infrared photon. The entire Paschen series lies in the infrared portion of the electromagnetic spectrum.

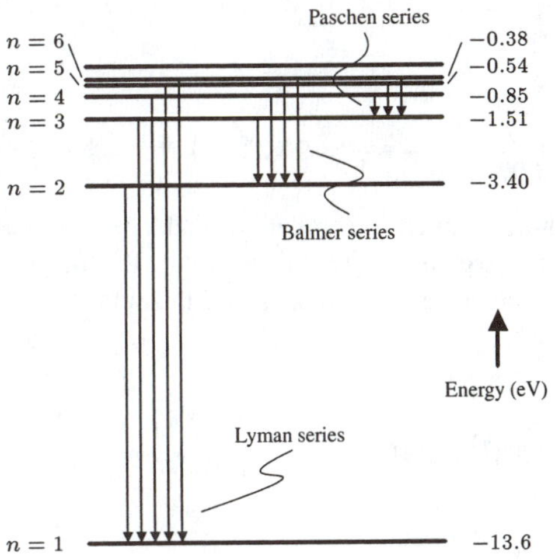

Figure 28.5. Transitions in the Lyman, Balmer, and Paschen series in hydrogen.

To summarize, Bohr's theory of the hydrogen atom correctly predicted the size of the atom, the allowed energy levels, and the transitions between levels. The success of this model gave credibility to the quantum nature of the atom and the quantization of electromagnetic radiation as photons. In Section 28.3, we encounter other experimental evidence supporting the Bohr theory. However, in Section 28.4, we explain how the Bohr theory is incomplete and eventually had to be replaced by modern quantum theory. Nevertheless, Bohr's work stands as a great achievement in twentieth-century physics and was an important step toward our current understanding of atomic physics.

Example 28.1. Using the Bohr model, find the speed of the electron in the ground state and the first excited state of hydrogen.

Solution. From Equation (28.6), the speed of the electron in the ground state ($n = 1, r = a_0$) is

$$v = \frac{n\hbar}{rm} = \frac{\hbar}{a_0 m}.$$

Inserting numerical values,

$$v = \frac{1.055 \times 10^{-34}\ \text{J·s}}{(5.29 \times 10^{-11}\ \text{m})(9.11 \times 10^{-31}\ \text{kg})} = 2.19 \times 10^6\ \text{J·s(m·kg)} = 2.19 \times 10^6\ \text{m/s}.$$

In the first excited state, $n = 2$ and $r = n^2 a_0 = 4a_0$. Thus,

$$v = \frac{n\hbar}{rm} = \frac{2\hbar}{(4a_0)\,m} = \frac{\hbar}{2a_0 m} = \frac{1.055 \times 10^{-34}\ \text{J·s}}{(2)(5.29 \times 10^{-11}\ \text{m})(9.11 \times 10^{-31}\ \text{kg})} = 1.09 \times 10^6\ \text{m/s}.$$

These speeds are high but not unreasonably so, because they are still less than 1% the speed of light. It should be clear that the electron speed decreases as the quantum number n increases. △

Example 28.2. What is the wavelength of the photon for the series limit of the Lyman series?

Solution. This series limit corresponds to a transition from $n_u = \infty$ to $n_\ell = 1$. Then, becaus the limit of $1/n_u^2$ is zero as n_u approaches infinity,

$$\frac{1}{\lambda} = R\left(\frac{1}{n_\ell^2} - \frac{1}{n_u^2}\right) = \left(1.097 \times 10^7\ \text{m}^{-1}\right)\left(\frac{1}{1^2} - 0\right) = 1.097 \times 10^7\ \text{m}^{-1},$$

so

$$\lambda = \frac{1}{1.097 \times 10^7\ \text{m}^{-1}} = 9.12 \times 10^{-8}\ \text{m} = 91.2\ \text{nm}.$$

This is the shortest-wavelength photon that can be emitted from the hydrogen atom, because it represents the largest possible difference between energy levels. Earlier, we found that the transition from $n_u = 2$ to $n_\ell = 1$ produced a 122-nm photon. Therefore, the entire Lyman series falls in the narrow range from 91.2 nm to 122 nm. △

28.3 Experimental Verification

28.3.1 X Rays

Strictly speaking, Bohr's model cannot be applied to larger, multielectron atoms. This is because the interactions between the electrons have strengths (in terms of force or potential energy) comparable to the interaction between the nucleus and the electrons. It is impossible to take these interactions into account in a precise quantitative model, since all the electrons are moving so rapidly in their orbits. However, Bohr reasoned that in larger atoms, the electrons still must be placed in quantized levels, which can be given quantum numbers $n = 1, 2, 3, \ldots$. in order of increasing distance from the nucleus. Strong evidence for the quantization of electron levels is provided in the fact that atoms (and molecules) have characteristic optical spectra. The limitations mentioned above make it impossible to predict the observed wavelengths. However, Bohr's theory can be used to make good quantitative predictions about the wavelengths (or frequencies) of x-ray photons emitted by atoms.

There are two principal ways in which x rays can be produced. One method is known by the German word **bremsstrahlung** (literally, "braking radiation"). In this process, a high-energy electron loses some of its kinetic energy when it interacts with matter, for example, an atomic nucleus, as shown in Figure 28.6. The lost kinetic energy appears as the energy $h\nu$ of a photon. The photon's energy is just the difference between the initial and final electron energies, or

$$h\nu = E_i - E_f. \tag{28.15}$$

Quantum theory places no restriction on the amount of energy that can be lost in this process. Therefore, there is no restriction on the photon frequency, and when a large number of high-energy electrons are involved in the bremsstrahlung process, a *continuous* x-ray spectrum is observed.

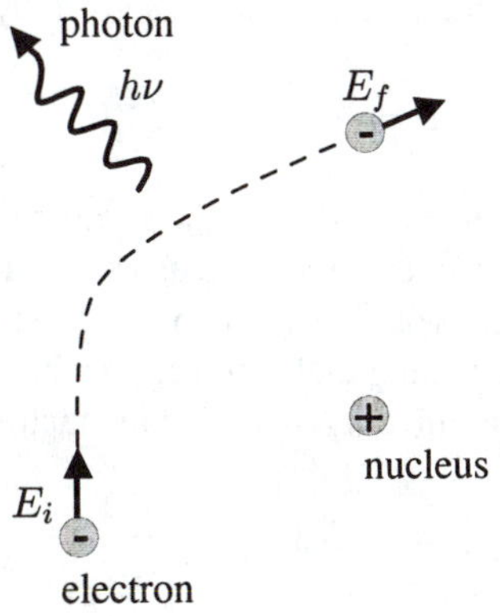

Figure 28.6. High-energy electron interacting with an atomic nucleus.

There is a second way in which x rays can be produced when high-energy electrons interact with matter. As mentioned above, electrons in an atom fill the various quantum levels $n = 1, 2, 3, \ldots$. (also called **orbitals**). As in the Bohr hydrogen atom, the higher orbitals have higher energies. If an incoming electron has sufficient energy to remove an electron from a lower orbital (say $n = 1$), the atom is not stable. An electron from a higher orbital falls to the $n = 1$ orbital, with the energy difference between the two levels appearing as a photon emitted by the atom. In larger atoms, the energies involved are such that the emitted photon is typically in the x-ray portion of the electromagnetic spectrum. X rays produced in this manner are called **characteristic x rays**. For a particular atom, there exists a characteristic x-ray spectrum similar to the optical/infrared/ultraviolet spectra in hydrogen. Only particular wavelengths are found in a characteristic spectrum, because electronic transitions can be made only from one quantum level to another.

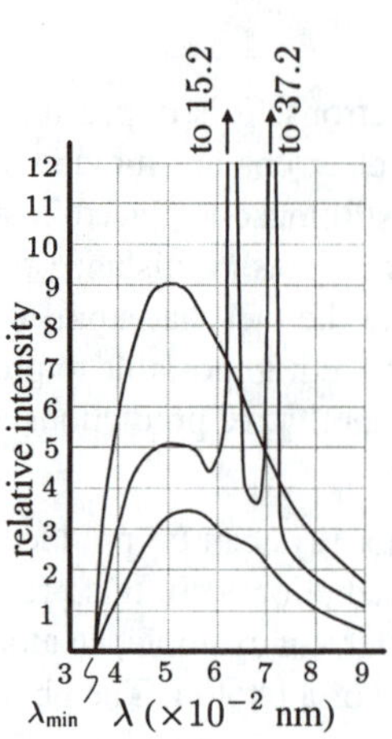

Figure 28.7. The experimental x-ray spectra of several materials. The three graphs are, from bottom to top, Cr, Mo, and W. From C.T. Ulrey, Physical Review **11**, 405 (1918).

When electrons are fired into a metal target, we often observe bremsstrahlung and characteristic x rays simultaneously. In Figure 28.7, we show the experimental x-ray spectra of several materials. The sharp peaks correspond to the characteristic x rays, and the broad peaks to bremsstrahlung.

In 1913–14, the young English experimentalist Henry G. J. Moseley (1887–1915) did the painstaking work of obtaining the characteristic x-ray spectra of numerous elements in the periodic table. He found that, for each element, the frequency ν of one characteristic x ray is given by the empirical formula

$$\nu = \frac{3cR}{4}(Z-1)^2 \tag{28.16}$$

where Z is the atomic number of the element and R is the Rydberg constant from the Bohr theory. The atomic number Z equals the number of protons in the nucleus, so the nucleus has a net charge $+Ze$.

Moseley's empirical result is consistent with Bohr's theory of the hydrogen atom. First, we note (see the problems at the end of the chapter) that for a single-electron atom with nuclear charge $+Ze$, Bohr's predicted wavelengths [Equation (28.14)] can be modified by including the factor Z^2:

$$\frac{1}{\lambda} = Z^2 R\left(\frac{1}{n_\ell^2} - \frac{1}{n_u^2}\right). \tag{28.17}$$

Because Moseley's formula is in terms of the photon frequency rather than wavelength, let's first rewrite Equation (28.17) in terms of the frequency, using the fact that for electromagnetic radiation $1/\lambda = \nu/c$. Thus,

$$\frac{\nu}{c} = Z^2 R\left(\frac{1}{n_\ell^2} - \frac{1}{n_u^2}\right)$$

or

$$\nu = Z^2 cR\left(\frac{1}{n_\ell^2} - \frac{1}{n_u^2}\right). \tag{28.18}$$

For an electronic transition with $n_u = 2$ and $n_\ell = 1$, the expression $1/n_\ell^2 - 1/n_u^2 = 3/4$, which brings us very close to Moseley's formula. Remember that the factor Z^2 has to be added for a *single-electron* atom. In a multielectron atom that produces x rays, the $n = 1$ orbital normally has two electrons. When one electron is knocked out of this orbital, an x ray is produced when another electron falls from the $n = 2$ orbital to fill that vacancy. But because one electron remains in the $n = 1$ orbital, an electron from the $n = 2$ level is attracted by a net charge of $+Ze$ (the nucleus) + $(-e)$ (the remaining electron) = $(Z - 1)e$. Therefore, when dealing with x-ray production, we need to replace Z in Equation (28.18) with $Z - 1$. With this change, we see that an x ray produced when an electron falls from $n = 2$ into the vacancy at $n = 1$ has a frequency of

$$\nu = (Z-1)^2 cR\left(\frac{1}{n_\ell^2} - \frac{1}{n_u^2}\right) = (Z-1)^2 cR\left(\frac{3}{4}\right) = \frac{3cR}{4}(Z-1)^2$$

in agreement with Moseley's experimental result.

Example 28.3. Find the theoretical value of the frequency of a photon produced when an electron is removed from the $n = 1$ orbital of molybdenum and an electron falls from the $n = 2$ orbital to fill the vacancy. Use your result to find the wavelength, and compare with the peaks in the characteristic x-ray spectrum in Figure 28.7.

Solution. Looking in the periodic table, we see $Z = 42$ for molybdenum. Therefore,

$$\nu = \frac{3cR}{4}(Z-1)^2 = \frac{3(2.997 \times 10^8 \text{m/s})(1.097 \times 10^7 \text{ m}^{-1})}{4}(41)^2 = 4.15 \times 10^{18} \text{ Hz}.$$

The wavelength of the photon is

$$\lambda = \frac{c}{\nu} = \frac{2.997 \times 10^8 \text{ m/s}}{4.15 \times 10^{18} \text{ s}^{-1}} = 7.2 \times 10^{-11} \text{ m}.$$

This computed wavelength is a good match to the second characteristic peak for the molybdenum spectrum in Figure 28.7. △

The frequency given by Equation (28.16) is just one possible x-ray frequency for an atom with atomic number Z, because as we have seen, that formula is good only for a transition from $n = 2$ to $n = 1$, and other transitions are possible. In the early days of x-ray spectroscopy, a nomenclature was developed for the different transitions, with the letters $K, L, M, \ldots.$ corresponding to the orbitals $n = 1, 2, 3, \ldots.$ (Figure 28.8). An x ray produced in a transition to the K ($n = 1$) orbital is called a $K_\alpha, K_\beta, K_\gamma, \ldots.$ x ray if the transition started on the $n = 2, 3, 4$, and so on, level as shown in the diagram. For example, a transition from $n = 2$ to $n = 1$ produces a K_α x ray, and a transition from $n = 3$ to $n = 1$ produces a K_β x ray.

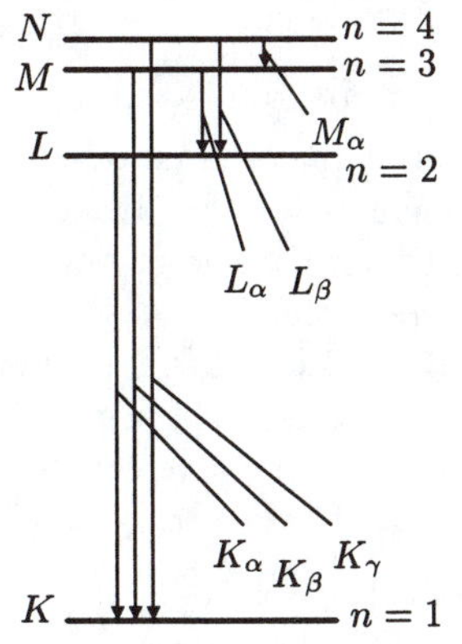

Figure 28.8. Nomenclature for x-ray transitions.

The Bohr theory can be used to predict the frequencies of other x-ray photons produced in atomic transitions. Let's consider the K_β x ray as an example, which involves a transition from $n = 3$ to $n = 1$. In this case, we have

$$\frac{1}{n_\ell^2} - \frac{1}{n_u^2} = \frac{1}{1^2} - \frac{1}{3^2} = \frac{8}{9}.$$

By the same reasoning as we used to find the frequency of the K_α x ray previously, the frequency of the K_β x ray is

$$\nu_{K_\beta} = \frac{8cR}{9}(Z-1)^2. \tag{28.19}$$

Example 28.4. Find the frequency and wavelength of a K_β photon produced in molybdenum, and compare the wavelength with the data in Figure 28.7.

Solution. The frequency is given by Equation (28.19). With $Z = 42$ for molybdenum,

$$\nu_{K_\beta} = \frac{8cR}{9}(Z-1)^2 = \frac{8(2.997 \times 10^8 \text{ m/s})(1.097 \times 10^7 \text{ m}^{-1})}{9}(41)^2 = 4.91 \times 10^{18} \text{ Hz}.$$

Then the wavelength is

$$\lambda_{K_\beta} = \frac{c}{\nu_{K_\beta}} = \frac{2.997 \times 10^8 \text{ m/s}}{4.91 \times 10^{18} \text{ s}^{-1}} = 6.10 \times 10^{-11} \text{ m}.$$

Again, this matches one of the experimental peaks very well. △

Moving on to consider the L_α x ray, the situation is not quite as simple. In its ground state, an atom (with ten or more electrons) has two electrons in the $n = 1$ orbital and eight more electrons in the $n = 2$ orbital. The L_α x ray occurs in a transition from the $n = 3$ to the $n = 2$ level. With one electron removed from the $n = 2$ level, you might expect an outer electron would "feel" a net charge of $(Z - 9)e$, with $+Ze$ coming from the nucleus and $-9e$ from the remaining $n = 1$ and $n = 2$ electrons. However, Moseley's experiments revealed that the net charge experienced by the electron falling from the $n = 3$ level to the $n = 2$ level was only about $(Z - 7.4)e$. The frequency of the L_α x ray is therefore given by

$$\nu_{L_\alpha} = \frac{5cR}{36}(Z - 7.4)^2$$

because $1/2^2 - 1/3^3 = 9/36 - 4/36 = 5/36$. This result (confirmed by experiment for many different elements) tells us that we should not think of the electron orbitals as being fixed circles with well-defined radii, as we assumed in our study of the hydrogen atom. Indeed, in modern quantum theory, the electrons themselves are not point particles. Like photons, electrons have both wave and particle properties and are spread throughout the atom in a probabilistic "cloud."

Moseley's work with x rays provided a quick justification and deeper understanding of Bohr's atomic theory, by confirming that all atoms (not just hydrogen) have quantized energy levels. Unfortunately, this was the end of Moseley's promising career. He was among the 9 million who perished fighting in World War I.

28.3.2 The Franck-Hertz Experiment

In this section, we describe briefly another experiment that confirmed the existence of quantized energy levels in atoms. In 1914, the German physicists James Franck (1882–1964) and Gustav Hertz (1887–1975)

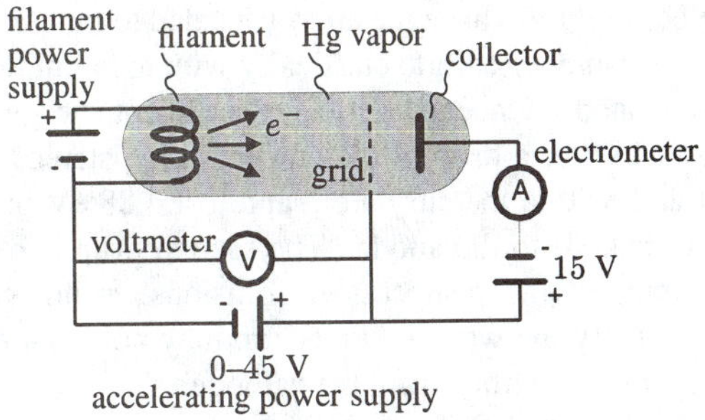

Figure 28.9. The Franck-Hertz experiment.

(the nephew of Heinrich Hertz), designed and performed an experiment with that purpose in mind. Rather than looking at atomic spectra, however, their goal was to show that in order for atoms in the ground state to reach an excited state, must absorb a definite amount of energy, corresponding to the difference between the quantum energy levels in the atom.

In the **Franck-Hertz** experiment, electrons are released from a metal cathode and attracted toward an anode at the other end of a glass tube (Figure 28.9). The tube is evacuated, except for a small amount of mercury vapor. (Mercury is typically used, but other elements, such as neon, can substitute.) The potential difference between the cathode and anode can be varied from zero to about 50 V. This means that an electron passing from the cathode to anode can achieve a kinetic energy of up to 50 eV, with the kinetic energy in electron-volts corresponding to the potential difference in volts. But what happens if an electron collides with a mercury atom on its way from the cathode to the anode? This is the crux of the experiment, and to understand what happens we show some typical results from a (modern) Franck-Hertz experiment in Figure 28.10. As the potential difference between cathode and anode is increased from zero, the electron current reaching the anode increases. This is what we would expect, because the force attracting the electrons toward the anode increases as the potential difference increases. However, as the potential difference is increased, at some point there is a sharp drop in the current, followed by another increase. The decrease and increase repeat as the potential difference is increased further.

The drops in the current occur at regular intervals of about 4.88 V. This can be explained by the fact that the difference between the energies of the ground state and first excited state in mercury is 4.88 eV. The

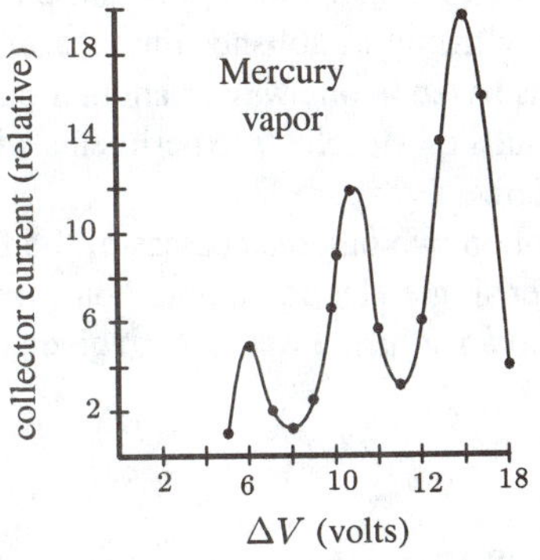

Figure 28.10. Anode current vs. accelerating potential difference in the Franck-Hertz experiment.

current rises initially, because electrons don't have the energy needed to raise a mercury atom from its ground state to its first excited state. They can only collide elastically with an atom, losing no kinetic energy in the process, so they can still reach the anode. Once they do have sufficient energy to raise a mercury atom from the ground state to the first excited state, a drop in the current is registered. Then the current rises again, because electrons that have collided with a mercury atom and lost 4.88 eV of kinetic energy in the process may still have enough energy to make it to the anode. The second drop is registered when electrons have enough energy to excite *two* mercury atoms in successive collisions. In this way, the process repeats (refer again to the graph in Figure 28.10). By the way, the first drop may not be at exactly 4.88 V, because it can take some additional energy to remove electrons from the cathode.

The explanation we have just given is plausible only if there are quantized energy levels in the mercury atom. The 4.88-eV difference between the ground state and first excited state cannot be predicted by the Bohr theory, but rather, it is experimentally determined by the Franck-Hertz experiment. Based on what you have already learned about atoms, you should not expect the excited mercury atoms to remain that way. Rather, they should spontaneously return to the ground state and in the process emit a photon of energy 4.88 eV. Franck and Hertz also realized this, and looked for photons with the right energy. The energy of a photon is related to the wavelength by

$$E = \frac{hc}{\lambda}. \tag{27.52}$$

Therefore, Franck and Hertz sought to detect electromagnetic radiation with a wavelength

$$\lambda = \frac{hc}{E} = \frac{(4.136 \times 10^{-15}\ \text{eV}\cdot\text{s})(2.997 \times 10^{8}\ \text{m/s})}{4.88\ \text{eV}} = 2.54 \times 10^{-7}\ \text{m} = 254\ \text{nm}.$$

This is in the ultraviolet part of the spectrum.. Sure enough, Franck and Hertz detected this radiation, but only after the potential difference between the anode and cathode was high enough to reach the first dip on the curve shown in Figure 28.10. This result is strong evidence in support of the existence of quantized energy levels in atoms.

28.4 The Bohr Model in Perspective

28.4.1 Successes and Limitations of the Theory

We have seen how Bohr's atomic theory can be used to make a number of predictions that can be verified experimentally. It gives the observed emission wavelengths and energy levels in hydrogen, for example. Perhaps more important, Bohr's theory firmly establishes the idea of quantized energy levels in atoms. A large amount of 20th century physics is based in one way or another on this idea. You have already seen how quantized energy levels explain the results of Moseley's experiments with characteristic x rays and the atomic excitations observed by Franck and Hertz.

Without going into detail, we mention two other successes of the Bohr theory. First, as noted in Section 28.3, the theory can be applied to other single-electron atoms (ions, really): He^{+}, Li^{++}, and so on. Because the nuclear charge is $+Ze$, this leads to an adjusted wavelength given by

$$\frac{1}{\lambda} = Z^2 R \left(\frac{1}{n_\ell^2} - \frac{1}{n_u^2} \right). \tag{28.17}$$

We have already seen how Equation (28.17) can be applied successfully to obtain the wavelengths of characteristic x rays. It can also be used to predict the observed emission spectra of single-electron ions such as He^{+}.

Another success worth mentioning is the accuracy that can be obtained if we eliminate the assumption that the proton is fixed. Recall that one of Bohr's assumptions is that the dynamics of the hydrogen atom are governed by classical mechanics when the atom is in a stationary state. Because the proton's mass is not infinite, both the proton and electron revolve around their common center of mass. This is a small correction, but an important one if precise measurements are made of the emission spectrum. It turns out that with this correction, the emission wavelengths are given by a formula similar to the one we found previously [Equation (28.14)] but with an adjusted Rydberg constant R_H. That is,

$$\frac{1}{\lambda} = R_H \left(\frac{1}{n_\ell^2} - \frac{1}{n_u^2} \right) \tag{28.20}$$

with

$$R_H = R \left(1 + \frac{m}{M} \right)^{-1}, \tag{28.21}$$

where m is the electron mass and M is the proton mass. The factor $(1 + m/M)^{-1}$ has a numerical value of about 0.99946. This factor is very close to one, but it is needed to bring Bohr's theory into better agreement with experiment when spectral lines are measured with precision. Further, this correction can be applied to any hydrogen isotope (e.g., deuterium) or hydrogenlike ion (for example, He^+), with M as the mass of the nucleus. The agreement between theory and experiment is then extremely good.

In spite of these successes, it soon became apparent that the Bohr theory was incomplete. As we have seen, it can be used to predict all the atomic states in single-electron atoms or ions (even though it can be applied to find the energies of x-ray transitions). Even in hydrogen, Bohr's theory does not account for the relative intensity of the spectral lines. This is because the theory says nothing about the relative probability of an atom being excited to a given state, or once in that state, to which lower state it may fall. The theory says nothing about how atoms join to form molecules. Another experimental fact not addressed by the Bohr theory was that in hydrogen (and other elements) each spectral line is actually a composite of several closely spaced lines. This is known as **fine structure** of the spectral lines.

The incompleteness in Bohr's theory led to a flurry of activity in the years following 1913, though with a significant reduction of activity during World War I. Finally, in 1925–1926, the German physicist Werner Heisenberg (1901–1976) and the Austrian physicist Erwin Schrödinger independently developed a more complete form of the quantum theory. The theories of Heisenberg and Schrödinger are mathematically equivalent, though they are formulated differently. The **Schrödinger wave equation** is still the basis for much of quantum theory today, and you will likely study it in your next course in physics.

28.4.2 Conclusion

Although beyond the scope of this course, Schrödinger's equation is in keeping with our theme of integrating physics and mathematics, because it is a second-order partial differential equation. Thus has it been for centuries, and so is it likely to remain: Physics and mathematics stand as independent disciplines, but they are closely related in many ways. In ancient times, the names of Pythagoras, Plato, and Archimedes stand out as pioneers in connecting mathematics with the natural world. During the scientific revolution of the seventeenth century, Galileo and Newton developed essential ties between mathematics and the physics of motion. You have seen how the physics of electromagnetism is intimately connected with vector calculus. In the twentieth century, the ties between quantum theory and abstract algebra and between general relativity and differential geometry have emerged. Other subfields of physics and mathematics continue to be connected in important ways, and there is no reason to believe this will not continue in the twenty-first century. As the mathematician and physicist Eugene Wigner wrote:

The language of mathematics reveals itself (to be) unreasonably effective in the natural sciences ... a wonderful gift which we neither understand nor deserve. We should be grateful for it and hope that it will remain valid in future research and that it will extend, for better or for worse, to our pleasure even though perhaps also to our bafflement, to wide branches of learning.

28.5 Problems

28.1 Background: Early Theories and Experiments

1. Use Balmer's formula to find all the wavelengths emitted by hydrogen in the visible part of the spectrum (400–700 nm).

2. Assume a hydrogen atom with a spherical nucleus of radius 1.2×10^{-15} m and an electron orbiting at a radius of 5.3×10^{-11} m. Taking the "atom" to be a sphere of radius equal to the radius of the electron's orbit, what is the ratio of the volume of the nucleus to the volume of the atom? (The fact that so much of the atom is "empty space" made the nuclear model difficult for scientists to believe at first.)

28.2 The Bohr Model

1. Find a general formula for the speed of the electron in the nth quantum state. Evaluate numerically for $n = 10$.

2. Physicists have recently made hydrogen atoms (called Rydberg atoms) with rather large values of n. (a) Find the radius of an electron's orbit if $n = 200$. (b) What is the wavelength of the photon emitted if a Rydberg atom makes a transition from $n = 200$ to $n = 199$? (c) Explain why Rydberg atoms are extremely unstable.

3. What is the wavelength of the photon emitted when a hydrogen atom makes a transition from $n = 7$ to $n = 2$? In what part of the electromagnetic spectrum is this?

Problems 4 and 5 deal with Bohr's *correspondence principle.*

4. Find the frequency of the photon emitted when a photon makes a transition from the $n = 100$ state to the $n = 99$ state. Compare your result with the orbital frequency (the reciprocal of the period) for the electron in the $n = 100$ orbit, showing that these two frequencies are quite close.

5. Show that in the limit as n approaches infinity, the frequency of a photon emitted when an atom makes a transition from the $n + 1$ level to the n level is equal to the orbital frequency of the electron in the n level. This is an illustration of Bohr's correspondence principle, which states that the quantum theory must reduce to the classical result in the limits where the two should agree. In this case, we have reached the classical limit, because the atom becomes of macroscopic size.

6. Find the difference between the radii of these adjacent Bohr orbits in hydrogen: (a) $n = 1$ and $n = 2$, (b) $n = 3$ and $n = 4$, (c) $n = 19$ and $n = 20$.

7. Find the kinetic energy and magnitude of the linear momentum and angular momentum of the electron in the ground state of hydrogen. Repeat for the $n = 5$ state.

8. A 150-kg satellite is in a circular orbit 230 km above Earth's surface. (a) What is the magnitude of the satellite's angular momentum? (b) Assuming that the satellite's angular momentum is quantized like the electron in hydrogen (with $L = n\hbar$), what is the quantum number n? (c) Based on this example, explain why we don't notice quantum effects in our everyday experience.

9. For the hydrogen spectral series with $n_\ell = 6$ find (a) The longest and shortest wavelength of an emitted photon, (b) the highest and lowest frequency of an emitted photon.

10. A 410-nm photon strikes a hydrogen atom. What is the highest possible energy state to which the atom can be raised if its initial state is (a) $n = 1$, (b) $n = 2$, (c) $n = 3$?

11. (a) Find the photon energies corresponding to the limits of the visible spectrum, with wavelengths of 400 nm and 700 nm. (b) Find the photon energies corresponding to each of the visible photons that can be emitted by hydrogen.

28.3 Experimental Verification

1. Find the frequency and wavelength of the K_α x ray from the following elements: (a) aluminum, (b) iron, (c) lead.

2. Repeat Problem 1 for the K_β x ray.

3. Repeat Problem 1 for the L_α x ray.

4. In this problem, you are to find an approximate range of x-ray energies by comparing a weak x ray (a L_α transition in a light metal) and a strong x ray (a K_β transition in a heavy metal). (a) Find the energy (in electron-volts) and wavelength of the L_α transition in aluminum. (b) Find the energy and wavelength of the K_β transition in uranium. (c) Compare the energies you found in (a) and (b).

5. Find the energy of the K_α transition in potassium. Which element has a K_a transition with twice this much energy?

6. Which element has a K_α photon frequency of 4.15 $\times 10^{18}$ Hz? Find the K_β and L_α frequencies for this element.

7. Find a general formula for the frequency of the K_γ x ray, as a function of the atomic number Z.

8. Consider a Franck-Hertz experiment with an atom other than mercury. What are the possible values of the difference between the ground state and first excited state if visible photons are to be observed from the atomic excitations?

28.4 The Bohr Model in Perspective

Problems 1–3 refer to electronic transitions in the single-electron He^+ ion.

1. Find the wavelength of the photon emitted in a transition from (a) $n = 2$ to $n = 1$, (b) $n = 3$ to $n = 2$. (c) Compare the results in both (a) and (b) with the wavelengths of the same transitions in hydrogen.

2. What is the range of possible wavelengths for the "Balmer series" with $n_\ell = 2$?

3. Describe all the *visible* (i.e., $\lambda = 400$–700 nm) photons that can be emitted by He^+ in any transition.

4. Use the corrected Rydberg constant to find the wavelength of the red Balmer line in hydrogen to five significant digits.

5. Repeat the preceding problem for *deuterium*, the isotope of hydrogen with one proton and one neutron in the nucleus. Compare the numerical value of the wavelength with the corresponding line in hydrogen.

APPENDIX A

Mathematics

ALGEBRA AND TRIGONOMETRY

Quadratic Formula

If $ax^2 + bx + c = 0$ then $x = \dfrac{-b \pm \sqrt{b^2 - 4ac}}{2a}$

Circumference, Area, Volume

Where $\pi \simeq 3.14159 \ldots$

circumference of circle	$2\pi r$
area of circle	πr^2
surface area of sphere	$4\pi r^2$
volume of sphere	$\frac{4}{3}\pi r^3$
area of triangle	$\frac{1}{2}bh$
volume of cylinder	$\pi r^2 \ell$

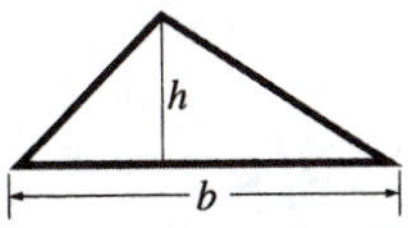

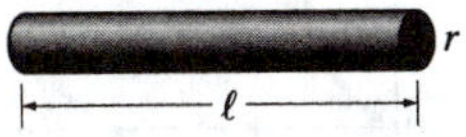

Trigonometry

definition of angle (in radians): $\theta = \dfrac{s}{r}$

2π radians in complete circle
1 radian $\simeq 57.3°$

Trigonometric Functions

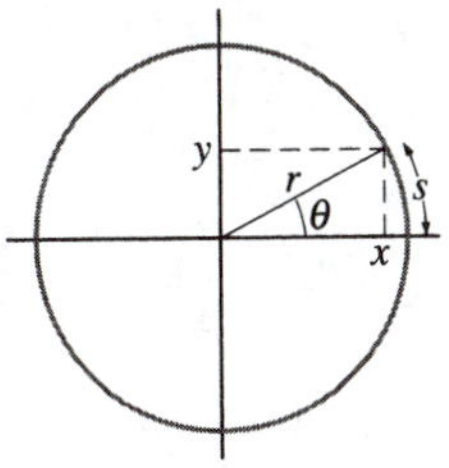

$$\sin\theta = \frac{y}{r}$$

$$\cos\theta = \frac{x}{r}$$

$$\tan\theta = \frac{\sin\theta}{\cos\theta} = \frac{y}{x}$$

Values at Selected Angles

$\theta \rightarrow$	0	$\frac{\pi}{6}$ (30°)	$\frac{\pi}{4}$ (45°)	$\frac{\pi}{3}$ (60°)	$\frac{\pi}{2}$ (90°)
$\sin\theta$	0	$\frac{1}{2}$	$\frac{\sqrt{2}}{2}$	$\frac{\sqrt{3}}{2}$	1
$\cos\theta$	1	$\frac{\sqrt{3}}{2}$	$\frac{\sqrt{2}}{2}$	$\frac{1}{2}$	0
$\tan\theta$	0	$\frac{\sqrt{3}}{3}$	1	$\sqrt{3}$	∞

Graphs of Trigonometric Functions

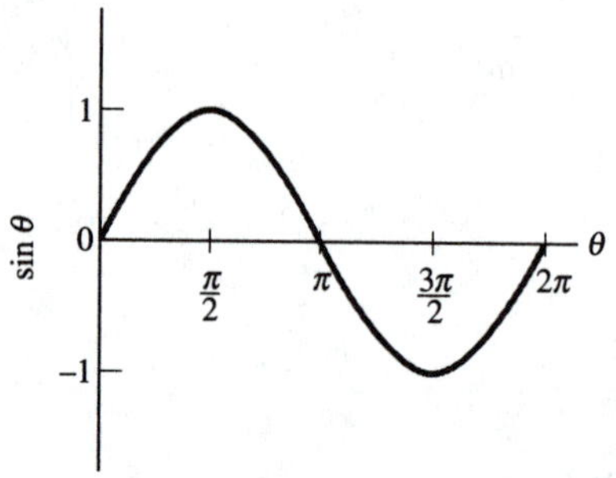

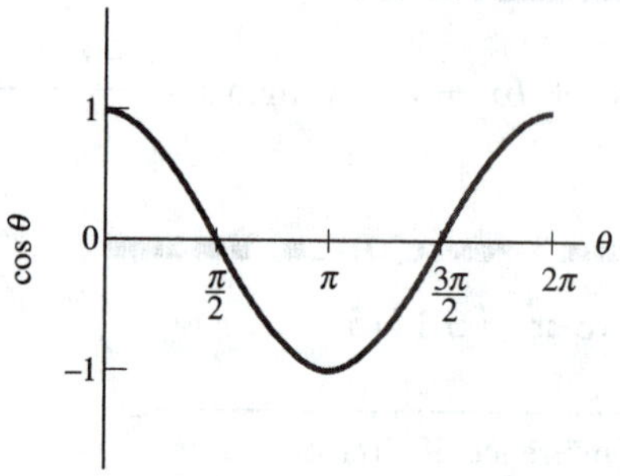

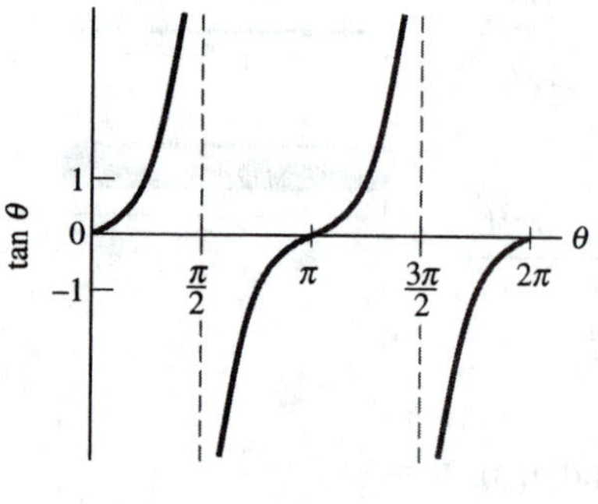

Trigonometric Identities

$$\sin(-\theta) = -\sin\theta$$

$$\cos(-\theta) = \cos\theta$$

$$\sin\left(\theta \pm \frac{\pi}{2}\right) = \pm\cos\theta$$

$$\cos\left(\theta \pm \frac{\pi}{2}\right) = \mp\sin\theta$$

$$\sin^2\theta + \cos^2\theta = 1$$

$$\sin 2\theta = 2\sin\theta\cos\theta$$

$$\cos 2\theta = \cos^2\theta - \sin^2\theta = 1 - 2\sin^2\theta = 2\cos^2\theta - 1$$

$$\sin(\alpha \pm \beta) = \sin\alpha\cos\beta \pm \cos\alpha\sin\beta$$

$$\cos(\alpha \pm \beta) = \cos\alpha\cos\beta \mp \sin\alpha\sin\beta$$

$$\sin\alpha \pm \sin\beta = 2\sin[\tfrac{1}{2}(\alpha \pm \beta)]\cos[\tfrac{1}{2}(\alpha \mp \beta)]$$

$$\cos\alpha + \cos\beta = 2\cos[\tfrac{1}{2}(\alpha + \beta)]\cos[\tfrac{1}{2}(\alpha - \beta)]$$

$$\cos\alpha - \cos\beta = -2\sin[\tfrac{1}{2}(\alpha + \beta)]\sin[\tfrac{1}{2}(\alpha - \beta)]$$

Laws of Cosines and Sines

Where A, B, C are the sides of an arbitrary triangle and α, β, γ the angles opposite those sides:

Law of cosines

$$C^2 = A^2 + B^2 - 2AB\cos\gamma$$

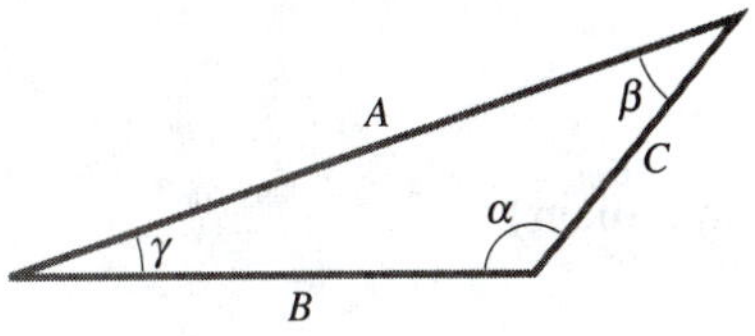

Law of sines

$$\frac{\sin\alpha}{A} = \frac{\sin\beta}{B} = \frac{\sin\gamma}{C}$$

Exponentials and Logarithms

Graphs

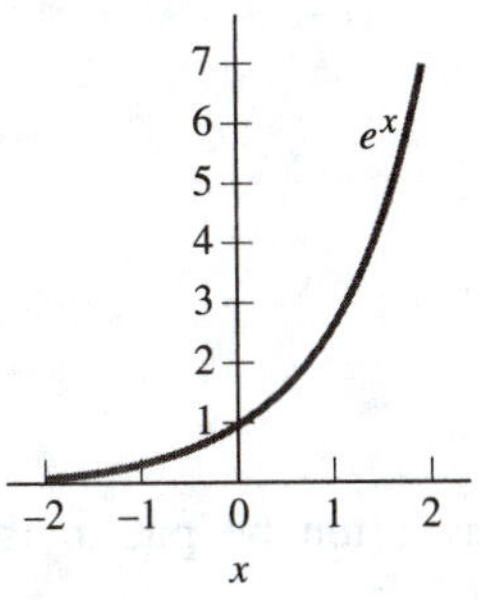

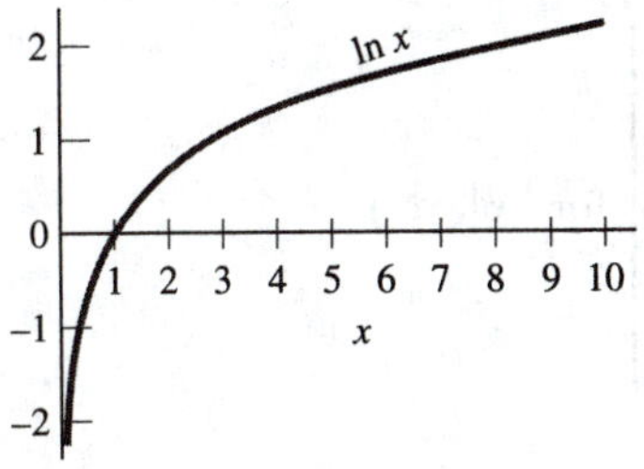

Exponential and Natural Logarithms Are Inverse Functions

$$e^{\ln x} = x, \quad \ln e^x = x \quad e = 2.71828\ldots.$$

Exponential and Logarithmic Identities

$a^x = e^{x \ln a}$ $\qquad \ln(xy) = \ln x + \ln y$

$a^x a^y = a^{x+y}$ $\qquad \ln\left(\frac{x}{y}\right) = \ln x - \ln y$

$(a^x)^y = a^{xy}$ $\qquad \ln\left(\frac{1}{x}\right) = -\ln x$

$\log x \equiv \log_{10} x = \ln(10) \ln x \simeq 2.3 \ln x$

Expansions and Approximations

Series Expansions of Functions

Note: $n! = n(n-1)(n-2)(n-3)\cdots(3)(2)(1)$

$$e^x = 1 + x + \frac{x^2}{2!} + \frac{x^3}{3!} + \cdots \quad \text{(exponential)}$$

$$\left.\begin{aligned} \sin x &= x - \frac{x^3}{3!} + \frac{x^5}{5!} - \cdots \quad \text{(sine)} \\ \cos x &= 1 - \frac{x^2}{2!} + \frac{x^4}{4!} - \cdots \quad \text{(cosine)} \end{aligned}\right\} (x \text{ in radians})$$

$$\ln(1 + x) = x - \frac{x^2}{2} + \frac{x^3}{3} - \cdots \quad \text{(natural logarithm)}$$

$$(1 + x)^p = 1 + px + \frac{p(p-1)}{2!}x^2 + \frac{p(p-1)(p-2)}{3!}x^3 + \cdots$$

(binomial, valid for $|x| < 1$)

Approximations

For $|x| \ll 1$, the first few terms in the series provide a good approximation; that is,

$$\left.\begin{aligned} e^x &\simeq 1 + x \\ \sin x &\simeq x \\ \cos x &\simeq 1 - \tfrac{1}{2}x^2 \\ \ln(1 + x) &\simeq x \\ (1 + x)^p &\simeq 1 + px \end{aligned}\right\} \text{for } |x| \ll 1$$

Expressions that do not have the forms shown may often be put in the appropriate form. For example:

$$\frac{1}{\sqrt{a^2 + y^2}} = \frac{1}{a\sqrt{1 + \dfrac{y^2}{a^2}}} = \frac{1}{a}\left(1 + \frac{y^2}{a^2}\right)^{-1/2}.$$

For $y^2 \ll a^2$, this may be approximated using the binomial expansion $(1 + x)^p \simeq 1 + px$, with $p = -\frac{1}{2}$ and $x = y^2/a^2$:

$$\frac{1}{a}\left(1 + \frac{y^2}{a^2}\right)^{-1/2} \simeq \frac{1}{a}\left(1 - \frac{1}{2}\frac{y^2}{a^2}\right).$$

Vector Algebra

Vector Products

$\mathbf{A} \cdot \mathbf{B} = AB\cos\theta$

$|\mathbf{A} \times \mathbf{B}| = AB\sin\theta$, with direction of $\mathbf{A}\times\mathbf{B}$ given by right-hand rule:

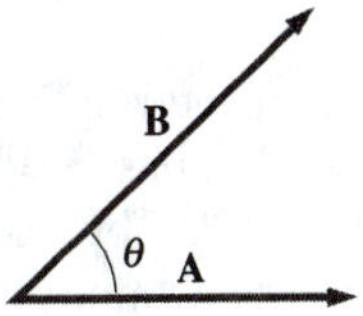

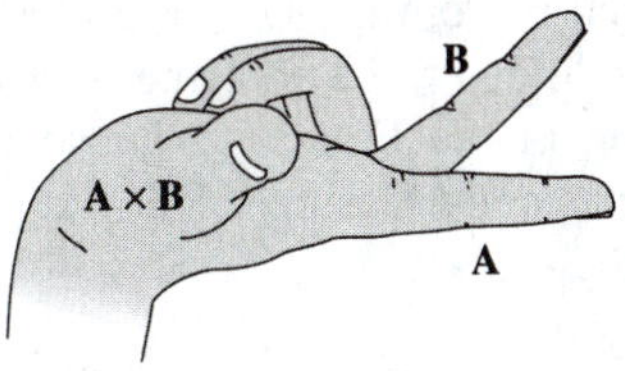

Unit Vector Notation

An arbitrary vector $\mathbf{A}$ may be written in terms of its components A_x, A_y, A_z and the unit vectors $\hat{\mathbf{i}}$, $\hat{\mathbf{j}}$, $\hat{\mathbf{k}}$ that have length 1 and lie along the x, y, z axes:

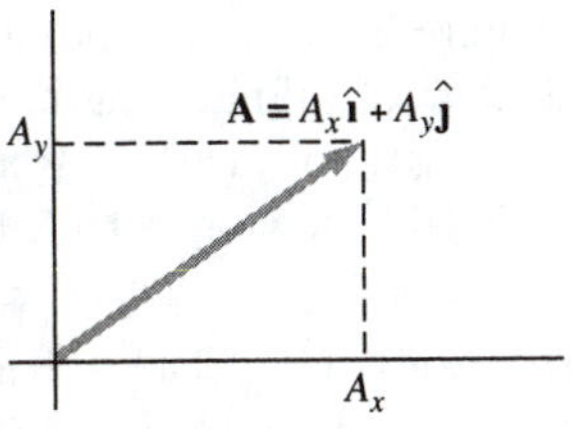

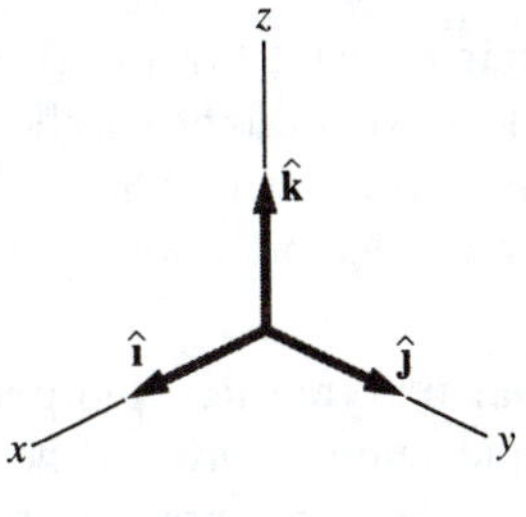

In unit vector notation, vector products become

$\mathbf{A} \cdot \mathbf{B} = A_xB_x + A_yB_y + A_zB_z$

$\mathbf{A} \times \mathbf{B} = (A_yB_z - A_zB_y)\hat{\mathbf{i}} + (A_zB_x - A_xB_z)\hat{\mathbf{j}} + (A_xB_y - A_yB_x)\hat{\mathbf{k}}$

Vector Identities

$\mathbf{A} \cdot \mathbf{B} = \mathbf{B} \cdot \mathbf{A}$

$\mathbf{A} \times \mathbf{B} = -\mathbf{B} \times \mathbf{A}$

$\mathbf{A} \cdot (\mathbf{B} \times \mathbf{C}) = \mathbf{B} \cdot (\mathbf{C} \times \mathbf{A}) = \mathbf{C} \cdot (\mathbf{A} \times \mathbf{B})$

$\mathbf{A} \times (\mathbf{B} \times \mathbf{C}) = (\mathbf{A} \cdot \mathbf{C})\mathbf{B} - (\mathbf{A} \cdot \mathbf{B})\mathbf{C}$

APPENDIX B

The International System of Units (SI)

This material is from the United States edition of the English translation of the sixth edition of "Le Système International d'Unités (SI)," the definitive publication in the French language issued in 1991 by the International Bureau of Weights and Measures (BIPM). The year the definition was adopted is given in parentheses.

unit of length (meter): The meter is the length of the path traveled by light in vacuum during a time interval of 1/299 792 458 of a second. (1983)

unit of mass (kilogram): The kilogram is the unit of mass; it is equal to the mass of the international prototype of the kilogram. (1889)

unit of time (second): The second is the duration of 9 192 631 770 periods of the radiation corresponding to the transition between the two hyperfine levels of the ground state of the cesium-133 atom. (1967)

unit of electric current (ampere): The ampere is that constant current which, if maintained in two straight parallel conductors of infinite length, of negligible circular cross section, and placed 1 meter apart in vacuum, would produce between these conductors a force equal to 2×10^{-7} newton per meter of length. (1948)

unit of thermodynamic temperature (kelvin): The kelvin, unit of thermodynamic temperature, is the fraction 1/273.16 of the thermodynamic temperature of the triple point of water. (1957) Also, the unit kelvin and its symbol K should be used to express an interval or a difference of temperature.

unit of amount of substance (mole): (1) The mole is the amount of substance of a system that contains as many elementary entities as there are atoms in 0.012 kilogram of carbon 12. (1971) (2) When the mole is used, the elementary entities must be specified and may be atoms, molecules, ions, electrons, other particles, or specified groups of such particles.

unit of luminous intensity (candela): The candela is the luminous intensity, in a given direction, of a source that emits monochromatic radiation of frequency 540×10^{12} hertz and that has a radiant intensity in that direction of (1/683) watt per steradian. (1979)

SI Base and Supplementary Units

QUANTITY	SI UNIT NAME	SI UNIT SYMBOL
Base Unit		
Length	meter	m
Mass	kilogram	kg
Time	second	s
Electric current	ampere	A
Thermodynamic temperature	kelvin	K
Amount of substance	mole	mol
Luminous intensity	candela	cd
Supplementary Units		
Plane angle	radian	rad
Solid angle	steradian	sr

SI Prefixes

FACTOR	PREFIX	SYMBOL
10^{24}	yotta	Y
10^{21}	zetta	Z
10^{18}	exa	E
10^{15}	peta	P
10^{12}	tera	T
10^{9}	giga	G
10^{6}	mega	M
10^{3}	kilo	k
10^{2}	hecto	h
10^{1}	deka	da
10^{0}	—	—
10^{-1}	deci	d
10^{-2}	centi	c
10^{-3}	milli	m
10^{-6}	micro	μ
10^{-9}	nano	n
10^{-12}	pico	p
10^{-15}	femto	f
10^{-18}	atto	a
10^{-21}	zepto	z
10^{-24}	yocto	y

Some SI Derived Units with Special Names

QUANTITY	SI UNIT NAME	SYMBOL	EXPRESSION IN TERMS OF OTHER UNITS	EXPRESSION IN TERMS OF SI BASE UNITS
Frequency	hertz	Hz		s^{-1}
Force	newton	N		$m \cdot kg \cdot s^{-2}$
Pressure, stress	pascal	Pa	N/m^2	$m^{-1} \cdot kg \cdot s^{-2}$
Energy, work, heat	joule	J	$N \cdot m$	$m^2 \cdot kg \cdot s^{-2}$
Power	watt	W	J/s	$m^2 \cdot kg \cdot s^{-3}$
Electric charge	coulomb	C		$s \cdot A$
Electric potential, potential difference, electromotive force	volt	V	J/C	$m^2 \cdot kg \cdot s^{-3} \cdot A^{-1}$
Capacitance	farad	F	C/V	$m^{-2} \cdot kg^{-1} \cdot s^4 \cdot A^2$
Electric resistance	ohm	Ω	V/A	$m^2 \cdot kg \cdot s^{-3} \cdot A^{-2}$
Magnetic flux	weber	Wb	$V \cdot s$	$m^2 \cdot kg \cdot s^{-2} \cdot A^{-1}$
Magnetic field	tesla	T	Wb/m^2	$kg \cdot s^{-2} \cdot A^{-1}$
Inductance	henry	H	Wb/A	$m^2 \cdot kg \cdot s^{-2} \cdot A^{-2}$
Radioactivity	becquerel	Bq	1 decay/s	s^{-1}
Absorbed radiation dose	gray	Gy	J/kg, 100 rad	$m^2 \cdot s^{-2}$
Radiation dose equivalent	sievert	Sv	J/kg, 100 rem	$m^2 \cdot s^{-2}$

APPENDIX C

Conversion Factors

The listings below give the SI equivalents of non-SI units. To convert from the units shown to SI, multiply by the factor given; to convert the other way, divide. For conversions within the SI system see table of SI prefixes in Appendix B, Chapter 1, or inside front cover. Conversions that are not exact by definition are given to, at most, 4 significant figures.

Length

1 inch (in.) = 0.0254 m
1 foot (ft) = 0.3048 m
1 yard (yd) = 0.9144 m
1 mile (mi) = 1609 m
1 nautical mile = 1852 m

1 angstrom (Å) = 10^{-10} m
1 light-year (ly) = 9.46×10^{15} m
1 astronomical unit (AU) = 1.5×10^{11} m
1 parsec = 3.09×10^{16} m
1 fermi = 10^{-15} m = 1 fm

Mass

1 slug = 14.59 kg
1 metric ton (tonne; T) = 1000 kg

1 unified mass unit (u) = 1.660×10^{-27} kg

Force units in the English system are sometimes used (incorrectly) for mass. The units given below are actually equal to the number of kilograms multiplied by g, the acceleration of gravity.

1 pound (lb) = weight of 0.454 kg
1 ton = 2000 lb = weight of 908 kg

1 ounce (oz) = weight of 0.02835 kg

Time

1 minute (min) = 60 s
1 hour (h) = 60 min = 3600 s

1 day (d) = 24 h = 86 400 s
1 year (y) = 365.2422 d* = 3.156×10^{7} s

* The length of the year changes very slowly with changes in Earth's orbital period.

Area

1 hectare (ha) = 10^4 m^2
1 square inch (in.^{2}) = 6.452×10^{-4} m^2
1 square foot (ft^2) = 9.290×10^{-2} m^2
1 acre = 4047 m^2
1 barn = 10^{-28} m^2
1 shed = 10^{-30} m^2

Volume

1 liter (L) = 1000 cm^3 = 10^{-3} m^3
1 cubic foot (ft^3) = 2.832×10^{-2} m^3
1 cubic inch (in.^{3}) = 1.639×10^{-5} m^3
1 fluid ounce = 1/128 gal = 2.957×10^{-5} m^3
1 barrel = 42 gal = 0.1590 m^3
1 gallon (U.S.; gal) = 3.785×10^{-3} m^3
1 gallon (British) = 4.546×10^{-3} m^3

Angle, Phase

1 degree (°) = $\pi/180$ rad = 1.745×10^{-2} rad
1 revolution (rev) = 360° = 2π rad
1 cycle = 360° = 2π rad

Speed, Velocity

1 km/h = (1/3.6) m/s = 0.2778 m/s
1 mi/h (mph) = 0.4470 m/s
1 ft/s = 0.3048 m/s
1 ly/y = 3.00×10^8 m/s

Angular Speed, Angular Velocity, Frequency, and Angular Frequency

1 rev/s = 2π rad/s = 6.283 rad/s (s^{-1})
1 Hz = 1 cycle/s = 2πs^{-1}
1 rev/min (rpm) = 0.1047 rad/s (s^{-1})

Force

1 dyne = 10^{-5} N
1 pound (lb) = 4.448 N

Pressure

1 dyne/cm^2 = 0.10 Pa
1 atmosphere (atm) = 1.013×10^5 Pa
1 torr = 1 mm Hg at 0°C = 133.3 Pa
1 bar = 10^5 Pa = 0.987 atm
1 lb/in.^{2} (psi) = 6.895×10^3 Pa
1 in. H_2O (60°F) = 248.8 Pa
1 in. Hg (60°F) = 3.377×10^3 Pa

Energy, Work, Heat

1 erg = 10^{-7} J
1 calorie* (cal) = 4.184 J
1 electron-volt (eV) = 1.602×10^{-19}J
1 foot-pound (ft·lb) = 1.356 J
1 Btu* = 1.054×10^3 J
1 kWh = 3.6×10^6 J
1 megaton (explosive yield; Mt) = 4.18×10^{15} J

* Values based on the thermochemical calorie; other definitions vary slightly.

Power

1 erg/s = 10^{-7} W
1 horsepower (hp) = 746 W
1 Btu/h (Btuh) = 0.293 W
1 ft·lb/s = 1.356 W

Magnetic Field

1 gauss (G) = 10^{-4} T
1 gamma (γ) = 10^{-9} T

Radiation

1 curie (ci) = 3.7×10^{10} Bq
1 rad = 10^{-2} Gy
1 rem = 10^{-2} Sv

Energy Content of Fuels

ENERGY SOURCE	ENERGY CONTENT
Coal	2.9×10^{7} J/kg = 7300 kWh/ton = 25×10^{6} Btu/ton
Oil	43×10^{6} J/kg = 39 kWh/gal = 1.3×10^{5} Btu/gal
Gasoline	44×10^{6} J/kg = 36 kWh/gal = 1.2×10^{5} Btu/gal
Natural gas	55×10^{6} J/kg = 30 kWh/100 ft^3 = 1000 Btu/ft^3
Uranium (fission)	
Normal abundance	5.8×10^{11} J/kg = 1.6×10^{5} kWh/kg
Pure U-235	8.2×10^{13} J/kg = 2.3×10^{7} kWh/kg
Hydrogen (fusion)	
Normal abundance	7×10^{11} J/kg = 3.0×10^{4} kWh/kg
Pure deuterium	3.3×10^{14} J/kg = 9.2×10^{7} kWh/kg
Water	1.2×10^{10} J/kg = 1.3×10^{4} kWh/gal = 340 gal gasoline/gal
H_2O	
100% conversion, matter to energy	9.0×10^{16} J/kg = 931 MeV/u = 2.5×10^{10} kWh/kg

APPENDIX D

The Elements

The atomic weights of stable elements reflect the abundances of different isotopes; values given here apply to elements as they exist naturally on Earth. For stable elements, parentheses express uncertainties in the last decimal place given. For elements with no stable isotopes (indicated in boldface), sets of most important isotopes are given. (Exceptions are the unstable elements thorium, protactinium, and uranium, for which atomic weights reflect natural abundances of long-lived isotopes.) See also periodic table inside back cover.

ATOMIC NUMBER	NAMES	SYMBOL	ATOMIC WEIGHT
1	Hydrogen	H	1.00794 (7)
2	Helium	He	4.002602 (2)
3	Lithium	Li	6.941 (2)
4	Beryllium	Be	9.012182 (3)
5	Boron	B	10.811 (5)
6	Carbon	C	12.011 (1)
7	Nitrogen	N	14.00674 (7)
8	Oxygen	O	15.9994 (3)
9	Fluorine	F	18.9984032 (9)
10	Neon	Ne	20.1797 (6)
11	Sodium (Natrium)	Na	22.989768 (6)
12	Magnesium	Mg	24.3050 (6)
13	Aluminum	Al	26.981539 (5)
14	Silicon	Si	28.0855 (3)
15	Phosphorus	P	30.973762 (4)
16	Sulfur	S	32.066 (6)
17	Chlorine	Cl	35.4527 (9)
18	Argon	Ar	39.948 (1)
19	Potassium (Kalium)	K	39.0983 (1)
20	Calcium	Ca	40.078 (4)
21	Scandium	Sc	44.955910 (9)
22	Titanium	Ti	47.88 (3)
23	Vanadium	V	50.9415 (1)
24	Chromium	Cr	51.9961 (6)
25	Manganese	Mn	54.93805 (1)
26	Iron	Fe	55.847 (3)
27	Cobalt	Co	58.93320 (1)

ATOMIC NUMBER	NAMES	SYMBOL	ATOMIC WEIGHT
28	Nickel	Ni	58.69 (1)
29	Copper	Cu	63.546 (3)
30	Zinc	Zn	65.39 (2)
31	Gallium	Ga	69.723 (1)
32	Germanium	Ge	72.61 (2)
33	Arsenic	As	74.92159 (2)
34	Selenium	Se	78.96 (3)
35	Bromine	Br	79.904 (1)
36	Krypton	Kr	83.80 (1)
37	Rubidium	Rb	85.4678 (3)
38	Strontium	Sr	87.62 (1)
39	Yttrium	Y	88.90585 (2)
40	Zirconium	Zr	91.224 (2)
41	Niobium	Nb	92.90638 (2)
42	Molybdenum	Mo	95.94 (1)
43	**Technetium**	**Tc**	**97, 98, 99**
44	Ruthenium	Ru	101.07 (2)
45	Rhodium	Rh	102.90550 (3)
46	Palladium	Pd	106.42 (1)
47	Silver	Ag	107.8682 (2)
48	Cadmium	Cd	112.411 (8)
49	Indium	In	114.82 (1)
50	Tin	Sn	118.710 (7)
51	Antimony (Stibium)	Sb	121.75 (3)
52	Tellurium	Te	127.60 (3)
53	Iodine	I	126.90447 (3)
54	Xenon	Xe	131.29 (2)
55	Cesium	Cs	132.90543 (5)
56	Barium	Ba	137.327 (7)
57	Lanthanum	La	138.9055 (2)
58	Cerium	Ce	140.115 (4)
59	Praseodymium	Pr	140.90765 (3)
60	Neodymium	Nd	144.24 (3)
61	**Promethium**	**Pm**	**145, 147**
62	Samarium	Sm	150.36 (3)
63	Europium	Eu	151.965 (9)
64	Gadolinium	Gd	157.25 (3)
65	Terbium	Tb	158.92534 (3)
66	Dysprosium	Dy	162.50 (3)
67	Holmium	Ho	164.93032 (3)
68	Erbium	Er	167.26 (3)
69	Thulium	Tm	168.93421 (3)
70	Ytterbium	Yb	173.04 (3)
71	Lutetium	Lu	174.967 (1)
72	Hafnium	Hf	178.49 (2)
73	Tantalum	Ta	180.9479 (1)
74	Tungsten (Wolfram)	W	183.85 (3)
75	Rhenium	Re	186.207 (1)
76	Osmium	Os	190.2 (1)
77	Iridium	Ir	192.22 (3)
78	Platinum	Pt	195.08 (3)
79	Gold	Au	196.96654 (3)
80	Mercury	Hg	200.59 (3)
81	Thallium	Tl	204.3833 (2)
82	Lead	Pb	207.2 (1)
83	Bismuth	Bi	208.98037 (3)

ATOMIC NUMBER	NAMES	SYMBOL	ATOMIC WEIGHT
84	Polonium	Po	209, 210
85	Astatine	At	210, 211
86	Radon	Rn	211, 220, 222
87	Francium	Fr	223
88	Radium	Ra	223, 224, 226, 228
89	Actinium	Ac	227
90	Thorium	Th	232.0381 (1)
91	Protactinium	Pa	231.03588 (2)
92	Uranium	U	238.0289 (1)
93	Neptunium	Np	237, 239
94	Plutonium	Pu	238, 239, 240, 241, 242, 244
95	Americium	Am	241, 243
96	Curium	Cm	243, 244, 245, 246, 247, 248
97	Berkelium	Bk	247, 249
98	Californium	Cf	249, 250, 251, 252
99	Einsteinium	Es	252
100	Fermium	Fm	257
101	Mendelevium	Md	255, 256, 258, 260
102	Nobelium	No	253, 254, 255, 259
103	Lawrencium	Lr	256, 258, 259, 261
104	Rutherfordium	Rf	257, 259, 260, 261
105	Dubnium	Db	260, 261, 262
106	Seaborgium	Sg	259, 260, 261, 263
107	Bohrium	Bh	261, 262
108	Hassium	Hs	264, 265
109	Meitnerium	Mt	266
110	—	—	269
111	—	—	272
112	—	—	277

APPENDIX E

Astrophysical Data

SUN, PLANETS, PRINCIPAL SATELLITES

BODY	MASS (10^{24} kg)	MEAN RADIUS (10^6 m EXCEPT AS NOTED)	SURFACE GRAVITY (m/s²)	ESCAPE SPEED (km/s)	SIDEREAL ROTATION PERIOD* (days)	MEAN DISTANCE FROM CENTRAL BODY† (10^6 km)	ORBITAL PERIOD	ORBITAL SPEED (km/s)
Sun	1.99×10^6	696	274	618	36 at poles 27 at equator	2.6×10^{11}	200 My	250
Mercury	0.330	2.44	3.70	4.25	58.6	57.6	88.0 d	48
Venus	4.87	6.05	8.87	10.4	−243	108	225 d	35
Earth	5.97	6.37	9.81	11.2	0.997	150	365.3 d	30
Moon	0.0735	1.74	1.62	2.38	27.3	0.385	27.3 d	1.0
Mars	0.642	3.38	3.74	5.03	1.03	228	1.88 y	24.1
Phobos	9.6×10^{-9}	9-13 km	0.001	0.008	0.32	9.4×10^{-3}	0.32 d	2.1
Deimos	2×10^{-9}	5-8 km	0.001	0.005	1.3	23×10^{-3}	1.3 d	1.3
Jupiter	1.90×10^3	69.1	26.5	60.6	0.414	778	11.9 y	13.0
Io	0.0889	1.82	1.8	2.6	1.77	0.422	1.77 d	17
Europa	0.478	1.57	1.3	2.0	3.55	0.671	3.55 d	14
Ganymede	0.148	2.63	1.4	2.7	7.15	1.07	7.15 d	11
Callisto	0.107	2.40	1.2	2.4	16.7	1.88	16.7 d	8.2
and 13 smaller satellites								
Saturn	569	56.8	11.8	36.6	0.438	1.43×10^3	29.5 y	9.65
Tethys	0.0007	0.53	0.2	0.4	1.89	0.294	1.89 d	11.3
Dione	0.00015	0.56	0.3	0.6	2.74	0.377	2.74 d	10.0
Rhea	0.0025	0.77	0.3	0.5	4.52	0.527	4.52 d	8.5
Titan	0.135	2.58	1.4	2.6	15.9	1.22	15.9 d	5.6
Iapetus	0.0019	0.73	0.2	0.6	79.3	3.56	79.3 d	3.3
and 12 smaller satellites								
Uranus	86.6	25.0	9.23	21.5	−0.65	2.87×10^3	84.1 y	6.79
Ariel	0.0013	0.58	0.3	0.4	2.52	0.19	2.52 d	5.5
Umbriel	0.0013	0.59	0.3	0.4	4.14	0.27	4.14 d	4.7
Titania	0.0018	0.81	0.2	0.5	8.70	0.44	8.70 d	3.7
Oberon	0.0017	0.78	0.2	0.5	13.5	0.58	13.5 d	3.1
and 11 smaller satellites								
Neptune	103	24.0	11.9	23.9	0.768	4.50×10^3	165 y	5.43
Triton	0.134	1.9	2.5	3.1	5.88	0.354	5.88 d	4.4
and 7 smaller satellites								
Pluto	0.015	1.2	0.4	1.2	−6.39	5.91×10^3	249 y	4.7
Charon	0.001	0.6			−6.39	0.02	6.39 d	0.2

*Negative rotation period indicates retrograde motion, in opposite sense from orbital motion. Periods are sidereal, meaning the time for the body to return to the same orientation relative to the distant stars rather than the Sun.

†Central body is galactic center for Sun, Sun for planets, and planet for satellites.

APPENDIX F

A Brief Table of Integrals

1. $\int u\,dv = uv - \int v\,du$

2. $\int a^u\,du = \frac{a^u}{\ln a} + C, \qquad a \neq 1, \qquad a > 0$

3. $\int \cos u\,du = \sin u + C$

4. $\int \sin u\,du = -\cos u + C$

5. $\int (ax+b)^n\,dx = \frac{(ax+b)^{n+1}}{a(n+1)} + C, \qquad n \neq -1$

6. $\int (ax+b)^{-1}\,dx = \frac{1}{a}\ln|ax+b| + C$

7. $\int x(ax+b)^n\,dx = \frac{(ax+b)^{n+1}}{a^2}\left[\frac{ax+b}{n+2} - \frac{b}{n+1}\right] + C, \qquad n \neq -1, -2$

8. $\int x(ax+b)^{-1}\,dx = \frac{x}{a} - \frac{b}{a^2}\ln|ax+b| + C$

9. $\int x(ax+b)^{-2}\,dx = \frac{1}{a^2}\left[\ln|ax+b| + \frac{b}{ax+b}\right] + C$

10. $\int \frac{dx}{x(ax+b)} = \frac{1}{b}\ln\left|\frac{x}{ax+b}\right| + C$

11. $\int (\sqrt{ax+b})^n\,dx = \frac{2}{a}\frac{(\sqrt{ax+b})^{n+2}}{n+2} + C, \qquad n \neq -2$

12. $\int \frac{\sqrt{ax+b}}{x}\,dx = 2\sqrt{ax+b} + b\int \frac{dx}{x\sqrt{ax+b}}$

13. (a) $\int \frac{dx}{x\sqrt{ax+b}} = \frac{2}{\sqrt{-b}}\tan^{-1}\sqrt{\frac{ax+b}{-b}} + C, \quad \text{if} \quad b < 0$

(b) $\int \frac{dx}{x\sqrt{ax+b}} = \frac{1}{\sqrt{b}}\ln\left|\frac{\sqrt{ax+b}-\sqrt{b}}{\sqrt{ax+b}+\sqrt{b}}\right| + C, \quad \text{if} \quad b > 0$

14. $\int \frac{\sqrt{ax+b}}{x^2}\,dx = -\frac{\sqrt{ax+b}}{x} + \frac{a}{2}\int \frac{dx}{x\sqrt{ax+b}} + C$

15. $\int \frac{dx}{x^2\sqrt{ax+b}} = -\frac{\sqrt{ax+b}}{bx} - \frac{a}{2b}\int \frac{dx}{x\sqrt{ax+b}} + C$

16. $\int \frac{dx}{a^2+x^2} = \frac{1}{a}\tan^{-1}\frac{x}{a} + C$

17. $\int \frac{dx}{(a^2+x^2)^2} = \frac{x}{2a^2(a^2+x^2)} + \frac{1}{2a^3}\tan^{-1}\frac{x}{a} + C$

18. $\int \frac{dx}{a^2-x^2} = \frac{1}{2a}\ln\left|\frac{x+a}{x-a}\right| + C$

19. $\int \frac{dx}{(a^2-x^2)^2} = \frac{x}{2a^2(a^2-x^2)} + \frac{1}{2a^2}\int \frac{dx}{a^2-x^2}$

20. $\int \frac{dx}{\sqrt{a^2+x^2}} = \sinh^{-1}\frac{x}{a} + C = \ln\left(x + \sqrt{a^2+x^2}\right) + C$

21. $\int \sqrt{a^2+x^2}\,dx = \frac{x}{2}\sqrt{a^2+x^2} + \frac{a^2}{2}\ln\left(x + \sqrt{a^2+x^2}\right) + C$

22. $\int x^2\sqrt{a^2+x^2}\,dx = \frac{x}{8}(a^2+2x^2)\sqrt{a^2+x^2} - \frac{a^4}{8}\ln\left(x + \sqrt{a^2+x^2}\right) + C$

23. $\int \frac{\sqrt{a^2+x^2}}{x}\,dx = \sqrt{a^2+x^2} - a\ln\left|\frac{a+\sqrt{a^2+x^2}}{x}\right| + C$

24. $\int \frac{\sqrt{a^2+x^2}}{x^2}\,dx = \ln\left(x + \sqrt{a^2+x^2}\right) - \frac{\sqrt{a^2+x^2}}{x} + C$

25. $\displaystyle\int \frac{x^2}{\sqrt{a^2+x^2}}\,dx = -\frac{a^2}{2}\ln\left(x+\sqrt{a^2+x^2}\right) + \frac{x\sqrt{a^2+x^2}}{2} + C$

26. $\displaystyle\int \frac{dx}{x\sqrt{a^2+x^2}} = -\frac{1}{a}\ln\left|\frac{a+\sqrt{a^2+x^2}}{x}\right| + C$

27. $\displaystyle\int \frac{dx}{x^2\sqrt{a^2+x^2}} = -\frac{\sqrt{a^2+x^2}}{a^2x} + C$

28. $\displaystyle\int \frac{dx}{\sqrt{a^2-x^2}} = \sin^{-1}\frac{x}{a} + C$

29. $\displaystyle\int \sqrt{a^2-x^2}\,dx = \frac{x}{2}\sqrt{a^2-x^2} + \frac{a^2}{2}\sin^{-1}\frac{x}{a} + C$

30. $\displaystyle\int x^2\sqrt{a^2-x^2}\,dx = \frac{a^4}{8}\sin^{-1}\frac{x}{a} - \frac{1}{8}x\sqrt{a^2-x^2}\,(a^2-2x^2) + C$

31. $\displaystyle\int \frac{\sqrt{a^2-x^2}}{x}\,dx = \sqrt{a^2-x^2} - a\ln\left|\frac{a+\sqrt{a^2-x^2}}{x}\right| + C$

32. $\displaystyle\int \frac{\sqrt{a^2-x^2}}{x^2}\,dx = -\sin^{-1}\frac{x}{a} - \frac{\sqrt{a^2-x^2}}{x} + C$

33. $\displaystyle\int \frac{x^2}{\sqrt{a^2-x^2}}\,dx = \frac{a^2}{2}\sin^{-1}\frac{x}{a} - \frac{1}{2}x\sqrt{a^2-x^2} + C$

34. $\displaystyle\int \frac{dx}{x\sqrt{a^2-x^2}} = -\frac{1}{a}\ln\left|\frac{a+\sqrt{a^2-x^2}}{x}\right| + C$

35. $\displaystyle\int \frac{dx}{x^2\sqrt{a^2-x^2}} = -\frac{\sqrt{a^2-x^2}}{a^2x} + C$

36. $\displaystyle\int \frac{dx}{\sqrt{x^2-a^2}} = \cosh^{-1}\frac{x}{a} + C = \ln\left|x+\sqrt{x^2-a^2}\right| + C$

37. $\displaystyle\int \sqrt{x^2-a^2}\,dx = \frac{x}{2}\sqrt{x^2-a^2} + \frac{a^2}{2}\ln\left|x+\sqrt{x^2-a^2}\right| + C$

38. $\displaystyle\int \left(\sqrt{x^2-a^2}\right)^n dx = \frac{x\left(\sqrt{x^2-a^2}\right)^n}{n+1} - \frac{na^2}{n+1}\int \left(\sqrt{x^2-a^2}\right)^{n-2} dx, \quad n \neq -1$

39. $\displaystyle\int \frac{dx}{\left(\sqrt{x^2-a^2}\right)^n} = \frac{x\left(\sqrt{x^2-a^2}\right)^{2-n}}{(2-n)a^2} - \frac{n-3}{(n-2)a^2}\int \frac{dx}{\left(\sqrt{x^2-a^2}\right)^{n-2}}, \quad n \neq 2$

40. $\displaystyle\int x\left(\sqrt{x^2-a^2}\right)^n dx = \frac{\left(\sqrt{x^2-a^2}\right)^{n+2}}{n+2} + C, \quad n \neq -2$

41. $\displaystyle\int x^2\sqrt{x^2-a^2}\,dx = \frac{x}{8}(2x^2-a^2)\sqrt{x^2-a^2} - \frac{a^4}{8}\ln\left|x+\sqrt{x^2-a^2}\right| + C$

42. $\displaystyle\int \frac{\sqrt{x^2-a^2}}{x}\,dx = \sqrt{x^2-a^2} - a\sec^{-1}\left|\frac{x}{a}\right| + C$

43. $\displaystyle\int \frac{\sqrt{x^2-a^2}}{x^2}\,dx = \ln\left|x+\sqrt{x^2-a^2}\right| - \frac{\sqrt{x^2-a^2}}{x} + C$

44. $\displaystyle\int \frac{x^2}{\sqrt{x^2-a^2}}\,dx = \frac{a^2}{2}\ln\left|x+\sqrt{x^2-a^2}\right| + \frac{x}{2}\sqrt{x^2-a^2} + C$

45. $\displaystyle\int \frac{dx}{x\sqrt{x^2-a^2}} = \frac{1}{a}\sec^{-1}\left|\frac{x}{a}\right| + C = \frac{1}{a}\cos^{-1}\left|\frac{a}{x}\right| + C$

46. $\displaystyle\int \frac{dx}{x^2\sqrt{x^2-a^2}} = \frac{\sqrt{x^2-a^2}}{a^2x} + C$

47. $\displaystyle\int \frac{dx}{\sqrt{2ax-x^2}} = \sin^{-1}\left(\frac{x-a}{a}\right) + C$

48. $\displaystyle\int \sqrt{2ax-x^2}\,dx = \frac{x-a}{2}\sqrt{2ax-x^2} + \frac{a^2}{2}\sin^{-1}\left(\frac{x-a}{a}\right) + C$

49. $\displaystyle\int \left(\sqrt{2ax-x^2}\right)^n dx = \frac{(x-a)\left(\sqrt{2ax-x^2}\right)^n}{n+1} + \frac{na^2}{n+1}\int \left(\sqrt{2ax-x^2}\right)^{n-2} dx$

50. $\displaystyle\int \frac{dx}{\left(\sqrt{2ax-x^2}\right)^n} = \frac{(x-a)\left(\sqrt{2ax-x^2}\right)^{2-n}}{(n-2)a^2} + \frac{(n-3)}{(n-2)a^2}\int \frac{dx}{\left(\sqrt{2ax-x^2}\right)^{n-2}}$

51. $\displaystyle\int x\sqrt{2ax-x^2}\,dx = \frac{(x+a)(2x-3a)\sqrt{2ax-x^2}}{6} + \frac{a^3}{2}\sin^{-1}\left(\frac{x-a}{a}\right) + C$

52. $\displaystyle\int \frac{\sqrt{2ax-x^2}}{x}\,dx = \sqrt{2ax-x^2} + a\sin^{-1}\left(\frac{x-a}{a}\right) + C$

53. $\displaystyle\int \frac{\sqrt{2ax-x^2}}{x^2}\,dx = -2\sqrt{\frac{2a-x}{x}} - \sin^{-1}\left(\frac{x-a}{a}\right) + C$

54. $\displaystyle\int \frac{x\,dx}{\sqrt{2ax - x^2}} = a \sin^{-1}\left(\frac{x - a}{a}\right) - \sqrt{2ax - x^2} + C$

55. $\displaystyle\int \frac{dx}{x\sqrt{2ax - x^2}} = -\frac{1}{a}\sqrt{\frac{2a - x}{x}} + C$

56. $\displaystyle\int \sin ax\,dx = -\frac{1}{a}\cos ax + C$

57. $\displaystyle\int \cos ax\,dx = \frac{1}{a}\sin ax + C$

58. $\displaystyle\int \sin^2 ax\,dx = \frac{x}{2} - \frac{\sin 2ax}{4a} + C$

59. $\displaystyle\int \cos^2 ax\,dx = \frac{x}{2} + \frac{\sin 2ax}{4a} + C$

60. $\displaystyle\int \sin^n ax\,dx = -\frac{\sin^{n-1} ax \cos ax}{na} + \frac{n - 1}{n}\int \sin^{n-2} ax\,dx$

61. $\displaystyle\int \cos^n ax\,dx = \frac{\cos^{n-1} ax \sin ax}{na} + \frac{n - 1}{n}\int \cos^{n-2} ax\,dx$

62. (a) $\displaystyle\int \sin ax \cos bx\,dx = -\frac{\cos (a + b)x}{2(a + b)} - \frac{\cos (a - b)x}{2(a - b)} + C, \quad a^2 \neq b^2$

(b) $\displaystyle\int \sin ax \sin bx\,dx = \frac{\sin (a - b)x}{2(a - b)} - \frac{\sin (a + b)x}{2(a + b)} + C, \quad a^2 \neq b^2$

(c) $\displaystyle\int \cos ax \cos bx\,dx = \frac{\sin (a - b)x}{2(a - b)} + \frac{\sin (a + b)x}{2(a + b)} + C, \quad a^2 \neq b^2$

63. $\displaystyle\int \sin ax \cos ax\,dx = -\frac{\cos 2ax}{4a} + C$

64. $\displaystyle\int \sin^n ax \cos ax\,dx = \frac{\sin^{n+1} ax}{(n + 1)a} + C, \quad n \neq -1$

65. $\displaystyle\int \frac{\cos ax}{\sin ax}\,dx = \frac{1}{a}\ln|\sin ax| + C$

66. $\displaystyle\int \cos^n ax \sin ax\,dx = -\frac{\cos^{n+1} ax}{(n + 1)a} + C, \quad n \neq -1$

67. $\displaystyle\int \frac{\sin ax}{\cos ax}\,dx = -\frac{1}{a}\ln|\cos ax| + C$

68. $\displaystyle\int \sin^n ax \cos^m ax\,dx = -\frac{\sin^{n-1} ax \cos^{m+1} ax}{a(m + n)} + \frac{n - 1}{m + n}\int \sin^{n-2} ax \cos^m ax\,dx, \quad n \neq -m$

(If $n = -m$, use No. 86.)

69. $\displaystyle\int \sin^n ax \cos^m ax\,dx = \frac{\sin^{n+1} ax \cos^{m-1} ax}{a(m + n)} + \frac{m + 1}{m + n}\int \sin^n ax \cos^{m-2} ax\,dx, \quad m \neq -n$

(If $m = -n$, use No. 87.)

70. $\displaystyle\int \frac{dx}{b + c \sin ax} = \frac{-2}{a\sqrt{b^2 - c^2}}\tan^{-1}\left[\sqrt{\frac{b - c}{b + c}}\tan\left(\frac{\pi}{4} - \frac{ax}{2}\right)\right] + C, \quad b^2 > c^2$

71. $\displaystyle\int \frac{dx}{b + c \sin ax} = \frac{-1}{a\sqrt{c^2 - b^2}}\ln\left|\frac{c + b \sin ax + \sqrt{c^2 - b^2}\cos ax}{b + c \sin ax}\right| + C, \quad b^2 < c^2$

72. $\displaystyle\int \frac{dx}{1 + \sin ax} = -\frac{1}{a}\tan\left(\frac{\pi}{4} - \frac{ax}{2}\right) + C$

73. $\displaystyle\int \frac{dx}{1 - \sin ax} = \frac{1}{a}\tan\left(\frac{\pi}{4} + \frac{ax}{2}\right) + C$

74. $\displaystyle\int \frac{dx}{b + c \cos ax} = \frac{2}{a\sqrt{b^2 - c^2}}\tan^{-1}\left[\sqrt{\frac{b - c}{b + c}}\tan\frac{ax}{2}\right] + C, \quad b^2 > c^2$

75. $\displaystyle\int \frac{dx}{b + c \cos ax} = \frac{1}{a\sqrt{c^2 - b^2}}\ln\left|\frac{c + b \cos ax + \sqrt{c^2 - b^2}\sin ax}{b + c \cos ax}\right| + C, \quad b^2 < c^2$

76. $\displaystyle\int \frac{dx}{1 + \cos ax} = \frac{1}{a}\tan\frac{ax}{2} + C$

77. $\displaystyle\int \frac{dx}{1 - \cos ax} = -\frac{1}{a}\cot\frac{ax}{2} + C$

78. $\displaystyle\int x \sin ax\,dx = \frac{1}{a^2}\sin ax - \frac{x}{a}\cos ax + C$

79. $\displaystyle\int x \cos ax\,dx = \frac{1}{a^2}\cos ax + \frac{x}{a}\sin ax + C$

80. $\int x^n \sin ax\,dx = -\frac{x^n}{a}\cos ax + \frac{n}{a}\int x^{n-1}\cos ax\,dx$

81. $\int x^n \cos ax\,dx = \frac{x^n}{a}\sin ax - \frac{n}{a}\int x^{n-1}\sin ax\,dx$

82. $\int \tan ax\,dx = \frac{1}{a}\ln|\sec ax| + C$

83. $\int \cot ax\,dx = \frac{1}{a}\ln|\sin ax| + C$

84. $\int \tan^2 ax\,dx = \frac{1}{a}\tan ax - x + C$

85. $\int \cot^2 ax\,dx = -\frac{1}{a}\cot ax - x + C$

86. $\int \tan^n ax\,dx = \frac{\tan^{n-1} ax}{a(n-1)} - \int \tan^{n-2} ax\,dx, \qquad n \neq 1$

87. $\int \cot^n ax\,dx = -\frac{\cot^{n-1} ax}{a(n-1)} - \int \cot^{n-2} ax\,dx, \qquad n \neq 1$

88. $\int \sec ax\,dx = \frac{1}{a}\ln|\sec ax + \tan ax| + C$

89. $\int \csc ax\,dx = -\frac{1}{a}\ln|\csc ax + \cot ax| + C$

90. $\int \sec^2 ax\,dx = \frac{1}{a}\tan ax + C$

91. $\int \csc^2 ax\,dx = -\frac{1}{a}\cot ax + C$

92. $\int \sec^n ax\,dx = \frac{\sec^{n-2} ax \tan ax}{a(n-1)} + \frac{n-2}{n-1}\int \sec^{n-2} ax\,dx, \qquad n \neq 1$

93. $\int \csc^n ax\,dx = -\frac{\csc^{n-2} ax \cot ax}{a(n-1)} + \frac{n-2}{n-1}\int \csc^{n-2} ax\,dx, \qquad n \neq 1$

94. $\int \sec^n ax \tan ax\,dx = \frac{\sec^n ax}{na} + C, \qquad n \neq 0$

95. $\int \csc^n ax \cot ax\,dx = -\frac{\csc^n ax}{na} + C, \qquad n \neq 0$

96. $\int \sin^{-1} ax\,dx = x\sin^{-1} ax + \frac{1}{a}\sqrt{1 - a^2x^2} + C$

97. $\int \cos^{-1} ax\,dx = x\cos^{-1} ax - \frac{1}{a}\sqrt{1 - a^2x^2} + C$

98. $\int \tan^{-1} ax\,dx = x\tan^{-1} ax - \frac{1}{2a}\ln(1 + a^2x^2) + C$

99. $\int x^n \sin^{-1} ax\,dx = \frac{x^{n+1}}{n+1}\sin^{-1} ax - \frac{a}{n+1}\int \frac{x^{n+1}\,dx}{\sqrt{1 - a^2x^2}}, \qquad n \neq -1$

100. $\int x^n \cos^{-1} ax\,dx = \frac{x^{n+1}}{n+1}\cos^{-1} ax + \frac{a}{n+1}\int \frac{x^{n+1}\,dx}{\sqrt{1 - a^2x^2}}, \qquad n \neq -1$

101. $\int x^n \tan^{-1} ax\,dx = \frac{x^{n+1}}{n+1}\tan^{-1} ax - \frac{a}{n+1}\int \frac{x^{n+1}\,dx}{1 + a^2x^2}, \qquad n \neq -1$

102. $\int e^{ax}\,dx = \frac{1}{a}e^{ax} + C$

103. $\int b^{ax}\,dx = \frac{1}{a}\frac{b^{ax}}{\ln b} + C, \qquad b > 0, \quad b \neq 1$

104. $\int xe^a\,dx = \frac{e^{ax}}{a^2}(ax - 1) + C$

105. $\int x^n e^{ax}\,dx = \frac{1}{a}x^n e^{ax} - \frac{n}{a}\int x^{n-1}e^{ax}\,dx$

106. $\int x^n b^{ax}\,dx = \frac{x^n b^{ax}}{a\ln b} - \frac{n}{a\ln b}\int x^{n-1}b^{ax}\,dx, \qquad b > 0, \quad b \neq 1$

107. $\int e^{ax}\sin bx\,dx = \frac{e^{ax}}{a^2 + b^2}(a\sin bx - b\cos bx) + C$

108. $\int e^{ax}\cos bx\,dx = \frac{e^{ax}}{a^2 + b^2}(a\cos bx + b\sin bx) + C$

109. $\int \ln a\,dx = x\ln ax - x + C$

110. $\int x^n(\ln ax)^m\,dx = \frac{x^{n+1}(\ln ax)^m}{n+1} - \frac{m}{n+1}\int x^n(\ln ax)^{m-1}\,dx, \qquad n \neq -1$

111. $\displaystyle\int x^{-1}(\ln ax)^m = \frac{(\ln ax)^{m+1}}{m+1} + C, \qquad m \neq -1$

112. $\displaystyle\int \frac{dx}{x \ln ax} = \ln|\ln ax| + C$

113. $\displaystyle\int \sinh ax\,dx = \frac{1}{a}\cosh ax + C$

114. $\displaystyle\int \cosh ax\,dx = \frac{1}{a}\sinh ax + C$

115. $\displaystyle\int \sinh^2 ax\,dx = \frac{\sinh 2ax}{4a} - \frac{x}{2} + C$

116. $\displaystyle\int \cosh^2 ax\,dx = \frac{\sinh 2ax}{4a} + \frac{x}{2} + C$

117. $\displaystyle\int \sinh^n ax\,dx = \frac{\sinh^{n-1} ax \cosh ax}{na} - \frac{n-1}{n}\int \sinh^{n-2} ax\,dx, \qquad n \neq 0$

118. $\displaystyle\int \cosh^n ax\,dx = \frac{\cosh^{n-1} ax \sinh ax}{na} + \frac{n-1}{n}\int \cosh^{n-2} ax\,dx, \qquad n \neq 0$

119. $\displaystyle\int x \sinh ax\,dx = \frac{x}{a}\cosh ax - \frac{1}{a^2}\sinh ax + C$

120. $\displaystyle\int x \cosh ax\,dx = \frac{x}{a}\sinh ax - \frac{1}{a^2}\cosh ax + C$

121. $\displaystyle\int x^n \sinh ax\,dx = \frac{x^n}{a}\cosh ax - \frac{n}{a}\int x^{n-1}\cosh ax\,dx$

122. $\displaystyle\int x^n \cosh ax\,dx = \frac{x^n}{a}\sinh ax - \frac{n}{a}\int x^{n-1}\sinh ax\,dx$

123. $\displaystyle\int \tanh ax\,dx = \frac{1}{a}\ln(\cosh ax) + C$

124. $\displaystyle\int \coth ax\,dx = \frac{1}{a}\ln|\sinh ax| + C$

125. $\displaystyle\int \tanh^2 ax\,dx = x - \frac{1}{a}\tanh ax + C$

126. $\displaystyle\int \coth^2 ax\,dx = x - \frac{1}{a}\coth ax + C$

127. $\displaystyle\int \tanh^n ax\,dx = -\frac{\tanh^{n-1} ax}{(n-1)a} + \int \tanh^{n-2} ax\,dx, \qquad n \neq 1$

128. $\displaystyle\int \coth^n ax\,dx = \frac{\coth^{n-1} ax}{(n-1)a} + \int \coth^{n-2} ax\,dx, \qquad n \neq 1$

129. $\displaystyle\int \operatorname{sech} ax\,dx = \frac{1}{a}\sin^{-1}(\tanh ax) + C$

130. $\displaystyle\int \operatorname{csch} ax\,dx = \frac{1}{a}\ln\left|\tanh\frac{ax}{2}\right| + C$

131. $\displaystyle\int \operatorname{sech}^2 ax\,dx = \frac{1}{a}\tanh ax + C$

132. $\displaystyle\int \operatorname{csch}^2 ax\,dx = -\frac{1}{a}\coth ax + C$

133. $\displaystyle\int \operatorname{sech}^n ax\,dx = \frac{\operatorname{sech}^{n-2} ax \tanh ax}{(n-1)a} + \frac{n-2}{n-1}\int \operatorname{sech}^{n-2} ax\,dx, \qquad n \neq 1$

134. $\displaystyle\int \operatorname{csch}^n ax\,dx = -\frac{\operatorname{csch}^{n-2} ax \coth ax}{(n-1)a} - \frac{n-2}{n-1}\int \operatorname{csch}^{n-2} ax\,dx, \qquad n \neq 1$

135. $\displaystyle\int \operatorname{sech}^n ax \tanh ax\,dx = -\frac{\operatorname{sech}^n ax}{na} + C, \qquad n \neq 0$

136. $\displaystyle\int \operatorname{csch}^n ax \coth ax\,dx = -\frac{\operatorname{csch}^n ax}{na} + C, \qquad n \neq 0$

137. $\displaystyle\int e^{ax}\sinh bx\,dx = \frac{e^{ax}}{2}\left[\frac{e^{bx}}{a+b} - \frac{e^{-bx}}{a-b}\right] + C, \qquad a^2 \neq b^2$

138. $\displaystyle\int e^{ax}\cosh bx\,dx = \frac{e^{ax}}{2}\left[\frac{e^{bx}}{a+b} + \frac{e^{-bx}}{a-b}\right] + C, \qquad a^2 \neq b^2$

139. $\displaystyle\int_0^\infty x^{n-1}e^{-x}\,dx = \Gamma(n) = (n-1)!, \qquad n > 0$

140. $\displaystyle\int_0^\infty e^{-ax^2}\,dx = \frac{1}{2}\sqrt{\frac{\pi}{a}}, \qquad a > 0$

141. $$\int_0^{\pi/2} \sin^n x\,dx = \int_0^{\pi/2} \cos^n x\,dx = \begin{cases} \dfrac{1\cdot 3\cdot 5\cdots(n-1)}{2\cdot 4\cdot 6\cdots n}\cdot\dfrac{\pi}{2}, & \text{if } n \text{ is an even integer} \geq 2 \\[2ex] \dfrac{2\cdot 4\cdot 6\cdots(n-1)}{3\cdot 5\cdot 7\cdots n}, & \text{if } n \text{ is an odd integer} \geq 3 \end{cases}$$

Answers to Selected Odd-Numbered Problems

Section 13.1

3. $z = (-5x - 4y + 46)/3$
5. $x - 3 = 0$ (no slope-intercept form)
7. $5x - 9y - 2z + 31 = 0$; $z = (5x - 9y + 31)/2$
9. $42x - 12y + 32z - 238 = 0$; $z = (-21x + 6y + 119)/16$
11. $(x + z - \pi/2)/\sqrt{2} = 0$ or $z = -x + \pi/2$
13. $1.44 \approx 82.6°$

Section 13.2

15. $R(v_0, \phi) = (v_0^2/g)\sin(2\phi)$
17. $A(b, h) = bh/2$

Section 13.3

5. (b) $z = u^2 - v^2$
   (c) hyperbolic paraboloid

Section 13.4

1. planes with normal $\langle -5, -2, 4\rangle$
3. spheres centered on the origin

Section 13.5

1. does not exist
3. does not exist
11. continuous at the origin

Section 14.1

1. $\frac{28}{3}$
3. $8 \ln 5$
5. 0
7. $-3e^4(e^4 - 1)/8$
9. 0.4035; 1.2680
11. 0.00661313; 0.0173766

Section 14.2

1. $441/10$
3. $3/35$
5. $(\pi \ln 2)/4$
7. $372/5$
9. $135/4$
11. $9207/20$
13. $4\sqrt{2} - 10\sqrt{5}$
15. $-11/2$
17. $38/3$
19. $3\pi/2$
21. $3/2$
23. $\pi/12$

Section 14.3

1. 1
3. 144
5. $-60 + 2 \ln 4 - 14 \ln 7 + 56 \ln 28 - 8 \ln 64$
7. $r = \sqrt{10}$; $\theta = 4.39$; $z = -3$
9. $r = 4.23$; $\theta = -0.594$; $z = -2.0$
11. $x = 1.30$; $y = 3.14$; $z = -1.3$
13. $x = -2.5$; $y = 4.33$; $z = 1.4$
15. solid region: $0 \le r \le R$, $0 \le \theta \le \pi, 0 \le z \le h$;
    curved side: $r = R, 0 \le \theta \le \pi$, $0 \le z \le h$;
    flat side: $0 \le r \le R, \theta = 0, \pi$, $0 \le z \le h$;
    top and bottom: $0 \le r \le R$, $0 \le \theta \le \pi, z = 0, h$
17. $\rho = \sqrt{19}$; $\phi = 2.33$; $\theta = 4.39$
19. $\rho = 5.22$; $\phi = 1.96$; $\theta = -0.594$
21. $x = 0$; $y = 3.5$; $z = 0$
23. $x = -11.44$; $y = -7.42$; $z = -6.24$
25. cylindrical coordinates, solid region: $0 \le r \le \frac{R}{h}z$, $0 \le \theta \le 2\pi, 0 \le z \le h$;
    curved side: $r = \frac{R}{h}z$, $0 \le \theta \le 2\pi$, $z = h$;
    top: $0 \le r \le R, 0 \le \theta \le 2\pi$, $z = h$.
    spherical coordinates, solid region: $0 \le \rho \le \frac{h}{\cos\phi}$, $0 \le \phi \le \tan^{-1}(R/h)$, $0 \le \theta \le 2\pi$;
    curved side: $0 \le \rho \le \sqrt{R^2 + h^2}$, $0 \le \phi \le \tan^{-1}(R/h)$, $0 \le \theta \le 2\pi$;
    top: $\rho = \frac{h}{\cos\phi}$, $0 \le \phi \le \tan^{-1}(R/h)$, $0 \le \theta \le 2\pi$
27. -162π
29. 6
31. $(31\pi^2)/10$
33. 8π

Section 14.4

1. $\langle 4/3, 8\rangle$ cm
3. on the symmetry axis a distance $3h/4$ from the vertex
5. 2238 g; on the symmetry axis 7.5 cm from each end
7. $X = 0.342$ cm
9. $(7MR^2)/5$
11. $(2MR^2)/3$
13. $(2M/5)(R_2^5 - R_1^5)/(R_2^3 - R_1^3)$
    (a) $(2MR^2)/5$ (b) $(2MR^2)/3$

Section 15.2

3. 75.5 y
5. $\sqrt{3}a/2$ from the center
7. 2.49×10^{11} m; 2.07×10^{11} m
9. $C = 8.84 \times 10^{12}$ m^3/s^2; $T = 6060$ s

Section 15.3

1. (a) 274 m/s^2 (b) 5.93×10^{-3} m/s^2
3. 2.0×10^{30} kg
7. $T = (\frac{2\pi^2 R^3}{GM})^{1/2} = 12.2$ y
9. $F = 9.8$ N; $a = 1.6 \times 10^{-24}$ m/s^2
11. (a) 2.66×10^{40} J·s
    (b) 2.925×10^4 m/s; 3.024×10^4 m/s
    (c) 8.01×10^{18} m^2 for both
13. $T^2 = \frac{4\pi^2 r^{n+1}}{GM}$

Section 15.4

1. (a) -5.31×10^{33} J
   (b) -1.96×10^{26} J
3. about 8 m/s
5. 1.77×10^{32} J; 990 m/s
7. 936 m/s
9. -1.27×10^{10} J

Section 15.5

3. $T = 2\pi\sqrt{\frac{R^3}{MG}}$

Section 15.6

1. 6.5×10^{-4} m/s^2
3. 42.1 km/s
5. 5.01 km/s
7. $r = 2R$
9. (a) 117.7 min, about 9 min longer
   (b) 1630 m/s
11. 1.23 m/s
13. 1690 km
15. (a) 0.689 (b) 1480 m/s

Section 15.7

1. 5.57×10^{-9} N

Section 16.1

1. 6.24×10^{18}; 6.24×10^{12}
3. (a) 2.86×10^{32} C
   (b) 3.50×10^{-33}
5. 96,500 C

Section 16.2

1. 410 N
3. electrostatic 6.14×10^{55} N; gravitational 1.99×10^{20} N; ratio $= 3.09 \times 10^{35}$
5. (a) 1.33 N (b) 3.63 μC; 1.89 N
7. 0.28 m (center-to-center)
9. 8.99×10^5 N

Section 16.3

1. 2.40×10^{-3} N, directed away from the opposite diagonal
3. between the two charges, $d\left(\frac{3}{2} - \frac{\sqrt{3}}{2}\right)$ away from the $3q$ charge; no
5. (a) $1.91kQ^2/d^2$ toward the opposite diagonal (b) 0
7. $kqQ/2L^2$ toward the rod
9. The force is in the x-direction with $F_x = -2kq_0C\left(-\frac{1}{\sqrt{5}} + \ln\left(\frac{1+\sqrt{5}}{2}\right)\right)$
11. $2kq_0\lambda/d$ in the $-y$-direction
13. $4kq_0Q/\pi R^2$ in the $-y$-direction
17. (a) $\frac{4kq_0Q}{R^2}\left(1 - \frac{d}{\sqrt{R^2+d^2}}\right)$ in the $-y$-direction

Section 17.1

1. $\vec{F}(\vec{r}) = k\hat{r}/r^3$
3. $\langle x, y\rangle$
5. $5(x^2+y^2)^{-3/2}\langle x, y\rangle$
7. $\sqrt{x^2+y^2}\langle x, y\rangle$
9. $\langle x, y, z\rangle$
11. $5(x^2+y^2+z^2)^{-3/2}\langle x, y, z\rangle$
13. $\sqrt{x^2+y^2+z^2}\langle x, y, z\rangle$

Section 17.2

1. (a) $1.15 \times 10^5\,\hat{k}$ N/C
   (b) $(1090\,\hat{\imath} + 564\,\hat{\jmath})$ N/C
   (c) $(-391\,\hat{\imath} - 1140\,\hat{\jmath} + 212\,\hat{k})$ N/C
3. (a) $(-269\,\hat{\imath} + 87.7\,\hat{\jmath})$ N/C
   (b) $(1.75 \times 10^{-6}\,\hat{\imath} - 5.70 \times 10^{-7}\,\hat{\jmath})$ N
5. (a) 6.34×10^{28} N/C
   (b) 1.02×10^{10} N straight up
   (c) 6.34×10^{22} N/C; 1.02×10^4 N straight up
7. 111 nC
9. $\frac{kQ}{(x+a)^2} + \frac{kQ}{(x-a)^2}$ in the $-x$-direction
11. $x > a$: $\frac{kQ}{(x+a)^2} + \frac{kQ}{(x-a)^2}$ in the $+x$-direction;
    $x < a$: $\frac{kQ}{(x+a)^2} + \frac{kQ}{(x-a)^2}$ in the $-x$-direction
13. (a) $\frac{2kQa}{((x-a)^2+a^2)^{3/2}} - \frac{2kQa}{((x+a)^2+a^2)^{3/2}}$ in the $-y$-direction
    (b) the field decreases in proportion to $1/x^4$
19. $\frac{-2kQ}{R\sqrt{L^2+4R^2}}\,\hat{\jmath}$
21. (a) $kC(\frac{2RL}{R^2-L^2} + \ln\frac{R-L}{R+L})\,\hat{\imath}$
    (b) $kC(-\frac{2RL}{R^2-L^2} + \ln\frac{R+L}{R-L})\,\hat{\imath}$
23. $2k\lambda/R$ toward the line of charge
25. (a) $2900\,\hat{\imath}$ N/C (b) $-2900\,\hat{\jmath}$ N/C
27. $z = -7.3$ m on the z-axis
29. (a) 2.259×10^5 N/C
    (b) 2.214×10^5 N/C
    (c) 2.214×10^5 N/C

Section 17.3

1. (a) $(3.77 \times 10^{-16}\,\hat{\imath} - 1.19 \times 10^{-15}\,\hat{\jmath})$ N
   (b) $(2.25 \times 10^{11}\,\hat{\imath} - 7.11 \times 10^{11}\,\hat{\jmath})$ m/s^2
3. 5.37×10^{-15} m, a factor of 3.87×10^{13} less than the deflection of the electric field
5. (a) $v_0\,\hat{\imath} + \frac{QE_0}{m}t\,\hat{k}$
   (b) $v_0t\,\hat{\imath} + \frac{QE_0}{2m}t^2\,\hat{k}$; a parabola
7. $(0.315\,\hat{\imath} + 3.15\,\hat{\jmath} - 0.417\,\hat{k})$ m/s; $(3.91\,\hat{\imath} + 13/7\,\hat{\jmath} - 1.04\,\hat{k})$ m
9. 6.09×10^{-10} C/m^2; upper
11. (a) 2.5×10^{-8} C·m
    (b) 1.77×10^{-3} N·m 1.77×10^{-3} J
13. $2\pi\sqrt{I/pE}$

Section 18.1

1. (a) $f_x(0,0) = -5.45$; $f_y(0,0) = 1.50$
   (b) $f_x(0.6, 0.8) = -4.80$; $f_y(0.6, 0.8) = 3.45$
   (c) $f_x(-0.4, 0.8) = -4.80$; $f_y(-0.4, 0.8) = 2.65$

(d) $f_x(-0.2,-0.6) = -5.90$; $f_y(-0.2,-0.6) = 2.85$

3. $f_x(x,y) = 9x^2 + 4y^4$; $f_y(x,y) = 16xy^3$; $f_x(-1,0) = 9$; $f_y(-1,0) = 0$
5. $f_x(x,y) = -4yx^{y-1}$; $f_y(x,y) = -4x^y \ln x$; $f_x(2,1) = -4$; $f_y(2,1) = -8\ln 2$
7. $f_x(x,y) = 6ye^y$; $f_y(x,y) = 6x(1+y)e^y$; $f_x(2,-1) = -6/e$; $f_y(2,-1) = 0$
9. $f_x(x,y) = -x(x^2+y^2)^{-3/2}$; $f_y(x,y) = -y(x^2+y^2)^{-3/2}$; $f_x(-1,1) = 1/2^{3/2}$; $f_y(-1,1) = -1/2^{3/2}$
11. $f_x(x,y,z) = 4yz + 6x$; $f_y(x,y,z) = 4xz - 2yz^3$; $f_z(x,y,z) = 4xy - 3y^2z^2$; $f_x(1,1,1) = 10$; $f_y(1,1,1) = 2$; $f_z(1,1,1) = 1$
13. $f_x(x,y,z) = -2(14y^2+z)^{-1}$; $f_y(x,y,z) = 56xy(14y^2+z)^{-2}$; $f_z(x,y,z) = 2x(14y^2+z)^2$; $f_x(5,4,3) = -2/227$; $f_y(5,4,3) = -1120/51529$; $f_z(5,4,3) = -10/51529$
15. $f_x(x,y,z) = -3\ln(yz)$; $f_y(x,y,z) = -3x/y$; $f_z(x,y,z) = -3x/z$; $f_x(2,2,1) = -3\ln 2$; $f_y(2,2,1) = -3$; $f_z(2,2,1) = -6$
17. $f_x(x,y,z) = yze^{xyz}$; $f_y(x,y,z) = xze^{xyz}$; $f_z(x,y,z) = xye^{xyz}$; $f_x(1,1,4) = 4e^4$; $f_y(1,1,4) = 4e^4$; $f_z(1,1,4) = e^4$
19. $4\sigma T^3$
21. $-\omega\cos(kx - \omega t)$
23. mv
25. $ikAe^{ikx} - ikBe^{-ikx}$
27. γ

Section 18.2

1. 3.114
3. $z = 36(x-2) - 12(y-3)$
5. $z = 6\ln 2 + \frac{3}{2}(x-2) + \frac{3}{2}(y-2)$
7. $z = e^2 + e^2(x-2) + e^2(y-1)$
9. $z = 1$
11. $z = -7 - \frac{7}{2}(x-1) + \frac{7}{2}(y-3)$
13. $f(x,y,z) \approx 6 + 6(x-1) + 3(y-2) + 2(z-3)$
15. $f(x,y,z) \approx -13 - 12(x-2) + (y+1) + 12(z+1)$
17. $f(x,y,z) \approx 1 + (x-1) + y$
19. $f(x,y,z) \approx 5^{1/4} + \frac{1}{(2)5^{1/4}}(x-1) + \frac{1}{5^{1/4}}(y-2) - \frac{1}{(2)5^{1/4}}(x+1)$
21. $(-\frac{1}{3}, -\frac{1}{6})$
23. (a) +0.94 m (b) +0.13 m (c) −0.94 m

Section 18.3

1. $\langle y, x\rangle$
3. $\langle y^2 - 2xy, 2xy - x^2\rangle$
5. $\langle yz, xz, xy\rangle$
7. $-(x^2+y^2+z^2)^{-3/2}\langle x,y,z\rangle$
9. (a) $\sqrt{5}$ (b) $\sqrt{5}$
11. $4\sqrt{2}$
13. $8\sqrt{2}$
15. $4\sqrt{2}$
17. $4\sqrt{3}$
19. $18/\sqrt{14}$
21. $2r\hat{r}$
23. $-\hat{r}/r^2$
25. $nr^{n-1}\hat{r}$
27. (d) (2.5, 4)

Section 18.4

1. $W = 1.15 \times 10^{-28}$ J; $\Delta U = -1.15 \times 10^{-28}$ J
3. -2.74×10^{-7} J
5. -3.22×10^{-19} J
7. -1.64×10^{-10} J
9. (a) acceleration along the $+z$-axis, but magnitude of acceleration decreases with distance (c) 4.24×10^{-3} m/s (d) 6.0×10^{-3} m/s
11. -4.71×10^{-10} V
13. no
17. (a) 1.67×10^4 V/m from positive toward the negative (c) 6.0 mm (d) uniforrm acceleration toward the lower potential (e) uniform acceleration toward the higher potential (f) -8.01×10^{-17} J
19. -1.14×10^{-5} V
21. 6.67 C

Section 18.5

1. $-4x\,\hat{\imath} - 4y\,\hat{\jmath}$
3. $\frac{2x}{(x^2+y^2)^2}\,\hat{\imath} + \frac{2y}{(x^2+y^2)^2}\,\hat{\jmath}$
5. $-6\cos(2x)\cos(y)\,\hat{\imath} - 3\sin(2x)\cos(y)\,\hat{\jmath}$
7. $(x^2+y^2+z^2)^{-3/2}(x\,\hat{\imath} + y\,\hat{\jmath} + z\,\hat{k})$
9. $-\frac{1}{x}\,\hat{\imath} - \frac{1}{y}\,\hat{\jmath} - \frac{1}{z}\,\hat{k}$
11. (a) −2060 V (b) 513 V (c) −32, 500 V
13. (a) $(-4x^3+4y)\,\hat{\imath} + (-4y^3+4x)\,\hat{\jmath}$ (c) $(-1,-1)$ local minumum; $(0,0)$ saddle; $(1,1)$ local maximum

Section 18.6

1. $-\frac{2kq}{\sqrt{2}a^2}\,\hat{\imath}$; $\frac{8kq}{9a^2}\,\hat{\imath}$
3. $2\pi k\sigma(\sqrt{R^2+z^2} - z)$
5. outside: kQ/r; inside: $\frac{kQ}{2R}(3 - \frac{r^2}{R^2})$
7. $2k\lambda/x$ away from the line
9. (a) $k\frac{Q}{L}\ln\frac{L+\sqrt{4x^2+L^2}}{-L+\sqrt{4x^2+L^2}}$ (b) $E_x = \frac{2kQ}{x\sqrt{4x^2+L^2}}$ (c) At great distances x the field decreases as $1/x^2$
11. (a) $\frac{k\lambda_0}{4}\left(L\sqrt{4x^2+L^2} + 2x^2\ln\left(\frac{-L+\sqrt{4x^2+L^2}}{L+\sqrt{4x^2+L^2}}\right)\right)$ (b) away from the charged wire

Section 18.7

1. $f_{xx}(x,y) = 0$; $f_{xy}(x,y) = f_{yx}(x,y) = 1$; $f_{yy}(x,y) = 0$
3. $f_{xx}(x,y) = -\frac{3}{x^2}$; $f_{xy}(x,y) = f_{yx}(x,y) = 0$; $f_{yy}(x,y) = -\frac{3}{y^2}$
5. $f_{xx}(x,y) = -y^2\sin(xy)$; $f_{xy}(x,y) = f_{yx}(x,y) = \cos(xy) - y\sin(xy)$; $f_{yy}(x,y) = -x^2\sin(xy)$

7. $f_{xx}(x,y,z) = 0$;
 $f_{xy}(x,y,z) = f_{yx}(x,y,z) = z$;
 $f_{yy}(x,y,z) = 0$;
 $f_{yz}(x,y,z) = f_{zy}(x,y,z) = x$;
 $f_{zz}(x,y,z) = 0$;
 $f_{xz}(x,y,z) = f_{zx}(x,y,z) = y$
9. $f_{xx}(x,y,z) = 8yz^3$;
 $f_{xy}(x,y,z) = f_{yx}(x,y,z) = 8xz^3 - 10yz^2$;
 $f_{yy}(x,y,z) = -10xz^2$;
 $f_{yz}(x,y,z) = f_{yz}(x,y,z) = 12x^2z^2 - 20xyz$;
 $f_{zz}(x,y,z) = 24x^2yz - 10xy^2$;
 $f_{xz}(x,y,z) = f_{zx}(x,y,z) = 24xyz^2 - 10y^2z$
11. $f_{xxx}(x,y) = 6y^2$;
 $f_{xxy}(x,y) = f_{xyx}(x,y) = f_{yxx}(x,y) = 12xy$;
 $f_{xyy}(x,y) = f_{yxy}(x,y) = f_{yyx}(x,y) = 6x^2$;
 $f_{yyy}(x,y) = 0$;
13. $f(x,y) \approx xy$
15. $f(x,y) \approx \ln 2 + \ln 2(x-1) + \frac{1}{2}(y-2) + \frac{1}{2}(x-1)(y-2) - \frac{1}{8}(y-2)^2$
17. $f(x,y) \approx \frac{\sqrt{2}}{4} + \frac{\sqrt{2}}{2}(x-\frac{\pi}{4}) - \frac{\sqrt{2}}{4}(y-\frac{\pi}{4}) + 2\sqrt{2}(x-\frac{\pi}{4})^2 - \frac{\sqrt{2}}{2}(x-\frac{\pi}{4})(y-\frac{\pi}{4}) - \frac{\sqrt{2}}{8}(y-\frac{\pi}{4})^2$
19. (b) $f(x,y,z) \approx 6 + 6(x-1) + 3(y-2) + 2(z-3) + 3(x-1)(y+1) + (y-2)(z-3) + 2(x-1)(z-3)$

Section 19.1

1. 5.61 cm
5. (a) 3.39×10^{-10} F
 (b) 3.37×10^{-10} F
7. (a) 2.07×10^{-7} C/m^2
 (b) 3.12×10^{-10} s
 (c) 1.34×10^{-8} s
9. (a) 10^{-7} C (b) 4.23×10^7 N/C
11. 1.1×10^{-7} s
13. (a) $R = 8.99$ mm (b) $R = 8.99$ m
 (c) $R = 8.99$ km
17. 9 μF
19. 1 μF
21. 2 μF
23. 550 nF in parallel
25. $\frac{60}{13}$ μF, $\frac{70}{9}$ μF, 10 μF, $\frac{100}{9}$ μF
27. 0.924 μF
29. 1.50 μF
31. 1.12×10^{-5} C
33. 1.43×10^{-7} J/m^3
35. (a) 358 V/m (b) 7.93×10^{-14} C
 (c) 0.179 V
37. (a) 6.13×10^{-3} J
 (b) 1.67×10^{-3} J

Section 19.2

1. (a) 8.87 pF (b) 106 pV (c) 2880 V
3. 787 μC
5. $(\kappa + 1)C_0/2$
7. 1.39×10^{-10} F

Section 19.3

1. 6.24×10^{15}
3. (a) 3.52×10^9 m/s^2
 (b) 1.42×10^{-12} s;
 3.55×10^{-15} m
5. (a) 0.656 Ω (b) 91.5 A;1.31 cm/s
7. 8.75×10^{-5} Ω
9. 19.2 W
11. (a) 21 Ω (b) 6.7 MΩ (c) 85 MΩ
13. 192 Ω
15. 230 J

Section 20.1

1. $f(x) = x$ on $(0,1]$
3. $f(x) = x$ on $(0,1)$
5. no local extrema;
 $f(0) = 0$ is global min;
 $f(2) = 4e^{-2}$ is global max
7. local min and global min at $(\frac{5}{2})^{1/3}$; no global max
9. 14.7 m
11. $r = (\frac{5a}{2b})^{1/3}$

Section 20.2

1. $(0,0)$ is stationary input; neither min nor max
3. $(1,4)$ is local minimizer
5. $(0,0)$ is saddle; $(\frac{125}{12}, -\frac{25}{6})$ is local minimizer
7. replace open disk by open ball
9. $(0,0,0)$ is local min
11. $(0,0,0)$ is saddle
13. all gradient vectors point away from minimizer
15. some gradient vectors point away from and some point toward saddle point
17. closed curves with maximizer in interior

Section 20.3

1. $(0,0)$ is saddle
3. $(1,4)$ is local min
5. $(0,0)$ is saddle; $(\frac{125}{12}, -\frac{25}{6})$ is local minimizer
7. $(0,0)$ is local min

Section 20.4

1. interior: $\{(x,y)|y > 0\}$,
 exterior: $\{(x,y)|y < 0\}$,
 boundary: $\{(x,y)|y = 0\}$;
 open, unbounded
3. interior: $\{(x,y)|\frac{x^2}{5^2} + \frac{y^2}{3^2} < 1\}$,
 exterior: $\{(x,y)|\frac{x^2}{5^2} + \frac{y^2}{3^2} > 1\}$,
 boundary: $\{(x,y)|\frac{x^2}{5^2} + \frac{y^2}{3^2} = 1\}$;
 closed, bounded
5. interior: empty,
 exterior: $\{(x,y)|x \neq 0\}$,
 boundary: $\{(x,y)|x = 0\}$;
 closed, unbounded
7. interior: $\{(x,y)|y < 0\}$,
 exterior: $\{(x,y)|y > 0\}$,
 boundary: $\{(x,y)|y = 0\}$;
 open, unbounded
9. min: $f(0,0) = 0$;
 max: $f(1,2) = 2$
11. min: $f(-2,0) = 5$;
 max: $f(2,0) = 21$
13. min: $f(-1,0,1) = -3$;
 max: $f(1,4,3) = 19$

Section 20.5

1. min: -6; max: 6
3. min: 9; max: 80
5. $(\frac{21}{29}, \frac{20}{29})$
7. $(\frac{18}{41}, -\frac{3}{41}, \frac{6}{41})$
9. $x = \sqrt{\frac{2t}{s+t}A}$; $y = \sqrt{\frac{s+t}{2t}A}$

Section 21.1

1. 1.94 Ω; 3.60 V
3. (a) 0.479 A (b) 22.94 W; 0.06 W
 (c) 23.0 W total

5. short circuit 10.9 W; with resistor 9.9×10^{-4} W
7. 125 s; 12 J

Section 21.2

3. (a) 82 Ω (b) 73.2 mA; 0.878 V; 1.098 V; 1.463 V; 2.561 V
5. (a) five in parallel (b) two in parallel connected with three others in series (c) two in parallel in a series with five in parallel (d) one resistor in series with a parallel combination of two plus a parallel combination of five
7. Each bulb is as bright as the reference bulb. Removing one does not change the brightness of the other two.
9. Each bulb is slightly dimmer than the reference bulb, and they all go out if one is removed.
11. B_1 is brighter than the reference bulb, and if it's removed, they all go out; B_2 and B_3 are slightly dimmer than the reference bulb, and if either is removed, B_1 and the remaining bulb are equally bright and as bright as the reference bulb.
13. 30 Ω
15. 75 V
17. 3.04 V; 0.46 V

Section 21.3

1. 0.125 A, 0.10 A, 0.225 A
3. 25 V, 35 V, 200 mA
5. $3R_0/4$

Section 21.4

1. 27.5 mΩ
3. 0.507 Ω; 6.92 mA
5. 0.346 Ω; 0.343 Ω
7. 200 kΩ

Section 21.5

3. (a) 3.4×10^{-11} C (b) 3.1×10^{-10} C (c) 1.5×10^{-9} C
5. (a) $\tau \ln 2$ (b) 1.23τ

Section 22.1

1. $\vec{r}(u) = \langle u+2, 0\rangle,\ u \in [-4, 0]$; $\vec{r}(u) = 2\langle \cos u, \sin u\rangle,\ u \in [0, \pi]$
3. $\vec{r}(y) = \langle \sqrt{1+y^2}, y\rangle,\ y \in \mathbb{R}$
5. $\vec{r}(u) = \langle 11u-3, 1\rangle,\ u \in [0, 1]$; $\vec{r}(u) = \langle 8, -3u+4\rangle\ ,\ u \in [1, 2]$
7. $\vec{r}(u) = \langle \cos u, 0, \sin u\rangle,$ $u \in [0, \frac{\pi}{2}]$
9. $\vec{r}(u) = \langle 3\cos u, 3\sin u, \frac{u}{\pi}\rangle,$ $u \in [0, 10\pi]$

Section 22.2

1. 2π
3. 0
5. 8
7. 2
9. $-\pi$
11. (a) $-\pi$ (b) $-\pi$ (c) results are equal

Section 22.3

1. $V(x,y) = \frac{1}{2}x^2 + xy - \frac{1}{2}y^2$
3. no potential function
5. $V(x,y) = \frac{1}{2}x^2 + \frac{1}{2}y^2 + \sin x$
7. $V(x,y) = \sin xy$
9. (a) 3 (b) 3 (c) 3 (d) results are equal
11. 22
13. $-\sin 2$

Section 22.4

1. 0
3. 1
5. 1/3
7. $-\pi$
9. the two line integrals have opposite values

Section 23.1

1. (a) 600 N·m²/C (b) 1040 N·m²/C
3. (a) 1.69×10^{10} J (b) 4.66×10^{6} W
5. 8470 N·m²/C
7. (a) 2820 N·m²/C
11. $z = -10$ cm face: 3.46×10^4 N·m²/C; $z = +10$ cm face: 1.06×10^4 N·m²/C; other four faces: 2.26×10^4 N·m²/C
17. (a) $\frac{\rho a^2}{2\pi\varepsilon_0 r}$ outward (b) $\frac{\rho r}{2\pi\varepsilon_0}$ outward
19. Q on solid sphere, $-Q$ on inner surface of shell, $+Q$ on outer surface of shell
21. Q on solid sphere, $-Q$ on inner surface of shell, $-2Q$ on outer surface of shell
23. (a) 0 (b) $\frac{\rho(r^2-b^2)}{2\varepsilon_0 r}$ outward (c) $\frac{\rho(c^2-b^2)}{2\varepsilon_0 r}$ outward
25. $\frac{\rho_0 R^4}{4\varepsilon_0 r^2}$ outward
27. $E = \frac{Ze}{4\pi\varepsilon_0 r^2} - \frac{Zer}{4\pi\varepsilon_0 a^3}$ (outward) for $r \leq a$, $E = 0$ for $r > a$

Section 23.2

1. $\vec{r}(\phi, \theta) =$ $2\langle \sin\phi\cos\theta, \sin\phi\sin\theta, \cos\phi\rangle$ with $(\phi, \theta) \in [0, \frac{\pi}{2}] \times [0, 2\pi]$
3. $\vec{r}(\phi, \theta) =$ $2\langle \sin\phi\cos\theta, \sin\phi\sin\theta, \cos\phi\rangle$ with $(\phi, \theta) \in [0, \pi] \times [0, \pi]$
5. $\vec{r}(x,y) = \langle x, y, x^2+y^2\rangle$ with $(x,y) \in \mathbb{R} \times \mathbb{R}$
7. $\vec{r}(\theta, z) = \langle 2\cos\theta, 2\sin\theta, z\rangle$ with $(\theta, z) \in [0, 2\pi] \times \mathbb{R}$.
9. $\vec{r}(\theta, z) = \langle 2z\cos\theta, 2z\sin\theta, z\rangle$ with $(\theta, z) \in [0, 2\pi] \times \mathbb{R}$.
11. $\|\vec{n}(\pi/2, \theta)\| = 1$; $\|\vec{n}(3\pi/4, \theta)\| = \sqrt{2}/2$
13. $\|\vec{n}(u,v)\| = 3 + \cos v$
15. (a) double cone (b) $\|\vec{n}(u,v)\| = \sqrt{2}v$

Section 23.3

1. $2\pi RH$
3. $\pi R\sqrt{R^2+H^2}$
5. 0
7. 4π
9. 1/4
11. $16\pi/3$

Section 24.2

1. $9.38 \times 10^{-5}\,\hat{\jmath}$ N
3. 0
5. $(-4.73\,\hat{\imath} + 3.94\,\hat{\jmath})$ N
7. $(-2.05 \times 10^{-3}\,\hat{\imath} - 7.68 \times 10^{-14}\,\hat{\jmath}$ $-2.57 \times 10^{-3}\,\hat{k})$ N

13. (a) $y = \frac{eEx^2}{2mv_0^2}$ (b) 1.07 mm

15. (a) 4.09×10^{-15} m (b) 5.57×10^{-11} N/C

19. (a) 3.39 mm (b) 1.85×10^{-6} m

21. 4.22×10^{-6} m

23. (a) 1.89×10^5 m/s (b) 0.0522 T

25. $B = 1$ T, $R = 32.2$ cm

27. $R = 2.81$ cm

29. 1.93×10^5 Hz and 1.96×10^5 Hz

31. turn off one field

Section 24.3

1. (a) IaB perpendicular to the wire for the two equal sides; $\sqrt{2}IaB$ perpendicular to the wire for the hypotenuse (b) 0
3. 0.336 N
5. $F_x = -IBR$; $F_y = IBR$
7. 1.87×10^{-2} T in the $+x$-direction

Section 24.4

3. (a) 1.16×10^{-3} N·m (b) 0 (c) 0
5. 6.56×10^{-2} A·m^2
7. 1.99 cm; 12
9. 4.68×10^{-5} kg·m^2
11. (a) 0 (b) $2\pi\sqrt{I/(\mu B)}$

Section 24.5

1. $n = IB/(ehV_H)$
3. (a) 5.86×10^{28} m^{-3} (b) 1
5. (a) 5.43×10^{-2} T (b) 5.43 T

Section 25.1

3. $B_z = \frac{\mu_0 I}{4\pi}\int_C \frac{R}{(R^2+z^2)^{3/2}}\,ds$
5. $B_z \approx \frac{\mu_0 I R^2}{2z^3}$
9. $B_z = \frac{\mu_0 I L}{2\pi y\sqrt{L^2+y^2}}$
13. $B_z = -\frac{\mu_0 I}{4\pi y_0}\left(\frac{2\sqrt{x_0^2+y_0^2}+x_0+y_0}{\sqrt{x_0^2+y_0^2}}\right)$
15. $B_z = \frac{\mu_0 I}{4R}$
17. $B_z = -\frac{\mu_0 I}{4R} - \frac{\mu_0 I}{2\pi R}$
19. $\frac{2\sqrt{2}\mu_0 I}{\pi L}$

Section 25.2

1. 3.22×10^{-5} T
3. (a) 1.18×10^{-2} T (b) straight line with constant velocity (c) circle of radius 8.85 mm
9. inside: 0; outside: $\frac{\mu_0 I_0}{2\pi r}$

Section 25.3

1. (a) $-3.87 \times 10^{-6}\,\hat{\jmath}$ T (b) 0 (c) $3.87 \times 10^{-6}\,\hat{\jmath}$ T
3. 2.10×10^{-6} N/m, attractive in 1 and repulsive in 2
5. $-1.55 \times 10^{-6}\,\hat{\jmath}$ T
7. $(-7.5\times10^{-7}\hat{\imath}-2.5\times10^{-7}\hat{\jmath})$ N/m; $(7.5\times10^{-7}\,\hat{\imath}-2.5\times10^{-7}\,\hat{\jmath})$ N/m; $(7.5\times10^{-7}\,\hat{\imath}+2.5\times10^{-7}\,\hat{\jmath})$ N/m
9. $-1.55 \times 10^{-6}\,\hat{\jmath}$ T
11. $(7.5\times10^{-7}\,\hat{\imath}-7.5\times10^{-7}\,\hat{\jmath})$ N/m; $(-7.5\times10^{-7}\hat{\imath}-7.5\times10^{-7}\hat{\jmath})$ N/m; $(-7.5\times10^{-7}\hat{\imath}+7.5\times10^{-7}\hat{\jmath})$ N/m

Section 25.5

3. (a) -6.52×10^{-2} A/m (b) -1.1×10^{-7} J/T
5. 4×10^{-6}
7. 2.0×10^{-3}

Section 25.6

1. $2x + \frac{2y}{(x^2+y^2)^2}$
3. $3y + 4z + 24xz^3$
5. $2xe^{x^2} + 1/y - xy/\sqrt{1-x^2y^2z^2}$
9. $2/r$
11. 0
13. $6x^2 + 6y^2 - 2\cos^2 x + 2\sin^2 x$
15. 6
17. $-2\sin x\cos y$

Section 25.7

1. $\langle -4y^2z, -4e^x, -x\rangle$
3. $\langle \frac{-6y}{(x^2+y^2+z^2)^2}, \frac{-6z}{(x^2+y^2+z^2)^2}, -2\rangle$
5. $\langle 0, 2z + \frac{1}{\sqrt{1-x^2}}, -8xe^{-x^2} - 2y\rangle$
7. 0
9. $-2x - 6xy$
13. $\vec{0}$
15. $\vec{0}$
21. $W = 0$ along both paths; conservative
23. $\langle 0, 0, 2\rangle$; not conservative

Section 26.1

7. 1.084×10^{-6} T·m^2; 1.084×10^{-5} T·m^2
9. 2.55 μA
11. Baa_0t
13. B^2L^2v/R in the direction of motion
15. 4.94×10^{-8} T·m^2
17. $\frac{\mu_0 I v}{2\pi}\ln(1 + \frac{a}{x_0})$
19. $\mu_0 n A I_0 \omega \sin(\omega t)$
21. (a) $\frac{20}{3}$ T·m^2 (b) 0 (c) -0.558 T·m^2
23. (a) 0.295 V (b) 44.2 kV
27. 267 V

Section 26.2

5. 1.97×10^{-6} H
7. 8.44 cm
9. (a) 1.76×10^{-5} s (b) 1.16×10^{-4} s (c) 3.85×10^{-4} s (d) 7.69×10^{-4} s
11. (a) 58.9 mA (b) 114 mA (c) $I_0e^{-Rt/L}$
13. 60 mA
15. 1.12×10^{-4} s
17. 1.63×10^{-5} C
19. (a) 1.24×10^{-5} s (b) 6.20×10^{-6} s
21. (a) J/m^3
23. (a) 2.31×10^{-2} T (b) 212.3 J/m^3; 4.96×10^{-2} J (c) 4.96×10^{-2} J
25. (a) 3.26 mH (b) 2.00×10^{-2} J (c) 11.1 J/m^3
27. overdamped
29. $(-Q_0 + A - A\beta t)e^{-\beta t}$
31. (a) critical damping (b) overdamping

Section 27.1

1. $-4/3$
7. 0

Section 27.2

1. both sides of the equation equal 0
5. 30

Section 27.4

5. (a) 1.83×10^{-7} T (b) 4.02 W/m^2 (c) 2.68×10^{-8} J/m^3

7. The direction is in the direction of current flow.
9. $\vec{E} = \langle E_0 \cos(ky + \omega t), 0, 0\rangle$; $\vec{B} = \langle 0, 0, B_0 \cos(ky + \omega t)\rangle$
11. 1.55×10^{22} J
13. 15 km
15. 5.83×10^{-7} m

Section 27.5

1. (a) 1.76×10^{25} (b) 2.72×10^{18} (c) 2.62×10^{11}
5. 1×10^{45}

Section 28.1

1. 656.2 nm; 486.1 nm; 434.0 nm; 410.1 nm

Section 28.2

1. $\hbar/(ma_0 n)$; 2.19×10^5 m/s
3. 397 nm, ultraviolet
7. ground state: $K = 13.6$ eV, $p = 1.99 \times 10^{-24}$ kg·m/s, $L = 1.05 \times 10^{-34}$ J·s; $n = 5$: $K = 0.544$ eV, $p = 3.99 \times 10^{-25}$ kg·m/s, $L = 5.28 \times 10^{-34}$ J·s
9. (a) 12.4 μm; 3.28 μm (b) 9.14×10^{13} Hz; 2.42×10^{13} Hz
11. (a) 3.10 eV; 1.77 eV (b) 1.89 eV; 2.55 eV; 2.86 eV; 3.02 eV

Section 28.3

1. (a) 3.55×10^{17} Hz; 8.44×10^{-10} m (b) 1.54×10^{18} Hz; 1.94×10^{-10} m (c) 1.6×10^{19} Hz; 1.85×10^{-11} m
3. (a) 2.56×10^{15} Hz; 2.09×10^{-8} m (b) 1.58×10^{17} Hz; 1.90×10^{-9} m (c) 2.54×10^{18} Hz; 1.18×10^{-10} m
5. 3.31 keV; $Z = 27$, cobalt
7. $(\frac{15cR}{16})(Z-1)^2$

Section 28.4

1. (a) 30.4 nm (b) 164 nm (c) a factor of four lower in both cases
5. 656.30 nm

Index